Impact of

GLOBAL CLIMATIC CHANGES

on

Photosynthesis and Plant Productivity

Impact of

GLOBAL CLIMATIC CHANGES

on

Photosynthesis and Plant Productivity

PROCEEDINGS

(of the Indo-US Workshop held on January 8-12, 1991 at New Delhi, India)

Organized by
Indian Council of Agricultural Research
and
Far Eastern Regional Research Office
of the United States Department of Agriculture

OXFORD & IBH PUBLISHING CO. PVT. LTD.
New Delhi Bombay Calcutta

EDITORS

Abrol, Y.P.
Division of Plant Physiology
Indian Agricultural Research Institute
New Delhi 110 012, India

Wattal, P.N.
Division of Plant Physiology
Indian Agricultural Research Institute
New Delhi 110 012, India

Gnanam, A.
University of Pondicherry
Pondicherry 605 014, India

Govindjee
Department of Biophysics and Plant Biology
University of Illinois
Urbana, Illinois 61801, USA

Ort, D.R.
USDA/ARS
Department of Plant Biology
University of Illinois
Urbana, Illinois 61801, USA

Teramura, A.H.
Department of Botany
University of Maryland
College Park
Md. 20742-5815, USA

ISBN 81-204-0614-1

Published in India by Mohan Primlani for Oxford & IBH Publishing Co. Pvt. Ltd., 66 Janpath, New Delhi 110 001, processed and printed from camera ready pages by Rekha Printers, Pvt. Ltd., A-102/1, Okhla Industrial Area, Phase II, New Delhi 110 020.

1-L1-7.5

FOREWORD

We live as guests of the green plants on earth. This is because green plants, through the process of photosynthesis, constitute the source of primary productivity. Naturally, the process of photosynthesis has evoked great interest among physiologists, breeders, biochemists, geneticists and agronomists. Recent studies on potential changes in climate as a result of the growing imbalance between atmospheric carbon emissions and carbon absorption have underlined the urgency of understanding the impact of high concentration of CO_2 and other greenhouse gases in the atmosphere on food security. Precipitation, temperature, sea levels and ultraviolet radiation can all be affected by changes in atmospheric CO_2 levels and in the ozone layer. Anticipatory research is, therefore, essential to anticipate the potential impact of climatic factors on photosynthesis.

Considerable data are already available on the impact of CO_2 on photosynthesis and photorespiration. However, photorespiration rates may decrease with an increase in atmospheric CO_2. Past studies tend to suggest that a higher CO_2 content in the atmosphere may be an advantage particularly in areas where water may be a limited factor. Unfortunately, most studies on direct CO_2 effects have involved CO_2 concentrations of 660 ppm. We need studies using levels ranging from 450 to 500 ppm. Also under field conditions, the anticipated gain may not be realised due to the complex nature of interactions involved among atmospheric CO_2 levels, temperature, precipitation and sunlight hours.

This publication is thus timely. It contains a wealth of information which will be valuable in crop modelling and in the agronomic and genetic improvement of productivity. In population-rich, but land-hungry countries like India, China and Bangladesh, the only pathway available for meeting the growing food needs is the improvement of productivity per day, and per units of land, water and energy. Much of the progress made so far in improving the yield of crops has come from a better partitioning of photo-synthates in favour of the economic part. We now need advances in improving total biomass production. We also need for each agro-ecological region of India a detailed analysis of the consequences of climatic change on food and drinking water security.

I am confident that the Proceedings of the US Workshop will be found extremely valuable by scientists and scholars interested in understanding methods of improving plant productivity both under normal and altered climatic conditions.

I congratulate Dr.Y.P. Abrol for getting this publication ready for dissemination both speedily and in a very readable manner.

M.S. SWAMINATHAN
Director
Centre for Research on Sustainable
Agriculture and Rural Development,
Madras

PREFACE

The biosphere, lithosphere, hydrosphere and atmosphere of the earth are in dynamic equilibrium. Human activities during the last century involving particularly landscape modification, resource exploitation and effluent flow have reached sufficient magnitude as to disturb the global eco-system. Higher temperatures, redistribution of rainfall, increased UV-B radiation due to stratospheric ozone depletion and increased levels of atmospheric CO_2 and other greenhouse gases may have profound but at the present state of our knowledge unpredictable effects on photosynthesis and plant productivity.

Interaction of US and Indian scientists for over a decade due to operation of US-India Fund supported co-ordinated project entitled "Photosynthesis and Crop Productivity under Tropical Environment" at some ICAR Institutes/Universities led to a suggestion by the US scientists to hold a workshop in India to develop a clear understanding of the climate changes at global level and to pool research efforts to develop strategies so as to ameliorate the basic environment in which we grow our crops. Further, it was felt that this will provide an opportunity to the international community of scientists to freely exchange ideas, share data and develop collaborative research efforts designed to understand the impact of global climatic changes.

In response to the above suggestion, the Indian Council of Agricultural Research constituted a steering committee under the Chairmanship of Dr. N.S. Randhawa, Director-General, ICAR with the following members: Dr. R.S. Paroda, Deputy Director General (CS); Dr. A.M. Michael, Director, IARI; Dr. R.C. Hedlund, Acting Director, FERRO; Dr. S.C. Adlakha, Agricultural Research Specialist, FERRO; Mr. J.C. Malhotra, Assistant Director General (PIU); and Dr. Y.P. Abrol, Head, Division of Plant Physiology, IARI, as *Member-Secretary.*

The workshop was inaugurated at the Indian Agricultural Research Institute Auditorium by the Deputy Prime Minister of India, Shri Devi Lal. During the function, a book entitled "Photosynthesis and plant productivity under tropical environment — a co-ordinated effort", summarizing the work done in the project was released. From the US side, His Excellency the US Ambassador to India, Mr. William Clark, Jr. also participated in the inaugural function. Professor M.G.K. Menon, M.P., Rajya Sabha, delivered the *Opening Address.* This was followed by an *Overview Lecture* on "Global Climate Change Scenario" by Dr. A.P. Mitra, Director General, CSIR and *Special Lecture* by Professor Fakhri A. Bazzaz, Harvard University, on "CO_2 and global climate change".

The five-day deliberations which included presentation by invited speakers, poster presentations and discussions by the working groups were held at Hyatt Regency, New Delhi. Experts from different fields of environmental sciences including Dr. T.P. Coohill (Atmospheric Physics), Dr. B.L. Deekshatulu (Remote Sensing), and Dr. D.R. Sikka (Tropical Meteorology) and a number of agro-climatologists were invited to share their views on possible hazards of climatic changes with experts in the fields of photosynthesis and plant productivity. This *Proceedings Volume* comprises papers presented/submitted for presentation and recommendations of the working groups regarding identification of future research priorities and areas where collaboration is desirable amongst the international community of scientists.

I wish to put it on record that the success of this workshop was entirely due to the active support by all members of the Organizing/Technical Programme Committees, Chairpersons and Rapporteurs of different sessions/working groups, staff of FERRO and the Division of Plant Physiology. My special thanks are to Drs. Aruna Sharma and S.N.Bhardwaj for their ungrudging help in attending to each and every detail of preparation of materials required for the workshop.

Thanks are due to ISRO and CSIR for supplying copies of "Geosphere and Biosphere Programme" and "Global Change — Indian contributions, 1980–1989", respectively, for distribution among the participants.

Division of Plant Physiology
IARI, New Delhi 110012, India
May, 1991

Y.P. ABROL
Organizing Secretary

National Organizing Committee

Paroda, R.S., New Delhi,
Chairman
Bose, Salil, New Delhi
Gnanam, A., Pondicherry
Goswami, N.N., New Delhi
Lahiri, A.N., Jodhpur
Maheshwari, S.C., New Delhi
Mohan Ram, H.Y., New Delhi
Rama Das, V.S., Palampur
Ramaiah, P.K. Bangalore
Sane, P.V., Lucknow
Singh, J.S., Varanasi
Singh R., Hisar
Sinha, S.K., New Delhi
Abrol, Y.P., New Delhi,
Organizing Secretary

U.S. Organizers

Ort, D.R.
USDA/ARS
Department of Plant Biology
University of Illinois
Urbana, Illinois

Govindjee
Department of Biophysics and Plant Biology
University of Illinois
Urbana, Illinois

Whitmarsh, J.
USDA/ARS
Department of Plant Biology
University of Illinois
Urbana, Illinois

Technical Programme Committee

Abrol, Y.P., New Delhi,
Chairman
Agarwal, P.K., New Delhi
Babu, C.R., New Delhi
Balasubramanian, V., Hyderabad
Bhardwaj, S.N., New Delhi
Das, T.M., Calcutta
Murty, K.S., Cuttack
Sengupta, U.K., New Delhi
Singhal, G.S., New Delhi
Udaya Kumar, M., Bangalore
Mohanty, P., New Delhi,
Member-Secretary

Local Organizing Committee

I.C.A.R.
Abrol, Y.P.
Bhardwaj, S.N.
Malhotra, J.C.
Nath, V.
Sharma, Aruna
Tomar, D.P.S.
Wattal, P.N.

FERRO
Thomas, J.H.
Hedlund, R.C.
Adlakha, S.C.

Co-ordination and Administrative Support

Nand Kishore
Dhingra, J.L.
Tuli, T.P.S.
Ramesh Chander
Meena, R.C.
Gupta, G.K.
Kapoor, Usha
Sharma, Anil
Chatwal, Rakesh

WORKING GROUPS

Session I:	**UV-B Radiation Effects**		
	Chairpersons:	A.H. Teramura, E. Wellman	
	Rapporteurs:	P.A. Kumar, Aruna Sharma	
Session II:	**CO_2 Enrichment/Balance**		
	Chairpersons:	N.E. Tolbert, V.S. Rama Das	
	Rapporteurs:	U.K. Sengupta, P. Raghuveer	
Session III:	**Photosynthesis and Environmental Stresses**		
	Chairpersons:	D.R. Ort, A. Gnanam	
	Rapporteurs:	M.C. Ghildiyal, T. Sudhakar Babu	
Session IV:	**Crop Modelling**		
	Chairpersons:	S.K. Sinha, J.D. Eastin	
	Rapporteurs:	B. Baldev, N.V.K. Chakravarty	
Session V:	**Agro and Natural Eco-systems**		
	Chairpersons:	R.S. Paroda, C.J. Nelson	
	Rapporteurs:	P.N. Wattal, Renu Khanna-Chopra	
Poster Session			
	Convenor:	D.C. Uprety	
Plenary Session			
	Chairperson:	R.S. Paroda	
	Convenor:	Y.P.Abrol	

CONTENTS

III. PHOTOSYNTHESIS AND ENVIRONMENTAL STRESS

IV. CROP MODELLING

V. AGRO AND NATURAL ECO-SYSTEMS

OVERVIEW: GLOBAL CHANGE AND INDIAN EXPERIENCE

A.P. Mitra

Director General, Council of Scientific and Industrial Research, Rafi Marg, New Delhi-110 001.

Present address: Bhatnagar Fellow, National Physical Laboratory Pusa Campus, New Delhi-110 012.

1. INTRODUCTION

In this overview lecture I present to you, instead of an overview of the general subject of global change - matters that have been described in different fora in considerable details - an outline of the role that Indian scientists have been playing in this area in the past and are continuing today and the plans for the future.

Past work that I will outline here includes:

i) the ozone network in India

ii) the Indian Middle Atmosphere Programme (IMAP)

I will also give you an account of the major facilities that have been developed over the years, including the period of the IMAP and finally an outline of the current activities and some results.

2. THE INDIAN MIDDLE ATMOSPHERE PROGRAMME (IMAP)

During the years since the IGY, the most important international programme from Indian point of view and also in the context of global change was an intensive multi-institutional and multi-disciplinary study of the middle atmosphere beginning on January 1, 1982, and continuing till March, 1989. This is the Indian Middle Atmosphere Programme (IMAP), a part of the International MAP activity. About 200 Indian scientists (at least 50 of whom were young and new entrants in this area) coming from 20 institutions were involved in this programme, operating in stations spread all over India (Fig.1).

The Indian effort was centered on specially organized campaigns with specific goals. These campaigns were multi-technique and multi-institutional. These were so organized that normal routine or standard measurements (often continuing over a long time) could also be integrated with these programmes with enhanced utility. Thus the effort was also to include input of scientists from small institutions.

The programme was divided into five broad disciplines: minor constituents and atmospheric chemistry, atmospheric dynamics, radiation, ionization including electrodynamics, modelling and theoretical studies, and four operational activities: balloons, rockets, satellite data and software and information cell. There was provision also for the establishment of centres. All these areas are critical for global change studies, especially those relating to major constituents, dynamics and radiation.

A total of 178 rockets were launched for the primary and subsequent phases of IMAP. This was a major effort of a magnitude we had not undertaken before and required careful planning for production of rockets in time, for payload fabrication and integration to be in phase. Of those large number of rockets a major fraction went for Ozone Intercomparison and Equatorial Dynamics Campaigns. The Ozone Intercomparison (March 23-31, 1983) Campaign and the Diurnal Variation Campaign (Dec.3-8, 1987), both carried out jointly with Soviet scientists, with payloads from both India and USSR, used up a total of 37 M-100 rockets fired within two periods of roughly a week each.

A national high altitude balloon facility operating at Hyderabad was available to the IMAP community. 100 kg (ionization and minor constituents) balloons were used. Special high altitude balloons were used during the equatorial wave campaign from a number of places.

Balloons were planned to be one of the major tools for study and a total of 33 balloons were to be deployed for special experiments. These were high altitude balloons and the expected minimum altitudes to be attained were between 25 and 35 km. These were in addition to those flown by India Meteorological Department till the end of 1984, launched 1000 high altitude balloons for temperature and wind measurements from different places in the country; about 50 ozonesonde flights (fortnightly) from Delhi, Pune and Trivandrum; about 400 radiometersonde flights from 8 stations.

A major aim was to evolve a first order reference middle atmosphere over India.

A reference atmosphere must include at least the following elements:

i) Concentrations of the main atmospheric constituents N_2, O_2 and the neutral temperature T.

ii) Distributions of the interacting minor constituents:

a) greenhouse molecules CO_2, CH_4, NO_x, CFMs

b) ozone

c) radicals OH, C10, NO

d) aerosols

iii) Electron and ion distributions; and preferably also the nature of the ions,

iv) Profiles of wind motions of different types

For the reference atmosphere, the least known and in the context of the global change problem, the most critical were the minor species. Amongst the many in the oxygen, hydrogen, nitrogen, carbon and chlorine families, not all could be measured in the short timeframe or with the techniques and facilities available in India or could be developed during IMAP. Nor was it considered necessary to measure all minor species in India. For example, CO_2 measurements at Mona Loa continuing for several decades are global representative values and can be accepted for Indian conditions as for others. The status is given in Table 1.

Table 1: Greenhouse molecules - Status of Indian observations

CO_2	Not measured routinely but flux from biomass burning by portable gas chromatograph (PGC)
CH_4	Yes. PGC (NPL) Profile: Balloons (Hyderabad)
O_3	Yes. Extensive: Groundbased, Balloons, Rockets
NO_x	NO in mesophere (Rockets)
	N_2O in Tropo/Strat (Balloons)
	NO_2 in Tropo/Strat (Laser heterodyning system)
CFM_s	Tropo/Strat in Hyderabad (Balloons)
H_2O	Radiosondes 6-8 km

LASER HETERODYNING	UPTO 35 Km
ROCKETS	UPPER STRAT/MESOSPHERE

The hierarchy of interest was:

Troposphere	<u>H_2O</u>, <u>O_3</u>, <u>CH_4</u>, <u>CFMs</u>, <u>N_2O</u>, OH
Stratosphere	<u>H_2O</u>, <u>O_3</u>, OH, NO, C10, <u>CH_4</u>, <u>CFMs</u>
Mesosphere	<u>NO</u>, <u>H_2O</u>, <u>O_3</u>, O

Those underlined were the ones for which measurements were undertaken. This meant introduction of new ground-based and a heavy dependence on balloons and rockets.

In the early 80's when the IMAP started, ozone measurements were well in hand with a long series of measurements with Dobson spectrophotometers and some balloon and rocket measurements, but there were limited amount of observations for other minor species. For nitric oxide measurements had been made with rocket-borne techniques for the mesophere but not for the stratosphere or the troposphere. Water vapour measurements through the radiosonde technique gave H_2O concentrations upto only 6-8 km. The newly introduced NPL technique which uses atmospheric emission around 22.3 GHz essentially measured water vapour content and the profile determination involved several assumptions and in any case was restricted to the lower troposphere. The only stratospheric H_2O measurements were those by the Soviet scientists who launched rocket-borne techniques in the late seventies over Thumba. However, the concentrations were believed to be too large.

During the IMAP we intended to correct this situation. A new high quality ground-based technique was introduced: the laser heterodyning system at NPL (described later) that could make realtime profile measurements of H_2O (in addition to O_3) upto about 25 km. Added to this were some estimates by Jayaraman and Subbaraya <u>et al</u> from attenuation of solar radiation with balloon-borne payloads carried in March 1988. As a result we find that H_2O mixing ratio profile is now on a firm footing.

There were successful measurements of CFMs, N_2O, CH_4 from Hyderabad through a joint Indo-German campaign. Scientists from PRL, Ahmedabad and MPAE, Lindau, had a composite payload including a cryogenic gas sampler from MPAE and a multichannel suntracking photometer provided by PRL. Trace gases collected and analysed were CO_2, CH_4, CO, O_3; halocarbons such as CFC-11, CFC-12, CFC-22, CFC-13B1, CFC-114, CFC-12B1, CFC-113, CH_3CCl_3 and CCl_4. As a result of the IMAP activities, representative profiles of several critical minor species have been derived. These are shown in Fig.2.

3. OZONE OVER INDIA

Of several minor constituents, measurements on ozone have been the most comprehensive in India and involving essentially all known techniques except satellite-borne measurements. New techniques such as laser heterodyning system and UV-B photometers were introduced during IMAP. The observational techniques include:

i) Dobson spectrophotometers
ii) Surface chemical ozonesondes
iii) Balloon measurements
iv) Rocket measurements at Thumba (by two Indian groups and also by jointly with USSR scientists)
v) Ground-based UV-B photometer network
vi) A laser heterodyning system at New Delhi; and
vii) Microwave radiometry at 110.836 GHz (limited measurements)

Dobson values are available in most cases from IGY onwards and the Delhi Dobson equipment was calibrated with world standard at International Intercomparison in Boulder in 1977. Dakshin Gangotri station is new and so far had limited operation. Balloon measurements are available from 1971. Intercomparison was made with six other countries at Hohenpeissenberg, FRG in January-February, 1970. Rocket measurements are relatively recent. UV-B photometry started in NPL several years ago and with its help a network was established under the umbrella of IMAP.

Details of the different set-up along with data availability are indicated below:

(i) Dobson Spectrophotometers

Srinagar (Dobson No.10)	34°05'N,74°50'E	1959 onwards
New Delhi (Dobson No.36)	28°38'N,77°13'E	1957 onwards
Varanasi (Dobson No.55)	25°18'N,80°01'E	1963 onwards
Ahmedabad (Dobson No.54)	23°0'N, 72°39'E	(earliest since 1951; initiated by Ramanathan)
Pune (Dobson No.39)	18°32'E, 73°51'E	1973 onwards

Kodaikanal (Dobson No.45)	10°14'N, 77°28'E	1953 onwards
Dakshin Gangotri	70°S, 11°E	1987 onwards

(ii) Surface Measurements

Pune	1972 onwards
Trivandrum	1972 onwards
New Delhi	1973 onwards
Kodaikanal	1978 onwards
Nagpur	1978 onwards

(iii) Balloon Measurements

Pune

Delhi

Trivandrum

Dakshin Gangotri

Notes: (a) Intercomparison with sondes from six other countries at Hohenpeissenberg, FRG. In January-February, 1970. The instrument produced slight positive deviations from the mean in the troposphere (15%) and corresponding low readings in the stratosphere (5%) - several improvements since.

(b) Measurements available from 1971 onwards. Feqquency of soundings: fortnightly. More frequent during campaigns.

(c) At Dakshin Gangotri, special flights in 1987, as part of the International Ozone Hole Campaign; flights continuing in 1988, 1989 and 1990.

(d) ECC ozonesondes used.

(iv) Rockets

Thumba

Notes: (a) Indo-USSR Intercomparison Campaign: March 23-31, 1983.

(b) Second Indo-USSR Campaign: December 3-8, 1987.

(v) UV-B Photometers

Delhi	First one to be set up; Data since November, 1978.
Waltair	set up during IMAP
Jodhpur	set up during IMAP
Mysore	set up during IMAP
Shilong	
Trivandrum	
Pune	

Notes: (a) Narrowband UV-B photometer at ground to measure directly the erythemally effective radiations with 3 interference filters of 10 nm half bandwidth; central wavelengths are:
280±10nm; 290±10nm and 310±10nm

(b) Observations obtained first in Delhi (NPL) and data available since November, 1978.

(vi) Laser Heterodyning System

Delhi (NPL)

Note: Uses tunable CO_2 waveguide laser. Ozone profiling 15-40 km. Capable of monitoring several other minor species: Water vapour (0-20 km) and NO_2.

(vii) Microwave Radiometry

Bangalore (RRI)

Notes: (a) Vivekanand and Arora used a mm wave superheterodyne receiver built for radio astronomy to obtain spectral line observations of atmospheric ozone at 110.836GHz.

(b) A new radiometer is now under construction at NPL for regular operation.

The distribution of Indian Ozone observing network with indications of the techniques used is given in Fig.3.

It is also important to note that Indian Dobson network is amongst the most dense in the tropical regions. Amongst 71 Dobson stations in the world, only 19 are within ±30°N,S and 9 within ±20°N,S. Of these 19, India has five.

The second major strength of the Indian Ozone programme is the multiplicity of techniques often used simultaneously or near simultaneously for profile determinations. In this context, the two rocket-balloon-groundbased intercomparison campaigns carried out at Thumba, as part of the IMAP during 1983 and 1987 are of special value.

1983 intercomparison data have been compiled and issued (1987) and a mean reference profile has been derived from these and other rocket and balloon measurements for the tropics and specifically for Trivandrum. For the latter, Subbaraya uses a total of 19 ozone profiles obtained during 1980 to 1984. In Fig.4 we show the spread in the values for the March intercomparison campaign, irrespective of the techniques used as well as profiles obtained by laser heterodyning system and the microwave radiometry techniques (the last two are at different times and different locations). One notices: (i) a very sharp rise in ozone concentration from the tropopause level, (ii) high variability at tropopause.

We have now a very substantial set of data on ozone for the Indian subcontinent. We can use this: (a) to provide a base reference profile, (b) to examine trends in ozone changes over the Indian subcontinent as others are doing for data elsewhere to examine if there has been any measureable effect so far on CFC injections, and (c) the nature and extent of the ozone "depression" that already exists over low latitudes as in India. This last aspect is one that one needs to emphasise at the outset. There is a low ozone belt over a good part of India. Any further decrease would have more impact than in mid-latitude stations.

4. MAJOR FACILITIES AVAILABLE FOR GLOBAL CHANGE STUDIES

4.1 Rocket Soundings

There are three rocket ranges; (Thumba close to geomagnetic equator, Sri Hari Kota range near Madras and at Balasore in Orissa) the rockets normally available for global change related studies are:

RH200	Meteorological payloads (Chaff)	89 km/10.3 kg at 85°
RH300	Scientific experiments	100 km/50 kg at 82°
RH300	Scientific experiments	150 km/60 kg at 82°

The locations of the three rocket ranges are indicated in Fig.

A picture of an IMAP payload carried in a rocket is shown in Fig.5.

4.2 Balloon Facilities at Hyderabad

A permanent Balloon Facility was set up at Hyderabad in 1971 by the Tata Institute of Fundamental Research. Subsequently this became a National Facility and Facility was widely used during the Indian Middle Atmosphere Programme. The current capability is fabrication and launching of balloons upto 175000 m^3, payloads upto 100 kg and flights upto 38 km altitude and durations of 10-12 hours. The facility has been available for sometime for participation of scientists from different parts of India on programmes that are examined by the Balloon Board. Joint programme with other countries have also been carried out (such as the ones on jointly organized between PRL and Max Planck Institute at Lindau, referenced earlier).

4.3 **Lidar and Laser Heterodyning System**

There were several new facilities set up during the IMAP period. The lidar facility is at Thumba on the magnetic equator designed, installed and operated by VSSC. It is a ruby lidar operating on 694.3 nm with a pulse energy of 10J, a pulse width of a micro-second and with essentially regular observations since October, 1986.

The laser heterodyning system located in Delhi, designed, fabricated and operated by the NPL, uses a tunable CO_2 laser. The experimental set up is shown in Fig.6. The equipment, in principle, can monitor the following absorption lines of gaseous species in 9.11 nm range:

CF_2C12	$926.7510cm^{-1}$	10-20 km
NH_3	966.2514	0-15
C_2H_4	974.7500	
H_2O (Vapour)	975.9400	0-20
O_3	1043.1775	15-40
CFC13	1084.6000	10-20

The heterodyning system has been used quite successfully to derive profiles in a real time sense for ozone, water vapour and nitrogen dioxide. The profiles were checked against other standard measurements such as balloon ozonesondes and the technique has been validated. An example is shown in Fig.7. This shows a run on ozone with near simultaneous balloonsonde measurement (both at Delhi).

4.4 **Indian Remote Sensing Satellite IRS-1A**

This Satellite launched on March 17, 1988 provides a major tool for Indian scientists for remote sensing of several areas of interest in global change: land use, forestry, water resources, marine resources and agricultural products. A diagramatic representation of the IRS system (courtesy: ISRO) is given in Fig.8. The satellite carries three cameras using CCDs as detectors. The satellite has two types of imaging sensors - one with a special resolution of 72.5 metres and the other with two separate imaging sensors with a special resolution of 36.25 metres each.

4.5 **UV-B Photometers**

To measure directly the solar ultraviolet-B radiations, instead of competing it indirectly through measurements of ozone. A network of UV-B filter photometers have been set up in India (outlined earlier) operating at 290, 300 and 310 nm. One unit was sent to the Antarctica to measure UV-B radiation en route and at the Indian Station, Maitri. Data are standardised in units of microwatt cm^{-2} nm^{-1} for radiation received at ground for different times and different solar zenith angles. The intensity of the received radiation depends on a number of environmental parameters: aerosol loading, cloud (in particular).

In addition, a network of multiwavelength radiometers has been set up under the supervision of the Trivandrum group.

5. CURRENT PROGRAMME AND SOME RESULTS

5.1 Reference ozonosphere over India and Special Characteristics

Since ozone has been measured very extensively over India in different locations, a Reference Ozonosphere for different seasonal conditions and for different locations can be built up. One example of the profile structures over Trivandrum (8°N), as obtained from simultaneous or near simultaneous use of different techniques, has already been given earlier. Since India covers a fairly large latitudinal range from equatorial latitudes to mid-latitude conditions, it is also of interest to see how the profiles shape changes with latitude. This diagram is shown in Fig.9 in which average profiles are compared for Delhi, Varanasi, Poona and Trivandrum.

No detailed trend analysis of the data, some existing for three decades or more, has been attempted seriously in India. In the WMO-NASA Report on International Ozone Trends Panel 1988, most of Indian Dobson Stations have not been included for trend analysis. The primary reason it is understood, is claimed to be lack of information on instrument calibration.

These comments, in our view are not well taken. It has been pointed out by the Indian Meteorological Department that when the International Ozone Commission introduced new absorption coefficients in 1958, these were brought into use in India from 1963. Corrections for the old data from 1957 to 1963 were not made, but an 8 per cent correction can be made for all data using the C-wavelength before 1963. Delhi data from 1971 to April 1974 should be deleted from long-term series, since Dobson No.36 at Delhi suffered some damage in 1971 and withdrawn from routine use in April 1974 whe No.112 was brought into operation. It seems to us that once these points, and those regarding calibration procedures used are kept in view, a reliable trend analysis is possible.

The next important thing to note is that a good part of India is sitting in a low ozone belt. This is clearly seen in Fig.10. Any further decrease would have more serious impact than in mid-latitude stations, because then, in addition to increase in UV-B flux, radiations below 290 nm, which are biologically more active will begin to penetrate to the ground. The diagrams include both summer and wintere values and the very large changes between these dosages are also worth noting.

The third important point concerns India's own observations in its Antarctic Station Dakshin Gangotri as a part of the International Antarctic Ozone Hole Campaign of 1987. The observations were made with balloons throughout the year and have been continued during the years 1988, 1989 and 1990. The 1987 October observations, when the depletion was at its maximum, are shown from a number of stations in the Antarctic (including the observations made in Dakshin Gangotri (Fig.12). It is important to remember that the Indian station (70°S, 11°E) is located on the fringe of the polar vortex and as the boundary of the vortex shifts from one day to another, the Indian station also shifts into the containment vessel and outside.

The fourth point is the direct measurement of the UV-B radiation at several wavelengths in the erthymcally active regions in different parts of India. Instead of calculating UV-B dosages from observations of ozone, use of these photometers gives us directly the flux reaching the ground. In one sense these equipments are unique. It is possible to generate UV-B dosages contours over India from the network that has been set up. The distribution of dosage over Delhi for different times of the year are shown in Fig.13 (after Srivastava et al).

Since CFCs have now been recognised to be the principal contributor to the depletion of ozone, view with which Indian scientists are in agreement, a policy decision undertaken in India is to undertake technology developments for CFCs alternatives. Two CSIR laboratories are currently engaged in this work: the National Chemical Laboratory at Poona and the Indian Institute of Chemical Technology. Along with this, parameters needed for design modifications such as compressor, heat exchangers, motors, insulational lubricants as also the question of toxicity of the substitutes are under invesdtigation at several CSIR laboratories.

In regard to R-22 which is being suggested as an immediate (but temporary) solution, Indian manufacturers which are now producing R-11 and R-22 are also in position to produce R-22. However, India have some hesitation in using R-22 as a replacement. The reasons are the following:

a) Although it has a lower ozone depletion potential (ODP), the dissociation rate is nevertheless appreciable. This has been also seen in the balloon measurements of R-22 concentrations over Hyderabad recently through a joint experiment between India and West Germany (Fig.14); and

b) Consequent changes in the components and the mechanical system for refrigeration are not considered worthwhile.

5.2 **Warming Scenario**

IPCC scenario for changes by 2030 for Southern Asia (5°to 30°N, 70° to 150°E) is outlined to be as follows:

"The Warming varies from 1 to 2°C throughout the year. Precipitation changes little in winter and generally increases throughout the region by 5 to 15 per cent in summer. Summer soil moisture incrreases by 5 to 10 per cent".

To provide a better resolution over the entire Indian sub-continent a large mass of meteorological data recorded for over 100 years are now being analysed in a number of Indian Institutions. Two special efforts are by the India Meteorological Department and a group at ISRO. The trend analysis from the India Meteorological Department of the annual temperature of India with five year running mean is shown in Fig.15. The variations are within the ± level but nevertheless a trend (of about 0.3°C over the 90 year period) is observable. The ISRO group (Datta and Chakravarty, personal communication) have taken 90 years of temperature and precipitation data over some 100 stations in India and have chosen from this 5 warmest years and 5 coldest years. The warmest years chosen were 1937, 1953, 1973, 1980 and 1983; the coldest years chosen were 1903, 1906, 1908, 1916 and 1964. The differences in average temperatures and in precipitation are then plotted. The contours of such differences are given in Figs. 16 and 17. The deviations are from longterm averages of 1951-1970. Global deviations for these two groups were: -0.41°C and +0.22°C. The corresponding values for India are: 0.24°C and +0.29°C (*i.e.* a warming of 0.53°C). Thus warming over India was nearly of the same magnitude as that for the above as a whole. One notices that except for a few localized regions of cooling, warming occurs over most part of India with varying magnitudes upto 0.8°C. In regard to rainfall, changes in monsoon rainfall are shown in Fig.17. We notice excess rainfall over a large part of India. Rainfall deficiencies are seen in South and South-East and in North-West regions.

5.3 Methane

The IPCC Document (1990) has made a somewhat sweeping statement relating to emission of methane from paddy fields in Asia: "one difficulty in obtaining accurate estimates is that almost 90% of the world's harvested area of rice paddies is in Asia and of these about 60% are in China and in India from which no detailed data are available". Detailed data do indeed exist for several parts of Asia and measurements in India have been going on for a number of years, principally by the National Physical Laboratory. A number of publications exist describing the details of the measurement techniques and the results of measurements from paddy fields under different conditions (Parashar *et al*, 1988; Saha *et al*, 1989; Parashar *et al*, 1990; Parashar *et al*, 1991). The measurement method involves the use of a gas chromatograph with FID. Air samples are sometimes analysed *in situ* with a portable gas chromatograph. Also, samples are collected in teflon bags and analysed with a microprocessor controlled gas chromatograph in the laboratory. Methane standards for comparison were prepared in the laboratory.

Methane flux rate was found to depend as indeed seen elsewhere on a number of parameters including: (a) whether the paddy field is water-logged or not, (b) on soil temperature, soil moisture and its pH value, (c) on the nature and content of fertilizers and (d) the phase of the going season.

Measurements in India shows a very wide range of values for the flux rate. Fluxes for alkaline soil could be as much as 100 times larger than the acidic soil and similar differences in value exist for water-logged and non-water-logged areas. The representative flux rates are:

Inundated rice fields : 3-10 $mg/m^2/h$

Wet soil with no stagnating water : 0.1-0. $mg/m^2/h$

Comparisons given by Parashar (private communication) with data taken from other countries are shown in Table 2.

Table 2: Comparison of India Observations with those elsewhere (after Parashar)

(In gm $m^{-2}d^{-1}$)

Indian Data		OECD Data
Acidic	$(0.5 - 3.2) \times 10^{-3}$	0.19 - 0.69 (China) (Av: 0.5)
Alkaline	$(5-75) \times 10^{-3}$	0.16 - 0.38 (Italy) (unfertilized)
LOWLAND: Acidic	$(5-75) \times 10^{-3}$	0.28 - 0.60 (Italy) (fertilized)
Alkaline	$(31-1530) \times 10^{-3}$	0.25 (California) (fertilized)

Total contribution is a product of three parameters: the flux rate (A), the area of the particular category (B) and the duration of the rice growing season (C). The total emission is then given by a product of these three.

The duration assumed in the Indian calculations is 90 days and the estimated acreage of water-logged paddy fields is taken to be 16 million hectares. With these values the current Indian estimate of methane emission from paddy fields in India is between 5 to 10 terragram per year in contrast with over 30 Tg/yr reported in some calculations elsewhere.

6. Concluding Remarks

It is clear that the currently existing facilities in India (IRS satellite, the three rocket ranges, Balloon Facility in Hyderabad; the LIDAR at Thumba, the Laser Heterodyning System etc.) provide excellent opportunities for global change programme. The National Natural Resources Management System (NNRMS), now being established under ISRO/DOS, which uses remote sensing data from IRS-1A, Landsat, SPOT and NOAA, data interpretation facilities at five Regional Remote Sensing Service Centres and 21 State Remote Sensing Centre will provide a major support system. IGBP-related information expected from NNRMS includes:

(i) Regular mapping of changes in forerst cover
(ii) Mapping of flooded areas
(iii) Land form changes
(iv) Components of hydrological cycle

Other satellite applications envisaged include: vegetation index derivation from NOAA-AVHRR; study of Thar desert covering parts of Rajasthan, Haryana, Punjab and Gujarat states using currently available aerial photographs from Landsat MSS/TM, IRS-1A, LISS-1 and II, SPOT-HRV, NOAA-AVHRR and also future missions like ERS-1 and EOS polar platform; sea surface temperature and chlorophyll concentrations using NOAA-AVHRR channels.

A new facility available to atmospheric scientists will be the MST (Mesospheric Stratospheric Tropospheric). radar now operating on ST mode in a place (13°N, 79°E) near Tirupati in Andhra Pradesh (Fig.18). The neighbouring rocket range at SHAR provides opportunities for intercomparison and complementarity. The radar operates at 53 MHz with peak power of 2.5 MW (and a peak power aperture products of 3 x $10^{10} Wm^{-2}$), an antenna array consisting of 1024 crores yagi elements.

The Radar was commissioned on ST mode on 29 October, 1990 and is available also to international community of scientists. The radar provides high resolution 3-d contours of dynamical parameters on a realtime basis. In addition, it allows a study of the tropopause characteristics with a resolution of 150m, and consequently provides an important tool for study of stratosphere - tropospheric exchange processes.

A major effort this year will be the "1991 Methane Campaign". This will comprise of networked and intercalibrated set of measurements in different rice-growing areas of India (Orissa, Assam, West Bengal, Bihar, Andhra Pradesh, Tamil Nadu, Kerala, Delhi, Haryana etc.) and new measurements of emissions from animals vis-a-vis nature of food intake.

REFERENCES

CSIR Programme on Global Change, Document No.1, CSIR, July 1990.

Geosphere-Biosphere Programmes: ISRO/DOS Plan, ISRO-SP-47-89, ISRO, October 1989.

Global Change: Indian Contributions, 1980-1990, CSIR Global Change Series, Vol.2(Revised), February 1991.

Ind. J. Radio & Space Phys., Special Issue on Indian Middle Atmosphere Programme, October & December, 1989.

Proc. National Workshop on Ozone (April 13-15, 1987), Science Academy, Ed. B.N. Srivastava, K.S. Zalpuri, National Physical Laboratory, New Delhi, 1988.

Indian Geosphere-Biosphere, Ed. T.N. Khoshoo and Manju Sharma, National Academy of Sciences, Vikas Publishing House Pvt. Ltd., 1991.

Monograph on Human Influences on Atmospheric Environment, A.P. Mitra, National Physical Laboratory, 1980.

IMAP FACILITIES AND INSTITUTIONS

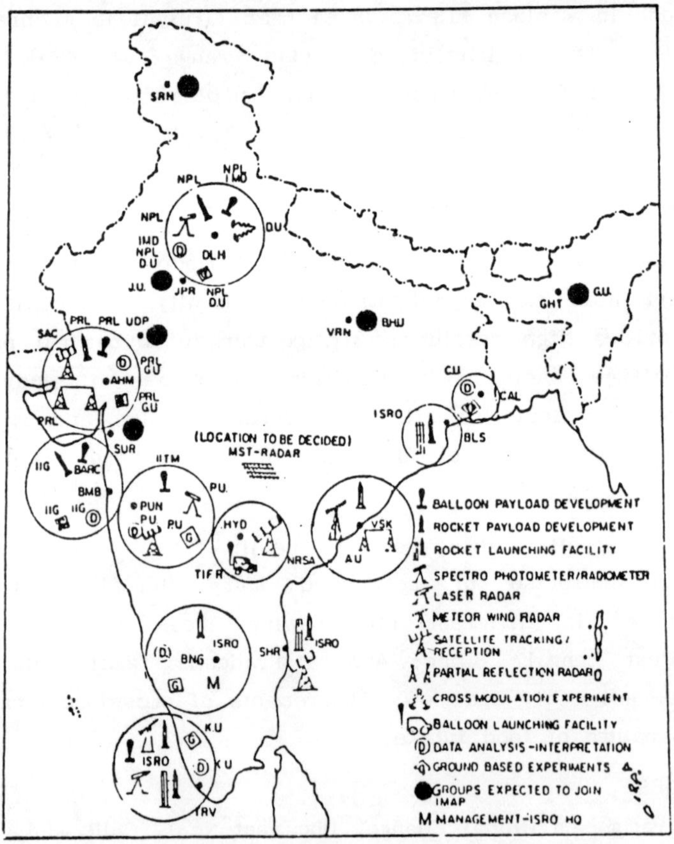

Fig. 1 Facilities and Institutions Participating in the Indian Middle Atmosphere Programme.

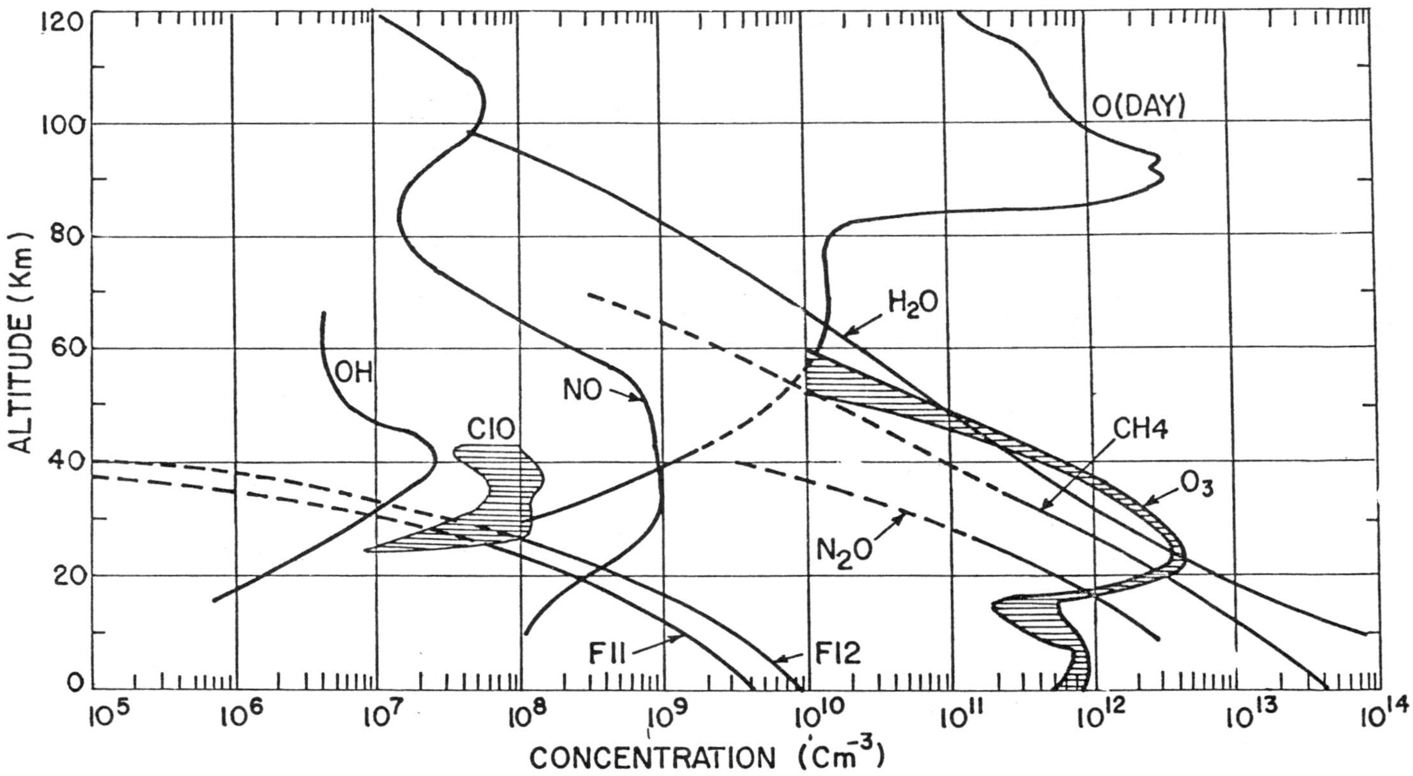

PROFILES OF IMPORTANT MINOR SPECIES

Fig. 2

INDIA METEOROLOGICAL DEPARTMENT

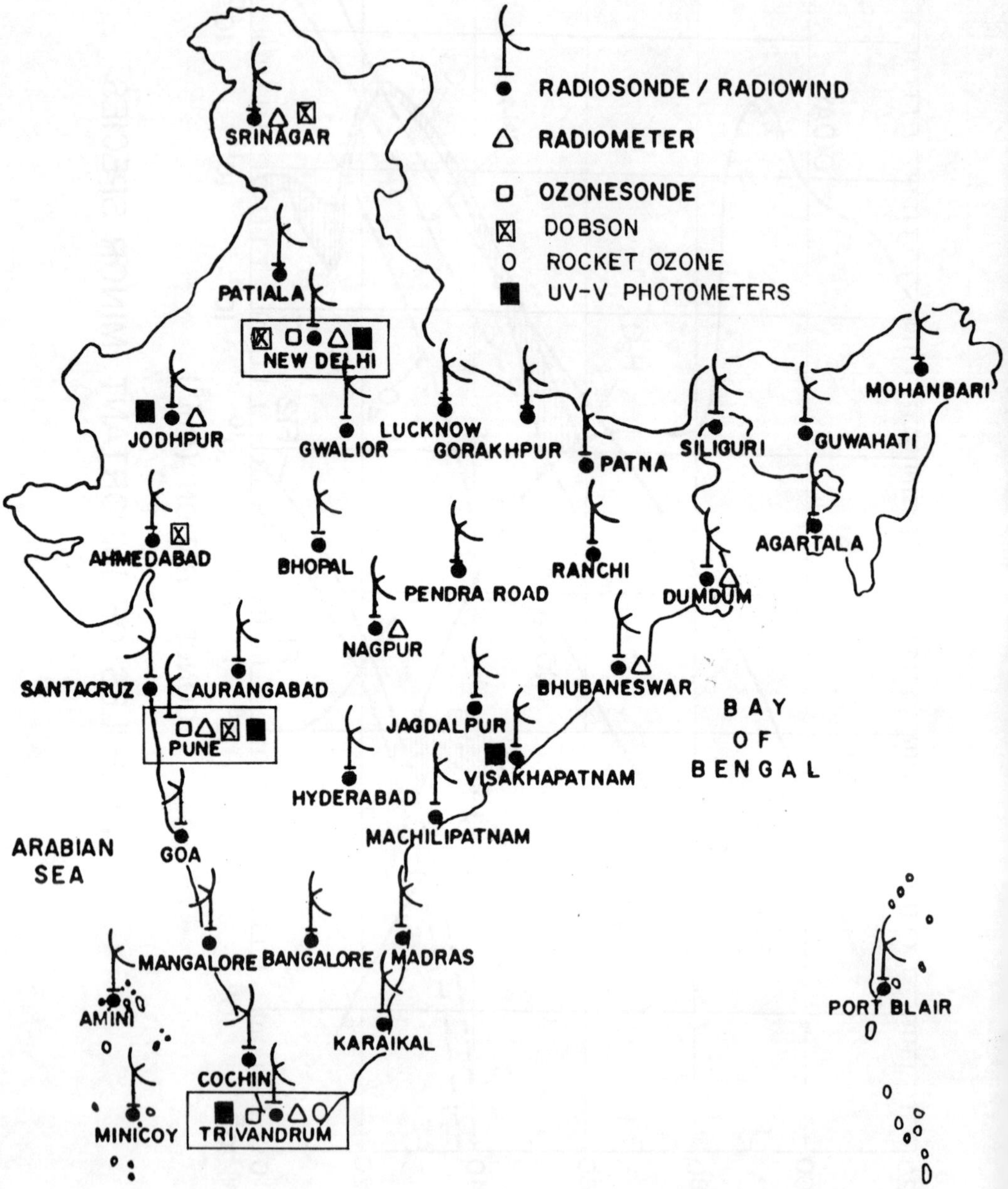

Fig.3

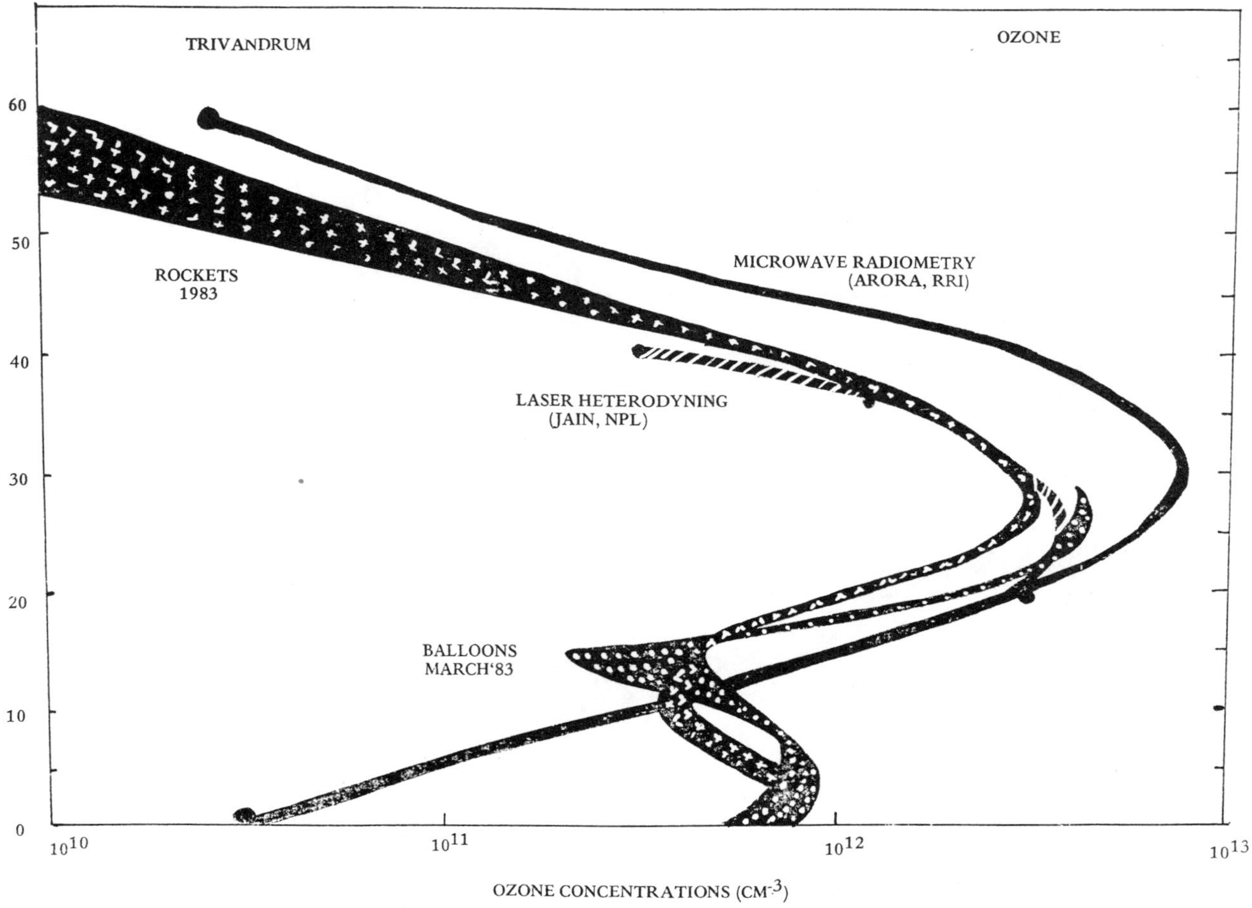
TRIVANDRUM
OZONE
ROCKETS
1983
MICROWAVE RADIOMETRY
(ARORA, RRI)
LASER HETERODYNING
(JAIN, NPL)
BALLOONS
MARCH'83
60
50
40
30
20
10
0
10^{10}
10^{11}
10^{12}
10^{13}
OZONE CONCENTRATIONS (CM^{-3})

Fig. 4

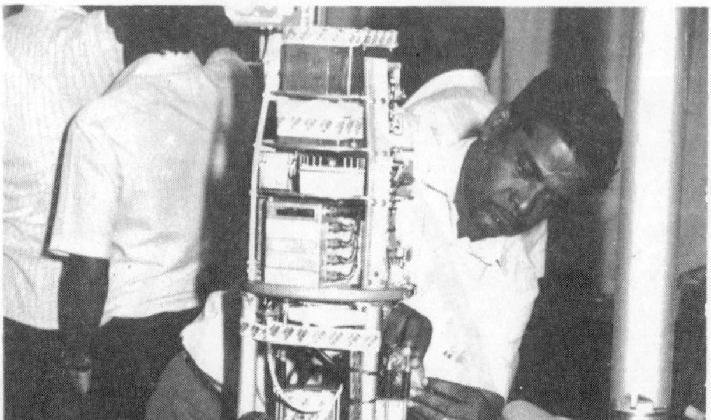

Fig. 5

Laser Heterodyne System developed by NPL for measuring vertical profiles of minor constituents.

Fig. 6

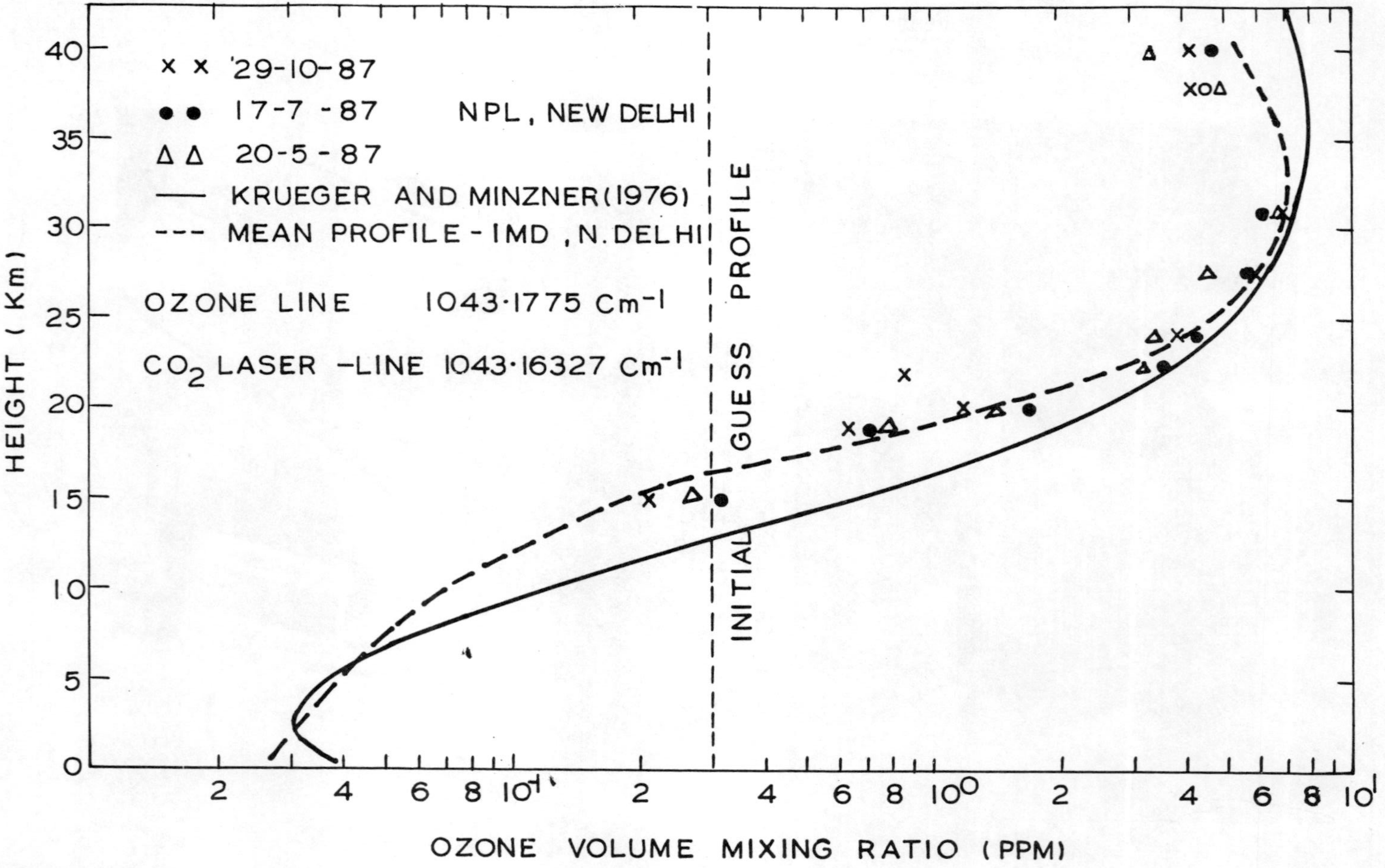

× × '29-10-87
● ● 17-7-87
NPL, NEW DELHI
Δ Δ 20-5-87
KRUEGER AND MINZNER (1976)
MEAN PROFILE - IMD, N. DELHI
OZONE LINE 1043·1775 Cm⁻¹
CO2 LASER -LINE 1043·16327 Cm⁻¹
INITIAL GUESS PROFILE
HEIGHT (Km)
OZONE VOLUME MIXING RATIO (PPM)

Fig. 7

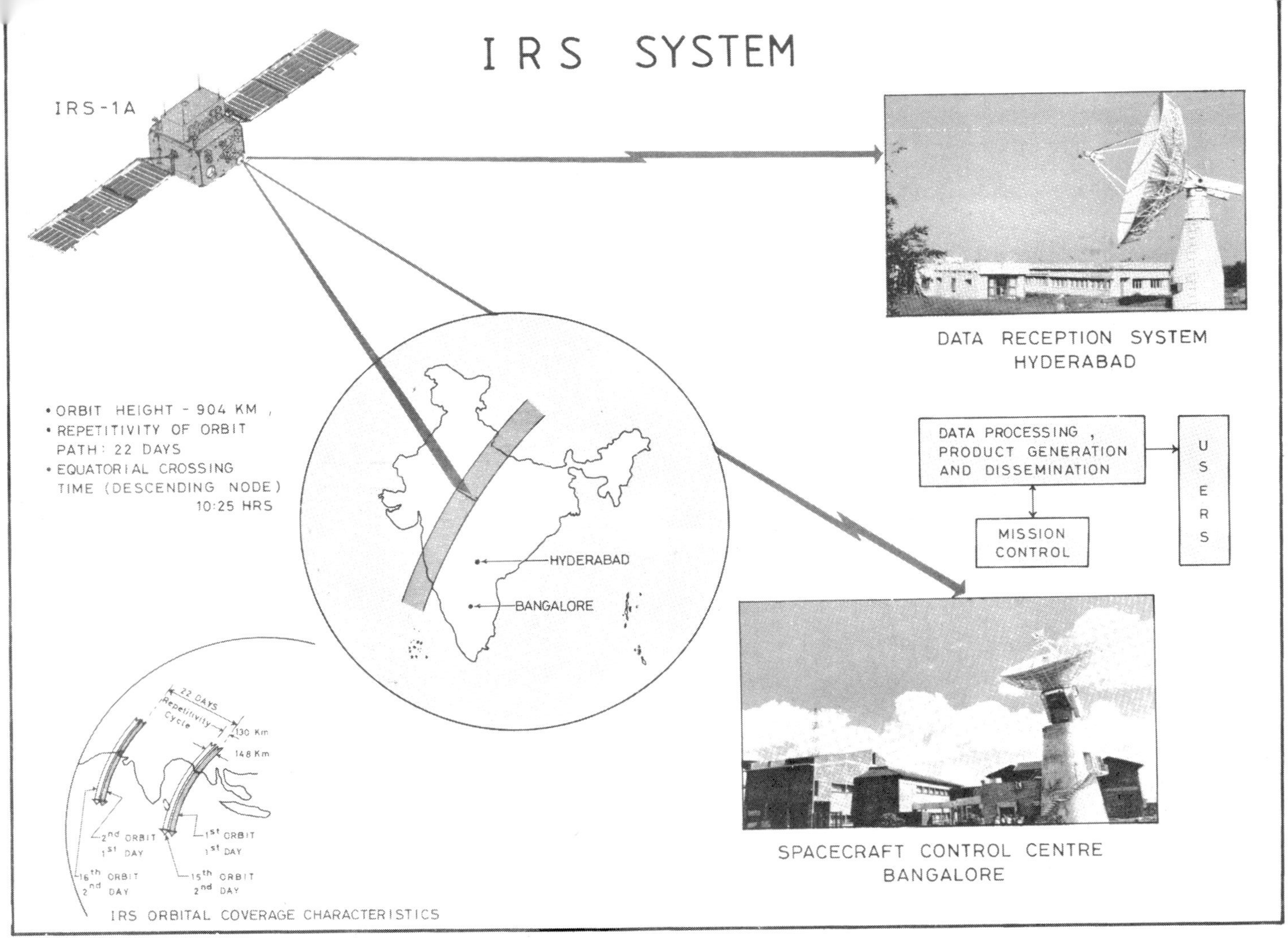
IRS SYSTEM
IRS-1A
• ORBIT HEIGHT - 904 KM ,
• REPETITIVITY OF ORBIT PATH: 22 DAYS
• EQUATORIAL CROSSING TIME (DESCENDING NODE) 10:25 HRS
HYDERABAD
BANGALORE
DATA RECEPTION SYSTEM HYDERABAD
DATA PROCESSING , PRODUCT GENERATION AND DISSEMINATION
USERS
MISSION CONTROL
SPACECRAFT CONTROL CENTRE BANGALORE
22 DAYS Repetitivity Cycle
130 Km
148 Km
2nd ORBIT 1st DAY
1st ORBIT 1st DAY
16th ORBIT 2nd DAY
15th ORBIT 2nd DAY
IRS ORBITAL COVERAGE CHARACTERISTICS

Fig. 8

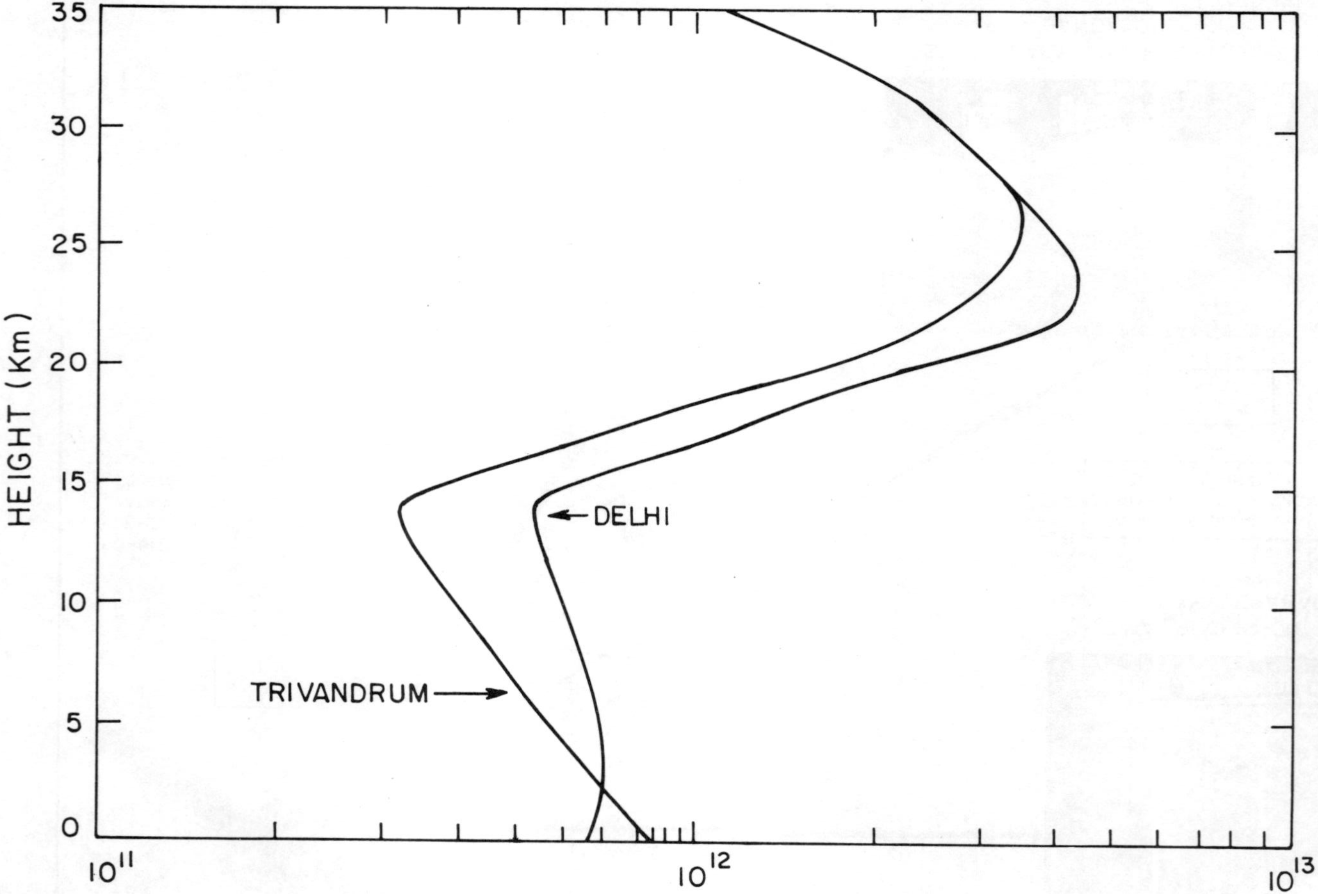

Fig.9 OZONE CONCENTRATION (cm^{-3})

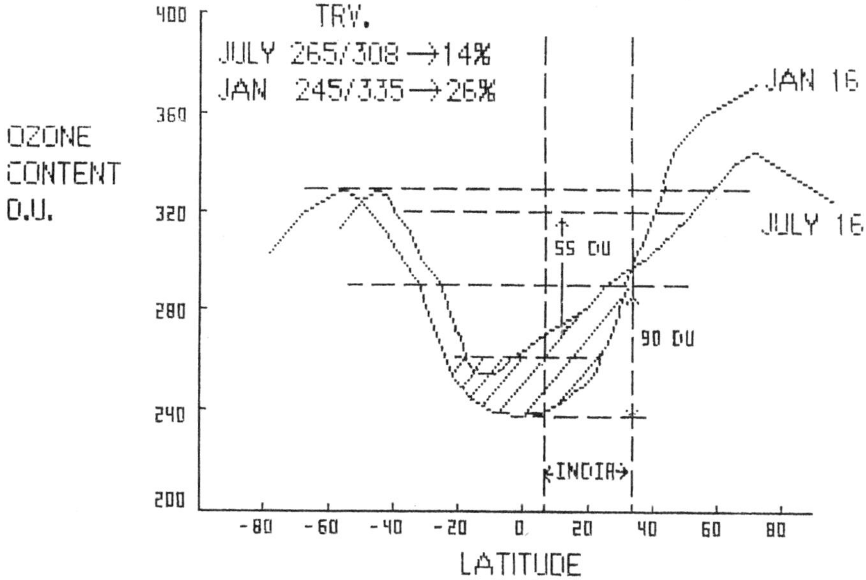

Fig. 10

UV RADIANCE

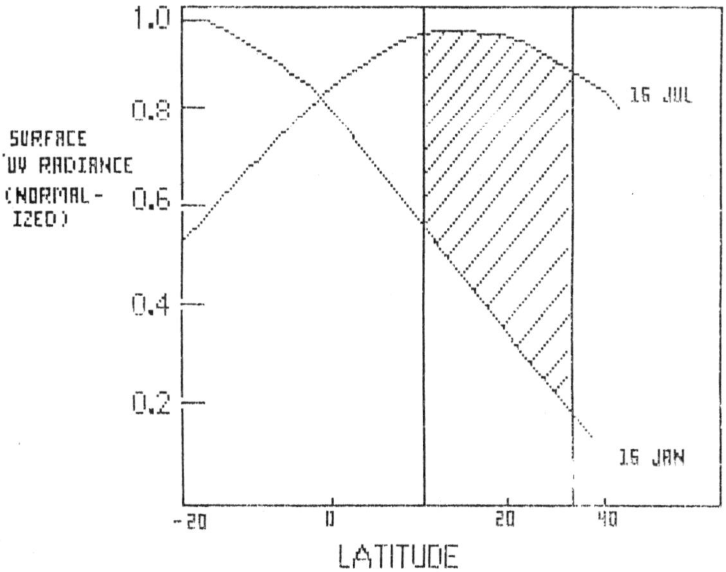

Fig. 11

Ozone Vertical Profiles

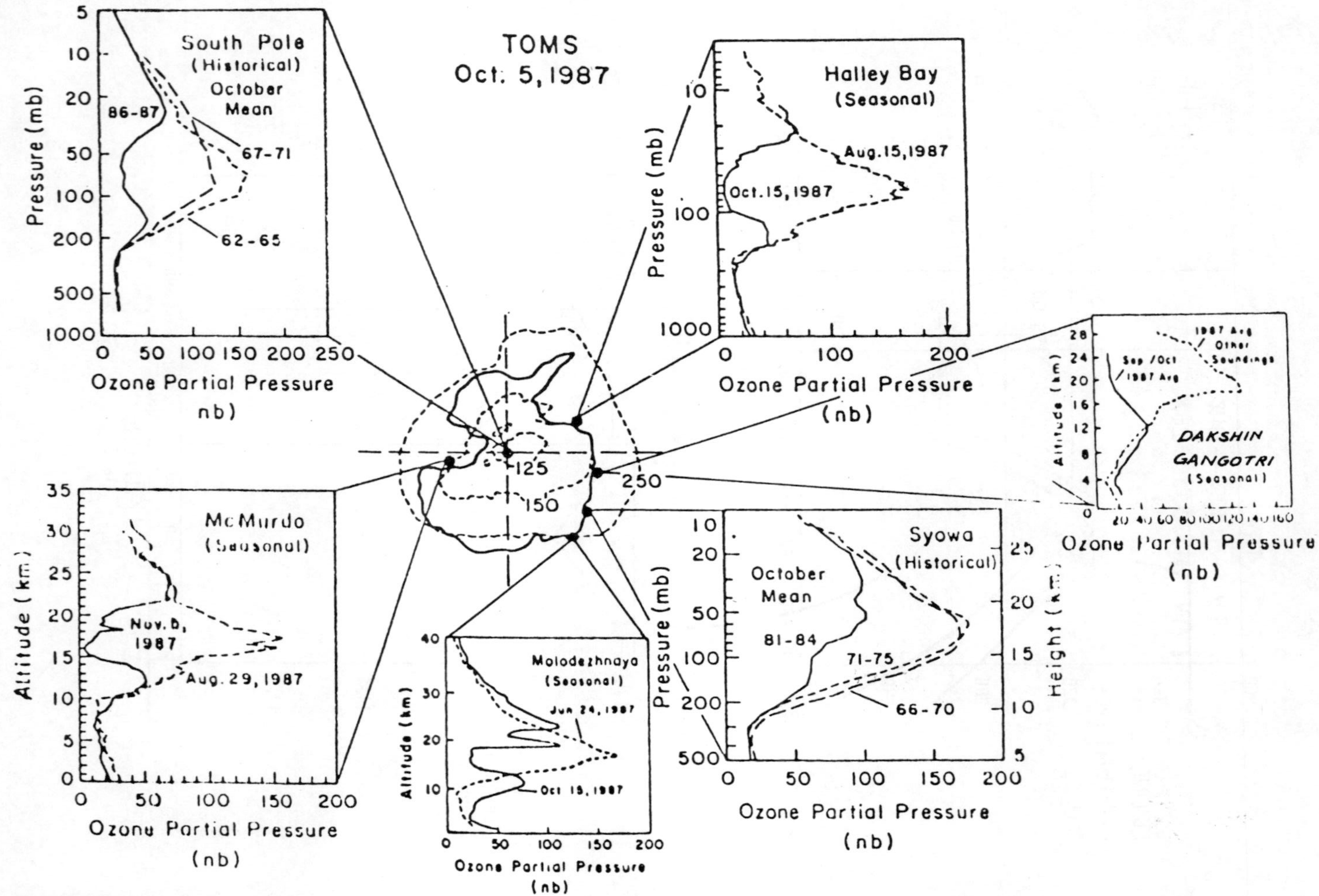

Fig. 12

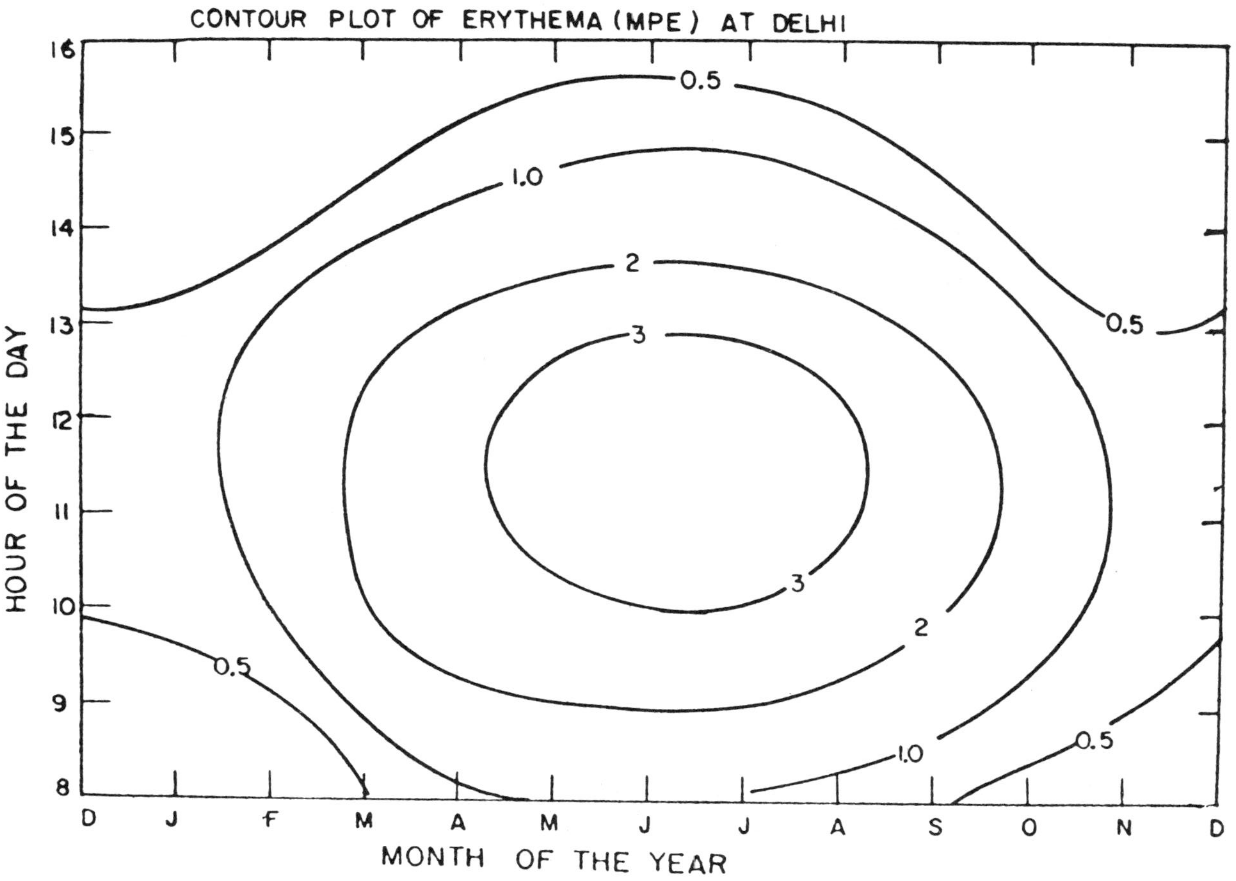

Fig. 13 (AFTER SRIVASTAVA ET AL)

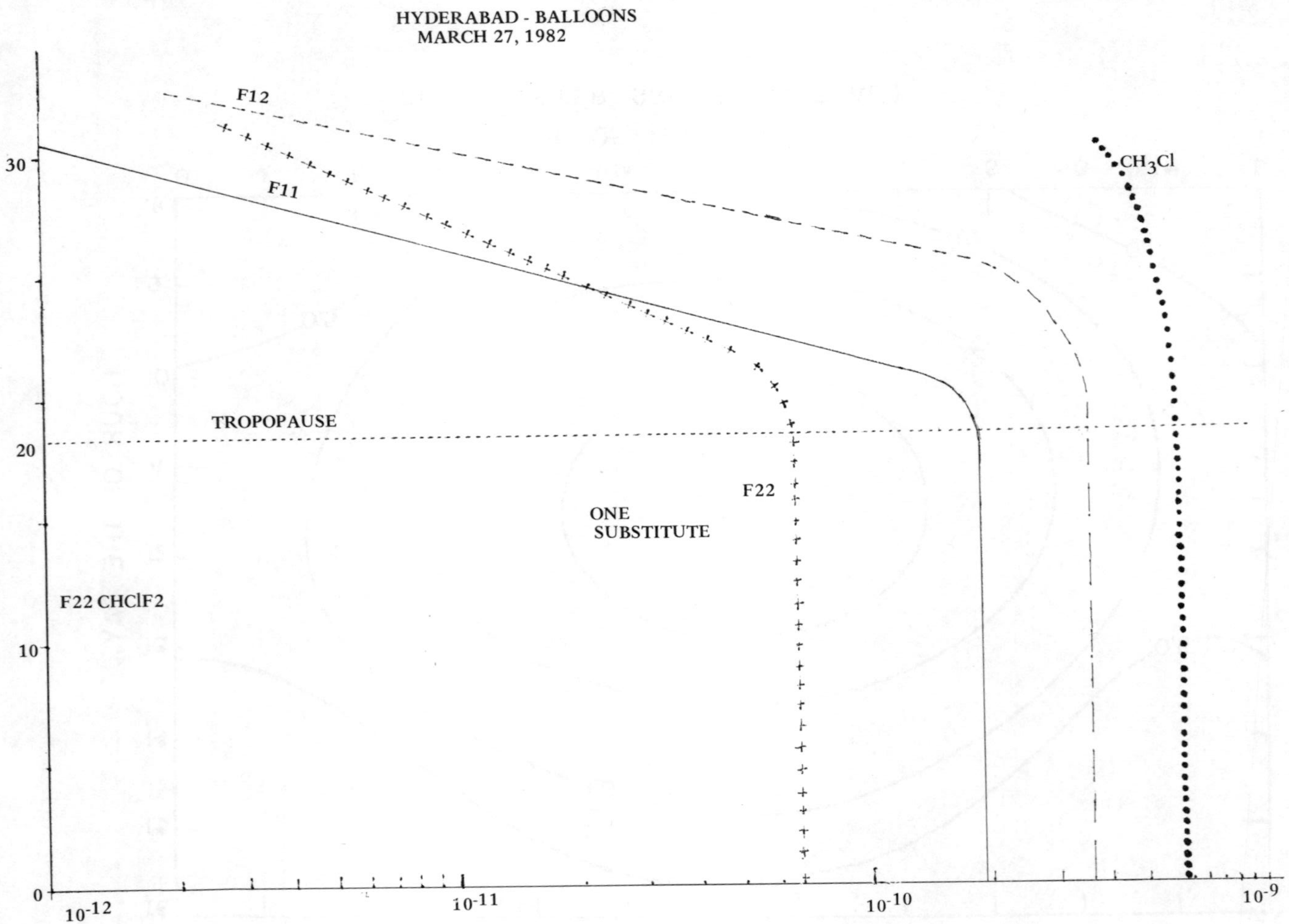

FIG. 14 DISTRIBUTIONS OF CFC-11, CFC-12 AND CFC-13 OVER HYDERABAD

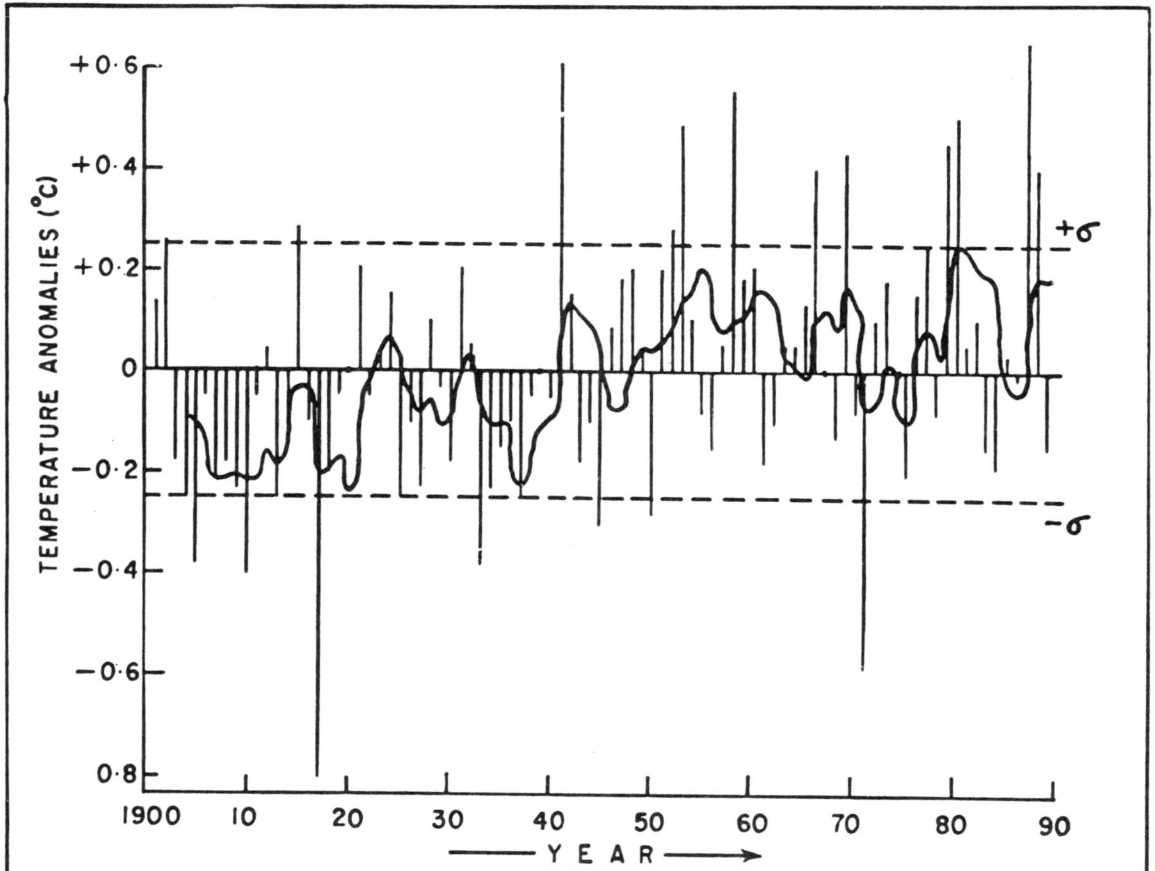

ANNUAL TEMPERATURE OF INDIA
(THICK LINE 5 YEAR RUNNING MEAN)

Fig. 15

SURFACE AIR TEMPERATURE (MAX.) DEVIATION OVER INDIA

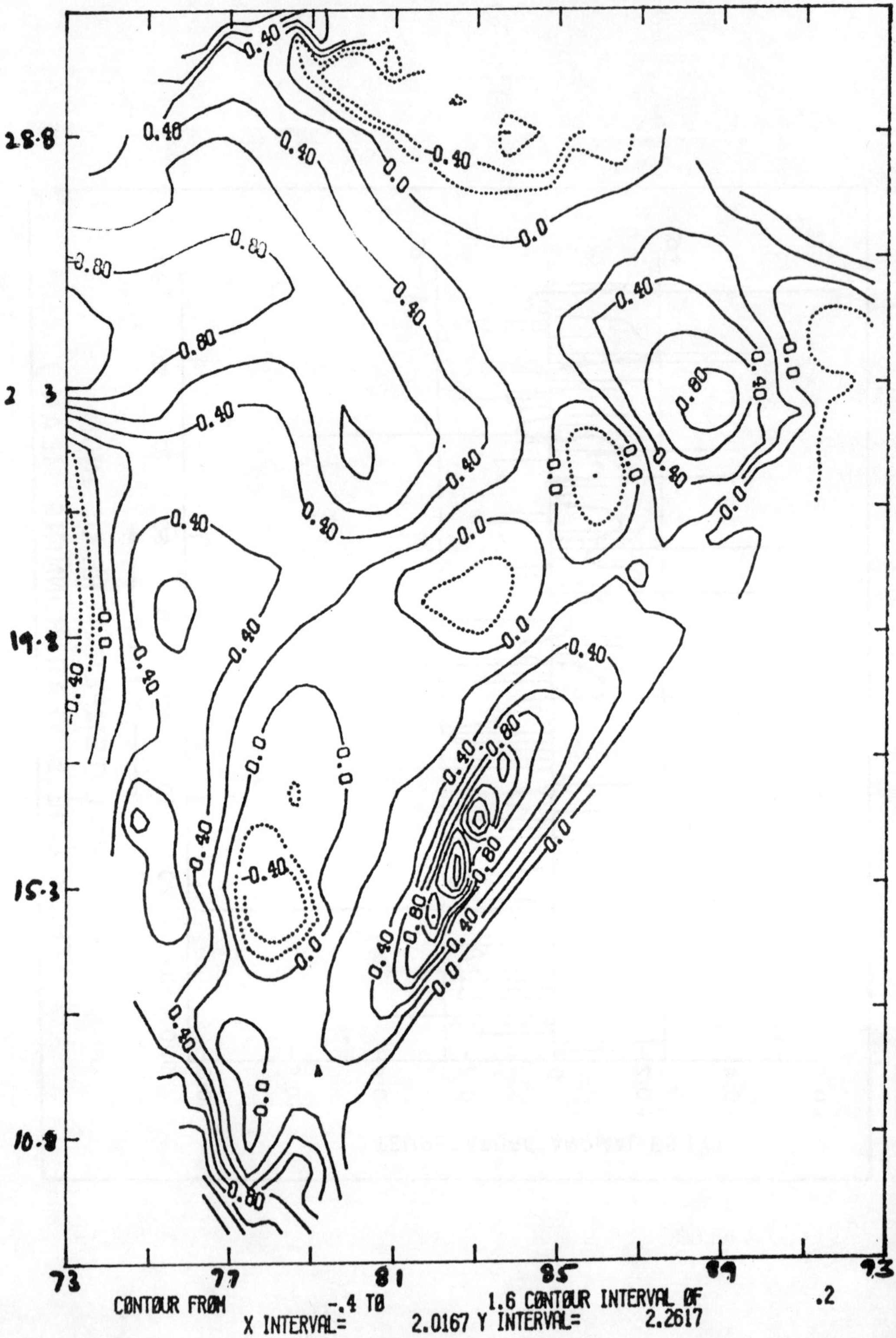

CONTOUR FROM -.4 TO 1.6 CONTOUR INTERVAL OF .2
X INTERVAL= 2.0167 Y INTERVAL= 2.2617

Fig. 16

MØNSØØN RAINFALL DEVIATIØN ØVER INDIA IN MMS

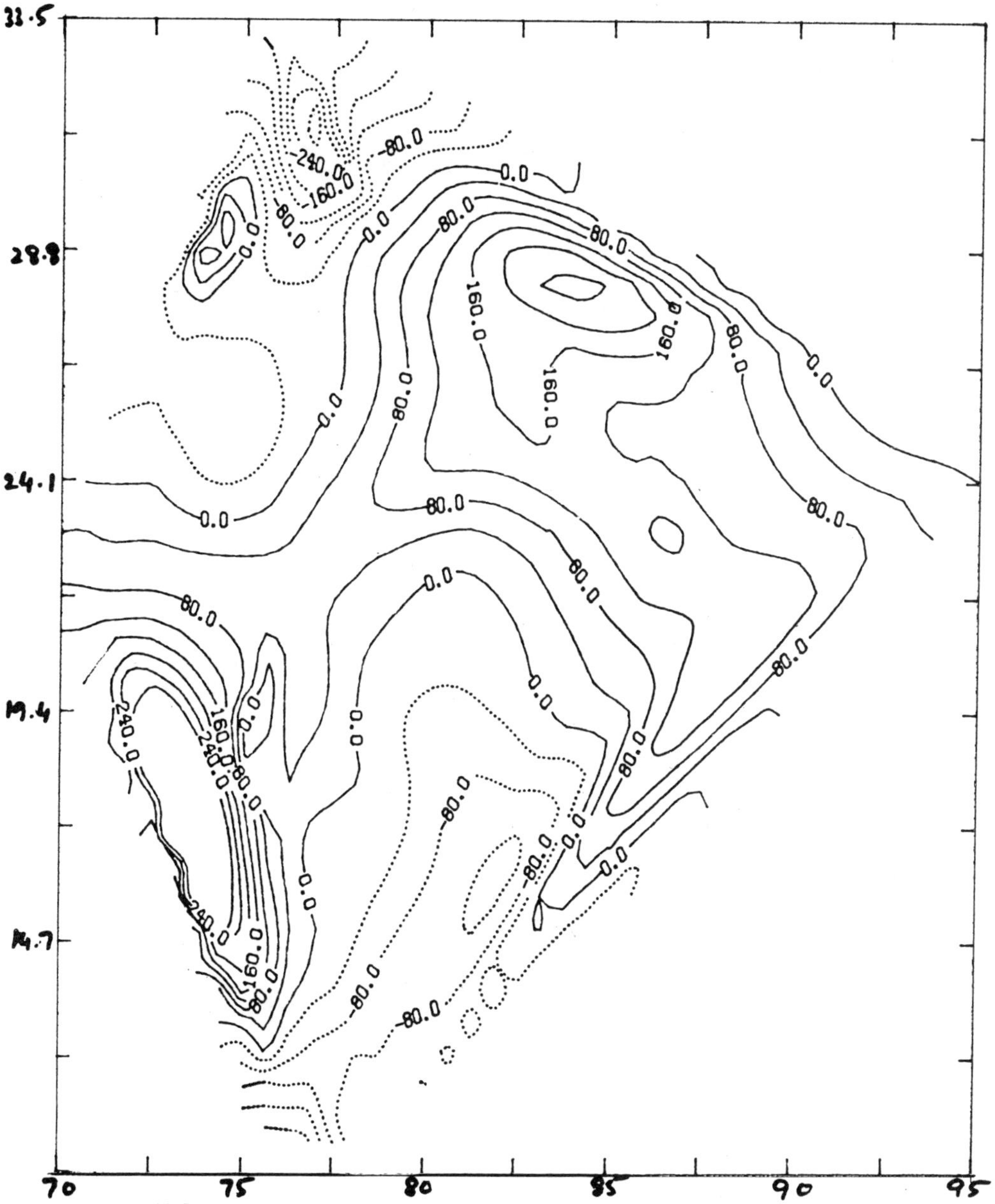

CØNTØUR FRØM -320 TØ 280 CØNTØUR INTERVAL ØF 40
X INTERVAL= 2.5 Y INTERVAL= 2.35

Fig.17

Fig. 18

I

UV-B RADIATION EFFECTS

STRATOSPHERIC OZONE DEPLETION AS IT AFFECTS LIFE ON EARTH – THE ROLE OF ULTRAVIOLET ACTION SPECTROSCOPY

Thomas P. Coohill
Department of Physics and Astronomy
Western Kentucky University
Bowling Green, Kentucky 42101, U.S.A.

ABSTRACT

It is widely agreed that a portion of the earth's protective stratospheric ozone layer is being depleted. The major effect of this ozone loss will be an increase in the amount of ultraviolet radiation (UV) reaching the biosphere. This increase will be completely contained within the UVB (290 nm - 320 nm). It is imperative that assessments be made of the effects of this additional UVB on living organisms. This requires a detailed knowledge of the UVB photobiology of these life forms. One analytical technique to aid in these approximations is the construction of UV action spectra for such important biological end-points as human skin cancer, cataracts, immune suppression; plant photosynthesis and crop yields; and aquatic organism responses to UVB, especially the phytoplankton. Combining these action spectra with the known solar spectrum (and estimates for various ozone depletion senarios) can give rise to a series of effectiveness spectra for these parameters. This manuscript gives a first approximation, rough estimate, for the effectiveness spectra for some of these bioresponses, and a series of crude temporary values for how a 10% ozone loss would affect the above end-points. These are not intended to masquerade as final answers, but, rather, to serve as beginning attempts for a process which should be continually refined. It is hoped that these estimates will be of some limited use to agencies, such as government and industry, that have to plan now for changes in human activities that might alter future atmospheric chemistry in a beneficial manner.

INTRODUCTION

By 1988, the scientific community had reached agreement that a portion of the earth's protective stratospheric ozone layer was being depleted, largely by man-made chemicals (1, 2, 3).This consensus was arrived at by the careful analysis of a substantial amount of data on the vertical ozone column, recorded by balloon, rocket, airplane, and other measurements. Earth- based wavelength measurements in the UVB (290-320 nm), largely conducted with "Dobson" meters, confirmed the former experiments (4, 5, 6). But the amount of stratospheric ozone loss above temperate areas, especially in the northern hemisphere, was not enough (less than 10%) to alarm the general public, even though (see below) it may have serious consequences for life on earth. It was the advent of the antarctic "ozone hole" that convinced the public, governmental agencies, and even skeptical scientists that the ozone layer was indeed being reduced, by as much as 50% in this case, and that this loss could be attributed, in large part, to human activities (3). The ozone hole (7), which appears only in the austral spring, was further observed to break up in the summer months and spin off depleted areas that caused severe temporary ozone losses above some major populated areas in the southern hemisphere (e.g. the reported 35% depletion above the city of Melbourne).

Although the list of chemical pollutants responsible for the loss of stratospheric ozone is long, one major group stands out, the chlorofluorocarbons (CFCs) invented in the 1930's (8). These CFCs are widely used in a variety of manufactured goods because of their stability, inertness, and long life. Thus many coolants, electronic solvents, foams, etc. contain CFCs which are slowly released into the atmosphere where they can exist for up to one hundred years. About thirty years after release these CFCs perculate into the stratosphere where, via a reaction involving short wavelength (λ) ultraviolet radiation (UV), they produce free chlorine which destroys ozone. Each CFC molecule acts in a catalytic fashion to destroy about 100,000 ozone molecules.

A rapid response to this world wide threat, considering the geopolitical situation produced the first Montreal Protocol in 1985, which addressed the problem directly by proposing limits on CFC production. These restraints have been made more stringent as additional data accumulates (9). It is hoped that this treaty will turn off the source of CFCs (and other offending gases) by the early twenty-first century.

The use of transition chemicals, eg. HCFCs, will allow us time to evolve to a series of chlorine free substitutes for CFCs that are both safe and energy efficient. However, even if all CFC usage were to stop by the turn of the millennium, the release of the CFCs already in the world, and their long lifetimes, means that the ozone layer will continue to be adversely affected until at least the mid-twenty-first century.

Ultraviolet Radiation and Living Forms

The immediate question that the stratospheric O_3 problem poses is "what will this mean for the world?" The most important direct consequence of this loss of ozone is an increase in the amount of a certain portion of the UV spectrum reaching the surface of the earth. This increase will be largely confined to the wavelength region 295-315 nm since shorter wavelengths are absorbed by other atmospheric components, while longer ones penetrate the current atmospheric column. This narrow region is a large fraction of the so-called UVB. Thus the effects of O_3 depletion can be assessed by estimating or measuring the additional consequences of increased UVB exposure. These consequences are largely biological. Hence, a broad understanding of the UVB photobiology of living organisms is essential, if we are to estimate the worldwide effect of O_3 depletion. What follows is a general attempt at such estimations and a description of the biological assay most applicable for addressing the extent of this

Abbreviations used in this manuscript:

UV- Ultraviolet radiation (190-380 nm).	AS - action spectrum (a)	λ - wavelength (s)
(Does not include the vacuum UV).	AAS - analytical AS	CFC - chlorofluorocarbon
UVA - UV of λ 320-380 nm	PAS - polychromatic AS	
UVB - UV of λ 290-320 nm	ES - effectiveness spectrum (a)	
UVC - UV of λ 190-290 nm		

problem. Future estimates will follow as the amount of data collected on these systems increases. A goal is to provide the scientific, governmental, and public communities with reasonable, scientifically based values that can be used for determining appropriate responses to this worldwide problem.

The vast majority of biological organisms evolved after the initial formation of the stratospheric O_3 layer. This layer provided for living forms an umbrella of protection from the deleterious effects of UV by absorbing heavily those UV wavelengths below 320 nm. It is probably not a coincidence that the single most important molecule in living cells, i.e. the genetic material DNA, has an absorption spectrum that peaks at 260 nm (well below 320 nm) and drops by three orders of magnitude at 320 nm. In addition, the molecules plastoquinone and plastoquinol, both important in photosynthesis, also absorb strongly at λ below 310 nm (figure 1). The percentage of UVB in the solar output reaching the top of the earth's atmosphere is less than 1.5% (20 Wm^{-2}); the amount reaching the earth's surface is less than 0.3% (2Wm^{-2}) due to the filtering effect of atmospheric chemicals. Even so, there is enough ambient UVB reaching the biosphere to produce some damage to cellular DNA. UV effects on DNA are widely reported and include such photochemical changes as pyrimidine dimers, 6-4 photoproducts, DNA-protein cross links, and lesions that can lead to single and double strand breaks (10, 11, 12). If left unrepaired, these lesions may lead to impairment, mutation, or even cell death. For these reasons, some organisms avoid exposure to high levels of ambient UV, or are in the case of humans advised to do so to prevent serious damage. The biochemical and physiological consequences of UV exposure to some biological systems are well characterized and reasonably well understood (13, 67, 14) when compared with other insults, eg. ionizing radiation (15).

Any analysis of biological responses to UV is complicated by a large number of variables (16). In addition, animals, especially humans, can avoid sunlight if they choose; terrestrial plants cannot. Some plants have evolved in brilliant sunlight and would be expected to have developed defensive mechanisms to cope with high levels of UV. Shielding of component cells and tissues in higher organisms, or shielding of single cells by other cells in a system, is difficult to estimate. Plant cells are often highly pigmented and hence especially able to protect centrally located target molecules such as DNA (17) from the harmful effects of UV. Pigmentation in animal cells varies. The wide variety of biological responses measured in plants also limits the degree to which comparisons between laboratories may be made (18). Plants are exposed to many stresses in addition to UV radiation (19, 20, 21). These include, but are not limited to, available nutrients, water stress, atmospheric composition. Any attempt at providing a general picture of a typical plant's responses to these varied parameters is at best difficult and at worst impossible. Most animal studies are confined to laboratory studies, often single cells or groups of cells in culture, a highly artificial situation (22). This paper attempts to provide a limited first approximation of the type of general data that may be useful in this regard. It follows from the plant work of Caldwell (18) and Tevini (23) and includes important results from their work and that of others (24, 25, 16, 26, 27, 28). Results reported for lower plants have already been compiled (29, 30). Data from animal, mostly mammalian cells in culture, are from a variety of sources, partially summarized by Coohill (22) and Coohill et al. (31).

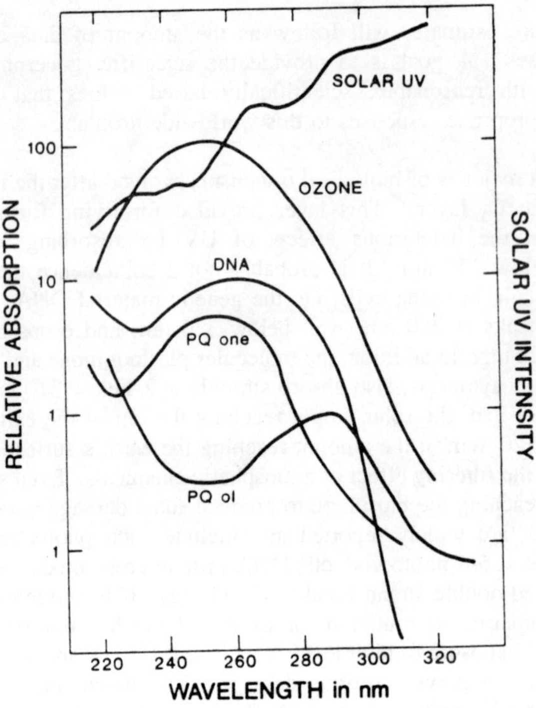

Figure 1. The umbrella-like protection from solar UV afforded to several important biomolecules by the absorption proporties of ozone. Solar UV at the top of the atmsophere. Absorption spectra for Ozone, DNA, Plastoquinone, and Plastoquinol.

Action Spectroscopy

Among the methods of obtaining and analyzing photobiological data, the technique known as Action Spectroscopy plays a central role in the initial characterization of bioresponses. Action spectroscopy is most simply defined as the measurement of a biological effect as a function of wavelength. Crude action spectra (AS) the term "action spectrum" was not coined until the nineteen forties (32) were first used in the nineteenth century to help identify chlorophyll as the chromophore most responsible for the growth of plants (33, 34, 35). In this century more sophisticated methods refined the analysis of AS such that it is now possible, in some instances, to make a reasonable determination of the molecule likely to contain the chromophore(s) responsible for the response being studied. It is this latter usage that has allowed such important identifications as the determination of DNA as the genetic material (36, 37). This was possible because of the experimental availability of small unicellular organisms (such as bacteria and fungi) for photobiological studies. Such systems meet a variety of rather stringent criteria (see below) for reliable analysis of what I will term an analytical action spectrum. However, it is still possible to glean some information about the photoresponses of organisms that do not meet these criteria, although it may be impossible to identify either the molecule(s) or the mechanism(s) involved in the process being studied. It may not be possible to circumvent these limitations if a complex process, such as plant "growth", is to be analyzed. Action spectra for this latter process may be nothing more than a determination of effect as a function of wavelength and remain uninterpretable as to the mechanism causing the effect. But even here, such action spectra can be useful in estimating response to ambient or changing light exposures.

Analytical Action Spectra (AAS)

If it is desirable and possible to do so, a carefully constructed AS can identify the absorbing chromophore. This can occur if the AS corresponds closely to the absorption spectrum of a molecule that can be shown to be affected by exposure to radiation in the λ region tested, and that, within reason, can be thought of as being involved in the metabolic process being studied. Whether this chromophore is directly responsible for the effect, or whether it has an energy level configuration that matches the energy of the incident photons which excite or alter the molecule and then, in turn, transfers the absorbed energy to a molecule that directly causes the effect, can sometimes be determined (13, 22, 31).

The following five fundamental conditions are necessary for the construction of an analytical action spectrum (for a more complete description see Jagger [13]). First, at no λ should more than half of the incident radiation be absorbed by the sample before every participating chromophore is exposed. Ideally, each chromophore should be exposed to the same number of photons in order to have the same probability of responding. This is never the case with biological samples. The higher the transmission of the sample, the more accurate the AAS can be. Thus, Gates (36) was able to obtain a relatively rigorous AAS, because he used the bacteria E. coli and S. aureus which have cellular diameters of less than 2 . He measured the absorption of a monolayer of these cells at 260 nm to be about 25% (36), saying in addition that a thin layer of such cells was "all but colorless in visible light, and so transparent that objects may be seen through it clearly and without distortion". Jagger (13) estimated the amount of absorption to the center of an E. coli cell, at this same λ, to be equal to or less than 15% (depending in the growth stage of the cell). Various measurements and calculations for mammalian cells show that more than half of the radiation at 260 nm will be absorbed before it reaches the center of a spherical cell (radius 10), while less than a third will be absorbed before it reaches the center of the nucleus of a flattened umbonate cell (38, 39). Even still, reasonable AAS for such important mammalian cell functions as survival, mutation, and viral induction have been reported (40, 41, 42, 43, 44, 45, 46). Significant absorption of UV radiation by multicellular organisms, such as plants, usually precludes an AAS, but useful AS have been constructed, even utilizing higher plants (24, 25, 16, 26, 47, 28).

The second criterion for an AAR is that scattering and absorption of radiation in front of the target chromophore should be either negligible, and thus satisfy the criterion of high transmission, or amenable to a "correction factor" for such shielding. For example, if the absorption and scattering properties of the membranes and cytoplasm, and their thicknesses between the incident beam and a centrally located target chromophore, such as nuclear DNA, are known, then it may be possible to subtract these effects and predict the intensity and wavelength distribution of the radiation reaching the target molecules (17, 39). Even better, if a target, such as adjacent pyrimidine molecules in DNA, is known to cause an effect, and if the consequent photoproduct (the pyrimidine dimer) can be measured, then an AS for dimer formation can be compared to an AS for cell killing. If they are similar, any observed shift in the peak of this spectrum compared to the absorption spectrum of isolated DNA (22) can be accounted for if knowledge of the target chromophore or of the absorption and scattering events before the radiation impinges on the chromophore is not detailed, then any correction factor is suspect. For those cells that can change their size (17) or shape (39) it is sometimes possible to alter the exposure to the cell surface in such a manner as to expose the center of the cell to the same fluence (17, 39). Again, such experimental modifications should be avoided if careful measurements of cellular dimensions are not available.

A third criterion demands that (for a given chromophore) the absorption spectrum in vitro, be identical to its absorption spectrum in vivo, in order to compare it to an AAS. In addition, the quantum yield, which is the probably of a photochemical change in a chromophore that has absorbed a photon, should be the same for all the λ tested in the AAR. Otherwise the effect can vary with λ even without a change in the absorption of radiation by the chromophore. Although this should be tested, it rarely is, but it is often a safe assumption for biomolecules (13).

The fourth criterion for an accurate AAS involves the shape of the fluence effect curves ("survival" curves if cell killing is being measured). It should be possible to multiply each separate λ survival curve by a suitable fluence modification factor so that the curves are reasonably superimposable upon one another. This is an easily tested and crucial factor in determining whether one is looking at the same mechanism of action throughout the λ region. This can often be complicated by the extent of the non responsive (shoulder) portion of the fluence effect curves, which in turn, may well be affected by extraneous factors, such as the amount of photochemical repair being accomplished during the assay period (13, 16).

The fifth, and often the easiest, criterion to test is whether the effect is the same regardless of the rate at which the exposure is given. In other words, does the efficiency of the process change, if a given amount of radiation is delivered to a sample in a short time at one λ, but due to experimental constraints, over a longer time period at a different λ? It is therefore imperative that each λ used be tested in fluence rate over as much of the range of fluence rates used at other λ as is possible. It is usually impossible to shorten each fluence rate to that of the briefest exposure, since sources do not usually emit at the same rate for each λ, so one normally extends the most effective λ to longer times. Often this reciprocity has to be tested over a fluence rate range of at least a factor of three.

Even if all of the above constraints are met, and they never are, there are experimental variables that can limit the reliability of any AS. These include, but are not limited to: the spectral purity of the radiation source; the accuracy of the dosimetry measurements; the placement of the dosimeter (ideally-at the sample); the presence of exogenous and/or endogenous "non-participating" chromophores; the ambient (even micro-environmental) conditions; the time in the life cycle (cellular) or growth cycle (developmental) of the exposed organism; and the physical state of the target molecule (eg., DNA extended or coiled in chromotion). In addition, cells often harbor sophisticated and efficient methods for repairing photobiological damage (13). Again, the extent of this repair during irradiation and before assay can significantly effect the measured response. All of these variables can given rise to errors in the interpretation of an AAS. Hence, "a particular action spectrum is a very specialized thing and may apply only under the conditions actually used to obtain it" (13).

Even given the above constraints, AAS have been produced that have pointed to the answer for crucial biological questions. Perhaps the most famous AAS is that of Gates (36) who showed (figure 2), for the first time, that nucleic acid (presumably DNA) was the target molecule for the death of bacterial cells (both E. coli and S. aureus). This was the first clear evidence that the genetic material was DNA, not, as was then widely believed, protein. Almost a decade later, even more tenable results for identification of the genetic material were the AAS that involved cellular mutation by UV. Knapp, Reuss, Risse, and Schreiber (48) using the spores of Sphaerocarpus Donnelli; Stader and Uber (49) using the pollen of maize; and Hollaender and Emmons (37) using the fungus Trichophyton metagrophytes, all showed that UV in the region about 265 nm was the most effective λ for cellular mutation. Hollaender and Emmons (37) went on to compare these results with those for cell killing in these organisms and concluded that "the 265 nm maximum coincides with the high absorption coefficient of nucleic acids near this wavelength". All of these studies and those of Gates (36) preceded the biochemical work of Avery (50) which is usually regarded as the first "clear" evidence that DNA was the genetic material.

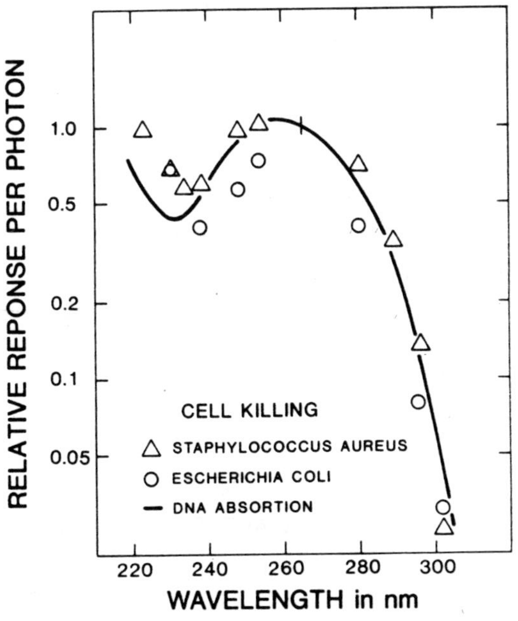

Figure 2. Analytical action spectrum for bacterial cell killing by UV radiation (Gates, 1930 [36]). This was the first clear evidence that DNA was the genetic material. The solid time is the absorption spectrum for DNA. The points are the quantum corrected efficiences for cell killing at the plotted wavelength.

Although AAS utilizing small cells flourished, it was difficult to extend these studies to larger (e.g. mammalian) cells because of the substantial absorption of UV by large cells and tissues. Early attempts by Mayer and Schreiber (51) failed to produce AS (figure 3) with the fine structure of those of Gates (36), because they employed hanging-drop mammalian tissue samples that were essentially opaque to radiation near the peak of DNA absorption (260 nm). But, with the advent of single cell mammalian culture techniques, and with the unique flattened geometry that mammalian cells assume when in monolayer culture (39) these studies began to appear in the later 1960's. The first AAS for mammalian cell killing (40) reported data similar to that of Gates (36) with bacteria, but with a peak effect shifted slightly toward longer λ (peak 270 nm). This gave rise to the belief that both nucleic acids (peak absorption 260 nm) and proteins (peak absorption 280 nm) might contribute to reproductive death in these cells. However, once measurements of the AS for pyrimidine dimer production were available (41, 44, 51, 52.), it could be shown that DNA alone was responsible for mammalian cell lethality (22).

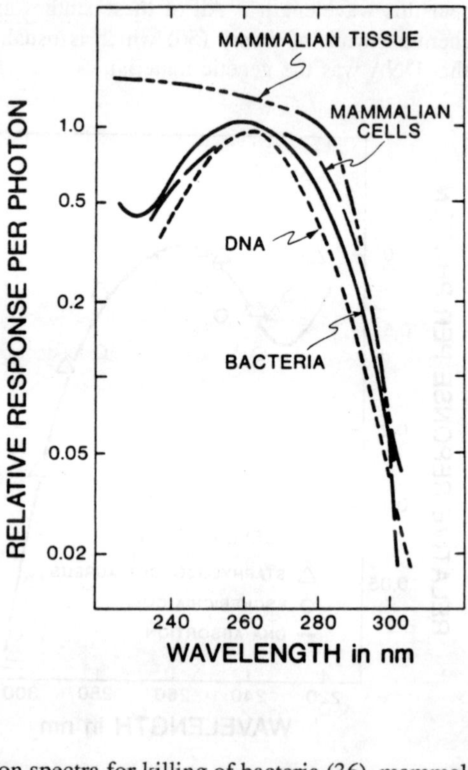

Figure 3. Action spectra for killing of bacteria (36), mammalian cells (22), and mammalian tissue (51). Also represented is the absorption spectrum for DNA. Note that the fine structure inherent in the AS for bacteria and mammalian single cells is absent in the case of mammalian tissue. This is because the latter is essentially opaque to UV (see text).

UV-A (320 - 380 nm) Action Spectra

It was widely thought that experimental work in the UVC and UVB (ie. λ range 190 -320 nm) could be extrapolated to predict photobiologial responses in the UVA (320 - 380 nm). This is not the case and UVA AS are much more complex than what had previously been predicted (53, 14, 31). Here even events such as cell mutation, which surely involves the genetic material, can be affected at fluences far below those necessary to affect DNA in these λ regions. It is thought that an intermediate molecule is involved that absorbs the incident UVA photon and transfers the effect to DNA (31). Hence, studies in the environmentally relevant UV wavelength region, ie those UV wavelengths that reach the earth's surface (290 - 380 nm), are being increasingly reported, and AS in this region do not lend themselves to ready interpretation (54, 55, 56, 57).

Polychromatic Action Spectra (PAS)

In addition, many authors, perhaps realizing the futility of attempting to construct AAS using multi-cellular organisms, have reported AS that employ polychromatic sources. These polychromatic AS (PAS) add to the complexity already inherent in the literature involving λ studies and bio-effects. These studies vary from irradiating the affected system with single UV wavelengths only (and then growing them under normal illumination, i.e. a "standard" AS), adding single wavelengths to a "white" light background, or generating a set of data using polychromatic sources that employ cutoff filters at successively shorter wavelengths. The former monochromatic system is accurate, and the preferred way for determining the target chromophore and the classical method for generating a true action spectrum, i.e. a plot of biological effect as a function of single-wavelength irradiations. However, this system is highly artificial and greatly removed from the natural setting. The polychromatic system is complex and tends to obscure individual chromophores, but it is closest to natural field conditions. A major advantage of using polychromatic radiation in the development of PAS is that interactions of biological responses to different wavelengths (usually unknown) can be empirically incorporated into the composite spectrum. For example, the involvement of photorepair systems or other cellular responses to longer wavelength radiation that might mitigate the damaging effect of shorter-wavelength radiation can be assessed without knowing about the nature or spectra for these repair and/or mitigating responses (13, 18, 58). While a PAS is useful because of its closer relation to the natural setting, it should be remembered that the monochromatic action spectra widely used for erythema (59) and for DNA (60) have proven useful in making general statements about biological responses to UV. The same would be true for more complex systems, such as plants.

Stratospheric Ozone Depletion and Action Spectroscopy

So the field of action spectroscopy is quite complex. However, once again, it is an important first analysis in determining bioresponses to the natural environment. For example, the known depletion of stratospheric ozone and the consequent increase in the amount of UVB (290 - 320 nm) radiation reaching the biosphere (4, 61), is of major concern to the worldwide community. Government agencies, the popular press, and others, are keenly aware of this environmental problem, and, as is correct, have turned to the photobiological community with the question "how will this affect life on earth?" One obvious method of providing at least limited estimates for these effects is to obtain AS for a given effect, and combine these with the known (or estimated) ambient solar radiation expected due to various ozone depletion scenarios. Such an effectiveness spectrum (ES, 62) can be used to give first approximation preliminary estimates for the effects of this increased UVB on the biosphere.

For example, one could combine the AS for cell erythema with both the known and an estimated depleted ozone layer, solar spectrum reaching the earth's surface to produce two ES and measure the increase in effect due to the latter (figure 4). Other well established AS could be used to produce similar ES that would allow reasonable estimates to be stated for the biological consequences of a decreased ozone layer. Preliminary attempts for some biological parameters, such as skin cancer, have already been made (63, 64, 65, 66, 67). For example, Henriksen, Dahlback, Larsen and Moan (65) predicted a 2% increase in overall skin cancer for each 1% depletion of stratospheric ozone. Other estimates for direct effects on human health are needed, e.g. cortical cataracts (68); some are being proposed, e.g. immune suppression (69).

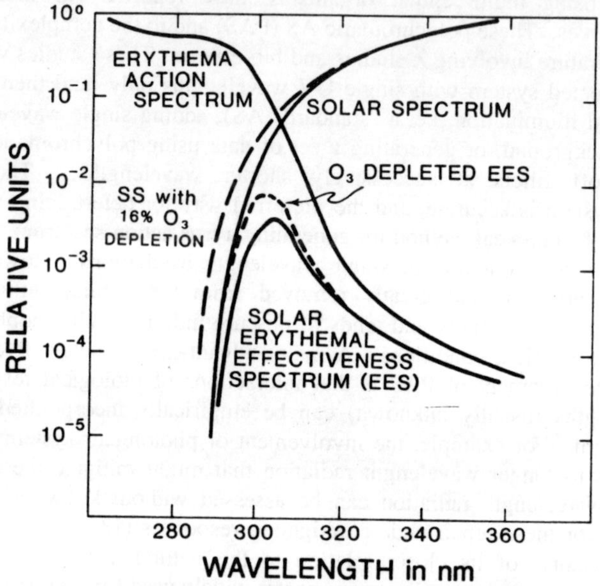

Figure 4. Erythema action spectrum (59) and solar effectiveness spectrum (14) for human skin. Added to the figure is the solar spectrum for a 16% O_3 depletion and the corresponding effectiveness spectrum for this additional UV radiation exposure.

But perhaps the most significant detrimental biological effects, even for human populations, will be those due to increased exposure to UVB of both terrestrial and aquatic organisms, especially plants. Data for UV effects on terrestrial higher plants is being collected, mainly for important crops, and, in some instances, forest species (70). Initial crop data show that several cultivars of some major crop species are sensitive to UVB. These include soybeans, beans, wheat, peas, rice, potato, squash, and cassava (71, 23, 72, 73, 74). Some studies show other cultivars of rice and wheat to be insensitive to UV (71). Other crops appear to be UV resistant, e.g. corn, peanuts, and cabbage (71). Estimates of the extent of this sensitivity are difficult to make even for such vague assays as "yield". But the results cited above seem to show that decreases in yield quantity and quality may well occur under enhanced UV exposure. Attempts to construct as ES from a generalized plant damage AS (18) or from a variety of end-points weighted for photosynthesis (75) allow a preliminary estimate of a 1% decrease for each 1% decrease in the ozone layer (figure 5). However, it must be remembered that no one AS represents the general responses for higher plants. If crop yields diminish, certain human populations already near starvation may be dramatically effected (76).

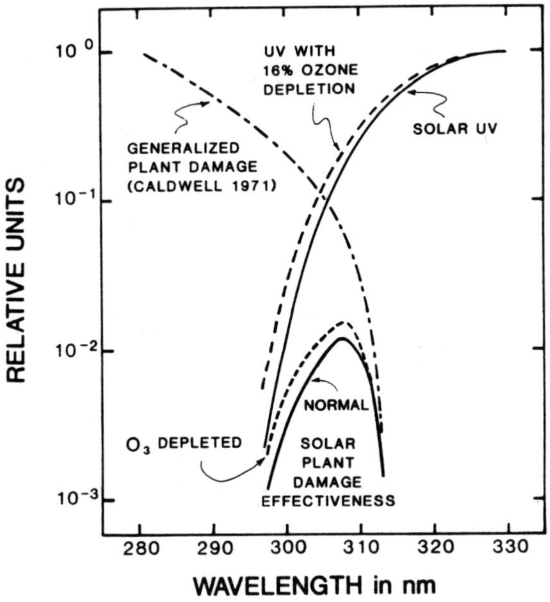

Figure 5. This figure (75) shows the increase in solar plant damage due to a 16% decrease in ozone. The area under the O_3 depleted effectiveness spectrum (dashed line) is 17% greater than that for a normal solar effectiveness spectrum. Data such as this, approximately translates into a 1% increase in plant damage for each 1% decrease in O_3.

All of the above ES show that the λ region 295 - 315 nm (ie., contained within the UVB) is the only region of the UV spectrum that will shown an increase in biological effect due to ozone depletion.

Some scientists think that the first noticeable effect of a depleted ozone layer might be a reduction in the phytoplankton community in the world's oceans (77, 78, 79, 80, 81). Laboratory results show that some phytoplankton already appear to be at, or near, their UV stress limit. Small increases in UV-B may dramatically lower their reproductive rate. In oceanic populations the situation is more complex since photoplankton may move to lower positions in the water column to avoid additional UV. Such movement would also lower their photosynthetic efficiency. In addition, different species may predominate in the new solar UV surroundings. Field studies, though difficult, are being attempted (82). Estimates for the effects of additional UV on total phytoplankton mass vary widely (83) from dramatic (ie., a 4% decrease for each 1% decrease in ozone (78, 79), to negligible (no decrease, (82), and even in some cases to an increase (83). At present, a crude estimate would be that the total phytoplankton population will decrease 1% for each 1% loss of ozone (84). Since plankton account for more than half of the earth's biomass and, hence, fix more than half of the atmospheric CO_2, any reduction in their population will have drastic effects on the aquatic food chain and could contribute significantly to the greenhouse effect.

Conclusions

Thus, accurate information is needed to assess the potential effects on living organisms caused by an increase in the ultraviolet radiation reaching the earth's surface due to the depletion of stratospheric ozone. Of immediate use would be a series of action spectra for general biological effect(s) that would typify responses to increased UVB. At minimum, this would require that data be collected at a variety of wavelengths that include, at least, the full UVB Range. Further extension of the system to longer wavelengths would be preferable, in part to estimate other cellular responses, including repair (85, 58, 56). Such a spectrum would provide biologists with a common starting point, similar to the erythemal action spectrum widely quoted as a typical response to UV for human skin (59). Because of the absorption properties of ozone and the solar UV spectrum, increases in the UV flux at the earth's surface would be concentrated in the UVB (61, 86). How these small increases in incident energy would affect living cells and ecosystems (87) is still at issue, but detrimental effects have been postulated (88, 23) and, in some cases, for example plants, reported (72, 19, 73, 88, 89, 90, 91).

So, once again, accurate AS are needed to answer key photobiological questions central to our understanding of the ecological consequences of human behavior. More AS have to be generated, especially for plants. Studies utilizing plant cells in culture might be able to identify the chromophores involved in plant responses to UV. Field studies utilizing intact plants and polychromatic sources added to the ambient background will be essential for realistic estimates of effect. From the above it is apparent that no single action spectrum or solar effectiveness spectrum can currently be considered to be general enough to predict the varied responses of all biological systems to wavelengths throughout the environmentally relevant UV range 290 to 380 nm. However, reasonable estimates can be made for plant damage by UVB and a portion of UVA (75). The usefulness of these spectra are limited and they will be replaced in the future by more accurate and common spectra as they are generated. The availability and use of numerous cell mutants has made a large contribution to the understanding of the responses of bacteria, fungi, and mammalian cells to UV. More plant mutant data, such as those reported by Galland (92), are also needed. If common assays and irradiation procedures are followed, then a generalized plant damage action spectrum may be forthcoming and will add to the body of animal cell data already generated. In addition, a representative AS for aquatic organisms, especially the phytoplankton, would be of immediate interest (92). If data are clearly listed in tabular form, comparisons will be facilitated. Of course, as is also the case for the widely used action spectrum for human skin erythema, no common spectrum will suffice to substitute for any specific spectrum for any given response. But the photobiological community will be asked to provide "a realistic action spectrum, developed under more natural conditions" (73), to allow government agencies (and others) to assume correct positions. Therefore, common attempts, though limited, may have widespread implications. Accordingly, table 1 is a first approximation rough estimate for the effects of a 10% depletion of stratospheric on some of the bioresponses mention above.

Table 1
Ozone Depletion and Life on Earth

Rough estimates of the effects on certain bioresponses from the additional UVB (290 - 320nm) reaching the biosphere due to a

10% depletion* of stratospheric ozone.

25%	increase in human non-melanoma skin cancers
15%	increase in human cortical cataracts
10%	increase in human immune system suppression+
10%	decrease in certain UV sensitive crop yields
10%	decrease in oceanic phytoplankton

* A value that could occur at high latitudes over inhabited areas by the mid XXI century.
\+ largely from data utilizing a murine model
estimates vary widely for these crucial organisms (see text)

REFERENCES

1. Frederick, J. (1990) Trends in atmospheric ozone and ultraviolet radiation: mechanisms and observations for the northern hemisphere. Photochem. Photobiol. **51**, 757-763.

2. Caldwell, M. M., S. Madronich, L. O. Björn and M. Ilyas (1989) Ozone reduction and increased solar ultraviolet radiation. In "Environmental effects panel report", United Nations Environmental Program. ISBN 92 807 1245 4.

3. Stolarski, R. S. (1988) The antarctic ozone hole. Sci. Amer. **258**, 30-36.

4. Lubin, D., J. E. Frederick, C. R. Both, T. Lucas and D. Nueschuler (1989) Measurements of enhanced springtime ultraviolet radiation from Palmer Station Antarctica. Geophys. Res. Lett. **16**, 783-785.

5. Federick, J. E., H. E. Snell and E. K. Haywood (1989) Solar ultraviolet radiation at the earth's surface. Photochem. Photobiol. **50**, 443-450.

6. World Meteorological Organization (1989) Scientific assessment of stratospheric ozone: 1989, Global Ozone Research and Monitoring Project, Geneva. In press.

7. Farman, J. C., B. G. Gardiner and J. D. Shanklin (1985) Large losses of total ozone in Anatarctica reveal seasonal ClO_x/NO_x interaction. Nature. **315**, 207-210.

8. Rowland, F. S. (1989). Chlorofluorocarbons and the depletionof stratospheric ozone. Amer. Sci. **77**, 36-45.

9. Montreal Protocol (1987), Substances that deplete the ozone layer: Final Act UNEP Na. 87-6106.

10. Peak, M. J. and J. G. Peak (1986b) DNA-to-protein crosslinks and backbone breaks caused by far- and near-ultraviolet, and visible radiations in mammalian cells. In Mechanisms of DNA Damage and Repair. Implications for Carcinogenesis and Risk Assessment (Edited by M. G. Simic, L. Grossman and A. C. Upton), pp. 193-202, Plenum Press, New York.

11. Mitchell, D. L. and R. S. Nairn (1989) The biology of the (6-4) photoproduct. Photochem. Photobiol. **49**, 805-819.

12. Ananthaswamy, H. N. and W. E. Pierceall (1990) Molecular mechanisms of ultraviolet radiation carcinogenesis. Photochem. Photobiol. **52**, 1119-1136.

13. Jagger, J. (1967) Introduction to Research in Ultraviolet Photobiology. Prentice-Hall, Englewood Cliffs, New Jersey.

14. Jagger, J. (1985) Solar-UV actions on living cells. Praeger, NY.

15. Elkind, M. M. and Whitmore, G. F., 1967, "Radiobiology of Cultured Mammalian Cells," Gordon and Breach, New York.

16. Caldwell, M. M., L.B. Camp, C. W. Warner and S. D. Flint (1986) Action spectra and their key role in assessing biological consequences of solar UV-B radiation. In R. C. Worrest and M. M. Caldwell (eds.) Stratospheric Ozone Reduction, Solar Ultraviolet Radiation and Plant Life. 87-111.

17. Coohill, T. P. and R. A. Deering (1969). Ultraviolet light inactivation and photoreactivation of Blastocladiella emersonii. Radiat. Res. **39**, 374-385.

18. Caldwell, M. M. (1971) Solar UV irradiation and the growth and development of higher plants. In Photophysiology Vol. 6, (Edited by A. C. Giese), pp. 131-177.

19. Teramura, A. H. (1983) Effects of ultraviolet-B radiationonthe growth and yield of crop plants. Physiol. Plant. **58**, 415-427.

20. Teramura, A.H. (1987) Current risks and uncertainties of stratospheric ozone depletion upon plants. In-Assessing the risks of trace gases that can modify the stratosphere. Vol. 8: Technical support documentation ozone depletion and plants. U.S.E.P.A., EPA 400/1-87/001H, 1-77.

21. Tevini, M., U. Thoma and Iwanzik (1983) Effects of enhanced UV-B radiation on germination, seedling growth, leaf anatomy and pigments of some crop plants. Z. Pflanzenphysiol. **109**, 435-448.

22. Coohill, T. P. (1984) Action spectra for mammalian cells in vitro. In Topics in Photomedicine (Edited by K. C. Smith), pp. 1-37, Plenum Press, New York.

23. Tevini, M. (1988) The effects of UV radiation on plants. J. Photochem. Photobiol. **2**, 401-403.

24. Bogenrieder, A. (1982) Action spectra for the depression of photosynthesis by UV irradiation in Lactu stativa L. and Rumex alpinus L. In Biologic Effects of UV-B Radiation (Edited by H. Baur, M. M. Caldwell, M. Tevini and R. C. Worrest), pp. 132-139. Proceedings of a workshop held in Munich-Neuherberg.

25. Bornman, J. F., L. O. Björn and H. E. Akerlund (1984) Action spectrum for inhibition by ultraviolet radiation of photosystem II activity in spinach thylakoids. Photobiochem. Photobiophys. **8**, 305-313.

26. Jones, L. W. and B. Kok (1966) Photoinhibition of chloroplast reactions. I. Kinetics and action spectra. Plant Physiol. **41**, 1037-1043.

27. Steinmüller, D. (1986) On the effect of ultraviolet radiation (UV-B) on lead surface structure and on the mode of action of cuticular lipid biosynthesis in some crop plants. Karler. Beitr. Entrw. Ökophysiol. **6**, 1-174. M. Tevini -ed.

28. Tevini, M. and D. Steinmüller (1988) Personal communication

29. Häder, D. P. (1988). Ecological consequences of photomovement in microorganisms. J. Photochem. Photobiol. B: **Biol. 1**, 385-414.

30. Nultsch, W., J. Pfau and M. Materna-Weide (1987) Fluence and wavelength dependence of photoinhibition in the brown alga Dictyota dichotoma. Mar. Ecol. Prog. Ser. **41**, 93-97.

31. Coohill, T. P., M. J. Peak and J. G. Peak (1987) The effects of the ultraviolet wavelengths of radiation present in sunlight on human cells in vitro. Photochem. Photobiol. **46**, 1043-1050.

32. Kleczkowski, A., 1972, Action spectra and their interpretation, In "Research Progress in Organic Biological and Medicinal Chemistry" (U. Gallo and L. Santamaria, ed.), Vol. 3 Part II, pp. 48-70.

33. Daubeny, C. (1836) On the action of light upon plants and plants upon the atmosphere. Phil. Trans. Roy. Soc. London, 149-179.

34. Engelmann, T. W. (1882) Ueber Sauerstoffausscheidung von Pfllanzensellen im Microspectrum. Bot. Zeit. **40**, 419-426.

35. Draper, J. W. (1884) Note on decomposition of carbonic acid by the leaves of plants under the influence of yellow light. Phil. Mag. **25**, 159-173.

36. Gates, F. L. (1930) A study of the bactericidal action of ultraviolet light. III. The absorption of ultraviolet light by bacteria. J. Gen. Physiol. **14**, 31-42.

37. Hollaender, A. and C. W. Emmons (1941) Wavelength dependence of mutation production in the ultraviolet, with special emphasis on fungi. Cold Spring Harbor Symp. Quant. Biol. **9**, 179-186.

38. Hatfield, J. M. R., L. Schulze and D. Ernet (1970). Measurement of the ultraviolet absorption in specific parts of both living and fixed mammalian cells, using a specially designed microspectrophotometer. Exptl. Cell Res. **59**, 484-486.

39. Coohill, T. P., Knauer, D. J., and Fry, D. G., (1979), The effects of changes in cell geometry on the sensitivity to ultraviolet radiation of mammalian cellular capacity, Photochem. Photobiol. **30**, 565-572.

40. Todd, P., Coohill, T. P., and Mahoney, J. A. (1968), Responses of cultured chinese hamster cells to ultraviolet light of different wavelengths, Radiat. Res. **35**, 390-400.

41. Rothman, R. H. and R. B. Setlow (1979) An action spectrum for cell killing and pyrimidine dimer formation in Chinese hamster V-79 cells. Photochem. Photobiol. **29**, 57-62.

42. Jacobson, E. D., Krell, K., and Dempsey, M. J., 1980, The Wavelength dependence of ultraviolet light-induced cell killing and mutagenesis in L5178Y mouse lymphoma cells, Photochem. Photobiol. **33**, 257-260.

43. Kantor, G. J., Sutherland, J. C., and Setlow, R. B., 1980, Action spectra for killing non-dividing normal human and xeroderma pigmentosum cells, Photochem. Photobiol. **31**, 459-464.

44. Doniger, J., Jacobson, E. D., Krell, K., and DiPaolo, J. A., 1981, Ultraviolet light action spectra for neoplastic transformation and lethality of syrian hamster embryo cells correlate with spectrum for pyrimidine dimer formation in cellular DNA, Proc. Natl. Acad. Scie. U.S.A. **78(4)**, 2378-2382.

45. Sutherland, B. M., Delihas, N. C., and Sutherland, J. C., 1981, Action spectra for ultraviolet light-induced transformation of human cells to anchorage-independent growth, Cancer Res. **41**, 2211-2214.

46. Coohill, T. P., Moore, S. P. Knauer, D. J., Fry, D. G., Eichenbreener, T.J., and Bockstahler, L. E., (1982), Action spectrum for the in vitro induction of simian virus 40 by ultraviolet radiation, Mutat. Res. **95**, 95-103.

47. Negash, L. (1987) Wavelength-dependence of stomatal closure by ultraviolet radiation in attached leaves of Eragrostis tef: action spectra under backgrounds. Plant Physiol. Biochem. **25**, 753-760.

48. Knapp, E., Reuss, A., Risse, O. and H. Schreiber (1939). Quantitative analyse der mutationsauslösenden wirkung monochromatischen UV-lichtes Naturviss. **27**, 304.

49. Stadler, L. J. and F. M. Uber (1941) Genetic effects of ultraviolet radiation in maize. IV. Comparison of monochromatic radiation. Genetics. **27**, 84-118.

50. Avery, O. T., C. M. MacLeod and M. McCarty (1944) Studies on the chemical nature of the substance inducing transformation of pneumococcal types. J. Exp. Med. **79**, 137-158.

51. Mayer, E. and H. Schreiber (1934) Die wellanangabhangiegkeit der ultraviolettwurkung auf gewebekuturen (Reinkulturen), Protoplasma. **21**, 34-61.

52. Rosenstein, B. S. and D. L. Mitchell (1987) Action spectra for the induction of pyrimidine photoproducts and cyclobutane pyrimidine dimers in normal human skin fibroblasts. Photochem. Photobiol. **45**, 775-780.

53. Keyse, S. M., S. H. Moss and D. J. G. Davies (1983) Action spectra for inactivation of normal and Xeroderma pigmentosum human skin fibroblasts by ultraviolet radiations. Photochem. Photobiol. **37**, 307-312.

54. Rundel, R. D. (1983) Action spectra and estimation of biologically effective UV radiation. Physiol. Plant. **58**, 360-366.

55. Peak, J. J. and J. G. Peak (1983) Use of action spectra for identifying molecular targets and mechanisms of action of solar ultraviolet light. Physiol. Plant. **58,** 360-366.

56. Peak, M. J. and J. G. Peak (1986) Molecular Photobiology of UVA. In The Biological Effects of UVA Radiation. (Edited by F. Urbach and R. W. Gange), pp. 42-56. Plenum Press, New York.

57. Tyrrell, R. M. and M. Pidoux (1987) Action spectra for human skin cells: estimates of the relative cytotoxicity of the middle ultraviolet, near ultraviolet and violet regions of sunlight on epidermal keratinocytes. Cancer Res. **47**, 1825-1829.

58. Beggs, C. J., A. Stolzer-Jehle and E. Wellmann (1985) Isoflavonoid formation as an indicator of UV-stress in bean (Phaseolus vulgaris L.) leaves. The significance of photorepair in assessing potential damage by increased solar UV-B radiation. Plant Physiol. **79**, 630-634.

59. Parrish, J. A., R. R. Anderson, F. Urbach, and D. Pitts (1978) UV-A: Biological Effects of Ultraviolet Radiation with Emphasis on Human Responses to Longwave Ultraviolet. Plenum Press, NY

60. Setlow, R. B. (1974) The wavelengths in sunlight effective in producing skin cancer: A theoretical analysis. Proc. Nat. Accad. Sci. USA. **71**, 3363-3366.

61. Frederick, J. E. and D. Lubin (1988) Possible long-term changes in biologically active ultraviolet radiation reaching the ground. Photochem. Photobiol. **47**, 571-578.

62. Turro, N. J. and Lamola, A. A. (1977), Photochemistry. In "The Science of Photobiology" (K. C. Smith, ed.), pp. 87-112, Plenum Press, New York.

63. Dahlback, A., T. Henriksen, S. H. H. Larsen and K. Stamnes (1989) Biological UV-doses and the effect of an ozone layer depletion. Photochem. Phtotobiol. **49**, 621-625.

64. Dahlback, A., and J. Moan (1990) Annual exposures to carcinogenic radiation form the sun at different latitudes and amplification factors related to ozone depletion. The use of different geometrical representations of the skin surface receiving the ultraviolet radiation. Photochem. Photobiol. **52**, 1025-1028.

65. Henriksen, T., A. Dahlback, S. H. H. Larsen and J. Moan (1990) Ultraviolet radiation and skin cancer. Effect of an ozone layer depletion. Photochem. Photobiol. **51**, 579-582.

66. Kelfkens, G., F. R. DeGruÿl and Jan Van der Leun (1990) Ozone depletion and increase in annual carcinogenic ultraviolet dose. Photochem. Photobiol. **52**, 819-823.

67. Urbach, F. (1989) Potential effects of altered solar ultraviolet radiation on human skin cancer. Photochem. Photobiol. **50**, 507-513.

68. Taylor, H. R. (1989) The biological effects of UV-B on the eye. Photochem. Photobiol. **50**, 489-492.

69. DeFabo, E. C., F. P. Noonan and J. E. Frederick (1990) Biologically effective doses of sunlight for immune suppression at various latitudes and their relationship to changes in stratospheric ozone. Photochem. Photobiol. **52**, 811-817.

70. Sullivan, J. H. and A. H. Teramura (1988) Effects of ultraviolet B irradiation on seedling growth in the Pinaceae. Amer. J. Botony. **75**, 225-230.

71. Tevini, J., A. H. Teramura, G. Kulandaivelu, M M. Caldwell, and L. O. Björn (1989) Terrrestial plants. In "Environmental effects panel report", United Nations Environmental Program. ISBN. **92,** 807 1245 4.

72. Teramura, A. H. and N. S. Murali (1986) Intraspecific differences in growth and yield of soybeans exposed to ultraviolet-B radiation under greenhouse and field conditions. Env. Exp. Bot. **26**, 89-95.

73. Teramura, A. H. and N. S. Murali (1986) Intraspecific differences in growth and yield of soybeans exposed to ultraviolet-B radiation under greenhouse and field conditions. Env. Exp. Bot. **26**, 89-95.

74. Teramura, A. H. personal communication.

75. Coohill, T. P. (1989) Ultraviolet action spectra (280 to 380 nm) and solar effectiveness spectra for higher plants. Photochem. Photobiol. **50**, 451-457.

76. Worrest, R. C. personal communication.

77. Bidigare, R. R. (1989) Potential effects of UV-B radiation on marine organisms of the southern ocean: distributions of photoplankton and krill during austral spring. Photochem. Phtotobiol. **50**, 469-477.

78. El-Sayed, S. Z. (1988) Fragile life under the ozone hole. Nat. Hist. **97**, 72-83.

79. El-Sayed, S. Z., F. C. Stephens, R. R. Bidigore and M. E. Ondrusek (1989) Potential effects of solar ultraviolet radiation on antartic phytoplankton. In "Effects of solar ultraviolet radiation on biogeochemical dynamics in aquatic environments. WHOI-90-09, 141-142.

80. Häder, D.P. (1986) Signal perception and amplification in photomovement of prokaryotes. Biophys. Acta. **864**, 107-122.

81. Häder, D. P., R. C. Worrest, and H. D. Kumar (1987) Aquatic ecosystems. In "Environmental effects panel report", United Nations Environmental Program. ISBN. 92 807 1245 4.

82. Mitchell, B. G., M. Vernet and O. Holm-Hansen (1989) Ultraviolet radiation in antarctic waters: Particulate absorption and effects on photosynthesis. In "Effects of solar ultraviolet radiation on biogiochemical dynamics in aquatic environments", WHOI-90-09, 135-136.

83. Morel, F. M. M. and N. M. Price (1989) Indirect effects of UV radiation on phytoplanton. In "Effects of solar ultraviolet radiationon biogiochemical dynamics in aquatic environments", WHOI-90-09, 110-112.

84. Hader, D. P. personal communication.

85. Caldwell, M. M. (1984) Effects of UV radiationon plants in the transition region to blue light. In-Blue Light Effects in Biological Systems. (Ed. H. Senger) 20-28.

86. Worrest, R. C. (1986) The effect of solar UV-B radiation on aquatic systems: an overview. In: Titus, J. G. (ed.) Effects of changes in stratospheric ozone and global climate. Vol. 1. Overview, U.S. Environmental Protection Agency and United Nations Environmental Program. 175-191.

87. Caldwell, M. M. (1968) Solar ultraviolet radiation as an ecological factor for alpine plants. Ecol. Monogr. **38**, 243-268.

88. Caldwell, J. M., W. G. Gold, G. Harris and C. W. Ashurst (1983) A modulated lamp system for solar UV-B (280 - 320 nm) supplementation studies in the field. Photochem. Photobiol. **37**, 479-485.

89. Tevini, M., W. Iwanzik and U. Thoma (1981) Some effects of enhanced UV-B radiation on the growth and composition of plants. Planta. **153**, 388-394.

90. Tevini, M., U. Thoma and Iwanzik (1983) Effects of enhanced UV-B radiation on germination, seedling growth, leaf anatomy and pigments of some crop plants. Z. Pflanzenphysiol. **109**, 435-448.

91. Tevini, M. and A. H. Teramura (1989) UV-B effects on terrestial plants. Photochem. Photobiol. **50**, 479-487.

92. Galland, P. (1983) Action spectra of photogeotropic equilibrium in phycomyces wild type and three behavioral mutants. Photochem. Photobiol. **37**, 221-228.

SOLAR ULTRAVIOLET-B RADIATION MEASUREMENT AT THE EARTH'S SURFACE: TECHNIQUES AND TRENDS

Ann R. Webb, Department of Meteorology, University of Reading, England.

ABSTRACT

An increase in solar ultraviolet-B radiation reaching the earth's surface due to stratospheric ozone depletion has potential impacts on much of the biosphere : human health, natural ecosystems, both aquatic and terrestrial, and agriculture. To understand the implications of a change in ultraviolet-B climate the effects of ultraviolet-B radiation on different life forms must be understood, and analysed with respect to a knowledge of the present levels of ultraviolet-B radiation currently tolerated, and the expected changes in those levels. Both the mechanisms and end results of ultraviolet-B effects on plants are poorly understood, complicated by interaction with other stresses, differing plant responses, and a history of research based on various unrealistic radiation regimes. The strong wavelength dependency of both the solar spectrum and the plant action spectra make questionable the relevance of results obtained with artificial radiation sources when extrapolated to natural environmental radiation. The solar ultraviolet-B radiation has great natural variability and has been measured in a number of ways by both broadband techniques and more recently by instruments with high wavelength resolution, providing spectral data. There is an increasing demand for spectral data to facilitate the planning and analysis of ultraviolet-B impact experiments that are relevant to what may be experienced in the environment. Identifying radiation changes and their cause will also be helped by spectral measurements as ozone depletion would lead to greatest proportional increase at short wavelengths, but because of the great intensity change across the ultraviolet-B waveband this signal may be hidden in an all-encompassing broadband measurement. The current magnitudes of ultraviolet-B changes with season, and at different wavelengths, are illustrated with spectral data taken at Reading, England (51.5°N).

INTRODUCTION

The potential for damage due to a global change in ultraviolet-B radiation (UVB, 280-320nm) has received widespread general recognition and an international agreement to reduce anthropogenic gases which lead to increased levels of UVB. The Montreal Protocol calls for a drastic reduction in the use of chlorofluorcarbons (CFCs) which, although benign in the troposphere, release chlorine compounds in the stratosphere and destroy ozone very effectively through a series of catalytic chain reactions (1). Ozone in itself is not beneficial to lifeforms (2) : in the troposphere it damages plant life and adversely effects the respiratory system of mammals, being fatal in high doses. However, its remote presence in the stratosphere is essential to our well-being. Ozone is a strong absorber of the short wavelength ($\lambda < 280$ nm) solar radiation incident on the earth's atmosphere. All UVC (200-280nm) is absorbed by oxygen and ozone in the upper atmosphere, none of this highly energetic and damaging radiation reaching the biosphere. As wavelength increases above 280nm the absorption coefficient of ozone decreases and a small amount of UVB radiation reaches the earth's surface. The intensity of UVB at the surface depends on the extraterrestrial intensity and pathlength through the atmosphere (functions of latitude and time), and the ozone encountered along that pathlength. The former are dependent on astronomical factors (the earth's rotation about the sun and its own axis) the latter, ozone concentration, is due to a delicate balance of photo and chemical reactions which is being perturbed. A reduction in atmospheric ozone will increase ambient levels of UVB radiation reaching the biosphere and all forms of life will be exposed to higher intensities and shorter wavelengths of radiation than previously. The greatest proportional increase in irradiation will occur at the shortest wavelengths, those associated with the most energetic and biologically active photons, and most strongly absorbed by ozone. The reaction initiated by such photons is generally detrimental to systems which have not evolved to protect themselves from such energy fluxes.

To assess the effects of changing irradiation on a given system one needs to understand the mechanism by which the radiation is affecting the system, the action spectrum of the response and the dose-response relationship. Over a period of time repair processes and/or adaptive mechanisms must also be considered and all these effects related to the changes in incident radiation.

UVB EFFECTS ON PLANTS

In the case of UVB radiation effects on plants, knowledge in all the above areas is limited and variable. The overall effectiveness of UVB varies between species and also between cultivars of the same species. Effects on sensitive plants include reduced growth (measured by leaf area, plant height, etc.) reduced photosynthetic activity (directly or indirectly) and flowering (3). Some sensitive plants show an adaptive defence, accumulating UV-absorbing compounds in the outer tissues to screen the inner layers of tissue - a mechanism similar to the tanning of human skin. No common action spectrum has been found to adequately express the response of plants in general to different wavelengths of UV radiation (4) and thus the interaction between responses of different plants must also be considered. Will UVB reduced growth seriously reduce the competitiveness of a plant in either a natural ecosystem or an agricultural situation (crop vs. weed) and will the balance of natural vegetation or the viability of a crop thus change? What role does UVB stress play when a crop is multiply stressed, for example by water shortage or pests? Is it a dominant factor or merely contributory with other stresses being more important?

The lack of comprehensive knowledge of UVB effects, mechanisms and interactions with other environmental factors is due in part to the difficulty of performing realistic research in this field (3). The problem of realistic irradiation. Plants need light to grow, therefore there is no such thing as a simple dark control in plant research. Plants are also sensitive not only to the total radiative energy available but also to the spectral distribution of that energy, given wavebands controlling or having a greater efficiency for the different growth processes. Thus it is the ratio of different wavebands that must be carefully controlled as well as overall energy when designing a radiation experiment on plants. For example, under low levels of visible light UVB sensitivity of plants increases (3). The reasons for this are not fully understood but it may be analogous to damage of the human eye : At low visible light levels the eye is doing the job it evolved for and making maximum use of the available light, allowing maximum radiation to enter the eye, thereby unwittingly subjecting itself to damage by the unexpectedly high levels of UVB. A plant unsaturated by photosynthetically active radiation may suffer through a corresponding evolutionary expectation that other radiation levels are also low, and defence mechanisms against UVB do not need to be fully developed. On the other hand high levels of visible light, and also UVA/blue light mediate photorepair mechanisms (5) and thus reduce the effects of UVB damage.

Much of the UVB-plant information currently available has been obtained using UV radiation sources to enhance the short wavelength end of an existing (or control) source, be it the artificial light in growth chambers or solar radiation in the field or glasshouses. However, the spectral distribution of these artificial sources is not the same as that of the sun and their relative effectiveness has to be assessed by using a biological weighting function. The DNA-action spectrum (6) or the generalised plant action spectrum of Caldwell (7) is often used but neither of these is fully representative of the response of full plants of different species (4). Reducing high levels of ambient UVB in a realistic manner has been achieved by placing ozone containing cuvettes over growth chambers situated at low latitudes to give high (ambient) and low (ozone reduced) levels of UVB to corresponding plants (8). More such realistic research needs to be done to better assess the effects of UVB on plants in field conditions.

The importance of providing realistic irradiation to investigations of plant response to UVB, both in terms of energy and spectral distribution has been outlined above. What then is realistic for solar UVB radiation? What are current ambient levels of UVB to which each species is adapted in its natural environment? How does UVB change with latitude and season? If ozone depletion was equivalent to changing the UVB climate of 40°N to that currently experienced at 10°N what would this mean for spectral intensity distribution? Will such UVB climate changes really occur and will they be recognised if they do? The remainder of this paper will discuss the techniques for measuring solar UVB radiation and describe some of the natural variability in ambient UVB levels.

MEASUREMENT TECHNIQUES

Broadband Sensors

There are two basic categories of UVB measurements : broadband and spectral. The most widely available are broadband measurements which indicate the total irradiation across the UVB waveband and are often weighted by the sensors response function to give an 'equivalent' or 'effective' measure of the radiation. The sensor response is frequently matched, or approximated to, a biological response e.g. the erythema action spectrum as in the Robertson-Berger meter (9), and the radiation is then measured in terms of its overall effectiveness for producing erythema. This may be expressed as sunburn units or Wm^{-2} erythemally effective radiation. The principle is the same as that used for other wavebands e.g. a photopic measure of visible radiation rather than the absolute energy in the visible waveband. However, in the UVB, tailoring the response of the instrument by choice of detector and/or filters is both difficult and critical. Figure 1 shows the generalised action spectra for erythema (10), plants (7) and DNA (6), the response spectrum of the Robertson-Berger meter (9) and the shape of the solar UVB spectrum.

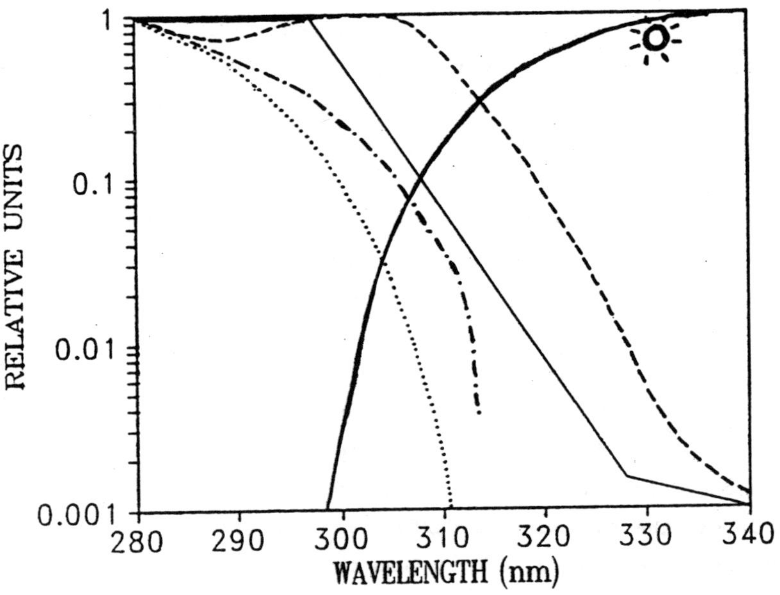

Figure 1: The generalised action spectra normalised at 280nm for erythema (——), plants (—•—) and DNA (•••••), the response spectrum of the Robertson-Berger meter (- - - -), and the shape of the solar spectrum (—☼—).

Because of the steep and opposing slopes of the solar spectrum and biological responses across the UVB waveband any mismatch of the instrument response and the biological effect it purports to represent could greatly reduce the useful information to be gained from the measurements. This is particularly so when the instrument response extends to longer wavelengths than the biological response. As the spectral intensity at for example 320 nm is 2 order of magnitude greater than that at 300 nm any detection of biologically ineffective radiation at the longer wavelengths can mask changes at short wavelengths which are very small in energy terms but extremely important for their biological activity. This problem is compounded further if broadband measurements are used to compare radiation sources with different spectral outputs e.g. in relating artificial lights to sunlight. Nontheless broadband instruments have been, and are, frequently used for UVB measurements for a number of practical and technological reasons.

Broadband radiometers are relatively cheap, small and easy to operate. They can withstand long term exposure to environmental conditions, and with an electrical output can be used for continous monitoring. Long term solar UVB waveband measurements have been made in this way in the United States, 1974-85 (11) and Switzerland, 1981-89 (12) but interpretation of the results is hampered by the spectral bandwidth of the instrument. The study in the U.S. used Robertson-Berger meters sited at airports and indicated a decrease in UVB over the study period. This does not correspond to ozone depletion theory (13) and it has been suggested that increased low level pollution has reduced overall radiation levels such that ozone induced changes at the shortest most susceptible wavelengths could be hidden by the stronger absolute signal at longer wavelengths. Alternatively, or included in the general pollution, there may have been an increase in tropospheric ozone which could offset a small decrease in the stratosphere. Tropospheric ozone is a more effective absorber of solar UVB (14) because at low levels the radiation is diffuse and its scattered pathlength through the ozone layer is longer than the pathlength of the direct beam radiation through the stratospheric ozone layer. The chance of the UVB photons encountering an ozone molecule and being absorbed are thus greater in the troposphere than the stratosphere for equivalent concentrations of ozone.

The measurements in Switzerland were made at an altitude of 3576 m, above the pollution level, and analysed with respect to total incoming solar radiation (which was assumed not to change). They show a small increase in UVB radiation over the 9 years of measurement.

Increase or decrease, extrapolating these results to effects upon the biosphere is difficult with no further spectral detail, for reasons already discussed. Techniques to by-pass this problem and measure a biological response directly have been used in particular cases e.g. the first stage of cutaneous vitamin D_3 synthesis (the photoconversion of 7-dehydrocholesterol to previtamin D_3 by radiation of wavelengths < 315 nm) has been studied in different climates using a solution of 7-dehydrocholesterol exposed to sunlight in a quartz test tube (15), but such methods are necessarily effect-specific and suffer the same problems of extrapolation to other effects as the broadband radiometers. Another alternative to radiometers is the personal dosimeter for UVB radiation e.g. polysulphone film badges (16). These give a measure of the total UVB dose received while the badge is being worn, weighted "erythemally" (though the film may be calibrated in absolute energy units (17)). Dosimeter badges have the advantage of being small and easily worn (or attached) to people (or plants) in many different situations. They have been used to study, amongst other things, occupational exposure to UVB (18), the effects of age and life-style (19), the exposure of different body sites (20) and climate effects (21). Widespread use of dosimeter badges in epidemilogical studies prove their chief advantage; volunteers to wear a backpack radiometer for a year would be few. Small size and weight also allows the badges to be sited in delicate positions e.g. on the face, or the leaf of a plant.

The Need For Spectral Data

While all these broadband UVB sensors have recognised practical advantages, the requirement for UVB monitors with greater spectral resolution is increasing. The general questions 'Is ozone being depleted and UVB likely to increase?' and 'Will this affect life forms?' have been replaced by more searching enquiries : 'How will ozone changes in the stratosphere and troposphere influence UVB at ground level?' and 'What are the mechanisms of action of UVB radiation and the specific wavelengths involved; will they be significantly affected by ozone change?' Only be answering these questions can the need and the means to offset potential damage be understood. Political moves to prevent further anthropogenic ozone depletion are underway but success is not guaranteed, and for the immediate future CFCs continue to accumulate. Nothing further can be done to prevent increases in ambient UVB : increasing tropospheric ozone is undesirable because of its own detrimental effects. However, knowing the mechanisms of UVB damage, their action spectra, dose-response relationship, and the current levels of UVB radiation tolerated may allow some precautions to be taken against UVB damage. Constructive interference is most likely, and most easily achieved, where there is control and economic incentive - in agriculture. Understanding the mechanisms of UVB action and the tolerances of different crop species and cultivars could lead to the breeding of more UV tolerant strains, or with knowledge of the UVB climate in different regions a choice of the most suitable cultivars (a UV-sensitive cultivar of soybean is the currently preferred cultivar in the U.S. (3)).

Spectral Measurement

All this prospective understanding and adaptation depends upon spectral knowledge of the solar (or artificial) UVB radiation. Yet there are very few such measurements and no long term comprehensive data sets. The most extensive study to date is the work of Berner (22) who made spectral measurements at Davos, Switzerland, from 1958-61, investigating spectral distribution and its relation to season, ozone and other factors such as cloud. The work was conducted at altitude (cf. the later work of Blumthaler) away from the influence of pollution. Since that definitive work there has been very little information available, through a combination of technical constraints, lack of demand and cost.

Good solar spectral UVB measurements require high performance from an instrument in a number of areas. The shape of the solar spectrum across the UVB waveband imposes a need for good wavelength resolution and precision, excellent stray light rejection, high sensitivity and a dynamic range of several orders of magnitude. Such criteria are met by spectroradiometers employing double gratings to separate the radiation into its component wavelengths and then a suitable detector (often a photomultiplier tube) to measure spectral intensities. Laboratory based spectroradiometers have long been capable of providing the performance standards sought, but not when moved outside and exposed to a hositle and changing environment. Developing a compact, portable spectroradiometer capable of operating reliably outdoors was more of a challenge. Today there are a number of commercially available instruments capable of this task, plus other instruments developed by individual research laboratories, from which data is beginning to appear in press (23-26), but no long-term data is available, nor is there good areal coverage - measurements have been made at isolated sites. However as both instrument technology and interest in UVB data increases this situation is expected to improve. A different technique for detecting spectra is currently employed in instruments used for measuring gas concentrations (inc. ozone) in the atmosphere by taking the ratio of two wavelengths of radiation, one absorbed by the gas and one not. Such instruments use a diode array to take a 'snapshot' of the radiation spectrum which, after being split into its component wavelength, is allowed to fall on an array of diodes which simultaneously record the intensity across the spectrum. Some of these instruments (e.g. the Brewer spectrophotometer when modified) can detect UVB wavelengths, but they have still to prove that they do not suffer from stray light detection when being used for absolute energy measurements. With such positive proof instruments already routinely used for monitoring atmospheric constitutents could also provide UVB data and greatly increase the current sparse sources of information.

Measurements now in progress will form the basis for future comparisons of long-term trends in solar UVB. The expected differences are small, and besides the practical challenge of maintaining performance and calibration over a long period, it will be many years before a trend in UVB might be detected beyond the natural variability. Ground level UVB at any one location varies cyclically both daily and seasonally due to the motion of earth and sun. Imposed on this regular change is the natural seasonal pattern of ozone change (27), which varies from year to year even without interference from man. There are also changes in solar activity which alter the radiation reaching the atmosphere, and within the troposphere there are the random occurrences of cloud, haze, aerosol etc., all of which can attenuate UVB (28). To construct a general UVB climatology for a region takes many years : to identify change even longer.

SOLAR UVB IN SOUTH-EAST ENGLAND

Meanwhile, the data that is currently available provides information on the levels of UVB that can be tolerated by plants as they are grown today, and gives a scale against which to match and assess the UVB treatment supplied in experimental situations, allowing more meaningful interpretation of results. To illustrate the natural variations in the UVB spectrum measured at one location there follows the results of measurements made at Reading (51.5°N) since August 1989. The instrument used was an Optronics 742 spectroradiometer, scanning the waveband 280-400 nm and sampling every 2 nm. The data shown here concentrates on the UVB waveband. Measurements were all made under cloudless skies, thus removing a major atmospheric variable, cloud, and providing a measure of the maximum radiation that might be expected for a given time. No other criteria have been used to select data, any other changes in atmospheric constitutents are inherent in the measurements.

Fig.2 shows two solar spectra measured at noon (1200 UT) on the 29th November, 1989 and 11th July, 1990. Note how both the short wavelength limit and the intensity of the whole spectrum change with season, with intensity changes becoming increasingly more dramatic toward the shortest wavelengths. As ozone absorption is most effective at short wavelengths the longer pathlength and higher ozone concentrations of the winter months both serve to cause disproportionate increases in attenuation across the UVB waveband, accentuating the seasonal changes at short wavelengths.

This point is reiterated in Fig.3, which shows the daily change of (a) radiation intensity at a wavelength of 300 nm, and (b) the total energy in the UVB waveband obtained by summing the monochromatic intensities at all wavelengths $\leq$ 320 nm. In each case daily change is illustrated for two entirely cloudless days: July and November. Note that the changes are greatest at the short wavelength on both a daily and on an annual basis. The differences for the entire UVB waveband are less dramatic and indicate how the signal from the most biologically active end of the spectrum can be subdued in broadband measurements by the greater intensity of radiation at long wavelengths. Another important feature to notice is the daily distribution of UVB, the majority of it is concentrated in the two hours either side of noon. This may be important in planning experimental radiation regimes, as even at elevated UVB levels the natural irradiation of plants would be confined to a few hours in the middle of the day. There are then periods

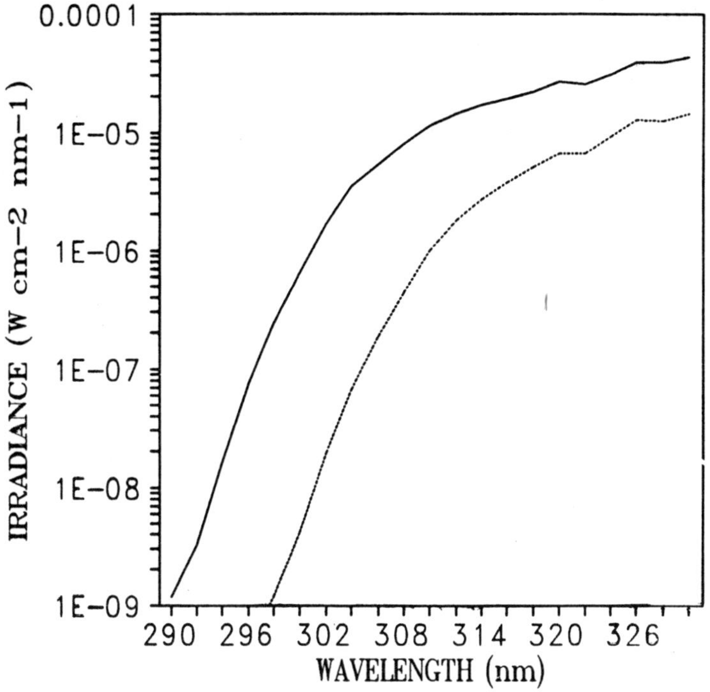

Figure 2: The solar UVB spectrum measured at noon in Reading on 11 July (———), and 29 November (------).

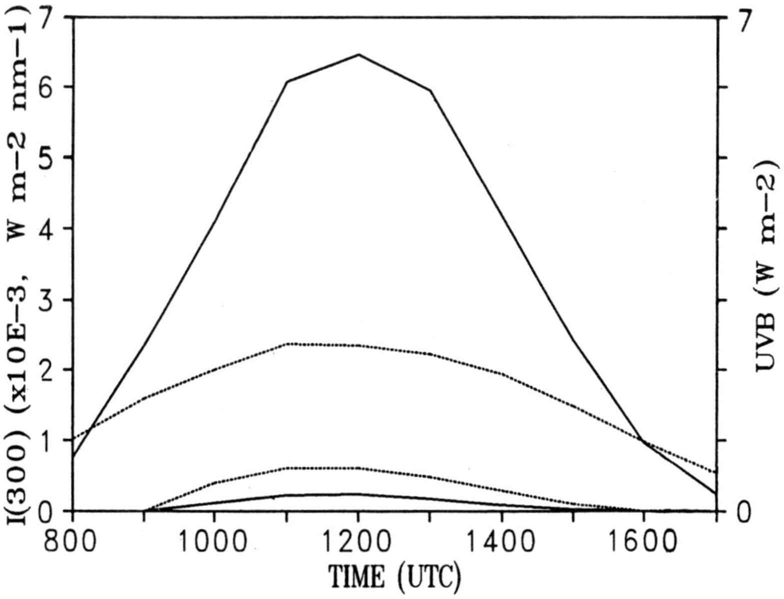

Figure 3: The diurnal change of solar radiation intensity at λ = 300nm (———) and over the complete UVB waveband (--------) on cloudless days in July and November.

of illumination with longer wavelengths of radiation (UVA and visible), separated by a dark period, during which repair mechanisms may be active, before the next period of exposure to UVB. To expose plants to UVB continuously or in a simple light-dark cycle may hinder repair processes and indicate a lower tolerance to UVB than might be the case in field conditions.

It is also important to realise that in the UVB much of the incident radiation in the environment is diffuse. This is due to scattering in the atmosphere, where Rayleigh scattering is much more efficient at short wavelengths. Thus 72% of the UVB waveband is diffuse at noon in July, this figure increasing to 95% in November. The corresponding figures for a wavelength of 300 nm are 76% and 99%. In comparison, the diffuse component of total solar radiation in Reading on a clear day is about 20%. A plant, or part of a plant, in visual shadow may therefore still be exposed to a significant amount of UVB.

CONCLUSION

It is possible that there will be future changes in the global UVB climate due to anthropogenic gases depleting the stratospheric ozone layer. The magnitude of these changes is difficult to predict because of uncertainty about future gas emissions and the effect of changing concentrations on the complex chemistry of the atmosphere. There is also tropospheric pollution to be considered; increased ozone in the troposphere could offset stratospheric depletion. While the extent of UVB change is uncertain, it is widely expected that increases in this region of the solar spectrum will have a net detrimental effect on the biosphere. By understanding, in advance, how the mechanisms of UVB damage work and what the threshold or tolerance limits are for fatal damage, avoidance measures can be taken. In the area of plant research and agricultural strategy more work is required to understand how UVB affects plants, what other factors increase the potential for damage, or enable repair processes, and which species and cultivars are most susceptible to UVB damage. This research must be conducted in a realistic radiation environment, using data available on current UVB levels to design experiments that imitate as closely as possible the present and possible future exposure regimes that field crops would experience. It is important that not only total UVB energy but also the spectral details of radiation regimes are monitored if results are to be correctly interpreted, and these results extrapolated to predict the effects of UVB climate change.

Acknowledgements

The measurement programme at Reading is supported by the Department of the Environment PECD 7/12/18.

REFERENCES

1. German Bundestag. ed. (1989). *Protecting the Earth's Atmosphere. An International Challenge.* Interim Report of the Study Commission of the 11th German Bundestag. Bonner Universitats - Buchdruckerei pp.126-161.
2. Schneider, T., (1989). *Atmospheric Ozone Research and its Policy Implications.* Elsevier Science Publishers, Amsterdam.
3. Tevini, M. and Teramura, A.H. (1989). UV-B effects on terrestrial plants. *Photochem. Photobiol.* 50(4): 479-487.
4. Coohill, P. (1989). Ultraviolet action spectra (280 to 380nm) and solar effectiveness spectra for higher plants. *Photochem. Photobiol.* 50(4): 451-457.
5. Wellman, E. (1989). Lichtabhangige Schutzmechanismen gegen UV-B Schaden bei Pflanzen. In: Lauggener Sèminarbeitrage 3-88 II. Fronz Ruttner Symposium, Munchen.
6. Setlow, R.B. (1974). The wavelength in sunlight effective in producing skin cancer: a theoretical analysis. *Proc. Natl. Acad. Sci.* 71: 3363-3365.
7. Caldwell, M.M. (1971). Solar UV irradiation and the growth and development of higher plants. In: *Photophysiology,* Giese, ed., Vol. VI, Academic Press, New York, pp.131-268.
8. Tevini, M., Grusemann, P. and Pieser, G. (1988). Assessment of UV-B stress by Chlorophyll fluorescence analysis. In: *Applications of Chlorophyll fluorescence,* Lichtenhaler ed. Kluwer, Dordrecht, pp.229-238.
9. Berger, D.S. (1987). Fluctuations and trends in environmental UV loads. In: Passchier and Bossnajakovic (eds). *Human exposure to ultraviolet radiation. Risks and regulations.* Elsevier Science Amsterdam, pp.213-221.
10. McKinley, A.F. and Diffey, B.L. (1987). A reference action spectrum for ultra-violet induced erythema in human skin. In: *Human Exposure to Ultraviolet Radiation: Risks and Regulations.* Passchier and Bossnajakovic eds., Elsevier, Amsterdam, pp.83-87.

11. Scotto, J., Cotton, G., Urbach, F., Berger, D. and Fears, T., (1988). Biologically effective ultraviolet radiation: Surface measurements in the United States, 1974 to 1985. *Science,* 239: 762-764.

12. Blumthaler, M. and Ambach, W., (1990). Indications of increasing solar ultraviolet-B radiation flux in Alpine regions. *Science,* 248: 206-208.

13. Watson, R.T., Prather, M.J. and Kurylo, M.J., (1988). Present state of knowledge of the upper atmosphere 1988: An assessment report. NASA reference publication 1208, Washington D.C.

14. Bruhl, C. and Crutzen, P.J. (1989). On the disproportionate role of tropospheric ozone as a filter against solar UV-B radiation. *Geophys. Res. Lett.* 16(7): 703-706.

15. Webb, A.R., Kline, L. and Holick, M.F. (1988). Influences of season and latitude on the cutaneous synthesis of vitamin D_3: Exposure to winter sunlight in Boston and Edmonton will not promote vitamin D_3 synthesis in human skin. *J. Clin. Endocrinol. and Metabol.* 67(2): 373-378.

16. Davis, A., Deane, G.H.W. and Diffey, B.L. (1976). Possible dosimeter for ultraviolet radiation. *Nature,* 261: 169-170.

17. Webb, A.R., (1985). Solar ultraviolet radiation and vitamin D synthesis in man. PhD. Thesis, Nottingham University, England.

18. Larko, O. and Diffeyt, B.L. (1983). Natural UVB radiation received by people with outdoor, indoor and mixed occupations and UVB treatment of psoriasis. *Clin. Exp. Dermatol.* 8: 279-295.

19. Corless, D., Gupta, S.P., Sattar, D.A., Switala, S. and Boucher, B.J. (1979). Vitamin D status of residents of an old people's home and long stay patients. *Gerontology,* 25:350-355.

20. Holman, C.D.J., Gibson, I.M., Stephenson, M. and Armstrong, B.K. (1983). Ultraviolet irradiation of human body sites in relation to occupation and outdoor activity: field studies using personal UVR dosimeters. *Clin. Exp. Dermoatol.* 8: 269-277.

21. Diffey, B.L., Larko, O. and Swanbeck,G. (1982). UV-B doses received during different outdoor activities and UVB treatment of psoriasis. *Br. J. Dermatol.* 106: 33-41.

22a. Bener, P. (1960). Investigation of the spectral intensity of ultraviolet sky and sun+sky radiation (between 297.5mμ and 370mμ) under different conditions of cloudless weather at 1590m.a.s.l. Contract AF 61(052)-54. Technical Summary Report No.1. Physikalisch-Meteorologisches Observatorium Davos, Davos-Platz,

Switzerland.

22b. Bener, P. (1963). The diurnal and annual variations of the spectral intensity of ultraviolet sky and global radiation on cloudless days at Davos 1590m.a.s.l. Contract AF 61(052)-618 Technical Note No.2 Physikalisch-Meteorologisches Observatorium Davos, Davos Platz, Switzerland.

23. Webb, A.R. and Campbell, G.S. (1987). Clear sky ultraviolet-B radiation in the English Midlands: a comparison of measured and modelled values. *Theor. Appl. Climatol*, 38: 79-84.

24. Henriksen, K., Stamnes, K., Volden, G. and Falk, E.S., (1989). Ultraviolet radiation at high latitudes and the risk of skin cancer. *Photodermatology*, 6: 110-117.

25. Stamnes, K., Henriksen, K. and Ostensen, P., (1988). Simultaneous measurements of UV radiation and total ozone. *Geophys. Res. Lett.* 15: 784-787.

26. Webb, A.R., (1991). Ultraviolet radition: the missing link in the ozone debate. Norsk. Geologisk Tidsskrift (in press).

27. Dahlback, A., Henriksen, T., Larsen, S.H.H. and Stamnes, K. (1989). Biological UV-doses and the effect of an ozone layer depletion. *Photochem. Photobiol.* 49(5): 621-625.

28. Frederick, J.E., Snell, H.E. and Haywood, E.K. (1989). Solar ultraviolet radiation at the earth's surface. *Photochem. Photobiol.* 50(8): 443-450.

EFFECTS OF UV-B AND VISIBLE LIGHT ON THE REACTION MECHANISM OF PHOTOSYNTHETIC WATER CLEAVAGE

G. Renger and H.-J. Eckert

Max-Volmer-Institut für Biophysikalische und Physikalische Chemie, Technische Universität Berlin, Straße des 17. Juni 135, D-1000 Berlin 12, Germany

ABSTRACT

Deleterious effects of UV and visible light on photosynthetic water cleavage in PS II membrane fragments from spinach were analyzed. Comparative measurements of (a) 830 nm absorption changes reflecting the turnover of P680, (b) oxygen evolution, and (c) activity of DCIP reduction in the absence and presence of exogenous electron donors, led to the conclusion that different target sites exist in PS II which are susceptible to UV and visible light stress. The mechanism of destruction was found to depend on the functional integrity of the PS II complex and the light quality. Visible light predominantly affects the stabilization of the primary charge separation in PS II membrane fragments when the catalytic site of water oxidation is intact; whereas, in systems deprived of their oxygen evolution capacity, the functional connection between P680 and Y_Z (Tyr-161 of polypeptide D1) is primarily interrupted with a rather high quantum yield. On the other hand, with UV-B the catalytic site of water oxidation is the most sensitive target in PS II membrane fragments, while non-photochemical quenchers of excitation energy are generated as well. We conclude that the deleterious effects of light stress cannot be ascribed to a single target site and a unique photochemical destruction mechanism.

INTRODUCTION

The interaction of solar radiation with photosynthetic organisms exerts multiple effects of functional relevance. Beyond its indispensable role as driving force of photosynthesis, many regulatory functions of light are known that are especially relevant for the development of the photosynthetic apparatus (1-3). Cyanobacteria, red algae and green plants have developed a variety of suitable adaptation mechanisms at the exciton level in order to cope with different environmental illumination conditions (for a recent review see ref. (4)). However, these processes have a limited capacity. At high light intensities harmful effects arise which can lead to serious damage and the eventual death of the organism. The light intensity is the product of photon flux density (PFD) and the energy of the impinging photons (hv). At first glance, it appears reasonable to assume that the latter parameter plays a key role in the mechanism of destruction, because the photon energy determines the target molecules which can be attacked by light. Therefore, at shorter wavelength the number and type of target sites are expected to increase. In respect to the light-induced impairment of the photosynthetic apparatus, current research activities focus on two types of effects: (a) destruction by UV-B in relation to problems caused by a decrease of the protective ozone layer in the stratosphere due to man-made air pollutants (5,6) and (b) photo-inhibition as a consequence of irradiation with visible light at high photon flux densities (7,8). Accordingly, both phenomena are worth analyzing in terms of their underlying mechanisms.

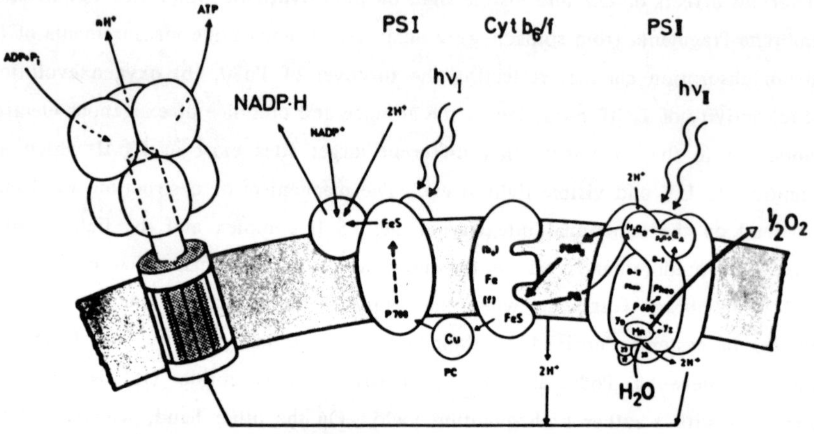

Fig. 1 *General scheme of the photosynthetic electron transport from water to* $NADP^+$ *coupled with ATP synthesis via proton fluxes (for details see text).*

The photosynthetic apparatus of all oxygen-evolving organisms (cyanobacteria, algae and higher plants) performs two types of different reaction sequences: (a) the light-induced water cleavage into dioxygen and metabolically bound hydrogen in the form of NADPH coupled with ATP synthesis. This process comprises the cooperation of at least four integral membrane processes; i.e., the reaction center complexes of PS I and PS II, the Cytochrome b_6/f complex and the ATP-synthase (see fig. 1). (b) The dark reactions of enzyme catalyzed CO_2 fixation into carbohydrates with NADPH and ATP as hydrogen and free energy source, respectively.

Numerous studies within the last three decades unambiguously showed that among the different complexes of sequence (a) the reaction pattern of PS II is most susceptible to deleterious effects either by UV or by visible light (9-14). The PS II complex is the site of the key steps of photosynthetic water cleavage. The overall reaction sequence of this complex is comprised of three different parts (15-18).

The first part is the generation of strongly oxidizing and moderately reducing redox equivalents in a π-electron mediated photoredox chemistry, taking place in the picosecond time domain. A special chlorophyll-a containing component, referred to as P680, ejects with a time constant of about 3 ps (19) an electron from its first excited singlet state to pheophytin a (Pheo) as primary acceptor. Subsequent electron transfer from $Pheo^-$ to a special noncovalent-bound plastoquinone, Q_A, leads with τ ~ 300 ps (20,21) to a radical pair $P680^+PheoQ_A^-$ that is sufficiently stabilized for further redox reactions.

The second part is the cooperative reaction of four oxidizing redox equivalents with two water molecules, giving rise to dioxygen and the release of four protons into the inner space (lumen) of the thylakoids. This reaction takes place via a sequence of four univalent redox steps at a manganese-containing catalytic site within the time domain of 50 μs to 1 ms (15-18,22).

The third part is the cooperative reaction of two reducing redox equivalents with plastoquinone. This reaction takes place at a special plastoquinone-binding site (Q_B site), located in polypeptide D1 of the PS II complex (23,24). Plastoquinone at the Q_B site becomes reduced via two sequential univalent redox steps with Q_A^- as electron donor. The half-life time of these two electron transfer steps is of the order of 100-400 μs (25,26). Only the plastosemiquinone form is comparatively strongly bound to the Q_B site. After double reduction and protonation, the plastoquinol exhibits a rather weak affinity and readily exchanges to plastoquinone, thereby completing the reaction cycle (for a review see (26)).

Detailed mechanistic studies on UV-B effects or photoinhibition are aimed at unraveling (i) the target reaction which becomes primarily affected by a photodynamic attack and (ii) mode of the modification of a specific redox component (and/or its protein matrix), giving rise to this effect. In order to address these problems, suitable methods are required. The application of sensitive spectroscopic techniques with high time resolution, however, has an inherently serious problem. As the optical and functional properties of leaves and also of thylakoid preparations do not permit a thorough analysis of the PS II reaction pattern, the desired experiments can be performed only with isolated PS II preparations. Therefore, a detailed mechanistic study cannot be performed with whole leaves, and in vitro studies with isolated systems are necessary. Two basically different approaches have been introduced. Both of them have their specific advantages and shortcomings. With the first method, whole plants are exposed to deleterious irradiation. After this treatment PS II membranes were isolated and used for mechanistic studies. At first glance, this procedure seems to provide more relevant information, because all deleterious or protective effects due to the modification of target molecules outside the photosynthetic apparatus contribute to the mode and rate of impairment of PS II. On the other hand, the PS II complex in modified leaves could be more susceptible to degradation during the isolation of PS II membranes and even smaller complexes so that secondary effects would mask the in vivo situation.

In the second approach, isolated preparations are illuminated at harmful light intensities and/or wavelengths. In this case, only effects due to direct interaction with the photosynthetic electron transport are monitored without interference from other processes. Although this method might not reflect the in vivo situation, it provides valuable information on the susceptibility of the isolated system to photodynamic degradation. Therefore, this approach is often used for mechanistic studies (11,14,28-30).

In this report the mechanisms of degradation by UV-B and by photoinhibition are analyzed in PS II membrane fragments from spinach.

MATERIALS AND METHODS

Oxygen-evolving PS II membrane fragments were isolated from spinach by solubilization of thylakoids (prepared as outlined in ref. 31) with Triton X-100 according to the procedure described in ref. (32), with some modifications as in ref. (33). Samples selectively deprived of their oxygen-evolving capacity were obtained by standard Tris-washing procedure (34). The exposure to UV-B irradiation and to visible light was performed while stirring the samples in Petri dishes in an ice bath (for details see refs. (11) and (14), respectively). Fluorescence decay curves were measured with a single photon counting technique described recently (35).

Laser flash-induced absorption changes at 830 nm, reflecting the turnover of P680, were monitored with a single beam flash photometer as outlined in ref. (36). For other experimental details see ref. 14.

O_2 evolution was measured with a Clark-type electrode and the DCIP reduction activity was spectrophotometrically monitored at 590 nm (14).

RESULTS AND DISCUSSION

I. Exciton Dynamics and Antenna Size

Absorption of photons in the visible or shorter wavelength region leads first to the formation of excited singlet states in the antenna. Most of them are transformed into chemical free energy or dissipated via fluorescence emission or radiationless decay. Some of them, however, can lead to photodynamic destruction depending on wavelength and light intensity. Therefore, the question arises as to what extent the antenna itself can be affected. Measurements of fluorescence induction curves in samples exposed to UV-B irradiation or high intensity visible light led to the conclusion that modifications can occur at the level of the exciton dynamics (37,38). In order to address this point, experiments were performed to analyze the exciton life time and the effective antenna size in light treated samples.

(A) Effects of UV-B on fluorescence life time

Fig. 2 shows typical fluorescence decay curves at 690 nm of control and UV-B treated PS II membrane fragments at room temperature in the absence of exogenous electron acceptors. The data reveal two interesting phenomena: the fluorescence decay exhibits multi-phase kinetics and exposure to UV-B irradiation, giving rise to a significant decrease of electron transport activity (vide infra), accelerates the fast phase(s) and retards the slow decay components. Additional experiments under

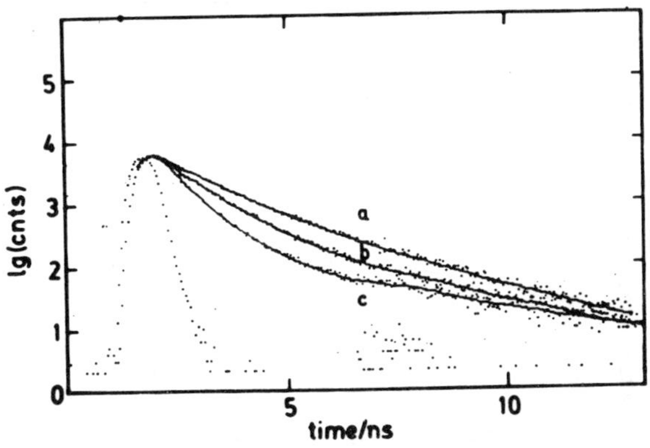

Fig. 2 *Fluorescence decay at 690 nm of control (a) and UV-B-treated (b,c) PS II membrane fragments at room temperature in the absence of exogenous redox components. UB-B irradiation time 30 min (b) and 90 min (c). Optical bandwidth for the emission measurements $\Delta\lambda = \pm 10$ nm, for other details see ref. 35.*

different experimental conditions (addition of exogenous oxidants and reductants, room temperature and 77 K, monitoring wavelength at 686 or 694 nm) unambiguously showed that the average lifetime of the fluorescence decay decreased with progressive UV-B irradiation, regardless of the functional state of the PS II reaction centers and the emission temperature or wavelength (35). In general, the complex decay curves of chlorophyll-a fluorescence in plants or thylakoids and isolated PS II preparations can be described either within the framework of model-based decay associated spectral analysis (for a recent review see ref. (39)) or by using lifetime distribution functions (40) reflecting heterogeneities of the pigment environment. In respect to the effects caused by UV-B, both types of numerical data fitting basically lead to the same conclusion; i.e., UV-B creates additional quenchers of excitons (35). The nature of these quenchers remains to be clarified.

(B) Effects of photoinhibition on the functional antenna size

The formation of dissipative sinks raises questions about the possibility of an interruption of exciton migration from the antenna to PS II reaction centers by selective photodynamic degradation of chlorophyll-a molecules in the immediate neighborhood of P680. This type of P680 disconnection from the antenna was recently proposed as a mechanism of photoinhibition (41). In order to check this idea, the extent of light-induced $P680^+$ formation was determined as a function of

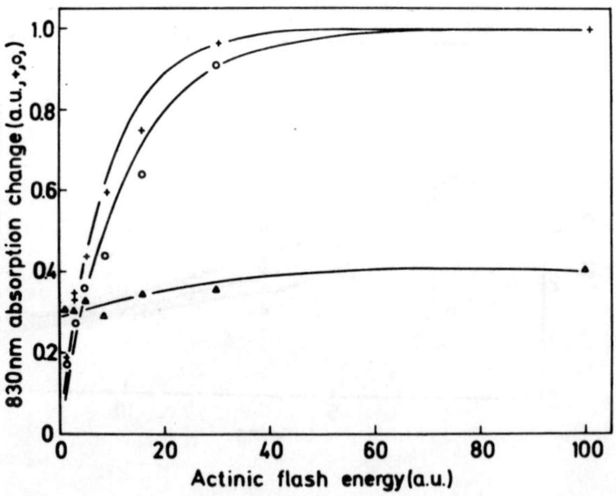

Fig. 3 *Total initial amplitude of the 830 nm absorption change as a function of the energy of actinic laser flash in Tris-washed PS II membrane fragments before (crosses) and after 30 min of photoinhibition (open circles) at 1700 W/m^2. For a comparison, the ratio of the amplitudes calculated from these data is also depicted (open triangles). For other details see ref. 14.*

the actinic laser flash intensity in Tris-washed PS II membrane fragments. The data obtained by measuring flash-induced absorption changes at 830 nm (vide infra, see fig. 8) are depicted in fig. 3. At first glance, no significant modifications of the saturation curve are observed. An analysis within the model of separate photosynthetic units and a quantum yield of PS II charge separation of about one indicates that the effective antenna size decreases in this particular photoinhibition assay (30 min irradiation with white light 1700 W/m^2, see ref. (14)) by about 25%; whereas, the capacity for stable charge separation becomes diminished by 75% under the same conditions. This result does not support the idea of a specific effect on exciton migration from the antenna to P680. The comparatively small decrease of the effective PS II antenna size is easier explained by different susceptibilities of α- and ß centers to photoinhibition (42). However, our data do not completely rule out the possibility that a fraction of PS II centers could become totally disconnected from the antenna. This problem will be addressed in future studies.

II. Functional Reaction Pattern of PS II

Information about the reaction pattern of PS II can be obtained by comparative measurements of 830 nm absorption changes, reflecting the turnover of P680 and of oxygen evolution. The transient 830 nm absorption change in the time domain of a few nanoseconds is the superposition of the dynamics of Chl-a singlet states, the formation of the primary radical pair $P680^{+}Pheo^{-}$ and either the electron transfer from $Pheo^{-}$ to Q_A in open reaction centers or the recombination of $P680^{+}Pheo^{-}$ in centers with Q_A^{-} kept reduced. As neither Q_A nor Q_A^{-} do significantly absorb in the near infrared and the state $P680^{+}PheoQ_A^{-}$ is formed within 1 ns, the initial amplitude detected at 830 nm with a time resolution of a few nanoseconds is a measure of the extent of $P680^{+}$ formation, provided that the yield of triplets is negligibly small. The relaxation of this absorption change reflects the $P680^{+}$ reduction kinetics. In PS II complexes functionally competent in oxygen evolution $P680^{+}$ becomes reduced predominantly by redox component Y_Z recently identified as Tyr-161 of polypeptide D1 in Synechoccystis sp. PCC 6803 (43,44). This electron transfer is dominated by kinetics of about 30 ns and 250 ns (45-48). Selective elimination of the oxygen-evolution capacity usually causes a significant retardation of the electron transfer from Y_Z to $P680^{+}$ (45,49,50). For example, in Tris-washed or trypsin-treated samples this reaction exhibits a pH-dependent half time of 2-40 μs (51,52). If the redox active tyrosine stays in its oxidized form Y_Z^{ox}, $P680^{+}$ becomes reduced by Q_A^{-} with a half lifetime of 100-200 μs (49,51,53).

P680 turnover characteristics in O_2-evolving PS II membrane fragments after exposure to UV-B and high intensity white light

Fig. 4 shows absorption changes at 830 nm induced by repetitive laser flash excitation in PS II membrane fragments exposed to different times of UV-B irradiation at 90 W/m². In the control sample, the relaxation is dominated by ns kinetics which are completely transformed into μs kinetics after selective destruction

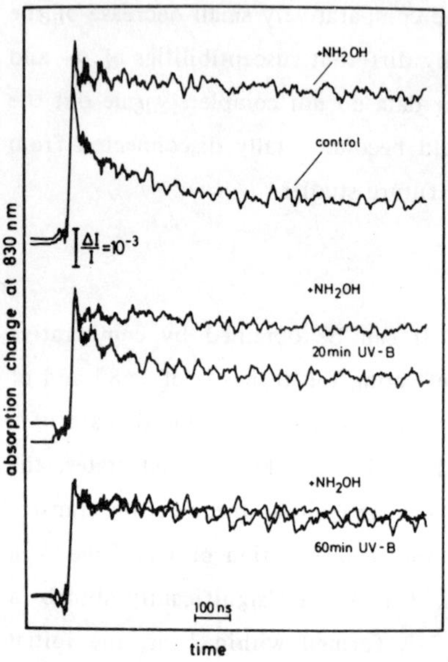

Fig. 4
Absorption changes at 830 nm as a function of time in PS II membrane fragments exposed to different times, t_{UV-B}, of UV-B irradiation at 90 W/m². Incubation with 3 mM NH_2OH after light treatment as indicated in the figure. The experiments were performed at a chlorophyll concentration of 50 μM in the presence of 1 mM $K_3[Fe(CN)_6]$; time between flashes 500 ms, other experimental conditions as described in Materials and Methods.

of the water-oxidizing system by incubation with 3 mM NH_2OH. Increasing exposure of the samples to UV-B before the measurements eliminates the relaxation in the ns time domain, while the detected initial amplitude and the absorption changes measured in the presence of 3 mM NH_2OH are much less affected. These findings suggest that UV-B irradiation primarily attacks the catalytic site of water oxidation. On the left side of fig. 5, the detected initial amplitude, the extent of the ns kinetics, the oxygen evolution capacity, and the activity of DCIP reduction with H_2O as donor are depicted as a function of UV-B irradiation time. The data reveal that the extent of the ns kinetics and the almost parallel decline of the O_2 evolution are markedly more susceptible to UV-B pre-treatment than the total extent of $P680^+$ formation. Likewise, the PS II electron transport with an exogenous PS II electron donor (DPC) becomes less affected (11).

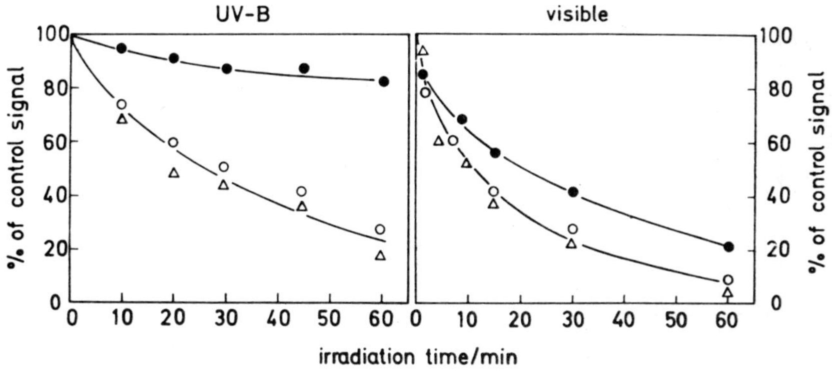

Fig. 5 *Normalized total initial amplitude,* ΔA^{total} *(closed circles), normalized extent of the ns kinetics,* ΔA^{ns} *(open circles) and oxygen evolution (open triangles) in PS II membrane fragments as a function of irradiation time with UV-B of 90* W/m^2 *(left side) or visible light of 1700* W/m^2 *(right side). All quantities are normalized to the respective value of the non-irradiated control sample.*

Analogous experiments were performed with PS II membrane fragments preilluminated with visible light at an intensity of 1700 W/m^2. The data obtained are summarized in fig. 5, right side. A comparison with the left side of fig. 5 reveals a significant qualitative difference. The decline of the detected initial amplitude due to progressing photoinhibition of the sample does not deviate much from that of the extent of the ns kinetics and the O_2 evolution. To illustrate this effect more clearly, in fig. 6 the normalized extent of the μs kinetics is compared as a function of exposure time to UV-B and visible light. In the former case an increase is observed, while in the latter case a decrease occurs. Based on the data of figs. 4-6, it is inferred that in isolated PS II membrane fragments the catalytic site of water oxidation is the primary target for damage by UV-B; whereas, photoinhibition by visible light predominantly eliminates the capability to perform a stable charge separation in PS II.

The deletion of the functional integrity of water oxidation by UV-B raises the question whether this effect comprises the blockage of a specific redox transition as in the case of Cl^- or Ca^{2+} depletion (55,56). To answer this question 830 nm-absorption changes were measured in dark-adapted samples, illuminated by a train of

....

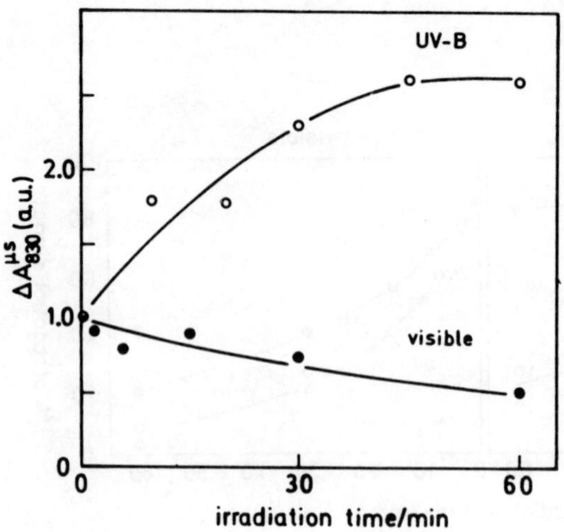

Fig. 6 *Normalized extent of the μs kinetics as a function of irradiation with UV-B at 90 W/m² (open circles) or visible light at 1700 W/m² (closed circles) of PS II membrane fragments. The extent of the μs kinetics is normalized to the respective value of the non-irradiated control sample.*

laser flashes. In order to improve the signal/noise ratio, the experiments were performed at a reduced time resolution. Under these conditions, the fast ns kinetics of $P680^+$ reduction escape detection. Therefore, in the control samples, the detected absorption changes as a function of flash number exhibit a marked oscillation with a maximum in the 3rd flash (see fig. 7), due to the dependence of the electron transfer rate from Y_Z to $P680^+$ on the redox state S_i of the catalytic site of water oxidation (48,57,58). In contrast to the control, the absorption changes in UV-B treated

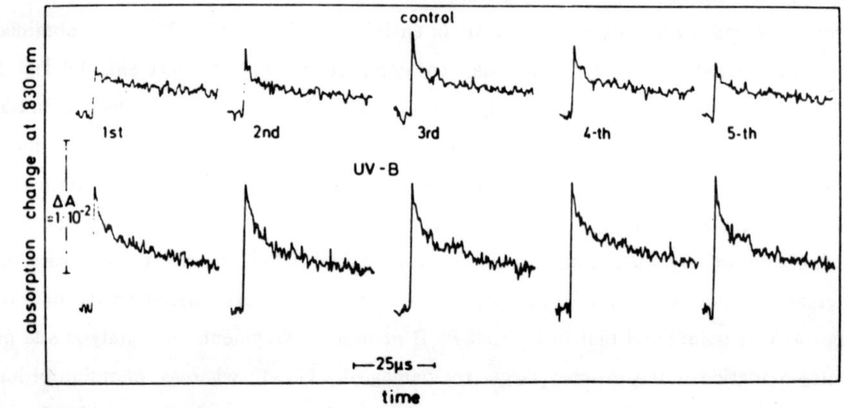

Fig. 7 *Absorption changes at 830 nm induced by a train of laser flashes in dark-control samples (top) or PS II membrane fragments irradiated for 30 min with UV-B at 90 W/m² before 5 min dark adaptation (bottom). For experimental details see ref. 11.*

samples hardly reveal any dependence on the flash number. This finding clearly shows that modifications caused by UV-B irradiation do not block a specific redox transition $S_i \rightarrow S_{i+1}$ in the water-oxidizing system. Interestingly, the pattern also indicates that Y_Z^{ox} becomes reduced to a significant extent during the dark time between the flashes. A closer inspection of the data reveals that a significant fraction of PS II exhibits a relaxation which is slower than the $P680^+$ reduction by Y_Z in preparations selectively deprived of their oxygen-evolution capacity. The origin of this effect is not yet completely determined, but it might be related to a different susceptibility of the P680-Y_Z connection in samples with and without functionally competent water oxidation.

P680 turnover characteristics of Tris-washed PS II membrane fragments after exposure to high intensity white light

Recent reports in the literature provided evidence for different mechanisms of photoinhibition, depending on the functional integrity of the catalytic site of water oxidation. To clarify this point, experiments were performed with Tris-washed PS II membrane fragments. Typical traces of 830 nm-absorption changes, induced by repetitive flash excitation, are depicted in fig. 8. In the control an almost mono-exponential relaxation is observed with a half lifetime of about 10 μs, which is typical for the rate of $P680^+$ reduction by Y_Z (51,52). This finding implies that Y_Z^{ox} becomes reduced during the dark time between the flashes. In samples

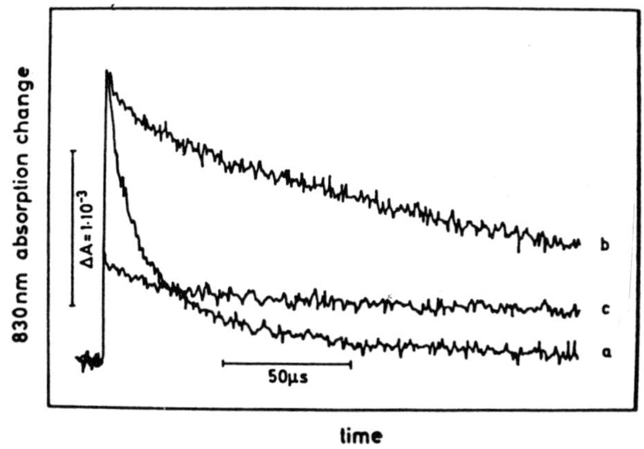

Fig. 8 *Time course of absorption changes at 830 nm induced by repetitive laser flashes in Tris-washed PS II membrane fragments: (a) control, (b) 1 min irradiation and (c) 30 min irradiation with white light at 1700 W/m². For experimental details see ref. 14.*

preilluminated for only 1 min with white light at an intensity of 1700 W/m^2 the $P680^+$ reduction is drastically retarded without any effect on the extent of $P680^+$ formation. Further photo-inhibitory treatment additionally caused a marked decrease of the initial amplitude similar to that of PS II membrane fragments with intact water oxidizing system (see fig. 5, right side). The results of fig. 8 can be interpreted with the assumption that the functional connection between P680 and Y_Z is very sensitive to photoinhibition in PS II membrane fragments deprived of an intact catalytic site of water oxidation. In this case, $P680^+$ cannot be reduced by Y_Z and undergoes a back reaction with Q_A^-. Alternatively, under repetitive flash excitation, the P680 recovery with Q_A^- as reductant can be also explained if Y_Z^{ox} is formed after the first flash and then stays oxidized. The latter interpretation, however, is much less likely for two reasons: (1) PS II electron donors like NH_2OH or $MnCl_2$, which are known to reduce Y_Z^{ox} at sufficient rates (59,60), do not affect the relaxation kinetics, and (2) the $P680^+$ reduction is about twice as slow ($t_{1/2} \approx$ 250 μs) as in PS II reaction centers with Y_Z^{ox} kept oxidized. The latter effect is probably due to an electrostatic effect of Y_Z^{ox} and the electron transfer rate from Q_A^- to $P680^+$ (60,61).

P680 turnover characteristics of PS II membrane fragments illuminated in the presence of 3 mM NH_2OH

Recently, it was shown that excitation of PS II membrane fragments with about 50 single turnover flashes in the presence of NH_2OH at concentrations leading, in the dark, to complete elimination of the oxygen-evolution capacity, causes a marked retardation of the $P680^+$ reduction kinetics (61). Illumination in the presence of NH_2OH was inferred to disconnect P680 from Y_Z as electron donor (66). Therefore, the slower kinetics, which resemble those of Tris-washed PS II membrane fragments after exposure to strong continuous white light, reflect the reduction of $P680^+$ by Q_A^- (instead by Y_Z) as confirmed by comparative measurements of absorption changes at 270 nm and 830 nm (61). A selective blockage of the electron transport from Y_Z to $P680^+$ should give rise to the formation of a comparatively stable $P680^+$ radical, provided that Q_A^- can be reoxidized rapidly enough in order to compete with the back reaction between Q_A^- and $P680^+$. A fast electron transfer from Q_A^- can be achieved with the oxidized form of the endogeneous high-spin nonheme iron that is located between Q_A and Q_B (62,63). Therefore, under these conditions, the lifetime of $P680^+$ should markedly increase. To check this idea, experiments were performed under the following assay conditions: PS II membrane fragments deprived of their oxygen-evolution capacity by incubation with 3 mM NH_2OH were photoinhibited by a train of 100 flashes, which is sufficient to functionally disconnect Y_Z from P680 as shown in fig. 9. After one flash the $P680^+$ reduction

•••••

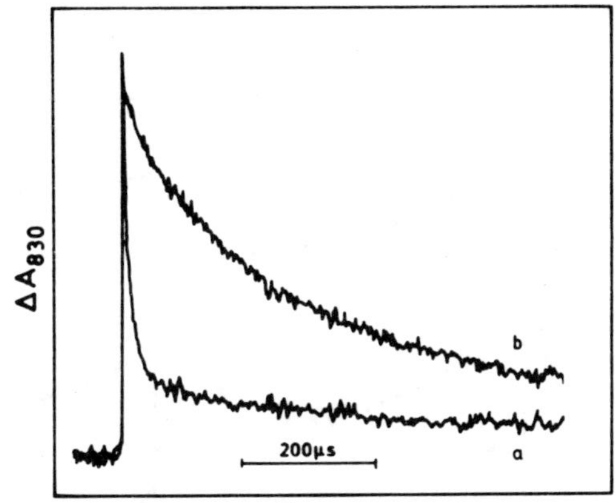

Fig. 9 *Time course of absorption changes at 830 nm induced by one laser flash in PS II membrane fragments treated with 3 mM NH_2OH in the dark (a) or in the presence of 100 laser flashes (b). Assay conditions: samples at 50 µM chlorophyll, 1 mM $K_3[Fe(CN)_6]$, 20 mM Mes, pH = 6.5, 10 mM NaCl, optical pathlength 1 cm, 16 signals were averaged at a repetition rate of 1 Hz.*

by Y_Z is observed, after 100 flashes the much slower electron transfer from Q_A^- (compare with fig. 8). After removal of NH_2OH by centrifugation and resuspension in buffer solution, the samples were incubated in the dark either in the absence or with 1 mM $K_3[Fe(CN)_6]$ which leads to the oxidation of Fe^{2+} (64,65). If samples without $K_3[Fe(CN)_6]$ were illuminated with a laser flash train, the $P680^+$ reduction kinetics are almost the same in each flash (see left traces in fig. 10), because the Q_A^- reoxidation by the acceptor side, even after the first flash, is not fast enough to compete with the $P680^+Q_A^-$ recombination. If one takes into account the electron transfer kinetics from Q_A^- to Q_B and Q_B^- (25,26) especially Q_B should efficiently compete with $P680^+$ as an oxidant of Q_A^-. This does not seem to be the case.Therefore, $P680^+$ probably retards the electron transfer from Q_A to Q_B. The mechanism (electrostatic) and the nature of this effect remain to be clarified (experiments addressing this problem are in progress). A markedly different phenomenon is observed in the presence of 1 mM $K_3[Fe(CN)_6]$. The relaxation kinetics of the 830 nm-absorption change after the first flash is much slower (order of tens of milliseconds; H.-J. Eckert, to be published) due to the rapid electron transfer from Q_A^- to Fe^{3+}. After the first flash, Fe^{2+} is formed and stays reduced, because the dark time between the flashes of the train is much shorter than the re-oxidation rate of Fe^{2+} by exogenous $K_3[Fe(CN)_6]$. Therefore, after the second and following flashes the relaxation kinetics of the 830 nm-absorption changes are

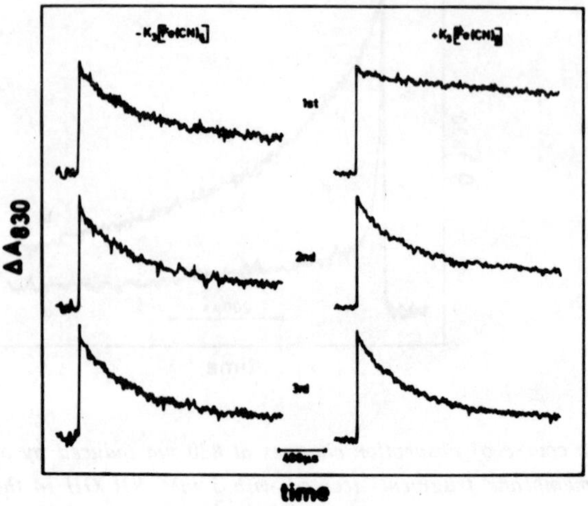

Fig. 10 *Time course of absorption changes at 830 nm induced by the 1st (top traces), 2nd (middle traces) or 3rd (bottom traces) laser flash of a train in PS II membrane fragments treated with 3 mM NH_2OH in the presence of 200 laser flashes followed by centrifugation and resuspension. The samples of the right side were preincubated with 1 mM $K_3[Fe(CN)_6]$ for 5 min in the dark before the measurements. Other conditions as in fig. 9, except for the omission of $K_3[Fe(CN)_6]$ in the assay medium of the traces at the left side.*

dominated by the recombination reaction between Q_A^- and $P680^+$. These findings demonstrate that under properly selected experimental conditions a comparatively long-lived $P680^+$ cation radical can be generated in PS II membrane fragments at room temperature. This technique will be used in future research for analysis of the $P680^+$ properties.

CONCLUSION

The present study revealed that the detailed mechanism of detrimental light effects in isolated PS II membrane fragments depends on both the radiation wavelength (visible versus UV-B light) and on the functional integrity of the catalytic site of photosynthetic water oxidation. Therefore, neither photoinhibiton nor UV-B induced PS II destruction can be ascribed to a unique mechanism with a single target site. This conclusion is corroborated by measurements of the action spectrum of the photodynamic effect (for a detailed discussion see ref. (11)).

ACKNOWLEDGEMENTS

The authors would like to thank S. Hohm-Veit for drawing the figures. The financial support by the Bundesministerium für Forschung und Technologie (0704546A), EPR-Sondervermögen (ERP 2603) and Deutsche Forschungsgemeinschaft (Sfb 312) is gratefully acknowledged.

REFERENCES

1. Senger, H. (1982) The effect of blue light on plants and microorganisms. Photochem. Photobiol. 35:911-920

2. Smith, H. (1982) Light quality, photo perception and plant strategy. Ann. Rev. Plant Physiol. 33:481-518

3. Shropshire, E. Jr. and Mohr,H. (Eds.)(1983) Encyclopedia of Plant Physiology, New Series, Vol. 16A and 16B, Springer, Heidelberg

4. Renger, G. (1991) Energy transfer and trapping in photosystem II. In: Topics in photosynthesis, Vol. 11 (Barber, J. ed.), Elsevier, Amsterdam (in press)

5. DeMore, W.B., Molina, M.J., Sander, S.P., Golden, D.M., Hampson, R.F., Kurylo, M.J., Howard, C.J. and Ravishankara (1987) Chemical kinetics and photochemical data for use in stratospheric modeling. In: Jet Propulsion Laboratory Publication 87-41, National Aeronautics and Space Administration, Pasadena, California

6. Caldwell, M.M., Robberecht, R., Nowak, R.S. and Billings, W.D. (1982) Differential photosynthetic inhibition by ultraviolet radiation in species from alpine life zone. Arctic and Alpine Res. 14: 195-202

7. Powles, S.G. (1984) Photoinhibition of photosynthesis induced by visible light. Annual Rev. Plant Physiol. 35, 15-44

8. Neale, P.J. (1987) Algal photoinhibition and photosynthesis in the aquatic environment. In: Topics in photosynthesis, Vol. 9 (Kyle, D.J., Osmond, C.B. and Arntzen, C.J., eds.), Elsevier, Amsterdam, pp. 39-65

9. Noorudeen, A.M. and Kulandaivelu, G. (1982) On the possible site of inhibition of photosynthetic electron transport by ultraviolet B (UV-B) radiation. Physiol. Plant. 55, 161-166

10. Bornmann, J.F. (1989) Target sites of UV-B radiation in photosynthesis of higher plants. J. Photochem. Photobiol. B:Biology 4:145-158

11. Renger, G., Völker, M., Eckert, H.J., Fromme, R., Hohm-Veit, S. and Gräber, P. (1989) On the mechanism of photosystem II deterioration by UV-B irradiation. Photochem. Photobiol. 49:97-105

12. Critchley, C. (1988) The molecular mechanism of photoinhibition in higher plants - facts and fiction. Aust. J. Plant Physiol. 15:27-41

13. Setlik, I., Allakhverdiev, S.I., Nebdal, L., Setlikova, E. and Klimov, V.V. (1990) Three types of photosystem II photo-inactivation. Photosynth. Res.23:39-48

14. Eckert, H.J., Geiken, B., Bernarding, J., Napiwotzki, A., Eichler, H.-J. and Renger, G. (1991) Two sites of photoinhibition of the electron transfer in oxygen-evolving and Tris-treated PS II membrane fragments from spinach. Photosynth. Res. (in press)

15. Babcock, G.T. (1987) The photosynthetic oxygen evolving process. In: Amesz, J. (ed.) New comprehensive biochemistry, Vol. 15, pp. 125-158, Photosynthesis, Elsevier, Amsterdam

16. Renger, G. (1987) Biological exploitation of solar energy by photosynthetic water splitting. Angew. Chem. Int. Ed. 26:643-660

17. Rutherford, A.W. (1989) Photosystem II, the water-splitting enzyme. Trends Biochem. Sci. 14:227-232

18. Hansson, Ö., Wydrzynski, T. (1990) Current perceptions of photosystem II. Photosynth. Res. 23:131-162

19. Wasielewski, M.R., Johnson, D.G., Seibert, M. and Govindjee (1989) Determination of the primary charge separation rate in isolated photosystem II reaction centers with 500 fs time resolution. Proc. Natl. Acad. Sci. USA 86, 524-528

20. Nuijs, A.M., van Gorkom, H.J., Plijter, J.J. and Duysens, L.M.N. (1986) Primary charge separation and excitation of chlorophyll a in Photosystem II particles from spinach as studied by picosecond absorbance-difference spectroscopy. Biochem. Biophys. Acta 848:167-175

21. Eckert, H.J., Wiese, N., Bernarding, J., Eichler,H.-J. and Renger, G. (1988) Analysis of the electron transfer from $Pheo^-$ to Q_A in PS II membrane fragments from spinach by time resolved 325 nm absorption changes in the picosecond domain. FEBS Lett. 240:153-158

22. Renger, G. and Weiss, W. (1986) Functional and structural aspects of photosynthetic water oxidation. Biochem. Soc. Trans. 14:17-20

23. Michel, H., Deisenhofer, J. (1988) Relevance of the photosynthetic reaction center from purple bacteria to the structure of photosystem II. Biochemistry 27:1-7

24. Trebst, A. (1986) The topology of the plastoquinone and herbicide binding peptides of photosystem II in the thylakoid membrane. Z. Naturforsch. 41c:240-245

25. Robinson, H.H. and Crofts, A.R. (1983) Kinetics of the oxidation reduction reactions of the photosystem II quinone acceptor complex and the pathway for deactivation. FEBS Lett. 153:221-226

26. Weiss, W. and Renger, G. (1984) Analysis of the system II reaction pattern by UV-absorption changes in Tris-washed chloroplasts. In: Advances Photosynth. Res. (Sybesma, C. ed.), Vol. I, Martinus Nijhoff, The Hague, pp. 167-170

27. Crofts, A.R., Wraight, C.A. (1983) The electrochemical domain of photosynthesis. Biochim. Biophys. Acta 726:149-185

28. Iwanzik, W., Tevini, M., Dohnt, G., Voss, M., Weiss, W., Gräber, P. and Renger, G. (1983) Action of UV-B radiation on photosynthetic primary reactions in spinach chloroplasts. Physiol. Plant 58:401-407

29. Styring, S., Virgin, I., Ehrenberg, A. and Andersson, B. (1990) Strong light photoinhibition of electron transport in Photosystem II. Impairment of the function of the first quinone acceptor, Q_A. Biochim. Biophys. Acta 1015:269-278

30. Blubough, D.J. and Cheniae, G.M. (1990) Kinetics of photoinhibition in hydroxylamine-extracted Photosystem II membranes: Relevance to photoactivation and sites of electron donation. Biochemistry 29:5109-5118

31. Winget, G.D., Izawa, S. and Good, N.E. (1965) Stoichiometry of photophosphorylation. Biochim. Biophys. Res. Commun. 21:438-443

32. Berthold, O.A., Babcock, G.T. and Yocum, C.F. (1981) A highly resolved, oxygen-evolving photosystem II preparation from spinach thylakoid membranes. FEBS Lett. 134:231-234

33. Völker, M., Ono, T., Inoue, Y. and Renger, G. (1985) Effect of trypsin on PS II particles. Correlation between Hill-activity. Mn-abundance and peptide pattern. Biochim. Biophys. Acta 806:25-34

34. Yamashita, T. and Butler, W.L.(1968) Photooxidation by photosystem II of Tris-washed chloroplasts. Plant Physiol. 43:1978-1984

35. Renger, G., Rettig, W. and Gräber, P. (1991) The effect of UV-B irradiation on the lifetimes of singlet excitons in isolated PS II membrane fragments from spinach. J. Photochem. Photobiol. B:Biology (in press)

36. Eckert, H.-J., Wydrzynski, T. and Renger, G. (1988) The effect of diisothiocyanostilbene-2,2'-disulfonic acid (DIDS), trifluo-perazine and lauroylcholinechloride on P-680$^+$ reduction and oxygen evolution. Biochim. Biophys. Acta 932:240-249

37. Okada, M., Kitajima, M. and Butler, W.L. (1976) Inhibition of Photosystem I and Photosystem II in chloroplasts by UV radiation. Plant & Cell Physiol. 17:35-43

38. Renger, G., Voss, M., Gräber, P. and Schulze, A., Effect of UV irradiation on different partial reactions of the primary processes of photosynthesis, in R.C. Worrest and M.M. Caldwell (eds.), Stratospheric Ozone Reduction, Solar Ultraviolet Radiation and Plant Life, Springer, Berlin, 1986, pp. 171-184

39. Holzwarth, A.R. (1989) Application of ultrafast laser spectroscopy for the study of biological systems. Quart. Rev. Biophys. 22:239-326

40. van de Ven, M., Govindjee, Preston, C., Seibert, M. and Gratton, E. (1990) Chlorophyll a fluorescence lifetime distributions in open and closed photosystem II reaction center preparations. Biochim. Biophys. Acta 1015:173-179

41. Thompson, L.K. and Brudvig, C.W. (1988) Cytochrome b-559 may function to protect photosystem II from photoinhibition. Biochemistry 27:6653-6658

42. Mäennpää, P., Andersson, B. and Sundby, C. (1987) Difference in sensitivity to photoinhibition between otosystem II in the appressed and non-appressed thylakoid region. FEBS Lett. 215:31-36

43. Debus, R.J., Barry, B.A., Sithole, I., Babcock, G.T., McIntosh, L. (1988) Directed mutagenesis indicates that the donor to P680$^+$ in photosystem II is tyrosine-161 of the D1 polypeptide. Biochemistry 27:9071-9074

44. Metz, J.G., Nixon, P.J., Rögner, M., Brudvig, G.W., Diner, B.A. (1989) Directed alteration of the D1 poly-peptide of photosystem II evidence that tyrosine-161 is the redox component, Z, connecting the oxygen-evolving complex to the primary electron donor, P680. Biochemistry 28:6960-6969

45. van Best, I. and Mathis, P. (1978) Kinetics of the reduction of the oxidized primary electron donor of photosystem II in spinach chloroplasts and in Chlorella cells in the microsecond and nanosecond time ranges following flash excitation. Biochim. Biophys. Acta 503:178-188

46. Brettel, K. and Witt, H.T.(1983) Reaction kinetics of the photo-oxidized chlorophyll a_{II} in chloroplasts measured in the nanosecond range at 837 nm under repetitive flash excitation. Photobiochem. Photobiophys. 6:253-260

47. Eckert, H.-J., Renger, G. and Witt, H.T. (1984) Reduction kinetics of the photo-oxidized chlorophyll a_{II}^+. FEBS Lett. 167:316-320

48. Eckert, H.-J. and Renger, G. (1988) Temperature dependence of $P680^+$ reduction in O_2-evolving PS II membrane fragments at different redox states S_i of the water-oxidizing system. FEBS Lett. 236:425-431

49. Gläser, M., Wolff, Ch. and Renger, G. (1976) Indirect evidence for a very fast recovery of chlorophyll a_{II} in spinach chloroplasts. Z. Naturforsch. 32c:712-721

50. Conjeaud, H., Mathis, P. and Paillotin, G. (1979) Primary and secondary electron donors in photosystem II of chloroplasts. Biochim. Biophys. Acta 546:280-291

51. Conjeaud, H. and Mathis, P. (1980) The effect of pH on the reduction kinetics of P-680 in Tris-treated chloroplasts. Biochim. Biophys. Acta 590:353-359

52. Renger, G., Völker, M. and Weiss, W. (1984) Studies on the nature of the water-oxidizing enzyme. I. The effect of trypsin on the system II reaction pattern in inside-out thylakoids. Biochim. Biophys. Acta 766:582-591

53. Renger, G. and Wolff, C. (1976) The existence of a high photochemical turnover rate at the reaction centers of system II in Tris-washed chloroplasts. Biochim. Biophys. Acta 423:610-614

54. Ono, T., Zimmermann, J.L., Inoue, Y. and Rutherford,A.W. (1986) EPR evidence for a modified S-state transition in chloride-depleted photosystem II. Biochim. Biophys. Acta 851:193-201

55. Boussac, A. and Rutherford, A.W. (1988) Nature of the inhibition of the oxygen-evolving enzyme of photosystem II induced by NaCl washing and reversed by the addition of Ca^{2+} or Sr^{2+}. Biochemistry 27:3476-3483

56. Ono, T. and Inoue, Y. (1990) A marked upshift in threshold temperature for the S_1 to S_2 transition induced by low pH treatment of PS II membranes. Biochim. Biophys. Acta 1015:373-377

57. Brettel, K., Schlodder, E. and Witt, H.T. (1984) Dependence of the $Chl\text{-}a_{II}^+$ reduction kinetics on the S state of the oxygen-evolving system. In: Biggins, J. (ed.) Progress photosynthesis research, Vol. I, Martinus Nijhoff, Dordrecht, pp. 295-298

58. Renger, G., Eckert, H.-J. and Völker,M. (1989) Studies on the electron transfer from Tyr-161 of polypeptide D1 to $P\text{-}680^+$ in PS II membrane fragments from spinach. Photosynth. Res. 22:247-256

59. Babcock, G.T. and Sauer, K. (1975) Two electron donation sites for exogenous reductants in chloroplast photosystem II. Biochim. Biophys. Acta 396:48-62

60. Ford, R.C. and Evans, M.C.W. (1985) Electron transfer to P-680$^+$ in active and inhibited photosystem II fractions from higher plants. Biochim. Biophys. Acta 807:1-9

61. Weiss, W. and Renger, G.(1986) Studies on the nature of the water-oxidizing system. II. On the functional connection between the reaction-center complex and the water-oxidizing enzyme system Y in photosystem II. Biochim. Biophys. Acta 850:173-183

62. Petrouleas, V. and Diner, B.A. (1986) Identification of Q_{400} a high potential acceptor of photosystem II with the iron of the quinone-iron acceptor complex. Biochim. Biophys. Acta 849:193-202

63. Diner, B.A. and Petrouleas, V. (1988) Q_{400} the non-heme iron of photosystem II iron-quinone complex. A spectroscopic probe of quinone and inhibitor binding to the reaction center. Biochim. Biophys. Acta 895:107-125

64. Wraight, C.A. (1985) Modulation of herbicide binding by the state of Q_{400}, an endogeneous component of photosystem II. Biochim. Biophys. Acta 809:320-330

65. Renger, G., Wacker, U. and Völker, M. (1987) Studies on the protolytic reactions coupled with waer cleavage in photosystem II membrane fragments from Spinach. Photosynth. Res. 13:167-189

66. Ghanotakis, D.F. and Babcock, G.T. (1983) Hydorxylamine as an inhibitor between Z and P680 in photosystem II. FEBS Lett. 153:231-234

INTERACTION OF HERBICIDE AND ULTRAVIOLET-B RADIATION ON THE PHOTOSYNTHETIC APPARATUS

G. KULANDAIVELU AND K. ANNAMALAINATHAN

School of Biological Sciences
Madurai Kamaraj University, Madurai 625 021, India.

ABSTRACT

When wheat (Triticum vulgare L.) seeds were treated with dichlorophenyl dimethylurea (DCMU) at 50 to 200 μM and allowed to germinate the growth characteristics of the seedlings did not undergo much changes except for a small loss in pigment level. The photosynthetic electron transport activity showed a loss of 50 and 60% at 50 and 200 μM concentrations, respectively. Chloroplasts isolated from such DCMU grown seedlings showed greater resistance to added DCMU concentration suggesting a possible modification of the 32 KDa herbicide binding protein (HBP). UV-B treatment decreased the photosynthetic electron transport, particularly the photosystem II (PSII) activity. Such decrease was less in DCMU grown seedlings indicating the same site of action of DCMU and UV-B radiation. Similarly PSII core complex modified by UV-B radiation was found to be less sensitive to added DCMU. Polypeptide analysis of PSII particles from DCMU grown seedlings revealed stable 32 KDa and LHCII components.

INTRODUCTION

Action of intense ultraviolet-B (UV-B, 280-320 nm) on the photosynthetic apparatus has been studied by several workers (1-5). It has been widely accepted that UV-B radiation affects primarily the PSII electron transport, and the PSI reaction is fairly stable (6). Two major sites in PSII reaction have been shown to be sensitive to UV-B radiation viz., the O_2 evolving site (5,7) and the P680 reaction centre complex (3). Recently it has been found that exposure of Vigna chloroplasts to short time of UV-B radiation induce certain polypeptides, which play a role in protection against UV-B damage (8). This suggests that the thylakoids undergo rapid reorganization in order to protect the senitive components against UV-B damage.

DCMU and other related herbicides are shown to act on the 32 KDa herbicide binding protein (HBP, also known as D_1 and Q_B protein) (9,10), which is one of the three reaction centre proteins (11,12). The 32 KDa protein is rapidly turned over as a function of the visible light intensity and the herbicides diuron and atrazine not only block the electron transport through PSII but also inhibit 32 KDa protein degradation (13,14) by diplacing a liganded quinone (Q_B) from the protein (15). In recent years this D_1 protein has received considerable attention because it is a major product of the chloroplast protein synthesizing machinery (16,17). Since herbicide treatment brings about a modification in this protein it would be of much interest to study whether such altered PSII core organization results in any change in the sensitivity to UV-B radiation as the latter has also been shown to act at the same site.

To study this, the seeds were pre-treated with DCMU at optimal concentrations to produce partial modification (incomplete inhibition of electron transport reaction) of the HBP and then following the response to various doses of UV-B radiation.

MATERIALS AND METHODS

Plant Materials

Triticum vulgare L. seedlings were raised on moist filter papers in petridishes under controlled environment in growth chambers (Forma Scientific, USA). Temperature was maintained at $25 \pm 0.2^{o}C$. Illumination at the seedling surface was 10 $W.m^{-2}$ and a 12 h light/12 h dark cycle was maintained. For all the experiments 10 day-old seedlings were used.

DCMU Treatment

Various concentration of DCMU was prepared from a stock and added to the seeds initially. Thereafter, only tap water was added periodicaly to keep the filter paper in a moistened condition.

UV-B Treatment

Treatment of seedlings

Ten day - old seedlings were brought under enhanced UV-B condition in growth chamber. Enhanced UV-B radiation was provided by three 20 W cool fluorescent lamps and one Philips 20 W/12 sunlamp (Philips, Holland). The irradiance of UV-B at the surface of seedlings was 1 $W.m^{-2}$.

Treatment of chloroplasts

Chloroplasts at a final chlorophyll concentration of 500 µg ml^{-1} in Tris-HCl buffer, pH 7.8 was uniformly spread in a thermostated irradiation chamber (25 mm diameter and 3 mm depth). Temperature during the treatment was maintained at 5 ± 0.2°C using a water bath circulator. Irradiance at the sample surface was 5 $W.m^{-2}$. Radiation below 280 nm was removed by using a Schott WG 290 filter.

Estimation of Pigments

The photosynthetic pigments were extracted in 80% acetone and the concentration of chlorophylls and carotenoids was estimated according to the procedures of Arnon (18) and Mackinney (19), respectively.

Isolation of Chloroplast and PSII Particles

Intact and broken (Type II) chloroplasts were isolated following the method of Reeves and Hall (20). PSII membrane fragments were prepared from thylakoid membrane by solubilization with Triton X-100 according to the procedure of Kuwabara and Murata (21).

Measurement of Photsynthetic Electron Transport Activities

The rate of whole chain electron tranpsort (H_2O --> MV) and PSI ($DCIPH_2$ --> MV) mediated electron transport in isolated chloroplasts were measured as O_2 uptake using Hansatech O_2 electrode for whole chain electron transport, the reaction mixtue in a final volume of 1 ml contained 20 mM Tris-HCl, pH 7.8, 5 mM $MgCl_2$, 10 mM NaCl, 100 mM sorbitol, 1 mM MV, 0.1 mM sodium azide, 5 mM NH_4Cl and chloroplasts equivalent to 20 μg Chl/ml. For PSI measurements the reaction mixture contained 20 mM Tris-HCl, pH 7.8,2 mM sodium ascorbate, 5 mM NH_4Cl, 0.1 mM sodium azide, 50 uM DCPIP, 5 uM DCMU and 1 mM MV and chloroplasts equivalent to 10 μg Chl/ml.

The rate of PSII electron transport (H_2O --> DCPIP) was determined spectrophotometrically. Actinic light (>640 nm) was provided at an irradiance of 150 $W.m^{-2}$. The photomultiplier was shielded by a blue (Corning CS 4-96) filter.

Measurement of Chl a Flourescence

Chl a fluorescence transients (fast kinetic) was followed in intact chloroplasts after excitation with broad band light (400-620 nm, Corning CS4-96) at an irradiance of 100 $W.m^{-2}$. The photomultiplier (Hamamatzu R376) placed at 90° to the excitaion beam was protected by an interference filter (λ max 690 nm). The signal was triggered with the help of an electric shutter with an opening time of 10 ms. The final chlorophyll concentration was 2 μg/ml.

Fluorescence Excitation Spectra

Room temperature fluorescence excitation spectra of chloroplast preparations were followed using a Hitachi MPF4 spectrofluorimeter. The spectra were corrected for the differences in excitation monochromator and photomultiplier response. The chlorophyll concentration was 5 μg/ml.

Polyacrylamide Gel Eelctrophoresis

Thylakoid polypeptide analysis was made using a 8-18% SDS-polyacrylamide gradient gel electrophoresis system, according to the procedure of Lamemmli (22).

RESULTS AND DISCUSSION

With the primary objective to study the interaction of UV-B with herbicides (DCMU) on the photosynthetic reaction, as a first step, the effect of DCMU on growth characteristics was studied. In order to find out the optimum concentration which alters the pigment composition to a minimum extent but produce significant changes in the photosynthetic electron transport reactions, wheat seeds were treated with DCMU in the concentration range of 50 to 200 μM and then allowed to germinate. Changes in the pigment level in control and DCMU treated seedlings are shown in Table 1. The total chlorophyll level showed a

Table 1.

Effect of various concentration of DCMU on the pigment composition of wheat seedlings. Leaves were harvested after 12 days for pigment analysis. Values represent average of three experiments.

Concentration of DCMU	Chl a	Chl b	Total Chl	carotenoids	Chl a/b
	mg / g fresh weight				
Control	1.30	0.51	1.81	0.52	2.55
50 μM	1.10	0.54	1.64	0.48	2.04
100 μM	1.08	0.55	1.63	0.48	1.96
200 μM	1.01	0.56	1.57	0.45	1.80

loss of about 20% with 200 μM DCMU while low concentration (50 μM) caused only marginal change. Very low concentration did not produce any significant change. The loss of chlorophyll was found to be mainly caused by the changes in Chl a level. Further, a 10% increase in this level was observed in 200 μM DCMU treated seedlings. Such differential change in Chl a and Chl b has resulted in progressive decrease in the Chl a/b ratio. The level of carotenoids also had decreased in DCMU treated seedlings. Absence of any change in the pigments indicate that the light harvesting system is not very much affected even with high DCMU concentration.

The photosynthetic electron transport systems particularly the PSII mediated activity showed progressive decrease with increasing concentration of DCMU. Both the overall and PSII activity showed approximately similar changes in the loss of activity with different DCMU concentrations (Table 2). Even after treatment with 200 uM DCMU,

Table 2.

Effect of various concentration of DCMU on the photosynthetic partial reactions of wheat chloroplasts. Concentration of MV and DCPIP was 1 and 0.05 mM, respectively. Values represent average of three individual measurements. Figures in parantheses indicate percentage of respective controls.

Concentration of DCMU	Whole chain electron transport H_2O---> MV	PSI $DCPIPH_2$ --->MV	PSII H_2O--->DCPIP
	μmol O_2 mg $chl^{-1}.h^{-1}$	μmol DCPIP mg $chl^{-1}.h^{-1}$	
Control	125 (100)	384 (100)	169 (100)
50 μM	68 (54.4)	418 (108.9)	117 (69.2)
100 μM	49 (39.2)	421 (109.6)	70 (41.4)
200 μM	46 (36.8)	421 (109.6)	64 (37.9)

as much as 36% of control PSII activity was found while with 50 μM DCMU it was more than 50%. This indicates that <u>in vivo</u> treatment produced partial modification of the electron transport intermediates, particularly the 32 KDa HBP. Under <u>in vitro</u> condition DCMU even at 1 μM brings about complete inhibition of electron transport activity.

The fact that in vivo DCMU treatment causes only partial modification of the HBP is also evidenced by the fluorescence measurements. Chloroplasts isolatd from control wheat seedlings showed the typical fast fluorescence induction with high level of variable part and addition of 5 μM DCMU completely removed the varibale part and also increased the fluorescence level (Fig.1). Chloroplasts isolated from 50 and 100 μM DCMU treated seedlings

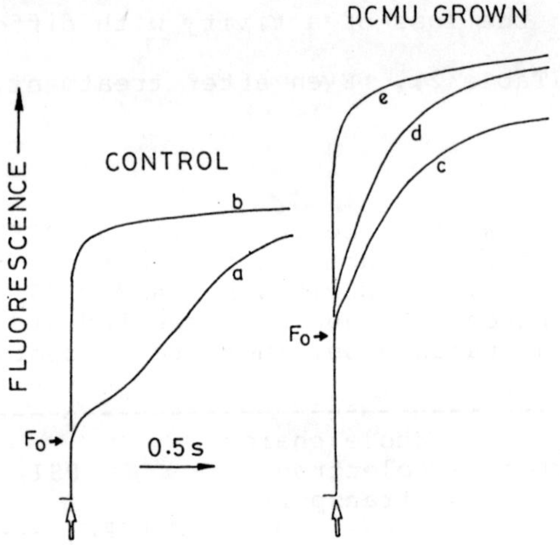

Fig. 1.
Typical kinetics of Chl a fluorescence induction in wheat chloroplasts isolated from control (a,b) and DCMU grown (c-e) seedlings. a. control (untreated), b. +5 μM DCMU, c. 50 μM DCMU grown, d. 100 μM DCMU grown, e. 50 μM DCMU grown + 5 μM DCMU. Other conditions as under materials and methods.

also showed varable fluorescence although a typical dip and slow increase were not present. Increasing concentration of DCMU treatment had proportionately enhanced the Fp level and also reduced the t1/2 for variable fluorescence induction. Nevertheless these chloroplasts responded to the added DCMU indicating that the DCMU action sites in HBP are still open.

Fluorescence excitation spectra of chloroplasts isolated from control and DCMU grown seedlings were followed to check for any organizational differences in the pigment protein complexes. Although the fluorescence excitation spectrum of chloroplasts form DCMU treated plants differed quantitatively over the control, no significant qualitative difference was found (Fig.2). This again support the data of pigment analysis.

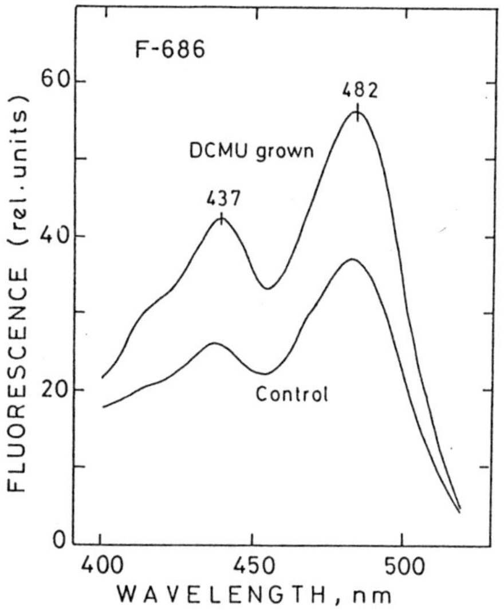

Fig. 2.
Room temperature excitation spectra of chloroplasts isolated from control and 50 uM DCMU grown wheat seedlings. Spectra were followed for F-686. Chlorophyll concentration of both samples was 5 µg/ml.

Fluorescence transients measurements in chloroplasts in the presence and absence of DCMU indicates that in DCMU grown plants although the HBP is partially modified still responds to further addition of DCMU. To check this further and also find out the optimal inhibiting concentration of DCMU in control and DCMU grown seedlings, rate of PSII electron transport was followed in the presence of varying concentration of DCMU. A plot of the ratio of control rate of PSII (Vo) and rate at specific DCMU concentration (V) against DCMU concentration indicates the requirement of almost two-fold higher concentration in DCMU grown seedlings as compared to control (Fig.3). Such a plot was shown to reveal the concentrations for 50% inhibition at the Vo/V ratio of 2 (23). Higher DCMU requirement in DCMU grown seedlings indicates a possible modification of the HBP (14,24).

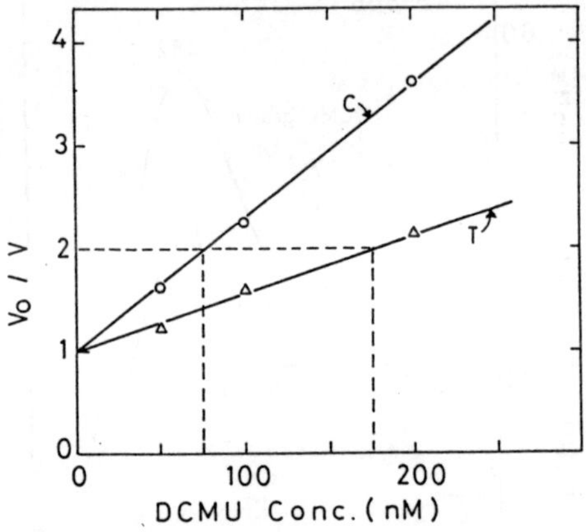

Fig. 3.
Ratio of control PSII rate and rate of specific DCMU concentration (v) against DCMU concentration for chloroplasts isolated from control and 50 μM DCMU grown seedlings.

From the above data, it is clear that DCMU treatment at 50 uM concentration brings about partial loss of electron transport and also produced significant modification of the PSII core complex particularly the HBP. To check the interaction of UV-B radiation on the DCMU induced changes, as a first step, the behaviour of various electron transport reactions to UV-B treatment was followed in chloroplasts isolated from control and DCMU grown seedlings. As shown by many workers (4,25,26), the PSI activity was less sensitive than PSII. As much as 90% loss of overall and PSII activity was seen in 60 min UV-B treated control chloroplasts (Fig.4). Chloroplasts

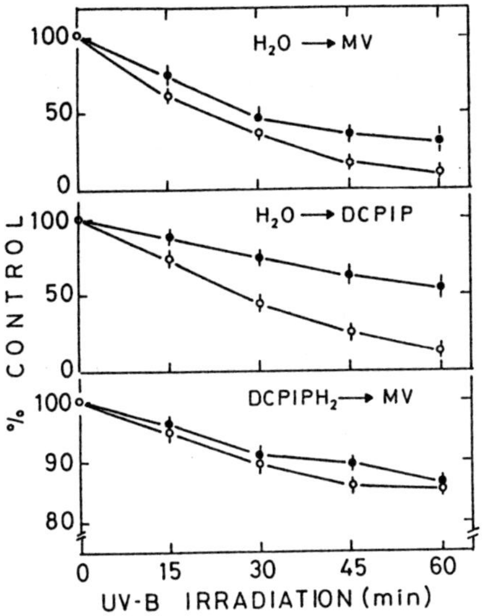

Fig. 4.
Changes in overall and partial photochemical activities in chloroplasts isolated from control (o o) and 50 uM DCMU (o o) grown wheat seedlings as a function of time of UV-B irradiation. The 100% values are H_2O --> MV,82; $DCPIPH_2$ --> MV, 315 umol O_2.mg $Chl^{-1}.h^{-1}$, H_2O -->DCPIP, 162 umol DCPIP mgChl^{-1}. h^{-1}.

isolated from DCMU grown seedlings showed resistance to UV-B treatment and approximatly 50% of initial activity was noticed even in 60 min UV-B treated sample. Noorudeen and Kulandaivelu (3), based on various electron transport activities and fluorescence measurements have concluded that UV-B irradiation on Amaranthus chloroplasts acts primarily at the PSII reaction centre complex. Changes in the reaction centre complex could be either at the level of D_1 or D_2 or on both proteins. Above experiments in wheat indicates i) modification of HBP and other PSII core complex proteins decreased the UV-B sensitivity and ii)

UV-B treatment besides affecting the PSII core complex also affects the H_2O oxidation site. In many plants UV-B radiation was shown to inactivate the O_2 evolving site by depleting the 33 or 23 KDa polypeptides (8).

To confirm the possible action of UV-B and also DCMU on O_2 evolution site PSII activity was followed with artificial electron donors. DPC and NH_2OH were used as electron donors. DPC was known to donate electron directly to PSII reaction centre while NH_2OH at a site between Z1 and Z2 (27). In control chloroplasts UV-B treatment progressively decreased the PSII activity. As much as 64% loss of PSII activity was noticed when H_2O was used as electron donor (Table 3). However, in the presence of artificial electron donor the extent of UV-B inhibition was low. Among the two electron donors tried DPC was found to be more effective than NH_2OH. This is in agreement with the previous report (3). DPC --> DCPIP reaction was reduced only by 30% after 45 min of UV-B treatment. This indicates that UV-B irradiation of wheat chloroplasts affects both the O_2 evolving site and also the PSII core complex.

Measurements of PSII activity with H_2O and other artificial electron donors in chloroplasts isolated from DCMU grown seedling showed interesting results. In these chloroplasts the H_2O --> DCPIP reaction was found to be inhibited to lesser degree as compared to the control chloroplasts. The extent of inhibition was approximately 30% indicating that in the DCMU grown seedlings due to stabilization of the HBP and PSII core complex, UV-B radiation had poor effect at this site, however, produced normal change at the

Table 3.

Restoration of PS II mediated Hill activity by the addition of electron donors like DPC and NH_2OH in UV -B treated and heat inactivated chloroplasts of control and 50 μM DCMU grown wheat seedlings. Figures in parentheses indicate the percentage of respective control activity. The water oxidizing complex was heat inactivated by incubating the chloroplasts at 45 °C for 5 minutes.

Treatment	PS II activity H_2O--->DCPIP	 DPC---> DCPIP	 NH_2OH-->DCPIP
	μmol of DCPIP reduced mg $chl^{-1}.h^{-1}$		
Control	159.0 (100)	149.0 (100)	128.0 (100)
+ 30 min UV-B (<u>in vitro</u>)	69.0 (43.4)	113.0 (75.8)	73.0 (57.0)
+ 45 min UV-B	58.0 (36.5)	105.0 (70.5)	71.0 (55.5)
50 μM DCMU grown	92.0 (100)	87.0 (100)	87.0 (100)
+ 30 min UV-B	71.0 (77.2)	92.0 (105.7)	72.5 (83.3)
+ 45 min UV-B	63.0 (68.5)	84.0 (96.6)	70.4 (81.0)

O_2 evolving site. This conclusion is supported by the rate of DCPIP reduction observed with artificial electron donors DPC and NH_2OH, where DPC mediated reaction showed only marginal loss.

The competitive action of UV-B and DCMU on the HBP PSII core was also checked by following the effect of DCMU at different concentration in UV-B treated chloroplasts. In control chloroplasts, DCMU, with increasing concentration inhibited the PSII activity proportionately and as much as 70% inhibition had occured at 100 nM DCMU. UV-B irradiation reduced the control activity level, however, decreased the extent of DCMU inhibition. Renger *et al.* (5) investigated modification of the D1/D2 complex by UV-B radiation and they have found that UV-B irradiation markedly reduces the number of ^{14}C-atrazine binding sites.

All these experiments indicate that DCMU pre treatment produced modification in the HBP-PSII core thereby reduce the sensitivity UV-B radiation and UV-B radiation affects both the O_2 evolving complex and the PSII core in wheat chloroplasts which is different from those reported for <u>Amaranthus</u> (3) and <u>Vigna sinensis</u> (8).

Above facts are supported by direct evidences obtained by the analysis of polypeptide composition of the PSII and chloroplasts. Fig.5 shows the polypeptide profiles of the PSII particles isolated from control and DCMU grown wheat

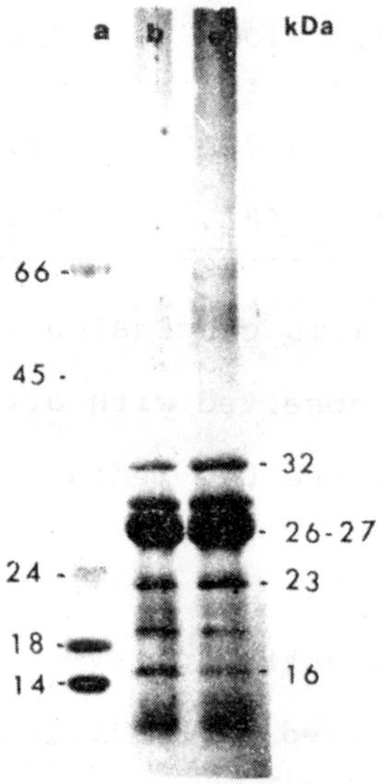

Fig. 5.

Typical SDS-PAGE electrophoretic pattern of thylakoid proteins of control and 50 μM DCMU grown wheat seedlings. a. marker proteins; b. control; c. 50 μM DCMU grown.

seedlings. In DCMU grown seedling although the level 16 and 19 KDa polypeptides showed a decrease, considerable increase was observed in the level of 32-33, 26-27 and 23 KDa polypeptides. The 32 KDa polypeptide showed large increase in its level. Similarly the 26-27 KDa, chlorophyll containing polypeptides also increased indicating possible modification in the PSII core complex.

ACKNOWLEDGEMENTS: We thank the Department of Science and Technology, Govt. of India for financial support. Senior Research Fellowship awarded to one of us (K.A) by CSIR is gratefully acknowledged.

REFERENCES

1. Brandle, J.R., Campbell,W.F., Sosson,W.B. and Caldwell, M.M. (1977). Net Photosynthesis, electron transport capacity, and ultrastructure of Pisum sativum L. Exposed to ultraviolet-B radiation. **Plant Physiol.** 60: 165-168.

2. Van, T.K., Garrard, L. and West, S.H. (1977). Effects of 298 nm radiation of photosynthetic reactions of leaf discs and chloroplast preparations of some crop species. **Environ. Exp. Bot.** 17: 107-112.

3. Noorudeen, A.M. and Kulandaivelu, G. (1982). On the possible site of inhibition of photosynthetic electron transport by UV-B radiation. **Physiol. Plant.** 55: 161-166.

4. Iwanzik, W., Tevini, M., Dohnt, G., Voss, M., Weiss, W, Graber, O. and Renger, G. (1983). Action of UV-B radiation on photosynthetic primary reactions in spinach chloroplasts. **Physiol. Plant.** 58: 401-407.

5. Renger, G., Volker, M., Eckert, H.J., Fromme, R., Hohn-Veit, S. and Graber, P. (1989). On the mechanism of photosystem II deterioration by UV-B irradiation. **Photochem. Photobiol.** 49: 97-105.

6. Bornman, J.F. (1989). Target sites of UV-B radiation in photosynthesis of Higher plants. **Photochem.** Photobiol. 4: 145-158.

7. Bornman, J.F., Björn, L.O. and Åkerlund, H-E. (1984). Action spectrum for inhibition by ultraviolet radiation of photosystem II activity in spinach thylakoids. **Photobiochem. Photobiophys.** 8: 305-313.

8. Nedunchezhian, N. and Kulandaivelu, G.(1990). Evidences for the ultraviolet-B (280-320 nm) radiation induced structural reorganization and damage of photosystem II polypeptides in isolated chloroplasts. **Physiol. Plant.** (in press).

9. Mullet, J.E. and Arntzen, C.J. (1981). Identification of 32-34 kilodalton polypeptide as a herbicide receptor protein in photosystem II. **Biochim. Biophys. Acta** 635: 236-248.

10. Pfister, K., Steinback, K.E., Gardner, B. and Arntzen, C.J. (1981). Photoaffinity labelling of an herbicide receptor protein in chloroplast membranes. **Proc. Natl. Acad. Sci.** 78: 984-985.

11. Nanba, O. and Satoh, K. (1987). Isolation of a photosystem II reaction centre consisting of D_1 and D_2 polypeptides and cytochrome b559. **Proc. Natl. Acad. Sci.** USA. 84: 109-112.

12. Chapman, D.J., Gounaris, K. and Barber, J. (1989). The D1/D2 cytochrome b559 photosystem two reaction centre from Pisum sativum L. Isolation, characterization and damage by light. **Photosynthetica** 23:411-426.

13. Mattoo, A.K., Pick, U., Holffman-Falk, H. and Edelman, M. (1981). The rapidly metabolized 32000-dalton polypeptide of the chloroplast is the `proteinaceous shield' regulating photosystem II electron transport and mediating diuron herbicide sensitivity. **Proc. Natl. Acad. Sci.** 78: 1572-1576.

14. Gaba, V., Marder, J.B., Greenberg, B.M., Mattoo, A.K. and Edelman, M. (1987). Degradation of 32 KD herbicide binding protein in far red light. **Plant Physiol.** 84: 348-352.

15. Vermaas, W.F.J., Arntzen, C.J., Gu, L.O and Vu, C.A. (1983). Interaction of herbicides and azoquinones at photosystem II binding in thylakoid membranes. **Biochim. Biophys. Acta** 723: 266-275.

16. Marder, J.B., Mattoo, A.K. and Edelman, M. (1986). Identification and characterization of the psb A gene product the 32 KDa chloroplast membrane protein. **Methods Enzymol.** 118: 384-396.

17. Mattoo, A.K., Marder, J.B. and Edelman, M. (1989). Dynamics of photosystem II reaction center. **Cell** 56: 241-246.

18. Arnon, D.I. (1949). Copper enzymes in isolated choroplasts. Polyphenoxidase in Beta vulgaris. **Plant Physiol** 24: 1-15.

19. Mackinney, G. (1941). Absorption of light by chlorophyll solutions. **J. Biol. Chem.** 140: 315-322.

20. Reeves, S.G. and Hall, D.O. (1973). The Stoichiometry (ATP/2e ratio) of non-cyclic photophosphorylation in isolated spinach chloroplasts. **Biochim. Biphys. Acta** 314: 66-78.

21. Kuwabara, T. and Murata, N. (1982). Inactivation of photosynthetic oxygen evolution and concomitant release of three polypeptides in the photosystem II particles spinach chloroplasts. **Plant Cell Physiol.** 23: 533-539.

22. Laemmli, U.K. (1970). Cleavage of structural proteins during the assembly of the head of bacteriophage T4. **Nature** 227: 680-685.

23. Galloway, R.E. and Mets, L. (1982). Non-Mendelian inheritance of 3-(3,4-Dichlorophenyl)-1,1-dimethylurea resistant thylakoid membrane properties in chlamydomonas. **Plant Physiol.** 70: 1673-1677.

24. Mattoo, A.K., Hoffman-Falk, H., Marder, J.B. and Edelman, M. (1984). Regulation of protein metabolsim. Coupling of photosynthetic electron transport to in vivo degradation of the rapidly metabolized 32-Kilodalton protein of the chloroplasts membrane. **Proc. Natl. Acad. Sci.** USA 81:1380-1384.

25. Kulandaivelu, G. and Noorudeen, A.M. (1983). Comparative study of the action of UV-C and UV-B radiation on photosynthetic electron transport. **Physiol. Plant.** 58: 389-394.

26. Renger, G., Graber, P., Döhnt, G., Hagemann, R., Weiss, W. and Voss, R. (1982). The effect of UV irradiation of primary process of photosynthesis. In: **Biological effects of UV-B radiation,** (Bauer, H., Caldwell, M.M., Tevini, M. and Worrest, R.C. eds). Gesellschaft für Strahlen und umweltforschung mbH, Munchen pp. 110-116.

27. Wydrzynski, R. and Govindjee (1975). A new site of bicarbonate effect in photosystem II of photosynthesis. Evidence from chlorophyll fluorescence transients in spinach chloroplasts. **Biochim. Biophys. Acta** 387: 403-408.

UV-B EFFECTS ON THE DEVELOPMENT OF PHOTO-SYNTHETIC APPARATUS, GROWTH AND PRODUCTIVITY OF HIGHER PLANTS

Yu.E. Giller

Institute of Plant Physiology and Biophysics of the Academy of Sciences of Tajik SSR, Dushanbe, USSR

Abstract - In laboratory and field experiments having used erythemal lamps the influence of zone B ultraviolet irradiation (UV-B, 280-320 nm) on photosynthetic apparatus of young cotton, barley and arabidopsis plants (single irradiation), triticale, pea and cotton chloroplasts pigment system, growth and development (irradiation of etiolated seedlings during greening) as well as on chloroplasts pigment system, growth and productivity of cotton and soybean (irradiation during vegetation of plants in field conditions) has been studied. An idea of relationship between plants resistance to UV-B and their taxonomic state has been confirmed. It has been shown that response feature of a chloroplasts pigment system to UV-B irradiation depends on the age of leaves: young ones have deep, but reversible damages, old leaves have only quantitative and not reversible damages. There has been revealed a significant role played by visible light intensity in the resistance of photosynthetic apparatus pigment system formation and plants growth function to UV-B irradiation. Field experiments have shown that in similar conditions the reduction of cotton and soybean plants productivity is much the same due to the increase of UV-B intensity.

INTRODUCTION

Due to anthropogenic atmosphere polution resulting in depletion of the stratospheric ozone layer the irradiation of the Earth's surface by ultraviolet (UV) radiation is increasing. The most changes are taking place within the spectral range of 290-320 nm (middle-wave UV irradiation - zone B UV radiation - UV-B) /I,2/.Revealing plants response to an increasing UV-B intensity is one of the most important trends in the evaluation of ecological and economic consequences of the ozone layer reduction. It is already known that UV-B mostly affects negatively practically on all physiological reactions of a plant organism /3-6/. In view of the mentioned above the study of an affect of increased UV-B compared with natural intensity on the production process of plants has acquired a great theoretical and applied significance. The results of such investigations are necessary to forecast the consequences of ozone lauer reduction as related to changes in productivity of Earth's vegetable life.

The main component of the plants production process is photosynthesis. The known by now UV irradiation action spectra on photosynthetic reaction either enable to admit the existence of a considerable maximum over the UV-B short-wave limit /7,8/, or they have a maximum of 270-290 nm with a marked shoulder of 320-330 nm /9,10/. The maximum is within the DNA, RNA and protein absorption range, and probably it is to a greatest extent due to UV-inactivation of photocatalitic centres of photosynthesis, especially those of photosystem II centres /1,8,11-14/. The shoulder of the UV-B action spectrum on photosynthesis might be associated with the absorption of UV-B by chloroplasts pigments, since a chlorophyll and some carotenoids have specific absorption bands within the range of 290-320 nm. The intensity of the latter is always 2-4 times weaker than those of main bands /15/. In addition recent calculations have shown that it is precisely within range of 310-330 nm the solar effectiveness spectra of the plants damage have (depending on the action spectra used) sufficiently wide maxima /8/. Thus the study of UV-B effects on the chloroplasts pigment system is significant not only for the evaluation of photosynthetic apparatus changes, indicative of development and functional wholeness of which is the state of pigment system /16-18/, but as well as for the understanding of mechanisms and consequences of UV-B interaction with one more group of acceptors in a plant cell.

Another important component of plant production process is a growth function. It is already known that the increase of UV-B intensity as a rule affects negatively on the growth and development of higher plants, but the magnitude of the effects depends on taxonomic state and cultivation conditions /6,19,20/. This suggests the importance of information about response of growth function of various species of cultived plants to UV-B in different growing conditions and different stages of ontogenesis.

In this communication the results of investigation of our laboratory on the UV-B effects on the chloroplasts pigment system and functional activity, on plants growth and their productivity have summarized. The main attention has been concentrated on the study of cotton plant, most important agricultural plant of the Central Asiatic region of the USSR.

MATERIALS AND METHODS

Cultivating of Plants

A commercial variety of cotton 108-F (Gossypium hirsutum L.) has been grown on sand in a complete darkness at 26±2°C during 5 days (to obtain etiolated seedlings), in the soil at 24±1°C under light intensity of 35.5 W/m^2 with photoperiod of 14 hrs and in the field according to standard agrotechnological practice.

Barley (Hordeum vulgare L.) has been grown in the soil at 24±I°C under light intensity of 35.5 W/m^2 with photoperiod of I4 hrs.

Arabidopsis (Arabidopsis thaliana L. Heynh.) has been grown in the soil at 20±I°C under light intensity of 42.6 W/m^2 with photoperiod of I8 hrs.

Pea (Pisum sativum L.) and triticale (Triticale) have been grown on a filter-paper moistened with nutrient solution in a complete darkness at 24±I°C.

Soybean (Glicine hispida Maxim.) has been grown in the field according to standard agricultural practice.

UV-B Irradiation of Plants

Erithemal lamps EL-30 were used as an radiation source. Energy distribution within irradiation spectrum 250-4I0 nm constituted (%): 3(250-280 nm), 50(280-320 nm) and 47(320-4I0 nm). Within the UV-B range the spectral characteristics of lamps and the Sun qualitatively differed small /20/.

Single irradiation of plants was carried out at UV-B intensity of 3-4 W/m^2 during I hr (cotton) and 2 hrs(barley, or 9.3 W/m^2 during 2-3 hrs (arabidopsis) and visible light intensity and temperature as which these plants were being cultivated.

During experiments etiolated plants were in luminostates : UV-B - 8, 3 or I.5 W/m^2 under the visible light intensity of 35.5 or I7.8 W/m^2 with the photoperiod of I4 hrs (cotton), or under a continuous lighting (triticale and pea) at temperature ranges of 22-29°C (in the daytime) and I5-I8°C (at night).

In field experiments the plants were exposed to UV-B daily from I0 to I6 oclock by local time (during the Sun's apogee). The intensity of artifical UV-B was 0.6I and 0.I6 W/m^2 (cotton) and 0.56 W/m^2 (soybean). Considering that natural UV-B intensity at the Dushanbe latitude at the Sun's apogee is approximately equal to 3 W/m^2, the additional UV-B intensity constituted 20 and 5% for cotton and I9% for soybean.

In all investigations controls were from the same plants, but without UV-B irradiation.

Analyses

Plastid pigments concentration in leaves were determined by using spectrophotometry of acetone extracts from leaf samples of specified mass and area /2I,22/. Chlorophyll native state changes were estimated by low-temperature fluorescence spectra of leaves /I6/.

Photosynthetic apparatus functional activity (electron transfer rate, photosystem activity) was evaluated by conventional methods based on the photochemical activity recording and "light - dark" absorption differential spectra of isolated chloroplasts and kinetics of leaves fluorescence /I9,23/.

Growth and productivity parameters were measured on model plants the number of which in every experiment was sufficient for obtaining statistically reliable data.

RESULTS AND DISCUSSION

Single Irradiation Effects on Photosynthetic Apparatus

It was established /24/ that during three days after irradiation the chlorophyll content in young cotton leaves calculated on green mass and particularly on dry mass increased and calculated on a leaf area unit it practically did not change, while carotenoids concentration increased independent on the means of calculation. In the similar conditions concentration of all plastid pigments in barley leaves calculated on mass did not change, but calculated on area reduced by 25-30%. No variations were observed in control leaves (Table I). In both types under study the drying of UV-B irradiated leaves was observed. That fact with no pigment quantity change in cells should have brought about an apparent increase in their concentration in a leaf especially calculated on green mass. The data obtained show that UV-B effect speed up pigment distruction process in barley to a greater extent than in cotton.

Leaves low-temperature fluorescence spectral variations under the UV-B affect were within the intensity increase band with their maxima of 685 and 695 nm which might have been associated with a partial distruction of the chlorophyll native form system. By this index too cotton plant appeared to be more stable compared with barley.

Single UV-B irradiation suppressed functional acti-

Table I

UV-B effect on plastid pigment content in plant leaves

Index	Controls			Experiment		
	Chlorophyll		Carotenoids	Chlorophyll		Carotenoids
	a	b		a	b	
			Cotton plant			
mg/g green mass	I.0±0.I	0.5±0.I	0.2±.02	I.6±0.3	0.4±0.I	0.5±0.I
mg/g dry mass	7.0±0.6	3.5±0.7	I.7±0.5	I0.±0.6	3.0±0.4	3.0±0.4
mg/dm^2	2.0±0.2	I.0±0.I	0.5±0.I	2.7±0.6	0.7±0.I	0.8±0.I
			Barley			
mg/g green mass	0.6±.04	0.2±.05	0.3±.0I	0.4±.03	0.2±.0I	0.3±0.I
mg/g dry mass	5.3±0.6	2.0±0.6	2.5±0.6	4.7±0.2	2.0±0.5	3.0±0.7
mg/dm^2	0.8±.05	0.3±.08	0.4±.02	0.5±.04	0.2±.02	0.3±.02

vity of cotton photosynthetic apparatus. In particular, electron-transfer rate in chloroplasts decreased by 25%

and photosystem II response centres activity reduced considerably /I9,23/, while no marked changes of photosystem I activity were observed /23/. This fact complies with the idea of its strong resistance to UV-B /II/.

In the experiments on arabidopsis the single UV-B irradiation affect on the photosynthetic apparatus pigment system depending on the plant age was studied /25/. As shown in Fig.IA, chlorophyll a concentration in rosette leaves during their development in the process of plant onthogenesis changed markedly. By the 20-th day this index reached its peak, then it decreased and from 22 to 26 days remained practically steady. Every time plants irradiation by UV-B caused reduction of pigment concentration in leaves, while differences between the variants "experiment" and "controls" keeping reliable at

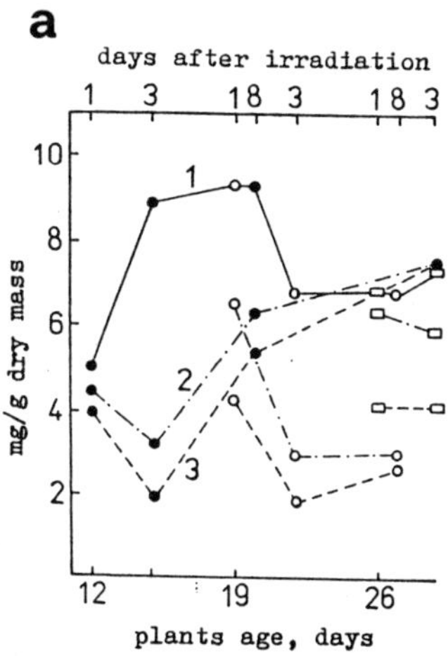

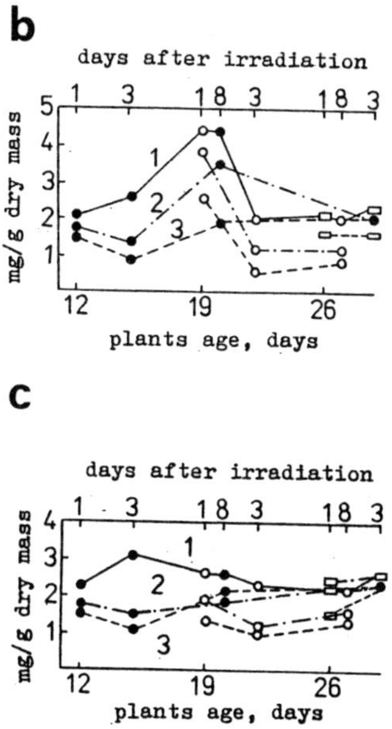

Figure I. UV-B affect on plastid pigments content in arabidopsis leaves
Notations : A - chlorophyll a, B - chlorophyll b, C - carotenoids; I - controls, 2 - UV-B irradiation 2 hrs (24 kJ/m^2), 3 - UV-B irradiation 3 hrs (36 kJ/m^2); ● - I2-day-old plants, ○ - I9-day-old plants, ▭ - 26-day-old plants

99% confidential interval.

Response character of the chloroplasts pigment system to UV-B action depended significantly on the plant's age. Thus irradiation of I2-day-old seedlings did not only reduce chlorophyll a concentration in their leaves, but it resulted in a reverse course (compared with controls) of time dependence of the index under study, since the plants of "experiment" variant by 22-26 days did not already differ from those of controls, i.e. the damages were reversible. UV-B irradiation of I9 and 26-day-old plants caused only quantitative, but irreversible changes (Fig.IA).

The doze dependence of the effects observed was rather trivial. As doze increased, changes increased too.

Practically similar data , but with less quantitative differences between the experiment variants were obtained for chlorophyll b (Fig.IB). This result agrees well with a known idea about more strong than chlorophyll a links of this pigment with membrane proteins /26/ and as a consequence its stronger resistance to stress effects.

Plants irradiation by UV-B decreased the carotenoids content in arabidopsis leaves. Changes were in a whole the same as considered for a chlorophyll, but they did not depend on irradiation time duration (Fig.IC).

The experimental results on a single irradiation confirmed a notion about the relationship between the chloroplasts pigment system response to UV-B irradiation and the plant variety affiliation /6/. Dependence nature of this response on the leaves age was established. It was shown that functional activity of a photosynthetic apparatus of cotton plant is damaged by UV-B earlier than its pigment system which is more resistant in cotton than in barley.

UV-B Effects on the Development of Chloroplasts Pigment System

As may be inferred from data in Fig.2 /25/ in the course of triticale seedlings greening under UV-B irradiation there was accumulated less chlorophyll a in leaves

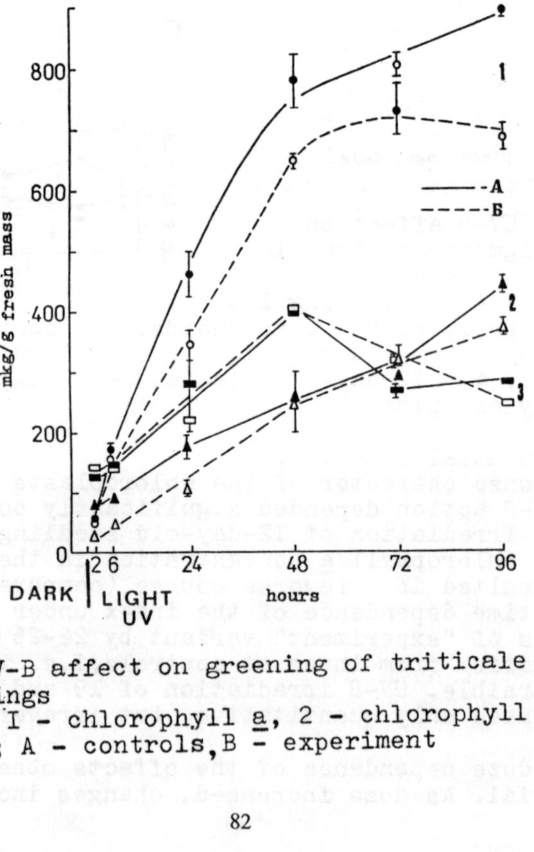

Figure 2. UV-B affect on greening of triticale postetiolated seedlings
Notations : I - chlorophyll a, 2 - chlorophyll b, 3 - carotenoids; A - controls, B - experiment

than in controls. Differences between chlorophyll a concentration were not great and they were reliable only during first days of greening. As to carotenoids concentration no significant differences were observed.

During greening stage in the conditions of increased UV-B intensity of pea etiolated seedlings (Fig.3) within first days of plants growth in light the differences from the controls were not great, but during 6 to 24 hrs the process of chlorophyll a accumulation speeded up. Even in the second day the shlorophyll and carotenoids accumulation in irradiated plants markedly slowed down /25/.

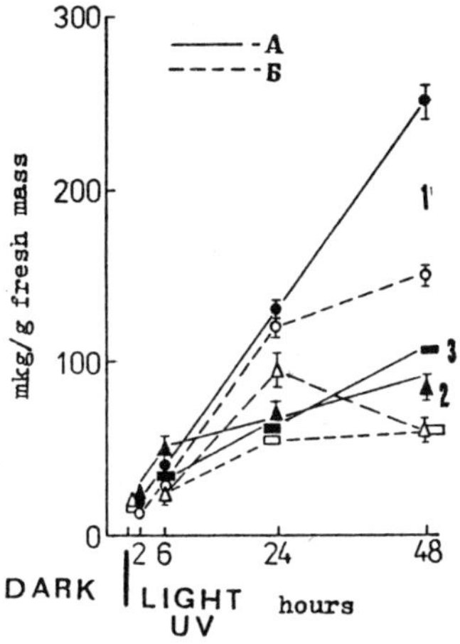

Figure 3. UV-B affect on greening of pea postetiolated seedlings
Notations : I - chlorophyll a, 2 - chlorophyll b, 3 - carotenoids; A - controls, B - experiment

In the experiments with postetiolated cotton plants /27/ three UV-B intensities were tested. As shown in Fig.4, under a greater visible light intensity among intensities tested UV-B irradiation practically did not affect on a chlorophyll a accumulation in postetiolated cotton seedlings. Under a less lighting a chlorophyll a accumulated more in cotyledon leaves of a "control"variant (at the expense of a longer period of linear growth of its concentration), but under UV-B effect beginning from 3-4 days of greening markedly slowed down. Similar fact was observed in the experiments with triticale too (Fig.2).

Just the same occurrence was observed for chlorophyll b, but with quantitatively less effect (Fig.5), similar to experiments with arabidopsis (Fig.I). In addition by the end of the experiment under a minimum intensity among UV-B tested ones there was observed a certain tendency towards chlorophyll b accumulation stimulation. In the experiments with pea similar fact was revealed in the beginning of greening (Fig.3).

The data in Fig.6 suggests that kinetics of carote-

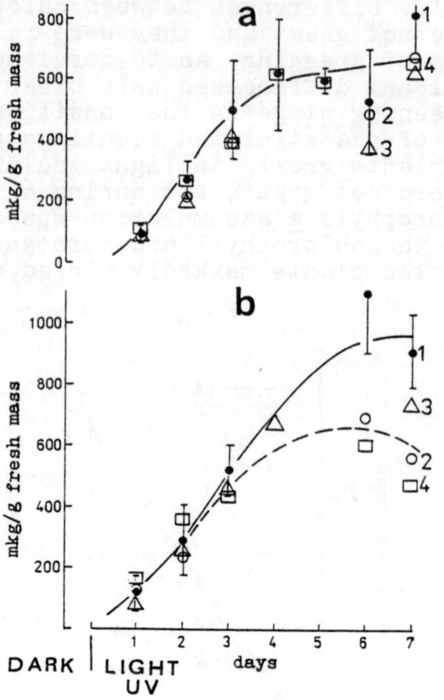

Figure 4. UV-B affect on chlorophyll _a_ accumulation in postetiolated cotton seedlings
Notations : I - controls, 2 - UV-B, 8 W/m^2, 3 - UV-B, 3 W/m^2, 4 - UV-B, I.5 W/m^2; A - 35.5 W/m^2, B - I7.8 W/m^2

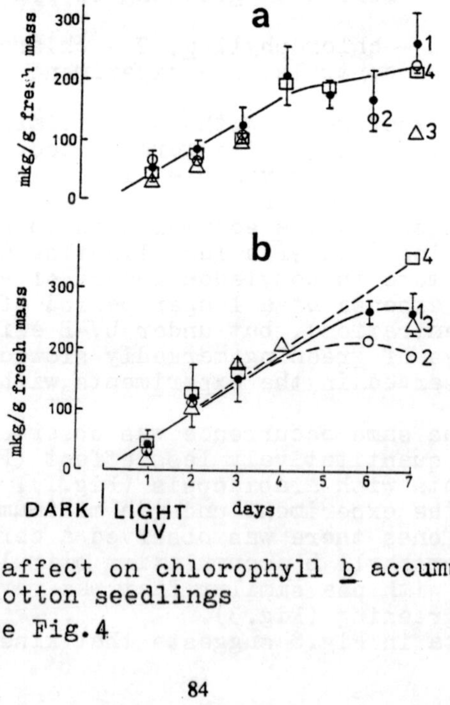

Figure 5. UV-B affect on chlorophyll _b_ accumulation in postetiolated cotton seedlings
Notations : see Fig.4

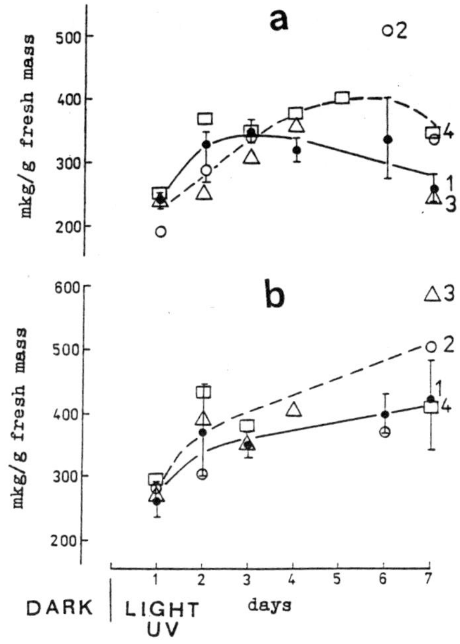

Figure 6. UV-B affect on carotenoids accumulation in postetiolated cotton seedlings
Notations : see Fig.4

noids accumulation in the course of greening postetiolated cotton seedlings depends on a visible light intensity. Having reached the value of 350 mkg/g during three days, the pigment concentration in cotyledon leaves under a greater visible light intensity slightly decreased (similar to experiments with triticale - Fig.2), and under a less one it increased up to 400 mkg/g.

UV-B irradiation stimulated carotenoids accumulation but the character of the effect also depended on lighting. Under a greater lighting a pigment accumulation rate in leaves reduced, but the process proceeded longer and a maximum carotenoids concentration was achived on the fifth day. At a less lighting a pigment accumulation rate merely increased. The effects observed did not depend on UV-B intensity. Carotenoids accumulation in leaves stimulated by UV-B was previously observed in cucumber seedlings /6/ and at a single irradiation of cotton plant (Table I).

The experimental results on postetiolated seedlings showed a high sensitivity of chloroplasts pigment system formation to UV-B. In the above experiments as in the case with single irradiation experiments of young plants a relatively high (compared with other plants studied) photosynthetic apparatus pigment system resistance of cotton to UV-B damage effect.

The most important result of the above experiments was the discovery of a significant role plaued by a visible light intensity in forming photosynthetic apparatus resistance to UV-B. Such a phenomenon was previously revealed while studying UV-B effect on photosynthetic intensity /28/, plant growth and its productivity /20,29/. What is the reason for this effect - either it is only due to photoreactivation of genome distructions /30/, or again and again physiological processes depending on light /20,28,29/ - is still not clear. Nevertheless the assumption that one of the reasons for the above could be the increase of plastid pigments concentration in leaves, our data (Figs.4-6) confirm as to carotenoids only.

Effect on Growth and Development of Cotton Seedlings

In these experiments UV-B affect on cotton plant in the conditions imitating the onset of plant autotrophic development since its coming out of soil was investigated: postetiolated seedlings growing in light were studied /3I/. In cotton-growing areas of the USSR located in the northen borders of this culture's distribution the initial vegetation stages of cotton plant proceed in the least favourable environment, thus increasing the importance of information on another ecological stress potentially conceivable in this period. Moreover, in the experiments with soybean it was established that the most sensitive to damage were the young leaves /32/.

During experiments (7-8 days) control seedlings were developing normally. After 2-3 days a bronze-brown "tan" began to apper on "experiment" variant leaves which were more pronounced at great UV-B intensities. By the end of the experiment stems of these plants became thin and dry and cotyledon leaves shed. In such conditions the postetiolated pea seedlings perished too /25/.

Fig.7 demonstrates seedlings growth dynamics in relative units the choise of which was prompted by necessity to have a growth function activity index independent on temperature /3I/. It is obvious that UV-B at all light modes tested suppressed cotton plant growth at a very early stage of development. This result on cotton plant compliments similar earlier data obtained for soybean, cucumber, barley, and pine tree /6,20,33/.

Visible light intensity considerably modified UV-B effect on the growth of postetiolated cotton seedlings. If under the most intensive light among tested the growth of "experiment" variant plants were completely absent (8 W/m^2) or it slowed down only during the first 2-3 days, then under a less lighting intensity such plants in general were growing only during the first day (Fig.7).

These experiments showed that cotyledon dry mass leaves in 7 days after lighting began did not depend on visible light intensity, and under UV-B affect at a less lighting tested decreased approximately by 20% /3I/.

Water-supply of "experiment" variant leaves decreased proportionally to UV-B intensity compared with controls, but independent of visible light intensity(Fig.8). Leaves drying had been mentioned previously under UV-B irradiation of barley /20/ and in our experiments at a single irradiation of young cotton plants and barley /24/.

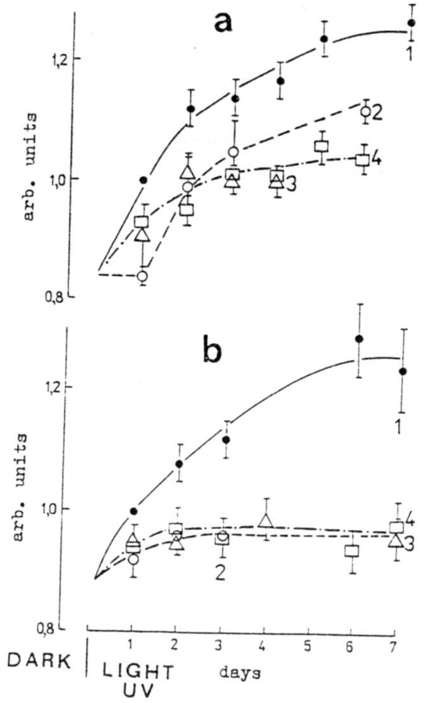

Figure 7. UV-B affect on postetiolated cotton seedlings growth
Notations : see Fig.4

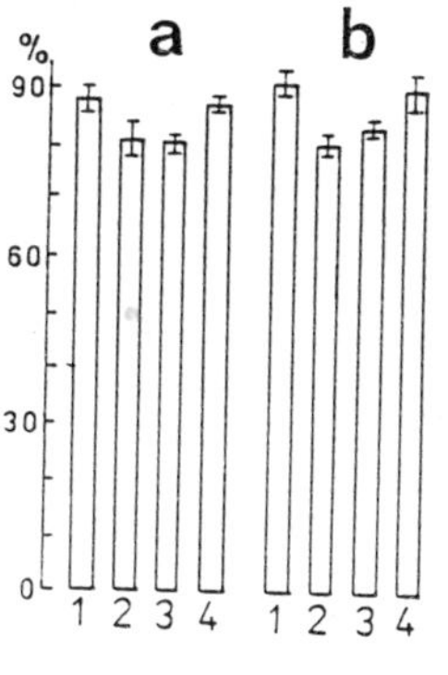

Figure 8. UV-B affect on water-supply of cotyledon leaves of postetiolated cotton seedlings
Notations : see Fig.4

Thus, in the experiments with cotton plants the well-known notions /20,28,29/ were confirmed as to the dependence of plant resistance to UV-B on visible light intensity against the background of which UV-B act. It was shown that cotton plant growth function sensitivity to UV-B is significantly decreased even under comparatively visible light intensities. It is probably this fact that could account for (along with the absence of control) the circumstances in development when growing this plant in field conditions. Moreover, in spring at the Dushanbe latitude natural Sun light exceeds the light of the order under which the increase of cotton plant growth function resistance to UV-B was observed.

Recent experiments with sunflower have revealed possibilities for a common mechanism of plant growth suppression by UV-B through direct photoconversion of indolil-acetic acid with consequent formation of oxidated products including growth inhibitors /6/. This fact did not add anything to already known notions about the principles of visible light protective effect /28-30/. Nevertheless the results of our experiments with cotton plant do not comply with an assumption about the relationship between increasing resistance to UV-B and more intensive plant development under greater visible light intensities /29/.

Affect on Cotton and Soybean Photosynthetic Apparatus, Growth and Productivity in Field Conditions

In the conditions of field experiments (two-year experimental data)/25/ no reliable changes of pigment concentration in plastids of cotton plants irradiated by UV-B compared with controls have been observed. Soybean in the same conditions in different years have demonstrated both reducing pigments content in leaves and a tendency for its increase (Table 2). Differences revealed have been probably caused by a different sensitivity of cotton plant and soybean to UV-B. Soybean among other things belongs plants particularly sensitive to above effect /32/.

Table 2

UV-B affect on plastid pigment content in soybean leaves

Index	Controls			Experiment		
	Chlorophyll		Carotenoids	Chlorophyll		Carotenoids
	a	b		a	b	
			Flowering, I988			
mg/g green mass	0.9±.05	.25±.0I	0.5±.02	0.7±.03	.24±.02	0.4±0.I
mg/dm^2	I.2±.06	.35±.0I	0.7±.02	I.0±.04	.33±.03	0.5±.02
			Fruiting, I988			
mg/g green mass	I.2±.03	0.6±.03	0.6±.02	0,8±.04	0.4±.02	0.4±.02
mg/dm^2	2.0±.07	I.0±.06	I.0±.04	I.5±.08	0.7±.06	0.8±.05
			Fruiting, I989			
mg/g green mass	I.3±.09	0.4±.02	0.5±.04	I.5±0.I	0.6±.05	0.6±.03
mg/dm^2	2.0±0.I	0.7±.04	0.8±.05	2.7±0.3	I.0±0.I	I.0±.06

Table 3

UV-B affect on cotton plant growth and yield

Index	Controls	Experiment	
		0.16 W/m^2	0.61 W/m^2
Plant growth, cm	104±9	93±10	80±10
Bolls number per plant, piece	9.5±1	7.9±0.7	8±3
Yield per plant, g	63±6	45±5	48±7
Raw cotton mass per boll, g	6.6±0.1	5.7±0.1	6.4±0.1

Table 4

UV-B affect on soybean growth and yield

Index	Controls	Experiments
Plant growth, cm	71±5	48±4
Pods number per plant, piece	120±15	85±13
Yield per plant, g	24±4	19±4
Mass of 100 seeds, g	9.2±0.2	11.6±0.4

From Tables 3 and 4 it follows that an additional UV-B irradiation during vegetation has resulted in a reliable suppression of growth and reducing cotton plant and soybean productivity. At the same time irradiated plants have reduced their fruiting elements (bolls in cotton plants and pods in soybean), but a mass of one fruiting element either has insignificantly reduced (cotton), or even increased (soybean). Similar effect on increasing seed mass (while reducing number of seeds) in barley irradiated by UV-B compared with controls has been observed earlier /20/.

At approximately equal UV-B intensity growth suppression was more prominent in soybeans - 32% (in cotton plants - 24%). Yield reduction in the above conditions was practically the same: cotton plant - 23%, soybean - 20% (Tables 3 and 4).

In the Introduction to one of recent reviews on the problem of a biological effect of UV-B /34/ it is said that only for soybean during six years of field investigations yield reduction was relatively demonstrated when UV-B intensity was increased according to 25% reduction of the ozone layer. This yield reduction for some varieties was 20-25%, while for others in the same conditions the yield increased by I0-22% /6/. The data for other plants are extremely controdictory, and for cotton plant no data are available at all /6/. The results of our investigations discussed above so far do not enable to quantitatively evaluate cotton yield loss depending on the degree of the ozone layer reduction. But the comparision of the above data (Tables 3 and 4) with the results of experiments on soybean /6/ enables to a sufficient degree of probability to consider that in the same conditions of UV-B intensity the reduction of cotton plant productivity (at least the variety under study) will be of the same order as of the soybean varieties sensitive to UV-B.

CONCLUSION

The experimental data under consideration most convincingly indicate that such an environmental factor like visible light intensity has a powerful effect on plants resistance to UV-B. Probably not only the absence of visible damage on a photosynthetic apparatus and growth function of plants in natural conditions is explained by the above factor which must be expected based on the experimental results under complete excluding UV-B, but as well as to a considerable extent by the very possibility of harvest formation itself. Particular importance of light factor as to plants response to UV-B irradiation is emphasized by another circumstance that the reducing of lighting increases sensitivity, while other factors reduce it. The study of molecular mechanisms of a modifying effect of visible light on the consequence of plants irradiation by UV-B seems now absolutely necessary.

Our experiments have shown that cotton plant in judged by chloroplasts pigment system response is relatively resistive to UV-B. On the other hand in field conditions its productivity under UV-B influence is reduced the same manner as for soybean which is considered to be a particularly sensitive plant. In spite of that rather discouraging result it seems sufficiently promising to further researches into the physiological tests on plants resistance to UV-B. It is hardly possible to solve the problem of plants selection for resistance to the consequences of global climatic changes of the Earth having not taken into consideration the mentioned above effect.

REFERENCES

I. Caldwell, M.M. (I984). Effect of UV radiation on plants in the transition region to blue light. In: Blue Light Effects in Biological Systems, H. Senger ed., Springer,

Berlin, pp. 20-28.

2. Frederick, J.E., Snell, H.E. and Haywood E.K. (I989). Solar ultraviolet radiation at the Earth's surface. Photochem. and Photobiol. 50: 443-450.

3. Dubrov,A.P. (I968). Genetical and Physiological Effects of Ultraviolet Radiation Action on Higher Plants. Nauka, Moscow, pp 56-I37. (In Russian).

4. Caldwell, M.M. (I968). Solar ultraviolet radiation as an ecological factor for alpine plants. Ecol. Monogr. 38: 243-258.

5. Caldwell, M.M., Gold, W.W., Harris, G. and Ashurst, C. W. (I983). A modulated lamp system for solar UV-B (280-320 nm) supplementation studies in the field. Photochem. and Photobiol. 37: 479-485.

6. Tevini, M. and Teramura, A.H. (I989). UV-effects on terestrial plants. Photochem. and Photobiol. 50: 479-487.

7. Bell, L.N. and Merinova, G.L. (I96I). Effects of UV radiation dose and wave-length on Chlorella photosynthesis. Biophysics (Moscow). 6: I59-I64. (In Russian).

8. Coohill, T.P. (I989). Ultraviolet action spectra (280-320 nm) and solar effectiveness spectra for higher plants. Photochem. and Photobiol. 50: 45I-457.

9. Nasyrov, Yu.S. (I966). Photosynthesis of Plants of Tajikistan Vertical Zones and Means of Rising its Productivity. Doctoral Thesis, Donish, Dushanbe, pp. 35-37. (In Russian).

I0. Bornman, J.F., Björn, L.O. and Åkerlung, H.E.(I984). Action spectrum for inhibition by ultraviolet radiation of photosystem II activity in spiach thylakoids. Photobiochem. and Photobiophys. 8: 305-3I3.

II. Noorudeen, A.M. and Kulandaivelu, G. (I982). On the possible site of inhibition of photosynthetic electron transport by ultraviolet-B (UV-B) radiation. Physiol. Plantarum. 55: I6I-I66.

I2. Iwanzik, W., Tevini, M., Dohnt, G., Voss, M., Gräber, P. and Renger, G. Action of UV-B radiation on photosynthetic primary reactions in spinach chloroplasts. Physiol. Plantarum. 58: 40I-407.

I3. Tevini, M. and Pfister, K. (I985). Inhibition of photosystem II by UV-B radiation. Z. Naturforsch. 40 C: I29-I33.

I4. Renger, G.,Völker, M., Eckert, H.J., Fromme, R., Hohm-Veit, S. and Gräber, P. (I989). On the mechanisms of photosystem II deterioration by UV-B irradiation. Photochem. and Photobiol. 49: 97-I05.

I5. Rabinowitch, E. (I953). Photosynthesis. Vol.2. Izd. Inostr. Lit. Moscow, pp. 9-80. (In Russian).

I6. Krasnovsky, A.A. (I974). Light Energy Conversion in the Photosynthesis. Molecular Mechanisms, Nauka, Moscow, pp. I-64. (In Russian).

17. Shiryaev, A.I. (1978). Submicroscopic and Macromolecular Organization of Chloroplasts, Naukova Dumka, Kiev, pp. 1-64. (In Russian).

18. Krinsky, N. (1971). Function. In: Carotenoids, O. Osler, ed., Birkhäuser Verlag, Basel, pp. 669-716.

19. Usmanov, P.D., Mednik, I.G., Lipkind, B.I. and Giller, Yu.E. (1987). Genotypic specificity in plant responses to middle-wave ultraviolet radiation. Plant Physiol. (Moscow). 34: 721-729. (In Russian).

20. Kanash, E.V., Savin, V.N. and Osipov, Yu.A. (1988). Growth and productivity in barley exposed to ultraviolet radiation at different stages of plant development. Plant Physiol. (Moscow). 35: 726-733. (In Russian).

21. Wettstein, D. (1957). Chlorophyll-letale und der submikroskopische formwechsel der plastiden. Experimental Cell Res. 12: 424-506.

22. Shlyk, A.A. (1968). Spectroscopic determination of chlorophyll a and b. Biochemistry (Moscow). 33: 275-285. (In Russian).

23. Giller, Yu.E. and Lipkind, B.I. (1988). UV-B effects on photosynthetic apparatus of cotton plants. Transact. of the Acad. of Sci. of Tajik SSR, Dep. of Biol. Sci. N3(112): 31-36. (In Russian).

24. Scherbakova, I.Yu., Lipkind, B.I., Karieva, F.A. and Giller, Yu.E. (1989). UV-B effects on chloroplasts pigment system of cotton and barley plants. Proceed. of Acad. of Sci. of Tajik SSR. 32: 856-859. (In Russian).

25. Giller, Yu.E., Scherbakova, I.Yu., Lipkind, B.I., Karieva, F.A. and Shishkin, V.A. (1991). UV-B effects on photosynthetic apparatus and productivity of higher plants. Cosmic. Biol. and Air-cosmic Medicine. In press (In Russian).

26. Shlyk, A.A. and Prudnikova, I.V. (1965). Dark biosynthesis of chlorophyll in homogenate. Proceed. of Acad. of Sci. of USSR. 160: 720-723. (In Russian).

27. Giller, Yu.E., Karieva, F.A., Lipkind, B.I. and Scherbakova, I.Yu. (1991). UV-B effects on the development of pigment system of cotton chloroplasts. Proceed. of Acad. of Sci. of Tajik SSR. 34: In press (In Russian).

28. Warner, C.W. and Caldwell, M.M. (1983). Influence of photon flux density in the 400-700 nm waveband on inhibition of photosynthesis by UV-B (280-320 nm) irradiation in soybean leaves: separation of indirect and immediate effects. Photochem. and Photobiol. 38: 341-346.

29. Mirecki, R.M. and Teramura, A.H. (1984). Effects of ultraviolet-B irradiance on soybean - V. The dependence of plant sensitivity on the photosynthetic photon flux dencity during and after leaf expansion. Plant Physiol. 74: 475-480.

30. Wellmann, E. (1983). UV radiation: definitions, characteristics and general effects. In: Encyclopedia of Plant Physiology, New Series, Photomorphogenesis, Vol. 16B, W. Shropshire and H. Mohr, eds., Springer, Berlin, pp. 745-756.

31. Giller, Yu.E., Karieva, F.A., Lipkind, B.I. and Scherbakova, I.Yu. (1991). UV-B effects on growth of cotton seedlings. Proceed. of Acad. of Sci. of Tajik SSR. 34: In press (In Russian).

32. Teramura, A.H. and Caldwell, M.M. (1981). Effects of ultraviolet-B irradiance on soybean - IV. Leaf ontogeny as a factor in evaluating ultraviolet-B irradiance effects on net photosynthesis. Amer. J. Bot. 68: 934-941.

33. Sullivan, J.H. and Teramura A.H. (1988). Effects of ultraviolet-B irradiation on seedling growth in the Pinaceae. Amer. J. Bot. 75: 225-230.

34. Urbach, F. (1989). The biological effects of increased ultraviolet radiation: an update. Introduction. Photochem. and Photobiol. 50: 439-441.

U.V. INHIBITION OF THE DISSOLVED INORGANIC CARBON CONCENTRATING MECHANISM(S) AND ALTERNATIVE RESPIRATION IN UNICELLULAR GREEN ALGAE

Arun Goyal and N.E. Tolbert, Department of Biochemistry, Michigan State University, East Lansing, MI 48824, USA.

ABSTRACT :

Because of low levels of CO_2 in air, plants and algae have processes for concentrating CO_2 internally to increase net photosynthesis and decrease photorespiration. Unicellular green algae and cyanobacteria develop dissolved inorganic carbon concentrating mechanism (often referred to as "DIC-pumps") when grown or adapted to air levels of CO_2 for 6-24 hours in the light. These DIC-pumps are suppressed when elevated levels of CO_2 (1-5%) are provided during growth. The induction or development of the pump is inhibited by the low levels of UV-B (280-320 nm, 2 $\mu W.cm^{-2}$) but not by UV-A. A preliminary action spectra indicated maximum inhibition around 290 nm. These low UV treatments do not inhibit photosynthetic CO_2 fixation with excess CO_2, but UV-B prevent an adaptive change in $K_{0.5}$ (DIC). Dissolved inorganic carbon concentrating mechanism and alternative, cyanide-resistant respiration, are both inhibited by SHAM, an inhibitor of the alternative respiration. Thus these two processes may be related. UV-A but not UV-B inhibited the capacity of alternative respiration in leaves and algae. A working model is UV-B inhibition of the formation of DIC pump enzymes and UV-A inhibition of the alternative respiration.

INTRODUCTION :

Atmospheric CO_2 levels are controlled by the amount of photosynthesis, respiration and photorespiration by plants and algae, using the bifunctional enzyme Rubisco (1). Because levels of ambient CO_2 are insufficient for rapid growth, processes are present for increasing the CO_2 concentration in the cell, specifically at the site of Rubisco to increase the rate of photosynthetic CO_2 fixation and to suppress photorespiration. The two processes are a C_4 organic acid cycle in macroalgae and in the bundle sheath cells of C_4 plants and in CAM plants; and mechanisms for concentrating dissolved inorganic carbon (DIC or CO_2 and HCO_3^-) by systems that are present in unicellular green algae and cyanobacteria.

Dissolved Inorganic Carbon Concentrating Mechanism(s): Unicellular green algae and cyanobacteria, when grown with or adapted to air levels of CO_2 in the light, develop dissolved inorganic carbon concentrating mechanisms, or "DIC-pumps" (2). These pumps need to be biochemically characterized. Depending on the microalgae and the pH of the growth medium during DIC accumulation, either CO_2 or HCO_3^- may be imported (2,3). The CO_2 pump is usually characterized by the presence of an extracellular carbonic anhydrase (CA) to increase the rate of extracellular HCO_3^- conversion to CO_2 (4). Intracellular isozymes of CA (5), and Vanadate sensitive plasmalemma type ATPase-HCO_3^- transporter at the chloroplast envelop are also probably involved (6). For bicarbonate uptake from the medium, an ATPase -HCO_3^- transporter at the plasmalemma of the outer cell membrane of some algae functions at alkaline pH to transport HCO_3^- from the medium (7). The DIC pumps are repressed by high CO_2 (1-5% v/v), but induced in the light within about 6-24 hours after changing high CO_2 grown cells to ambient air levels of CO_2. The $K_{0.5}$(DIC) levels for the DIC pumps may be about 1 μM, well below the $K_m(CO_2)$ of >26 to 50μM for Rubisco (1).

Alternative Respiration : The ubiquitous alternative pathway of respiration is found in plant systems and in algae but the role of the alternative oxidase is not understood (8). Algae grown with low CO_2 to induce DIC-pumps, always have a large capacity for the alternative oxidase. A decrease in the alternative oxidase in algae occurs during log phase of growth on high CO_2 (9). Because both the DIC-pump and the alternative oxidase are inhibited by SHAM it is postulated that two processes may be related (10).

This report on dissolved carbon concentrating mechanism(s) and on alternative respiration deals their inhibition by low dosages of ultraviolet radiation, UV-A and B in unicellular green algae. Alternative respiration in pea seedlings is also inhibited by UV-A.

METHODS :

Algae and Growth Conditions (3,6,9,10) : *Chlamydomonas reinhardtii* 137 cells were grown photoautotrophically in minimal medium at 26 ± 2C with continuous shaking and bubbling with air enriched with 3-5% CO_2 . The photon flux density was 150 $\mu E.m^{-2}.s^{-1}$ with a light /dark regime of 16/8 hours. The pH of algal growth medium started around 7 and drifted to near 5 during growth of the culture. While in the exponential growth phase and one day before the use, cultures were diluted with fresh growth media and aerated only by shaking in air for 24 hours to develop DIC pumps. Cultures were harvested by centrifugation (1000g x 10 min); the cell pellet was washed once with the fresh growth medium (pH 6.8), and centrifugation was repeated. Finally the cell pellet was resuspended in a small volume of assay buffer, kept on ice, and used within 2 h. Chlorophyll was estimated on an aliquot after ethanol extraction.

Pea Plants (11): Seeds of *Pisum sativum L.* cv. Alaska were from Atlee Burpee Seed Company, Warminster, PA. seeds were washed several times with tap water, soaked overnight in distilled water, and sown in 1:1:1 mixture of peat moss, topsoil, and vermiculite. Plants were raised in green house and watered with Hoagland nutrient solutions. Three weeks old plants were used for control and UV-A treatment.

Measurement of Photosynthetic O_2 Evolution and $K_{0.5}$(DIC) (3,7,10): Photosynthetic DIC dependent oxygen evolution was measured with an oxygen electrode in 2 ml of 25 mM buffers of various pH. The light intensity was 800 $\mu E.m^{-2}.s^{-1}$. In order to keep DIC content low, all buffers were prepared daily in freshly distilled water and sparged continuously with nitrogen until use. The $K_{0.5}$(DIC) values were determined by measuring the rate of oxygen evolution and slope of the progress curve (3,6).

Inorganic Carbon Uptake : DIC accumulation by algal cells were estimated by silicone oil filtration technique (3,6). Reactions in microfuge tubes were done at designated pH. After illumination for 1 min, DIC uptake was initiated by the addition of 5 μL of $NaH^{14}CO_3$ (final concentration, 150 μM) and the incubation was terminated after 15 to 60 s by turning on the microcentrifuge to remove the algae.

UV Treatment for Algal Cells (Goyal and Tolbert, Unpublished) : For UV-B experiments, algal cell suspensions were exposed by two procedures. 1: Algal cultures stirred in an open beaker were exposed from a distance of 4 feet to a long UV handheld lamp; 2: Cell suspensions in thermostated (25°C) 3 ml quartz cuvettes in a Gilford spectrophotometer were exposed to various wavelengths of UV-B light between 280 to 320nm. The slit of the UV lamp was fully open and the suspensions were continuously bubbled with air. In addition to UV, all high-CO_2 grown cells were exposed simultaneously overhead to white light (500 $\mu E.m^{-2}.s^{-1}$) from a 300 W reflector lamp to activate DIC concentrating mechanisms. As measured by a Laser Power meter, the UV intensities from a long wave length handheld lamp was about 2 $\mu W.cm^{-2}$, from the spectrophotometer at 300 nm 0.2 $\mu W.cm^{-2}$, and from the short wavelength UV-C handheld lamp was 0.25 $\mu W.cm^{-2}$).

UV Light Treatment and Recovery of Alternative Oxidase in Pea Plants (Goyal and Tolbert, Unpublished): Each tray of 50-75 pea seedlings were exposed to UV-A by a hand-held UV lamp (Model UVL-21, Blay-Ray Lamp, Longwave UV-366 nm, Ultra-Violet Products, Inc., San Gabriel, CA) by placing the lamp 1 meter above the plants for 8 h. The UV light had an intensity of 2$\mu W.cm^{-2}$. White light of 500 $\mu E.m^{-2}.s^{-1}$ was also provided to the plants from an incandescent 300 W reflector lamp filtered through a solution of 1 M $CuSO_4$. For the control plants, only white light was provided. The rate of respiration and capacities for respiration were measured before and after the UV treatments and recovery experiments. Part of plants were also held for recovery from the UV, and respiration rates were run at 20, and 48 h after the exposure and removal of UV treatment. During the recovery a 10 h light and 14 h dark cycle was used, and plants were watered every 12 h.

Measurement of Respiration (Goyal and Tolbert unpublished): Dark respiration and capacities of Cyt-*c* and alternative respiration in unicellular green algae were measured as described in Goyal and Tolbert (9). Dark oxygen uptake by leaf disks were measured polarographically with a Rank Brothers oxygen electrode at 25°C in a total volume of 3 ml of 5 mM potassium phosphate buffer (pH 7.5) or water. Twenty leaf disks (90 mg fresh weight) were punched out of leaves immediately before the assay, equilibrated for 5 to 8 min in the buffer in the cuvette with continuous stirring. Afterwards the rate of respiration, v_T was determined. After about 8 min either 1 mM KCN or 5 mM SHAM was added to the reaction vial, and oxygen uptake measurements were recorded for the next 5 min, after which the other inhibitor, SHAM or KCN was added.

Calculations For the Capacity and the Rate of Respiration (9) :

v has been designated for the measured rate of respiration and V for the "capacity". v_T, velocity of total respiration; v_{CN}, respiration as oxygen uptake that was inhibited by cyanide; v_{SHAM}, amount of inhibition of oxygen uptake by SHAM; v_{res}, velocity of residual respiration; V_{cyt}, capacity for the Cyt-c pathway; V_{alt}, capacity for the alternative pathway.

V_{cyt}, capacity for the Cyt-c pathway is measured as the rate of oxygen uptake that was inhibited by 1-2 mM CN in the presence of 5mM SHAM; V_{alt}, capacity for the alternative pathway is measured as the rate of oxygen uptake that was inhibited by 5 mM SHAM in the presence of 1-2 mM CN.

Theoretically, if both pathways were operating to their full capacities,

$$v_T = V_{cyt} + V_{alt} + v_{res}$$

Normally either one or both pathways presumably did not operate at full capacity for the measured total velocity, v_T, was always less than the sum of

$$V_{cyt} + V_{alt} + v_{res}$$

RESULTS :

When green algae are grown on high CO_2 (5% v/v in air) they do not have DIC-pumps, however when high CO_2 grown cells are transferred to low, air levels of CO_2 in the presence of light, within 6-24 hours they form a DIC-pump, which is usually characterized by :

1. Extracellular carbonic anhydrase to facilitate conversion of HCO_3^- to CO_2, and isoforms for intracellular carbonic anhydrase. DIC pumps in turn are repressed by high CO_2. The $K_{0.5}$ (DIC) for the CO_2 pump is about 1-5μM, well below the K_m(CO_2) of 26-50μM for rubisco (1).
2. DIC-pumps need photosynthetic or respiratory energy to drive the DIC pump(6). Light is needed for the induction of the CO_2 pump, but the plasmalemma bicarbonate transporter can be partially formed in the dark, and seems to use ATP from respiration. For the induction of the chloroplasts ATPase-HCO_3 transporter, light is required (7).
3. During induction of the DIC-pump at least four polypeptides are formed (12).
4. Intact chloroplast from air grown cells probably have an ATPase -bicarbonate transporter but not from the CO_2 cells (6).

5. Some algae which transport bicarbonate have a bicarbonate transporter at the outer membrane (7).

6. A low dosage of UV-B inhibits induction of the DIC pump (Goyal and Tolbert, Unpublished)

Our current general model for algal DIC pump is described below (Goyal and Tolbert, Unpublished) . This model includes two different pathways, and both may or may not be present in the same algae. In one case the "CO_2 pump", usually enhances the replenishment of CO_2 from HCO_3^-, as CO_2 enters the cell to be concentrated by a HCO_3^- transporter at the chloroplast envelope and converted back in the chloroplasts to CO_2 as substrate for Rubisco. In other case, the "HCO_3^- pump", a plasmalemma ATPase transporter is proposed for HCO_3^- uptake, particularly in alkaline media, as well as the chloroplast transporter.

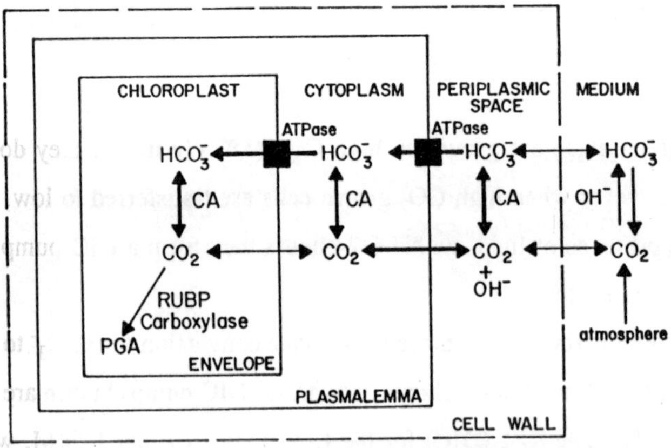

A proposed model for inorganic carbon accumulation by unicellular green algae.

When *Chlamydomonas* or *Dunaliella* cells were grown on high CO_2, about 200 μM bicarbonate was needed for V_{max} photosynthesis (Fig 2A). The external DIC concentration for half-maximal CO_2-dependent oxygen evolution [$K_{0.5}$ (DIC)] was about 75 μM. Because these algae are absorbing CO_2, the $K_{0.5}$(DIC) values varies with pH of the test medium. However when high CO_2 grown cells were adapted in light with air levels of CO_2 to develop the DIC-pump, the $K_{0.5}$(DIC) dropped to about 1-5μM (Fig 2B) (3,6). Salicylhydroxamic acid (SHAM), an inhibitor of the alternative respiration in plant mitochondria, inhibited DIC-pumps (Fig 2C) (10). This inhibition of the DIC-pumps can be relieved by high concentrations of DIC. The SHAM inhibition of the DIC-pump did not occur when higher concentration of the DIC (1 mM) were used, where a DIC pump was not necessary, thus confirming that the inhibition was specifically due to the CO_2 transport process and not to photosynthesis (10). When high CO_2 grown cells were adapted in low CO_2 to form the DIC-pump in the presence of added low levels of the UV-B, the UV-B inhibited the formation of the DIC-pump, and cells behaved as if they were high CO_2 grown cells (Fig 2D) (Goyal and Tolbert, Unpublished). Both UV-B and UV-C inhibited the induction of the DIC-pump, where the $K_{0.5}$(DIC) for UV treated cells remained at 75 μM, similar to high CO_2 grown cells

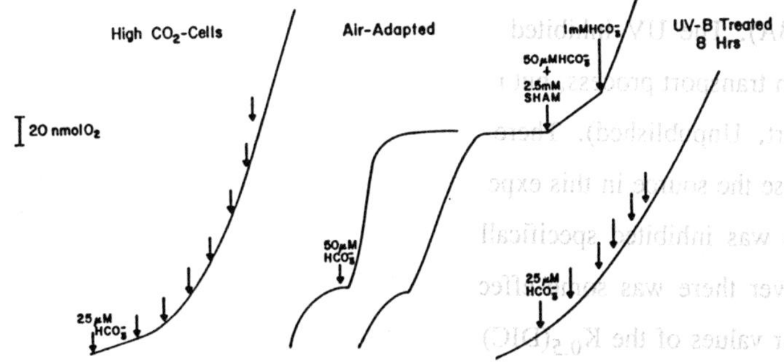

Figure 2. Photosynthetic oxygen evolution by *Chlamydomonas reinhardtii*. A. High CO2 grown cells; B. Air adapted cells; C. Inhibition of the pump by SHAM; D. Inhibition of the induction of the pump by UV-B (Goyal and Tolbert, Unpublished)

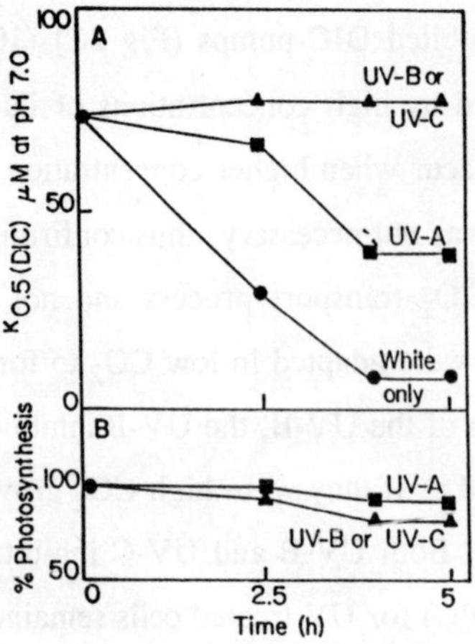

Figure 3. Effect of UV-A, B, or C on the A. induction of the DIC-pump; B. photosynthesis, in high CO_2 grown *Chlamydomonas*. (Goyal and Tolbert, Unpublished)

in case of UV-A treated cells the $K_{0.5}$(DIC) decreased to about 40 μM but not to 5 μM as in the case of the white light only, where the DIC pump was fully activated (Fig 3A). The UV inhibited specifically the induction of the DIC-pump and the carbon transport process, but not photosynthesis with high CO_2 (Fig 3B) (Goyal and Tolbert, Unpublished). There was some effect of UV-A long wavelength light, because the source in this experiment was not monochromatic. The induction of the pump was inhibited specifically by the UV-B and UV-C but not by the UV-A, however there was some effect of UV-A on the induction of the pump, such as higher values of the $K_{0.5}$(DIC) than to the control.

The air adapted cells can concentrate within 1 min up to 2 mM DIC from low levels of DIC extracellular (Fig 4A) (3,6). However this process of DIC accumulation is inhibited by the treatment by UV-B, UV-C and 5 mM SHAM, and to the some extent by UV-A (Fig 4B), this inhibition was relived by higher concentration of DIC, where a DIC-pump is not required. This confirms that the carbon transport process was specifically inhibited. By using silicone oil centrifugation technique (6,10), it is established that the reduced photosynthetic rates were due to the reduced carbon transport and accumulation of carbon.

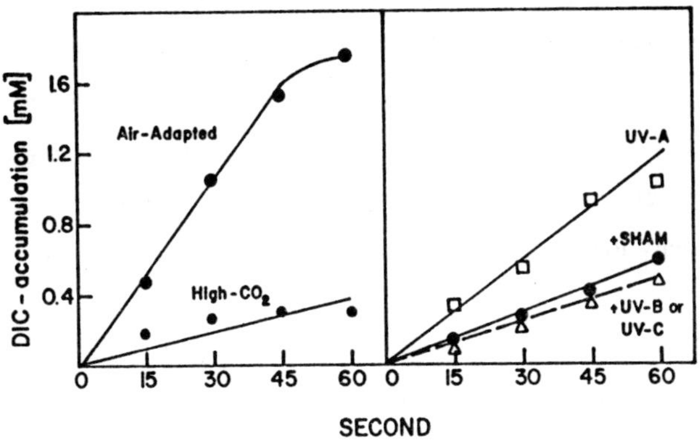

Figure 4. Intracellular accumulation of the DIC by *Chlamydomonas reinhardtii* and inhibition by UV-B and SHAM.

There are reports in the literature where mixtures of UV-A and UV-B produced less inhibitory effects than only one UV. It is argued that in the presence of other wavelength UV lesser effect is due to induction of repair mechanism. At this time we do not know the mechanisms involved in the inhibition of the induction of the DIC-pumps by UV-B. We have not considered UV-C because it affects nucleic acids and other quinone displacing processes. Miyachi's lab in Tokyo has also reported a failure to form CA of the CO_2 in similar experiments with 280 nm light (13).

UV-A Inhibition of the Alternative Respiration in Plants and Algae :

Practically all plant mitochondria isolated so far show a cyanide-resistant respiration, which is often termed as alternative respiration (8). CN inhibits cytochrome *c* oxidase; alternative respiration or cyanide resistant electron transport system consist of a branch point from the conventional electron transport system at reduced ubiquinone with a ubiquinone oxidase, the alternative oxidase, which is distinct from cytochrome oxidase. The cyanide resistant electron pathway is inhibited by SHAM, propyl galate,and disulfiram.

Cultures growing with high CO_2 lost their capacity for alternative respiration during the log phase of exponential growth (9). During 48 h of growth the capacity of the alternative respiration reached minimal levels (Fig 5) to about 10% of the maximal alternative respiration. After 48 h the capacity of the alternative oxidase progressively resumed as the cultures entered late log phase of growth. On the other hand the capacity of Cyt c respiration did not changed. In contrast to growth on high CO_2 cells, cells growing with air levels of CO_2 which formed DIC pumps, did not lose their capacity for the alternative respiration (Fig 5).

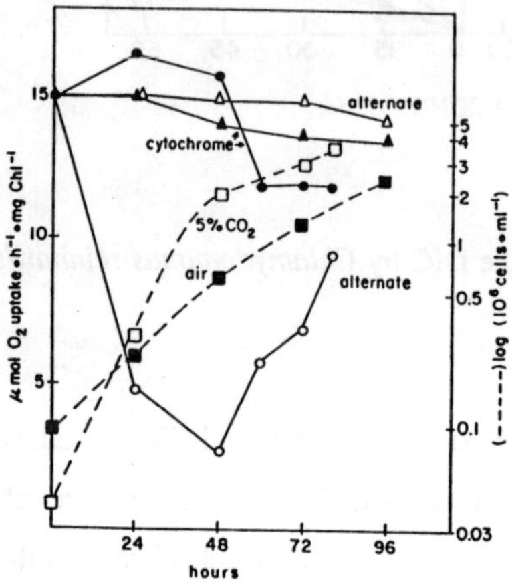

Figure 5: Variation of Cyt *c* and alterative respiration by cultures of *Chylamydomonas* growing with air or high CO_2, solid line, respiration and dashed line for cell density. O, V_{alt}, ●, V_{cyt} for cells grown on 5% CO_2; Δ, V_{alt}, ▲, V_{cyt} for cells grown on air.

The capacity of the alternative respiration in these *Chlamydomonas* decreased when cells were exposed to UV-A but not to UV-B or C (Fig 6). The alternative oxidase capacity in these low CO_2 cells recovered within 24 hours after removal of the UV-A (Goyal and Tolbert, Unpublished).

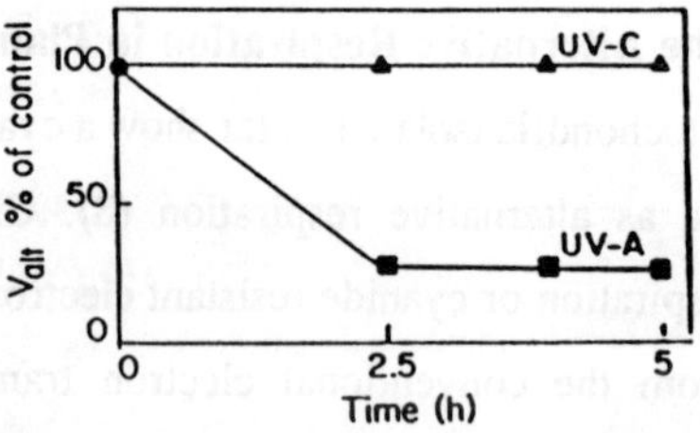

Figure 6. Effect of UV on the capacity of the alternative oxidase in *Chlamydomonas* .

Table I. Effect of UV-A on the capacity of alternative pathway (V_{alt}) and cytochrome pathway (V_{cyt}) in leaf disks of pea cv. Alaska. Values are expressed as a percentage of total dark respiration (v_T). Results are mean ± SE, n = 3 to 6. v_T is expressed as $\mu mol.O_2.mg^{-1}$ fresh wt. h^{-1}.

	v_T $\mu mol.O_2.g^{-1}$fresh wt.h^{-1}	V_{cyt} % of v_T	V_{alt}
Zerotime Control (No UV-A)	42 ± 2.5	60	30
After 6-8 hr exposure to UV-A	37 ± 0.5	56	13
20 hr after UV treatment and recovery in light			
Control (No UV-A)	42 ± 0	60	28
UV treated plants	35 ± 0	70	21
48 hr after UV treatment and recovery in light			
Control (No UV-A)	44 ± 0	64	29
UV treated plants	40 ± 0	66	24

Similar effects of UV-A on the capacity of the alternative oxidase were observed in higher plants, Table 1 shows the effect of UV-A exposure on the capacity of alternative respiration in pea leaves and it's recovery after removal of the UV-A radiation.

DISCUSSION :

UV-B specifically inhibits the formation or induction of the dissolved inorganic carbon concentrating mechanism in algae. When a water column is exposed to increasing UV-B radiations due to the depletion of the ozone layer, it may reduce the level or efficiency of DIC-pumps *in vivo*, which in turn would cause reduced photosynthesis, increased rate of photorespiration and in the long run cause an increase in the atmospheric CO_2. On the other hand the UV-B probably does not affect the capacity of the alternative respiration, if it is already present. However UV-A may effect photosynthetic rates, if alternative respiration for ion gradient involves DIC pumps. These results are preliminary, but the observed phenomenon are reproducible and significant.

REFERENCES:

1. Tolbert, N.E. (1991). Photosynthetic Carbon Metabolism and the Atmospheric CO_2 and O_2. In: *Encyclopedia of Earth System Science*, Academic Press (in press).
2. Lucas, W.J. and Berry, J.A. (1985). *Inorganic Carbon Uptake by Aquatic Photosynthetic Organisms.* American Society of Plant Physiologists, Rockville, MD.
3. Moroney, J.V. and Tolbert, N.E. (1985). Inorganic Carbon Uptake by *Chlamydomonas reinhardtii. Plant Physiol.* 77: 253-258.
4. Coleman, J.R., Berry, J.A., Togasaki, R.K., and Grossman, A.R. (1984). Identification of Extracellular Carbonic Anhydrase of *Chlamydomonas reinhardtii. Plant Physiol.* 76: 472-477.
5. Moroney, J.V., Togasaki, R.K., Husic, H.D., and Tolbert, N.E. (1987). Evidence that an Internal Carbonic Anhydrase is Present in 5% CO_2 Grown and Air Grown *Chlamydomonas. Plant Physiol.* 84:757-761.
6. Goyal, A. and Tolbert, N.E. (1989). Uptake of Inorganic Carbon by Isolated Chloroplasts for Air Adapted *Dunaliella. Plant Physiol.* 89: 1264-1269.
7. Thielmann, J., Tolbert, N.E., Goyal, A, and Senger, H. (1990). Two systems for Concentrating CO_2 and Bicarbonate During Photosynthesis by *Scenedesmus. Plant Physiol.* 92: 622-629.
8. Dry, I.B., Bryce, J.H., and Wiskich, J.T. (1987). Regulation of Mitochondrial Respiration. In: *The Biochemistry of Plants.* DD Davis ed.; Vol 11, Academic Press, NY 213-252.
9. Goyal, A. and Tolbert, N.E. (1989). Variations in the Alternative Oxidase in *Chlamydomonas* in Air or High CO_2. *Plant Physiol.* 89: 958-962.
10. Goyal, A., Tolbert, N.E. (1990). Salicylhydroxamic Acid (SHAM) Inhibition of the Dissolved Inorganic Carbon Concentrating Process in Unicellular Green Algae. *Plant Physiol.* 92: 630-636.
11. Goyal, A., Hiser, C.H., Tolbert, N.E., and McIntosh, L. (1991). Progress No. 9 Cultivar of Pea Has Alternative Respiratory Capacity and Normal Glycine Metabolism. *Plant Cell Physiol.* 32: (In Press).

12. Manuel, L.J. and Moroney, J.V. (1988), Inorganic Carbon Accumulation by *Chlamydomonas reinhardtii*. New Proteins are Made During Adaptation to Low CO_2. *Plant Physiol.* 88: 491-496.

13. Dionisio, M.L., Tsuzuki, M., and Miyachi, S. (1989). Blue Light Induction of Carbonic Anhydrase Activity in *Chlamydomonas reinhardtii*. *Plant Cell Physiol.* 30: 215-219.

EFFECTS OF ULTRAVIOLET-B RADIATION ON GROWTH, PIGMENTATION $NaH^{14}CO_3$ UPTAKE AND NITROGEN METABOLISM IN *NOSTOC MUSCORUM*

RashmiTyagi, H.D. Kumar, Deepak Vyas and Ashok Kumar

Department of Botany, Banaras Hindu University, Varanasi-221005

ABSTRACT

The impact of ultraviolet-B (UV-B) treatment on growth, N_2 fixation and $^{14}CO_2$ uptake in a nitrogen fixing cyanobacterium Nostoc muscorum has been studied. Continuous exposure to UV-B (5.5 W/m^2) for 37 min caused 50% survival of the organism whereas 100-min exposure caused complete loss of survival. UV-B drastically decreased pigment content, especially of phycocyanin. Exposure of actively N_2-fixing culture suspension to UV-B instantly inhibited nitrogenase activity and there was complete inactivation of nitrogenase in 20 min. Some inhibition of nitrate reductase activity was also observed following UV-B exposure but complete inactivation did not occur even after 100 min. Furthermore, induction of nitrate reductase was not affected by UV-B treatment. $^{14}CO_2$ uptake was almost completely abolished under UV-B light. These findings suggest that UV-B radiation exerts damaging effect on Nostoc muscorum, particularly on its photosynthesis and N_2 fixation.

INTRODUCTION

Ultraviolet (<400 nm) is an integral part of solar radiation. Based on wavelength, ultraviolet radiation is subdivided into three parts, namely UV-A (315-400 nm), UV-B (280-315 nm) and UV-C (<280 nm). Amongst these, the UV-B radiation is severely affected by reduction in stratospheric ozone. Scotto et al. (1) demonstrated significant variations in solar UV-B flux with latitude, seasonal changes and sun angles. The influence of solar UV on biological and other phenomena depends not only on such factors as ozone layer thickness and sun angle, but also on the nature of the biological effects themselves. The damaging effects of enhanced UV-B radiation resulting from a possible destruction of the stratospheric ozone layer by man-made chemicals, such as chlorofluorocarbons, have been demonstrated by several research groups during the last 15 years (2, 3, 4, 5, 6).

The impact of increasing UV-B radiation on biological systems has been investigated only recently. In higher plants, reductions of leaf area, fresh and dry weight, lipid content and of photosynthetic activity have been reported in a number of UV-B sensitive plant species (6, 7, 8). It has been also reported that UV-B radiation affects many metabolic processes, pigmentation, and community composition of biological systems other than higher plants (9). Few, if any, studies have been made on cyanobacteria, a group that occupies an important place in both aquatic and terrestrial food webs (10,11). This is unfortunate because cyanobacteria constitute phylogenetically the oldest group of O_2-evolving photosynthetic prokaryotes and it seems probable that during their evolutionary history, they had faced more intense ambient solar UV radiation than other organisms. Cyanobacteria grow in diverse habitats ranging from hot springs to the arctic and therefore they are expected to face differential levels of UV-B. In addition to these properties some of these organisms fix atmospheric N_2 and enrich the fertility of paddy fields and other soils; consequently, any impact of UV-B radiation on nitrogen-fixing cyanobacteria would be expected to affect the productivity in such habitats. Newton et al. (11) showed that the nitrogen-fixing enzyme in cyanobacteria is more sensitive to UV-B than are other metabolic processes. However, detailed investigations of UV-B effects on cyanobacteria are needed to reveal the sensitivity of various metabolic processes.

The purpose of the present work was to study the impact of UV-B radiation on growth, N_2 fixation and $^{14}CO_2$ uptake in the nitrogen-fixing, rice-field cyanobacterium Nostoc muscorum. Our findings suggest that UV-B radiation is generally inhibitory for several physiological processes and particularly to the photosynthetic components.

MATERIALS AND METHODS

Test Organism and Growth Conditions

The filamentous, heterocystous, N_2-fixing cyanobacterium (blue-green alga) Nostoc mucorum Ag. ex Born. et Flah. (ATCC 27893) was used in the present study. The culture was obtained through the kind courtesy of Dr. Robert Haselkorn (Univerity of Chicago, USA). This alga was selected mainly for its faster growth rate and ability to form discrete colonies on agar plates. The modified Chu-10 medium of Safferman and Morris (12) was routinely employed as the basal culture medium. Unless otherwise stated, cultures were grown in Erlenmeyer flasks containing 100 ml of medium lacking any combined nitrogen source, at 27±2°C in a culture room. Cultures were illuminated with daylight fluorescent tubes (14 W/m^2= 2400 lux) for 14 h/d.

Source and Mode of UV-B Treatment

The source of UV-B radiation was a Fotodyne, Inc., USA lamp CAT No. 3-4408 giving its main output at 312.67 nm. The desired radiation dose was obtained by adjusting the distance between the UV-B light source and the sample. Irradiations were performed in 75-mm sterile petri dishes with the lids removed during irradiation, and containing 10 ml of exponential phase culture suspensions. The suspension was gently stirred during irradiation. For determining per cent survival, 0.05 ml aliquots were withdrawn at known time intervals and plated on agar plates. Per cent survival was scored by colony counts and plotted semilogarithmically. Similarly, samples were withdrawn at desired intervals for measuring growth, pigment content, $^{14}CO_2$ uptake, nitrogenase, and nitrate reductase activities. Three replicate experiments were performed for each parameter. A separate control receiving fluorescent light (14 W/m^2) was kept for each experiment.

Nitrogenase Activity

Nitrogenase activity was measured by the acetylene reduction technique (13). Unless otherwise stated, 2-ml UV-B treated cultures were taken in 7-ml vacutainer tubes (Becton Dickinson, Rutherford, New Jersey, USA) and acetylene was injected by a hypodermic syringe to attain a final concentration of 10%. The tubes were incubated in fluorescent light (14 W/m^2) at 27°C. The ethylene formed was determined at intervals lasting upto 3 h in a CIC (Baroda) gas chromatograph fitted with a porapak R column and flame ionization detector.

Nitrate Reductase Activity

In vivo nitrate reductase activity was estimated by the method of Camm and Stein (14). Nitrite formed was determined by the diazocoupling method of Lowe and Evans (15). The absorbance of pink colour was estimated at 540 nm in a Bausch and Lomb Spectronic 20 colorimeter.

Uptake of $NaH^{14}CO_3$

The uptake of $NaH^{14}CO_3$ was estimated as per the method of Kumar *et al*. (16). 10 ml of culture suspension supplemented with 50 µl of $NaH^{14}CO_3$ (specific activity 0.026 µCi/ml) was exposed to UV-B/fluorescent light for different, known time periods. At desired intervals, 1 ml aliquots were withdrawn and transferred into scintillation vials containing 0.1 ml of 50% acetic acid. The resulting suspension was bubbled with air for 3 min whereafter 10 ml scintillation cocktail (Bray's solution) was added. Samples were counted in a Beckman-LS-7000 liquid scintillation counter.

Extraction and Estimation of Pigments

Chlorophyll *a* and carotenoids were extracted and measured in acetone according to Myers and Kratz (17). Phycocyanin was extracted in 2.5 mM potassium phosphate buffer (pH 7.0) after repeated freezing and thawing, and measured as per the method of Brody and Brody (18).

Chemicals

Sulfanilamide and N-1-naphthylethylene diamine dihydrochloride were purchased from Sigma Chemical Co., St. Louis, USA. $NaH^{14}CO_3$ (specific activity 5 µCi/ml) was obtained from BARC, Bombay.

RESULTS

Cultures of Nostoc muscorum were exposed to 2.5, 5.5 and 10.5 W/m^2 doses of UV-B for different time periods in an

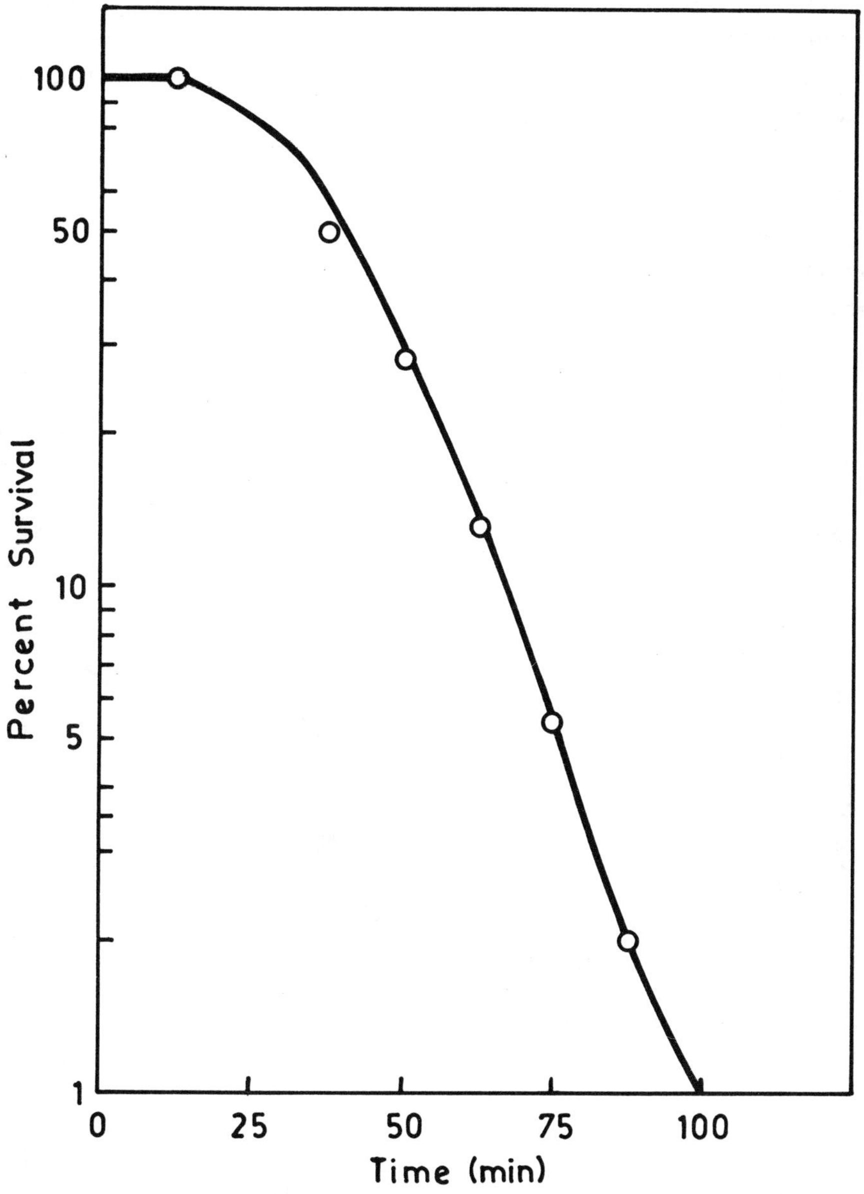

Figure 1. UV-B survival of Nostoc muscorum. Exponentially grown culture was exposed to 5.5 W/m^2 dose of UV-B.

attempt to select a mid-inhibitory survival dose. UV-B

treatment at 2.5 W/m^2 for 37 min did not affect growth significantly but 10.5 W/m^2 elicited complete killing. There was only 50% killing at 5.5 W/m^2 following 37 min treatment and therefore this dose was used in all further experiments. (This dose corresponded to a distance of 30 cm between the UV-B tube and the petri dish). Fig. 1 shows the survival curve, based on colony counts, after UV-B treatment for different durations. The survival was not affected upto 25 min of treatment but thereafter there was an abrupt decrease in per cent survival. There was 50% survival after 37 min treatment, and complete killing occurred after 100 min exposure.

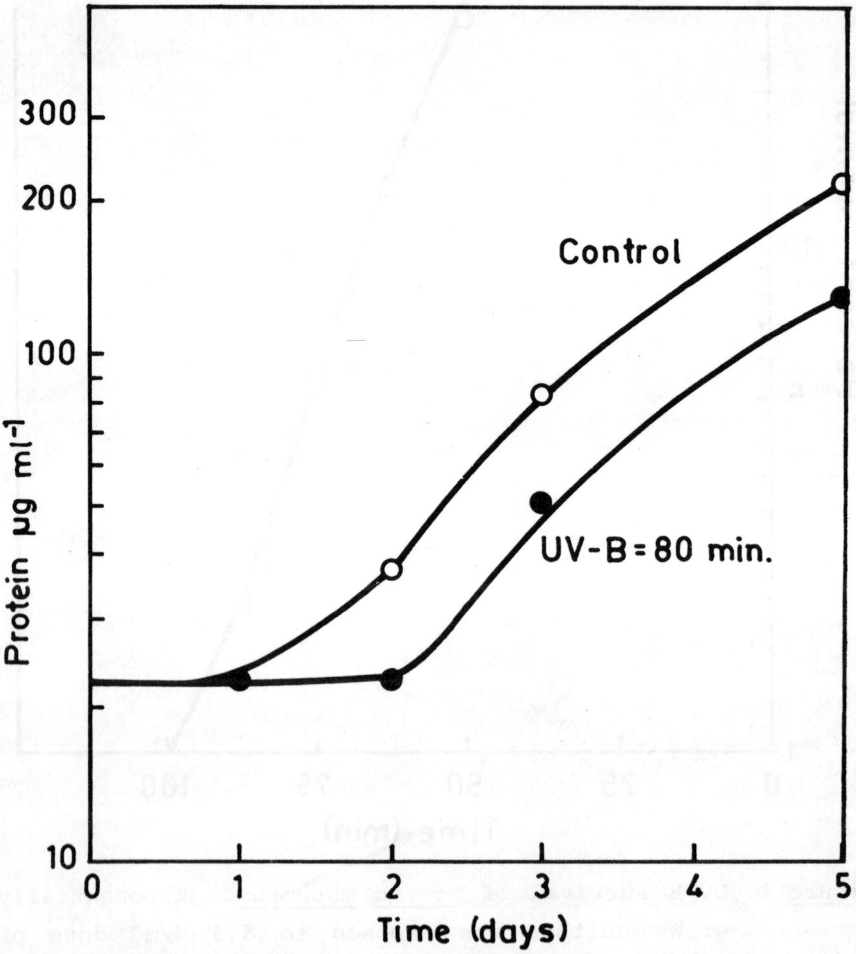

Figure 2. Growth behaviour of N. muscorum after UV-B treatment. Dose of UV-B was the same as in Figure 1.

Fig. 2 shows the growth pattern in liquid medium after 80 min of UV-B treatment. The UV-B treated cells showed a lag of 2 days after which there was slow but gradual growth. The growth was 40% lesser in UV-B treated culture in comparison to that in untreated control. Once it became apparent that UV-B treatment causes inhibition of growth, an attempt was made to understand the mechanism involved in the inhibition of growth.

Table-1 shows the content of photosynthetic pigments in cultures treated with UV-B for different durations. It is apparent that 30 min treatment did not affect carotenoids, chlorophyll a, or phycocyanin content significantly; but 80-min treatment caused significant decrease. Amongst the pigments, the content of phycocyanin was most severely affected.

Table 1. Effect of UV-B radiation (5.5 W/m^2) on photosynthetic pigment content of *Nostoc muscorum*[a].

Photosynthetic pigments		UV-B exposure duration (min)		
		0[b]	30	80
Chlorophyll a	$\mu g\ ml^{-1}$	0.97	0.79	0.18
	%	100.00	81.40	18.50
Phycocyanin	$\mu g\ ml^{-1}$	37.0	30.00	4.00
	%	100.00	81.08	10.80
Carotenoids	$\mu g\ ml^{-1}$	0.60	0.55	0.01
	%	100.00	91.60	20.00

a. Actively growing culture of *N. muscorum* was exposed to UV-B for different time periods and thereafter transferred to fluorescent light for growth. Pigment was extracted after 3 days incubation.

b. The value in untreated control culture was 100% as observed at 0 min.

Knowing that UV-B radiation inhibits growth and pigmentation, we became interested to study its effect on N_2 fixation. Fig. 3 shows that nitrogenase activity is affected instantly and dramatically in UV-B treated

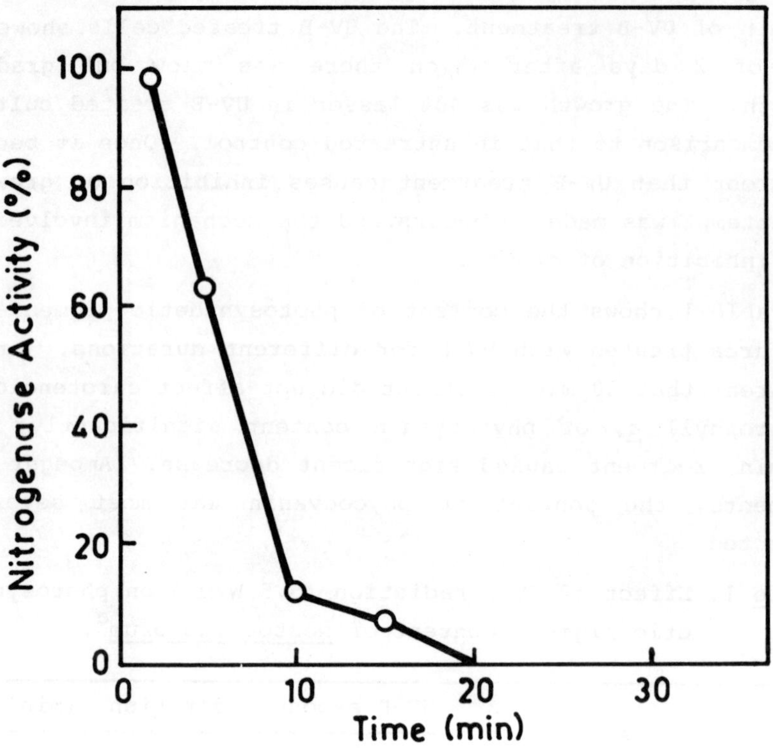

Figure 3. Effect of UV-B on nitrogenase activity. Actively N_2-fixing culture was exposed to UV-B (5.5 W/m^2) for different time period and thereafter C_2H_2 reduction assay was made under fluorescent light.

cultures. Exposure of cultures to UV-B for as low as 5 min resulted in significant (75%) decrease in C_2H_2 reduction. There was complete loss of C_2H_2 reduction in cultures exposed for 20 min to UV-B. As both nitrogenase and nitrate reductase are molybdoenzymes whose activity depends on supply of ATP and suitable reductant, it was thought proper to test the effect of UV-B on the activity of nitrate reductase. Accordingly, we tested the effect of UV-B on the activity as well as induction of nitrate reductase. Like nitrogenase, nitrate reductase activity was also affected instantly following UV-B treatment but, interestingly, there was no complete inactivation even after 80-min of UV-B treatment (Fig. 4). In fact a significant amount of activity (about 25%) was still

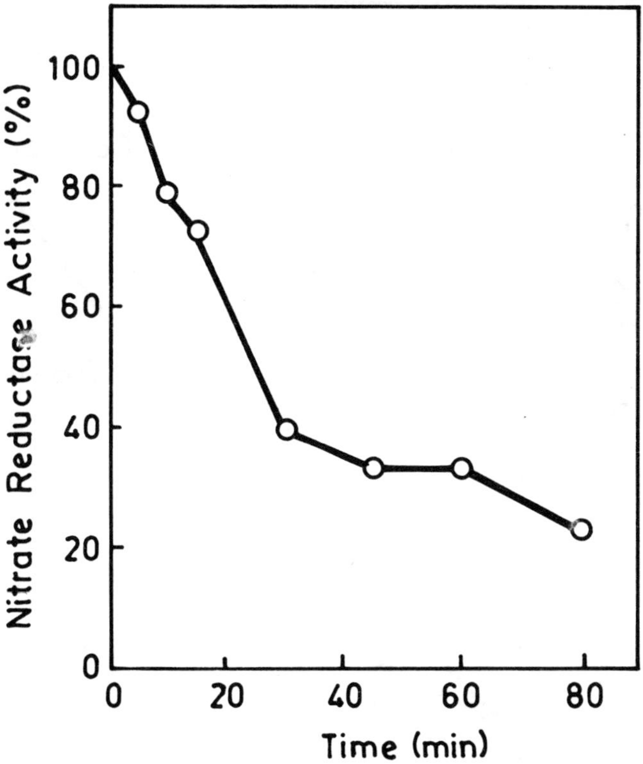

Figure 4. Nitrate reductase activity in UV-B treated N. muscorum. NO_3^--grown culture was used in this experiment.

discernible in 80-min UV-B treated cultures. Experiments on induction of nitrate reductase showed no alteration in the induction period following UV-B treatment (Fig. 5). However, there was marked difference between nitrate reductase activities of UV-B treated and untreated cultures in terms of total activity for a period lasting upto 72 h (Fig. 5).

As the processes of N_2 fixation and nitrate reduction are dependent on photosynthesis, it was imperative to test the effect of UV-B on $^{14}CO_2$ uptake. It is seen from Fig. 6 that exposure of cultures to UV-B causes complete loss of $^{14}CO_2$ uptake capacity.

DISCUSSION

The above results show that N. muscorum is sensitive to UV-B radiation. The effect of UV-B does not appear to be

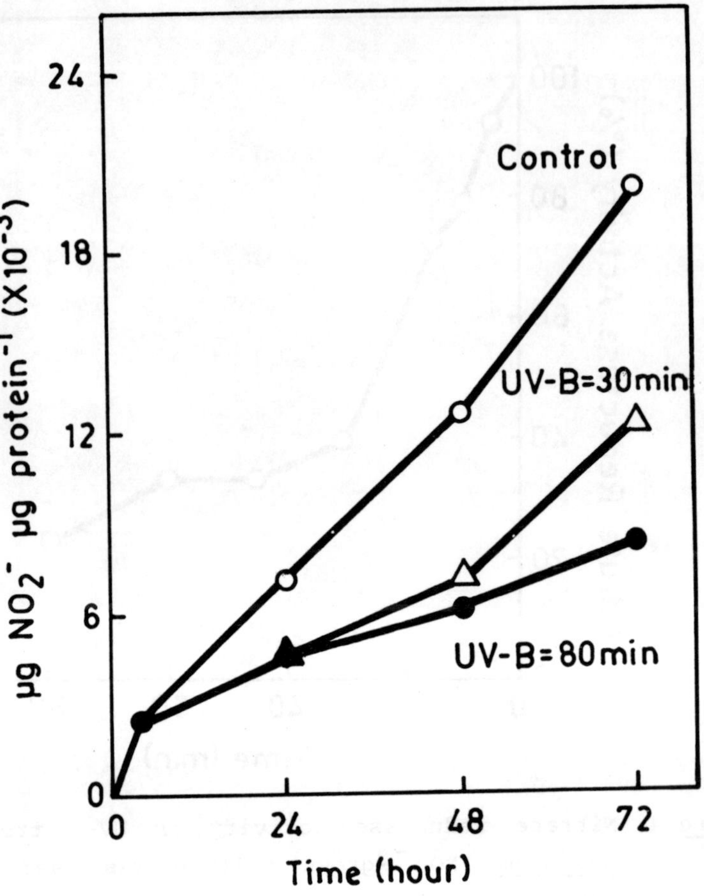

Figure 5. Effect of UV-B treatment on the induction of nitrate reductase activity. N_2-fixing culture was transferred into NO_3^--containing (5 mM) medium and thereafter UV-B treatment was given for 30 and 80 min. UV-B treated cultures were incubated in fluorescent light and activity was measured at indicated times.

specific for a particular metabolic process; rather it influences a number of physiological processes. Inhibition of growth and killing of cyanobacteria by UV-B radiation have been reported by Döhler et al. (10), Newton et al. (11) and Häder et al. (19). It has been demonstrated that at sublethal exposure levels, UV-B produces intracellular damage and affects growth and the endogenous rhythms found in many microorganisms. Our findings on killing of N.

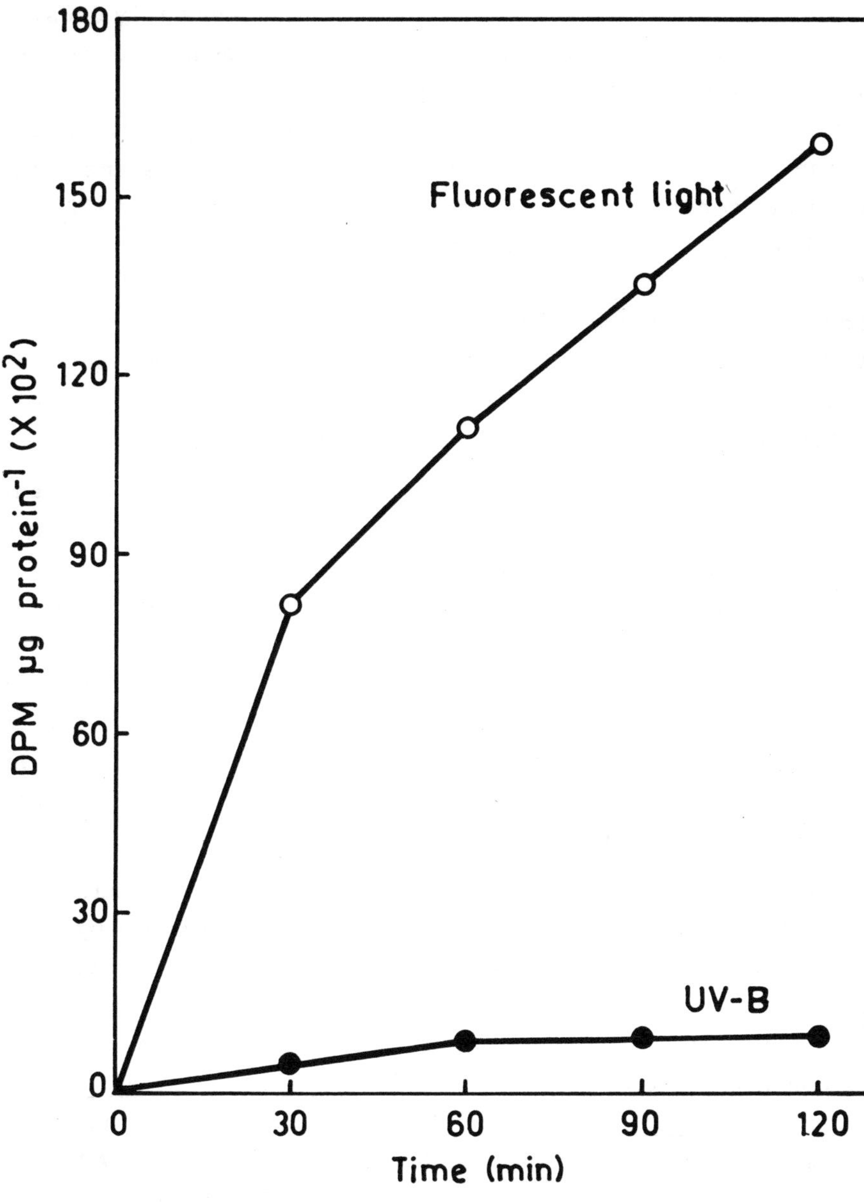

Figure 6. $^{14}CO_2$ uptake by N. muscorum under UV-B and fluorescent light.

muscorum following UV-B exposure are in agreement with the above findings. The killing of cells following exposure to UV-B may result from damage to cellular proteins. It has been also reported that higher doses of UV-B affect the cellular membranes' permeability to important chemicals and induce irreversible damage to proteins, eventually causing death (20). Furthermore, DNA has been thought to be the primary UV-B target in certain microorganisms (21). This conclusion stems primarily from the similarity between the action and absorption spectra. Overall, it seems that cellular constituents absorbing light between 280-320 nm are destroyed by UV-B radiation, accounting for the death of the cells. The slow growth observed in liquid culture might be due to the presence of a few UV-B resistant cells. It is also possible that damaged macromolecules such as proteins may be repaired by the dark or light repair mechanisms, thus permitting growth after a lag.

Our observation of drastic reduction in pigment content is consistent with the findings of Häder _et al._ (22) and Döhler _et al._ (10) who reported pigment bleaching following exposure to sunlight or UV-B. Döhler _et al._ (10) reported a close correlation in the decrease of phycocyanin with the level of UV-B radiation. The reduction in the content of photosynthetic pigments indicates that the protein moieties of pigments are the primary target of UV-B effect. As phycocyanin is a chromoprotein, the UV-B radiation has drastic effect on this pigment.

Our data of C_2H_2 reduction activity suggest that most probably nitrogenase enzyme complex is inactivated by UV-B treatment. This is based on the finding that inhibition of C_2H_2 reduction occurred within 5 min of UV-B treatment and there was no activity after 20 min. Almost similar findings have been obtained with _Anabaena flos-aquae_ and _Anabaena azollae_ (11). Furthermore, the loss of activity may also be due to impairment in the supply of ATP and reductant which are essential for nitrogenase activity. However, any imbalance in the supply of ATP and reductant would not be expected to block nitrogenase activity instantly because nitrogenase activity is not lost abruptly, even in the dark or in DCMU-treated cultures. It should also be remembered that growth is not severely affected by 20-min UV-B treatment and hence the observed effect appears to be solely due to inactivation of nitrogenase polypeptide.

The data of nitrate reductase activity also favour the latter conclusion because appreciable amount of nitrate reductase activity was observed in UV-B treated culture. If UV-B immediately blocks the supply of ATP and reductant, one would expect nitrate reductase to behave in the same way as does nitrogenase. It therefore appears likely that nitrogenase is specifically inactivated by UV-B in N_2-fixing cyanobacteria. The continued expression of nitrate reductase even after long exposure to UV-B might be due to the supply of some preformed or residual ATP and reductant to the enzyme molecule. Likewise, failure of UV-B to inhibit induction of nitrate reductase appears to be due to the presence of bulk of constituents required for the synthesis of nitrate reductase polypeptide in the cell itself.

Depression of photosynthetic CO_2 fixation by UV-B radiation may be explained by a diminution of supply of ATP and $NADPH_2$. It has been demonstrated by many workers that UV-B inhibition of ATP synthesis represents an effect on the photosystem II reaction centre (8, 23).

ACKNOWLEDGEMENTS

This work has been supported by a research project from the Department of Environment, Government of India (14/28/89-MAB/RE). Experimental work has been done by RT and DV.

REFERENCES

1. Scotto, J., Cotton, G., Urbach, F., Berger, D. and Fears, T. (1988). Biologically effective ultraviolet radiation: Surface measurements in the United States, 1974 to 1985. Science 239: 762-764.

2. Biggs, R.H., Kossuth, S.U. and Teramura, A.H. (1981). Response of 19 cultivars of soybeans to ultraviolet-B irradiance. Physiol. Plant. 53: 19-26.

3. Flint, S.D. and Caldwell, M.M. (1984a). Comparative sensitivity of binucleate and trinucleate pollen to ultraviolet radiation: A theoretical perspective, In: Stratospheric Ozone Reduction, Solar Ultraviolet Radiation and Plant Life, R.C. Worrest and M.M. Caldwell (eds.), Springer-Verlag, Berlin.

4. Gold, W.G. and Caldwell, M.M. (1983). The effects of ultraviolet-B radiation on plant competition in terrestrial ecosystems. Physiol. Plant. 58: 435-444.

5. Renger, G., Völker, M., Eckert, H.J., Fromme, R., Hohm-Veit and Gräber, P. (1989). On the mechanism of photosystem II deterioration by UV-B irradiation. Photochem. Photobiol. 49: 97-105.

6. Teramura, A.H. (1983). Effects of ultraviolet-B radiation on the growth and yield of crop plants. Physiol. Plant. 58: 415-427.

7. Caldwell, M.M. (1981). Plant response to solar ultraviolet radiation, pp. 167-197, In: Encyclopaedia of Plant Physiology, Vol. 12A, Physiological Plant Ecology I, O.L. Lange, P.S. Nobel, C.B. Osmond and H. Ziegler (eds.), Springer-Verlag, Berlin.

8. Iwanzik, W., Tevini, M., Dohnt, G., Voss, M., Weiss, W., Gräber, P. and Renger, G. (1983). Action of UV-B radiation on photosynthetic primary reactions in spinach chloroplasts. Physiol. Plant. 58: 401-407.

9. Worrest, R.C. (1986). The effect of solar UV-B radiation on aquatic systems: An overview, pp. 175-191, In: Effects of Changes in Stratospheric Ozone and Global Climate, Vol. 1: Overview, J.G. Titus (ed.), United Nations Environ. Prog. and U.S. Environ. Protec. Agency.

10. Döhler, G., Biermann, I. and Zink, J. (1986). Impact of UV-B radiation on photosynthetic assimilation of ^{14}C-bicarbonate and inorganic ^{15}N-compounds by cyanobacteria. Z. Naturforsch. 41c: 426-432.

11. Newton, J.W., Tyler, D.D. and Slodki, M.E. (1979). Effect of ultraviolet-B (280 to 320 nm) radiation on blue-green algae (cyanobacteria), possible biological indicators of stratospheric ozone depletion. Appl. Environ. Microbiol. 37: 1134-1141.

12. Safferman, R.S. and Morris, M.E. (1964). Growth characteristics of the blue-green algal virus LPP-1. J. Bacteriol. 88: 771-773.

13. Stewart, W.D.P., Fitzgerald, G.P. and Burris, R.H. (1968). Acetylene reduction by nitrogen-fixing blue-green algae. Arch. Mikrobiol. 62: 336-348.

14. Camm, E.L. and Stein, J.R. (1974). Some aspects of the nitrogen metabolism of *Nodularia spumigena* (Cyanophyceae). Can. J. Bot. 52: 719-726.

15. Lowe, R.H. and Evans, H.J. (1964). Preparation and some properties of a soluble nitrate reductase from *Rhizobium japonicum*. Biochim. Biophys. Acta 85: 377-389.

16. Kumar, A., Tabita, F.R. and Van Baalen, C. (1982). Isolation and characterization of heterocysts from *Anabaena* sp. strain CA. Arch. Microbiol. 133: 103-109.

17. Myers, J. and Kratz, W.A. (1955). Relationship between pigment content and photosynthetic characteristics in a blue-green alga. J. Gen. Physiol. 39: 11-21.

18. Brody, S.S. and Brody, M. (1961). A quantitative assay for the number of chromophores on a chromoprotein: its application to phycoerythrin and phycocyanin. Biochim. Biophys. Acta 53: 348-352.

19. Häder, D.-P., Watanabe, M. and Furuya, M. (1986). Inhibition of motility in the cyanobacterium, *Phormidium uncinatum*, by solar and monochromatic UV irradiation. Plant Cell Physiol. 27: 887-894.

20. Döhler, G. (1984). Effect of UV-B radiation on the marine diatoms *Lauderia annulata* and *Thalassiosira rotula* grown in different salinities. Marine Biol. 83: 247-253.

21. Yammamoto, K.M., Satake, M., Shinagawa, H. and Fujiwara, Y. (1983). Amelioration of the ultraviolet sensitivity of an *Escherichia coli rec* A mutant in the dark by photoreactivating enzyme. Mol. Gen. Genet. 190: 511-515.

22. Häder, D.-P., Rhiel, E. and Wehrmeyer, W. (1988). Ecological consequences of photomovement and photobleaching in the marine flagellate *Cryptomonas mucilata*. FEMS Microbiol. Ecol. 53: 9-18.

23. Kulandaivelu, G. and Noorudeen, A.M. (1983). Comparative study of the action of ultraviolet-C and ultraviolet-B radiation on photosynthetic electron transport. Physiol. Plant. 58: 389-394.

THE RESPONSE OF PLANTS TO ULTRAVIOLET-B RADIATION AND METAL STRESS

Janet F. Bornman and Sylvain Dubé

Department of Plant Physiology, University of Lund, Box 7007,
S-220 07 Lund, Sweden

ABSTRACT

Increasing deposits of heavy metals and other toxic pollutants are being found in soils in many parts of the world as a result of anthropogenic pollution and addition of phosphate fertilizers. In addition, the threat of a concomitant decrease in the ozone layer with subsequent increase in UV radiation will conceivably place plant and other life under increasing stress. Consequently the assessment of multiple effects is very important in obtaining a more meaningful and realistic view of the current changes in the environment.

Picea abies was grown from seed and exposed to UV-B radiation simulating a 5% decrease in the ozone layer at Lund, Sweden (55.7 °N, 13.4 °E). Analyses were carried out after 10 weeks of treatment. The combination of UV-B radiation and Cd^{2+} resulted in a reduction of needle dry weight and plant height, although in the latter case Cd^{2+} alone reduced seedling height most. CO_2 assimilation rates also declined. In contrast to the treatments with UV or Cd^{2+} alone, the combination treatment also altered the kinetics of the variable fluorescence, namely in the F_M, or second maximum region. For comparative purposes plants of *Brassica campestris* L. cv. Tove, grown under similar conditions, were also included in some of the analyses.

Key words - *Brassica campestris*, cadmium, CO_2 assimilation, chlorophyll fluorescence induction, *Picea abies*, Norway spruce, ultraviolet radiation.

MATERIALS AND METHODS

Material and Growth Conditions

Seeds of Norway spruce (*Picea abies* L. Karst) were sown in vermiculite and germinated in a greenhouse. After 6 weeks they were transplanted to pots and placed in a growth chamber under either control or treatment conditions. Background light (WL) in all cases was 500 $\mu mol\ m^{-2}\ s^{-1}$ PAR (400-700 nm, Philips HPLR 400 W lamps). Half the plants received UV-B radiation simulating a 5% decrease in stratospheric ozone calculated (Björn and Murphy 1985) for Lund, Sweden (55.7 °N, 13.4 °E), for a cloudless day (15 July) with aerosol zero. The biologically effective radiation was 6.17 $kJ\ m^{-2}\ day^{-1}$, computed according to Caldwell (1971). A cut-off Plexiglas filter (FBL.2458, Röhm GMBH, Chemische fabrik, Germany; 3 mm in thickness) removed nearly all wavelengths below 280 nm. Between 275-280 nm, radiation corresponding to 0.3% of that measured between 250-320 nm was found.

The plants under both light conditions were divided into two groups, one of which received 3 $\mu M\ CdCl_2$ in the nutrient medium (pH 4.5; pH 5.9 without cadmium) twice a week. Relative humidity was ca 50%. The photoperiod was 16 h; day/night temperature of $22 \pm 2/19 \pm 1°$ C.

Plants were measured after 10 weeks of treatment. The first 7 cm of the main shoot were used for all measurements in order to eliminate large differences in sample size. *Brassica* plants were similarily treated except that they were maintained under treatment conditions for 2 weeks and supplied with 15 $\mu M\ CdCl_2$ twice a week.

Variable Chlorophyll Fluorescence

An integrating sphere interfaced with a data acquisition system and control software (essentially according to Dubé and Vidaver 1990) was used to monitor variable chlorophyll fluorescence (F_{var}). Both *Picea abies* and *Brassica campestris* were used for comparative purposes.

Photosynthetic Rates

Gas exchange measurements were conducted on plant material using a portable infrared gas analyser (Analytical Development Co., type LCA 2).

Pigment Analyses

Chlorophyll

Chlorophyll was determined according to Inskeep and Bloom (1985). Eighteen needles per treatment were macerated in DMF (N,N-dimethylformamide). The extraction solution was measured spectrophotometrically using an AMINCO DW-2 UV/VIS spectrophotometer (Silver Spring, ML, USA) in the split beam mode.

UV-B screening pigments

Eighteen needles per treatment were ground in methanol-H_2O-HCl (79:20:1 v/v; Robberecht and Caldwell 1986) in quartz sand using a pestle and mortar. Extracts were centrifuged (1 600 g, 10 min) and the supernatants scanned in an AMINCO DW-2 UV/VIS spectrophotometer (Silver Spring, ML, USA).

Growth Parameters

Plant height was measured in cm from soil level to the base of the apical bud of the main shoot, and heights averaged for each treatment.
Twenty needles per plant (upper 7 cm of the main shoot) were taken for fresh and dry weight determinations.

RESULTS AND DISCUSSION

After 10 weeks, plants treated with cadmium alone gave a 32% decrease in plant height, while those exposed to UV-B and cadmium showed a 25% decrease relative to controls. Plants exposed to UV-B alone gave a 21% decrease (Fig. 1.).

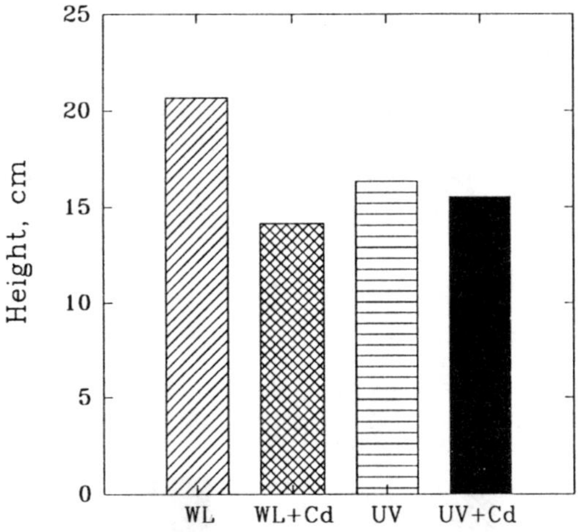

Fig. 1. Height in cm of spruce seedlings after 10 weeks of treatment.

Dry weight was decreased most both for cadmium treatment alone as well as for cadmium in combination with UV radiation (Fig. 2).

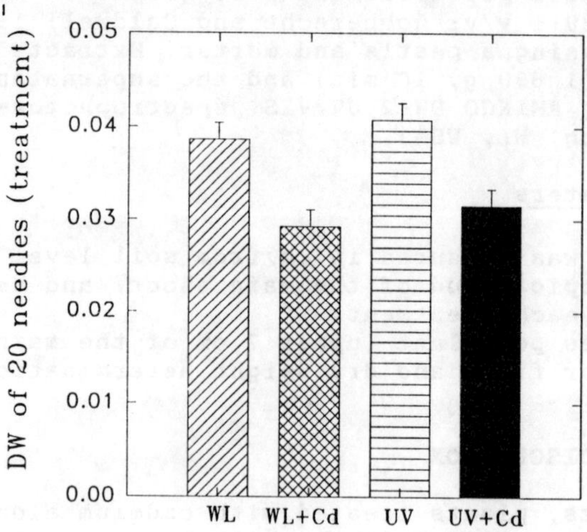

Fig. 2. Dry weight of spruce needles after 10 weeks of treatment.

The total Chl content in cadmium treated plants showed a 22% increase as compared to controls (Fig. 3). This increase is rather surprising, since numerous reports have shown that cadmium often decreases Chl content, probably by impairing precursor synthesis (e.g. Stobart et al. (1985) or by decreasing important ions such as iron (Haghiri 1973).

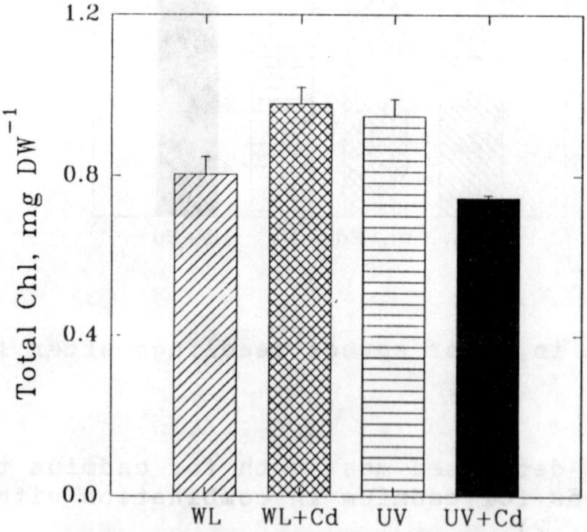

Fig. 3. Total Chl (mg DW^{-1}) of spruce needles after 10 weeks of treatment.

In contrast, Chl content in those plants treated with UV and cadmium was decreased by 7% compared to controls. This decrease was due to a larger reduction in Chl *b*, which may indicate a change in the size of the antenna pigments. UV-B screening pigments were increased most from plants exposed to UV plus cadmium (results not shown). The reason for this is not clear.

Rates of apparent photosynthesis, normalized to needle dry weight, decreased by 33% in UV plus cadmium treated plants (Fig. 4). UV-B treated plants also showed a decrease compared to controls, while cadmium alone resulted in increased rates.

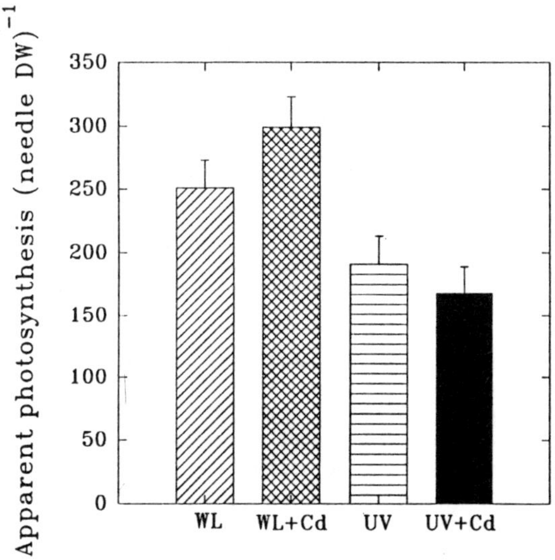

Fig. 4. Apparent photosynthesis of spruce seedlings after 10 weeks of treatment. Values are expressed per needle dry weight.

It appears that the reduction in rates was mainly due to the effect of UV radiation, since differences between the combination treatment and UV alone were small (Fig. 4).

For plants exposed to UV-B radiation and cadmium, the kinetics of the variable fluorescence were altered in the F_M (second maximum) region for both *Brassica* and *Picea*. Results for *Brassica* are shown in Fig. 5a, although similar changes were found for *Picea*.

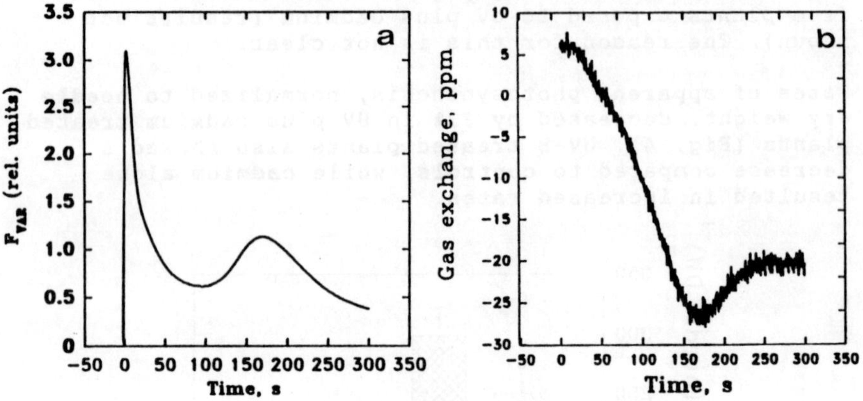

Fig. 5. a) Fluorescence induction curve of *Brassica campestris* leaves after 2 weeks of treatment with simultaneous exposure to UV-radiation and cadmium supply. **b)** A trace from the same leaf sample as in **b)** showing CO_2 gas exchange rates. Note the delay in the shoulder (second peak after ca 250 s) as compared with the fluorescence transient of **a)** (second peak after ca 170 s).

Fluorescence transients are usually characterised by typical OIDPSMT phases (Govindjee and Papageorgiou 1971). However, after UV and cadmium treatment, the second maximum, or F_M was no longer discernible. Instead, a distinct rise was seen after the S-stage (Fig. 5a), accompanied by a similar, although delayed, rise in the curves of the CO_2 gas exchange (Fig. 5b). Greger and Ögren (1989) reported that for sugar beet seedlings, the kinetics of the slow phases of the induction curve also lacked certain oscillations, which may represent regulatory reactions of carbon metabolism (Walker et al. 1983).

CONCLUSIONS

It is possible that the regulatory processes of the carbon reduction cycle were altered by the combination of UV radiation and cadmium. Changes usually attributed to cadmium alone (e.g. as far as Chl content and CO_2 assimilation rates were concerned) were not manifested in the present study, probably as a result of the low concentrations of cadmium administered as compared to other investigations. The interesting outcome of this study was that in combination with UV radiation, many of these cadmium effects became evident, pointing to a multiple stress phenomenon.

REFERENCES

Björn, L.O. and Murphy, T.M. (1985). Computer calculation of solar ultraviolet radiation at ground level. *Physiol. Vég.* 23:555-561.

Caldwell, M.M. (1971). Solar UV irradiation and the growth an ddevelopment of higher plants. In: *Photobiology,* A.C. Giese, ed., Vol. 6, pp. 131-177. Academic Press, New York.

Dubé, S. and Vidaver, W. (1990). An integrating fluorometer data acquisition system. *Plant Physiol. Biochem.* 28: 539-546.

Govindjee and Papageorgiou, G. 1971. Chlorophyll fluorescence and photosynthesis: Fluorescence transients. In: *Photophysiology,* A.C. Giese, ed., Vol. VI, pp. 2-45. Academic Press.

Greger, M. and Lindberg, S. (1987). Effects of Cd^{2+} and EDTA on young sugar beets (*Beta vulgaris*). II. Net uptake and distribution of Mg^{2+}, Ca^{2+} and Fe^{2+}/Fe^{3+}. *Physiol. Plant.* 69: 81-86.

Greger, M. and Ögren, E. 1989. Direct and indirect effects of cadmium on photosynthesis in sugar beet (*Beta vulgaris*). Doctoral Thesis. ISBN 91-7146-717-3.

Haghiri, F. (1973). Cadmium uptake by plants. *J. Environ. Qual.* 2: 93-96.

Inskeep, W.P and Bloom, P.R. (1985). Extinction coefficients of chlorophyll a and b in N,N-dimethylformamide and 80% acetone. *Plant Physiol.* 77: 483-485.

Robberecht and Caldwell, M.M (1986). Leaf UV optical properties of *Rumex obtusifolius* L. in regard to a protective mechanism against solar UV-B radiation injury. In: *Stratospheric Ozone Reduction, Solar Ultraviolet Radiation and Plant Life*, R.C. Worrest and M.M. Caldwell, eds, Vol. G8, pp. 251-259. Springer-Verlag, NATO ASI Series. ISBN 3-540-13875-7.

Stobart, A.K., Griffiths, W.T., Ameen-Bukhari, I. and Sherwood, R.P. (1985). The effect of Cd^{2+} on the biosynthesis of chlorophyll in leaves of barley. *Physiol. Plant.* 63: 293-298.

Walker, D.A., Horton, P., Sivak, M.N. and Quick, W.P. (1983). Antiparallel relationship between O_2 evolution and slow fluorescence induction kinetics. *Photobiochem. Photobiophys.* 5: 35-39.

PHYSIOLOGICAL AND MORPHOLOGICAL RESPONSES OF SNAPBEAN PLANTS TO OZONE STRESS AS INFLUENCED BY PRETREATMENT WITH UV-B RADIATION

Madhoolika Agrawal[1], Shashi B. Agrawal[1] Donald T. Krizek[2], George F. Kramer[2], Edward H. Lee[2], Roman M. Mirecki[2], and Randy A. Rowland[2], [1]Department of Botany, Banaras Hindu University, Varanasi, India and [2]USDA Climate Stress Laboratory, ARS, Beltsville, MD 20705, USA

ABSTRACT

Snapbean (Phaseolus vulgaris L. cv. 'Bush Blue Lake 290') plants were grown under enhanced UV-B radiation for 21 days in a growth chamber at 11.7 kJ m^{-2} d^{-1} of biologically effective ultraviolet radiation ($UV\text{-}B_{BE}$) (normalized at 300 nm) and 300 µmol m^{-2} s^{-1} of photosynthetic photon flux prior to ozone (O_3) (3 h at 0.25 µmol mol^{-1}) to determine the influence of UV-B preconditioning on O_3 sensitivity. This level of $UV\text{-}B_{BE}$ represents approximately a 20% decrease in stratospheric O_3 for clear sky conditions at Beltsville, MD (39°N) on June 21.

Plants grown from seed under enhanced UV-B radiation showed marginal cupping of the primary and trifoliate leaves, marginal chlorosis, and marked reductions in fresh weight, dry weight, and leaf area as compared to those grown in the absence of UV-B radiation. UV-B treated plants subjected to O_3 showed greater injury of the first trifoliate and less injury of the primary leaf after 48 h than non-irradiated plants.

Chlorophyll concentration was decreased in UV-B treated plants as determined by SPAD chlorophyll meter readings and measurements of leaf discs extracted in 80% acetone. The chlorophyll a/b ratio was decreased by both UV-B and O_3 treatment. Flavonoid absorbance at 300 nm, however, was decreased by UV-B treatment.

UV-B irradiation caused nearly a two-fold reduction in photosynthetic rate but had no significant effect on stomatal conductance of the first trifoliate leaf. In contrast, O_3 fumigation not only decreased photosynthetic rate but caused a marked reduction in stomatal conductance.

Measurements of chlorophyll fluorescence kinetics revealed an increase in area under the fluorescence curve under elevated UV-B radiation and a decrease in half-rise time under O_3 stress, suggesting an increase in PS II activity.

These findings demonstrate the need to conduct studies on interactions of UV-B radiation and other environmental stress factors in order to develop realistic assessments and valid models of projected changes in global climate.

INTRODUCTION

There is increased concern regarding the adverse biological effects of stratospheric ozone reduction and the projected increase in biologically effective ultraviolet-B ($UV\text{-}B_{BE}$) radiation reaching the earth's surface (1-5). There is also growing concern about the possible direct and indirect effects of increased levels of trace gases such as carbon dioxide, nitrous oxides (N_2O), methane (CH_4), and tropospheric ozone (O_3) (2,6).

Numerous studies have been conducted on the effects of UV-B radiation on plant growth and development (5,7-11). Extensive literature has also been published on the effects of ozone (12-20).

Extensive research has been conducted to determine the physiological and biochemical basis for adaptation to UV-B radiation (9,11,21) and O_3 pollution (15,17-19,22-23). Flavonoids have been shown to serve as important UV-screening pigments (4,24-27).

To date, relatively little is known about the combined effects of UV-B radiation and air pollutants (2). Global climate change is likely to result in both positive and negative environmental effects of plants.

Thus, additional data on environmental interactions are needed in order to develop a meaningful assessment of projected increases in UV-B radiation and global climate changes on agricultural production. Such information will also be required in order to develop improved models of global climate change.

The objective of the present study was to determine the influence of UV-B pretreatment on the response of plants to O_3 fumigation under controlled environment conditions.

MATERIALS AND METHODS

Plant Material

Seeds of *Phaseolus vulgaris* L. 'Bush Blue Lake 290', an O_3 sensitive cultivar (13) were planted in 12.5 cm plastic pots containing a peat-vermiculite potting mix (Jiffy Mix, Jiffy Products of America, Inc., West Chicago, IL). Upon emergence, the seedlings were selected for uniformity and thinned to two plants per pot. The plants were fertilized once a week with 200 ml Peters 20-20-20 fertilizer (Peter's Fertilizer Products, W.R. Grace & Co., Fogelsville, PA, USA). The remaining days, they received only distilled water.

UV Pretreatment

Prior to O_3 fumigation, the plants were grown for 21 days from seed in two microprocessor-controlled plant growth chambers (Model M-13, Environmental Growth Chamber Inc., Chagrin Falls, OH, USA) under the following environmental conditions: 25/20°C day/night temperature, 70% relative humidity, 300 µmol m^{-2} s^{-1} of photosynthetic photon flux (PPF) provided by 1500 mA cool white fluorescent lamps and supplemental incandescent lamps, and a 16 h photoperiod.

One group of plants was grown under enhanced UV-B radiation (referred to as +UV). A second group of plants was grown in the absence of UV-B radiation (referred to as -UV). Each growth chamber was equipped with six Q-Panel (Cleveland, OH, USA) UVB-313 lamps covered with either 0.13 mm (5 mil) cellulose diacetate to transmit UV-B (cut off ca 292 nm) or with 0.13 mm (5 mil) polyester (Cadillac Plastics, Baltimore, MD, USA) to exclude UV-B (cut off ca 318 nm).

The +UV plants were exposed to an irradiance which simulated the amount of $UV\text{-}B_{BE}$ that would reach the ground at Beltsville, MD (39° N), under clear skies on June 21, assuming approximately a 20% reduction in stratospheric ozone. This exposure was based on calculations using a UV-dosage model (28) and a generalized plant action spectrum (1) normalized at 300 nm. This dose was 11.7 kJ m^{-2} d^{-1}. Filters were changed every 4-5 days. Daily adjustments in UV-B irradiance were made by means of a dimming system based on measurements made with a portable radiometer calibrated against a spectroradiometer (29). Plants were irradiated for 6 h daily (1100-1700 h) midway through the photoperiod.

Ozone Treatment

On day 22, the +UV and -UV treated plants were transferred to a controlled environment chamber at 25°C where they were equilibrated for 1 h before O_3 fumigation. The PPF level during fumigation was

300 µmol m^{-2} s^{-1} of PPF (provided by 1500 mA cool white fluorescent and GE 52 watt supplemental incandescent lamps), and the relative humidity was 70-80%. Following equilibration, the plants were fumigated for 3 h at 0 or 0.25 µmol mol^{-1} of O_3. The O_3 concentration was monitored with a TECO model 49 (Thermo Electron Corporation, Hopkinton, MA, USA) instrument. Upon completion of the O_3 treatment, measurements were taken on selected plants for growth and physiological response while others were returned to their former growth chambers without UV treatment to score UV and O_3 injury.

Extraction and Determination of Pigments

Chlorophyll was extracted from leaf discs with 80% acetone and determined according to Arnon (30). Flavonoids were extracted from leaf discs (1 cm^2 area) in ethanol-acetic acid (99:1 v/v) according to Flint *et al.* (26). The absorbance was read on a Shimadzu Model 160A UV/VIS Recording Spectrophotometer (Columbia, MD, USA). In situ measurements of chlorophyll concentration were also determined by using a Minolta SPAD model 501 chlorophyll meter.

Growth Measurements

Measurements of growth responses were determined on day 9, 15, and 22, but data are only presented for day 22. Plant height was measured from the cotyledonary node to the base of the uppermost leaf. Leaf areas of the primary leaves and trifoliate leaves were determined with a LI-COR LI-3000 Leaf Area Meter (Lincoln, NE, USA). Fresh and dry weights were determined separately for each leaf but only pooled data are reported here. After determining fresh weights, the samples were dried in a forced draft oven at 70°C for 48 h before dry weights were obtained.

Measurements of Gas Exchange

Measurements of photosynthetic rate stomatal conductance were determined on the terminal leaflet of the first trifoliate leaf from the base of the plant by means of a LI-COR Model LI-6200 Portable Photosynthesis System (Lincoln, NE, USA).

Chlorophyll Fluorescence Induction

Measurements of chlorophyll fluorescence induction kinetics were made at room temperature (23±2°C) using a Model MF-1 portable chlorophyll fluorometer (Univ. of Missouri, Columbia, MO, USA), which was connected to a personal computer equipped with hardware and software to convert the computer to a digital oscilloscope (Rapid Systems Inc., Seattle, WA, USA). The adaxial leaf surface was irradiated with a red light emitting diode and the fluorescence signal collected from the same surface. Leaves of all plants were dark-adapted for 6 min before obtaining measurements. Data are presented for variable fluorescence (F_v) (F_{max} - F_o), (F_v/F_o), area under the curve obtained during a period of 1800 ms, and half-rise time (from F_o to F_{max}).

Statistical Analysis

SAS (SAS Institute, Carry, NC, USA) was used to conduct an analysis of variance on all growth and physiological parameters. Because of space limitations, F values are only shown for selected variables.

RESULTS

Leaf Injury

Visual injury from UV-B irradiation was observed on the first trifoliate leaf within 15 days of treatment. This was manifested as severe inhibition of leaf enlargement, marginal cupping of the leaves, and marginal chlorosis. UV-B treated plants subjected to O_3 treatment showed greater O_3 injury on the first trifoliate but less O_3 injury on the primary leaf at 48 h after fumigation than -UV plants.

Growth Measurements

The most striking growth response of snapbean plants to +UV treatment after 3 weeks exposure was a two-fold reduction in fresh and dry weights and more than a two-fold reduction in leaf area as compared to controls (-UV) (Tables 1, 2, 3, 4, and 5). Specific leaf weights (SLW) were increased by UV-B exposure (Tables 5, 6). After 9 and 15 days of UV-B treatment, there was a significant reduction in shoot elongation (data not shown), but by 21 days, this effect was largely lost (Tables 4, 5). UV-B irradiation also increased the percent dry weight (DW) of the leaves (Tables 4, 5). There was a significant UV x O_3 interaction in SLW and % DW after 3 weeks of UV treatment and 3 h of O_3 exposure (Table 5).

Table 1. Fresh Weights of 'BBL 290' Snapbean Plants after 21 Days of UV and 3 Hours of O_3 Treatments

UV Treatment	O_3 conc. $\mu mol\ mol^{-1}$	Fresh Weight (g)		
		T	T+L+P	W
-	0	5.6 $\pm$ 0.6	10.0 $\pm$ 0.9	14.6 $\pm$ 1.4
-	0.25	4.9 $\pm$ 0.5	7.9 $\pm$ 1.2	11.5 $\pm$ 1.5
+	0	3.3 $\pm$ 0.1	5.6 $\pm$ 0.2	8.9 $\pm$ 0.5
+	0.25	2.8 $\pm$ 0.7	5.4 $\pm$ 1.0	8.6 $\pm$ 1.6

T = Trifoliates; L = Laterals; P = Primary Leaves;
W = Leaves and Stems.

Table 2. Dry Weights of 'BBL 290' Snapbean Plants after 21 Days of UV and 3 Hours of O_3 Treatments

UV Treat-ment	O_3 conc. $\mu mol\ mol^{-1}$	Dry Weight (g) T	T+L+P	W
-	0	0.88 $\pm$ 0.08	1.38 $\pm$ 0.11	1.80 $\pm$ 0.15
-	0.25	0.73 $\pm$ 0.07	1.10 $\pm$ 0.12	1.40 $\pm$ 0.15
+	0	0.42 $\pm$ 0.02	0.68 $\pm$ 0.03	0.94 $\pm$ 0.05
-	0.25	0.39 $\pm$ 0.09	0.72 $\pm$ 0.12	1.02 $\pm$ 0.16

T = Trifoliates; L = Laterals; P = Primary Leaves;
W = Leaves and Stems.

Table 3. Leaf Area of 'BBL 290' Snapbean Plants after 21 Days of UV and 3 Hours of O_3 Treatments

UV Treat-ment	O_3 conc. $\mu mol\ mol^{-1}$	Leaf Area (cm^2) T	L	T+L+P
-	0	281.8 + 27.5	8.3 + 3.3	445.8 + 36.5
-	0.25	247.9 + 24.8	4.7 + 2.6	364.7 + 42.7
+	0	111.6 + 5.7	2.2 + 2.2	187.6 + 8.1
+	0.25	92.2 + 20.3	2.5 + 2.5	213.0 + 58.0

T = Trifoliate; L = Laterals; P = Primary Leaves.

Table 4. SLW, % DW of Leaves, and Height of 'BBL 290' Snapbean after 21 Days of UV and 3 Hours of O_3 Treatments

UV Treat-ment	O_3 conc. $\mu mol\ mol^{-1}$	Ht cm	SLW $mg\ cm^{-2}$	% DW
-	0	11.4 $\pm$ 2.1	3.12 $\pm$ 0.08	15.9 $\pm$ 0.6
-	0.25	8.4 $\pm$ 0.7	2.93 $\pm$ 0.00	14.8 $\pm$ 0.1
+	0	7.3 $\pm$ 0.0	3.79 $\pm$ 0.04	12.8 $\pm$ 0.2
+	0.25	7.6 $\pm$ 0.5	4.21 $\pm$ 0.17	14.1 $\pm$ 0.7

Table 5. F Value for Growth Responses to UV-B and O_3 Treatments

Parameter	UV	O_3	UV x O_3
FW - Tops	0.0017**	0.2272	0.3157
DW - Tops	0.0019**	0.2780	0.1207
Leaf Area Tops	0.0001***	0.2694	0.7086
Height	0.0638	0.2513	0.1871
SLW	0.0001***	0.2834	0.0136*
% Dry Weight	0.0031**	0.8662	0.0329*

*** : $P < 0.001$; ** : $P < 0.05$; * : $P < 0.01$.

Measurements of Gas Exchange

Photosynthetic rates in the first trifoliate were decreased by both long-term UV-B pretreatment and a single 3 h O_3 exposure (Tables 6 and 7). Ozone treatment had a greater suppressive effect on photosynthesis of 'Bush Blue Lake 290' snapbean plants than did UV-B enhancement. When bean plants were subjected to a combination of the two environmental stresses, a synergistic effect was obtained (Tables 6 and 7). Stomatal behaviour in UV-B irradiated snapbean plants was unaffected by UV-B exposure while O_3 exposure caused a marked reduction in stomatal conductance (Tables 6 and 7).

Table 6. Photosynthetic Rates (Pn) and Stomatal Conductance (Cs) of First Trifoliate of 'BBL 290' Snapbean after 21 Days of UV and 3 Hours of O_3 Treatments

UV Treatment	O_3 conc. $\mu mol\ mol^{-1}$	Pn rate $\mu mol\ m^{-2}\ s^{-1}$	Cs $cm\ s^{-1}$
-	0	3.63 $\pm$ 0.33	1.65 $\pm$ 0.40
-	0.25	1.08 $\pm$ 0.15	0.52 $\pm$ 0.12
+	0	2.09 $\pm$ 0.34	1.18 $\pm$ 0.16
+	0.25	0.31 $\pm$ 0.51	0.35 $\pm$ 0.01

Table 7. F Value for Gas Exchange Responses to UV-B and O_3 Treatments

Parameter	UV	O_3	UV x O_3
Pn rate	0.0117*	0.0003***	0.3412
Cs	0.1953	0.0023**	0.5246

*** : P < 0.001; ** : P < 0.05; * : P < 0.01.

Photosynthetic Pigments

Exposure of snapbean plants to increased UV-B irradiance caused a reduction in chlorophyll a concentration on a fresh weight (FW) and DW basis (Tables 8 and 10) but had no appreciable effect on chlorophyll b concentration (data not shown). The chlorophyll a/b ratio was reduced by UV-B enhancement (Tables 8 and 10) largely reflecting the change in chlorophyll a concentration. Ozone fumigation also caused a reduction in chlorophyll a/b ratio (Tables 8 and 10). A reduction in total chlorophyll as measured with a SPAD chlorophyll meter was also obtained.

Table 8. Chlorophyll Conc. in First Trifoliate of 'BBL 290' Snapbean after 21 Days of UV and 3 Hours of O_3 Treatments

UV Treat-ment	O_3 conc. μmol mol^{-1}	Chl a conc. (μg ml^{-1}) FW basis	Chl a conc. (μg ml^{-1}) DW basis	Chl a/b ratio
-	0	1.7 ± 0.1	17.1 ± 0.6	3.2 ± 0.1
-	0.25	1.5 ± 0.1	15.5 ± 0.7	3.0 ± 0.1
+	0	1.5 ± 0.0	15.0 ± 0.4	2.9 ± 0.0
+	0.25	1.4 ± 0.1	14.1 ± 0.7	2.7 ± 0.0

Flavonoid Content

Under the relatively low PPF conditions used in this experiment, UV-B enhancement for 3 weeks caused a reduction in UV-B screening pigments. Flavonoid absorbance at 300 nm was decreased by half in UV-B irradiated plants as compared to controls (Tables 9 and 10). This was true on both a FW (data not shown) and a DW basis.

Chlorophyll Fluorescence Kinetics

Under increased UV-B radiation, there was an increase in the area under the fluorescence curve, but no change in variable fluorescence or half rise time. Under O_3 stress there was a decrease in half rise time but no change in variable fluorescence or area under the fluorescence curve (Tables 11 and 12).

Table 9. Flavonoid Absorbance of First Trifoliate of 'BBL 290' Snapbean after 21 Days of UV and 3 Hours of O_3 Treatments

UV Treatment	O_3 conc. $\mu mol\ mol^{-1}$	Flavonoid absorbance at 300 nm		
		Area basis abs cm^{-2}	FW basis abs mg^{-1}	DW basis abs mg^{-1}
-	0	2.9 ± 0.3	0.14 ± 0.02	1.3 ± 0.2
-	0.25	3.1 ± 0.3	0.15 ± 0.02	1.3 ± 0.1
+	0	1.5 ± 0.4	0.06 ± 0.01	0.6 ± 0.2
+	0.25	2.7 ± 0.7	0.13 ± 0.03	1.0 ± 0.2

Table 10. F Value for Pigment Responses to UV and O_3 Treatments

Parameter	UV	O_3	UV x O_3
Chl a conc. DW basis	0.0218*	0.0706	0.6237
Chl a/b	0.0002***	0.0008***	0.9720
Flavonoid Abs 300 nm DW basis	0.0295*	0.1926	0.3295

*** : $P < 0.001$; * : $P < 0.01$.

Table 11. Chlorophyll Fluorescence Kinetics of 'BBL 290' Snapbean Plants after 21 Days of UV and 3 Hours of O_3 Treatments

UV Treatment	O_3 conc. $\mu mol\ mol^{-1}$	F_v	Fl. area[z] x 10^5	F_v/F_o	Half-rise time ms
-	0	96 ± 15	4.85 ± 0.26	0.45 ± 0.07	78 ± 2
-	0.25	65 ± 9	4.77 ± 0.11	0.29 ± 0.04	49 ± 3
+	0	67 ± 7	4.95 ± 0.04	0.28 ± 0.04	68 ± 3
+	0.25	85 ± 1	5.37 ± 0.00	0.35 ± 0.01	47 ± 2

[z]Area expressed in mv per 1800 ms.

Table 12. F Value for Chlorophyll Fluorescence Kinetics

Parameter	UV	O_3	UV x O_3
F_v	0.6712	0.5377	0.0365*
Area	0.0385*	0.2611	0.1182
F_v/F_o	0.2602	0.3601	0.0482*
Half-rise time	0.2501	0.0001***	0.5323

*** : P < 0.001;   * : P < 0.01.

DISCUSSION

Subjecting bean plants to increased UV-B radiation for 21 days from the time of seeding resulted in a marked inhibition of biomass production and leaf enlargement. Shoot elongation, however, was not significantly reduced by increased UV-B radiation. These results are consistent with those reported by Cen and Bornman (9) for Phaseolus vulgaris (cv. 'Stella') under conditions of low PPF (250 μmol m^{-2} s^{-1}). Under moderate (500 μmol m^{-2} s^{-1}) and high (700 μmol m^{-2} s^{-1}) PPF, however, these workers failed to obtain a decrease in fresh or dry weight. In contrast to our findings and those of Cen and Bornman (9) obtained under growth chamber conditions, Dumpert and Knacker (31) found that increased UV-B radiation under greenhouse and field conditions caused a decrease in plant height in P. vulgaris as well as a reduction in leaf area and fresh and dry weights.

Leaf thickness was increased by +UV treatment as evidenced by the increase in specific leaf weight (Table 4). Bean plants given +UV also tended to be less succulent as reflected in the higher percent DW of the leaves as compared to -UV treated plants (Table 4). A similar increase in leaf thickness has been reported for P. vulgaris (cv. 'Stella') (9) and for other species (5,11,27). It is difficult to explain the interaction between UV-B treatment and O_3 treatment in terms of SLW and percent DW (Table 4). Since growth measurements were made directly after fumigation, it is not surprising that O_3 treatment had no significant effect on growth responses.

UV-B irradiation and O_3 fumigation applied separately each caused a decrease in photosynthetic rate, but the inhibitory effects of O_3 were greater than those of UV-B irradiation (Table 6). When 'Bush Blue Lake 290' snapbean plants were exposed to both UV-B and O_3 stress, there was a synergistic effect. Contrary to findings of other workers (5,8,11), UV-B irradiation had no significant effect on stomatal behavior. This was in contrast to O_3 treatment which caused a marked reduction in stomatal conductance (Tables 6 and 7). The fact that stomatal behavior was unaffected by UV-B irradiation but was greatly altered by O_3 fumigation suggests a different mode of action for plant response to these two stress.

The decrease in flavonoid absorbance was unexpected and suggests that flavonoid synthesis itself may have been damaged by UV-B irradiation, possibly by damage to the epidermal layer where flavonoids are typically found (9,32). Alternatively, it is also possible that for P. vulgaris, a PPF level of 300 μmol m^{-2} s^{-1} is too low to induce flavonoid synthesis. Cen and Bornman (9) found in their study that even 500 μmol m^{-2} s^{-1} might have been too low to support flavonoid synthesis in P. vulgaris (cv. 'Stella'). Since flavonoids are known to provide a protective function as an optical screen to UV-B irradiation (11,24, 26,33), it is clear that threshold levels of PPF required for flavonoid induction must be determined for each species and/or cultivars if meaningful studies on UV-B irradiation are to be conducted under controlled environment conditions.

Interpretation of the data on chlorophyll fluorescence kinetics after 21 days of UV-B irradiation and 3 h of O_3 exposure depended on the parameter used. On the basis of increased area under the fluorescence curve, +UV treatment appeared to result in an increase in PS II activity. Similarly, on the basis of reduction in half-rise time, O_3 treatment also appeared to cause an increase in PS II activity (Tables 11 and 12). These results were unexpected and further measurements will be needed to confirm these results (Tables 11 and 12). Our findings differ from those reported by Cen and Bornman (9) for P. vulgaris L. (cv. 'Stella') exposed to 6.17 kJ m^{-2} d^{-1} of $UV\text{-}B_{BE}$ and other investigators (21,34).

Our findings demonstrate the difficulty of extrapolating the results of UV-B studies conducted under relatively low PPF conditions (typically available to most growth chamber users) to natural field conditions. They also illustrate the complexity of conducting UV-B research. It is clear that if realistic assessments of UV-B effects are to be made, further studies will need to be conducted on interactions of UV-B radiation and other environmental stress factors.

REFERENCES

1. Caldwell, M.M. (1971). Solar UV irradiation and the growth and development of higher plants. In: Photophysiology, Giese, A.C. ed., Vol.6, Academic Press, New York, pp. 131-177.

2. Krupa, S.V. and Kickert, R.N. (1989). The greenhouse effect: Impacts of Ultraviolet-B (UV-B) radiation, carbon dioxide (CO_2), and ozone (O_3) on vegetation. Environ. Pollut. 61: 263-393.

3. National Aeronautics and Space Administration (NASA) (1988). Executive summary of the ozone trends panel. NASA, Washington, D.C.

4. Robberecht, R. (1989). Environmental photobiology. In: The Science of Photobiology, 2nd Ed., Smith, K.C. ed., Plenum, New York, pp. 135-154.

5. Teramura, A.H. (1983). Effects of ultraviolet-B radiation on the growth and yield of crop plants. Physiol. Plant. 58: 415-427.

6. Krizek, D.T. (1989). The changing atmosphere: Implications for agriculture. Proc. Agr. Res. Inst. 37: 21-34.

7. Basiouny, F.M. (1982). Effects of UV-B irradiation on growth and development of different vegetable crops. Proc. Fla. State Hort. Soc. 95: 356-359.

8. Bennett, J.H. (1981). Photosynthesis and gas diffusion in leaves of selected crop plants exposed to ultraviolet-B radiation. J. Env. Qual. 10: 271-275.

9. Cen, Y-P. and Bornman, J.F. (1990). The response of bean plants to UV-B radiation under different irradiances of background visible light. J. Expt. Bot. 41: 1489-1495.

10. Krizek, D.T. (1975). Influence of ultraviolet radiation on germination and early seedling growth. Physiol. Plant. 34(3): 182-186.

11. Tevini, M. and Teramura, A.H. (1989). UV-B effects on terrestrial plants. Photochem. Photobiol. 50: 479-487.

12. Amiro, B.D., Gillespie, T.J. and Thurtell, G.W. (1978). Injury response of Phaseolus vulgaris to ozone flux density. Atmos. Environ. 18(6): 1207-1215.

13. Bennett, J.H., Lee, E.H. and Heggestad, H.E. (1984). Biochemical aspects of ozone and oxyradicals: superoxide dismutase. In: Gaseous Air Pollutants and Plant Metabolism, Koziol, M.J. and Whatley, F.R. eds., Butterworths, London, UK, pp. 413-424.

14. Dugger, W.M., Taylor, O.C., Cardiff, E. and Thompson, C.R. (1962). Stomatal action in plants as related to damage from photochemical oxidants. Plant Physiol. 37: 487-491.

15. Heagle, A.S. (1989). Ozone and crop yields. Ann. Rev. Phytopathol. 27: 397-423.

16. Krupa, S.V. and Nosal, M. (1989). Effects of ozone on agricultural crops. In: Atmospheric Ozone Research and Its Policy Implications, Schneider, T., Lee, S.D., Wolters, J.R. and Grant, L.D. eds., Elsevier, Amsterdam, The Netherlands, pp. 29-38.

17. Lee, E.H. and Bennett, J.H. (1982). Superoxide dismutase - A possible protective enzyme against ozone injury in snapbeans (Phaseolus vulgaris L.). Plant Physiol. 69: 1444-1449.

18. Lee, E.H., Jersey, J.A., Gifford, C. and Bennett, J.H. (1984). Differential ozone tolerance in soybean and snapbeans: analysis of ascorbic acid in O_3-susceptible and O_3-resistant cultivars by high-performance liquid chromatography. Environ. Expt. Bot. 24(4): 331-341.

19. Olszyk, D.M. and Tibbitts, T.W. (1981). Stomatal response and leaf injury of Pisum sativum L. with SO_2 and O_3 exposures. Plant Physiol. 67: 539-544.

20. Reinert, R.A., Heggestad, H.E. and Heck, W.W. (1982). Response and genetic modification of plants for tolerance to air pollutants. In: Breeding Plants for Less Favorable Environments, Christiansen, M.C. ed., John Wiley & Sons, New York, NY, pp. 259-292.

21. Tevini, M. and Pfister, K. (1985). Inhibition of photosystem II by UV-B radiation. Zeitscrift fur Naturforschung 40: 129-122.

22. Bennett, J.H., Lee, E.H. and Heggestad, H.E. (1990). Inhibition of photosynthesis and leaf conductance interactions induced by SO_2, NO_2 and SO_2+NO_2. Atm. Environ. 24A: 557-562.

23. Heath, R.L. (1988). Biochemical mechanisms of pollutant stress. In: Proceedings International Conference on Assessment of Crop Loss from Air Pollutants, Heck, W.W., Taylor, O.C. and Tingey, D.T. eds., Elsevier Applied Science, London, UK, pp. 259-286.

24. Beggs, C.J., Schneider-Ziebert, U. and Wellman, E. (1986). UV-B radiation and adaptive mechanism. In: Stratospheric Ozone Reduction, Solar Ultraviolet-Radiation and Plant Life, Worrest, R.C. and Caldwell, M.M. eds., Vol.G8, Springer-Verlag, NATO Series, pp. 235-250.

25. Beggs, C.J., Stolzer-Jehle, A. and Wellman, E. (1985). Isoflavonoid formation as an indicator of UV stress in bean (Phaseolus vulgaris L.) leaves: The significance of photorepair in assessing potential damage by increased solar UV-B radiation. Plant Physiol. 79(3): 630-634.

26. Flint, S.D., Jordan, P.W. and Caldwell, M.M. (1985). Plant protective response to enhanced UV-B radiation under field conditions: Leaf optical properties and photosynthesis. Photochem. Photobiol. 41: 95-99.

27. Wellman, E. (1976). Specific ultraviolet effects in plant morphogenesis. Photochem. Photobiol. 24: 659-660.

28. Green, A.E.S., Cross, K.R. and Smith, L.A. (1980). Improved analytical characterization of ultraviolet skylight. Photochem. Photobiol. 31: 59-65.

29. Krizek, D.T., Mirecki, R.M., Philbeck, R.B. and Booker, F.L. (1990). An improved lamp/filter system for use in UV enhancement studies under greenhouse and field conditions. In: Air Pollution Reports. 26, Payer, H.D. et al. eds., Com. Eur. Com. pp. 271-278.

30. Arnon, D.I. (1949). Copper enzymes in isolated chloroplasts. Polyphenoloxidase in Beta vulgaris. Plant Physiol. 24: 1-15.

31. Dumpert, K. and Knacker, T. (1985). A comparison of the effects of enhanced UV-B radiation on some crop plants exposed to greenhouse and field conditions. Biochemie und Physiologie der Pflanzen. 180: 599-602.

32. Bornman, J.F. Bjorn, L.O. and Olsson, E. (1986). Effects of ultraviolet radiation on fluorescence kinetics in isolated thylakoids and intact leaves. In: Stratospheric Ozone Reduction, Solar Ultraviolet Radiation and Plant Life, Worrest, R.C. and Caldwell, M.M. eds., Vol. G8, Springer-Verlag, NATO Series, pp. 185-197

33. Caldwell, M.M., Robberecht, R. and Flint, S.D. (1983). Internal filters: Prospects for UV-acclimation in higher plants. Physiol. Plant. 58: 445-450.

34. Bornman, J.F., Bjorn, L.O. and Akerlund, H.-E. (1984). Action spectrum for inhibition by ultraviolet radiation of photosystem II activity in spinach thylakoids. Photobiochem. and Photobiophys. 8: 305-313.

FIELD STUDIES OF UV-B RADIATION EFFECTS ON PLANTS: CASE HISTORIES OF SOYBEAN AND LOBLOLLY PINE

A. H. Teramura and J. H. Sullivan

Department of Botany, University of Maryland, College Park, Maryland, 20742 U.S.A.

ABSTRACT

By far, the bulk of our understanding of the impacts of UV-B radiation comes from studies conducted in artificially controlled environments (1). Because environmental conditions within growth chambers or greenhouses are unlike those found in nature, plant responses under such conditions may neither quantitatively nor qualitatively resemble field responses. For instance, it is now widely known that plants grown in growth chambers appear to be more sensitive to a given UV dose than field-grown plants (2). The basis for this difference in sensitivity comes from the fact that in artificial environments (growth chambers and greenhouses) a single factor is generally manipulated, while all other factors are either kept constant or are optimized for growth. Such single-factor stresses are rarely experienced by plants outdoors. Instead, under actual conditions, plants would commonly experience simultaneous, multiple stresses. For example, plants receive their maximum daily UV-B irradiance during the period of maximum air temperatures, visible irradiance, and evaporative demand for water. Unlike plants in growth chambers where nutrient solutions may be applied daily, most native plants and many agricultural crops grow in soils that are low or deficient in nutrients. In addition to these differences in physical factors, artificial environments almost always exclude biotic factors, such as the interactions between other plants, insects, diseases, etc. Finally, inherent limitations on size of controlled-environment facilities make it impractical to conduct studies on crop yields, which must ideally be conducted in carefully designed field studies.

Weighed against these shortcomings of controlled-environment studies are the enormous complexities associated with field studies. Here, daily fluctuations in environmental factors are superimposed upon longer-scale seasonal and annual fluctuations making interpretation extremely difficult and necessitating multi-year experimental designs. Both temporal and spatial variability often result in inconsistencies in plant responses between one year and the next. Experimental field plot studies will be described for a crop plant, soybean (Glycine max L.), and a forest tree species, loblolly pine (Pinus taeda L.). In these field studies ambient levels of solar UV-B radiation have been supplemented by artificial UV-B radiation provided from fluorescent sunlamps. The effectiveness of UV-B radiation on physiological processes, growth, and yield will be described for these plants grown under a full solar UV spectrum and in combination with drought.

INTRODUCTION

The effects of ultraviolet-B (UV-B, between 280 and 320 nm) radiation have now been studied on approximately 300 species and varieties of plants. Of those studied, nearly one-half showed physiological damage and/or growth reductions in response to UV-B radiation (3). Some of the most sensitive families include the Fabaceae, Cucurbitaceae and Brassicaceae. General physiological effects of UV-B radiation include reductions in photosynthetic carbon assimilation, alterations in stomatal function, phytohormone activity and foliar chemistry (4). These responses are often then manifested in reductions in dry matter accumulation, height, leaf area and seed yield. The bulk of information, however, on the impacts of UV-B radiation comes from studies conducted in artificially controlled environments (1). Unfortunately, since environmental conditions within growth chambers or greenhouses are unlike those found in nature, plant responses under such conditions may neither quantitatively not qualitatively resemble field responses. For instance it is now widely known that plants grown in growth chambers may be severely damaged by UV-B irradiances which simulate current ambient solar UV-B fluxes. The basis for this difference in sensitivity comes from the fact that in artificial environments (growth chambers and greenhouses) a single factor is generally manipulated while all other factors are either kept constant or are optimized for growth. Such single-factor stresses are rarely experienced by plants outdoors. Instead, under actual conditions, plants would commonly experience simultaneous, multiple stresses. For example, plants receive their maximum daily UV-B irradiance during the period of maximum air and soil temperatures, visible irradiance and evaporative demand for water.

A significant difference between growth chamber or greenhouse environments and field conditions lies in the visible irradiance or PFD (Photosynthetic Photon Fluence) present. It has long been observed that leaves grown in full sunlight often differ anatomically, morphologically and chemically from shade leaves. Therefore, these responses to sunlight may provide protection from UV-B radiation and the absence of these responses may contribute to the high sensitivity of plants grown inside controlled environments. Teramura and Murali (2), demonstrated that four of five soybean cultivars tested were roughly twice as sensitive to UV-B, in terms of alterations in vegetative characteristics, when grown under greenhouse compared to field conditions. In other studies

(5,6), the importance of visible irradiance, or the ratio of PPF to UV-B, in modifying the response to UV-B was clearly established. For example, Mirecki and Teramura (6) found significant reductions of photosynthesis in response to UV-B radiation when the background PPF was 800 μmol $m^{-2}s^{-1}$ but no reductions at a background PPF of 1400 μmol $m^{-2}s^{-1}$.

In addition to the above complications of the artificial vs. the field environment, some other constraints of growth chamber or greenhouse studies reduce their usefulness in assessing the consequences of an increase in UV-B radiation. For example, artificial environments almost always exclude biotic factors such as interactions between other plants, insects, diseases, etc. and such studies also face inherent limitations on size and available space and thus are generally not suitable for yield analyses. Therefore, for all these reasons field validation studies are essential to the realistic assessment of the impacts of increasing solar UV-B radiation.

The enormous complexity of field studies must be weighed against the shortcomings of controlled environment studies. In the field, the daily fluctuations in environmental factors are superimposed upon longer-scale seasonal and annual fluctuations. This makes the interpretation of results extremely difficult. Furthermore, this temporal and spatial variability may result in inconsistent responses between one year and the next and necessitates the implementation of multiple year experimental designs. To date only about 20 field studies have been conducted on the response of plants to UV-B radiation and of these, only 5 were continued for more than a single growing season. Only two of these 5 were multi-year studies lasting for more that two consecutive growing seasons. One of these studies grew soybean (Glycine max. (L.) Merr.) for six years and the other grew loblolly pine (Pinus taeda L.) for four years.

FIELD VALIDATION STUDIES: SOYBEAN

Previous studies have shown that a wide range of intraspecific variability exists in soybean with respect to sensitivity to UV-B radiation. A summary of studies conducted on soybean (2,7,8) shows that 26 of 41 cultivars tested were sensitive (growth reductions exceeding 5%) to UV-B radiation in field or greenhouse studies (Table 1). The only field study, however, undertaken over more than two seasons was conducted at the University of Maryland and evaluated two soybean cultivars over a six-year period from 1981 to 1986 (9).

Two soybean (Glycine max (L) Merr.) cultivars were chosen for study based upon preliminary greenhouse trials for UV sensitivity and planted into the field (2). Based upon overall growth performance, Essex was found to be sensitive while Williams was tolerant to UV-B radiation. Field experiments were conducted during May through October of 1981 to 1986 at the Agricultural Research Center, USDA, Beltsville, Maryland, U.S.A.

Supplemental UV-B radiation was supplied by filtered Westinghouse FS-40 sunlamps oriented perpendicular to the planted rows (rows oriented in an East-West direction) and suspended above the plants. Lamps were filtered either with 0.13 mm thick cellulose diacetate (transmission down to 290 nm) for supplemental UV-B radiation or 0.13 mm Mylar Type S plastic films (absorbs all radiation below 320 nm) as a control. The radiation filtered through the cellulose diacetate supplied a weighted daily supplemental irradiance of either 3.0 or 5.1 effective kJ m^{-2} UV-B_{BE} using the generalized plant response action spectrum (10) normalized to 300 nm. Therefore, plants beneath these lamps received supplemental doses in addition to ambient levels of UV-B radiation. These increased levels of UV-B radiation (supplemental + ambient) were similar to those which would be received at College Park, Maryland, U.S.A. (39°N) with anticipated 16 and 25% stratospheric ozone reductions during a cloudless day on the summer solstice (11). The weighted irradiance of Mylar filtered lamps was 0, so plants beneath these lamps received only ambient levels of UV-B (8.5 effective kJ m^{-2} UV-B_{BE} on the summer solstice). Different UV-B doses were obtained by varying the distance between the lamps and the top of the plants (a distance of 0.75 and 1.0 m for 5.1 and 3.0 effective kJ m^{-2}, respectively). This was checked and maintained weekly as the plants grew. Cellulose diacetate filters were pre-solarized for 8h and changed weekly to ensure uniformity of UV-B transmission. The FS-40 sunlamps were preburnt and matched for desired spectral irradiance, prior to the initiation of the experiment each year. Spectral irradiance beneath the lamps as measured with an Optronics Model 742 spectroradiometer equipped with a double monochromator with dual holographic grating and interfaced with a Hewlett Packard 85 printing calculator. The spectroradiometer was calibrated using a National Institute of Standards and Technology (NIST) 1000 W tungsten halogen lamp and wavelength alignment checked against known mercury emission lines using a Hg Arc lamp.

The results of this 6-year field study demonstrate intraspecific differences in UV-B sensitivity in soybean. In Essex, the sensitive cultivar, reductions in photosynthetic carbon assimilation were manifested ultimately in reduced vegetative biomass and seed yield (9,12,13). In Williams, however, neither photosynthetic capacity nor growth were affected by supplemental UV-B radiation. In Essex, a simulated 25% ozone reduction reduced yield by 19-25% during 4 of the 6 years (Table 2).

The annual variation in responsiveness to UV-B raises the question as to the potential interaction of UV-B with other environmental factors and supports the need for multi-year studies. For example, if the results from only 1983 were available then Williams rather than Essex would have been considered the sensitive cultivar. Parallel field studies (13,14) have shown that UV-B effects may be modified by water availability. In these two field studies, when conditions of drought were imposed on soybean, no reductions in photosynthetic, growth or yield were observed in response to supplemental UV-B radiation. Likewise, the 1983 and 1984 field seasons were characterized with prolonged periods of drought and in these seasons their was little or no UV-B radiation effect on yield in Essex (Table 2).

In another series of studies (15,16) in which levels of phosphorous were manipulated, similar results were obtained in that the effectiveness of UV-B in altering plant physiology or growth was diminished when nutrient stress was imposed. In both the cases of drought and nutrient limitations the masking of UV-B effects appeared to arise from the plants biochemical and morphological response to the primary stress (e.g. drought or limiting P). These responses included growth reductions, increases in specific leaf weight (SLW) and an accumulation of foliar UV-absorbing compounds.

In similar fashion the observed inter- and intra-specific variation in sensitivity to UV-B may be dependent in part on these same responses to UV-B. One primary means of defense against incoming UV-B radiation would be to prevent the radiation from reaching sensitive internal targets such as chloroplasts or nuclei (17). Incoming irradiance striking the leaf surface may be reflected, absorbed by the epidermis or transmitted into the leaf mesophyll. Increased leaf thickness (as indicated by an increase in SLW) could increase the path length of the irradiance to the mesophyll. The well-known sun-leaf anatomy, for example, is considered an acclimation to high levels of visible irradiance but it is possible that UV-B radiation might also contribute to this response.

Another general response to drought, nutrient limitations and high irradiances (visible and UV) is the accumulation of flavonoids in epidermal tissues. These compounds absorb strongly in the UV-A and UV-B regions and their accumulation in the epidermis has been shown to reduce epidermal transmittance of UV (18). The increase in epidermal concentrations of these compounds in response to drought or nutrient deficiency may serendipitously protect the plant from UV-B radiation. Also, the ability to accumulate these compounds in response to UV-B varies greatly among species and among cultivars of a given species. Some of the observed sensitivity differences to UV-B radiation may be due in part to this response. It has now been established that the synthesis of several of the key enzymes in the flavonoid biosynthetic pathway are induced by UV-B (19). Visible irradiance and UV-A irradiation are also apparently important in this regard. For example it was observed nearly four decades ago that flavonoid concentrations were greater in plants grown in high light conditions and that flavonoid levels drop rapidly in response to cloud or shade conditions (20). However, the genetic basis of flavonoid accumulation in response to UV-B radiation remains poorly understood and some plants do not appear to be able to increase flavonoids in response to UV-B.

The variation among cultivars of soybean in terms of UV-B sensitivity seems to parallel that of the interspecific variability observed in all other species combined. The intraspecific variation and interactive response to co-occurring stresses may be due to gene-level responses such as the induction of specific genes in the flavonoid biosynthetic pathway and other as of yet undetermined responses. However, at the present it seems likely that the productivity of at least some soybean cultivars would be substantially reduced if ozone depletion continues. Adequate assessments of potential damage depend upon a better understanding of how the response to UV-B might be modified by other environmental changes such as elevated CO_2 concentrations, increased temperature and modified water availability. In addition, further uncertainties exist as to whether insect or pathogen damage may be affected by UV-B and as to how changes in solar irradiance might alter competitive interactions between species.

FIELD VALIDATION STUDIES: LOBLOLLY PINE

In contrast to crops, few studies have been undertaken on woody perennials (trees), which account for about two-thirds of global net primary productivity and occupy as much as one-third of the land area of the United States (21). Only three studies have been completed on trees under field conditions and one of these examined the effects of exclusion of solar UV-B radiation (22). Exclusion of naturally occurring UV-B radiation increased the growth of four broadleaf species. In the first study to supply supplemental UV-B irradiation to tree seedlings in the field, supplemental UV-B radiation had no effects on growth in either Engelmann spruce or lodgepole pine (23). Three additional studies conducted under either growth chamber or greenhouse conditions have demonstrated deleterious effects of UV-B radiation on tree growth and physiology (24,25,26). However, none of these studies were carried out for more than a single growing season, so the long-term effects of UV-B radiation on trees is unknown.

The examination of UV-B radiation effects on perennial species provides a unique opportunity to observe more subtle responses to protracted UV exposure which are impossible to investigate in annual species. For instance, it is presently unknown whether UV repair mechanisms can mitigate UV damage during the dormant period when ambient levels of solar UV are at their seasonal minimum. Likewise, we have no information on whether the extensive physiological changes which accompany tissue hardening prior to entering the dormant period modify its sensitivity to subsequent UV exposure. The range of responses observed both inter- and intra-specifically suggests that extrapolations between annual and perennial species may not be feasible. The strict adherence to the assumption that UV affects all plants similarly could lead to substantial errors in the assessment of the potential impacts of increases in UV-B radiation, when considered on a global basis. Therefore, some direct field validation experiments on key forest species are essential before realistic estimates of this nature can be made.

Due to their economic importance and widespread global distribution, conifers have been selected for study in over half of the studies of UV-B radiation effects on trees. Some 15 species of conifers have been tested to date for susceptibility to UV-B radiation. Of these, 7 were deleteriously affected, 5 were resistant, and 3 were favored by UV-B radiation (Table 3). In a recent study by Sullivan and Teramura (25) 4 of the 10 species tested were susceptible to UV-B in terms of reductions in height or total biomass production.

Loblolly pine was one of the most susceptible with reductions of biomass and height of 40 and 16%, respectively. Loblolly pine is the leading commercial species in the southeastern United States. It is found at elevations below about 300 m and is the dominant species of the region, occupying nearly 16% of the total acreage over its range (27). The southern pine forests account for two-thirds of the United States pulp producing capacity. Nearly all of the annual harvest is sawtimber, which is used primarily for plywood as well as naval stores, poles, and pilings (28). Therefore, there would be enormous economic consequences should increasing levels of UV-B substantially reduce loblolly pine productivity.

In the first multi-year field study on UV-B effects on trees, loblolly pine was grown for three successive years under ambient and ambient plus supplemental UV-B irradiances. The irradiation methodology was similar to that described previously for soybean and the supplemental irradiances simulated that anticipated with a 16 or 25% ozone depletion over College Park, Maryland. The objectives of this study were to determine if the physiology or growth of loblolly pine was affected by UV-B under field conditions and if annual differences in microclimate would modify this effectiveness.

Plants from seven seed sources were grown for one season and total plant biomass was reduced in 4 and 5 of these seed sources at simulated 16 and 25% ozone depletions, respectively. The overall reductions for all sources combined were approximately 7 and 14% for the 16 and 25% ozone depletions, respectively. In order to determine whether sensitivity or tolerance was maintained over subsequent years, plants from two selected seed sources were irradiated for two additional years. One source was initially characterized as sensitive and the other as tolerant to UV-B radiation.

During the second and third growing seasons measurements of photosynthetic capacity revealed that light utilization (apparent quantum efficiency) and the response to internal CO_2 concentrations was reduced by UV-B radiation. Unlike soybean and previous greenhouse studies on loblolly pine, there was little additional accumulation of flavonoids in response to UV-B radiation. However, the absolute levels of UV-absorbing compounds in loblolly pine was comparable to that observed in other species. Possibly the response was saturated in full sunlight so that no additional increases were stimulated by supplemental UV-B radiation. The nature of physiological responses (photosynthesis and pigment accumulation) appeared to be partially dependent upon needle age. Additional studies are ongoing to elucidate the particular age dependence of the response to UV-B. Canopy modeling efforts then might be useful in attempts to explain UV-B effects on dry matter accumulation.

The reductions in photosynthetic carbon assimilation apparently lead to continued growth reductions over the second and third years of the study. At the conclusion of the second year, biomass of plants from both seed sources was reduced (14%) at the higher UV-B irradiance. Reduction in biomass of approximately 17% were observed for both seed sources and at both supplemental irradiances after three years (Table 4). The total reductions in biomass were split roughly equally between root and shoot after three years.

The reductions in biomass (including stem diameter and volume), suggest that wood or pulp production might be affected both quantitatively or qualitatively if significant ozone depletion continues. An decrease in plant height coupled with a trend toward an increase in the number of lateral branches, could reduce the wood quality for use as sawtimber as well as decrease pulp quantity. However, the extrapolation of data from even three years growth to adult tree productivity is difficult. Therefore, continued studies over extended periods should be considered on loblolly pine and other conifers as well as hardwoods, for which almost no information exists. The reductions observed in seedling growth may also contribute to changes in the competitive ability of plants both in natural ecosystems and in commercial forestry operations.

CONCLUSIONS AND AREAS FOR FUTURE RESEARCH

Two multi-year field have demonstrated significant reductions in yield or production in soybean and loblolly pine. However, the large range of variability in terms of sensitivity of UV-B makes it difficult to generalize with respect to regional-scale alterations in net productivity. Another reason for this is that we know very little about potential indirect effect of UV-B radiation. Changes in competitive ability or resistance to insects or pathogens (of trees or crop species) may ultimately prove more important than direct UV-B effects on growth, in terms of alterations in natural community dynamics or commercial productivity. Orth et al. (29) recently demonstrated an increase in disease severity of cucumber plants exposed to UV-B radiation prior to fungal inoculation. These would be important areas of research to pursue in the future.

LITERATURE CITED

1. Tevini, M. and A.H. Teramura. 1989. UV-B effects on terrestrial plants. Photochem. Photobiol. 40:479-487.

2. Teramura, A.H. and Murali, N.S. (1986). Intraspecific differences in growth and yield of soybean exposed to ultraviolet-B radiation under greenhouse and field conditions. Environ. and Exp. Bot. 26:89-95.

3. Teramura, A.H. (1987). Ozone depletion and plants. In: Assessing the risks of trace gases that can modify the stratosphere, Vol. VIII. U.S. E.P.A. Report, Washington, D.C.

4. Teramura, A.H. (1983). Effects of ultraviolet-B irradiances on soybean. Plant Physiol. 65:483-488.

5. Warner, C.W. and Caldwell, M.M. (1983). Influence of photon flux density in the 400-700 nm waveband on inhibition of photosynthesis by UV-B (280-320 nm) irradiation in soybean leaves: separation of indirect and immediate effects. Photochem. Photobiol. 38:341-346.

6. Mirecki, R.M. and Teramura, A.H. (1984). Effects of ultraviolet-B irradiance on soybean. The dependence of plant sensitivity on the photosynthetic photon flux directly during and after leaf expansion. Plant Physiol. 74:475-480.

7. Biggs, R.H., Kossuth, S.V. and Teramura, A.H. (1981). Response of 19 cultivars of soybeans to ultraviolet-B irradiance. Physiol. Plant. 53:19-26.

8. Lydon, J., Teramura, A.H., and Summers, E.G. (1986). Effects of ultraviolet-B radiation on the growth and productivity of field-grown soybean. In: Stratospheric ozone reduction, solar ultraviolet radiation, and plant life. R.C. Worrest and M.M. Caldwell eds. Springer-Verlag, pp. 313-325.

9. Teramura, A.H., Sullivan, J.H. and Lydon, J. (1990). Effects of UV-B radiation on soybean yield and seed quality: a 6-year field study. Physiol. Plant. 80:5-11.

10. Caldwell, M.M. (1971). Solar irradiation and the growth and development of higher plants. In: Photophysiol. Vol. VI, A.C. Giese ed., Academic Press, pp. 131-177.

11. Green, A.E.S., Cross, K.R. and Smith, L.A. (1980). Improved analytical characterization of ultraviolet skylight. Photochem. Photobiol. 31:59-65.

12. Murali, N.S. and Teramura, A.H. (1986). Effects of supplemental ultraviolet-B radiation on the growth and physiology of field-grown soybean. Env. Exp. Bot. 26:233-242.

13. Sullivan, J.H. and Teramura, A.H. (1990). Field study of the interaction between solar ultraviolet-B radiation and drought on photosynthesis and growth in soybean. Plant Physiol. 92:141-146.

14. Murali, N.S. and Teramura, A.H. (1986). Effectiveness of UV-B radiation on the growth and physiology of field-grown soybean modified by water stress. Photochem. Photobiol. 44:215-220.

15. Murali, N.S. and Teramura, A.H. (1985). Effects of ultraviolet-B irradiance on soybean. VI. Influence of phosphorus nutrition on growth and flavonoid content. Physiol. Plant. 63:413-416.

16. Murali, N.S. and Teramura, A.H. (1985). Effects of ultraviolet-B irradiance on soybean. VII. Biomass and concentration and uptake of nutrients at varying P supply. J. Plant Nutr. 8:177-182.

17. Beggs, C.J., Schneider-Ziebert, R. and Wellman, E. (1986). UV-B radiation and adaptive mechanisms in plants. In: R.C. Worrest and M.M. Caldwell eds. Stratospheric Ozone Reduction, Solar ultraviolet radiation and plant life. Springer-Verlag. pp. 235-250.

18. Robberecht, R. and Caldwell, M.M. (1978). Leaf epidermal transmittance of ultraviolet radiation and its implication for plant sensitivity to ultraviolet-radiation induced injury. Oecologia 32:277-287.

19. Ohl, S., Hahlbrock, K. and Schafer, E. (1989). A stable blue-light-derived signal modulates ultraviolet-light-induced activation of the chalcone-synthase gene in cultured parsley cells. Planta 177:228-236.

20. Lautenschlager-Fleury, D. (1955). Uber die ultraviolettdurchlassig keit von blattepidermen. Ber. Schweiz Bot. Ges. 65:343-386.

21. Whittaker, R.H. (1975). Communities and Ecosystems. MacMillan Co., New York.

22. Bogenrieder, A. and Klein, R. (1978). Die abhangigkeit der UV-empfindlichkeit von der lichtqualitat bel der aufzucht (Lactuca sativa L.). Angew. Botanik 52:283-293.

23. Kaufmann, M.R. (1978). The effect of ultraviolet (UV-B) radiation on Engelmann spruce and lodgepole pine seedlings. UV-B Biological and Climatic Effects Research (BACER). Final Report EPA-IAG-D6-0168. EPA, Washington, DC.

24. Kossuth, S.V. and Biggs, R.H. (1981). Ultraviolet-B radiation effects on early seedling growth of Pinaceae species. Can. J. For. Res. 11:243-248.

25. Sullivan, J.H. and Teramura, A.H. (1988). Effects of ultraviolet irradiation on seedling growth in the pinaceae. Amer. J. Bot. 75:225-230.

26. Sullivan, J.H. and Teramura, A.H. (1989). The effects of ultraviolet-B radiation on loblolly pine: 1. Growth, photosynthesis and pigment production in greenhouse-grown saplings. Physiol. Plant. 77:202-207.

27. Harlow, W.M. and E.S. Harbar (1969). Textbook of Dendrology. McGraw-Hill, pp. 87-90.

28. Walker, L.C. (1980). The southern pine region. In: Regional Silviculture of the United States. J.W. Barrett ed., John Wiley and Sons, p. 231-276.

29. Orth, A.R., Teramura, A.H. and Sisler, H.D. (1990). Effects of ultraviolet-B radiation on fungal disease development in Cucumis sativus. Amer. J. Bot. 77:1188-1192.

Table 1. Summary of UV-B Radiation Effects on Soybean Cultivars

Sensitive[1]	Tolerant[2]	Favored[3]
Acadian	Americana	Cobb
Altona	Biloxi	*Forrest
*Bay	Bossier	Ware
Calland	*Centennial	
Crawford	Davis	
Cumberland	Douglas	
Cutler	Hood	
Desoto	Hutton	
Elf	Pixie	
*Essex	Roanoke	
Hamilton	Wye	
Hardee	York	
*James		
Jupiter		
Kent		
Miles		
Mineira		
Otootan		
Pella		
Pickett		
Santa Maria		
Seminole		
Shore		
Sprite		
Will		
*Williams		

[1]Sensitive = reductions exceeding 5%
[2]Tolerant = effects 5% or less
[3]Favored = enhancement exceeding 5%

*Cultivars included in more than one study

Table 2. Summary of UV-B radiation effects on soybean yield. A 25% ozone depletion was simulated over College Park, Maryland U.S.A.

Year	% Change in Yield	
	Essex	Williams
1981	-25	+22
1982	-23	+14
1983*	+ 6	-11
1984*	- 7	+10
1985	-20	+ 4
1986	-19	+ 6

*Years with prolonged drought

Table 3. Summary of effects of UV-B radiation on growth of 15 coniferous species based upon change in biomass accumulation and height

Species	Sensitivity Rating*
Abies concolor	favored
Abies fraseri	favored
Abies procera	sensitive
Picea engelmannii	favored
Picea glauca	resistant
Pinus contorta	sensitive
Pinus edulus	resistant
Pinus elliottii	sensitive
Pinus nigra	resistant
Pinus ponderosa	sensitive
Pinus resinosa	sensitive
Pinus strobus	resistant
Pinus sylvestris	sensitive
Pinus taeda	sensitive
Pseudotsuga menziesii	resistant

*Sensitive = reductions exceeding 5% of controls
Resistant = ± 5% of controls
Favored = increases exceeding 5% of controls

Table 4. Summary of effects of UV-B radiation on growth of loblolly pine after three years of supplemental UV-B radiation.

Seed Source[1]	Year	UV-B Treatment ($kJ\ m^{-2}$ UV-B)	Dry Weight (g) Root	Stem	Needle	Total
		0	3.8	1.8	3.7	9.3
	1	3.1	4.2	1.9	4.0	10.1
		5.0	4.8	2.0	4.2	11.0
		0	80	45	75	210
Wicomico	2	3.1	80	52	75	207
County MD		5.0	73	44	63	180
		0	268	335	363	966
	3	3.1	226	286	318	830
		5.0	228	252	301	781
		0	5.9	2.2	5.1	13.2
	1	3.1	5.6	2.4	5.7	13.7
		5.0	4.3	1.9	3.9	10.1
		0	84	52	83	219
Virginia	2	3.1	80	55	78	213
		5.0	71	42	72	185
		0	277	331	366	974
	3	3.1	226	280	304	810
		5.0	219	274	312	805

[1]Seed sources were either an open pollinated stand in Wicomico County, Maryland, or a stand in northern Virginia.

CONTRIBUTION TO NEW DELHI - WORKSHOP

A COMPARISON OF THE GROWTH, PHOTOSYNTHESIS AND TRANSPIRATION OF WHEAT AND MAIZE IN RESPONSE TO ENHANCED ULTRAVIOLET-B RADIATION

By: Jelte Rozema, Jos van de Staay, Vergilio Costa, José Torres Pereira*, Rob Broekman, Ger Lenssen and Martin Stroetenga

Department of Ecology and Ecotoxicology, Faculty of Biology, Vrije Universiteit, De Boelelaan 1087, 1081 HV Amsterdam, The Netherlands.

* Department of Biology, Universidade de Tras os Montes e Alto Douro, 5001 Vila Real Codex, Portugal.

ABSTRACT

A comparison has been made between the growth response of the C_3 crop wheat (*Triticum aestivum*) and the C4 crop maize (*Zea mays*) to enhanced Ultraviolet-B radiation. Plant were grown in pots with garden soil in a greenhouse at a relatively low light intensity (200 - 250 μ Einstein $m^{-2}.sec^{-1}$). The relative growth rate of both plant species was depressed by enhanced UV-B. The decrease of the RGR of wheat by elevated UV-B was 32 - 47% of the control, and for maize the decrease the RGR was 22-39% of the control treatment. The reduction of RGR with enhanced UV-B was associated with a marked decline of the Net Assimilation Rate in both wheat and maize. The Leaf Area Ratio and the Specific Leaf Area were not or only slightly affected by increased UV-B. There was no significant change of the Leaf Weight Ratio with increased UV-B in the two species. Gas exchange measurements indicate that the rate of net photosynthesis of wheat is depressed by increased UV-B and to a less extent of maize, while transpiration is not significantly influenced by increased UV-B radiation in both species.

The combined effect of increased atmospheric CO_2 and elevated UV-B radiation, which form both part of global climate change on plant growth and photosynthesis is discussed.

Increases of plant growth and productivity in response to global atmospheric carbon dioxide increase may be greatly reduced when UV-B radiation is enhanced simultaneously. However only a few experimental data are available on the combined effect of CO_2 enrichment and rising UV-B radiation.

INTRODUCTION

There is increasing evidence for continued gradual increase of atmospheric carbon dioxide and doubling of the present level of 350 $\mu l\ l^{-1}$ CO_2 is expected to occur around the year 2050. At the same time solar UV-B radiation is expected to increase due to depletion of stratospheric ozone. Emissions of chlorofluorocarbons, nitrous oxide and methane are considered to cause breakdown of ozone produced in the stratosphere.

Although uncertainties exist relating to the details of the chemical processes involved, the breakdown of ozone in the stratosphere and the consequences are now extensively studied. With partial depletion of stratospheric ozone increased levels of solar UV-B radiation occur at the surface of the earth.

The effects of atmospheric carbon dioxide and enhanced solar UV-B as separate components of global change have been studied extensively. More recently the responses of plants to a combination of CO_2 enrichment and elevated UV-B radiation have also been examined (Rozema et al. 1990, Teramura et al. 1990).

In an earlier studies (Staay et al., 1990; Rozema et al., 1990) we reported adverse effects of enhanced UV-B radiation on the growth of the salt marsh species *Aster tripolium* and *Spartina anglica.* Also it appeared that *Aster tripolium* was more sensitive to increased UV-B radiation than *Spartina anglica.* It has been generally reported that great differences exist between plant species in their sensitivity to enhanced UV-B (Teramura

(1983), Caldwell *et al.* (1989).

In the present paper, we compare the growth and gas exchange of the C_3 cereal crop wheat and the C_4 cereal crop maize in response to increased UV-B. The present study has been performed in the greenhouse, but currently studies of plant responses to enhanced UV-B are also performed outdoors in an experimental field with natural solar UV-B as a background.

MATERIALS AND METHODS

For wheat (*Triticum aestivum* L.) cultivar Obelisk, and for maize (*Zea mays* L.) cultivar Limagrain 5 was used. Seeds were germinated in garden soil that was frequently moistened with demineralized water. Germination occurred within a week and seedlings of about 7 cm height were transferred to black plastic (polyethylene) pots 1800 cm^3) filled with garden soil (Calceolaria, Jongkind, Aalsmeer). Three seedlings were planted per pot. Plants were grown in a greenhouse with Philips HPI/T lamps, with a photon flux density (PAR) 200 - 250 μ Einstein m^{-2} sec^{-1}. UV-B radiation was varied using Philips TL/2/40 tubes wrapped in cellulose acetate foil, at 1.0 (control) and 2.0 Wm^{-2} (enhanced) W m^{-2}.UV-B radiation was measured using a UVX radiometer with a UV-X-31 sensor (Van de Staay et al., 1990). Plants were harvest after two weeks (n = 10) and after five weeks (n = 10).

Leaf area was measured with a Li-3100 area meter (Li. Corp., Inc., Lincoln, Nebraska, USA). Photosynthesis and transpiration was measured of individual leaves with the Parkinson Leaf Chamber of a portable ADC-LCA3 system (The analytical Development Company Ltd., Haddesdon, Herts. UK) at a photon flux density of about 250 μ Einstein $m^{-2}sec^{-1}$ (PAR). Details can be obtained from a more detailed report that is prepared (Staay et al., 1991).

Table 1. Relative growth rate and growth parameters NAR, LAR, LWR and SLA of wheat and maize in response to enhanced UV-B. Means of ten replications with standard deviation. The percentage indicates the change of plants grown with enhanced UV-B compared to the control plants. The data refer to the first harvest.

UV-B radiation ($W\ m^{-2}$)	RGR $mg\ g^{-1}$ day	NAR $g\ m^{-1}$ day	LAR $m^{-2}g^{-1}$ day	LWR $g\ g^{-1}$ day	SLA m^2g^{-1} day	% H_2O
Wheat (Triticum aestivum)						
Control	57.6	2.56	0.022	0.655	0.034	90.2
1.0	± 8.6	± 0.48	± 0.0027	± 0.028	± 0.0025	± 0.38
Enhanced	30.5	1.34	0.022	0.661	0.033	89.3
2.0	± 4.4	± 0.47	± 0.0023	± 0.026	± 0.0028	± 0.53
	- 48%	- 49%	0%	+ 1%	+ 3%	- 1%
Maize (Zea mays)						
Control	37.3	1.49	0.025	0.56	0.044	93.1
1.0	± 11.1	± 0.42	± 0.0027	± 0.036	± 0.0032	± 0.50
Enhanced	22.6	1.09	0.020	0.50	0.041	92.0
2.0	± 4.4	± 0.24	± 0.0023	± 0.049	± 0.0041	± 0.61
	- 40%	- 27%	- 20%	- 4%	- 7%	- 1.8%

RESULTS

The mean relative growth rate of wheat and maize is decreased with enhanced UV-B radiation. The reduction of the RGR with enhanced UV-B was up to 50% for wheat and up to 40% for maize (Table 1 and 2). The Net Assimilation Rate (NAR) of both plant species was markedly affected by increased UV-B radiation. This effect was most obvious in wheat. At the first harvest, the Leaf Area Ratio (LAR) and the Specific Leaf Area (SLA) were not significantly or only slightly affected by enhanced UV-B radiation. At the second harvest LAR of wheat and maize was 41% and 23% depressed with enhanced UV-B.

At the first harvest the Leaf Weight Ratio (LWR) showed no shift with the UV-B treatment. The LWR of wheat is somewhat higher than that of maize. At the second harvest both for wheat and maize LWR is increased markedly with enhanced UV-B radiation. The content of water of wheat and maize was not influenced by the UV-B treatment. The water content of wheat is slightly lower than that of maize. Net photosynthesis of both wheat and maize measured April 19, 1990 was markedly depressed by enhanced UV-B radiation (Table 3). Measurement of net photosynthesis, May 11, 1990 and July 6, 1990 confirmed the inhibition by enhanced UV-B in wheat. For wheat the reduction of P_n by enhanced UV-B measured in May and July is less than in April. For maize repeated measurements of P_n show variation with increased UV-B but do not show a clear trend.

Transpiration rates of maize leaves tend to be lower than those of wheat. In both species there is no significant reduction of the rate of transpiration with enhanced UV-B. For wheat transpiration measured in April, May and July does not vary greatly. For maize there is greater variation of the transpiration rate with the UV-B treatment. Measurements of the transpiration rate in maize in April and July demonstrate increased evaporation with a higher level of UV-B radiation.

Table 2. Relative growth rate and growth rate and growth parameters NAR, LAR, LWR and SLA of wheat and maize in response to enhanced UV-B radiation. Means of ten replications with standard deviation. The percentage indicates the change of plants grown with enhanced UV-B compared to the control plants. The data refer to the second harvest.

UV-B radiation ($W\ m^{-2}$)	RGR mg g^{-1} day	NAR g m^{-2} day^{-1}	LAR $m^2 g^{-1}$	LWR g g^{-1}	SLA $m^2 g^{-1}$	% H_2O
Wheat (Triticum aestivum)						
Control	71.4	4.10	0.017	0.464	0.037	88.5
1.0	± 4.3	± 0.54	± 0.0019	± 0.031	± 0.0026	± 0.58
Enhanced	48.8	2.04	0.024	0.625	0.039	86.7
2.0	± 3.2	± 0.30	± 0.0026	± 0.027	± 0.0037	± 0.85
	- 33%	- 51%	+ 41%	+ 34%	+ 5%	- 2%
Maize (Zea mays)						
Control	69.9	3.26	0.021	0.476	0.045	90.3
1.0	± 4.3	± 0.33	± 0.0011	± 0.020	± 0.0024	± 0.52
Enhanced	55.0	2.11	0.026	0.499	0.052	91.5
2.0	± 4.6	± 0.24	± 0.0017	± 0.031	± 0.0037	± 0.30
	- 22%	- 36%	+ 23%	+ 55%	+ 15%	+ 1%

Table 3 Net rate of photosynthesis (P_n, $\mu mol\ CO_2\ m^{-2}sec^{-1}$) and transpiration (E, $mmol\ H_2O\ m^{-2}sec^{-1}$) of leaves of wheat and maize in response to enhanced UV-B radiation. Average values of nine replications with standard deviation. Measurement were conducted April 19 (1), May 11 (2) and July 6 (3, after the second harvest).

UV-B ($W\ m^{-2}$)	Wheat (Triticum aestivum)					
	1		2		3	
	Pn	E	Pn	E	Pn	E
Control	3.46	7.89	5.72	7.40	6.72	6.68
1.0	± 2.52	± 1.65	± 2.04	± 1.32	± 1.04	± 0.63
enhanced	1.35	7.20	4.66	8.13	4.69	6.65
2.0	± 1.00	± 1.01	± 1.55	± 1.88	± 1.08	± 1.15
	- 61%	- 8%	- 19%	+ 9%	- 30%	- 1%

UVB ($W\ m^{-2}$)	Maize (Zea mays)					
	1		2		3	
	Pn	E	Pn	E	Pn	E
Control	5.32	1.79	5.38	2.01	5.70	1.70
1.0	± 2.15	± 0.52	± 1.69	± 0.55	± 1.35	± 0.29
enhanced	1.48	3.19	3.80	1.65	7.98	2.24
2.0	± 1.09	± 0.43	± 1.22	± 0.33	± 1.47	± 0.19
	- 82%	+ 78%	- 44%	- 18%	+ 40%	+ 31%

DISCUSSION

In a series of other papers (Rozema et al., 1990, et al. 1991, Staaij et al. 1992), we report on the combined effect of increased atmospheric CO_2 and elevated UV-B radiation on plant species. There are marked effects of atmospheric CO_2 and elevated UV-B on the plant species tested *viz.* tomato, bean, *Aster tripolium* and *Elymus athericus*. Growth increases caused by atmospheric carbon dioxide enrichment is reduced when increased CO_2 is combined with elevated UV-B. Teramura et al. (1990) also report that plant responses to increased atmospheric CO_2 are modified by enhanced UV-B radiation. For the species tested the effects of increased CO_2 and UV-B seem to be additive. This may be expected based on the different mode of action of CO_2 enrichment and elevated CO_2. Increased atmospheric CO_2 stimulates carboxylation in C_3 plants with raised net photosynthesis and an increased of net assimilation rate as a result. Elevated UV-B may reduce photosynthesis by disturbing PSII reactions and by damage to photosynthetic pigments.

In earlier studies we examined the effects of carbon dioxide enrichment on C_3 en C_4 plant species of Dutch salt marshes. There was improved growth and increased photosynthesis in the C_3 plant species *Puccinellia maritima*, *Scirpus maritimus* and *Elymus pycnanthus* (= *E. athericus*), but not in the C_4 halophyte *Spartina anglica* (Rozema *et al.* 1991, Lenssen & Rozema, 1990, Lenssen *et al.*, 1992. Enhanced atmospheric carbon dioxide causes reduced transpiration due to partial stomatal closure both in the C_3 and C_4 plant species examined. Accordingly the water potential of the plant increased with elevated CO_2. The results presented here demonstrate adverse effects of increased ultraviolet-B radiation on two important monocotyledonous cereal crops, wheat and maize. Similar reduction of growth of wheat and maize by elevated UV-B has been reported earlier. In most of the studies reported,.maize appears to be less sensitive to enhanced UV-B radiation than wheat (Teramura 1983). Sensitivity of plants to UV-B radiation seems to be dependent also on various environmental factors such as water and nutrient availability. Plants cultivated in the field show a higher production of UV-B absorbing pigments than in the greenhouse (Beyschlag *et al.*, 1988). The growth inhibition by increased UV-B in the maize cultivar studied namely cultivar Limagrain 5 was generally less than in the wheat cultivar Obelisk.

The values of the mean RGR are relatively low for wheat and maize. This may be related to the photon flux density (PAR) in the greenhouse, that was between 200 and 250 µ Einstein $m^{-2}.sec^{-1}$. Net photosynthesis of both crops will not be light saturated at this light intensity.

The analysis of the mean relative growth rate and the growth parameters may partially reveal the way in which inhibition of growth by enhanced UV-B in these two grasses occurs. Growth reduction appears not to be primarily related to the the morphological component of the growth, that is the Leaf Area Ratio and more in particular the Specific Leaf Area. Enhanced UV-B radiation did not influence the allocation of biomass to the aerial or belowground parts, since there is no effect of UV-B on the Leaf Weight Ratio at the first harvest. The Net Assimilation Rate appears to be much affected by enhanced UV-B in wheat and maize. This indicates that one or several phases of the processes occurring from carbon assimilation to increase of plant biomass will be inhibited or disturbed by enhanced UV-B.

The measurements of leaf photosynthesis support this and demonstrate a drop of the net photosynthetic rate in both wheat and maize with raised UV-B. This can not directly be related to stomatal closure since in the case of the strongest reduction of P_n, the transpiration rate appears not to be affected by UV-B or even increased. It is well-known that enhanced UV-B may disturb the photosystem II processes of photosynthesis. Measurements of photosynthesis and transpiration in a later phase of the growth experiment show some variation. It seems that the reduction of P_n with elevated UV-B at the end of the growth period is less marked than at the start of the experiment. When cellular components that absorb UV-B radiation are involved in adjustment to enhanced UV-B, the gradual build up of UV-B absorbing components may explain this change of the plant response to enhanced UV-B radiation with time. The relatively low photon flux density may not only have reduced the mean relative growth rate, it may also have influenced the photosynthetic response of wheat and maize to enhanced UV-B (cf. Cen & Boruman, 1990). At photosynthetically Active Radiation (PAR)-saturated and ambient CO_2, Beyschlag et al. (1988) did not find effects of enhanced UV-B irradiation on photosynthetic characteristics of wheat. Growth of plant cultivated under a high photon flux density appears to be less reduced by elevated UV-B than plants grown at low light intensity.

In a more extensive paper (Staay et al., 1991, Staaij et al., 1992) details on growth and physiological parameters (chlorophyll content, UV-B absorbing pigments) are presented. Transpiration rates of the maize leaves show some variation with time. When enhanced UV-B radiation disturbs the control mechanism of stomatal opening and closure both reduced and increased transpiration rates may be expected.

Growth of maize cv. Limagrain 5 seems to be less sensitive to enhanced UV-B than growth of the wheat cultivar Obelisk. This difference may relate to the more tropical or subtropical origin of maize and the fact that wheat the cultivar may have been evolved from wheat populations adapted to climatic conditions of the moderate zones. Since no precise details are known of the evolutionary history of maize and wheat cultivars used, the above hypothesis needs to be checked carefully. In further studies the analyses of growth of these crops in response to elevated UV-B is extended and not only greenhouse studies but also field studies will be conducted. In addition, it is attempted to explain the response of plants to UV-B in terms of development and distribution of UV-B absorbing compounds.

REFERENCES

Beyschlag, W., Barnes, P.W., Flunt, S.D. & Caldwell, M.M. (1988). Enhanced UV-B irradiation has no effect on photosynthetic characteristics of wheat (*Triticum aestivum* L.) and wild oat (*Avena fatua* L.) under greenhouse and field conditions. Photosynthetica 22: 516-525.

Caldwell, M.M., Teramura, A.H. & Tevini, M.M. (1989). The changing solar ultraviolet climate and the ecological consequences for higher plants. Tree 4: 363-367.

Cen, Y.P. & Bornman, J.F. 1990. The response of bean plants to UV-B radiation under different irradiances of background visible light. J. Exp. Bot. 41: 1489-1495.

Lenssen, G.M. & Rozema, J. (1990). The effect of atmospheric CO_2 enrichment and salinity on growth, photosynthesis and water relations of some salt marsh species. In: Goudriaan, J., Keulen, H. van, & Laar, H.H. van (eds.). The greenhouse effect and primary productivity in European agro-ecosystems. Pudoc, Wageningen, 64-67.

Lenssen, G.M., Lamers, J., Stroetenga, M. & Rozema, J. (1992). Interactive effects of CO_2 enrichment, salinity and flooding on growth of C_3 (*Elymus athericus*) AND C_4 (*Spartina anglica*) salt marsh species. In: Rozema, J., Lambers, H., Geyn, S. van de & Cambridge, M.L. (eds). CO_2 and Biosphere. Kluwer Academic Publishers, Dordrecht, in press.

Rozema, J., Dorel, F., Janissen, R., Lenssen, G.M., Broekman, R.A., Arp, W. & Drake, B.G. (1991). Effects of elevated atmospheric carbon dioxide on growth, photosynthesis and water relations of salt marsh grass species. Aq. Bot. 39: 45-55.

Rozema, J., Lenssen, G.M. & Staay, J.W.M. van de. (1990). The combined effect of increased atmospheric CO_2 and UV-B radiation on some agricultural and salt marsh species. In: J. Goudriaan, H. van Keulen & H.H. van Laar (eds.). The greenhouse effect and primary productivity in European agro-ecosystems, pp 68-71, Pudoc, Wageningen.

Rozema, J., Lenssen, G.M., Arp, W.J., Staaij, J.W.M. van de. (1991). Global cxhange, the impact of the greenhouse effect (atmospheric CO_2 enrichment) and the increased UV-B radiation on terrestrial plants. In: J. Rozema & J.A.C. Verkleij (eds.). Ecological responses to environmental stresses. pp. 220-231. Kluwer Academic Publishers, Dordrecht.

Staaij, J.W.H. van de, Lenssen, G.M., Stroetenga, M. & Rozema, J. (1992). The combined effect of elevated CO_2 and UV-B radiation on growth characteristics of *Elymus athericus*. In: J. Rozema, H. Lambers, S. van de Geyn, & M.L. Cambridge (eds.) CO_2 and Biosphere. Kluwer Academic Publishers, Dordrecht. In press.

Staaij, J.W.M. van de, Costa, V., Rozema, J., J.M.G. Torres Pereira, Broekman, R.A. & Stroetenga, M. (1991). Analysis of growth and physiology of wheat and maize in response to enhanced UV-B (in preparation).

Staaij, J.W.M. van de, Rozema, J. & Stroetenga, M. (1990). Expected changes in Dutch coastal vegetation resulting from enhanced levels of solar UV-B. In: J.J. Beukema, W.J. Wolff & J.J.W.M. Brouns (eds.). Expected effects of climatic change on marine coastal ecosystems. pp. 211-217. Kluwer Academic Publishers, Dordrecht.

Teramura, A.H. (1983). Effects of ultra violet B radiation on the growth and yield of crop plants. Physiol. Plant. 58: 415-427.

Teramura, A.H., Sullivan, J.H. & Ziska, L.H. (1990). Interaction of elevated ultraviolet-B radiation and CO_2 onm productivity and photosynthesis characteristics in wheat, rice and soybean. Plant Physiol. 94: 470-475.

Tevini, M.M. & Teramura, A.H. (1989). UV-B effects on terrestrial plants. Photochem. Photobiol. 50: 479-487.

EFFECT OF SOLAR UV-B RADIATION ON GROWTH OF MUNGBEAN (*VIGNA RADIATA* L. WILCZEK) PLANTS

Aruna Sharma, Y.P. Abrol and U.K. Sengupta
Division of Plant Physiology, Indian Agricultural Research Institute, New Delhi 110012, India

The stratospheric ozone layer which is the primary attenuator of solar ultraviolet radiation is gradually being depleted due to increased man-made pollutants. Enhancement in UV-B region (280-320 nm) is of particular interest in view of its adverse effects on biological systems (5). Further, it is likely that the problem will be more serious for countries near the equator as they already receive greater amount of UV-B radiation due to sun's angle for penetration of ultraviolet wave band through the ozone layer.

A number of investigations have shown that UV-B radiation affects the morphology and anatomy of the plants (1,4,11,12). Changes in physiological and biochemical processes have also been studied (6,9,10). Most of the information has been obtained from artificially controlled environment using UV-B lamps in growth chambers or green houses. Due to manipulation in a single factor, plants tend to show increased sensitivity to UV-B. The shortcomings of these types of experiments are discussed in this proceedings (13).

In the present communication, the effect of filtering off of UV-B in the natural light on growth of mungbean (Vigna radiata L. Wilczek) is reported.

Mungbean plants cv. PS 16 were grown in the field in July 1990 under optimal conditions of growth. During this period, sky under Delhi conditions remains partly cloudy but receives bright sunshine due to clearing off of the atmospheric dust particles by the monsoon rain. A thin polyester sheet was erected above the plot in order to exclude solar UV-B radiation reaching the plants. The polyester film filtered off all the radiation in the spectral zone of 280-320 nm and allowed only 23 per cent of 320 nm through the filter (Figure 1). The visible radiation (PAR) received at the ground level through the filter was about 86 per cent of the unfiltered radiation.

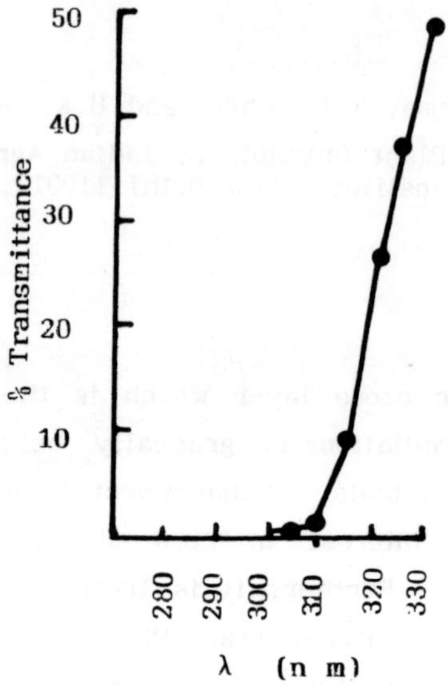

Figure 1. Percentage transmittance of UV wave band through polyester filter.

The mungbean cultivar used in this study is a short duration crop and takes two months to mature. The day length during its growth period (July-August) in this region is about 13 hours. Total sunshine hours are, however, curtailed by cloudy and overcast sky. During the experimental period, the crop experienced only 392 hours of sunshine as against 830 hours on the basis of day length. The plants received brighter solar radiation on clear days because some of the atmospheric pollutants were cleared off by monsoon rain.

In general, the growth of plants under the UV-B filter was faster and the leaves were wide apart from each other. Under the normal sun light, leaves had a tendency to remain closer to each other at noon time. It suggests that plants may have some protective mechanism against direct solar radiation containing UV-B wave band. The exclusion of UV-B from natural radiation caused significant increase in plant height. The differences were quite narrow at seedling stage but increased with maturity. Exclusion of UV-B did not alter the leaf number per plant indicating that increased plant height was due to internodal elongation. Considerable effect of UV-B exclusion was noticed on leaf area per plant. This was due to increase in the size of the individual leaves. The increase in leaf area and stem length was accompanied by respective increase in their dry weight (Figure 2). This indicates that exclusion not only caused elongation but also increased the growth in terms of dry matter production by the plants.

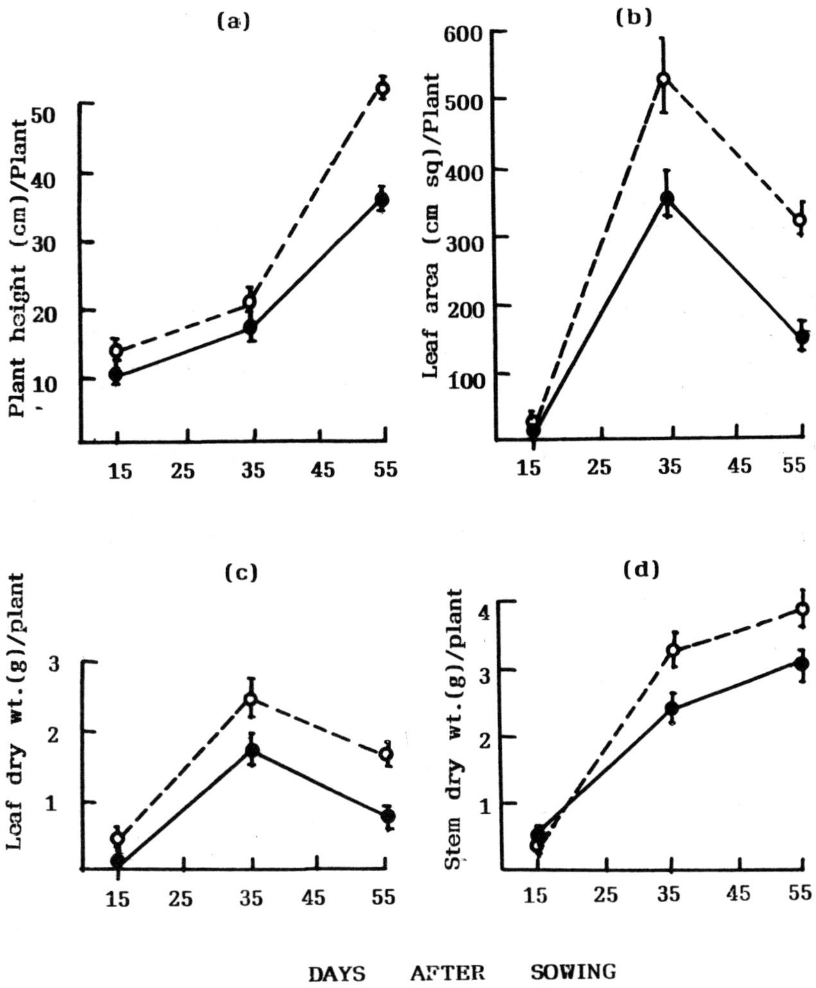

Figure 2. Effect of exclusion of UV-B on (a) plant height (cm), (b) leaf area (cm sq/plant), (c) leaf dry weight (g/plant), (d) stem dry weight (g/plant)

o________o full sun light. o------o full sunlight(-UV-B)

Elimination of UV-B also caused significant alteration in reproductive parameters of the plants. There was increase in pod and seed number per plant as well as their dry weight (Table 1). It is likely that the effect of UV-B is on photosynthesis (6,8) which in turn affects dry matter production as well as partitioning to different plant organs.

Table 1. Effect of exclusion of solar UV-B radiation (per plant) on reproductive parameters in mungbean cv. PS16.

Treatment	Pod number	Pod weight (g)	Seed number	Seed weight (g)
Full sunlight	3.80 ±0.24	0.70 ±0.09	25.25 ± 2.61	0.57 ±0.10
Full sunlight	10.28 ± 1.65	1.38 ±0.30	69.20 ± 7.16	1.86 ±0.21

The use of supplemental UV-B light to study the effect of enhanced UV-B radiation under green house and growth chamber on different crop species have shown similar type of response on growth of plants. In some sensitive crop plants, leaf expansion was reduced due to increased UV-B radiation (2,3,14,15). Teramura (12) reviewed the response of increased UV-B light and has shown that overall growth processes of plants are inhibited by it. Similar effects on growth parameters in the present study under natural environments indicate that the UV-B wave band present in the current level of solar radiation has already increased to cause stunting of plants and decrease other growth parameters. Thus, the plants under atmospheric conditions of Delhi are already growing under above optimal conditions of UV-B radiation.

REFERENCES

1. Brandle, J.R., Campbell, W.F., Sisson, W.B. and Caldwell, M.M. (1977). Net photosynthesis, electron transport capacity and ultra structure of Pisum sativum L. exposed to ultraviolet B radiation. Plant Physiol. 60: 165-169.

2. Biggs, R.H. and Kossuth, S.V. (1978). Effects of ultraviolet-B radiation enhancements under field conditions. In UV-B Biological and Climatic Effects Research (BACER). Final Report.

3. Briggs, R.H., Kossuth, S.V. and Teramura, A.H. (1981). Response of 19 cultivars of soybeans to ultraviolet-B radiation. Plant Physiol. 53: 19-26.

4. Caldwell, M.M. (1971). Solar UV-radiation and the growth and development of higher plants. pp. 131-171. In:A.C. Giese ed. Photophysiology vol. 6, Academic Press, New York.

5. Coohill, T.P. (1991). Stratospheric ozone-depletion as it affects life on earth - the role of ultraviolet action spectroscopy. In: Proceedings Indo-US Workshop: Global climatic changes - Photosynthesis and Plant Productivity, Jan. 8-12, New Delhi, India

6. Kulandaivelu, G., Maragatham, S. and Nedunchezhian, N. (1989). On the possible control of ultraviolet-B induced response in growth and photosynthetic activities in plants. Physiol. Plant. 76: 398-404.

7. Owens, O.H. and Krizek, D.T. (1980). Multiple effects of UV radiation (265-330 nm) on fungal spore emergence. Photochem. Photobiol. 32: 41-49.

8. Murali, N.S. and Teramura, A.H. (1986). Effect of supplemental ultraviolet-B radiation on growth and physiology of field grown soybeans. Environ. Exp. Bot. 26: 233-242.

9. Sullivan, J.H. and Teramura, A.H. (1989). The effects of ultraviolet-B radiation on loblolly pine.I. Growth, photosynthesis and pigment production in green house grown saplings. Physiol. Plant. 77: 2022-207.

10. Sullivan, J.H. and Teramura, A.H. (1990). Field study of the interaction between solar ultraviolet-B radiation and drought on photosynthesis and growth in soybean. Plant Physiol. 92: 141-146.

11. Teramura, A.H. (1980). Effects of ultraviolet-B irradiance on soybean. I. Importance of photosynthetically active radiation in evaluating ultraviolet-B irradiance effects on soybean and wheat growth. Physiol Plant. 48: 333-339.

12. Teramura, A.H. (1983). Effects of ultraviolet-B radiation on the growth and yield of crop plants. Physiol. Plant. 58: 415-427.

13. Teramura, A.H. and Sullivan, J.H. (1991). Field studies on UV-B radiation effects on plants; Case histories of soybean and loblolly pine. In: Proceedings Indo-US Workshop: Global climatic changes - Photosynthesis and Plant Productivity, Jan. 8-12, New Delhi, India

14. Tevini, M. and Teramura, A.H. (1989). UV-B effects on terrestrial plants. Photochem. Photobiol. 40: 479-487.

15. Tevini, M., Iwanzik, W. and Thoma, U. (1981). Some effects of enhanced UV-B irradiation on the growth and composition of plants. Planta. 153: 388-394.

II

CO_2 ENRICHMENT/BALANCE

INITIAL GROWTH AND ONTOGENY OF BIGLEAF MAPLE (*ACER MACROPHYLLUM*) IN AN ENRICHED CARBON-DIOXIDE ENVIRONMENT

John D. Bailey[1], Robert K. Dixon[2] and Robin Rose[3]
[1]Mantech Environmental Technology, US EPA - Corvallis, OR 97333
[2]US Environmental Protection Agency, Corvallis, OR 97333
[3]Nursery Technology Coop., Oregon St. Univ., Corvallis, OR 97331

ABSTRACT

A controlled-environment experiment was initiated to evaluate the influence of CO_2 enrichment on the growth and ontogeny of bigleaf maple (Acer macrophyllum). Development of seedlings was monitored from seed germination through the first five months of ontogeny in growth chambers containing 350 (ambient), 575 and 700 ppm CO_2. Seedling shoot elongation, leaf and branch expansion, and foliage retention were altered by ambient CO_2 conditions. These preliminary morphological responses are consistent with reports from previous experiments which employed tree seedlings pre-conditioned in ambient CO_2 conditions. Alterations in seedling crown and leaf morphology due to CO_2 enrichment may influence future adaptive responses of tree species to environmental stress agents.

The information in this document has been funded wholly by the U.S. Environmental Protection Agency. It has been subjected to the Agency's peer and administrative review, and it has been approved for publication as an EPA document. Mention of trade names or commercial products does not constitute endorsement or recommendation for use.

INTRODUCTION

An increase in the concentration of atmospheric carbon dioxide (CO_2) may significantly affect tree anatomy and physiology and substantially alter ontogeny (1,2,3,4,5,6). Potential direct and indirect effects of long-term CO_2 enrichment on juvenile and mature tree carbon assimilation and allocation is virtually unknown (7,8). Moreover, current knowledge of CO_2 effects on mature trees and forest ecosystems is based almost exclusively on short-term assessments of tree seedlings pre-conditioned in ambient CO_2 conditions (9). These experiments provide little information regarding initial seedling ontogeny, morphology or eco-physiological processes in an enriched CO_2 environment.

Previous research demonstrated consistent changes in conifer and broadleaf tree seedling morphology and physiology in response to short-term doses of elevated atmospheric CO_2 (1,3,9). Stimulation of carbon assimilation rates resulting from higher internal CO_2 concentrations in the leaves of plants in an enriched-CO_2 environment (all other things being equal) lead to:

1) increased stem and foliar biomass, surface area, number of branches and foliage retention (1,2,3,4,5,6,9,10,11,12,13,14);

2) greater root biomass and volume (4,5,13,14,15,17);

3) increased bud size and weight (15);

4) greater production and storage of carbon metabolites (17,18,19); and,

5) stimulation of root symbiotic associations (20,21,22).

The temporal and ontogenic responses of tree seedlings to CO_2 enrichment vary significantly by species and site conditions (9,12,23). Significant attributes of these responses will be reviewed.

Temporal Responses

Eco-physiological changes in tree seedlings in response to CO_2 enrichment occur within hours to days, and plants can acclimate to higher CO_2 concentrations in several weeks. For example, Arctic plants under field conditions (23) and temperate plants in growth chambers (2,5) have both shown acclimation responses within a single growing season. Acclimation of seedlings within growth chambers, however, may be attributed to other factors such as reduced carbon sink strength in a limited rooting environment (19,20). Acclimation responses vary widely by site and plant genotype as suggested by *in situ* assessment of coastal marsh grasses in an enriched CO_2 environment (12,23).

A long-term assessment of juvenile tree response to CO_2 enrichment was reported by Surano et al. (7) and Houpis et al. (8). Both reports summarize a 2.5 year experiment with ponderosa pine (Pinus ponderosa saplings grown in field chambers. Surano et al. (7) reported a general decrease in tissue heat tolerance (i.e., the upper thermal limits of photosynthetic apparatus) with increased CO_2 concentrations even though the optimum temperature for photosynthesis increased. Saplings became chlorotic, and growth rates eventually decreased (relative to control treatments) after early stimulation of height and diameter growth. Based on these results, Surano et al. (7) and Houpis et al. (8) proposed that pine saplings reach a new homeostasis in higher CO_2 concentrations, losing short-term growth benefits as environmental stress develops. However, these results are from one study conducted under one set of conditions on one sapling per treatment.

Ontogeny

Carbon dioxide enrichment may alter the cycle of growth and dormancy in trees, potentially altering adaptive responses to environmental stress (3,9,10). Enrichment with CO_2 may directly increase foliage retention or delay bud-set, and/or indirectly alter a seedling's cold-hardiness or bud dormancy (9). In a short-term study by Rogers et al. (1), loblolly pine (Pinus taeda) seedlings grown in an enriched CO_2 environment continued active shoot and needle growth after trees in ambient chambers became dormant. Commercial greenhouse facilities have for decades altered ambient CO_2 conditions to influence seedling growth and dormancy patterns (24). Elevated leaf temperatures observed in CO_2 enrichment experiments (7), may also influence temperature control of dormancy cycles.

An enriched CO_2 environment may alter dormancy cycles through changes in the tissue biochemistry in seedlings, particularly carbohydrate and starch metabolism (3,7,19). Plants metabolize and store photosynthate derived compounds to provide energy for maintenance respiration during dormancy and growth following bud-break (9). Oberbauer et al. (15) observed starch accumulation in leaf chloroplasts of tropical tree species subjected to CO_2 enrichment. Campagna and Margolis (5) measured starch accumulation in the needles of black spruce (Picea mariana) at elevated CO_2 concentrations. The effect of these CO_2 stimulated biochemical responses (e.g., starch accumulation) on tree dormancy and growth cycles is unknown, as is the effect on subsequent ontogeny.

The influence of CO_2 enrichment on initial seedling ontogeny, morphology and canopy eco-physiological processes is unknown. Moreover, previous morphological and physiological assessments of shade tolerant and intolerant species have employed seedlings or saplings pre-conditioned in ambient CO_2 conditions (9). The response of shade-tolerant tree species (eg, Acer) to elevated CO_2 has not been thoroughly examined. The objective of this preliminary experiment is to evaluate selected morphological responses of bigleaf maple (Acer macrophyllum) seedlings, during the initial five months of ontogeny, to CO_2 enrichment.

METHODS

The experiment was conducted in three EGC M-11 (4.2 meters^2) growth chambers, which regulate light, temperature and humidity. Three CO_2 treatments were evaluated: 350 (ambient), 575 and 700 ppm by volume. Chamber CO_2 concentration was monitored with a LICOR 6251 infra-red gas analyzer and 0.5 to 5.0-sec pulse injections of CO_2 were added as needed to maintain treatment levels. Enrichment with CO_2 was maintained $\pm$ 20 ppm of target concentrations. Sample and injection time, CO_2 concentration, temperature and light level were recorded at 4-minute intervals for each chamber. Seedlings were rotated among chambers monthly to account for any undetected differences in temperature, light quantity or light quality among chambers.

Bigleaf maple seed were stratified for 90 days at 2°C at which time radicals were emerging from the viable seed (25). Germinants were planted in Hahn-192 Styroblocks and placed in their respective CO_2 treatments. Chambers conditions were initially cool (9°C) with low light (200 $\mu E\ m^{-2} s^{-1}$). Ambient conditions were altered gradually: 500-1000 $\mu E\ m^{-2}s^{-1}$depending on plant height, 25°C days and 19°C nights, 14 hour days, and a constant 70% humidity. After one month seedlings were transplanted to 4x14-Treepots (10x10x35 cm) and grown for another four months. Seedlings were watered as needed and fertilized (NPK + micronutrients) weekly during the experiment. Plant density was held constant among treatments.

Seedling morphology was assessed during the experiment to record trends in height, root collar diameter, and number of leaves. After five months 24 seedlings from each treatment were harvested. Seedlings were dissected into three segments: 1) upper segment, extending from the last fully expanded leaves to the meristem (mean length, 12 cm); 2) lower section, from the cotyledon scar to the first fully-expanded, non-senescent leaves (mean length, 37 cm); and 3) middle section, between lower and upper sections (mean length, 34 cm). Basal diameter to 0.01 mm, length to 0.1 cm, total number of leaves (leaf scar count) and leaves retained, and leaf and stem color using a Munsell color guide (26) were measured for each stem section. Total seedling height, number of leaves per seedling, and number of leaves retained on the shoot were calculated. The total number of branches were recorded for the entire shoot. All data were subjected to analysis of variance ($P=0.05$).

RESULTS AND DISCUSSION

Height and diameter growth patterns of the bigleaf maple seedlings grown in an enriched CO_2 environment are shown in Figure 1. Enrichment with CO_2 increased mean root collar diameters 27%. Height of seedlings in the CO_2 enrichment treatments were not significantly different from those grown in ambient conditions. However, average internodal length of the mid-stem ranged from 10.5 cm in ambient conditions to 17.2 in 700 ppm CO_2 treatment (Table 1). Previous reports, with plants preconditioned in ambient CO_2, revealed total height and root collar diameter of broadleaf tree seedlings subjected to short-term CO_2 enrichment did not necessarily increase (2,7,17).

Seedling leaf morphology was significantly different among CO_2 treatments (Table 1). Plants in ambient CO_2 produced 20% more leaves over the five-month experiment than those in the 700 ppm CO_2 treatment. However, seedlings in the 575 and 700 ppm CO_2 treatments retained more leaves on lower and middle sections of the crown relative those growing in ambient CO_2. Tinus (24), Tolley and Strain (2,11), Higginbotham et al. (14), and Oberbauer et al. (15) previously reported significant increases in tree seedling leaf area following short-term exposure to an enriched CO_2 environment. Retention of leaves by seedlings in an elevated CO_2 environment was reported earlier by Tolley and Strain (2) and Norby et al. (17), in Liquidambar and Quercus, respectively.

Seedling branch morphology differed significantly among elevated CO_2 treatments (Table 1). The total number of branches per seedling was significantly increased by exposure to CO_2. Sionit et al. (4) observed a two-fold increase in number of branches in sweetgum (Liquidambar styraciflua) after one growing season in a 650 ppm CO_2 environment. In the current experiment, seedlings in the 700 ppm treatment had relatively large crowns and a proliferation of secondary branches. These data suggest CO_2 enrichment stimulates the proliferation and development of lateral buds (9).

Seedlings exposed to 700 ppm CO_2 were distinctly more yellow (2.5GY7/8, hue, value, and chroma) relative to seedlings grown in 350 ppm (5GY6/8) and 575 ppm treatments (7.5GY5/8, green). Leaves of seedlings grown in 700 ppm also exhibited distinctive basipetal curling. Leaf curling reduced total foliage area of seedlings in the 700 ppm treatment relative to those plants grown in ambient conditions. The cause of leaf discoloration and curling is not fully known but has been reported previously (7,8).

CONCLUSIONS

Elevated CO_2 concentration significantly altered the initial growth and morphology of bigleaf maple seedlings. Seedling height, root collar diameter, number of branches, leaf area and retention, and leaf color differed among CO_2 treatments. To our knowledge, this is one of the first experiments to measure a CO_2 response in tree seedlings grown from seed. Previous short-term experiments with conifer and broadleaf tree species (eg, 2,5,17,23) employed seedlings or saplings pre-conditioned in ambient CO_2 conditions. Long-term experiments are necessary to validate these short-term ontogenic and morphological responses.

The preliminary responses observed in this experiment with bigleaf maple suggest that crown morphology and leaf canopy retention are altered by ambient CO_2. Curtis et al (23) observed stimulation of plant tillering in an estuarine marsh grass community subject to CO_2 enrichment. Tree species which respond to high CO_2 conditions by altering leaf and branch bud development and leaf retention may be subject to frost damage in temperate regions (9). In contrast, within tropical latitudes, tree species with large and robust canopies (photosynthetic capacity) in an enriched CO_2 environment may gain a competitive advantage over species with small canopies (15). Long-term, detailed assessments of whole-tree morphology and physiology are required to elucidate the implications of an enriched CO_2 atmosphere for woody plant species (12)

ACKNOWLEDGMENTS

The USDA Forest Service, Forest Sciences Laboratory in Corvallis, OR, USA, provided bigleaf maple seed and assistance with plant care during the course of this experiment. The authors thank Christian Andersen, Richard Norby, Joe Zaerr and John Zasada for constructive review comments.

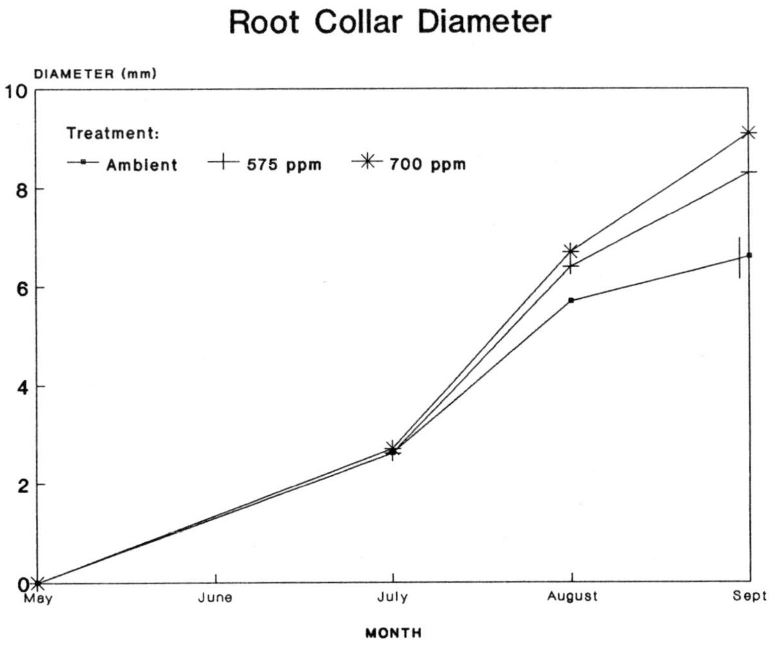

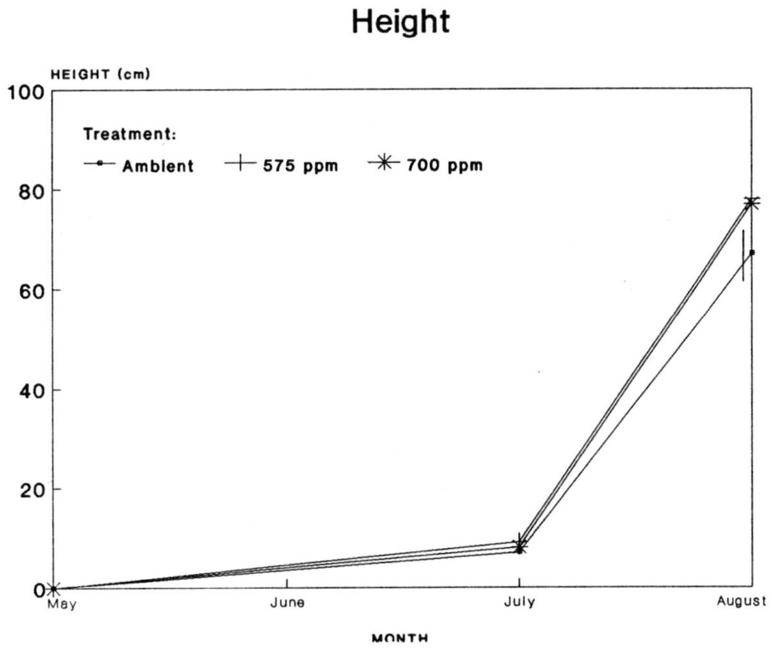

Figure 1. Root collar diameter and height growth patterns of bigleaf maple grown in ambient (350), 575 and 700 ppm CO_2 treatments after five months. Standard error of the mean indicated by lateral bar.

Table 1. Branch and leaf morphology of bigleaf maple seedlings grown in CO_2 enriched environments.

	CO_2 Treatment				
	350 ppm[a]	575 ppm	700 ppm	p-value[b]	R^2 [c]
Seedling Sections					
upper					
leaves (#)	11[d]	10	8	.0001	0.38
mean internodal length (cm)	2.3	2.9	3.5	.0001	0.41
middle					
leaves (#)	7	5	5	.0140	0.25
mean leaf size (cm^2)	76	101	93	.0001	0.42
mean internodal length (cm)	10.5	14.9	17.2	.0001	0.64
lower					
leaves (#)	17	15	15	.0001	0.40
leaves retained (%)	32	45	47	--	--
mean internodal length (cm)	4.5	4.4	3.0	.0001	0.46
Whole Seedling					
root collar diameter (cm)	6.6	8.3	9.1	.0001	0.68
leaves (#)	34	30	28	.0001	0.52
leaves retained (#)	22	22	20	.1817	0.13
branches (#)	1	5	6	.0001	0.67

[a] ambient treatment.
[b] p-values for ANOVA tests on treatment significance.
[c] R^2 for the regression of the variable against treatment level.
[d] n=24.

LITERATURE CITED

1. Rogers, H.H., G.E. Bingham, J.D. Cure, J.M. Smith, and K.A. Surano. 1983a. Responses of selected plant species to elevated carbon dioxide in the field. *J. Environ. Qual.* 12(4):569-574.

2. Tolley, L.C., and B.R. Strain. 1984a. Effects of CO_2 enrichment and water stress on growth of *Liquidambar styraciflua* and *Pinus taeda* seedlings. *Can. J. Bot.* 62:2135-2139.

3. Norby, R.J., J. Pastor, and J.M. Melillo. 1986b. Carbon-nitrogen interactions in CO_2-enriched white oak: Physiological and long-term perspectives. *Tree Physiol.* 2:233-241.

4. Sionit, N., B.R. Strain, H. Hellmers, G.H. Riechers, and C.H. Jaeger. 1985. Long-term atmospheric CO_2 enrichment affects the growth and development of *Liquidambar styraciflua* and *Pinus taeda* seedlings. *Can. J. For. Res.* 15:468-471.

5. Campagna, M.A., and H.A. Margolis. 1989. Influence of short-term atmospheric CO_2 enrichment on growth, allocation patterns, and biochemistry of black spruce seedlings at different stages of development. *Can. J. For. Res.* 19:773-782.

6. Funsch, R.W., R.H. Mattson, and G.R. Mowry. 1970. CO_2-supplemented atmosphere increases growth of *Pinus strobus* seedlings. *Forest Science* 16:459-460.

7. Surano, K.A., P.F. Daley, J.L.J. Houpis, J.H. Shinn, J.A. Helms, R.J. Palassou, and M.P. Costella. 1986. Growth and physiological responses of *Pinus ponderosa* Dougl. ex P. Laws. to long-term elevated CO_2 concentrations. *Tree Physiol.* 2:243-259.

8. Houpis, J.L.J., K.A. Surano, S. Cowles, and J.H. Shinn. 1988. Chlorophyll and carotenoid concentrations in two varieties of *Pinus ponderosa* seedlings subjected to long-term elevated carbon dioxide. *Tree Physiol.* 4:187-193.

9. Jarvis, P.G. 1989. Atmospheric carbon dioxide and forests. Phil. Trans. R. Soc. Lond. B 324:369-392.

10. Canham, A.E., and W.J. McCavish. 1981. Some effects of CO_2, daylength and nutrition on the growth of young forest tree plants: In the seedling stage. *Forestry* 54:169-182.

11. Rogers, H.H., J.F. Thomas, and G.E. Bingham. 1983b. Response of agronomic and forest species to elevated atmospheric carbon dioxide. *Science* 220:428-429.

12. Curtis, P.S., B.G. Drake, P.W. Leadley, W.J. Arp, and D.F. Whigham. 1989. Growth and senescence in plant communities exposed to elevated CO_2 concentrations on an estuarine marsh. *Oecologia* 78:20-26.

13. Tolley, L.C., and B.R. Strain. 1984b. Effects of CO_2 enrichment on growth of *Liquidambar styraciflua* and *Pinus taeda* seedlings under different irradiance levels. *Can. J. For. Res.* 14:343-350.

14. Higginbotham, K.O., J.M. Mayo, S. L'Hirondelle, and D.K. Krystofiak. 1985. Physiological ecology of lodgepole pine (*Pinus contorta*) in an enriched CO_2 environment. *Can. J. For. Res.* 15:417-421.

15. Oberbauer, S.F., B.R. Strain, and N. Fetcher. 1985. Effect of CO_2-enrichment on seedling physiology and growth of two tropical tree species. *Physiol. Plant.* 65:352-356.

16. Brown, K., and K.O. Higginbotham. 1986. Effects of carbon dioxide enrichment and nitrogen supply on growth of boreal tree seedlings. *Tree Physiol.* 2:223-232.

17. Norby, R.J., E.G. O'Neill, and R.J. Luxmoore. 1986a. Effects of atmospheric CO_2 enrichment on the growth and mineral nutrition of *Quercus alba* seedlings in nutrient-poor soil. *Plant Physiol.* 82:83-89.

18. Hollinger, D.Y. 1987. Gas exchange and dry matter allocation responses to elevation of atmospheric CO_2 concentration in seedlings of three tree species. *Tree Physiol.* 3:193-202.

19. Norby, R.J., E.G. O'Neill, W.G. Hood, and R.J. Luxmoore. 1987. Carbon allocation, root exudation and mycorrhizal colonization of *Pinus echinata* seedlings grown under CO_2 enrichment. *Tree Physiol.* 3:203-210.

20. Norby, R.J. 1987. Nodulation and nitrogenase activity in nitrogen-fixing woody plants stimulated by CO_2 enrichment of the atmosphere. *Physiol. Plant* 71:77-82.

21. O'Neill, E.G., R.J. Luxmoore, and R.J. Norby. 1987b. Increases in mycorrhizal colonization and seedling growth in *Pinus echinata* and *Quercus alba* in an enriched CO_2 atmosphere. *Can. J. For. Res.* 17:878-883.

22. Arnone, J.A., and J.C. Gordon. In press. Effect of nodulation, nitrogen fixation and CO_2 enrichment on the physiology, growth and dry mass allocation of *Alnus rubra* seedlings. *New Phytol.*

23. Tissue, D.T., and W.C. Oechel. 1987. Response of *Eriophorum vaginatum* to elevated CO_2 and temperature in the Alaskan tussock tundra. *Ecology* 68:401-410.

24. Tinus, R.W. 1972. CO_2 enriched atmosphere speeds growth of ponderosa pine and blue spruce seedlings. *Tree Planters' Notes* 23:12-18.

25. Zasada, J., J. Tappeiner, and T. Max. 1990. Viability of bigleaf maple seeds ofter storage. *West. J. Applied Forestry.* 5:52-55.

26. *Munsell Color Charts for Plant Tissues*. 1977. Kollmorgen Corp., Baltimore, MD, USA.

CO_2 ENRICHMENT RESPONSES OF CHRYSANTHEMUM, CUCUMBER AND TOMATO: PHOTOSYNTHESIS, GROWTH, NUTRIENT CONCENTRATIONS AND YIELD

M.M. Peet[1], D. H. Willits[2], K. E. Tripp[1], W. K. Kroen[1], D.M. Pharr[1], M.A. Depa[1] and P.V. Nelson[1]; [1]Dept. of Hort. Sci.,[2] Dept. of Bio. & Agr. Eng., North Carolina State Univ., Raleigh, N.C. 27695-7609.

ABSTRACT

Yield increases from CO_2 enrichment varied from 54% in cucumber fruit weight, to 37% in chrysanthemum dry weight and 20% in tomato fruit weight. To determine the basis for response differences, photosynthesis was modeled [5] in chrysanthemum and photosynthesis, growth and leaf concentrations of starch, nutrients and carbon were measured in tomato. CO_2-enriched chrysanthemum leaves were more efficient than ambient-CO_2 grown leaves at irradiances below 400 μmol(photons) $m^{-2}s^{-1}$ because of greater photosynthetic efficiencies. At higher irradiances, ambient CO_2-grown leaves were more efficient because of greater CO_2 conductance. Measured at the same CO_2 concentration, ambient leaves had higher photosynthetic rates than enriched leaves because of their greater CO_2 conductance, but dry weights and *in situ* photosynthetic rates were still higher in enriched plants. Cucumber yield data appeared to fit this model, but CO_2-enriched tomatoes were much less responsive to CO_2 increases. In tomatoes, response to CO_2 enrichment did not increase with increasing irradiance and short daily periods of enrichment did not increase yield, suggesting that the CO_2 conductances declined significantly. Lower leaves of CO_2-enriched tomato plants were inrolled, chlorotic and purpled. We thought this was because of carbohydrate accumulation, possibly associated with feedback inhibition of photosynthesis. We concluded, however, after a 3-year study, that deformation severity was not directly related to foliar starch concentrations. We found more starch in CO_2-enriched leaves, but the pattern of accumulation did not appear to account for changes in deformation through development and with source-sink treatments. Severity of leaf deformity was also not correlated with low leaf photosynthetic rates. *In situ* photosynthetic rates were only slightly higher in enriched plants and did not differ with genotype. Severity of leaf deformation was correlated with high fruit yields and high C/N ratios, but low root weights and low foliar concentrations of K and N. We suggest that in tomatoes CO_2-enrichment increases sink strength more than source strength in tomatoes. During fruit development, carbohydrates may be partitioned to fruit at the expense of roots, leading to higher yields, but late-season nutrient stress.

INTRODUCTION

Rising global CO_2 levels are expected to increase growth of plants significantly in both managed and unmanaged ecosystems. Agricultural yields, for example, have been predicted to rise by as much as 33% [7]. It is increasingly observed, however, that actual responses to CO_2 enrichment can differ greatly with the species used, the portion of the crop's life cycle in which enrichment is given and the irradiance, temperatures and nutritional levels provided to the plant during enrichment. For example, photosynthesis and growth of seedlings of alfalfa [2] and tobacco [14] are much more responsive to high CO_2 than are these processes in mature plants.

CO_2 enrichment for extended periods of time can, in fact, visibly deform leaves of cotton, tomato, clover, soybean and *Desmodium* [3]. Leaf injury has also been observed in cucumber [6,15] and gerbera [15]. Foliar effects are particularly pronounced in tomato [8,9]. In the following paper, we summarize CO_2 enrichment research which took place from 1981-1989. The data is presented as case studies on chrysanthemums, cucumbers and tomatoes. These case studies illustrate the difficulty of generalizing plant responses to elevated CO_2 concentrations. For tomatoes, the yield response was explored in greater detail through physiological studies, including an in-depth study of the high CO_2-induced leaf injury response. As part of this study, we investigated the hypothesis proposed by Madsen [8] and others [3] that leaf injury in CO_2-enriched tomato is caused by excess foliar starch.

MATERIALS AND METHODS

<u>All Plants:</u> All experiments were conducted in double polyethylene-covered greenhouses. For these experiments 2 large (6.7 m x 12.2 m) and 4 small (5.2 m x 6.1 m) greenhouses were used. CO_2 enrichment was by liquid CO_2 with application controlled by computer. Greenhouse temperatures were regulated by computer at preset levels, with heating provided by natural gas fired unit heaters with polytube heat kits attached. The non-enriched houses were cooled by two-speed fans and evaporative pads.

In addition to these cooling systems, which require venting to the outside, in three CO_2-enriched houses, cooling was applied via 3.1 m wide x 5.4 m long x 1.8 m high rockstorages attached to the houses by insulated ducts. More specific details of the operation of the rockstorage system are contained in Willits and Peet [16].
Environmental variables in all compartments were monitored by both the control

computer and a datalogger. Outside air dry bulb temperature, outside dew point, solar radiation received on a horizontal surface outside the compartment (from an Eppley black and white pyranometer), inside dry bulb temperature of each compartment, inside wet bulb temperature and the CO_2 levels in the houses were recorded by datalogger.

Chrysanthemums: Rooted cuttings of chrysanthemums were grown hydroponically in an aerated, modified Hoagland's solution for 3 5-week experiments. Experimental procedures are described in Depa [4]. Two CO_2 levels (350 and 675 μmol mol^{-1}) and 4 irradiance levels [550,875,1100 and 1400 μmol(photons)$m^{-2}s^{-1}$] were applied in all seasons except summer. Summertime CO_2 enrichment could not be maintained in this experiment because of the need to ventilate the greenhouses. The small greenhouses were used for all chrysanthemum research. Nutrient solutions were changed weekly, at which time dry weights and leaf areas of all plant parts were determined.

Photosynthesis and transpiration were measured weekly using a plexiglass cuvette that enclosed an entire plant. Cuvette size was increased as plant size increased. The bottom portion of the cuvette contained a room temperature nutrient solution. Volume of this root cuvette also increased with plant size. Each photosynthetic measurement was conducted at the same irradiance and CO_2 concentration as that under which the plant was grown. Steady-state photosynthesis was measured with an Anarad infrared gas analyzer. During the 10-15 minutes required to obtain a reading, leaf and air temperatures in the cuvette and leaf temperature outside the cuvette were compared using copper-constantin thermocouples to ensure that cuvette temperatures did not rise more than 2^oC over those of outside leaves. Chamber temperatures were controlled by circulating air through a heat exchanger connected to an ice bath. Flow rates were adjusted, if necessary, to maintain cuvette leaf temperatures. The time between successive readings was 20-30 minutes. The analyzer used was sensitive to changes in reference CO_2 concentration, and so drifted significantly as greenhouse CO_2 levels fluctuated. To remove this drift as a source of error, the cuvette was disconnected from the system and a zero reading was made before and after taking photosynthesis readings. These zero readings were later subtracted from the reading with the plant in the chamber.

Cucumbers: Plants were grown as fall and spring crops. Details of production are given in Peet and Willits [12].All plants were grown in the small greenhouses in upright 18.93 l plastic bags filled with a 1:1 mixture of soilless media and pine bark mix. Individual emitters delivered water and nutrient solution to the plants 2-3 times daily, delivering 1-3 l/day. CO_2 regimes applied in the conventional houses with limited daily enrichment period(because of the necessity to vent the houses for cooling) were: ambient, 1000, 3000 and 5000 μmol mol^{-1} . In the rockstorage houses with an average of 83% more daily enrichment time than the conventionally ventilated houses, CO_2 regimes were: ambient,

600, 1000 and 1200 μmol mol^{-1} . A total of 4 different cultivars were used. Fruit were harvested 2-3 times weekly and fruit fresh weights and numbers recorded. Plants in these greenhouses were enriched an average of 10 hrs daily, representing 70% of daylight hours in the fall and spring and all daylight hours in the winter.

Tomato Yield studies in 1981-1986: Tomatoes were grown in the large greenhouses described in Willits and Peet [16] and Peet and Willits [10,11].

Physiological studies in 1987 and 1988: Two cultivars of greenhouse tomato, 'Michigan-Ohio' and 'Laura' were seeded and grown to transplant size in a glasshouse without CO_2 enrichment. Seedlings were transplanted into 18.93 or 9.46 l upright black polyethylene bags with holes on the sides for drainage (Hydro-Gardens, Colorado Springs, Colo.). Mix and drip irrigation procedures were similar to the cucumbers, but the nutrient solution concentrations were those described in Abbott et al [1]. Every 2-3 weeks, foliar samples and fertilizer solutions were analyzed for nutrient content.

Two large greenhouses were used in 1987 and four small greenhouses in 1988. One large and two small greenhouses were enriched to 1000 μmol mol^{-1} CO_2 using liquid CO_2 in rockstorage-connected greenhouses, with the remaining houses at ambient CO_2 concentrations (approximately 350 μmol mol^{-1}). In 1987, day and night heating began at 21^oC and 16^oC respectively. Cooling began at 25^oC, 27^oC and 28^oC for low, high vent and evaporative cooling pads, respectively. In the rockstorage houses, however, the rockstorage was used for cooling for as long as possible before CO_2 enrichment was terminated and venting to the outside began. In 1988, greenhouse temperatures were the same as in 1987 until 20 weeks plants age when minimum night temperatures were increased by 4^oC in 1 enriched and 1 ambient house.

Both cultivars were grown for 20 weeks (1987) or 18 weeks (1988) in treatments. The growing apex was removed when the sixth flower cluster had set at least one fruit. All axillary leaf growth was removed weekly. In 1987, treatments imposed on half the plants in each CO_2/night temperature regime were: reduced bag size and pruning all fruit clusters to the 2 largest fruit when both fruit had diameters of at least 1 cm. Each treatment was replicated randomly three times in each greenhouse except for CO_2 concentrations which were not replicated. In 1988, CO_2 concentrations were replicated once and all other treatments 4 times.

Plants were rated weekly for severity of leaf deformation on the lower, middle and upper thirds (1987) or lower and upper half (1988) of the canopies. Ratings were made independently by two individuals. Deformation occurred largely in the lower third of the canopy so foliar starch concentrations and carbon exchange rates (CER's) were also measured in the lower canopy.

In 1987, lower canopy foliage was sampled at 0900 and 1600 h for carbohydrate analysis at 9,13, 19 and 23 weeks plant age. Leaf punches were taken from the terminal leaflet of each of the 5 plants in the treatments, kept briefly on ice, then freeze-dried and stored desiccated until analysis. Samples were randomly chosen from 2 of the 3 replicates, resulting in 64 samples per sample date and time. Results of one sampling for each treatment group are therefore the means of 4 samples each made up of 5 leaf disks per sample.

In 1988, lower canopy leaves were sampled at 12, 17 and 23 weeks plant age. Procedures were similar to those in 1987, except that samples were taken at 0600, 1200 and 1800 h. In both years, leaf disks were ground in 3 mls of 80% ethanol using a Brinkman Polytron Homogenizer. Samples were then extracted 3 times in 3 mls 80% ethanol in tubes immersed in a boiling water bath. Ethanol insoluble material was pelleted by centrifugation, resuspended in 1 ml of 0.2 M KOH and gelatinized by immersion of the tubes in a boiling water bath for 30 min. After cooling, the solution was neutralized with 0.6 ml of 1 M acetic acid. Three mls of dilute, dialyzed α-amyloglucosidase from *Aspergillus oryzae* were added to each sample to digest the starch. The amylogucosidase had been previously dialyzed overnight against 50mM acetate buffer (pH 4.5) at 4°C. The dialyzed enzyme was then centrifuged and then diluted 1:24 before digestion for 1 h at 55°C. Digestion was stopped by immersion of the tubes in a boiling water bath for 5 mins. Samples were centrifuged to remove any remaining insoluble material. The supernatants were made up to 10 mls with distilled water and stored frozen. Following digestion, glucose concentration was determined using an enzyme link assay [13].

CER was measured using an LCA-2 portable leaf chamber analyzer system (Analytical Development Corporation, Andover, MA). All readings were taken between 0900 and 1500 h on sunny days. Readings were taken on unshaded terminal leaflets of leaves in the lower third of the canopy on the same date and on the same leaves sampled for starch. Sample size was 1 leaf on 6-10 different plants.

Physiological study in 1989: Procedures were as described above, except that 6 additional genotypes of tomato were grown for a total of 8 genotypes. Additional data taken on these plants were: foliar nutrient concentrations, C:N ratios, and end-of-season fresh weights. Photosynthetic measurements were taken on unshaded terminal leaflets of 1 leaf in the upper, middle and lower third of the canopies of 3 plants of each of 4 of the 8 genotypes in each house. The genotypes sampled were selected to represent the widest range of deformation response.

Leaf, stem, root and fruit fresh weights were measured at 23 weeks plant age on each of 4 randomly chosen plants of each genotype in each house. Leaves and stems were harvested and weighed . Root balls from the same plants were removed from the bags, washed in water, squeezed to remove excess water and weighed. Fruit fresh weights were calculated as cumulative total fruit weight harvested over the entire season on an individual plant basis. All fruit were harvested at the pink stage. Plant leaf, stem, root and fruit fresh weights were averaged by genotypes and CO_2 treatment. Foliage was sampled for nutritional analysis at 1000 h at 10,11,13,14,15,17,19, 21 and 23 weeks plant age. The terminal leaflet and 1 subtending leaflet were removed from representative leaves in the lower third of the canopy on 3 plants of each genotypes in each house. All leaflets for a given genotype in each house were pooled into one sample. Dried samples were processed through a Wiley Mill 20 mesh sieve, then dry ashed at 500°C to determine K, P,Ca,Mg, Mn, and Fe content . The resulting ash was dehydrated in 6N HCL and redissolved in 0.6N HCl. Atomic absorption spectrophotometry was used to analyze for all elements except P, N and C. Tissue P was analyzed spectrophotometrically . C and N were analyzed with a Perkin Elmer 2400 CHN Elemental Analyzer. K,P,Ca, N and Mg concentrations are reported as percent dry weight. Mn and Fe concentrations are reported as $\mu mol\ mol^{-1}$ dry weight. Nutrient concentrations were corrected for starch by subtracting mg starch in a given leaf sample from the total mgs of sample dry weight and then recalculating the nutrient concentrations. All nutrient concentrations reported have been corrected for starch, except when N concentrations were used in calculating C/N ratios.

RESULTS AND DISCUSSION

<u>Chrysanthemum Study:</u> Plants grown at the high CO_2 concentration had higher relative growth rates (0.0975 compared to 0.0903 $g\ g^{-1} day^{-1}$) and higher dry weights (data not shown). The source of these higher growth rates and dry weights were higher photosynthetic rates in enriched plants. Since our photosynthetic data was collected at a naturally fluctuating range of irradiances, we modelled the photosynthetic response based on the rectangular hyperbola model of France and Thornley [5]. We investigated several other photosynthetic models, but this seemed to best fit our data. The initial slope of the response curve is equal to the constant α, which is known as the photosynthetic efficiency. The other coefficient used in this equation is τ, which is referred to as the CO_2 conductance. The equation for net photosynthesis is:

$$Pn = \frac{\alpha I \,.\, \tau C}{\alpha I + \tau C} \qquad \text{(eq.1)}$$

where I is irradiance (PAR) and C is the CO_2 concentration. Photosynthetic efficiency was found to be significantly affected by time of day for the ambient-grown plants. Thus, for ambient plants, a = a + b[time], where b has a negative slope because of an afternoon decline in photosynthetic rates in plants in the ambient houses. There was also evidence of a decline in enriched plants, but there were not insufficient measurements from the enriched houses to determine if the decline was significant. We could not tell from our data whether this afternoon decline represented stomatal or non-stomatal factors because transpiration readings were extremely variable.

Values of α and τ differed for ambient CO_2-grown and CO_2-enriched plants. Photosynthetic efficiencies were greater in enriched plants (.153 compared to .092), but CO_2 conductances were greater in ambient-grown plants (.075 compared to .045). We used France and Thornley's equations with our data to produce an irradiance response curve (Fig. 1).

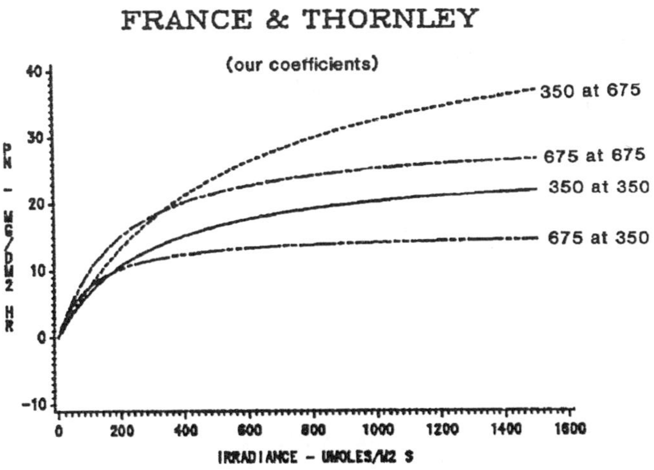

Fig. 1. Rectangular hyperbola model of individual leaf photosynthesis as a function of irradiance from France and Thornley [5] using our coefficients for CO_2 enriched and ambient CO_2 grown plants at measured at both 350 and 675 μmol mol^{-1}.

As seen on this figure, at irradiances below 400 μmol (photons) $m^{-2}s^{-1}$, CO_2-enriched plants are more efficient,because of their higher photosynthetic efficiencies. At higher irradiances, however, ambient-grown plants were more efficient because of their greater CO_2 conductance. The contrast is particularly great for the plants grown at 350 μmol mol^{-1}, but measured at 675 μmol mol^{-1} CO_2 and 1600 μmol(photons) $m^{-2}s^{-1}$. These plants have about a third higher photosynthetic rates than plants grown and measured at 675 μmol mol^{-1}. Plants grown at 675 and measured at 350 μmol mol^{-1} CO_2 have only a third the photosynthetic rates at 1600 μmol(photons) $m^{-2}s^{-1}$ of ambient-grown plants, measured at 675 μmol mol^{-1} CO_2. Under growth conditions, (675 plants at 675 and 350 plants at 350 µmol mol^{-1} CO_2), net photosynthesis is still slightly higher (about a sixth) in enriched compared to non-enriched plants, accounting for the greater growth of CO_2-enriched chrysanthemums. If CO_2 conductances had not declined in enriched plants, however, we would have predicted a much greater growth increase-almost a doubling at high irradiances.

Cucumbers:

Cucumbers responded very well to CO_2 enrichment, showing up to a 54% yield increase to some treatments. As might be predicted on the basis of the chrysanthemum photosynthesis data, the greatest response to high CO_2 in terms of plant weight gain per day came at high irradiances (Fig. 2).

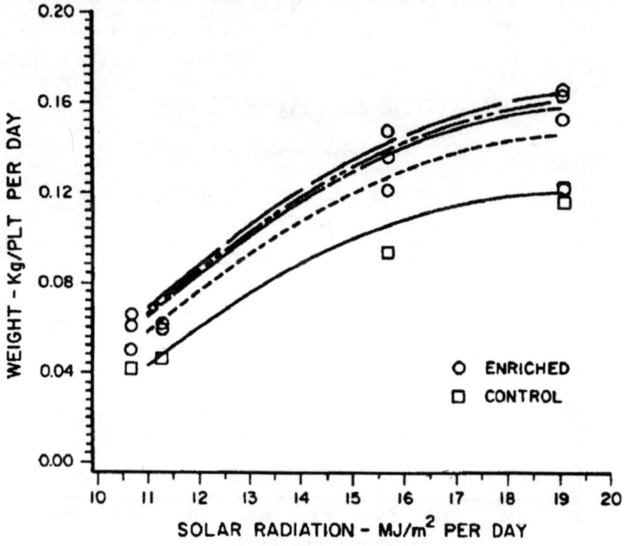

Fig. 2. Fruit weight vs. solar radiation for cucumbers. symbols represent seasonal averages. The solid line is Eq. 3 from [16] for no daily enrichment. The broken lines are Eq. 3 [16] for enrichment at 5000; 20,000; 10,000; and 15,000 µmol mol^{-1} h, respectively, bottom to top.

The decline in CO_2 conductance seen in the CO_2-enriched chrysanthemums was not a major limitation to CO_2 response. The longer each day that the crop could be enriched, the greater the response to high CO_2 (Fig. 3).

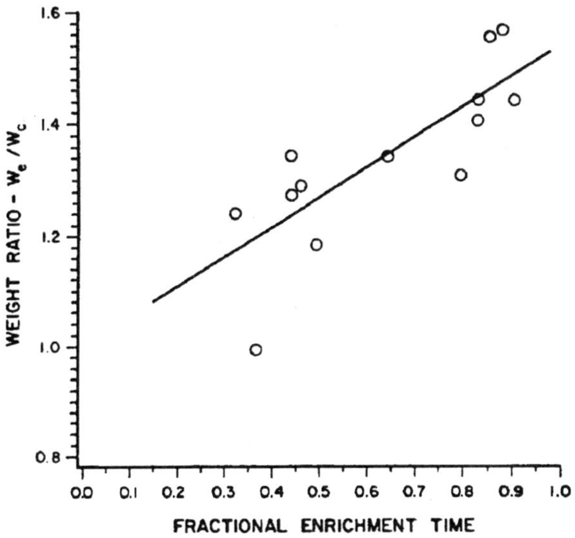

Fig. 3. Weight ratio of yield in enriched divided by yield in non-enriched cucumbers. Symbols represent seasonal averages of enriched treatments. The solid line is Eq. 7 [16].Fractional enrichment time is the % of daylight hours in which it was possible to CO_2 enrich the plants.

There was, however, an optimum CO_2 concentration for each length of daily enrichment. Enrichment at above optimal levels or for above optimal lengths at any daily enrichment period, resulted in reduced response to high CO_2. The best fitting relationship predicted maximum yields for cucumber at any combination of hours of enrichment and CO_2 concentration resulting in a cross-product of 14,400 μmol mol^{-1} CO_2x h. For example, this can be accomplished by applying a concentration of 1000 μmol mol^{-1} CO_2 for 14.4 h, 3000 μmol mol^{-1} for 4.8 h or 5000 μmol mol^{-1} CO_2 for 2.9 h.

Tomatoes

Yield studies in 1981-1986: The yield response of tomatoes to CO_2 was much less than cucumbers. Tomatoes also differed strikingly in their response to irradiance and duration of the enrichment period. While cucumbers had a greater response to CO_2-enrichment at high irradiance (Fig. 2), , as predicted by the model [5], tomato yields had a relatively flat response to irradiance (Fig.4).

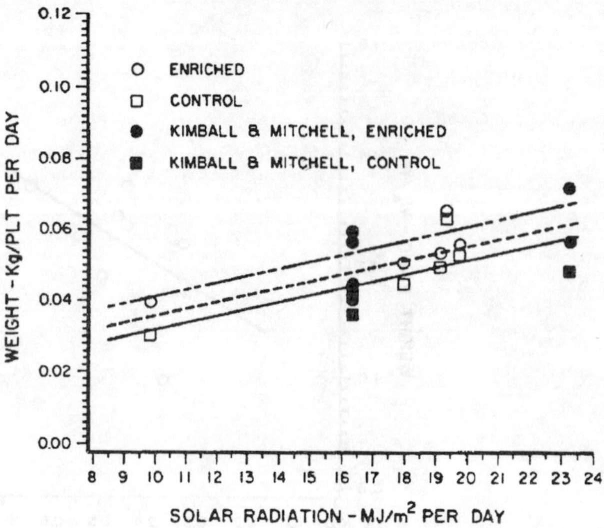

Fig. 4. Fruit weight vs. solar radiation for tomatoes. Symbols represent seasonal averages. The solid line is eq. 8 [16] with CO_2 concentration of 350 µmol mol-1. The broken lines are eq. 8 with CO_2 concentrations of 700 and 1200 µmol mol-1, respectively, bottom to top.

Tomatoes also required enrichment for a very long percentage of the day before significant effects on fruit weight were seen (Fig.5).

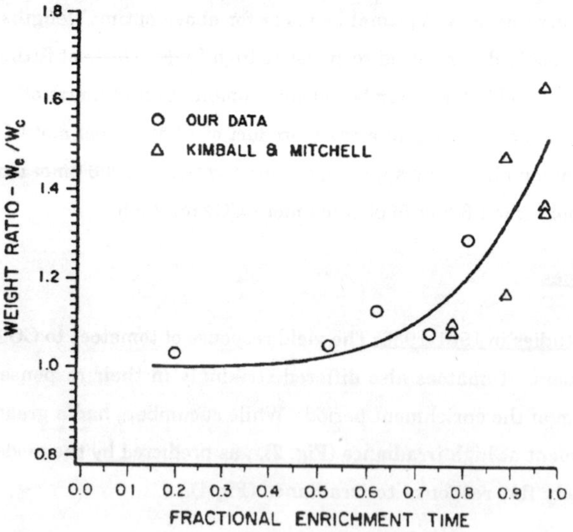

Fig. 5. Weight ratio of yield in enriched divided by yield in non-enriched cucumbers. Symbols represent seasonal averages. The solid line is eq. 14 [16].

This contrasts dramatically with the linear increase in cucumber fruit weight with duration of enrichment (Fig. 3). Possibly in tomatoes, the CO_2 conductance of enriched plants was sufficiently depressed that very long periods of high CO_2 were required to compensate for decreased conductances at low CO_2.

Physiological studies in 1987 and 1988: These studies were undertaken to explain the relative lack of yield response of tomatoes to high CO_2 concentrations noted above. We had observed in the 1981-1986 yield studies that CO_2 enrichment dramatically increased the severity of leaf deformation (purpling, chlorosis, curling) in tomatoes. It seemed likely that this deformation would reduce plant photosynthetic capacity and studies were initiated in 1987 to find the basis for this deformation. Deformation increased with CO2 enrichment in 1987 and 1988 and leaf starch was also seen to increase. The time course of development of leaf injury was very different from changes in foliar starch, however. Foliar injury was low until plants were 9 weeks old (4 weeks of CO_2 enrichment). From week 9 to week 13, foliar deformation symptoms (leaf curling, purpling and chlorosis) increased dramatically, but they then remained constant at week 13 levels for the 10 remaining weeks before the crop was harvested. Starch, on the other hand, decreased dramatically from week 9 to week 13, after which time it either was unchanged or increased. Differences between cultivars and between the pruning, bag size and night temperature treatments in leaf deformation (where they were found) also were not related to differences in leaf starch concentrations (Fig. 6).

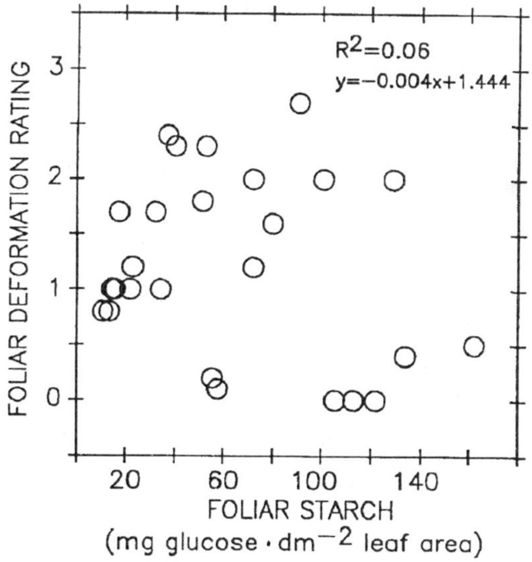

Fig. 6. Relationship between mean foliar starch concentration (mg glucose dm^{-2} leaf area) and mean foliar deformation severity of 2 cultivars of greenhouse tomato grown at ambient and 1000 $\mu mol\ mol^{-1}$ CO_2 in 1987 and 1988.

Photosynthetic rates were low in all leaves measured, especially after week 9. *In situ* photosynthetic rates in 1987 were not greatly different in enriched and ambient-grown plants. In 1988, there was some evidence of higher *in situ* rates of enriched plants , but again, rates were low. With these relatively small differences between enriched and non-enriched plants at growth CO_2 concentrations, and presumably lower CO_2 conductances in CO_2-enriched plants at ambient CO_2 concentrations, it is not surprising that tomatoes must be enriched for long periods of time in order to show a yield response to high CO_2, as previously seen in the yield studies (Fig.5).

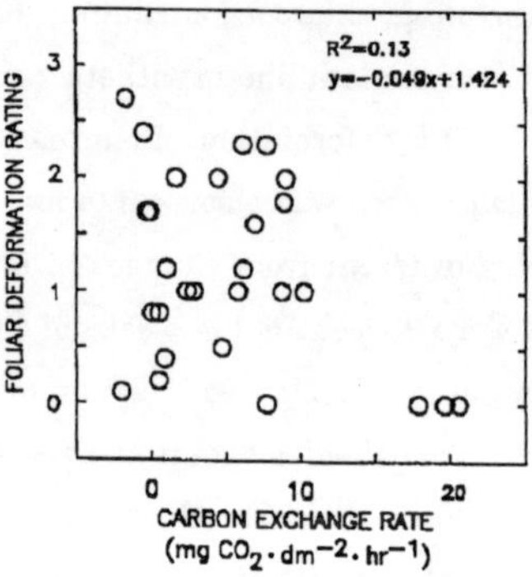

Fig. 7. Relationship between mean carbon exchange rates and mean foliar deformation severity in both CO_2-enriched and ambient CO_2-grown plants in 1987 and 1988.

Although we had hypothesized that deformed leaves would have lower photosynthetic rates, there appeared to be no relationship between the degree of deformity of the leaf and CER (Fig.7). It is necessary to point out, however, that CER was measured in enriched leaves at 675 $\mu mol\ mol^{-1}$ CO_2 and in ambient leaves at 350 $\mu mol\ mol^{-1}$. Thus, under conditions leading to greater deformation, i.e. CO_2 enrichment, CER was measured at a higher CO_2 concentration, possibly obscuring any negative relationship between deformation and CER. Examining the time course of development of deformation symptoms and changes in CER does not suggest a causal relationship of deformation in depressing CER, however. Deformation increased progressively in 1988 from week 9 to week to week 23, while CER was low throughout the measurement period. In 1987, deformation increased from none in week 9 to a high level in week 13 which was maintained until the end of the experiment. In contrast, CER in 1987 decreased from week 9 to 13, but then decreased again from week 19 to 23.

There was also little or no relation between leaf starch and CER (data not shown). All the leaves measured were in the lower canopy, however and CER was minimal. Examination of upper canopy leaves might have shown different relationships. For example, Yelle et al [17] observed seasonal increases in upper canopy starch in tomatoes throughout the season whereas lower canopy starch was declining. Since most of the deformation takes place in the lower canopy, however, lower canopy data should be the most relevant for our study.

Physiological studies in 1989: 1987 and 1988 data had shown very low CER rates in the lower canopy, but upper and middle canopy CER was not measured. In 1989, CERs were measured in all three regions of the canopy. CERs were highest in the upper canopy, low in the middle canopy and lowest in the lower canopy (data not shown). At 13, but not 19, weeks plant age, CERs were significantly greater in CO_2 enriched plants compared to ambient CO_2 grown plants. At 19 weeks, CERs were actually negative in both CO_2 treatments in the middle and lower canopy region. There were no significant differences between genotypes on either date and no apparent relationship with deformation (data not shown).

Across the 8 genotypes included in this study, CO_2 enrichment increased both yield and leaf deformation. Genotypes differed significantly in mean yield and deformation (data not shown). Regression analysis showed a significant, positive relationship between mean foliar deformation severity and mean total fruit yield, such that the highest yielding genotypes showed the greatest foliar deformation severity (Fig. 8).

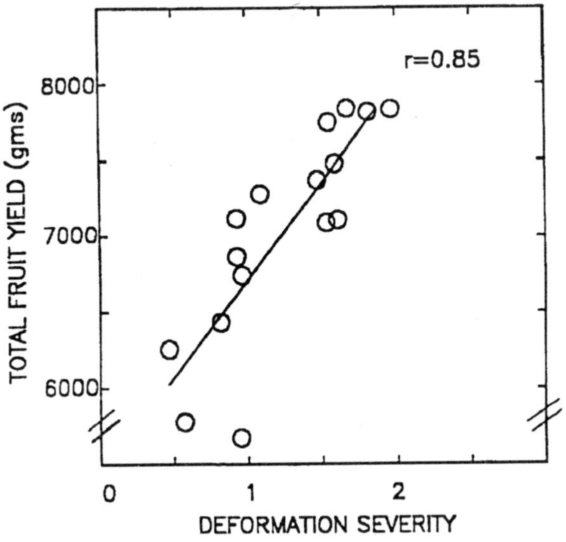

Fig. 8. Linear regression relationship between mean foliar deformation severity (0=none, 1=low, 2=moderate, 3=severe) and mean seasonal total fruit yield (g per plant) of 8 genotypes of tomato grown at ambient (ca. 350 and 1000 $\mu mol\ mol^{-1}\ CO_2$) for 16 weeks. Each point is the mean of 12 plants.

Total plant biomass increased only slightly with enrichment, with fruit weights increasing and root weights decreasing. These two factors; root weight and fruit weight;

were, in fact significantly negatively correlated, with an r^2 value of 0.72. There was no relationship between fruit fresh weight and either leaf (Fig. 9) or stem fresh weights (data not shown).

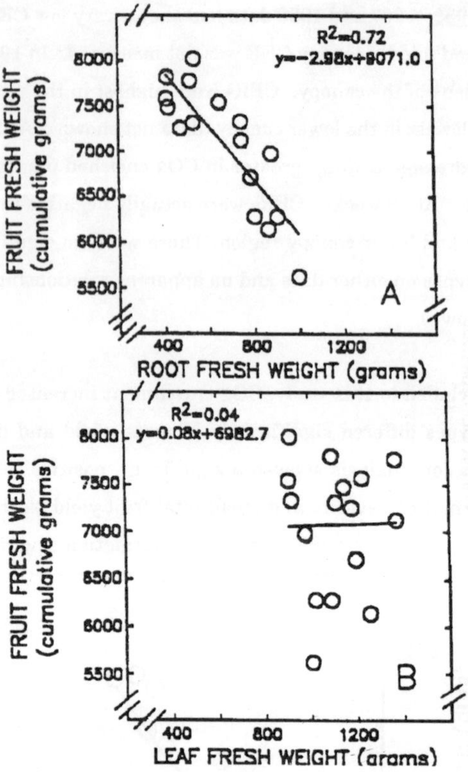

Fig. 9. Linear regression relationship between fruit fresh weight (mean cumulative yield, grams per plant), root(A), and leaf (B) fresh weights (grams per plant) of 8 genotypes of 23 week old tomato plants grown at ambient (ca 350 or 1000 µmol mol^{-1} CO_2 for 16 weeks. Each point is the mean of 8 plants for 1989 data.

Foliar K and N concentrations were both significantly reduced by CO_2 enrichment . For both nutrients, the time course of decrease in the foliage over the season was very similar to the time-course of increase in leaf deformation (Fig. 10 for K). When weekly ratings of foliar deformation were regressed against weekly leaf potassium and nitrogen concentrations, the r values were -0.65 and -0.66, respectively (p=.001). Mn concentrations were also significantly correlated with deformation ratings using seasonal, but not weekly, genotype means. Starch correction reduced the regression coefficient of deformation against all elements except potassium, indicating that the apparent correlation between low foliar nutrients and deformation in CO_2-enriched plants is at least partially a starch dilution effect. For example, the regression value for uncorrected nitrogen was 0.72 and the corrected value 0.66. For potassium, however, the correlation was equally strong with and without starch correction, suggesting that the effect of low foliar potassium values in causing leaf deformation in CO_2-enriched plants may be direct. A direct effect of K was also suggested by a preliminary study (Fig. 11) in which KH_2PO_4 was sprayed on the foliage of tomato plants during the period from 13 to 20 weeks plant age in which deformation symptoms normally appear. This foliar spray significantly reduced the deformation observed throughout the entire period Additional evidence for a direct role for K is the outward appearance of the deformed leaves which resembles K deficiency.

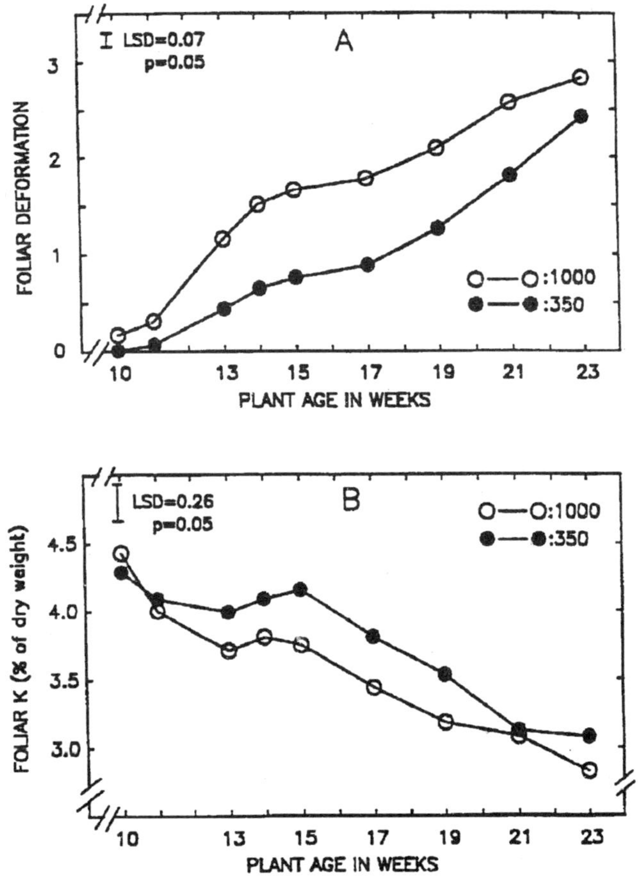

Fig. 10. Mean foliar deformation (A) and leaf K concentration (B) during growth at 350 (O) or 1000 (●) µmol mol^{-1} CO_2. Each deformation point represents the mean of 12 plants from each of 8 cultivars. Each K concentration point is the mean of 16 samples.

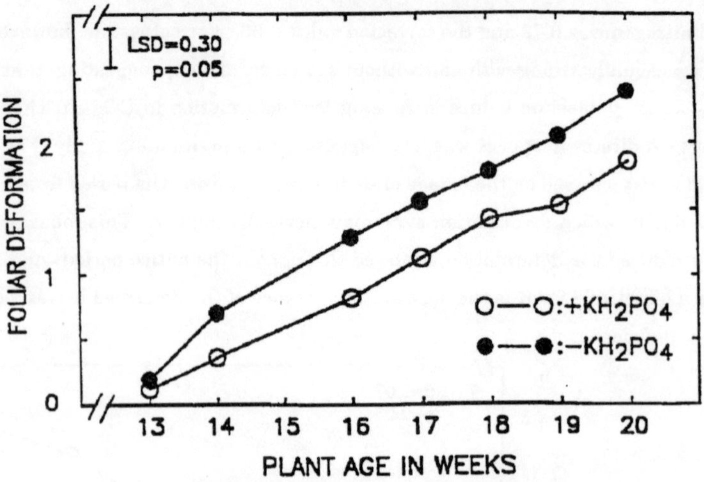

Fig. 11. Mean foliar deformation severity (0=none, 1=low, 2=moderate, 3=severe) of tomato foliage grown at ambient or 1000 mmol $mol^{-1} CO_2$ and either treated or untreated with 7 mM KH_2PO_4 foliar spray. Each point is the mean of ratings on 11 leaves in the lower third of the canopy.

Leaf carbon was high in CO_2-enriched plants, as expected, but was also high in high-yielding cultivars of both enriched and non-enriched plants (Fig. 12). The relationship was particularly strong for the enriched plants. Because non-starch corrected N was lower in CO_2-enriched and high-yielding cultivars, C:N ratios (N not starch-corrected) were significant and positively correlated with fruit yields (Fig. 12)

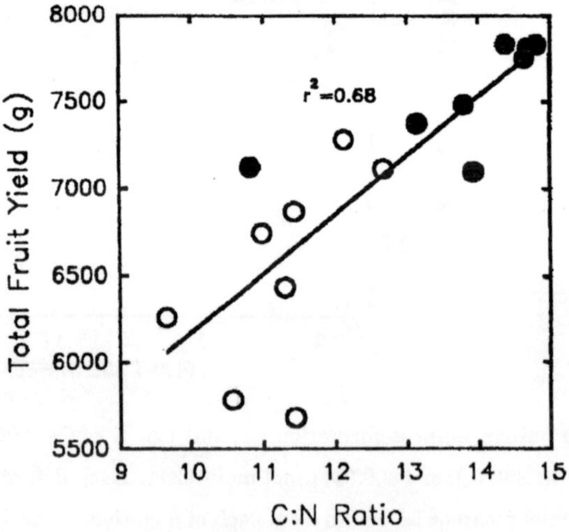

Fig. 12. Regression of cumulative fruit yield against C:N ratios (N not corrected for starch) during 16 weeks of growth at 350 or 1000 µmol mol^{-1} CO_2. Each symbol represents the mean of 12 plants. Open circles represent ambient-grown plants, closed circles, high CO_2-grown plants.

For the same reasons, i.e. higher leaf carbon and lower (uncorrected) leaf nitrogen, C:N were also positively correlated with leaf deformation(r^2=0.89)(Fig.13).

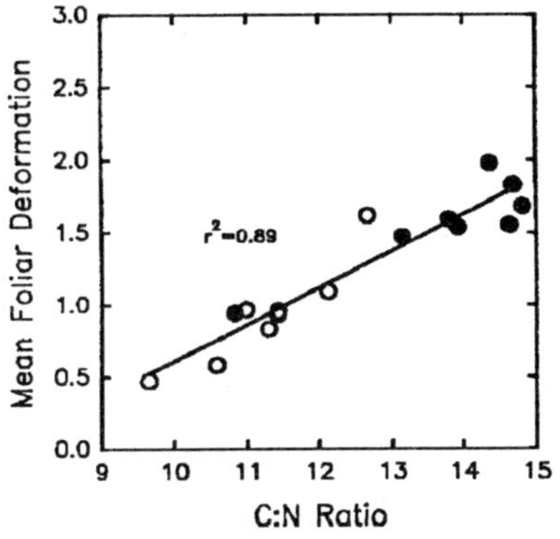

Fig. 13. Regression of deformation against C:N ratios (N not corrected for starch) during 16 weeks of growth at 350 or 1000 µmol mol^{-1} CO_2. Each symbol represents the mean of 12 plants. Open circles represent ambient-grown plants, closed circles, high CO_2-grown plants.

CONCLUSIONS

Chrysanthemums, cucumbers and tomatoes had very different responses to CO_2 enrichment. In chrysanthemums, dry weight, growth rate and photosynthetic responses to CO_2 enrichment were statistically significant for most of the experiments and were similar to the average given by Kimball [7]. Photosynthetic studies showed that CO_2 conductance was reduced in enriched plants, thus decreasing the potential gain in photosynthesis at high CO_2, and resulting in much less efficient leaves of CO_2-enriched plants at ambient CO_2 concentrations. Photosynthetic efficiency was increased in CO_2 enriched plants, however, making them more efficient at low irradiances than ambient CO_2-grown plants.

In cucumbers, yield response to CO_2 was greater than that predicted by Kimball [7]. Plants had a linear response to increasing daily hours of CO_2 enrichment. This suggests that CO_2 conductance was not greatly reduced in leaves of enriched plants, or they would have done poorly on days when enrichment was only possible for a few hours. They were also able to respond more to high CO_2 at high irradiance, as would have also been predicted based on the higher photosynthetic efficiencies of CO_2 enriched plants seen in chrysanthemum.

Tomatoes represented an extreme case of lack of response to high CO_2. Modeling data showed that unless they could be enriched for virtually the entire day, there was little yield benefit to CO_2 enrichment. Presumably this was because CO_2 conductance was decreased so much in enriched plants, that on days with partial enrichment, rates were depressed more by low CO_2 than they were increased by high CO_2 concentrations. Although we did not have as many irradiance data points for tomato, it appeared that yield response to CO_2 was actually less in tomatoes at high irradiance, or at least no greater.

We investigated the hypothesis that starch buildup in CO_2-enriched tomato leaves depressed photosynthesis and caused the characteristic leaf chlorosis, purpling and twisted we had observed we had observed consistently in these plants. Depressed photosynthetic rates in deformed leaves would then limit the response to high CO_2 concentrations. No such direct relationship between deformation, starch and CER could be found, however, to account for the lack of yield response. The time course of changes in foliar starch did not correspond with changes in deformation. The starch concentration of individual leaves was also not correlated with the degree of deformation of the leaves, except that CO_2 enriched leaves were on the average, more deformed and had higher starch concentrations compared to ambient CO_2-grown leaves. We were not able to relate CER in any of the canopy regions to differences in yields or deformation. Because CERs did not differ between genotypes, the observed significant genotypic differences in deformation and yield response could not be explained by differences in CER.

Examining the degree of deformation in 8 cultivars we were, however, able to make the following correlations. Across genotypes and CO_2 treatments, the most severe deformation was correlated with high fruit yield, but low root biomass. Root fresh weight was inversely proportional to fruit yield while stem and leaf fresh weights were not related to fruit yield. A high degree of deformation was also associated with low foliar K and N, but high C/N ratios.

The strong relationship between high fruit yield and low root weight suggested that CO_2 enrichment increased fruit sink strength, possibly through a direct effect on pollination and seed set. In a preliminary study, higher seed numbers per fruit were found in CO_2-enriched tomato fruit. This higher fruit sink strength may lead to a increase in partitioning to fruit at the expense of roots. The change in partitioning could reduce root weight at elevated CO_2 since CERs were not greatly increased by CO_2 enrichment and did not differ between cultivars. Reduced root /shoot ratios could have secondary effects on plant growth and development, such as reduced nutrient and water supply ,which may be involved in the deformation response. Mobile elements such as K and N may be withdrawn from lower leaves to support fruit growth, if adequate supplies are not available from the roots. This was suggested by the positive correlations of foliar K and

N with deformation, but the negative correlations of foliar concentrations of these elements with yield.

REFERENCES

1. Abbott, J.D., M.M. Peet, D.H. Willits, D.C. Sanders and Gough, R.E. (1986). Effects of irrigation frequency and scheduling on fruit production and radial fruit cracking in greenhouse tomatoes in soil beds and in a soil-less medium in bags. *Sci. Hortic.* 28: 209-217.

2. Baysdorfer, C. and Bassham, J.A. (1985). Photosynthate supply and utilization in alfalfa. A developmental shift from a source to a sink limitation of photosynthesis. *Plant Physiol.* 77:313-317.

3. Delucia, E.H.,T.W. Sasek and Strain, B.R. (1985). Photosynthetic inhibition after long-term exposure to elevated levels of atmospheric CO_2. *Photosyn. Res.* 7:175-184.

4. Depa, M.A. (1987). Predicting nutrient accumulation in chrysanthemum as influenced by levels of carbon dioxide and irradiance. M.S. thesis. North Carolina State University, Raleigh, N.C.

5. France, J. and Thornley, J. H. M. (1984). Mathematical models in agriculture. Butterworths. London. p. 119.

6. Ito, T. (1978). Physiological aspects of CO_2-enrichment to cucumber plants grown in greenhouses. *Acta Hortic.* 87:139-146.

7. Kimball, B.A. (1982). Carbon dioxide and agricultural yield: an assemblage and analysis of 430 prior observations. Water Conservation Laboratory Report 11. Water Conservation Laboratory, ARS/USDA, Phoenix, Arizona.

8. Madsen, E. (1974). Effect of CO_2 concentration on growth and fruit production of tomato plants. *Acta Agr. Scandinavica* 24:242-246.

9. Nagoaka, M., K. Takahashi, K. Arai, T. Hanada and Yoshioka, H. (1974). Effects of light intensity, night temperature and CO_2 concentration on the growth and yield of glasshouse tomato. Bull. Veg. Orn. Crops Res. Sta.:A 6:105.

10. Peet, M.M and Willits, D.H., (1984a). CO_2 enrichment of greenhouse tomatoes using a closed-loop heat storage: effects of cultivar and nitrogen. *Sci. Hortic.* 24:21-32.

11. Peet, M.M. and Willits, D.H. (1984b). Source/sink factors affecting response of greenhouse tomatoes to CO_2 enrichment. *Acta. Hortic.* 162:121-127.

12. Peet, M.M. and Willits, D.H. (1987). Greenhouse CO_2 enrichment alternatives: effects of increasing concentration or duration of enrichment on cucumber yields. *J. Amer Soc. Hort. Sci.* 112:236-241.

13. Robbins, N.S. and Pharr, D. M. (1987). Regulation of photosynthetic carbon metabolism in cucumber by light intensity and photosynthetic period. *Plant Physiology* 85:592-597.

14. Thomas, J.F., C.D. Raper, Jr., C.E. Anderson and Downs, R.J. (1975). Growth of young tobacco plants as affected by carbon dioxide and nutrient variables. *Agron. J.* 67:685-689.

15. Van Berkel, N. (1981). Three physiological disorders in glasshouse cultivation. *Acta Hortic.* 119:77-90.

16. Willits, D.H. and Peet, M.M. (1989). Predicting yield responses to different greenhouse CO_2 enrichment schemes: cucumbers and tomatoes. *Agricultural and Forest Meteorology* 44:275-293.

17. Yelle, S., R.C. Beeson, Jr., M. J. Trudel and Gosselin, A. (1989). Acclimation of two tomato species to high atmospheric CO_2. I. Sugar and starch concentrations. *Plant Physiol.* 90:1465-1475.

A PHYCOLOGICAL APPROACH TO ALLEVIATING THE GREENHOUSE PROBLEM

H.D.KUMAR, ASHOK KUMAR and L.K.SHARMA
Department of Botany, P.O. Box 5014, Banaras Hindu University, Varanasi-221005

ABSTRACT

One of the foremost environmental problems of global concern today is the greenhouse effect resulting mainly from the increasing concentration of CO_2 in the atmosphere. Terrestrial vegetation, forests, and soda lakes have attracted attention as sinks for some of the CO_2, but another equally important sink, namely, alkaline/saline wastelands appears to have been largely neglected. In this report we discuss the significance of such lands in countering, to some extent, the build-up of CO_2 in the atmosphere, thereby partly alleviating the greenhouse problem. We have surveyed a few "usar" soils and conducted some CO_2-enrichment experiments; these have revealed that these lands may act as a significant auto-sink for trapping some of the atmospheric CO_2. In a preliminary study we have observed that alkaline/saline soils show significant drop in pH following CO_2 absorption even in the presence of ambient concentration of CO_2; the pH fall is greater in the presence of CO_2-enriched atmosphere. Furthermore the algal flora of alkaline/saline soil appears to be another sink for trapping the CO_2. Our studies have demonstrated that the alkalophilic algae of usar lands can accumulate large amounts of inorganic carbon. Such accumulation appears to be due to the presence of the enzyme carbonic anhydrase. The strains having carbonic anhydrase on their outer membrane appear to be more efficient in trapping CO_2. Because of the high affinity for CO_2 and absence of photorespiration, these algae are suitable candidates for trapping CO_2. It is felt that by manipulating the carbonic anhydrase in plants other than algae, the trapping of CO_2 may be considerably increased. Although these alkaline/saline lands in India or elsewhere are treated as "wastelands", it is highly likely that they may prove to be a strong natural sink for CO_2. There is a need to propagate alkalophilic algae or other plants in these lands to fix CO_2 at high level.

INTRODUCTION

Global warming from increasing emissions of greenhouse gases has become a major scientific and political issue during the past decade. There is a fast growing and serious international concern at man's influence on the global climate. The greenhouse effect and climatic modifications are regulated by the interactions between tropospheric and stratospheric processes (1). There appears to have been an increase in tropospheric CO_2, N_2O, chlorofluorocarbons and CH_4 in recent years. All these gases (including N_2) accumulate in the atmosphere and act like glass in a greenhouse, letting in the warming rays, but inhibiting the escape of infra-red rays. Because CO_2 and other trace gases produced by industrial and agricultural processes absorb thermal radiation emitted by the earth's surface, recent studies have predicted that the increasing concentrations of these gases in the atmosphere will result in significant changes in climate which in turn may produce substantial changes in the location of the agricultural zones and shorelines (2).

Amongst all the greenhouse gases, relatively more work has been done on the increasing trend of CO_2, its source and its effect on the global temperature (3-5). Recently, it has been demonstrated that the amount of CO_2 in the atmosphere is steadily increasing by almost 1.5 ppm per year (6). Using average values, it has been estimated that global air temperature appears to have increased by roughly 0.7% over the past 140 years. Many workers have predicted that CO_2 concentrations will reach 600 ppm sometime between 2030 and 2080 A.D. It has also been projected that if CO_2 concentration were to double from 300 to 600 ppm, the earth's surface temperature would eventually warm up by somewhere between 3.5 and 5°C (4). A few workers have projected that an increase in global temperature of several degrees celsius will cause sea level to rise by 0.5 to 1.5 metres generally in the next 50 to 100 years; such a rise would endanger coastal settlements, estuarine ecosystems and the quality of coastal freshwater resources. Natural events such as intense drought, forest fires, intense hurricanes and flooding that have been hapenning for over a decade now clearly attest to the fact that the greenhouse effect is here. Unfortunately, no serious attempt has as yet been made to control the increasing level of atmospheric CO_2. Nature has created man and man is proving dangerous to mother nature. Despite our vast knowledge and technology development, one has to believe that most natural phenomena are subject to negation by natural laws/ processes. This dictum applies no less to the greenhouse problem and has prompted us to ask the question: "can we seek ways/means from nature to balancing the level of atmospheric CO_2?" In this paper we describe a few abiological and biological approaches to alleviating the greenhouse effect with special reference to the significance of microalgal communities in fixing CO_2.

TRAPPING OF CO_2 BY ALKALINE SOILS

The total land area of India is about 329 million hectares in which alkaline/saline lands occupy 64.65 lakh hectares (Fig. 1). (There are also a number of alkaline/ soda lakes and ponds.) These lands are characterized by impermeability, extreme hardness, and occasional presence of undesirable salts. They are locally known as 'usar' in Uttar Pradesh. The pH value of these soils often reaches up to 11, although the pH varies with depth. These lands also contain high levels of sodium. Their calcium content is generally low. These soils are usually infertile and are treated as wastelands. Their

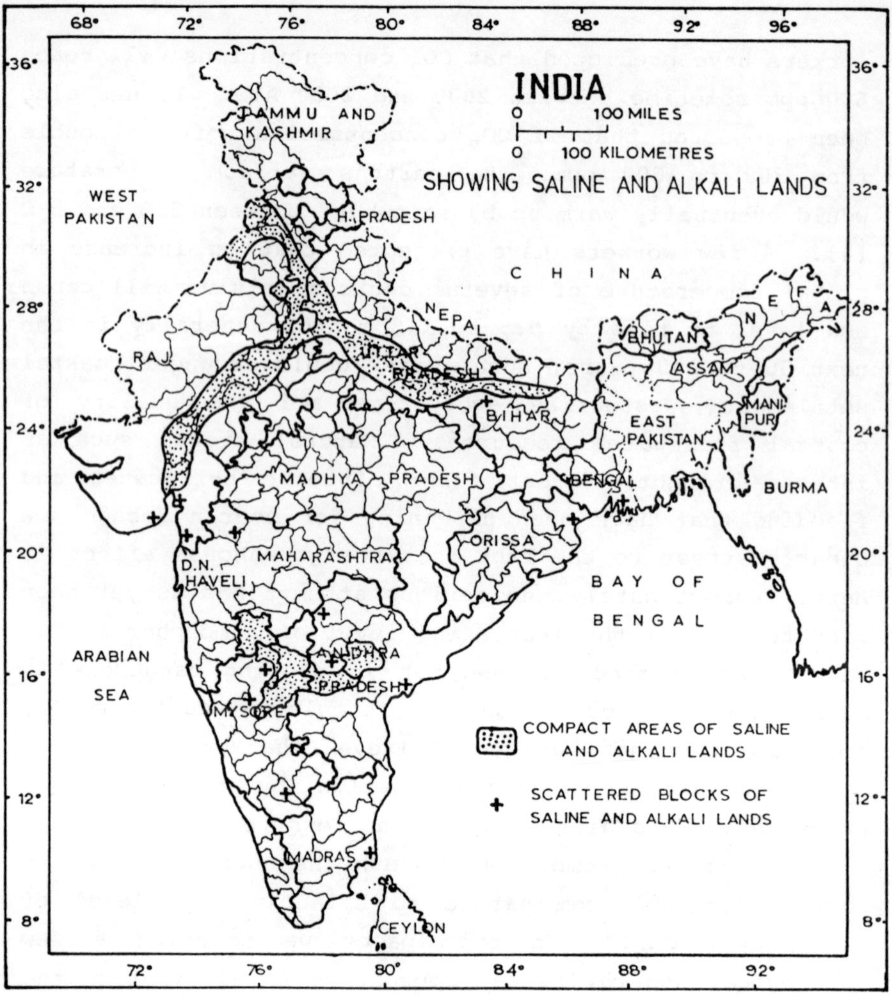

Fig. 1: Map showing distribution of alkaline/saline lands in India.

physico-chemical properties show fluctuations due to rainfall and microbial decomposition. A number of such lands have been transformed into fertile lands either by natural processes spontaneously or by man-made amendments (7). There are reports that reclamation of usar soils does occur; however, the real cause is not known (8). We feel that the fall in pH of reclaimed soils is an important factor occurring naturally.

In our study we have observed that usar soils have the capacity to absorb CO_2 resulting in a lowering of their pH. A suspension of usar soil having initial pH of

10.5 changed its pH to 10.02 upon continuous stirring for 36 hours. No such change in pH occurred in normal soil or double-distilled water, following stirring for the same period. That absorption of CO_2 by usar soils does occur sensu stricto was further evident from experiments where CO_2-enriched air was employed. Suspensions of usar soils were placed in airtight glass jars containing 5% CO_2 in air; they showed drop in pH from 10.5 to 8.15. This drop in pH remained constant for several days suggesting that CO_2 does not escape from the soil. The absorbed CO_2 is probably converted into Na_2CO_3, as these soils contain high level of sodium base. In the absence of periodical data, especially of pH, it is hard to speculate whether changes in pH have taken place in the existing usar soils which cover many thousands of hectares of land all over India. There is a need to test this phenomenon in natural conditions over a long period of time.

The above findings and our hypothesis are in agreement with earlier reports where trapping of atmospheric CO_2 by freshwater bodies has been shown (9). It has been demonstrated that the CO_2 trapping ability of lakes increases with increasing alkalinity and pH in saline/alkaline lakes. With the increase in alkalinity there is increased formation of $CaCO_3/Na_2CO_3$, some of which, however, tends to get deposited in the lake sediment.

DO ALKALOPHILIC ALGAE HAVE A ROLE IN TRAPPING CO_2?

A number of blue-green algae grow luxuriantly in alkaline/saline soils (8). The abundant algal growth produces striking and beneficial changes in such soils. The algal strains reported are mainly some species of *Microcoleus*, *Scytonema*, *Porphyrosiphon*, *Camptylonema*, *Cylindrospermum*, *Anabaena*, *Nostoc* and *Aulosira*. These algae form a thick mat on the soil surface during the rainy season (July to September) and during the retreating monsoon (December-January). Singh (8) made detailed investigation, both under field and laboratory conditions, with a view to understanding the role of these algae in the reclamation of usar soils. He found marked decrease in pH value, and there was a considerable improvement in the percentage of organic matter, total nitrogen, exchangeable calcium and water-holding capacity of the soil. Due to algal growth, the pH fell from 9.5 to 7.5, and the content of soluble CO_2 increased from 0.23 to 8.1 ppm. The changes in pH and CO_2 content of these soils following algal growth clearly point to the role of blue-green algae in trapping CO_2. We also have performed simple experiments with usar soil, double distilled water, pond water and a few algae collected from usar soils. From Table 1, it is seen that usar soils amended with algal mats are more efficient in absorbing CO_2 than the soils devoid of algae.

Table 1. Effect of algal supplementation on pH of usar*.

Condition/supplementation	Initial pH	Final pH after 80 hrs
Usar soil + double distilled water	10.55	9.94
Usar soil + DD water + algae	10.55	9.70
Usar soil + pond water (having pH 10)	10.55	9.85
Usar soil + pond water + algae	10.55	9.65

* 100 gm soil suspended in 200 ml of double distilled water or pond water, and usar algae (Aulosira, Nostoc, Anabaena etc.) were added. Initial pH was recorded and adjusted to 10.55 if required.

IS THE DROP OF pH A CONTINUOUS AND PERMANENT PHENOMENON IN USAR SOILS?

From our findings as well as tentative postulations, it appears that usar soils show gradual decrease in pH following CO_2 absorption. If this process continues, it would ultimately be expected to show considerable change in pH of all alkaline/saline soils and may even lead to complete reclamation of all such soils. However, due to operation of certain physico-chemical events, the soils regain their original pH and alkalinity and thus they maintain the original character. This is an important factor if these soils are going to act as a strong sink for absorbing CO_2.

METABOLIC PROCESSES INVOLVED IN THE TRAPPING OF CO_2

The notion that these algae are indeed involved in lowering the pH and increasing the soluble CO_2 content of alkaline/saline soils, apparently suggests that they possess unique metabolic processes/enzymes to absorb and metabolize the CO_2/HCO_3^-. It has been demonstrated that for microalgae (including cyanobacteria), the affinity for CO_2 in photosynthesis is much higher in cells grown under ordinary air (low-CO_2 cells) than in those grown under air enriched with 1-5% CO_2 (10). Based on the presence and location of carbonic anhydrase (CA), two groups of algae have been recognised: one having CA on the cell surface utilize HCO_3^- and the other having CA inside the cell utilize free CO_2 in photosynthesis. It has been clearly demonstrated that HCO_3^- is utilized after it has been converted to CO_2 by the CA on the cell surface (11). Therefore the actual species of inorganic carbon which crosses the cell membrane of these algae also appears to be free CO_2 (Fig. 2). Furthermore the existence of an active transport system for HCO_3^- has been suggested for cyanobacteria such as <u>Anabaena variabilis</u> (12).

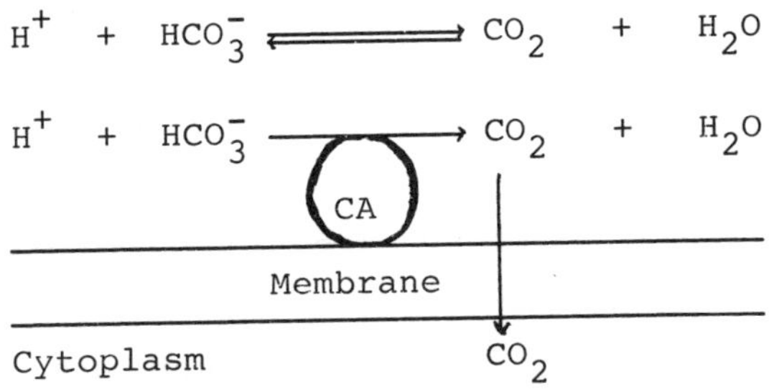

Fig. 2. Role of carbonic anhydrase in algal cells in utilization of HCO_3^- (15).

Accumulation of dissolved inorganic carbon (DIC) has been reported in a number of algae belonging to the classes Cyanophyceae, Chlorophyceae and Xanthophyceae (11). Reports are available which suggest that low-CO_2

cells of some species can accumulate DIC upto concentration 1000 times higher than that in the medium (12, see also 11).

Furthermore, the low-CO_2 cells of many microalgae have been shown to utilize DIC as efficiently as terrestrial C_4 plants. The presence of CA thus is required for lowering the apparent K_m (CO_2) for photosynthesis. The high DIC concentration inside the cell would obviously favour the rate of carboxylation reaction via RuBP carboxylase, and hence the lower apparent K_m (CO_2) value for photosynthesis. Enhancement of photosynthesis has been reported, even in higher plants, with the addition of CA. Shiraiwa and Miyachi (13) demonstrated multi-fold stimulation of $^{14}CO_2$ fixation by intact chloroplasts of spinach and _Bryopsis maxima_ in the presence of exogenously added CA. CA most probably causes accelerated conversion of HCO_3^- to CO_2, thereby enhancing the rate of $^{14}CO_2$ fixation.

The presence of both Na^+ ions and high pH in alkali soils is favourable for the high CO_2 accumulation within the microalgal cells. Abe _et al_. (14) reported that in the presence of Na^+ (30 meq) the rate of increase in internal DIC from added CO_2 increased in parallel with the increase in pH value from 7 to 9 whereas the same rate from added HCO_3^- did not change with change in pH. In the absence of Na^+ both rates decreased with increase in pH. Furthermore not only HCO_3^- but also CO_2 transport was stimulated by Na^+ and the enhancing effect of Na^+ on the transport of inorganic carbon was higher at higher pH in case of _Anabaena variabilis_.

Besides the above, the microalgae are also very efficient in CO_2 utilization. Kaplan _et al_. (12) observed high photosynthetic affinity to CO_2 in _A. variabilis_, in fact one of the highest reported for any organism, including C_4 species of higher plants.

The improved efficiency for CO_2 fixation observed appears to be a consequence of the fact that both low and high CO_2 cells accumulate inorganic carbon within the cells to concentrations which exceed that present in the medium (12). This permits more efficient activity of the RuBP carboxylase.

Coleman and Colman (16) demonstrated, by using a silicone fluid filtration technique, that the rate of accumulation of DIC inside the cells of the blue-green alga Coccochloris peniocystis was relatively constant over a wide pH range. At external DIC concentration of 0.56 to 0.89 mM, the internal concentration after 30 sec of illumination was greater than 3.5 mM over the pH range of 7.0 to 10.0. They also reported that at an alkaline pH where the intracellular pH is less than the external pH, the cell must rely upon active transport of HCO_3 and not a passive flux of CO_2 if inorganic carbon accumulation is to occur (17).

Though most of the algae present in alkaline/saline soils act as a significant sink by absorbing CO_2 for photosynthesis, N_2-fixing cyanobacteria appear to be even more important. The energetics of N_2-fixation in general show that this process requires large amount of reductant and ATP. The reductant in N_2-fixing cyanobacteria is supplied directly through photosynthesis. Thus the processes of N_2-fixation and CO_2 fixation in these organisms are interlinked. The property of N_2 fixation in a way forces these cyanobacteria to fix more CO_2 so as to drive nitrogenase reaction efficiently. The implication is that whereas green algae and other non-N_2-fixing algae fix CO_2 for simply meeting the requirement of less energy-consuming metabolic processes, some extra CO_2 may be fixed by N_2-fixing cyanobacteria to continue the process of N_2-fixation. Obviously the N_2-fixing cyanobacteria of alkaline/saline soils appear to play a key role in maintaining the nitrogen budget as well as fixing more CO_2 from the atmosphere.

It appears likely that CA is responsible for suppression of photorespiration in the microalgal cells. The CO_2/O_2 ratio inside the cells is one of the important factors which determine the relative ratio of photosynthetic CO_2 fixation to photorespiration. If some cooperation between CA and the CO_2 concentration mechanism brings about enhancement of CO_2 transport for RuBP carboxylase, it is possible that the presence of both factors suppresses photorespiration in cyanobacteria by increasing the CO_2/O_2 ratio in the cyanobacterial cytoplasm or the algal chloroplast stroma.

CA-deficient mutants and low-CO_2 cells of the green alga Chlamydomonas reinhardtii which had been treated with a CA inhibitor showed higher rates of glycolate production and higher O_2 inhibition. This indicates that CA may have a definite role in suppression of photorespiration, and that the CO_2 concentrating mechanism cannot work efficiently in isolation.

The overall assumption is that the algae having CA activity either inside or outside the cell membrane may play an important role in fixing CO_2 and augmenting the rate of photosynthesis.

REGENERATION OF ALKALINITY IN SOILS

In nature, there is some regeneration of alkalinity in usar soils. In areas having restricted underground drainage, salts which come through irrigation water or run-off fail to escape. The evaporation of water from the surface further facilitates concentration of salt in the soil. Capillary action during the summer months brings them to the surface where they form a white efflorescent crust. The downward movement of salts is much less than their upward movement with the result that salts tend to accumulate in high concentrations at or near the surface. These saline soils gradually deteriorate into alkali soils. Sodium salts enter the clay complex and form sodium clay by displacement of calcium. The above events go on continuously although the degree of alkalinity varies from year to year. Thus, no doubt these soils do absorb CO_2 and show decrease in pH but ultimately they regain the original characters in due course of time, by themselves, through physico-chemical processes.

Finally there are no reliable quantitative estimates of the extent to which wastelands or their algal cover absorb/fix CO_2 per unit area per unit time. This kind of data is badly needed and should be subjected to computer analysis for proper extrapolations. Global sink rates of CO_2 need to take the wasteland factor into account.

ACKNOWLEDGEMENTS

This work has been supported by research projects from the Department of Environment and Department of Science and Technology, Government of India. (Coordinated programme on the biology of alkalophilic microbes.)

REFERENCES

1. Wuebbles, D.J., Grant, K.E., Connell, P.S., and Penner, J.E. (1989). The role of atmospheric chemistry in climate change. JAPCA. 39: 22-8.

2. Krupa, S.V. and Kickert, R.N. (1989). The greenhouse effect: Impacts of ultraviolet-B (UV-B) radiation, carbon dioxide (CO_2) and ozone (O_3) on vegetation. Environmental Pollution 61: 263-393.

3. Detwiler, R.P. and Hall, C.A.S. (1988). Tropical forests and the global carbon cycle. Science 239: 42-47.

4. Schneider, S.H. (1989). The greenhouse effect: science and policy. Science 243: 771-781.

5. Kuo, C., Lindberg, C. and Thomson, D.J. (1990). Coherence established between atmospheric carbon dioxide and global temperature. Nature 343: 709-714.

6. Siegenthaler, U. (1990). El Nino and atmospheric CO_2. Nature 345: 295-96.

7. Bjorkman, O. (1989). Some viewpoints on photosynthetic response and adaptation to environmental stress. In "Photosynthesis". Alan R. Liss, Inc., pp. 45-58.

8. Singh, R.N. (1961). Role of Blue-green algae in Nitrogen economy of Indian Agriculture. Ind. Council of Agric. Res., New Delhi. pp. 83.

9. Lerman, A. and Stumm, W. (1989). CO_2 storage and alkalinity trends in lakes. Wat. Res. 23: 139-146.

10. Hogetsu, D. and Miyachi, S. (1977). Effects of CO_2 concentration during growth on subsequent photosynthetic CO_2 fixation in Chlorella. Plant Cell Physiol. 18: 347-352.

11. Aizawa, K. and Miyachi, S. (1986). Carbonic anhydrase and CO_2 concentration mechanisms in microalgae and cyanobacteria. FEMS Microbiology Reviews 39: 215-233.

12. Kaplan, A., Badger, M.R. and Berry, J.A. (1980). Photosynthesis and intracellular inorganic carbon pool in the blue-green alga Anabaena variabilis in response to external CO_2 concentration. Planta 149: 219-226.

13. Shiraiwa, Y. and Miyachi, S. (1978). Forms of inorganic carbon utilized for photosynthesis across the chloroplast membrane. FEBS Lett. 95: 207-210.

14. Abe, T., Tsuzuki, M., Kaplan, A., and Miyachi, S. (1987). Transport and fixation of inorganic carbon during photosynthesis in the cells of Anabaena grown under ordinary air. II. Effect of sodium concentration during growth on the induction of active transport system for inorganic carbon. Plant Cell Physiol. 28: 671-677.

15. Miyachi, S., Tsuzuki, M., and Yagawa, Y. (1985). Carbonic anhydrase in various microalgae. In: Inorganic carbon uptake by aquatic photosynthetic organisms. W.J. Lucas and J.A. Berry (eds.). pp. 145-154.

16. Coleman, J.R. and Colman, B. (1981). Inorganic carbon accumulation and photosynthesis in blue-green algae as a function of external pH. Plant Physiol. 67: 917-921.

17. Badger, M.R., Kaplan, A. and Berry, J.A. (1978). A mechanism for concentrating CO_2 in Chlamydomonas reinhardtii and Anabaena variabilis and its role in photosynthetic CO_2 fixation. Carnegie Inst. Year Book 77: 251-261.

ASSESSMENT OF LIMITATIONS FOR PHOTOSYNTHETIC RATE IN PLANTS ACCLIMATIZED TO ELEVATED CO_2 AND TEMPERATURES

M.UDAYA KUMAR and T.G.PRASAD

Department of Crop Physiology,
University of Agricultural Sciences,
GKVK Campus, BANGALORE 560065

To enhance the photosynthetic rate in plant species, it is essential to asses the limitations imposed for assimilation rates (A) by different photosynthetic traits and the genetic control of this traits.

The relative importance of these limitations vary depending upon the environmental factors. Since, the CO_2 is the major substrate, the predicted increase in global CO_2 may become an important environmental variable altering the limitations which may affect the potential carbon exchange rate of the species. Therefore, the major objective is to examine the relative limitations to 'A' at elevated CO_2 and the temperatures.

Any change in the ambient CO_2 concentration will have profound effect on photosynthetic characteristics. Not only the effect of elevated CO_2 on gas exchange rates, but also, the comprehensive status of the problem with regards to its effect on water use efficiency, leaf expansion, canopy development, allocation of photosysnthetic products, seedling growth rates and reproductive aspects have been critically reviewed by Leeman, (1983) and Eamus and Jarvis, (1989).

The increase in growth rates due to CO_2-fertilization has been unequivocally proven in several container studies in controlled experiments and in open top chamber studies under field conditions. Increase in biomass could be due to increase in assimilation rate or canopy photosynthesis or decrease in photorespiration. However, the response has been shown to be species specific. One of the reasons for the differential response is lack of maintenance of higher assimilation rates in long term experiments.

The plants are always CO_2 limited for photosynthesis. In all short term experiments when the leaves were exposed to elevated CO_2 levels for a short period, increase in photosynthetic rate has been clearly demonstrated. In long term experiments when the plants were continuously exposed to enriched CO_2 levels, the results have been contradictory.

Increase in assimilation rate at higher CO_2 concentration in long term experiment has been observed in several plant species by Havelka et al. (1984), Ocurson et al. (1984) Campbell (1987, 1988), Thomas and Harvey (1985), Porter et al. (1988), Nijs et al. (1989a, 1989b). In some crop species increased assimilation rates were maintained althrough the growth period and in such cases, a significant increase in growth rates were recorded. One of the factors attributed was an enhanced sink number and activity when plants were exposed to elevated CO_2 levels.

In contrast, a decrease in photosynthetic rate in plants grown at higher CO_2 concentration and measured at growth CO_2 concentration has been reported in water hyacinth (Spancer and Bowees, 1986), in cotton (Sivak et al., 1983), in cucumber (Peet et al., 1986), in desmoodium (Wulf and Strain, 1987) in a few forest species (Oberbaur et al., 1986) in Trifolium (Cave et al., 1981) in pinus seedling (Conroy et al., 1986a).

The absolute reduction in 'A' in plants exposed to elevated CO_2 for a long period is caused by the following factors.

1. End product inhibition - Accumulation of end products - starch and sucrose
2. Limitation due to inadequate Pi recycling.
3. Decrease in content and activation of RuBisCO.

End product: End product inhibition of photosynthesis has been attributed as a major reason (Azconbit and Osmond, 1983). Starch accumulation has been stated as a major reason for inhibition in several studies (Peet et al. (1986), Spencer and Bowees (1983) Wulf and Strain (1982) Cave et al. (1981) Oberbeur et al. (1986).

Several reasons are attributed for the observed decrease in 'A' due to starch accumulation. A reduction in chlorophyll content and grana formation due to starch accumulation has been reported by Wolf and Strain (1982), Cave et al. (1981). The other inhibitory effects of starch accumulation was reduction in light reaching the chloroplast (Waraned and Wilson, 1986) or increase in diffusive pathway length or interference with CO_2 transfer due to large starch grains. Direct damage to the chloroplast caused by starch granules has been clearly demonstrated by Nofziger and Koller (1976). Physical distortion of chloroplast due to starch grains are commonly observed in plants grown at high elevated CO_2 (Caves et al., 1981; Mauney et al., 1979) leading to suggestion that the decline in 'A' at high CO_2 results from either feed back inhibition or physical damage to chloroplast (Delusia et al., 1985).

Pi recycling: Inadequate recycling of inorganic phosphate limits the asimilation rate at elevated CO_2 levels even in short term experiments (Stitt, 1986). When the capacity to regenerate Pi declines the rate of electron transport also declines (Sharkey et al.,1989). It is well established that the rate of RuBP production and consumption and rate of triose phosphate utilization has been the major factor determining the extent of limitation imposed by inadequate Pi recycling. Therefore lack of utilization of end products, starch and sucrose leading to their accumulation rermains a major factor for inadequate Pi recycling.

RuBISCO

In plants aclamatized to high CO_2 concentrations in long term experiments, a decrease in RuBISCO activity has been shown in number of crops (Downton et al., 1987; Peet et al., 1986; Yelle et al., 1989). Wether, this is due to rerduction in specific activity of the enzyme or due to reduction in RuBISCO protein is not known. However, indirect measurement of RuBISCO protein by binding with [^{14}C] carboxyarabinitol biphosphate indicated that there was loss of RuBISCO protein in cabbage at elevated CO_2 (Sage, 1989). Not only RuBISCO content but also a decrease in its activation state also has been shown in plants acclamatized to high CO_2 (Sege et al., 1989).

The decrease in RuBISCO content in plants grown at elevated CO_2 has been attributed to reallocation of N to the other enzyme systems because carboxylation is favoured through high substrate levels. Under such conditions, the rate of utilization of end products may be higher and inorganic Pi recycling may not impose any limitations. But it is not clear why a reduction in the state of activation was observed with a reduction in the RuBISCO content.

It can be concluded that at high elevated CO_2 concentration the lack of maintenance of higher assimilation rate is due to reduction in Pi recycling caused by end product accumulation and also decrease in RuBISCO content and its activation state. Such a response was also seen in plants with rerstricted growth of sink. There is increasing evidence that maintenance of active sink for assimilates is necessary for the effective stimulation of photosynthesis at elevated CO_2 concentration.

Approach

The crop plants will be benefited to the predicted increase in global CO_2 concentration only when the observed short term responses of asimilation rates were maintained continuously. Though a number of factors discussed above has been attributed as major reasons, there is need to focus our attention on the following three aspects.

(1) Is really the sink limitation a constraint?

The maintenance of greater sink size is implicated as a prerequisite to maintain and continue the short term effect in assimilation rate. It is not always true that genotypes or species with higher sink capacity always maintins high assimilation rates. Therefore more intricate approaches needs to be adopted to assess the importance of sink size and its growth rates on assimilation rate at high substrate levels.

(2) Are there any specific constraints for sucrose biosysnthesis and its export?

Increased accumulation of starch is a common phenomena at elevated CO_2 levels. This could be due to constraint, in sucrose synthesis and export. Though the feedback control on the rate of utilization of triose phosphate is clearly demonstrated, the primary factors responsible for this is not understood so far.

(3) Is the observed decrease in RuBISCO content, activation and change in kinetic characteristics is a dirերct or indirect effect of elevated CO_2 concentration?

In non-responsive species at elevated CO_2 concentration, a deffenite reduction in RuBISCO activity has beend shown. Though the activation state of the enzyme is affected, te major factor atributed to the decreased activity seems to be due to reduction in RuBISCO protein content. However, the adoptive significance of lesser allocation of protein to this enzyme is not clear. Moreover, a substantial decrease in RuBISCO content is not associated with high assimilation rate nor showed an increase in activation state.

At higher substrate level more than Km (CO_2), specific activity, K(cat) assumes importance. Therefore species/ genotypes with high specific activity may be more benefited at high CO_2 levels.

There is a need to estimate the direct and indirect effect of elevated CO_2 concentration on RuBISCO protein content and kinetic characteristics. Whether a brief exposure to elevated CO_2 levels can alter the kinetic characters and the factors which influences the regulation of differential allocation of proteins to RuBISCO with change in CO_2 levels need greater indepth studies.

Temperature

The predicted changes in CO_2 is also associated with increase in global warming though the magnitude of change is debatable. Increase in temperature affects gas exchange traits and also alters dark respiraton, and growth rates of sink. Elevated temperature may have a direct effect on (a) specificity factor - photorespiration, (b) sucrose metabolism - its rate of synthesis and export and (c) indirect effect on gas exchange traits, through enhanced growth rates of sink.

At elevated temperature decrease in specificity factor and thus enhanced CO_2 loss through photorespiration has been well documented (Brooks and Farquhar, 1985; Jordon and Ogren, 1984). At higher CO_2 the changes that occur in photorespiratory when specificity factor known to decrease at high temperature could be species specific and needs to be investigated.

One of the reasons attributed for enhanced assimilation rates at higher temperature and CO_2 is due to decrease in limitations imposed by inorganic Pi recycling. The increase in sucrose metabolism has been implicated for observed decrease in limitations of Pi recycling (Stitt, 1986).

Approach

In long term experiments there is need to examine the combined effect of temperature and CO_2 to answer the following three aspects.

(1) Is the beneficial effect of elevated CO_2 is nullified by decrease in specificity factor?

(2) Is enhanced sucrose metabolism and thus high Pi recycling is maintained in long term experiments?

(3) What would be the indirect effect of enhanced growth rates of sink on assimilation rate?

These aspects have greater relevance in tropics with high temperature and light intensity.

REASONS FOR DECREASE IN 'A' IN LONG TERM CO_2 ENRICHMENT CONDITIONS

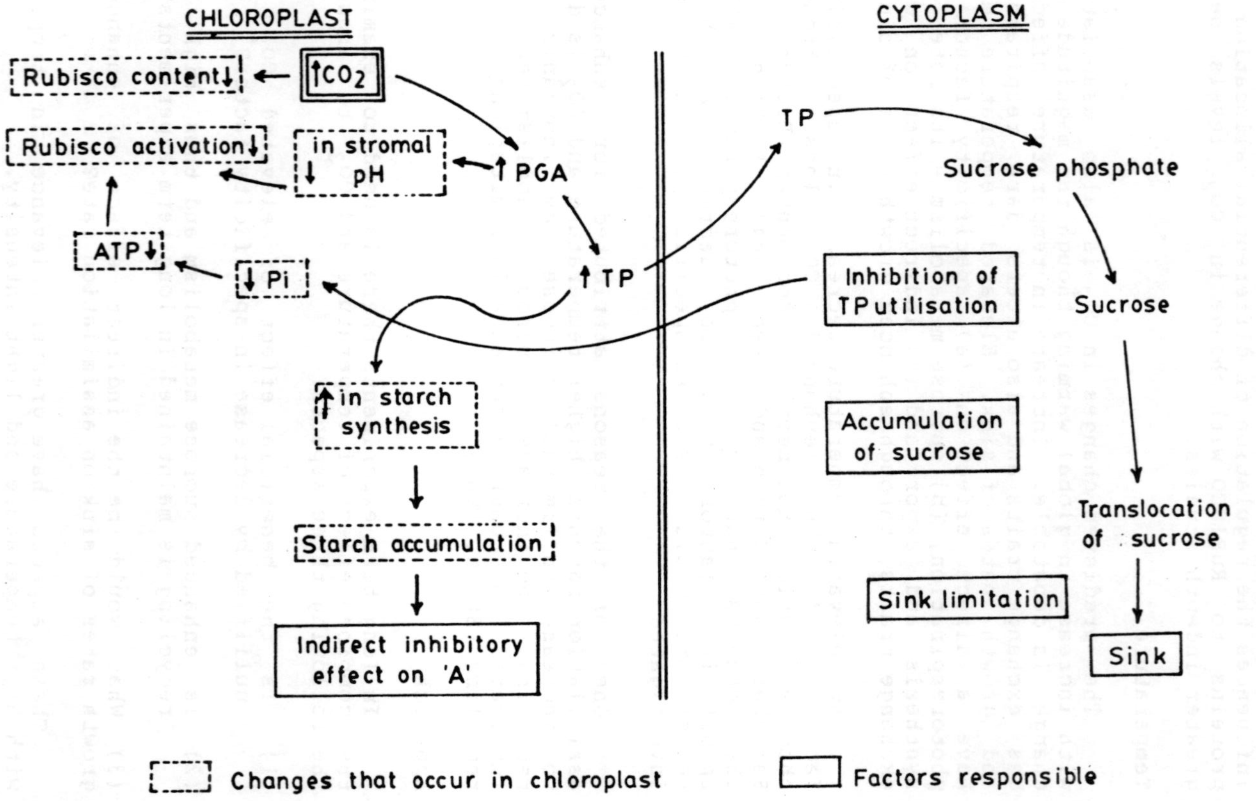

REFERENCES

Azcon-Bito, J. and Osmond, C.B., 1983. Plant Physiol,71: 574-581.

Brooks, A. and Farquhar, G.D., 1985, Planta, 165: 397

Cambell, W.L., Allen, L.H.Jr. and Bowes, G., 1988. Plant Physiol., 88: 669-688.

Campbell, W.J., Allen, L.H. and Bowes, G., 1987. Progress in Photosynthesis Research, IV: 253-256.

Cave, G., Tolley, L.C. and Strain, B.R., 1981. Physiol Plant, 51: 171-174.

Conroy, J., Barlow, E.W.R. and Bavage., 1986. Annal. Bot., 57: 165-177.

Downton, W.J.S., Grant, W.J.R. and Loveys, B.K., 1987. Aust. J. Plant Physiol, 14: 493-5Ø1.

Eamus, D. and Jarvis, P.G., 1989. Advances In Ecological Research, 19: 1-55.

Havelka, U.D., Ackerson, R.C., Boyle, M.G. and Ulitenbach, V.A., 1984. Crop Sci, 24: 1146-115Ø.

Jordon, D.B. and Ogren, W.L., 1984, Planta, 161: 3Ø8.

Lemon, E.R., 1983, AAAS Selected Symposium. pp.26-32.

Mauney, J.R., Guinn, G., Fry, K.E. and Hesketh, J.D., 1979. Photosynthetica, 13: 26Ø-266.

Nijs, I., Impens, I. and Beheaegha, T., 1989. Planta, 177: 312-32Ø.

Oberbauer, S.F., Sionit, N., Hastings. and Oechel, W.C., 1986. Can. J. Bot, 64: 2993.

Peet, M.M., Huber, S.C. and Patterson, D.T., 1986. Plant Physiol, 8Ø:63-67.

Porter, M., Pot, S. and Lambers, H., 1988. Physiol. Planta, 73: 312-32Ø.

Sage, R.F., Sharkey, T.D. and Seeman, J.R., 1989. Plant Physiol, 89:59Ø-596.

Sharkey, T.D., Berry, J.A. and Sage, R.F., 1989. Planta In press.

Spencer, W. and Bowes, G., 1986. Plant Physiol, 82: 528-533.

Stitt, M., 1986. Plant Physiol, 81: 1115-1122.

Stitt, M. and Quick, P., 1989. Physiol. Plant, 77: 633-641.

Thomas, J.F. and Harvey, C.N., 1983. Bot. Gaz, 144: 303-309.

Wulf, R.D. and Strain, B.R., 1982. Can. J. Bot, 60: 1084-1091.

Yelle, S., Beeson, R.C., Trudel, M.J. and Gosselin, A., 1986. Plant Physiol., 90: 1465-1472.

THE RESPONSE OF LEAF PHOTOSYNTHESIS TO ELEVATED CO_2

A. Laisk and V. Oja, Institute of Astrophysics and Atmospheric Physics, Estonian Acad. Sci., 202444, Tõravere, Estonia, U.S.S.R.

ABSTRACT

Plant responses to increasing CO_2 levels may be classified as short-term and long-term. Long-term responses are more complex as they include ecosystem- and organism-level responses also, while short-term responses may be understood considering leaf-level processes only. The active CO_2 concentration for photosynthesis is that at the active sites of the carboxylation enzyme. It is lower than ambient due to the concentration drop on the leaf diffusion resistances. Under atmospheric conditions RuBP carboxylase in C_3 plants operates at CO_2 concentrations below the $K_m(CO_2)$. Under saturating CO_2 the rate is limited by the supply of the CO_2 acceptor ribulosebisphosphate (RuBP). The latter is resynthesized in the Carbon Reduction Cycle (CRC). The rate of RuBP resynthesis is determined by: (1) concentrations of CRC intermediates, (2) CRC enzyme activities, and (3) levels of energy cofactors NADPH and ATP. Maximum concentration of CRC intermediates is determined by the inorganic phosphate (P_i) pool in chloroplasts which is kept constant. An increase in intermediate organic phosphates causes a decrease in free P_i. CRC enzymes are activated by the ferredoxin(Fd)-thioredoxin system which needs higher levels of reduced Fd for more enzyme activation. NADPH level is efficiently stabilized by the allosteric inhibition of the Fd-NADP reductase by NADPH. ATP/ADP ratio is determined by the proton gradient and free P_i. Photosystem II efficiency is controlled by the intrathylakoid pH. The maximum possible rate of leaf photosynthesis at saturating CO_2 and light levels is achieved as a compromise between the controversial conditions. High ATP levels are necessary to support high CO_2 assimilation rates. At high assimilation rates the concentrations of CRC intermediates increase and that of free phosphate decreases. Accumulation of organic phosphates is emphasized when sucrose synthesis is restricted. In order to maintain high ATP with decreasing P_i, transthylakoid pH difference increases, the intrathylakoid pH, consequently, decreases. Decreasing intrathylacoid pH causes downregulation of the photosystem II efficiency which decreases the electron transport rate. In a steady state, the rate of electron transport stoichiometrically corresponds to the rate of ATP synthesis at an ATP/ADP ratio which is sufficiently high to drive the carbon metabolism at a rate equal to the electron transport rate. Concurrently, the level of free phosphate is sufficiently low to activate sucrose synthesis to the level corresponding to the carbon assimilation and electron transport rates. In spite of almost a complete rate-control by PSII there is still enough reduced Fd to guarantee the necessary activation state of the CRC enzymes.

INTRODUCTION

Continuing deforestation and fossil fuel burning is shifting the carbon dioxide budget of the atmosphere towards an increase in CO_2 concentration. How does the biosphere respond to the increasing CO_2 level? It is a complicated question since, along with the direct effect of CO_2 on assimilation, possible indirect influences must also be considered. Even the direct effect of CO_2 on plants depends on the presence or absence of other growth-limiting factors besides the supply of carbohydrates. A short-term response of leaf photosynthesis to CO_2 is the easiest to analyse and even here we shall meet a number of complicated interactions. Nevertheless, a correct understanding of the primary effects of elevated CO_2 on leaf photosynthesis is a prerequisite for understanding what may happen at the plant, canopy, ecosystem and biosphere levels.

Carbon dioxide is a chemical substrate for photosynthesis while light is an energy cofactor driving the assimilation process. It is evident that at high light intensities the effects caused by elevated CO_2 concentration are more pronounced than at low light intensities since light determines a certain possible rate of photosynthesis which cannot be exceeded by improving the substrate supply. Temperature places a limit for the reaction rates, therefore, at low temperatures there will be little effect of increasing both, light intensity and CO_2 concentration. Limited supply of minerals will limit the synthesis of photosynthetic enzymes which is somewhat similar to the effect of low temperature. Low mineral supply, as well as low temperature or water stress may cause growth restrictions in the whole plant which sooner orater will be signaled to leaves and their photosynthesis regulated down, despite CO_2 concentration or light intensity. Evidently, elevated CO_2 has quite a limited chance to increase photosynthesis. But this conclusion is based on a simplified scheme of separated limiting factors assuming that when one factor is limiting, others cannot release the limitation. Reality is more complicated. CO_2 may release limitations caused, for example, by water deficit or low mineral supplies. It is also important to consider numerous regulatory phenomena which tend to stabilize the operation conditions of the photosynthetic machinery, so that under the wide variations of environmental factors the status of intermediate and energy metabolites (phosphate esters, ATP, NADPH etc.) varies as little as possible.

ACTIVE CO_2 CONCENTRATION. THE ROLE OF STOMATA.

In C_4 plants and in some algae CO_2 is actively transported to the CO_2-binding enzyme ribulosebisphosphate carboxylase (RuBPC-ase). In these plants the active CO_2 concentration may be higher than in the external air and the kinetics of CO_2 assimilation with respect to CO_2 are largely determined by the CO_2 transporting mechanism. In C_3 plants CO_2 is transported by diffusion and, therefore, during photosynthesis its concentration at the active sites of the carboxylation enzyme is lower than in the external air. There are several diffusion resistances on the pathway of CO_2, like leaf boundary layer, stomata, intercellular air spaces, cell walls and chloroplast stroma. Most of them are approximately constant for a given leaf in a short timescale, except stomata. Stomata are the entrance doors for the leaf gas exchange. With water deficiency, water economy is of primary importance and then stomata control their apertures according to the allowed rate of transpiration. If water is available, stomata regulate their apertures according to the demand of the leaf for CO_2. For instance, they tend to close at low light intensities at the normal atmospheric CO_2 concentration, at elevated CO_2 they close even during high light intensities. In both cases the aim of the regulation is to keep the intracellular CO_2 concentration more or less constant, nearly limiting for photosynthesis despite its surplus in the ambient air. Does it mean that stomatal closure tends to neutralize the effect of elevated ambient CO_2 by stabilizing the intracellular CO_2 concentration? The answer is yes if the photosynthetic machinery is not able to handle the unexpectedly high assimilation rates; but the answer is no if the machinery has a capacity that exceeds the rate at normal atmospheric CO_2 [25]. In the latter case the photosynthetic rate will increase to a value determined by the maximum capacity of the machinery. As explained above, this state may be temporary if the plant cannot utilize the increased rate of carbohydrate synthesis, or more steady if the growth processes have a correspondingly high capacity.

As we see, the maximum internal capacity of the photosynthetic machinery is an important parameter determining the response of the leaf photosynthesis to elevated CO_2.

THE ROLE OF RuBP CARBOXYLASE

RuBPC-ase catalyses the reaction of a 5-carbon sugar bisphosphate ribulosebisphosphate (RuBP) with CO_2 to form two molecules of 3-phosphoglycerate (PGA). The reaction is practically irreversible. Therefore, the rate of CO_2 assimilation in C_3 plants is always determined by the kinetic properties of the RuBP carboxylase on one hand, and CO_2 and RuBP concentrations at the reaction sites on the other hand. Accordingly, factors determining the CO_2 uptake rate may be divided into two groups: (a) kinetic factors controlling the concentrations of substrates, and (b) regulatory factors controlling the enzyme properties.

RuBPC-ase is activated by CO_2 and Mg^{2+} by carbamylation [22] and in the absence of 2-carboxyarabinitol 1-phosphate the carbamylation state mainly determines the relative activity of the enzyme [4]. In the activated state its K_m for CO_2 is about 10 - 15 µM [50] and about 15 - 30 µM for RuBP [23, 51], values up to 100 - 150 µM have been reported, [32]. The number of operating active sites is controlled by means of a tightly binding competitive inhibitor, 2-carboxyarabinitol 1-phosphate [17]. An ATP-dependent enzyme RuBPC-ase activase is involved in regulation of the activity of RuBPC-ase [30, 31]. A specific feature of the carboxylase s that the concentration of its active sites in chloroplast stroma is very high, 4 - 6 mM [15]. At low CO_2 levels high concentrations of RuBP up to 10 -20 mM may accumulate in the chloroplasts [2] which are believed to saturate the carboxylase with respect to this substrate. At RuBP-saturation the kinetic curve for CO_2 of the carboxylase is a Michaelis-Menten type rectangular hyperbola with half-saturation at 10-15 µM and practical saturation at about 40-60 µM. Atmospheric CO_2 concentration of 330 µl/l corresponds to about 13.5 µM in the gas phase at 25°C. Due to stomatal and other diffusion resistances the concentration in the leaf intercellulars is about 7 - 9 µM when stomata are normally open. At 25°C the CO_2 solubility constant in water is 0.83. There is also an additional small diffusion resistance in cell walls and chloroplast stroma. The resulting CO_2 concentration at the carboxylation sites is about 5 - 7 µM (5.2 µM in cotton mesophyll cells averaged over the vegetation period, [33]). Therefore, under normal atmospheric CO_2 concentration the carboxylase in leaves is functioning below its K_m for CO_2. In an assumption that RuBP remains saturating for the carboxylase at any CO_2 concentration, the rate of carboxylation might be increased by four to six times, increasing the external CO_2 concentration to 60 - 80 µM (1700 - 2000 µl l^{-1}).

Unfortunately, the assumption of constant RuBP is not true. Figure 1 shows an experiment carried out with a sunflower leaf in a rapid-response gas exchange measurement system [26]. The equipment has a response time of 2.2 seconds which allows us to monitor rapid transients in the

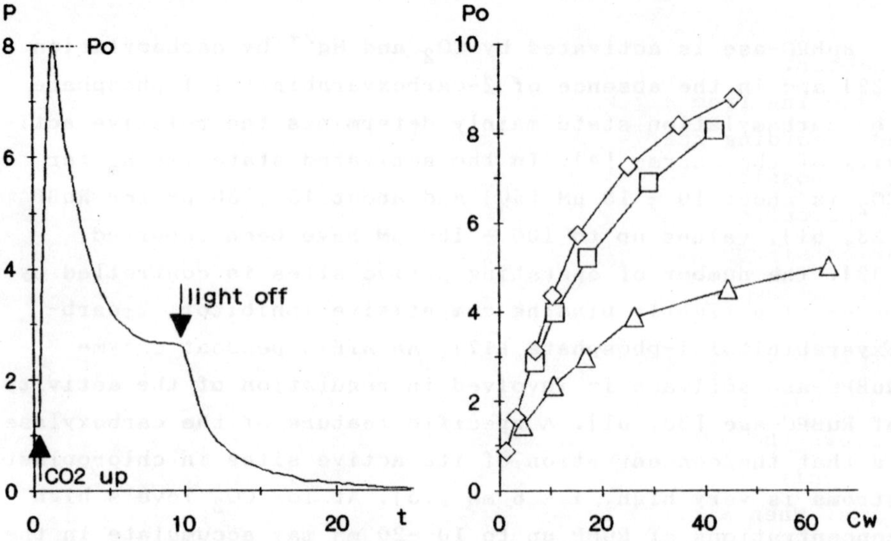

Fig. 1 (left). The recording of CO_2 uptake rate P (nmol cm^{-2} s^{-1}) in a sunflower leaf after a transition from 227 to 1907 µl l^{-1} CO_2 in N_2. Light I = 34 mW cm^{-2}, leaf temperature t_l=22°, time t in s.

Fig. 2 (right). Initial rates of CO_2 uptake after a transition from a low to different high CO_2 concentrations P_o (Fig. 1) as functions of the intracellular CO_2 concentration C_w (nmol cm^{-3}) at the maximum P_o. The three curves were measured at different temperatures: diamonds, t_l=33°, V_m=12.9 nmol cm^{-2} s^{-1}, K_m=21 nmol cm^{-3}; squares, t_l=24°, V_m=11.7, K_m=20.7; triangles t_l=14°, V_m=5.9, K_m=14.5.

leaf gas exchange. In this experiment the photosynthetic rate of a sunflower leaf was stabilized at a limiting CO_2 concentration at 1% O_2. Under these conditions the RuBP pool was big enough to saturate the carboxylase. Now the CO_2 concentration was rapidly increased to 2000 µl l^{-1}. Immediately a high CO_2 uptake rate was recorded which decreased in about 5 seconds to a much lower value (Fig. 1). A small part of the CO_2 uptake burst was caused by the solubilization of CO_2 in the chloroplast stroma [27], but the major part of it reflected the rapid carboxylation of RuBP. Evidently, the RuBP pool was not stable but rapidly decreased at the high CO_2. Switching the light off we could see a postillumination CO_2 uptake which reflected the remaining pool of the CO_2

acceptor.

Jumping from a low to different higher CO_2 concentrations and recording the initial rates of CO_2 uptake as in Fig. 1 it is possible to obtain clean kinetic CO_2 response curves of the carboxylase at a constant high RuBP level. Such curves are shown in Fig. 2. They are rectangular hyperbolae with V_m and K_m depending on temperature. The maximum rates V_m are much higher than the maximum steady photosynthetic rates of the leaf at the same temperatures are. This experiment shows us that the steady CO_2 uptake rate of the leaf follows the kinetic curve of the carboxylase at saturating RuBP only in a range of very low CO_2 (and O_2) concentrations. When a certain carboxylation rate is achieved, the stable RuBP concentration cannot be maintained but the higher the CO_2 concentration,the more it declines. At high CO_2 concentrations the rate of RuBP supply to the carboxylase is rate-determining, the RuBP concentration adjusts authomatically to the level which guarantees the equality of the carboxylation rate to the rate of RuBP synthesis [19, 20].

RuBP RESYNTHESIS IN THE CARBON REDUCTION CYCLE (CRC)

In the carboxylation reaction CO_2 is bound to a 5-carbon sugar bisphosphate and, as a result, two 3-carbon organic acid phosphates are formed. In CRC the 3-carbon organic acids are at first reduced to the carbohydrate level with the help of energy-rich compounds ATP and NADPH generated in the light reactions of photosynthesis. Then the net gain of carbon is exported from the chloroplasts to produce sucrose and the rest is rearranged to form RuBP again. Evidently, one from the six carbons reduced has to be exported and five used for RuBP resynthesis in order to guarantee stable operation of the cycle. As a 3-carbon moiety triose phosphate (T3P) is the smallest unit available for export, an integer stoichiometry is obtained after three turns of the cycle:

$$\underset{\text{RuBP}}{3\text{x}5} + \underset{CO_2}{3\text{x}1} = \underset{\text{PGA}}{6\text{x}3} \xrightarrow{\text{ATP,NADPH}} \underset{\text{T3P}}{6\text{x}3} - \underset{\text{export}}{1\text{x}3} = 5\text{x}3 \xrightarrow[\text{RuBP resyn.}]{\text{ATP}} \underset{\text{RuBP}}{3\text{x}5}$$

RuBP resynthesis proceeds via a network of reactions [6] in the last step of which ATP is used to convert ribulose monophosphate into RuBP. The rate of RuBP resynthesis depends on: (a) the concentrations of metabolites themselves as reaction substrates, (b) the activities of the enzymes catalysing the reactions, and (c) the concentrations of energy cofactors ATP and NADPH.

<u>CRC Intermediates.</u> All the CRC intermediates are phosphorylated sugars or organic acids with either one or two phosphate groups bound to their carbon skeletons. Therefore, the total concentration of CRC intermediates cannot exceed the total concentration of inorganic phosphate in the chloroplast. In a short timescale the latter may be considered a constant. There is a special enzyme in the chloroplast membrane, a phosphate translocator [49], which is responsible for the export of triose phosphate (and PGA, to some extent) from the chloroplast to the cytosol. But it functions like a merchant selling triose phosphate for a free inorganic phosphate (actually, it exchanges P_i, T3P and PGA^{2-} mutually in both directions). As a result of this, the total amounts of phosphate in the chloroplast and in the cytosol remain constant (the exchange of P_i between cytosol and vacuole is slow). This places an upper limit to the concentrations of photosynthetic intermediates in both compartments of the cell. It also opens a principle possibility to control the photosynthetic rate of leaves by the plant.

In photosynthesis, phosphate turns over between free and sugar-bound forms:

```
RuBP + CO2-->PGA -----> T3P -----------> RuBP
  Pi + ADP ------>ATP ---|----- ATP
  |                      |
  L----------------------|---------------------- Pi
Chloroplast              |                       |
=================================================
Cytosol                  |         SPS           |
                     T3P --> HeP --> Sucrose + Pi
```

In chloroplast P_i is bound to ADP forming ATP in photophosphorylation. The CRC is driven with the help of ATP and phosphate is transferred to CRC intermediates. Triosephosphates T3P are exported to the cytosol where, as a result of a series of transformations via hexosephosphates HeP, sucrose is formed and phosphate is released. From the cytosol P_i returns into the chloroplast in exchange of a triosephosphate. When the sucrose phosphate synthase (SPS) activity is downregulated [13, 14] and the rate of sucrose synthesis is low (for example, when the growth is suppressed and the plant cannot utilize all the sucrose produced in leaves), P_i is released at a restricted rate and at the same rate it can be imported into chloroplasts, incorporated into ATP and used in the CRC. It must be emphasized that the control of photosynthesis by sucrose synthesis is not absolute. When sucrose synthesis is restricted, the levels of CRC intermediates rise since export is suppressed. As a result of this, a pathway of starch synthesis in chloroplasts is activated and storage of carbohydrates continues. However, when the rate of photosynthesis increases as a result of an elevated CO_2 concentration, the CRC intermediates build up and the level of free phosphate, correspondingly, decreases [42]. In parallel with this the carbamylation of the RuBP carboxylase decreases [5] which allows higher RuBP concentrations to build up and store more P_i [41]. This leads to a decline in an equilibrium ATP/ADP ratio which is important to consider as a factor determining the maximum rate of photosynthesis. At maximum rates of photosynthesis, the CRC intermediates are at the maximum and free phosphate is at the minimum, influencing the rate of assimilation through the conditions for ATP synthesis (transthylakoid proton gradient) and the level of ATP.

CRC Enzymes. Several CRC enzymes catalysing PGA reduction and RuBP regeneration are controlled by reducing power generated in the light reactions of photosynthesis which act through the ferredoxin-thioredoxin enzyme control system [1, 3]. CRC enzymes glyceraldehyde-dehydrogenase, fructosebisphosphatase, sedoheptulosebisphosphatase and phosphoribulokinase plus not CRC enzymes NADP-malatedehydrogenase and ATPase (coupling factor), are shown to be activated in the light. In the inactive form these enzyme proteins contain thiol groups (- S - S -) which are reduced into sulphydryl groups (-SH SH-) during activation. The kinetics of activation is modulated by the presence of several CRC intermediates [37]. Reducing electrons come from ferredoxin, which is the primary output compound of the light reactions of photosynthesis, and are transferred to thioredoxins by the Fd-thioredoxin reductase FTR, and thioredoxins interact with enzymes:

```
Light     Light
  |         |
PSII--e-->PSI --> Fd⁻FNR------e----->NADPH
  |                 |                  |
  |                 |  RuBP+CO2-->PGA------->T3P------>RuBP
  e                FTR                 |               |
  |                 └---Thioredoxins----e-----------e
H2O                                     Activation
```

This scheme shows that reducing power is double-used in photosynthesis, both as a substrate and as a regulatory factor [11]. At low light intensities electrons do not move readily to PGA but at first considerable electron pressure has to be created before the enzymes become activated to the necessary extent. The same is true for the phosphorylation potential: considerable electrochemical and proton gradients are generated at low light because ATP synthesis and CO_2 assimilation are restricted by inactive enzymes. Quite clearly, such a regulatory system guarantees the build-up of relatively high redox and phosphorylation potentials in the cell already at low light intensities. In this way the conditions for cell metabolism are stabilized in a variable light environment. But in this case a question arises, what determines the rate of CO_2 assimilation at limiting light? It is not CO_2, NADPH or ATP because all these are in surplus. It is the CRC enzyme activation state. But the latter is automatically adjusted to a level which is just high enough to process the NADPH and ATP fluxes generated by light reactions at the given light intensity. Therefore, the photosynthetic rate still remains light-determined. A question arises whether the continous maintenance of the high metabolic resistance in the CRC causes an inevitable loss of photons, thereby decreasing the quantum efficiency of photosynthesis? Looking at the scheme above we see that ferredoxin needs to be more reduced in order to more activate the CRC enzymes. In principle, it causes also a higher reduction level of the PSI primary acceptor which may decrease the charge separation efficiency at PSI.

CRC enzymes are playing an important regulatory role at limiting light intensities, directed to stabilization of the energy cofactors in the cell. At high light and saturating CO_2 their activation level must be high corresponding to the high assimilation rate. Some kinetic data (oscillations of photosynthetic rate at saturating CO_2 concentrations) show that the maximum chemical capacities of the CRC enzymes still exceed the actual maximum metabolic fluxes and even at the maximum steady rates of photosynthesis the CRC enzymes still stay under the control.

Energetic Metabolites ATP and NADPH. These, especially ATP, are important factors determining the rate of CRC turnover at saturating CO_2 levels. The inability to generate higher ATP concentrations is one of the reasons which stop a further increase of the photosynthetic rate at saturating light and CO_2. At the same time the absolute levels of ATP may not be higher than at lower CO_2 uptake rates at lower CO_2 concentrations, since the metabolic resistance in CRC is now very low due to high activation state of the enzymes and high CO_2 concentration. According to experimental data, NADPH level is relatively stable, even somewhat increasing towards lower light intensities [45]. This is achieved by means of a very efficient control of the activity of ferredoxin-NADP reductase (FNR). This enzyme is extremely sensitively product-inhibited by the absolute concentration of NADPH [21]. Looking at the scheme above one can see that such a control is necessary to prevent overreduction of the NADP system due to the considerable difference in the standard redox potentials of Fd^-/Fd and NADPH/NADP systems (-414 and -320 mV correspondingly). Sensitive control of the enzyme keeps the NADP system moderately reduced despite variations in the redox state of Fd. Therefore, considering NADPH/NADP as one component of the driving force of the CRC, we see that it is relatively stabilized due to the feedback inhibition of the FNR. Evidently, the other component of the driving force, ATP/ADP ratio, must change parallel to changes in photosynthesis, of course, only as much as the metabolic resistance of the CRC enzymes may be considered constant. The latter tends to be regulated so that the phosphorylation state of the adenylate system varies less than expected [16].

LIGHT REACTIONS

In photosynthesis light energy is stored in two successive photoacts, Photosystem II (PSII) and Photosystem I (PSI) (Fig. 3). Electrons are donated to the PSII donor side (P680) in result of water oxidation, transferred from the PSII acceptor side (Q) to the PSI donor side (P700) with the help of two carriers, plastoquinone (PQ) and plastocyanine (PCN). Cytochrome b/f complex (Cyt b/f) facilitates the transfer of electrons from reduced PQ (PQH) to oxidized PCN. There are a few low-potential intermediates (A1 - A5) on the acceptor side of PSI through which electrons are rapidly transferred to ferredoxine (Fd). Ferredoxin-NADP reductase (FNR) catalyses the transfer of electrons from Fd to NADP. The whole process of electron transport occurs in the thylakoid membrane and the individual transfer reactions are specifically located either at the inner or the outer surfaces of the membrane. As a result of this electrons are going along an N-shaped way starting at the inner (thylakoid) and finishing at the outer (stroma) surface of the membrane.

Consequent to such a charge separation a difference in electric potentials over the membrane is generated, positive inside and negative outside. This energizes the membrane. Additionally, when plastoquinone receives two electrons, two protons from the stroma are bound to form plastohydrokinone (PQH_2). These protons are released inside the thylakoid when PQH is oxidized at the Cyt b/f complex. This creates a proton gradient on the membrane and increases its energization. The amount of protons appearing inside the thylakoid is relatively big. Counter ions Mg^{2+} serve to neutralize the charge but the proton gradient remains and it may be as high as 3.5 pH units.

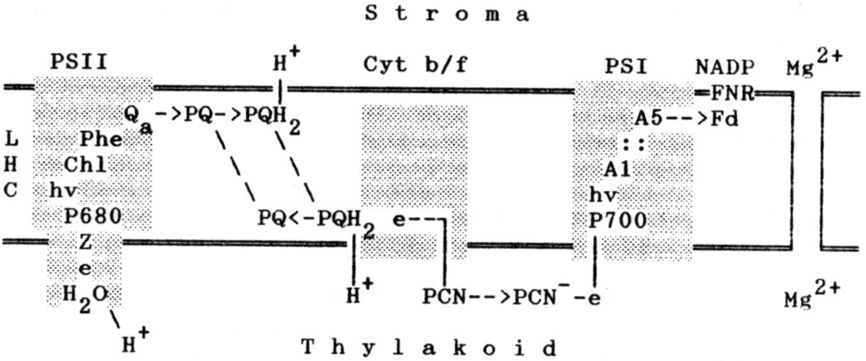

Fig. 3. A scheme of electron transport in photosynthesis [28]. Explanations in text.

Membrane energy, which is mostly in the form of proton gradient, is used for ATP synthesis. According to Mitchell, free energy of the proton gradient (and/or electrical potential) over the membrane can be used to shift the equilibrium of ATP hydrolysis towards ATP synthesis:

$$RT*\ln\left[\frac{ATP}{ADP*P_i}\right] = \Delta G_o + n*RT*\ln(H_{thyl}/H_{str}) \qquad (1)$$

from where

$$\frac{ATP}{ADP} = P_i*\exp\left[\frac{\Delta G_o}{RT} + 2.3026*n*(pH_{str} - pH_{thyl})\right] \qquad (2)$$

where ΔG_o is the standard free energy for ATP hydrolysis, RT is the multiplication of absolute temperature and gas constant and n is the number of protons whose energy is thermodynamically used for ATP synthesis when the protons are released via the coupling factor of ATP synthase.

From Eq.(2) it is possible to calculate the equilibrium ATP/ADP ratio as a function of P_i, ΔpH and n, the number of protons returned to stroma per P_i bound. According to the current information, the most likely value of n is 3 [46]. The ATP synthase is activated by the thioredoxin system but also by the proton pressure on the membrane. It is not clear whether it is activated to the extent that the ATP/ADP ratio in chloroplast would become close to the equilibrium determined by Eq.(2). Evidently, Eq.(2) gives an upper limit for the ATP/ADP ratio under given conditions.

An important feature of the photosynthetic machinery is that it supports a linear flow of electrons from water to NADP and further to CO_2 with a stoichiometry $4e/CO_2$. The same flow transports protons, that are necessary for ATP synthesis. According to the scheme in Fig. 3, four H^+ are released in water splitting and 4 H^+ are transported with plastoquinone per 4 electrons. Eight protons support the synthesis of 2.67 ATP-s, if there is no passive leakage across the membrane (but there is). The CRC needs 3 ATP-s per CO_2, plus 1/6 ATP per CO_2 in starch synthesis. Clearly, the linear electron flow produces ATP in an insufficient stoichiometry to support steady CO_2 uptake. ATP production will limit the rate of the CRC and at least 0.615 e per CO_2 must accumulate somewhere in the cell or to be captured by oxygen or some other oxidant.

The first compound where electrons can accumulate is ferredoxin. Here electrons branch between NADPH and other possible acceptors like plastoquinone, oxygen, thioredoxins, nitrite etc. The way how the electron flow is distributed between these acceptors seems to be one of the key questions in understanding the control processes in photosynthesis since at this point the equilibrium between ATP and NADP syntheses for the carbon assimilation, CRC enzyme activation, nitrite reduction and the consequent aminoacid synthesis rate and, finally, redox equivalent transport from the chloroplast to the cytosol are determined. Readjustments in the distribution of electron flow at the acceptor side of PSI are clearly seen in oscillations of photosynthesis.

UNDERSTANDING OSCILLATIONS MEANS UNDERSTANDING THE CONTROL OF PHOTOSYNTHESIS

For many years, oscillations in photosynthesis have commanded considerable interest, because in this process the rate-controlling regulatory systems are displaying their kinetics [18, 29]. Maximum rates of photosynthesis in the peaks of oscillations have been observed to be considerably

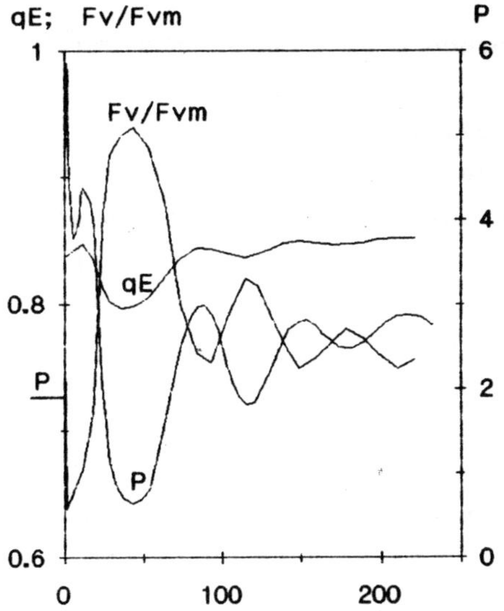

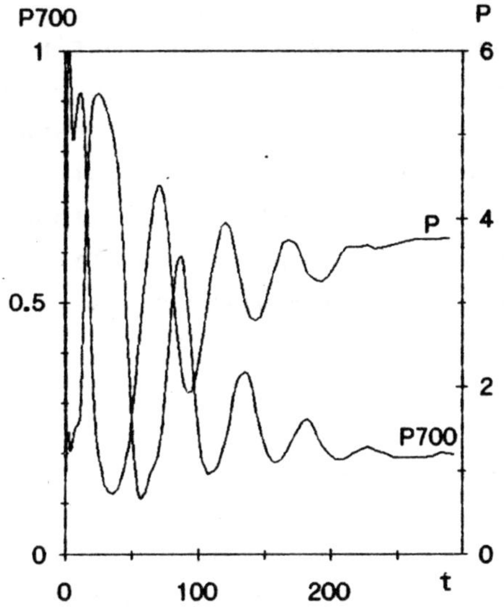

Fig. 4 (left). Simultaneous recordings of photosynthetic CO_2 uptake P (nmol cm^{-2} s^{-1}) and chlorophyll fluorescence parameters F_v/F_{vm} characterizing the reduction level of the primary acceptor Q_a and qE characterizing thylakoid energyzation in an <u>Helianthus</u> <u>annuus</u> leaf after a transition from 340 to 2000 µl l^{-1} CO_2, 21% O_2. Photon flux density 160 nmol cm^{-2} s^{-1}, time t, s.

Fig. 5 (right). Simultaneous recordings of photosynthetic CO_2 uptake P (nmol cm^{-2} s^{-1}) and the relative amount of nonoxidized P700 in a <u>Helianthus</u> <u>annuus</u> leaf after a transition similar to that in Fig. 4.

in excess of maximum steady state rates of light- and CO_2-saturated photosynthesis. Obviously, the potential of leaves for fast photosynthesis is not fully utilized under steady-state CO_2 and light saturation. An experiment with a sunflower leaf in which CO_2 was rapidly increased from 340 to 2000 µl l^{-1} is shown in Fig. 4. A deep minimum in photosynthesis occurs 45 seconds after the transition and the steady state is reached via a series of damped oscillations within about 200 seconds from the transition. Chlorophyll fluorescence was recorded simultaneously with the CO_2 uptake by means of a pulse-modulated fluorometer PAM 101 [39]. Variable fluorescence F_v and maximum variable fluorescence F_{vm} were measured using short dark and very high light pulses as is a common routine procedure [47]. A considerable reduction of the first acceptor quinone at PSII occurs in the first minimum of photosynthesis, as indicated by an increase in the F_v/F_{vm} ratio. The qE parameter which is believed to reflect the transthylakoid proton gradient and the absolute pH in the thylakoid lumen is decreasing, approaching a minimum slightly before the minimum in photosynthesis. These data indicate that a considerable reduction of the electron transport chain and de-energization of the thylakoid membranes occur in the first deep minimum of photosynthesis.

Oxidation of P700, the donor pigment of PSI can be measured monitoring the absorption of a leaf at 820 nm [40, 48]. The measurements can be calibrated assuming that P700 becomes completely reduced in the dark and completely oxidized with a far-red light which excites only PSI. Figure 5 shows simultaneous recordings of the CO_2 uptake and P700 in an experiment similar to that shown in Fig. 4. P700 was quite oxidised at the low CO_2 before the transition but it became almost completely reduced slightly before the CO_2 uptake rate reached its first minimum.

The interdependence between photosynthesis and nonoxidized P700 can be better visualized in a phase-portrait where the two variables are plotted against each other (Fig. 6). The points are plotted after 2 seconds and the direction of movement in time is shown by an arrow. A close inspection of the shape of the hysteresis curve in Fig. 6 enables us to recognize processes which follow approximately straight lines. One of the straight lines starts at the origin of the coordinates and ends at Pmax at the completely nonoxidized P700. In this experiment Pmax = 16 nmol $cm^{-2}s^{-1}$ which corresponds to 10-photon requirement per CO_2 in average for the both photosystems or 5 photons per CO_2 used in PSI in assumption of equal distribution of photons between the photosystems. This line corresponds to the complete control of PSI efficiency by the donor side. The proportion of photons used for charge separation is equal to the proportion of nonoxidized P700 when the experimental points are on this line. This happens when the rate of electron flow is completely controlled by PSII, therefore, this line will be

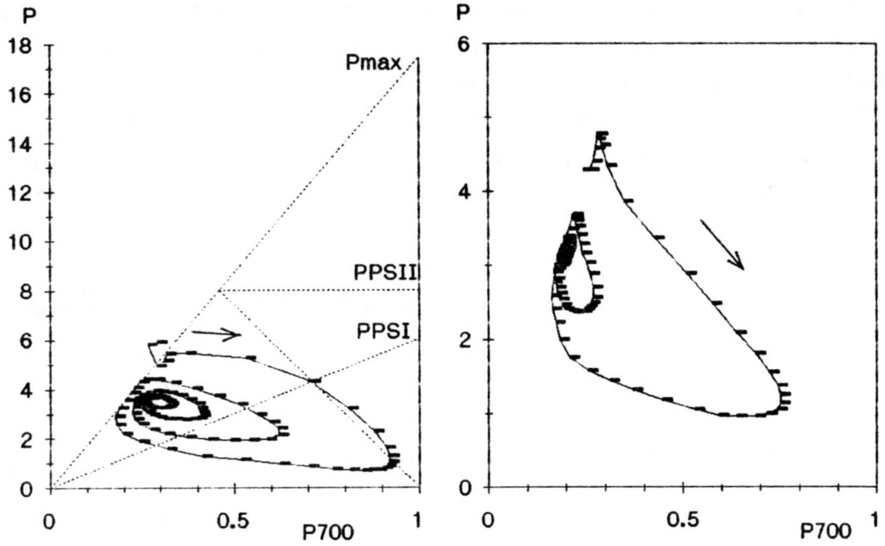

Fig. 6 (left). Phase-portrait of P against P700 from Fig. 5. After each 2 seconds the value of P was plotted against P700. For one of the points it is shown how the efficiencies of PSI and PSII are calculated: EPSII=PPSII/Pmax; EPSI=PPSI/Pmax.

Fig. 7 (right). Phase-portrait of P against P700 obtained from another experiment with a Helianthus annuus leaf which had weaker oscillations.

referred to as the PSII control line. PSI operates like a sucking pump whose input pressure is given by the nonoxidized P700. Evidently, the position of experimental points on this line shows us the depth of the PSII downregulation. In Fig. 6 the points are on the line when 20 to 35% of P700 are nonoxidized. The relative PSII efficiency (in an assumption of equal distribution of photons between the two photosystems) is of the same magnitude [10].

When photosynthesis is going through the troughs of oscillations the points in Fig. 6 depart to the right from the PSII control line. There seems to exist another linear process which causes the photosynthetic rate to decrease when P700 is more reduced than just necessary to remove the electrons donated by PSII. It may not be quite evident from Fig. 6 that the trajectories to the right from the PSII control line form a family of approximately straight lines ending up with zero photosynthesis at the completely nonoxidized P700, because there are simultaneous movements along both the lines during oscillations. It is more evident from Fig. 7, produced from another experiment where oscillations were less deep and more rapidly damped. For the points that lay to the right from the PSII control line in Fig. 6, P700 is more reduced than necessary to transport electrons donated by PSII, nevertheless, the electron transport rate through PSI (judged by the CO_2 uptake rate) is not greater. This suggests that a part of the PSI centres are blocked from the acceptor side. The possibility of PSI acceptor side reduction was shown also by Weis and Lechtenberg [48], especially at high CO_2 concentrations.

It is important that the rate of electron transport actually decreases approximately proportionally to the percentage of blocked PSI centres. This suggests that there exists a considerable channelling of electrons at the acceptor sides of PSI-s and between the two photosystems. If any PSII centre could donate electrons to any PSI centre, the rate of photosynthesis should not decline before P700 will be completely nonoxidized. Until there is still a sufficient number of nonblocked PSI centres present they could guarantee the necessary electron transport rate though the rest of the centres are blocked. For the case in Fig. 6, until there will be about 30% of PSI-s open on the acceptor side, the photosynthetic rate should stay constant. In this case 70% of the P700 would be nonoxidized because of the blocked acceptor side and 30% would be nonoxidized only on the donor side in order to perform an efficient electron transport. But experimental data do not agree with this model. The overall rate of electron transport through both the photosystems declines approximately proportionally to the percentage of blocked PSI-s. This suggests a model according to which electrons are restricted to 'channels' between certain pairs of PSII-PSI centres or, which is more likely, between relatively small groups of PSII-s and PSI-s. If a channel is blocked at the end it is switched off from the photosynthetic electron transport and electrons from the blocked channels cannot use neighbouring channels though those may be open.

The following scheme characterizes possible combinations which may occur in case of a simplified centre-to-centre channelling. In case of a grouped channelling the denotions OPen or CLosed for a PSII centre and Reduced or Oxidized for the Donor and Acceptor sides of a PSI centre will lose their absolute meanings and become averaged over the group. It is also not yet clear whether the PSII regulation mechanism operates on an on-off principle or PSII centres each can be regulated gradually.

```
hv  PSII      PSI
              DA
---> OP ---> RO --->NADP
---> OP ---> RR --->PQ, O2, NO2-, ThR
---> OP ---> OO
---> OP ---> OR
---> CL      OO
---> CL      OR
---> CL      RR
```

According to this scheme a successful charge separation at PSII occurs when it is OPen. A PSII may be CLosed because of the acceptor Q_a being reduced or the centre inactivated due to one of the thylakoid pH-dependent control mechanisms [9, 47]. After an electron is donated into the interphotosystem transport chain its fate depends on the state of the corresponding PSI centre(s). Centres with Reduced donor and Oxidized acceptor serve for photosynthetic electron transport to NADP. Because of the great redox potential difference between Fd^-/Fd and NADPH/NADP systems other acceptors are out of competition. It is suggested that a centre with Reduced donor and Reduced acceptor also can perform charge separation but the further electron transport is blocked at the ferredoxin-NADP-reductase (FNR) which is inhibited by reduced NADPH [21]. In these centres electrons accumulate at Fd creating a high negative redox potential. Under the thermodynamic pressure of this potential and NADPH being out of competition now, it appears possible to reduce plastoquinone PQ (cyclic transport), O_2 (pseudocyclic flow), nitrite and thioredoxins ThR. Therefore, this model assumes a 'division of labour': PSI centres with open FNR serve for photosynthesis without competition; only centres with inhibited

FNR (blocked by NADPH) can perform alternative reducing tasks, also without a competition by NADP. When the alternative channels are rate-saturated or blocked, electrons return to P700 via a back reaction from one of the primary acceptors of PSI.

The next two cases where PSII is open but P700 oxidized may occur transiently after a successful electron transfer by PSI when the intersystem electron transport is limiting and electrons do not arrive at P700 at a necessary rate. The last three cases refer to the situations when PSII is closed. These states accumulate in abundance when photosynthesis is completely PSII-controlled. In the scope of this model it is possible to follow the sequence of regulatory events at PSII and PSI during oscillations. Returning to Fig. 6 we see that immediately after the transition to the high CO_2 the system remains on the PSII control line as expected since at high CO_2 the metabolic resistance of the CRC decreases. During the initial two to four seconds after the transition PSII control releases a little and the points are moving upwards along the PSII control line. But then quite suddenly the points break to the right from the PSII control line indicating a developing acceptor side block at PSI. Assuming a linear decline in photosynthesis with the PSI acceptor side reduction we find that simultaneously with the developing overreduction at PSI the PSII control further releases. This is supported by the fluorescence data in Fig. 4 which show that the thylakoid energization-dependent quenching qE is decreasing while photosynthesis is dropping into the first trough. Dashed lines drawn through one of the points in Fig. 6 show how relative PSII and PSI efficiencies were calculated for every point. The relative efficiency EPSI is defined as the relative decrease in the photon yield of PSI caused by the acceptor side reduction. The relative efficiency EPSII is defined as the (hypothetical) photon yield of PSII in assumption of the full efficiency of PSI.

The relative efficiencies EPSII and EPSI are plotted against each other in Fig. 8. During the declining phase of oscillations the PSII control releases and simultaneously PSI becomes highly reduced losing more than 90% of its

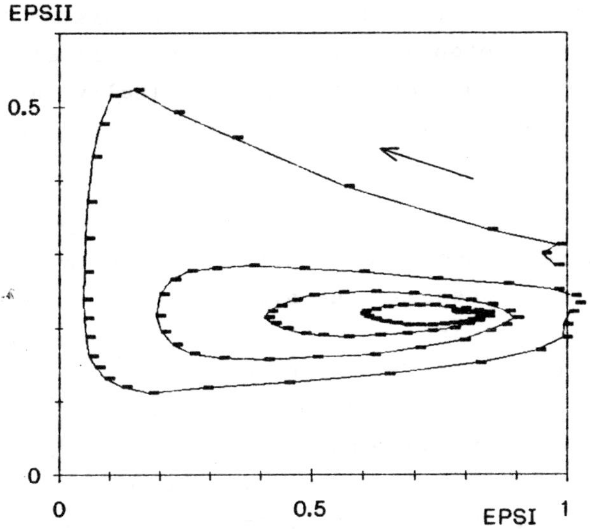

Fig. 8. Phase-portrait of the PSII efficiency EPSII against the PSI efficiency EPSI calculated from the data in Fig. 6 as given in the legend to Fig. 6.

efficiency for the photosynthetic electron transport. In the next phase, while PSI is still over-reduced, PSII gradually gains more control until it takes the complete control which enables the PSI over-reduction to disappear and the PSI efficiency to reach maximum again. This means that the decreasing trend in the proton gradient was reversed while PSI was overreduced. Evidently, either cyclic (this is, still, unlikely since the whole electrontransport chain was overreduced, see F_v/F_{vm} in Fig. 4) or pseudocyclic electron transport to O_2, or nitrite reduction helped to keep electrons flowing, or NADP-malate dehydrogenase was activated which accelerated the malate-oxaloacetate shuttle transport of electrons out from the chloroplast [38]. As a result of these processes, a great pH gradient was generated which produced considerable downregulation of the PSII centres, actually an overshoot. It is clearly seen from Fig. 6 that the points return onto the PSII control line at a much lower photosynthetic rate than that at which they branched off from the line. In this situation almost all FNR-s are open and almost all electrons are directed to NADPH supporting photosynthesis. But such a situation is unstable since the linear electron flow to NADP alone is unable to produce enough ATP to support photosynthesis. The accumulated ATP is gradually consumed, proton gradient released, PSII efficiency and the CO_2 uptake rate are increasing. Finally, unused NADPH begins to accumulate, it blocks FNR and the whole cycle begins again.

When the oscillations are over, a steady state photosynthetic rate is established as a result of a compromise between the control processes. The aim of these regulations is to establish a necessary ratio of NADPH/ATP syntheses and to activate the CRC enzymes to the necessary extent. When elctrons are transported to NADP only, the CO_2 uptake rate is determined by the corresponding ATP synthesis and NADPH accumulates, CRC enzymes are slowly deactivating. Accumulating NADPH blocks some FNR centres which causes electron accumulation in Fd at the blocked PSI centres. These centres start to reduce thioredoxins and, with their help, to activate the CRC enzymes. Simultaneously, NADP-malate dehydrogenase becomes activated which supports a more rapid transport of electrons to the cytosol where these can reach oxygen via the mitochondrial oxydative chains, coupled or alternative oxidases. This helps to produce some additional ATP in chloroplasts due to the linear electron flow not coupled to carbon reduction, and some in the cytosol. ATP can be produced also in result of cyclic and pseudocyclic electron flows and an additional flow for nitrite reduction, all of which take place at the closed centres. Therefore, establishment of the right proportion of the PSI centres with reduced acceptor is the main regulatory event which takes place when conditions for photosynthesis change, including changes in CO_2 concentration or light intensity. In the experiment in Fig. 6 the steady state point is a little to the right from the PSII control line indicating some permanent overreduction of the PSI centres. In the other experiment, Fig. 7, the steady over-reduction is almost not detectable. This does not mean that there were no PSI centres with reduced acceptor sides in this experiment. Rather, the rate of electron transport to the alternative acceptors was also PSII controlled.

But what detemines the maximum possible rate of photosynthesis at CO_2 and light saturation? The main driving forces for CO_2 assimilation are NADPH and ATP. NADPH cannot change in a wide range because its concentration determines the proportion of PSI centres with oxidised and reduced acceptor sides. Therefore, the main rate-determining driving force should be ATP and the main metabolic resistance at saturating CO_2 is a composite function of the CRC enzyme activities. As shown above, the ATP/ADP ratio is determined by transthylakoid pH difference and free phosphate concentration. To support a higher photosynthetic rate there must be a greater pH difference. At high photosynthetic rates which occur at elevated CO_2 concentrations rapid sucrose and starch synthesis rates are necessary. Control mechanisms for these pathways need low phosphate for their activation [12, 43]. Therefore, low free phosphate is a prerequisite for a high CO_2 uptake rate. At low P_i a greater pH difference is necessary to generate a high ATP/ADP ratio. On the other hand, the declining pH in the thylakoid lumens leads to inactivation of PSII centres and to a corresponding limitation in the electron transport rate. The following scheme illustrates the two simultaneous but controversial controls of photosynthesis linked to the end-product synthesis rates, both based on the thylakoid energization.

Scheme of the control limitation of photosynthesis

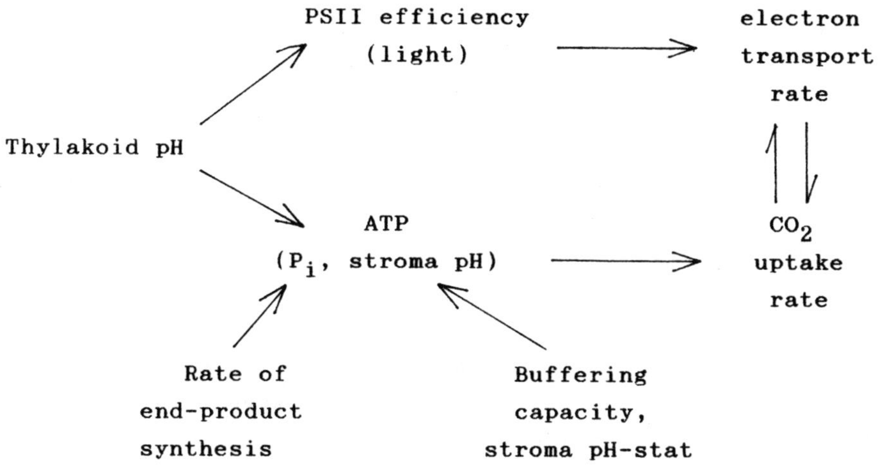

In this scheme, besides the free orthophosphate, also an

important role of the stroma pH is emphasized. It has been shown [48] that absolute pH in the thylakoid lumens controls the PSII efficiency. Therefore, a more alkaline pH in stroma would result in a greater pH difference and a higher ATP/ADP ratio at a given pH in the lumens. High buffering capacity and pH-stabilizing mechanisms in stroma help to support higher photosynthetic rates in leaves.

CONCLUSION AND DISCUSSION

The main conclusion is drawn from the kinetic studies of oscillations in photosynthesis. It is based on an experimental evidence that overreduction of the acceptor side of PSI causes a decrease in the CO_2 uptake rate approximately proportionally to the portion of the overreduced centres. This suggests that an overreduced centre cannot reduce NADP and the electrons fed to an overreduced centre cannot choose another centre which is not overreduced.

Weis and collaborators [21] have shown that reduced NADPH is an efficient inhibitor of ferredoxin-NADP reductase (FNR). We suppose that blocking of the PSI centres is a result of inhibition of the FNR by NADPH. Kinetic data show that the PSI centres are switched off one by one. Therefore, we have to conclude that the FNR active in photosynthetic NADP reduction is the membrane-bound enzyme attached to a PSI centre. In the scheme in Fig. 3 [28], Fd and FNR were shown as relatively free, not incorporated into the continous electron transfer chain at the acceptor side of the PSI centre. In Fig. 9 the model is changed structurally incorporating Fd and FNR into the PSI unit.

The acceptor side of PSI forms a continous 'tube' for electrons ending with NADP and, when open, the flow of electrons to NADP is out of competition. When the tube is closed (NADPH allosterically inhibits FNR), electrons accumulate at the bound Fd. Then a sufficiently high reducing potential is created which allows to reduce alternative acceptors (PQ, O_2, NO_2^-, thioredoxins). From our kinetic data it is impossible to say anything about the structural organization of the alternative reductases (Fd-plastoquinone reductase, FQR, Fd-nitrite reductase, FNiR, glutamate synthase, GOGAT and Fd-thioredoxin reductase, FTR). Robinson

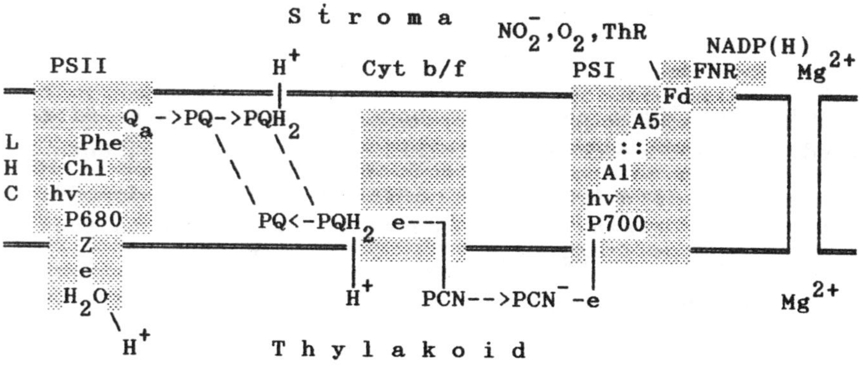

Fig. 9. A scheme of electron transport in photosynthesis. The difference compared to Fig. 3 is that Fd and FNR are incorporated as structural parts of the PSI centre complexes. These electron transport chains kinetically form more or less separated channels.

[34] reported that in his studies light-dependent CO_2 assimilation and NO_2^- reduction never suppressed each other, rather, there was some mutual activation between the two processes. He suggested that "there are individual linear electron transport systems which terminate on the reducing side of PSI by a specific reductase, i.e., either FNR, FTR, FNiR or GOGAT. Each terminal reductase would be equipped with its own bound set of Fd molecules. Ferredoxin which always would remain bound in place on the specific reductases, would recycle between the reduced and oxidized forms". The fact that nitrite photoreduction in vivo is inhibited by oxygen in soybean cell isolates [35] does not necessarily mean a competition between O_2 and NO_2^- reductions. Photorespiration which is active in the presence of oxygen seems to be an efficient ATP generator in whole cells [8]. At low O_2 this source must be replaced by an additional nonphotosynthetic electron flow which may be directed to NO_2^-.

There is also a more direct evidence in literature in support of our conclusion. Two molecular forms of FNR with molecular weights 33000 (FNR-S) and 75000 (FNR-L) have been isolated from spinach [44]. FNR-S is more tightly bound to the membrane than FNR-L. FNR-S is directly bound to PSI particles but FNR-L is also a functional constituent of the ferredoxin-NADP reductase complex. Absorbance transients have been observed in isolated thylakoids which could be identified to originate from the reduction of FNR to the semireduced FNR* by PSI. Generation of FNR* appears to be unpreturbed in the absence of Fd [7]. This is possible when a PSI centre can directly donate electrons to FNR structurally associated with the centre.

Our kinetic evidence also suggests that there must be a considerable channelling of the electron flow between PSII and PSI centres, so that only a limited number of PSII centres can donate electrons to a limited number of PSI centres. This is difficult to reconcile with the existing knowledge about the different number of PSI and PSII centres [24], and the spacial arrangement of the two photosystems, PSII-s in the granal and PSI-s in the stromal lamellae, free plastoquinone and plastocyanine pools interconnecting a number of the both photocentres. One of the possible compromises is that the channelled units are individual thylakoids. It is possible that all thylakoids are not equal in their performance. Something like stomatal patchiness may also exist on the thylakoid level. It seems that the grouping phenomenon is widespread on different levels of the natural hierarchy and there is no reason to expect that all thylakoid lamellae are absolutely similar.

What concerns the global importance of these conclusions, they once again emphasize the integrity of the plant as a living creature. Dependent on growth conditions and genotype, a long exposure of plants to elevated CO_2 may or may not cause an increase in the maximum CO_2 and light-saturated rate of photosynthesis [36]. Increased CO_2 concentration may cause positive shifts in the photosynthetic productivity on areas that are well supplied with growth substrates. Restrictions in photosynthesis caused by stomatal closure due to water stress may be released by high CO_2 but only if the growth rate of the plants is not drastically suppressed by the stress.

LITERATURE

1. Anderson, L.E. (1985). Light/dark modulation of enzyme activity in plants. In: Advances in botanical research, Callow, A.J., ed. Academic Press, New York, Vol. 12, pp. 1-46.

2. Badger, M.R., Sharkey T.D. and von Caemmerer, S. (1984). The relationship between steady gas exchange of bean leaves and levels of carbon-reduction-cycle intermediates. Planta, 160: 305-313.

3. Buchanan, B.B. (1984). The ferredoxin/thioredoxin system: A key element in the regulatory function of light in photosynthesis. Bio-Science, 34: 378-383.

4. Butz N.D. and Sharkey T.D. (1989) Activity ratios of ribulose-1,5-bisphosphate carboxylase accurately reflect carbamylation ratios. Plant Physiol. 89: 735-739.

5. von Caemmerer, S. and Edmondson, D.L. (1986). The relationship between steady-state gas exchange in vivo, RuP_2 carboxylase activity and some carbon reduction cycle intermediates in Raphanus sativus. Aust. J. Plant Physiol. 13: 669-688.

6. Edwards, G. and Walker, D.A. (1983). C_3, C_4: Mechanisms, and cellular and environmental regulation, of photosynthesis. Blackwell Scientific Publication.

7. Garab, G., Hong, H., Coughlan, S.J., Matthijs, H.C.P. and Hind, G. (1990). Absorbance transients of ferredoxin:$NADP^+$ reductase in isolated thylakoid membranes. In: Current research in photosynthesis, M. Baltscheffsky, ed., Kluwer Academic Publishers, the Netherlands, vol. II, pp. 667-670.

8. Gardeström, P. and Wigge B. (1988). Influence of photorespiration on ATP/ADP ratios in the chloroplasts, mitochondria and cytosol, studied by rapid fractionation of barley (Hordeum vulgare) protoplasts. Plant Physiol. 88: 69-76.

9. Genty, B., Briantais J.M. and Baker N.R. (1989). The relationship between quantum yield of photosynthetic electron transport and quenching of chlorophyll fluorescence.

Biochem. Biophys. Acta, 990: 87-92.

10. Harbinson J., Genty, B. and Baker, N.R. (1989). Relationship between the quantum efficiencies of photosystem I and II in pea leaves. Plant Physiol. 90: 1029-1034.

11. Harbinson, J., Genty, B. and Foyer, C.H. (1990). Relationship between photosynthetic electron transport and stromal enzyme activity in pea leaves. Toward an understanding of the nature of the photosynthetic control. Plant Physiol. 94: 545-553.

12. Heldt, H.W., Chon, C.J., Maronde, D., Herold, A., Stankovitch, Z.S., Walker, D.A., Kraminer, A., Kirk, M.R. and Heber, U. (1977). Role of orthophosphate and other factors in the regulation of starch formation in leaves and isolated chloroplasts. Plant Physiol. 59: 1146-1155.

13. Huber, S.C. and Huber, J.L.A. (1990). Regulation of spinach leaf sucrose-phosphate synthase by multisite phosphorylation. Current Topics in Biochemistry and Physiology, 9:

14. Huber, J.L.A., Huber, S.C. and Nielsen T.H. (1989). Protein phosphorylation as a mechanism for regulation of spinach leaf sucrose-phosphate synthase activity. Arch. Biochem. Biophys. 270: 681-690.

15. Jensen R.G.and Bahr J.T. (1977) Ribulose bisphosphate carboxylase/oxygenase. Annu. Rev. Plant Physiol. 28: 379-400.

16. Kobayashi, I., Inoue, I., Furuya, F., Shibata, K. and Heber, U. (1979). Regulation of adenylate levels in intact spinach chloroplasts. Planta, 147: 69-75.

17. Kobza J. and Seemann, J.R. (1988). Mechanisms for the light-dependent regulation of ribulose-1,5-bisphosphate carboxylase activity and photosynthesis in leaves. Proc. Natl. Acad. Sci. U.S.A. 85: 3815-3819.

18. Laisk, A. and Eichelmann, H. (1989). Towards understanding oscillations:a mathematical model of the biochemistry of photosynthesis. Phil. Trans. R. Soc. Lond. B 323: 369-384.

19. Laisk, A. and Oja V. (1974). Leaf photosynthesis under short pulses of carbon dioxide: carboxylation reaction in vivo. Fiziol. Rastenij, 121: 1123-1131 (in Russian).

20. Laisk, A. and Oja, V. (1976). Photosynthesis at light and CO_2 saturation is limited by the reactions of ribulose-bisphosphate resynthesis. Eesti NSV Teaduste Akadeemia Toimetised (Proc. of Estonian Acad. Sci.) Bioloogia, 25: 145-150 (in Russian).

21. Lechtenberg D., Voss, B. and Weis, E. (1990). Regulation of photosynthesis: photosynthetic control and thioredoxin-dependent enzyme regulation. In: Current research in photosynthesis, M. Baltscheffsky, ed., Kluwer Academic Publishers, the Netherlands, vol. IV, pp. 171-174.

22. Lorimer G.H., Badger M.R. and Andrews T.J. (1976). The activation of ribulose-1,5-bisphosphate carboxylase by carbon dioxide and Mg^{2+} ions. Equilibria, kinetics a suggested mechanism and physiological implications. Biochemistry, 15: 529-536.

23. Makino A., Mae T.and Ohira K. (1985). Enzymic properties of ribulose- 1,5-bisphosphate carboxylase/oxygenase purifed from rice leaves. Plant Physiol., 79: 57-61.

24. Melis A. 1989 Spectroscopic methods in photosynthesis: photosystem soichiometry and chlorophyll antenna size. Phil Trans. R. Soc. Lond. B 323, 227-239.

25. Moldau, H. and Sõber, J. (1988). Decrease in C_i/C_a-ratio under conditions favouring assimilate accumulation. Photosynthetica, 22: 9-19.

26. Oja, V. (1983). A Rapid-response gas exchange measurement system for studying the kinetics of leaf photosynthesis. Fiziol. Rastenij, 30: 1045-1052, (in Russian).

27. Oja, V., Laisk, A. and Heber U. (1986). Light-induced alkalinization of the chloroplast stroma in vivo as estimated from the CO_2 capacity of intact sunflower leaves. Biochim. Biophys. Acta 849: 355-365.

28. Ort, D.R. and Good, N. (1988). Textbooks ignore photosystem II-dependent ATP formation: is the Z scheme to blame.

Trends in Biochemical Sciences 13: 467-469.

29. Osmond C.B., Oja V., Laisk A. 1988 Regulation of carboxylation and photosynthetic oscillations during sun-shade acclimation in Helianthus annuus measured with a rapid-response gas exchange system. Aust. J. plant Physiol. 15, 239-252.

30. Portis, A.R. (1990). Rubisco activase. Biochim. Biophys. Acta 1015: 15-28.

31. Portis A.R., Salvucci M.E., Ogren W.L. 1986 Activation of ribulosebisphosphate carboxylase/oxygenase at physiological CO_2 and ribulosebisphosphate concentrations by rubisco activase. Plant Physiol. 82, 967-971.

32. Ranty B., Cavalie G. 1982 Purification and properties of ribulose 1,5-bisphosphate carboxylase from sunflower leaves. Planta, 155, 388-391.

33. Rasulov B.H., Laisk A.H. and Asrorov K.A. (1983). Photosynthesis and photorespiration in some cotton species during ontogeny. Fiziol. Rastenij 30: 637-645, (in Russian).

34. Robinson, M.R. (1988). The efficient allocation of photogenerated reductant for simultaneous CO_2 and NO_2^--photoassimilations in higher plants. In Report of the information exchange seminar on 'Photosynthetic productivity: biochemical and molecular genetic regulatory mechanisms', East-Westcenter, Honolulu.

35. Robinson, M.R. (1990). Nitrite photoreduction in vivo is inhibited by oxygen. Plant Physiol. 92: 862-865.

36. Sage, R.F., Sharkey, T.D. and Seemann, J.R. (1989). Acclimation of photosynthesis to elevated CO_2 in five C_3 species. Plant Physiol. 89: 590-596.

37. Scheibe, R. (1987). $NADP^+$-malate dehydrogenase in C_3 plants: regulation and role of a light-activated enzyme. Physiol. Plantarum, 71: 393-400.

38. Scheibe, R. and Stitt, M. (1988). Comparison of NADP-malate dehydrogenase activation, Q_A reduction and O_2 evolution in spinach leaves. Plant Physiol. Biochem. 26: 473-481.

39. Schreiber, U., Bilger, W. and Schliwa, U. (1986) Continuous recording of photochemical and non-photochemical chlorophyll fluorescence quenching with a new type of modulation fluorometer. Photosyn. Res. 10: 51-62.

40. Schreiber U., Neubauer C.and Klughammer C. (1989). Devices and methods for room-temperature fluorescence analysis. Phil. Trans. R. Soc. Lond. B 323: 241 - 251.

41. Sharkey T.D. (1990). Feedback limitation of photosynthesis and the physiological role of ribulose bisphosphate carboxylase carbamylation. Bot. Mag. Tokyo, Special issue 2: 3-10.

42. Sharkey, T.D. and Vanderveer, P. (1989). Stromal phosphate concentration is low during feedback limited photosynthesis. Plant Physiol. 91: 678-684.

43. Stitt, M. (1985). Fine control of sucrose synthesis by fructose-2,6-bisphosphate. In: Regulation of carbon partitioning in photosynthetic tissue. Heath, R.L. and Preiss J. ed-s. Waverley Press, Baltimore, pp. 109-126.

44. Shin M. (1990). Structure and function of ferredoxin-NADP reductase complex. In: Current research in photosynthesis, M. Baltscheffsky, ed., Kluwer Academic Publishers, the Netherlands, vol. II, pp. 659-662.

45. Takahama, U., Shimizu-Takahama, M. and Heber, U. (1981). The redox state of the NADP system in illuminated chloroplasts. Biochim. Biophys. Acta, 637: 530-539.

46. Tran-Anh T. and Rumberg B. (1987). Coupling mechanism between proton transport and ATP synthesis in chloroplasts. In: Progress in in photosynthesis research, vol. III, J. Biggins, ed. Dordrecht: Martinus Nijhoff, pp. 185-188.

47. Weis E. and Berry J. A. (1987). Quantum efficiency of photosystem II in relation to ' energy'-dependent quenching of chlorophyll fluorescence. Biochim. Biophys. Acta 894: 198 -208.

48. Weis E. and Lechtenberg D. (1989). Fluorescence analysis during steady-state photosynthesis. Phil. Trans. R. Soc. Lond. B 323: 253-268.

49. Werdan K.and Heldt, H.W. (1971). The phosphate translocator of spinach chloroplasts. In 'II nd International Congress on Photosynthesis', Stresa, 1337-1344.

50. Yeoh H.H., Badger M.R.and Watson L. (1980). Variarions in $K_m(CO_2)$ of ribulose bisphosphate carboxylase among grasses. Plant Physiol. 66: 1110-1112.

51. Yeoh H.H., Badger M.R.and Watson L. (1981). Variation in kinetic properties of ribulose-1,5-bisphosphate carboxylase among plants. Plant Physiol. 67: 1151-1155.

EFFECTS OF ELEVATED CO_2 CONCENTRATIONS ON CARBON METABOLISM AND RELATED PROCESSES IN LEAVES OF C_3 PLANTS

Randhir Singh
Department of Chemistry and Biochemistry,
Haryana Agricultural University,
Hisar - 125 004, India.

At canopy level, elevated CO_2 increases photosynthetic rates, leaf area, biomass and yield. Elevated CO_2 also reduces transpiration rate per unit leaf area, but not in proportion to reduction of stomatal conductance. With increasing leaf area and foliage temperature, photosynthetic water-use efficiency is increased. At cellular level, RuBP pool decreases with increasing CO_2. Organic phosphates increase at the expense of Pi, with chloroplastic PGA showing a marked increase. Hexose phosphates show an initial fall, followed by a marked increase. Sucrose synthesis is extremely sensitive to changes in CO_2 concentration. The [ATP]/[ADP] and the light scattering signal decline with increasing CO_2 pressure. These changes are generally interpreted as showing that the limiting element within the capacity for RuBP regeneration at high CO_2 pressure lies in the series of reactions between triose phosphates and RuBP, mainly because of mass- action ratios of RuBP regenerating reactions tending towards infinity at high CO_2 pressures.

The global atmospheric CO_2 concentration, currently estimated to be 350 umol mol^{-1},may double within the next century due to continued increase in fossil fuel use (1). Since CO_2 is the primary substrate for photosynthesis in green plants, it appears likely that an increase in CO_2 in the aerial environment would offer a great potential for increasing the growth of many agricultural crop species (2). Furthermore, because the photosynthetic capacity of C_3 crops is limited by the CO_2 level in the present atmosphere, an increase in CO_2 during growth might be expected to result in increased CO_2 fixation rates, and eventually to greater productivity (2,3).

By the early 1900s, leaf photosynthetic rates of certain species had been characterized as a function of several environmental variables such as CO_2 concentration, light and temperature. Over the years, many short term physiological studies of responses to CO_2 have been conducted. Also results of numerous earlier studies on the growth and yield responses of crops to elevated levels of CO_2 in enclosed environments have been compiled (4). In 1978, a workshop supported by the office of Carbon Dioxide Effects Research, OAES, U.S. Department of Energy, identified 31 plant responses to CO_2 enrichment (5). Some of the important responses identified included : increased photosynthesis, changed growth rate, decreased transpiration, decreased stomatal conductance, increased tolerance to atmospheric pollutants, increased leaf area, increased leaf dry weight/ leaf area etc. However, most of the responses reported as above had been obtained from experiments conducted in horticultural greenhouse. Hence, it was proposed that long term studies during all stages of the life cycle of plants were needed to answer questions on effects of long term global atmospheric CO_2 enrichment on vegetation. Furthermore, experiments are also needed to understand interactions between CO_2 and other environmental factors. A subsequent AAAS Symposium emphasized research needs at the carbon metabolism, leaf and other organ physiology, whole plant growth and development, terrestrial plant community, aquatic plant community and micro-organisms levels (6). One concern that emerged in this symposium was the uncertainity of how to make the linkage between the leaf physiological levels of response to CO_2 and the whole plant and community levels of response. These linkages are still being examined, researched,and coded into simulation models. A plan of research was developed by the U.S. Department of Energy, Carbon Dioxide Research Division to address these questions (7). Research within the framework of this plan has been underway for several years and many of the findings have been summarized by Strain and

Cure (8). However, systematic studies on these lines have not been conducted in this country. The objective of the present paper is not to give a comprehensive review but to discuss the recent data on crop and noncrop vegetation responses to elevated CO_2, both at canopy and cellular levels, by citing a few selected references.

Photosynthetic rates and carbon dioxide

Almost all recent studies show an increase of leaf photosynthetic rates when C_3 crop plants are exposed to long-term elevated CO_2 levels. A linear response of soybean leaves to increased CO_2 was reported when plants were grown at CO_2 concentration ranging from 340 to 910 umol mol^{-1} (9). Valle et al. (10) confirmed this type of response for soybean leaflets grown at, and exposed to two levels of CO_2, 330 umol mol^{-1} and 660 umol mol^{-1}. Their data showed a doubling of photosynthetic CER not only at high PPFD levels (PmaxL of 37 and 76 umol $CO_2 m^{-2} s^{-1}$ at CO_2 treatments of 330 and 660 umol mol^{-1}, respectively) but also across the whole range of PPFD levels. Leaf dark respiration rates were apparently higher for the 660 umol mol^{-1} treatment leaves than for the 330 umol mol^{-1} CO_2 treatment leaves. The light compensation point was reduced from 35 umol photon $m^{-2} s^{-1}$ at 330 umol mol^{-1} to 27 umol photom $m^{-2} s^{-1}$ at 660 umol mol^{-1}. The quantum yield mole CO_2 uptake, per mole photon intercepted, was 0.05 at 330 umol mol^{-1} and 0.09 at 660 umol mol^{-1}. Also leaves of soybean plants acclimated at 660 umol mol^{-1} CO_2 had higher CER across all ranges of external CO_2 concentration than leaves acclimated at 330 umol mol^{-1}. Likewise when leaf CER was plotted against intercellular CO_2 concentration, the leaves that were acclimated at 660 umol mol^{-1} CO_2 had higher CER than leaves acclimated at 330 umol mol^{-1} (11). In wheat also, the photosynthetic rate of the flag leaf of plants enriched to 1390 umol mol^{-1} was 50% higher than those of the plants grown at 332 umol mol^{-1} (12). However, in some other cases, plant photosynthetic responses have been observed to decrease following initial stimulation by elevated CO_2. Monoecious cucumbers exposed to continuous high levels of CO_2 showed reductions in photosynthetic rates and no increase in yield (13). This type of acclimation seems to be common in many ecosystem species where the initial stimulation of photosynthetic rates by elevated CO_2 decreases to the point where leaf or canopy photosynthetic rates become similar to those at ambient CO_2 concentrations.

The lack of response to high CO_2 is probably due to end-product feedback inhibition of photosynthesis due to imbalances of 'sink' for photoassimilate relatives to 'source' of photoassimilate. This phenomenon is usually manifested by accumulation of starch in chloroplast of leaves (14,15). Conversely, examples exist where a direct correlation between starch content and photosynthetic rate does not exist (16).

End-product feedback inhibition has been variously attributed to disruption of the grana of chloroplasts by starch grains (17), inhibition of RuPBcase activity by phosphorylated sugar via competition with RuBP for binding sites, some mechanism for inhibition of SPS, which would result in photosynthetic starch formation at the expense of photosynthetic sucrose formation (18) or some mechanism mediating the availability of phosphate at the chloroplast membrane. Regardless of the cause, the studies with soybean have indicated that the sink for photoassimilate (vegetative or reproductive growth) is always large enough for elevated CO_2 to manitain higher leaf photosynthetic rates (11,19) or whole canopy photosynthetic rates (20).

In cotton, field experiments have shown consistent large photosynthetic and yield responses to CO_2 (20). Rice canopy photosynthetic rate and yield responses to CO_2 increased to a concentration of 500 umol mol^{-1} but tended to level off at higher concentrations (21). Cure (22) summarised the effect of doubling CO_2 concentration on photosynthetic rates of nine crops, viz. wheat, barley, rice, maize, sorghum, soybean, alfalfa, cotton, potato and sweet potato, by comparing the percentage increase at 680 umol mol^{-1} above that at 340 umol mol^{-1} predicted by linear regression. She considered both short term CER (immediate response to CO_2 enrichment) as well as acclimated CER. Furthermore, she compared initial response net assimilation rate of CO_2 with long-term net assimilation rates. She computed weighted average increases of +52, +31, +23, and +18% for short-term CER, acclimated CER, initial net CO_2 assimilation rate, and long-term net assimilation rate, respectively. The computed weighted average increases for soybean were +78, +42, +35 and +23% for each type of measurement, respectively. The computed weighted average increases for soybean were +78, +42+35, and +23 for each type of measurement, respectively, for a doubling of CO_2 from 340 to 680 umol mol^{-1}. From these data, it is clear that as plants acclimate to higher CO_2, the leaf or whole plant photosynthetic CER will decrease. However, Valle et al. (10) found that leaves of soybean grown throughout their life cycle at 660 umol mol^{-1} CO_2 had higher photosynthetic CER capacity at all short-term levels of CO_2 from 90 to 990 umol mol^{-1}. Therefore, we should not generalize leaf and whole crop performance across a range of data sets. Acock et al. (23) concluded that soybean crop canopy photosynthetic rates could vary greatly throughout the growing season and that the variability appeared to be related to the requirement of the crop for photoassimilates.

Enzymatic responses to CO_2 enrichment

The research in this area has been conducted mainly on two enzymes, ribulose-1, 5-bisphosphate carboxylase (RuBPcase) and sucrose phosphate synthase (SPS). Vu et al. (24) reported that RuBPcase activities (on a per unit chlorophyll basis) of soybean leaf preparation with respect to 330 umol mol^{-1} were 92 to 95% for the 450 umol mol^{-1} treatment and 77 to 80% for the 800 umol mol^{-1} treatment for newly expanded leaves. However, the RuBPcase activity of older leaves showed smaller differences. The reactant RuBP concentrations were about 30 to 50% higher for leaves of soybean plants grown at 800 umol mol^{-1} than for plants grown at 330 umol mol^{-1}. Campbell et al. (11) found little differences in soybean leaf RuBPcase activity on a leaf area basis or soluble protein basis of plants grown over a range of CO_2 concentrations from 160 to 990 umol mol^{-1}. However, RuBPcase activity on a dry weight basis decreased across those respective CO_2 concentrations, presumably because of increased specific leaf weight (weight/area) attributable to increased leaf thickness and increased stored starch. RuBPcase protein was further shown to be a constant fraction (about 55%) of total soluble protein regardless of CO_2 treatment levels. Contrary to the findings of Vu et al. (24), Campbell et al. (11) found that RuBP levels in soybean leaves tended to decrease with increasing CO_2. No differences were found in RuBPcase activities of wheat leaves when expressed on a leaf area basis. RuBPcase activity in Nerium oleander was reduced significantly at 660 umol mol^{-1} with respect to 330 umol mol^{-1} when expressed on the basis of soluble protein, but the reduction was not statistically significant when expressed on a per-unit leaf area basis or chlorophyll basis. From the above discussion, it appears that elevated CO_2 raises the substrate RuBP slightly but reduces the in vitro activity of the enzyme RuBPcase slightly. However, neither of these responses would appear to be the primary cause of increased leaf photosynthesis rate of elevated CO_2.

Metabolite responses to CO_2 enrichment

Dietz and Heber (25,26) studied changes in the rate of CO_2 assimilation and in metabolite contents in the chloroplast compartment of spinach leaves with varying external CO_2 pressure. RuBP pool decreased with increasing CO_2 and remained high even at very low CO_2 pressures. Similar observations were made by Badger et al. (27) and by Stitt et al. (28). As the CO_2 pressure was increased, there was a general increase in organic phosphates at the expense of Pi, with chloroplastic PGA showing a marked increase. Hexose phosphates showed an initial fall, followed by a marked increase. The content of FBP remained relatively constant, although in another study it has been shown to fall in low CO_2 (28). The [ATP]/[ADP] and the light scattering signal were observed to decline with increasing CO_2 pressure. Dietz and Heber (25) interpreted these results as showing that that the limiting element within the capacity for RuBP regeneration at high CO_2 pressures lies in the series of reactions between triose phosphates and RuBP. They argued that electron transport does not limit RuBP regeneration becaue although [ATP]/[ADP] and [NADPH]/[NADP] decreased with increasing CO_2, pH (as measured by light scattering) stayed constant.

The site of this regulation between triose phosphate and RuBP remains elusive. Leegood and Walker (29) found CO_2 to have little influence on the activity of FBPase at below atmospheric CO_2 pressures, but it was inhibitory to redox regulation at above atmospheric CO_2 pressures. It, therefore, appears unlikely that redox modulation acts as a means of fine control over the flux through Calvin cycle at different CO_2 pressures. Since fru-6-P inhibits FBPase activity, its increased concentration with increasing CO_2 pressure would constitute a feasible mechanism for restricting FBPase activity. However, Dietz and Heber (25,26) argue from their data that FBPase does not limit at any CO_2 pressure and that regulation must lie in other reaction of RuBP regeneration, although the fact that the FBP content remains so constant over a wide range of CO_2 concentrations might equally well indicate that FBPase is regulated in such circumstances, even, if it cannot be shown to limit photosynthesis, since relatively large pool of FBP is unlikely to be maintained in the presence of active enzyme.

Stomatal conductance, transpiration and foliage temperature responses to elevated CO_2

Stomatal conductance, in general, decreases with increasing CO_2 concentration in most of the studies of about 50 species although a few cases of no response or very large response also have been reported (30,31). For some species, stomatal density may decrease with increasing CO_2 concentration (32). Some reports indicate no significant change (33) or increase in stomatal density with increasing CO_2 concentration (34). Morison (31) reported a reduction of about 40±5% in stomatal conductance for a doubling of CO_2 from 330 umol mol^{-1} based on 80 observations from the literature on 16 C_3 species. The mechanism of sensitivity of stomata to CO_2 remains unexplained.

Pallas (35) showed leaf transpiration rates of several species to decrease with increasing CO_2 concentration, whereas air temperature, humidity, irradiance, and flow rate were being maintained constant. Leaf temperature increases with decrease in transpiration by increased CO_2 concentration. All these factors will cause an increase in leaf vapour pressure through the increase in leaf temperature. This effect nullifies part of the direct reduction in transpiration expected due to a reduction in stomatal conductance.

Beladi et al. (36) developed a Soil-Plant-Atmosphere Model (SPAM) to determine canopy photosynthetic CER, transpiration plus evaporation from the soil surface (evapotranspiration or ET) etc. These models have been used to determine canopy photosynthetic CER at a number of CO_2 concentrations. The canopy ET decreases slightly with increasing CO_2 due to stomatal closure, and increases noticeably with increasing temperature-mediated VPD (vapour pressure deficit) at constant air vapour pressure. It is interesting to note that the decrease of ET due to a doubling of CO_2 concentration is approximately negated by an increase of ET due to a 3°C rise in air temperature.

The values of the foliage-to-air temperature differences were lower when the stomatal conductances were higher because of a greater degree of cooling associated with larger rates of transpiration. Soil surface moisture conditions may also affect the specific TF–TA values. Higher CO_2 world could increase foliage surface temperature by several degrees celsius. At a vapour pressure deficit of 4 kPa, the stomatal closure effect could cause foliage temperatures to be upto 4°C warmer at midday for an 800 umol mol^{-1} CO_2 world without any adjustments for atmospheric vapour pressure changes related to climate change or leaf area changes related to vegetation enrichment by CO_2. An increase in vapour pressure without any change in vapour pressure deficit would lead to an even greater increase in foliage temperature. Another scenario is that VPD would increase (mediated by global temperature rise) while atmospheric vapour pressure remains constant. This scenario would result in less increase of TF-TA. A more likely scenario is an increase in TA, a modest increase in atmospheric vapour pressure, and a partial closure of stomata, to produde a slight increase in TF-TA, with some increase in the absolute value of TF. The effect of higher TF-TA under raised global CO_2 levels might produce a local or regional warming, but the global effect would be small (37-39).

Water use and water-use efficiency in response to elevated CO_2

Plant canopy water use has been reported both to be reduced by CO_2 enrichment (40) and to be essentially unaffected by CO_2 (41). Although water use per unit leaf area decreases considerably with increasing CO_2 concentration due to stomatal closure, much of the reduction in transpiration that is expected from canopy is offset by larger LAI (41). Even if the LAI is similar, the canopy energy balance would cause foliage temperature to rise and prevent full expression of the reduction in stomatal conductance at elevated CO_2 on transpiration.

Water use efficiency (WUE) is defined herein as the ratio of photosynthetic CER to transpirational water loss. Although WUE is doubled (at equivalent leaf area) by a doubling of CO_2, most of the increase in C_3 plants comes from an increase in photosynthesis with only a small contribution from reduction in transpiration (10,41). High LAI canpies have lower WUE for a given CO_2 exposure level than low LAI canopies. Low LAI canopy (3.3) and the high LAI canopy (6.0) have almost the same canopy photosynthetic rates at equivalent CO_2 exposures, but the high LAI canopy has more transpiration which reduces its WUE.

Growth, biomass productivity and economic yield

The increased photosynthesis of plants translates eventually to increased biomass productivity and yield through individual plant growth processes. Acock and Allen (42) and Acock and Pasternak (43) have reviewed literature on these processes in detail. In general, higher CO_2 levels result in greater amounts of starch accumulation in leaves exposed to elevated levels of CO_2. Extra carbon in plant leaves induces by high CO_2 concentrations results in an increase in leaf size, number of branches (or tillers) and number of nodes along the branches which support leaves, and hence greater leaf area from these two effects. Leaves also increase in specific leaf weight (dry weight per unit leaf area) with increasing CO_2 enrichment. Allen et al. (44) predicted canopy photosynthetic rates biomass yield and seed yield to increase 53, 43 and 32%, respectively, for a doubling of CO_2 from 315 to 630 umol mol^{-1}. CO_2 enrichment of rice has been shown to increase dry weight, plant height, tillering and yield. Yoshida (45) reported optimum CO_2 concentration for growth and yield in rice to be between 1500 and 2000 umol mol^{-1}.

Interaction of enriched CO_2 with other environmental factors

Pearcy and Bjorkman (46) have reviewed leaf CER responses of a C_3 and a C_4 desert shurb to leaf temperature at 330 and 1000 umol mol^{-1} CO_2 at high light levels. They showed a low, flat temperature optimum of C_3 species with a maximum between 35 to 40°C when the CO_2 concentration was 330 umol mol^{-1},but a sharper temperature optimum about 2.5-fold larger near 44°C when the CO_2 concentration was 1000 umol mol^{-1}. In bigtooth aspen (C_3 species),peak CER increased 2.5-fold and the optimum temperature increased from 25 to 37°C when CO_2 level was raised from 320 to 1900 umol mol^{-1} (47). Presumably RuBP carboxylase- oxygenase affinity for CO_2 decreases with increasing temperature.

Whether increased temperature alongwith increased CO_2 also enhances overall growth or not over the long term depend strongly on germplasm and growth conditions. Okra, at ambient CO_2 concentration (350 umol mol^{-1}) and day night temperature below 26/20°C, did not survive, whereas the plants thrived at 20/14°C when CO_2 concentration was 450, 675 or 1000 umol mol^{-1}. Seed yield for soybean at 660 umol mol^{-1} decreased slightly with temperature over range of 23 to 33°C, but seed yield increased slightly with temperature at 330 umol mol^{-1} CO_2. Idso et al. (48) reported that the growth modification factor (i.e. the biomass growth increase ratio) for a 300 umol mol^{-1} elevation of CO_2 increased somewhat linearly (0.087 per °C) with increasing temperature over the range of 19 to 34°C for five C_3 species of plants. In fact, they reported a negative effect of elevated CO_2 for temperature less than 19°C. The findings with above crops indicate that whole-season, whole crop responses may be different from short-term, single leaf photosynthetic rate responses. Perhaps, many plants may thrive in cooler environments in higher global CO_2 atmospheres. Hence, quick predictions based on single leaf data may be misleading and that more research is needed to study long term CO_2 X temperature interactions on all plant processes.

Conclusions

From the discussion presented here, it is clear that the rate of photosynthesis is affected by many factors, particularly the availability of light, water and nutrients. It is not, however, very sensitive to temperature change. The rates of plant respiration and decay, on the other hand, do strongly depend on the temperature. A one-degree temperature change in either direction often alters rates of plant respiration by from 10 to 30%. Hence, the global warming will speed the decay of organic matter without appreciably changing the rate of photosynthesis. That will increase the release of CO_2 into atmosphere. A warming will also result in more methane, because methane is produced by respiration in regions where oxygen is not freely available, such as swamps, bogs and moist soils. In recent years, there has been a rise in the concentration of atmospheric methane of more than 1% per year. The increase is both rapid and significant because methane is 20 times as effective as CO_2 in trapping heat.

Though, an attempt was made here to discuss the direct effects of CO_2 on plant processes, less attention has been given to the question of interactions of CO_2 effects with potential climatic changes. Similarly, the effects of other green house grasses (methane, chlorofluorocarbons, nitrous oxide) in combination with CO_2 have not been identified. Hence, future research on direct effects of CO_2 on crops and ecosystems should also include interactions with the environmental variables (e.g. temperature, rainfall, vapour pressure deficit, and soil moisture) that may be involved with predicted future climatic changes.

References

1. White, R.M. (1990). The great climate debate. Vigyan 1: 18-29.
2. Allen, L.H., Boote, K.J., Jones, J.W., Jones, P.H., Valle, R.R., Acock, B., Rogers, H.H. and Dahlman, R.C. (1987). Response of vegetation to rising CO_2: Photosynthesis, biomass, and seed yield of soybean. Global Biogeochemical cycles 1: 1-14.
3. Downton, W.J.S., Bjorkman, O. and Pike,C.S. (1980). Consequences of increased atmosphere concentration of CO_2 for growth and photosynthesis of higher plants. In CO_2 and climate: Australian research. CSIRO Division of Atmospheric Physics. Aspendale, Victoria, Australian Academy of Sci., Canberra. G.I. Pearman ex., pp. 143-151.

4. Kimball, B.A. (1983). CO_2 and agricultural yield. An assemblage and analysis of 430 prior observations. Agron. J. 75: 779-788.
5. Strain, B.R. (1978). Report of the workshop on anticipated plant responses to global CO_2 enrichment. Dept. of Botany, Duke Univ., Durham, N.C.
6. Lemon, E.R. (1983). CO_2 and plants. The response of plants to rising levels of atmospheric CO_2. AAAS Selected Symp. 84. Westview Press, Boulder, CO.
7. Dahlman, R.C. and Strain, B.R. (1984). Vegetation response to CO_2 research plan. DOE/ER-0187, U.S. Dept. of Energy, Carbon Dioxide Research Div., Washington, DC.
8. Strain, B.R. and Cure, J.D. (1985). Direct effects of increasing CO_2 on vegetation. DOE/ER-0238, U.S. Dept. of Energy, Carbon Dioxide Research Div., Washington, DC.
9. Rogers, H.H., Bingham, G.E., Cure, J.D., Smith, J.M. and Surano, K.A. (1983). Responses of selected plant species to elevated CO_2 in the field. J. Environ. Qual. 12: 569-574.
10. Valle, R., Mishoe, J.W., Campbell, W.J., Jones, J.W. and Allen, L.H. (1985). Photosynthetic responses of Bragg soybean leaves adapted to different CO_2 environments. Crop Sci. 25: 333-339.
11. Campbell, W.J., Allen, L.H. and Bowes, G. (1988). Effects of CO_2 concentration on rubisco activity, amount and photosynthesis in soybean leaves. Plant Physiol. 88: 1310-1316.
12. Havelka, U.D., Wittenbach, V.A. and Boyle, M.G. (1984). CO_2 enrichment effects on wheat yield and physiology. Crop Sci. 24: 1146-1150.
13. Peet, M.M., Huber, S.C. and Patterson, D.T. (1986). Acclimation to high CO_2 in monoecious cucumbers II. Carbon exchange rates, enzyme activities, and starch and nutrient concentration. Plant Physiol. 80: 63-67.
14. Huber, S.C., Rogers, H.H. and Israel, D.W. (1984). Effects of CO_2 enrichment and photosynthate partitioning in soybean leaves. Physiol. Plant. 62: 95-101.
15. Vu, J.C.V., Allen, L.H. and Bowes, G. (1989). Leaf ultrastructure, carbohydrates, and protein of soybeans grown under CO_2 enrichment. Environ. Exp. Bot. 29: 141-147.
16. Potter, J.R. and Breen,P.J. (1980). Maintenance of high photosynthetic rates during the accumulation of high leaf starch levels in sunflower and soybean. Plant Physiol. 66: 528-531.

17. Cave, G., Tolley, L.C. and Strain, B.R. (1981). Effects of CO_2 enrichment on chlorophyll content and starch grain structure in Trifolium subterraneum leaves. Physiol. Plant. 51: 171-174.

18. Huber, S.C. (1983). Role of SPS in partitioning of carbon in leaves. Plant Physiol. 71: 818-821.

19. Smith, S.D., Strain, B.R. and Sharkey, T.D. (1987). Effects of CO_2 enrichment on four Great Basin grasses. Functional Ecol. 1: 139-143.

20. Jones, P., Allen, L.H., Jones, J.W. and Valle, R.R. (1985). Photosynthesis and transpiration responses of soybean canopies to short and long-term CO_2 treatments. Agron. J. 77: 119-126.

21. Baker, J.T., Allen, L.H., Boote, K.J., Rowland, A.J., Jones, J.W., Jones, P.H., Bowes, G. and Albrecht, S.L. (1988). Response of vegetation to CO_2. Plant Stress and Protection Research Unit, USDA-ARS, University of Florida.

22. Cure, J.D. (1985). CO_2 doubling response. A crop survey. In: Direct effects of increasing CO_2 on vegetation. DOE/ER-0238. U.S. Dept. of Energy. Carbon dioxide Res. Div., Washington, DC. pp. 99-116.

23. Acock, B., Reddy, V.R., Hodges, H.F., Baker, D.N. and McKinion, J.M. (1985). Photosynthetic response of soybean canopies to full-season carbon dioxide enrichment. Agron. J. 77: 942-947.

24. Vu, J.C.V., Allen, L.H. and Bowes, G. (1983). Effect of light and elevated atmospheric CO_2 on the RuBPcase activity and RuBP level of soybean leaves. Plant Physiol. 73: 729-734.

25. Dietz, K.J.and Heber, U. (1984). Rate limiting factors in leaf photosynthesis. I. Carbon fluxes in the Calvin cycle. Biochem. Biophys. Acta 767: 432-443.

26. Dietz, K.J. and Heber, U. (1984). Rate limiting factors in leaf photosynthesis.2. Electron transport. Biochem. Biophys. Acta 767: 444-450.

27. Badger, M.R., Sharkey, T.D. and Coemmerer, S.V. (1984). The relationship between steady-state gas exchange of bean leaves and the levels of carbon-reduction-cycle intermediates. Planta 160: 305-313.

28. Stitt, M., Wirtz, W. and Heldt, H.W. (1983). Regulation of sucrose synthesis by cytoplasmic fructose bisphosphatase and SPS during photosynthesis in varying light and CO_2. Plant Physiol. 72: 767-774.

29. Leegood, R.C. and Walker, D.A. (1982). Regulation of fructose-1,6-bisphosphatase in leaves. Planta 154: 449-456.

30. Morison, J.I.L. (1985). Sensitivity of stomata and water use efficiency to high CO_2. Plant Cell Environ. 8: 467-474.

31. Morison, J.I.L. (1987). Intracellular CO_2 concentration and stomatal response to CO_2. In: Stomatal Function, Zeiger et al., ed., Stanford Univ. Press, Stanford, CA, pp. 229-252.

32. Woodward, F.I. and Bazzaz, F.A. (1988). The response of stomatal density to CO_2 partial pressure. J. Exp. Bot. 39: 1771-1781.

33. Korner, C. (1988). Does global increase of CO_2 alter stomatal density? Flora 181: 253-257.

34. Thomas, J.F. and Harvey, C.N. (1983). Leaf anatomy of four species grown under continuous CO_2 enrichment. Bot. Gaz. 144: 303-309.

35. Pallas, J.E. (1985). Transpiration and stomatal opening with changes in CO_2 of the air. Science 144: 171-173.

36. Beladi, S.E., Allen, L.H., Campbell, W.J. and Jones, J.W. (1983). Simulation of crop canopy responses to CO_2 concentration, temperature and vapour pressure deficit. Agronomy abstracts, pp. 68.

37. Idso, S.B., Kimball, B.A. and Anderson, M.G. (1986). Foliage temperature increases in water hyacinth caused by atmospheric CO_2 enrichment. Arch. Meteorol. Geophys. Bioclimatol. Ser. B 36: 365-370.

38. Idso, S.B., Kimball, B.A., Anderson, M.G. and Mauney, J.R. (1987). Effects of atmospheric CO_2 enrichment on plant growth. The interactive role of air temperature. Agric. Ecosystems Environ. 20: 1-10.

39. Idso, S.B., Allen, S.G. and Choudhury, B.J. (1988). Problems with porometry: Measuring stomatal conductances of potentially transpiring plants. Agric. Forest Meteorol. 43: 49-58.

40. King, M.K. and Greer, D.H. (1986). Effects of CO_2 enrichment and soil water on maize. Agron. J. 78: 515-521.

41. Jones, P., Allen, L.H., Jones, J.W. and Valle, R.R. (1985). Photosynthesis and transpiration responses of soybean canopies to short and long-term CO_2 treatments. Agron. J. 77: 242-249.

42. Acock, B. and Allen, L.H. (1985). Crop responses to elevated CO_2 concentration. In: Direct effects of increasing CO_2 on vegetation, B.R. Strain and J.D. Cure, ed., DOE/ER-0238. U.S. Dept. of Energy, CO_2 Res. Div., Washington, DC. pp. 53-97.
43. Acock, B. and Pasternak, D. (1986). Effects of CO_2 concentration on composition, anatomy and morphology of plants. In: CO_2 enrichment of green house crops. Vol. II. Physiology, Yield and Economics, H.Z. Enoch and B.A. Kimball, ed., CRC Press, Boca Raton, FL, pp. 41-52.
44. Allen, L.H., Boote, K.J., Jones, J.W., Jones, P.H., Valle, R.R., Acock, B., Rogers, H.H. and Dahlman, R.C. (1987). Response of vegetation to rising CO_2: Photosynthesis, biomass, and seed yield of soybean. Global Biogeochemical Cycles 1: 1-14.
45. Yoshida, S. (1976). CO_2 and yield of rice. In: Climate and rice, IRRI, Los Banos, Philippines, pp. 211-221.
46. Pearcy, R.W. and Bjorkman, O. (1983). Physiological effects In: CO_2 and plants. The response of plants to rising levels of atmospheric CO_2, E.R. Lemon, ed., pp. 65-105.
47. Jurik, T.W., Webber, J.A.and Gates, D.M. (1984). Short term effects of CO_2 on gas exchange of leaves of bigtooth aspen in the field. Plant Physiol. 75: 1022-1025.
48. Idso, S.B., Kimball, B.A. and Mauney, J.R. (1987). Atmospheric CO_2 enrichment effects on cotton midday foliage temperature: Implications for plant water use and crop yield. Agron. J. 79: 667-672.
49. Cure, J.D., Rufty, T.W. and Israel, D.W. (1988). Phosphores stress effects on growth and seed yield responses of non-nodulated soybean to elevated CO_2. Agron. J. 80: 897-902.
50. Luxmore, R.J., O'Neil, E.G., Ellis, J.M. and Rogers, H.H. (1986). Nutrient uptake and growth responses of Virginia pine to elevated atmospheric CO_2. J. Environ. Qual. 15: 244-251.
51. Norby, R.J., O'Neill, E.G. and Luxmore, R.J. (1986). Effects of atmospheric CO_2 enrichment on the growth and mineral nutrition of Quercus alba seedlings in nutrient-poor soil. Plant Physiol. 82: 83-89.
52. Sionit, N., Mortensen, O.A., Strain, B.R. and Hellmers, H. (1981). Growth responses of wheat to CO_2 enrichment and different levels of mineral nutrition. Agron. J. 73: 1023-1027.

CARBON DIOXIDE ENRICHMENT, CLIMATIC CHANGE AND INDIAN AGRICULTURE: A PRELIMINARY ANALYSIS

D.P. SINGH, PHOOL SINGH, R.K. PANNU AND H.C. SHARMA

DIRECTORATE OF RESEARCH,
HARYANA AGRICULTURAL UNIVERSITY,
HISAR 125004, INDIA

ABSTRACT

Agriculture could be affected in two broad ways: by the direct effects of CO_2 on plants, and by warming effects of CO_2 and other greenhouse gases on the climate. A preliminary assessment of the potential effects of CO_2 enrichment and climate change suggests that higher CO_2 will have positive effects on growth and yield of major food crops grown under moderate to adequate supply of water through irrigation/rainfall in India. In these areas, the adverse effects of 1-2 °C rise in air temperature could be absorbed with the 5-10% increase in precipitation. For a doubling of CO_2 concentration, yield increases of 20-30% may be possible on about 70% area under rice and wheat. There is also possibility of improvement in the yield of cotton, groundnut and C4 species under new environmental conditions. In winter dryland crops, CO_2 enrichment and rise in temperature could probably result in depletion of soil water at rapid rate during vegetative phase of growth, leading to moisture stress during grain filling stage and in turn poor harvest index and low productivity. In North India, warming would offset some losses in yield by early pod set in winter grain legumes like chickpea and lentils, and by arresting frost damage in oilseed brassicas. A large proportion of an adverse effect on crop yield and production could be avoided or absorbed by suitable crop adjustments, developing proper genotypes, and adopting appropriate technology for the changed environmental conditions.

The coastal low lands and deltas could however be threatened by flooding, erosion and salt intrusion and thus are particularly at risk. Drainage may also be impeded as sea level rises, resulting in loss of yield or utlimately a change in land use of the agricultural areas. Salt water intrusion means that some crops may have to be abandoned or alternative varieties cultivated. In most respects, it would be wise to advance these issues for further research, rather than as findings.

The crop productivity could be affected in two broad ways by CO_2 enrichment, i.e. by the direct effects of CO_2 on plants, and by the effects of changes in climate. The burning of coal, gas, oil and the deforestation of large areas of land have caused a rise in the atmospheric CO_2 concentration from an estimated preindustraial level of about 265 ppmv (Parts per million by volume) to about 345 ppmv of air. Some scenarios predict that atmospheric CO_2 could double within 100 years. There is, however, major uncertainty about how quickly the CO_2 concentration will build up. Predictions of future concentrations depend, first on assumptions about the rates of consumption and mix of fossil fuels, and second on knowledge about the sources and sinks of carbon, i.e. the global carbon cycle. Added to CO_2 are the changes in the atmospheric concentration of other greenhouse gases (for example, CH_4, N_2O, Chlorofluorocarbons) which have the same radiative forcing as doubled CO_2. Thus, the effective doubling of CO_2 concentration will occur around the year 2030, if present omission trends continue [1]. The climate change caused by this effective doubling may be delayed owing to slow uptake of heat by the oceans and other factors. The results of most climate models suggest that a doubling of CO_2 will cause a global surface air warming from 1.5 to 4.5 °C from the so called greenhouse effect [1,46]. But the changes in climate will not follow step by step. There are also two major uncertainties about the climate change (when and where) which present major difficulties for agricultural impact analysis. Climatic changes will vary in direction and magnitude from one agricultural region to the next. Unfortunately, general circulation models (GCMs) can not yet provide reliable predictions of climatic change at regional scales appropriate for most impact analysis [25,46]. From the experimetnal observations and crop models, it is estimated that a doubling of atmospheric CO_2 concentration from 340 to 680 ppmv would increase the growth and yield of the major C3 crops by 10-50 per cent and of C4 crops by 0-10 per cent [1,2,46]. Crop impact analysis using crop climate models suggest that a 2 °C rise in air temperature in mid-latitude cereal regions may decrease yields by 3 to 17 per cent [46]. These results are, however, derived from leaf chambers, plant growth chambers, greenhouses and phytotrons in which environmental factors can be controlled with a high degree of precision. Evidence from actual field experiments is more limited due to difficulties in controlling and simulating actual field conditions [7,8,25]. Thus, it is unclear that how crops will fare in a CO_2 enriched world with competition from weeds, pests, diseases and other stresses. These lacks on vital information do not, however, preclude us in analysing the probable sensitivity of Indian agriculture to increased CO_2 concentrations and associated climatic changes. Such analysis is only tentative and aimed at to review the probable impact of CO_2 enrichment and climatic changes on photosynthesis, water use and productivity of important field crops grown in Indian sub-continent.

DIRECT EFFECTS OF CO_2 ENRICHMENT

Almost all recent studies have shown an increase of leaf photosyntheic rates when crop plants were exposed to long term elevated CO_2 levels [2,46]. This effect was most pronounced in C3 plants (e.g. wheat, soybean, rice, barley) as compared to C4 plants (e.g. maize, sugarcane, sorghum). From experimental observations, it was estimated that a doubling of atmospheric CO_2 concentration from 340 to 680 ppmv could increase the growth and yield of the major crops by 10-50 per cent [46]. Kimball [20] reviewed over 700 prior experiments involving effects of CO_2 on C3 agronomic plants and concluded that an atmosphere of 600 ppmv CO_2 of air would result in about one thrid increase in productivity. The plants with the C3 photosynthetic pathway are CO_2 limited and have higher photorespiration. In C3 plants, the CO_2 concentration increases directly the rate at which the carbon is fixed and also suppresses photorespiration. In contrast, C4 plants are largely CO_2 saturated at current ambient levels and also have negligible photorespiration. Therefore, higher CO_2 concentration has comparatively little effect in directly stimulating the photosynthetic response in C4 plants. For C4 species, the increase in yield would be less, 0-10 per cent [46].

At the crop level, several important feed backs can increase or reduce the effects of higher CO_2 on canopy photosynthesis over the duration of growth period [1,2,4,46]. One important feed back observed in many crop species was increased leaf area expansion which allowed greater interception of incident radiation and greater whole plant photosynthesis- a positive feed back. For 660 ppmv CO_2, photosynthesis rates were increased by 35, 25 and 10% in soybean, wheat and maize, respectively [1]. Cure [9] found that net photosynthesis of C3 crop species could be expected to increase by an average of 28% in an atmosphere of 680 ppmv CO_2. The proportion of N in the tissues at the final harvest was unaffected by elevated CO_2 concentration [3]. Therefore, the CO_2 enrichment functions mainly as a carbon nutrient for plants. These results are, however, derived largely from leaf chambers, green houses and phytotrons in which environmental factors can be controlled at desired level [25,46].

CO_2 enrichment also resulted in more number of tillers and branches which support the leaves and hence the greater leaf area from these two effects [2]. Self shading influenced the net assimilation rate at high CO_2 levels in soybean [15]. Evidence from field experiments carried out under 330 ppmv CO_2 concentration indicated that the incident photosynthetically active radiation (PAR) was not adequate to saturate the present day available canopy of most of the winter season crops [35,36,39,41,50] and wet monsoon season crops in North India [24,42]. Under controlled conditions, Singh et al [36,41] found light saturation of leaves at around 1600 $\mu E\ m^{-2}\ s^{-1}$ PAR at flowering stage in oilseed brassicas and chickpea. On the other hand, in the absence of environmental or physiological limitations, light response curves for leaves of C4 species did not saturate at 2400 $\mu E\ m^{-2}\ s^{-1}$ PAR intensity [21]. In North India, the incident PAR always remains below 1600 $\mu E\ m^{-2}\ s^{-1}$ in winter and wet monsoon seasons and below 2000 $uE\ m^{-2}\ s^{-1}$ during summer dry months [26,35,39]. Due to this sub-optimum available incident radiation, skipping one row after every two or three planted rows had rather beneficial effect in wheat [28], chickpea [37], oilseed brassicas [38] and cotton [43] in North India (Table 1). Under low light intensity, there was no difference in net photosynthesis rates of water-lily between 340 and 640 ppmv CO_2 concentration [4]. Therefore, the sub-optimal PAR at greater leaf area index with CO_2 enrichment will negate the positive response of elevated CO_2 on canopy photosynthesis and productivity to some extent under irrigated field conditions in India.

At leaf level elevated CO_2 reduced transpiration due to decline in stomatal conductance in most of over 50 species reviewed by Allen [2]. Thus the reduced transpiration and enhanced photosynthesis could result in higher water use efficiency (ratio between net CO_2 uptake for photosynthesis to water loss by transpiration). At the crop level, however, several important feed backs can modify the effects of higher CO_2 on plant growth over the duration of growth period. If vegetation LAI should increase with rising CO_2 then canopy ET is expected to increase due to the presence of more transpiring surface. Recent studies carried out on soybean [19] and wheat [7] have shown that the water transpired increased as the level of CO_2 increased. Water use efficiency with CO_2 enrichment increased in wheat [7] but decreased in soybean [2]. In soybean, the low LAI canopy (3.3) and high LAI canopy (6.0) had almost the same canopy photosynthetic rate at equivalent CO_2 exposure but the high LAI canopy had more transpiration which reduced its water use efficiency.

Table 1: Effect of crop geometry on yield of chickpea, mustard, wheat and cotton.

Crops	Crop geometry	Seed yield* (T/ha)
Chickpea [37]	30 cm row spacing	2.81
	Skipping one row after every two rows (30 + 60 cm)	2.91
Mustard [38]	30 cm row spacing	1.66
	Skipping one row after every two rows (30 + 60 cm)	2.22
Wheat [28]	22.5 cm row spacing	3.73
	Skipping one row after every three rows (22.5 + 67.5 cm)	3.80
Cotton [43]	75 cm row spacing	1.99
	120 cm gap between two rows planted 30 cm apart (30 + 120 cm)	2.93

* Seed cotton in case of cotton

Practically nothing is known about the consequences of increased ET with LAI at elevated CO_2 level on pattern of soil water use in plants under dryland sitatuions. In a recent study carried in boxes on winter wheat under high and low water level treatments (winter wheat planted in October was subjected to differential watering between April and June) provided some data on rooting pattern [8]. In the top soil layer (0-10 cm) drought stressed wheat grown with 825 ppmv CO_2 produced more roots than did adequately watered plants grown with 330 ppmv CO_2 concentration. Root weight remained almost constant below 50 cm soil depths for all the treatments. Authors did not, however, provide any data on water use pattern from different soil depths. Thus, it may be presumed that the increased ET (about 60 mm) at high CO_2 concentration in drought stressed wheat would have been supported by additional water extraction from top soil layers with improved root growth rather than more water absorption from the deeper layers of soil profile. It may have adverse influence on the productivity of those dryland crops which are grown entirely on conserved moisture in the profile before seeding [27,34]. In Indian context, this type of root system may be beneficial in those areas where little amount of rainfall is available at frequent interval to cause only mild stress in crop plants. The majority of dryland crops in India either have long spell of drought particularly during reproductive phase of growth in monsoon season or practically no rainfall during the entire growth period in winter season. On conserved soil moisture, the productivity was related to water use from sub-soil during reproductive phase of growth in oil seed brassicas [34], rice [23] and wheat [22]. Therefore, the over all productivity of real dryland winter season crops (grown predominately on stored soil water at planting) may decrease with CO_2 enrichment because of rapid early growth, higher ET during initial vegetative growth and exhaustion of soil water before grain filling, leading to forced maturity and poor harvest index. However, further research is needed to know more about the influence of elevated CO_2 on water use pattern, canopy expansion and partitioning of dry matter to grains under real dryland situations to draw any meaningful conclusion.

Other feed backs can also be negative. For instance, photosynthetic acclimation can result from an imbalance in the source/sink ratio: supply of photosynthate in the leaf due to an inadequate development of sink (or storage organs like grains) which can slow down photosynthesis at source. This type of acclimation seems to be common in many species where the initial stimulation of photosynthetic rate by elevated CO_2 decreases to the point where leaf or canopy photosynthetic rates become similar to those at ambient CO_2 concentrations [2,31,46,51]. No species exhibited an idealized acclimation response at the biochemical level and growth of three out of the five species (P. vulgaris, S. melongena, B. oleracea) was affected at high CO_2 level [31]. The weedy species (C. album) exhibited the most economical response to long term CO_2 enrichment because its rubisco content decreased yet the rate of photosynthesis at high CO_2 level increased. Thus the weedy species may enhance their ability to compete against the agricultural crop species with less effective acclimation responses. Evidently, there is no simple acclimation response to levels of CO_2; responses are likely to vary widely with species, environment, stage of growth and pretreatment conditions [2]. However, for most of the important crop species, the net effect of elevated CO_2 on plant growth and yield is positive [46].

EFFECTS OF TEMPERATURE

The second way in which agriculture could be affected is by changes in climate [46]. The surface temperature would increase from 1.5 to 4.5 °C for doubling of the pre-industrial CO_2 level [49]. Warming will probably be greater in the higher latitudes than in the tropics, and will be most pronounced during the winter half of the year. At higher latitudes the warming according to the models, will probably be at least twice the global average [16].

In North India for winter season crops like oilseed brassicas and chickpea, the response of canopy photosynthesis to PAR is temperature dependent [35,39]. At the current ambient CO_2 concentration the optimum temperatures for maximum rate of net canopy photosynthesis lie between 20-26 °C for winter season crops. During cooler months (December to first week of February), the canopy photosynthesis was always lower for a certain quanta of PAR during the first half than the second half of the day. Low temperatures also resulted in an ineffective flowering (pseudo and aborted flowers) in chickpeas [35] and frost damage in oilseed brassicas [10]. Therefore, the increase in air temperatures due to warming is expected to enhance the canopy photosynthesis and growth during the cool months in sensitive crops like oilseed brassicas and chickpea in North India. On the other hand, a reverse trend of air temperature on canopy photosynthesis was recorded on warm days, i.e. after the middle of February. Most of the winter cereal crops and pulses enter in the anthesis and grain filling stages of growth in February-March in North India. During this critical period of crop development, the elevated temperatures resulted in accelerated leaf senescence, decline in canopy photosynthesis and forced maturity in winter cereal crops [33,34,50].

INTERACTION OF CO2 AND TEMPERATURE

What is less clear, however, is how the expected rise in CO_2 and air temperature will interact and affect plant processes such as net photosynthesis and growth? Kimball [20] and Cure [9] reviewed the interactive effect of atmospheric CO_2 and temperature on plant growth. Kimball [20] reported that the percent increase in growth caused by CO_2 enhancement was about the same over the range of temperatures, and Cure [9] concluded that a doubling of CO_2 generally caused greater increase in growth at high temperature. Campbell et al [6] reported that a 10 °C difference in air temperature during the growth period of soybean caused only a 4 °C leaf temperature difference, which was insufficient to change the photosynthetic rate or rubisco activity in the canopies grown and measured at either 330 or 660 ppmv CO_2. However, Sionit et al [45] reported increased leaf photosythesis of soybean in response to CO_2 enrichment more at 22/16 °C than at 26/20°C. Backer et al [5] reported that seed yield of soybean at 660 ppmv CO_2 decreased slightly with temperature over range of 23 to 33 °C but it increased slightly with temperature at 330 ppmv CO_2. Contrarily, Idso et al [18] found negative effect of 300 ppmv elevated CO_2 for temperatures less than 19 C for five C3 species (carrot, radish, cotton, water-hycinth, water-fern) but a somewhat linear increase of growth at elevated CO_2 with increasing temperature over the range of 19 to 34 C for all the species. The plants of okra at ambient temperature did not survive at day/night temperatures blow 26/20 °C, whereas plants thrived at 20/14 °C when CO_2 concentration was 450, 675 and 1000 ppmv [44]. Conditions of cool temperature (20 °C) promoted leaf injury to CO_2 enriched plants of Phaseolus vulgaris [13]. The experience with okra and soybean as well as C4 grasses suggests that the whole crop responses of certain C3 plants may be different from short term, single leaf photosynthetic rate responses [2]. The comparison of whole crop, long term responses also suggests that quick predictions based on single leaf data may be misleading, and that more research is needed to study long term CO_2 and temperature interactions on all plant processes.

In many outdoor studies, increased vapour pressure deficit is caused by an increase in air temperature [2,4,19]. An increase of air temperature from 28 to 33 °C caused a 20% increased in soybean canopy transpiration at both 330 and 800 ppmv CO_2 and an increase in temperature from 28 to 35 °C caused 30% increase in canopy transpiration [19]. There was a linear increase in transpiration with increase in temperatures from 10 to 42 °C at 330 ppmv CO_2 in chickpeas [36] and oilseed rape [41] under controlled conditions. Under field conditions at ambient CO_2 level, there was a sudden rise in ET with temperatures above 20 °C in barley [50], chickpea, mustard and wheat (Fig.1), and above 32 °C in mungbean [24]. As stated earlier that if vegetation LAI increases with rising CO_2 concentration, then ET is expected to increase further with rise in CO_2 [2]. Higher temperatures could be deterimental to yields in most of the dryland crops in Indian sub-continent. In part this is because higher temperatures are associated with greater vapour pressure deficit and ET, and thus greater moisture stress, particularly during reproductive phase of growth in winter season crops grown on conserved moisture in soil profile. Because of more available water at the time of seeding of winter season rainfed crops, the CO_2 enrichment can lead to exhausten of limited water supply at rapid rate during initial vegetative phase of growth due to faster development of above ground canopy and thus soil may run out with water during critical reproductive phase of growth. Studies caried out in large containers also indicated that most of the crops unwatered after 30-35 days after seeding attained compensation point of photosynthesis at the time of anthesis (Table 2). Occurrence of moisture stress during the critical stages of flowering and grain filling may lead to drastic reduction in post-anthesis canopy photo-synthesis and economic yield (Fig. 2). The yield realization under such moisture stress conditions ranged from zero yield in chickpea [40] to 34% grain yield in mungbean [24]. Thus, an increase in LAI and vapour pressure deficit because of increase in temperature may lead to higher ET and thus less available water to plants in the soil profile. It is known that higher temperature accelerate plant development and shorten the growth period. This could lead to drought escape and realization of some yield in water deficit season, but could be detrimental to attain higher productivity in good rain fall years from dryland areas in general and irrigated areas in particular. It is estimated that a 2 °C increase in air temperature may decrease seed yields in mid-latitude cereal regions of North America and Europe by 3 to 17% [46], but no such estimates are available for Indian sub-continent.

Table 2: Plant water status for compensation point of canopy photosynthesis in important field crops.

Crops	Cultivars	Stage of growth	Leaf water potential (bars $\pm$ 2)
Wheat	WH-147	Anthesis	-42
Barley	BG-25	Anthesis	-46
Chickpea	H-355	Pod initiation	-30
Mustard	parkash	Flowering	-43
Pearl millet	BJ-104	Heading	-40
Cluster bean	FS-277	Pod initiation	-30
Green gram	T-44	Flowering	-16

CROP IMPACT ANALYSIS

There are large uncertainties about the rates at which CO_2 and other greenhouse gases will accumulate in the atmosphere. It is also not very certain that when and where it will be warm. Taking into accounts the projected concentrations of the greenhouse gases (with their associated uncertainties), there could possibly be equivalent doubling of CO_2 by as early as 2030 [49]. But changes in

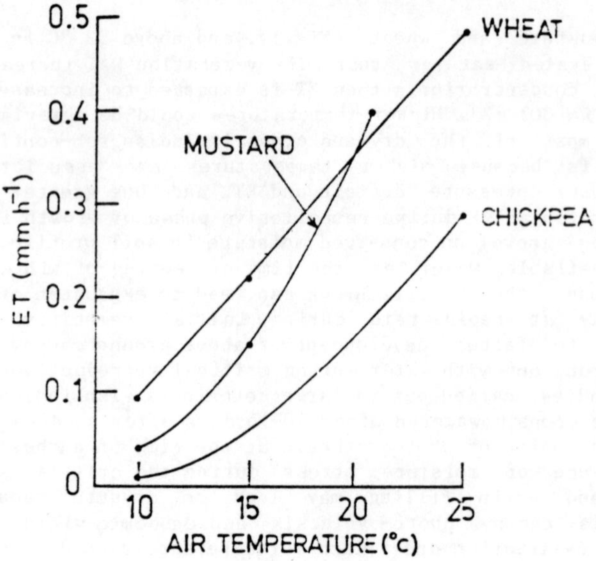

Fig. 1: Effect of air temperature on evapotranspiration (ET) in wheat ($Y = -0.168 + 0.026 X$; $R^2 = 0.87$), mustard ($Y = 0.118 - 0.028 X - 0.002 X^2$; $R^2 = 0.97$) and chickpea ($Y = - 0.0358 - 0.0045 X + 0.0007 X^2$; $R^2 = 0.83$). Diurnal data of clear days at anthesis have been used for curve fitting.

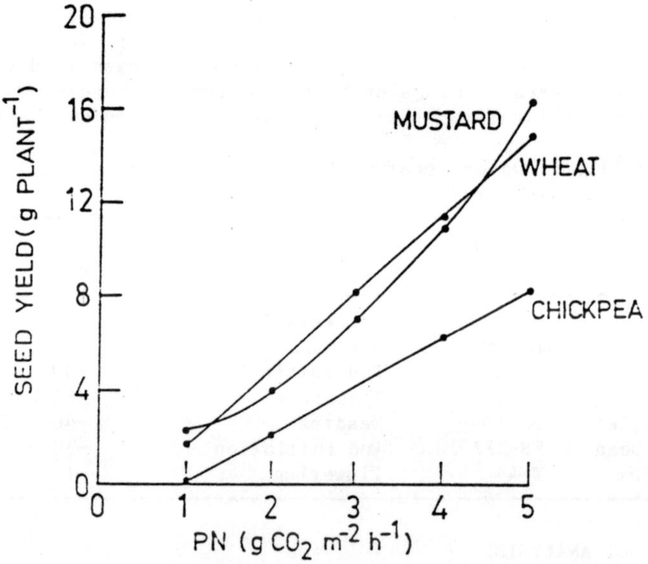

Fig. 2: Relationship between canopy net photosynthesis rates (ground area basis) and grain yield of mustard ($Y = 1.72 + 0.03 X + 0.58 X^2$; $R^2 = 0.90$), wheat ($Y = - 0.93 + 3.77 X$; $R^2 = 0.94$) and chickpea ($Y = 1.97 + 2.02 X$; $R^2 = 0.99$. Data of six irrigation treatments have been used for curve fitting.

climate will not follow step by step. Owing to the thermal intertia of the oceans, changes in global surface temperature will probably lag behind by several decades. Thus, the question of 'when' is clouded. The question of where is even more problematic. Climatic changes will vary in direction and magnitude from one agricultural region to the next. Unfortunately, GCMs cannot yet provide reliable predictions of climatic change at regional scales appropriate for most impact analysis. The current scientific consensus is that world should eventually warm by about 1.5 to 4.5 °C if the CO_2 concentration doubles [25,46,49]. There is general agreement that the higher latitudes will warm twice more than the lower latitudes, that the winter half of the year should warm more than the summer half and that on an average the world should receive more precipitation [47]. Lower latitudes (0-30 N) are expected to receive 5-10% more rainfall. Relative humidity will not change, and with clear sky, solar radiation is expected to decrease slightly by about 1% [17]. Humidity deficit, which is one of the factors governing the evapotranspiration will, however, increase by about 7% for each 1 °C rise in temperature if the relative humidity is unchanged. Temperature frequently plays a dominant role in the distribution and development of pests and diseases. It is expected that losses due to pests and diseases could increase as much as 15% at higher latitudes [25]. Scientific literature also suggests that sea level has risen by 10-20 cm over the last 100 years, an increase of 1-2 mm per year [47]. A rise in sea level also increases the vulnerability of coastal areas to flooding from storm surges. An example of this is the changes in Brahmputra delta, much of which is less than one meter above sea level.

The way in which climate scenarios are used for impact analysis, depends partly on what aspects of climatic changes are assumed to be most important for agriculture. The predominent view, which can be termed the slow change view considers the potential problem for agriculture to be slow, cummulative degradation or improvement of the growing environment. The problem for agriculture will be how to adapt to changes such as gradual regional warming and dessication. In contrast from a "shift-in-risk" view the potential problem of climatic change derives from a change in the frequency of the climatic extremes that are detrimental or beneficial to crop production on seasonal basis (drought, floods, frost etc). It has been demonstrated [46,48] that small changes in the mean climate can result in large changes in the frequency of extremes (assuming no change in the variance). Studies with this view point tend to choose short term, extreme values of climatic variables for impact analysis. In general, the few studies conducted using global agricultural models suggest that long term climatic effects on agriculture can to a large extent, be adjusted and accomodated by the system [46]. But it should be borne in mind that such results are only as good as the models themseleves [17]. Global agricultural models because of their complexity and the limitations of data are largely unvalidated. There is largely uncertainty as to whether agriculture would respond in a similar direction and magnitude in the real world.

For Indian sub-continent, it is expected that with the doubling of CO_2, there will be 1 to 2 °C rise in temperature, and 5 to 10% increase in precipation. In India, out of 142 million ha cultivated area, about 31.4% is irrigated [14]. The drylands, constiuting about 68.6% of arable land in the country, contribute around 45% of cereals and 75% oilseeds and grain legumes to the national food basket. Besides 31.4% irrigated area in India, about 28% area falls under assured rainfall conditions. Thus, about more than half of the cultivated area falls under moderate to adequate supply of water. The principal C3 crops occupy about 76.6% (117.8 million ha) and C4 species about 23.4% (36.0 million ha) of the 153.8 million ha gross area under major field crops (Table 3). Thus, the majority of crops which fall in C3 will respond favourably to CO_2 enhancement. Research results suggest that a doubling of CO_2 could increase the growth and yield of major C3 crops by 10-50%, provided all other factors remain optimum for growth [46]. For wheat and barley yield increases as much as 40% have been suggested [9]. Yoshino et al [52] reported that for a doubling of CO_2 atmosphere the rice yields in Japan are expected to increase by about 4% for quick maturing genotypes and 26% for late maturing genotypes. In soybean under high light conditions, seed yields are predicted to increase 32% for a doubling of CO_2 from 315 to 630 ppmv [2]. Although C4 plants in general are not expected to respond as much to increasing CO_2 levels as C3 plants [12], several other experiments have shown positive response of C4 grasses to CO_2 [30]. For C4 crops, the increase would be less, 0-10 per cent [46]. One should, however, bear in mind that such results are derived largely from the growth chambers and not from the farmer's field. How crops will fare in a CO_2 enriched world with competition from weeds, pests, diseases and other unforeseen adversities is a major uncertainty [25,46]. However, several studies suggest that for many species a doubling of CO_2 will lead to substantial increase in yield under field conditions [25].

Changes in CO_2 concentration and greenhouse gases are associated with changes in weather. Changes in weather could affect agriculture in number of ways, for example through changes in water use, length of growing season, geographical shifts in agricultural potential, incident of pests and changes in mean crop yield. Probably the most important consequences for agriculture would stem from a reduction in soil moisture due to higher rates of transpiration from plants and of evaporation from soil surfaces exposed to higher temperatures. Vapour pressure deficit which is one of the factors governing evapotranspiration will increase by about 7% for each 1 °C rise in temperature if the relative himidity is unchanged [17]. In soybean for each degree rise in temperature between 28 to 35 °C (both at 330 and 800 ppmv CO_2), ET increased by about 4% [19]. For a diurnal change in air temperature at ambient CO_2 level, the increases in ET with 1 °C rise in temperature were about 2.3% in barley (between 10-25 °C) [50], 3.2% in mustard (between 10-20 °C) [39], 2.4% in chickpea (between 10-25 °C) [35], 2.5% in wheat (between 10-25 °C; Fig. 1) and 2.5% in mungbean (between 30-40 °C; $ET(mm/h) = -0.707 + 0.027\ T\ (°C)$: $R^2 = 0.88$) [24]. Thus the rise in 1 to 2 °C air temperature might not have any adverse effect on soil water balance because of 5-10% increase in precipitation at lower altitudes (0-30 °N). There may be about 6-8% reduction in maturity duration for each oC rise in air temperature (Fig. 3). In wheat for each °C rise in air temperature, the reductions of about 5 days in flowering date [32] and 4 days from flowerig to maturity have been recorded in North India [11]. Temperature rise of 1-2 °C should increase the productivity of wheat in cool hilly region. In plains, however, wheat yields could be reduced with 1-2 °C elevated temperatures above mean temperature

Table 3: Cultivated and irrigated area of important C3 and C4 crops in India [14].

Crops	Area (m/ha)	Irrigated (%)
Rice (C3)	41.8	43.4
Wheat (C3)	24.0	77.3
Oilseeds (C3)	21.6	18.8
Pulses (C3)	23.2	9.8
Cotton (C3)	7.2	30.9
Sorghum (C4)	14.8	4.8
Pearl millet (C4)	12.0	5.7
Maize (C4)	5.9	20.8
Sugarcane (C4)	3.3	74.2

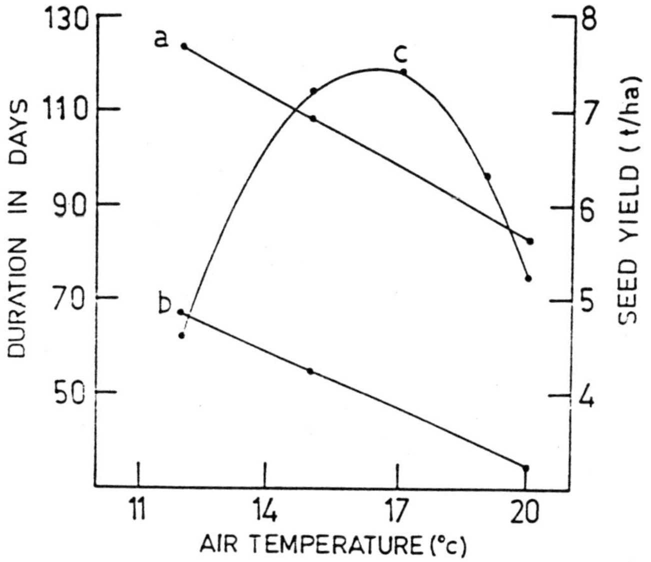

Fig. 3: Effect of air temperature on (a) duration of flowering (seeding to flowering [32]: $Y = 186.8 - 5.2\,X$; $R^2 = 0.92$), (b) grain filling (flowering to maturity [11]: $Y = 114.54 - 3.98X$; $R^2 = 0.95$), and (c) grain yield of wheat ([11]: $Y = -34.74 + 5.20\,X - 0.16\,X^2$; $R^2 = 0.99$).

of 17 °C during grain filling period (Fig.3) due to increased rate of senescence of flag leaf area and reduction in the grain filling duration [11]. These adverse effects of rise in temperature could be avoided by adjusting seeding time and developing suitable genotypes tolerant to higher temperature at seedling stage in irrigated wheat [33]. Warming, however, may cause some trouble to about 20% rainfed wheats grown in central India. This region will require either alternate crops or development of suitable genotypes tolerant to water stress and high temperature. This region, however, contribute relatively little to the overall production of wheat. The major wheat producing states of Uttar Pradesh, Punjab and Haryana grow 85 to 97% irrigated wheat. Therefore, it appears that the effects of slow change in climate with elevated temperatures of 1 to 2 °C could be absorbed by adjusting seeding time and developing suitable genotypes of wheat. Thus, the CO_2 enrichment will have positive effect (20-30%) on the productivity of rainfed wheat in hilly regions and irrigated wheat in plains of North India.

The rise in 1-2 °C in air temperature will be beneficial to majority of irrigated oilseeds and grain legumes grown in winter season in North India. It will help in early and more number of setting of pods by arresting pseudo-flowering, flower abortion and thus better reproductive growth of chickpea and lentil in North India. Warming will also provide safe guard against the frost damage in oilseed brassicas. The major area under winter season pulses and oilseeds is sown on stored soil moisture from monsoon rains. However, nothing is known about the effects of increased ET (due to increase in LAI and air temperature with CO_2 enrichment) on sesonal water use pattern in winter season rainfed crops. Seeding on conserved soil moisture with early vigrous growth in CO_2 enriched world may lead to relatively more water use during vegetative phase and thus exhaustion of soil water for reproductive phase of growth. This may cause forced maturity, poor harvest index and low productivity of dryland crops grown entirely on limited stored water in the soil profile. The effects of water deficit will further be aggravated due to relaively more warming between December and February months of winter season in South Asia [25]. One study carried out on winter wheat with limited water supply indicated about 8% (25 mm) and 18.5% (60 mm) increase in ET at 660 and 825 ppmv CO_2 over 340 ppmv CO_2 concentration [7]. CO_2 enrichment enhanced more root growth in top 50 cm soil profile [8]. Roots of wheat with CO_2 enrichment penetrated to the maximum depth of observation (176 cm) before roots of wheat grown under the ambient level of 340 ppmv CO_2. This study did not, however, provide any data on water use pattern during vegetative and reproductive phases of growth. There is a need to study the influence of elevated CO_2 and air temperature on water use pattern and partitioning of dry matter to roots, stems and grains under real dryland conditions.

Precipitaion is of prime importance to monsoon season crops, and models indicate an increase of 5-10% in the monsoon rainfall at lower latitudes. There are indications that under 1 °C warming and with precipitation increase of 100 mm in China, national yields of rice are estimated to increase by about 10 per cent [53]. Yields of rice are also expected to increase in Japan [52], but may decrease in other regions of Southeast Asia due to more rapid growth [25]. Late maturing varieties needing high temperatures for cutlivation would be more suited to exploit these new conditions. It has been estimated that the adoption of late mauring rice might increase yields by 26% with the temperature increase estimated for a doubling of CO_2 atmosphere [52]. Similar yields benefits are also expected in India where about 42% rice is irrigated and about 28% is grown under adequate rainfall conditions [14]. Even shortening of maturity duration of rice with global warming will be beneficial to maintain proper soil health by adjusting an additional leguminous crop in rice-wheat rotation in the states of Punjab, Haryana and Uttar Pradesh. Rao and Rajput [29] analysed the influence of weather parameters between 1978-1983 on rice yield at the Water Management Research Centres located throughout the country. The water surplus and water deficit regions when super imposed on yield values suggested that low rice yields were generally coincided with water surplus conditions and vice-versa. Multiple regression analysis further revealed that increasing vapour pressure deficit at the vegetatve phase showed a decrease in yield, probably due to increased respiration losses, where as more sunshine hours and increased vapour deficit (more dry conditions) were more conducive for proper grain filling and maturity of grains in rice. However, the representation of clouds and their radiative properties is a source of major uncertainty in modelling climatic changes [17]. In a warmer world the intertropical convergence zone would be likely to advance further northward into Africa and Asia [25]. If this occurs, then the total rainfall in the Sahel and India could increase. Rainfall should also be more intense in its occurrence and propogate flooding and erosion. In areas where increases in the intensity of rainfall occur, there may be more surface runoff, less percolation of water through the soil and less available soil moisture to plants. This will require changes in management to conserve more water in the soil profile and prevent soil erosion. In addition, increases in fertilizer use may be required in areas where greater rainfall is likely. Thus, it appears that the productivity of rice will increase significantly by adopting proper management strategies and genotypes in CO_2 enriched atmosphere in India.

Besides rice, the other important C3 species grown in India, are groundnut and cotton. These are mostly grown under moderate to adequate supply of water through rainfall and or irrigation. These crops are likely to be benefitted with CO_2 enrichment. The reduction in the length of growing season of cotton with rise in temperature will make this species even more suitable to double cropping system in the states of Punjab, Haryana and Uttar Pradesh. In these states, there is likelihood of some shift in area under groundnut from monsoon season to spring season with global warming. The other crops, which belong to C4 species such as sorghum, pearl millet, maize and surgarcane will be benefitted to some extent with CO_2 enrichment, and will be affected least by 1-2 ^{o}C rise in air temperature. There may be some shift in the area from monsoon season maize to winter maize in the northern States of India in case there is rise in temperature. The crops of pearl-millet and sorghum are mostly affected due to frequency, intensity and distribution of rainfall during monsoon season and they possess adequate drought escape and tolerance capacity. With the occurrence of more rainfall during monsoon season, the adverse effects of increased ET due to 1-2 ^{o}C rise in temperature will be absorbed by these species. Moreover, the Indian agriculture has shown its capacity to adjust such seasonal changes in the frequency of climatic extremes. India had experienced one of the worst drought of the century in 1987-88 in which about half of the area of the country was under severe category of drought [14]. It has adverse effect on animal population and their productivity, but resulted only little reduction in the production of food grains over previous year of 1986-87. Moreover, there is adequate development of technology, diversity in the cropping pattern and genetic variability between and within crop species to adapt the gradual changes or short term extreme-seasonal changes in climate. Also the shifts of crop potential with space and time because of change in climate could be adjusted due to availability of appropriate technology and inherent great diversity in Indian agriculture. Currently at national level only less than 50% yield of major crops is realized than what is obtained by progressive farmers or by experienced managers at commercial/research farms. Thus, there is a great scope to increase the productivity of major crops, if the present level of technology is made available for adoption to the farmers of the country.

FUTURE ACTIONS

The potential impact on agriculture arise from the direct effects of CO_2 enrichment on plants and from the effects of changes in climate. Several models predict that in the future higher CO_2 concentration and warm weather will be available to crop plants [25,46]. The uncertainties are however large, particularly with regard to changes in climate with space and time [17]. The results of controlled environment generally show that the higher CO_2 concentration will have positive effect on photosynthesis and productivity of major crop species. At present it is unclear whether the net effects of CO_2 and climate change on agriculture, particularly in low rain fall areas and coastal low lands, will be deterimental or beneficial [17,47]. Aspects which are important for agriculture, but about which there are great uncertainties and which need accurate assessment, include: regional and local variations in weather parameters, long term field studies involving the interactions of CO_2, temperature and water levels on allocation of dry matter in different under ground and above ground plant parts, soil water extraction pattern, ET, crop phenology, acclimation, nutrients requirement, and competition of crops with weeds, pests and diseases. There is also need to study the influence of interactions of rising CO_2 and temperatures with toxic air pollutants on plants. It seems desirable that for those regions that may be adversely affected by climate change, research priorities may be set on adapting species, varieties and production techniques to increased temperature, water demand and drought stress. The effects of climatic change on sea level rise and damage to natural ecosystems warrant accurate assessment and action plans to reduce the magnitude of such damages. It is likely to be several years before reliable predictions of global and particularly regional and local climate changes are available [17]. If, however, we are to improve our understanding of the significance of climatic change and its consequence for agriculture and mankind, it is utmost important to strengthen the current research into how agriculture can best adapt to, avoid or even profit from such changes.

REFERENCES

1. Adams, R.M., Rosenzweig, C., Pert, R.M., Ritchie,J.T., Mc Carl, B.A., Glyer, J.D., Curry, R.B., Jones, J.W., Boote, K.J. and Allen, L.H. Jr. (1990). Global climate change and US agriculture. Nature 345: 219-224.
2. Allen, L.H. Jr. (1990). Plant responses to rising carbon dioxide and potential interaction with air pollutants. J. Environ. Qual. 19: 15-34.
3. Allen, L.H. Jr., Vu, C.V., Valle, R.R., Boote, K.J. and Jones, P. (1988). Non-structural carbohydrates and nitrogen of soybean grown under carbon dioxide enrichment. Crop Sci. 28: 84-94.
4. Allen, S.G., Idso, S.B. and Kimball, B.A.(1990). Interactive effects of CO2 and environment on net photosynthesis of water lily. Agric. Ecosyst. Environ. 30: 81-88.
5. Baker, J.T., Allen, L.H. Jr., Boote, K.J., Jones, P. and Jones, W.J. (1989). Response of soybean to air temperature and carbon dioxide concentration. Crop Sci. 29: 98-105.
6. Campbell, W.J., Allen, L.H. Jr. and Bowes, G. (1990). Responses of soybean canopy photosynthesis to CO2 concentration, light and temperature. J. Expt. Bot. 41:427-433.
7. Chaudhuri, U.N., Kirkham, M.B. and Kanemasu, E.T. (1990). Carbon dioxide and water level effects on yield and water use of winter wheat. Agron. J. 82: 637-641.
8. Chaudhuri, U.N., Kirkham, M.B. and Kanemasu, E.T. (1990). Root growth of winter wheat under elevated carbon dioxide and drought. Crop Sci. 30:853-857.
9. Cure, J.D. (1985). Carbon dioxide doubling responses: a crop survey. In: Direct Effects of Increasing Carbon Dioxide on Vegetation, Strain, B.R. and Cure, J.D. eds. US Department of Energy, Washington, pp. 19-116.
10. Dhawan, A.K., Chhabra, M.L. and Yadava, T.P. (1983). Freezing injury in oilseed Brassica species. Annals Bot. 51: 673-678.
11. Dhiman, S.D., Singh, D.P.and Sharma, H.C. (1985). Grain growth of wheat as influenced by time of sowing and nitrogen fertilization. Haryana agric. Univ. J. Res. 15: 158-163.
12. Ehleringer, J. and Björkman, O. (1977). Quantum yield for CO2 uptake in C3 and C4 plants; dependent of temperature, CO2 and O2 uptake. Plant Physiol. 59:86-90.
13. Ehret, D.L. and Jolliffe, P.A. (1985). Leaf injury to bean plants grown in carbon dioxide enriched atmosphere. Can. J. Bot. 63: 2015-2020.
14. Fertilizer Statistics (1990-91). The Fertilizers Association of India, New Delhi.
15. Flint, E.P. and Patterson, D.T. (1983). Interference and temperature effects on growth in soybean (Glycine max) and associated C3 and C4 weeds. Weed Sci. 31:193-199.
16. Houghton, R.A. and Woodwell, G.M. (1989). Global climatic change. Scientific Amer. 260: 34-44.
17. Hume, C.J. and Cattle, H. (1990). The greenhouse effect-meteorological mechanisms and models. Outlook Agric. 19: 17-23.
18. Idso, S.B., Kimball, B.A., Anderson, M.G. and Mauney, J.R. (1987). Effects of atmospheric CO2 enrichment on plant growth: the interactive role of air temperature. Agric. Ecosyst.

Environ., 20: 1-10.

19. Jones, P., Allen, L.H. Jr., Jones, J.W. and Valle, R.R. (1985). Photosynthesis and transpiration responses of soybean canopies to short and long term CO2 treatments. Agron. J., 77: 119-126.

20. Kimball, B.A. (1985). Adaptation of vegetation and management practices to a higher carbon dioxide world. In: Direct Effects of Increasing Carbon Dioxide on Vegetation. Strain, B.R. and Cure, J.D., eds., U.S. Department of Energy, Washington, DC, pp. 185-204.

21. Ludlow, M.M. (1978). Light relations of pasture plants. In: Plant Water Relations in Pastures, Wilson, J.R. ed., CSIRO., Melbourne, pp. 35-49.

22. Morgan, J.M. and Condon, A.G. (1986). Water use, grain yield and osmoregulation in wheat. Aust. J. Plant Physiol. 13: 523-532.

23. O' Toole, J.C. and Chang T.T. (1979). Drought resistnce in cereals-rice a case study. In: Stress Physiology in Crop Plants .Mussell, H. and Staples, R.C. eds., Wiley Interscience, New York, pp. 373-405.

24. Pannu, R.K. and Singh, D.P. (1988). Influence of water deficits on morpho-physiological and yield behaviour of mungbean (Vigna radiata (L.) Wilczek). Proc. II. International Symposium on Mungbean, AVRDC, Taiwan, pp. 252-259.

25. Parry, M.L., Porter, J.H. and Carter, T.R. (1990). Climatic change and its implications for agriculture. Outlook Agric. 19: 9-15.

26. Phogat, B.S., Singh, D.P. and Singh, P. (1984). Response of cowpa (Vigna unguiculata (L.) Walp.) and mungbean (Vigna radiata (L.) Wilczek) to irrigation. II. Effects on CO2 exchange, radiation characteristics and growth. Irrig. Sci. 5: 61-72.

27. Passioura, J.B. (1986). Resistance to drought and salinity: avenues for improvement. Aust. J. Plant Physiol. 13: 191-201.

28. Prasad, K. and Rathi, K.S. (1987). Effect of method of sowing and fertilizer in wheat. Indian J. Agron. 32: 319-321.

29. Rao, G.G.S.N. and Rajput, R.K. (1986). Rice yield during kharif season as influenced by weather. Proc. Internal. Seminar on Water Management in Arid and Semiarid Zones. Organised by UNESCO and Department of Agric. Meteorol., Haryana Agric. Univ., Hisar, India, pp. 100-108.

30. Riechers, G. and Strain, B.R. (1988). Growth of blue grama (Boutelova gracilis)in response to atmospheric CO2 enrichment. Can. J. Bot. 166: 1570-1573.

31. Sage, R.F., Sharkey, T.D. and Seeman, J.R.(1989). Acclimation of photosynthesis to elevated CO2 in five C3 species. Plant Physiol. 89: 590-596.

32. Saini, A.D. and Nanda, R. (1987). Analysis of temperature and photo-periodic response to flowering in wheat. Indian J. Agric. Sci. 57: 351-359.

33. Singh, D.P., Behl, R.K., Chaudhary, B.D. and Singh, P.(1988) . Rainfed wheats in Indian sub-continent: an overview. Paper presented at International Symposium on Drought Resistance in Cereals. Organised by ICSU (Paris), Nov. 1988, Cairo, Egypt.

34. Singh, D.P., Chaudhary, B.D., Singh, P., Sharma, H.C. and Karwasra, S.P.S. (1990). ' Drought Tolerance in Oilseed

Brassicas and Chickpea. Directorate of Research, Haryana Agric. Univ., Hisar (India), pp. 60.

35. Singh, D.P., Peters, D.B., Singh, P., and Singh, M. (1987). Diurnal pattern of canopy photosynthesis, evapotranspiration and water use efficiency in chickpea (Cicer arietinum L.) under field conditions. Photosynthesis Research 11: 61-69.

36. Singh, D.P., Rawson, H.M. and Turner, N.C. (1982). Effect of different levels of radiation, temperature and humidity on the rate of photosynthesis, transpiration and water use efficiency of chickpea. Indian J. Plant Physiol. 25: 32-39.

37. Singh, D.P., Singh, J., Singh, P., Sharma, H.C. and Singh, M. (1982). Effect of crop geometry and irrigation management on growth, yield, evapotranspiration and water use efficiency of chickpea. Proc. IV. Afro-Asian Regional Conference, ICID, Nigeria, 1: 489-504.

38. Singh, D.P., Singh, P. and Sharma, H.C. (1985). Effect of crop geometry and irrigation management on yield, evapotranspiration and water use efficiency of mustard (Brassica juncea L.). Indian J. Agron. 30: 59-63.

39. Singh, D.P., Singh, P. and Sharma, H.C. (1986). Diurnal pattern of photosynthesis, evapotranspiration and water use efficiency in mustard at different growth phases under field conditions. Photosynthetica 20: 117-123.

40. Singh, D.P., Singh, P., Sharma, H.C. and Turner, N.C.(1987). Influence of water deficit on the water relations, canopy gas exchange and yield of chickpea (Cicer arietinum), Field Crops Res. 16: 231-241.

41. Singh, D.P., Turner, N.C. and Rawson, H.M. (1982). Effect of different levels of radiation, temperature and humidity on the rate of photosynthesis, transpiration and water use efficiency of oilseed rape. Biol. Plantarum 24: 132-135.

42. Singh, D.P., Singh, P., Yadav, S.K, Kumar,A. and Sharma, H.C.(1987). Diurnal pattern of plant water relations and canopy photosynthesis of maize, sorghum and pearl millet under field conditions. Indian J. Plant Physiol. 30:425-428.

43. Singh, P., Singh, P. and Singh, D.P. (1978). Seeding technology and irrigation water management in cotton. J. Agri. Engg. 5:175-180.

44. Sionit, N., Strain, B.R. and Beckford, R.A. (1981). Environmental control on the growth and yield of okra. I. Effect of temperature and CO2 enrichment at cool temperature. Crop Sci. 21: 885-888.

45. Sionit, N., Strain, B.R. and Flint, E.P.(1987). Interaction of temperature and CO2 enrichment on soybean: Photosynthesis and seed yield. Can. J. Plant Sci. 67: 629-636.

46. Warrick, R.A. (1990). Carbon dioxide, climatic change and agriculture. Geograph. J. 154: 221-233.

47. Warrick, R.A. and Barrow, E.M.(1990). Climatic and sea level change: A Perspective. Outlook Agric. 19: 5-8.

48. Wingley, T.M.L. (1985). Impact of extreme events. Nature 316: 106-107.

49. World Meteorological Organization (1985). Report of the international conference on the assessment of the role of carbon dioxide and other green house gases in climate variations and associated impacts, WMO, No. 661, Geneva.

50. Yadav, S.K., Singh, D.P., Singh, P. and Kumar, A. (1987). Diurnal pattern of photosynthesis, evapotranspiration and

water use efficiency in barley under field conditions. Indian J. Plant Physiol., 30: 233-238.

51. Yelle, S., Beeson, R.C.Jr., Trudel, W.J. and Goasselin, A.(1990). Duration of CO2 enrichment influences growth, yield and gas exchange of two tomato species. J. Amer. Hort. Sci. 115: 52-57.

52. Yoshino, M.M., Horie, T., Seino, H., Tsuji, H., Uchijima, T. and Uchijima, Z. (1988). Assessment in cool temperature and cold regions. In: The Impact of Climatic Variations on Agriculture. Parry, M.L., Carter, T.R. and Konijn, eds. Kluwer Dordrecht, Netherlands, pp. 725-863.

53. Zhang, J.C. (1989). The CO2 problem in climate and dryness in North China. Meteorol. Magz. 15: 3-8.

PHOTOSYNTHETIC PERFORMANCE UNDER ENRICHED LEVELS OF CARBON DIOXIDE: AN ASSESSMENT AND PREDICTIONS

V.S.Rama Das* and Sanjay Kumar
Council of Scientific and Industrial Research Complex, Palampur-176061 (H.P.) INDIA.
*Formerly at the School of Life Sciences, University of Hyderabad, Hyderabad-500 134, INDIA.

ABSTRACT:
The influences of higher than ambient levels of atmospheric carbon dioxide, the principal greenhouse gas, on plant photosynthesis are examined. The parameters of stomatal conductance, transpirational characteristics and of water use efficiency are considered. Higher carbon dioxide concentrations in the atmosphere are likely to alter the patterns of the partitioning of photosynthetically fixed carbon into different plant organs, starch/sucrose ratio and even phloem transport. Carbon dioxide enrichment might also affect the carbon allocation to dark respiration and other cellular metabolic processes. While looking into the effects of elevated carbon dioxide concentration on photosynthetic performance the differential responses of C_3, C_4 and of naturally occurring C_3/C_4 intermediate type plants are assessed in predicting whether the climate change would lead to modification of the evolutionarily stabilized pathways or to altered vegetational distribution.

INTRODUCTION:
Presently there is an increasing concern over the rising levels of major greenhouse gas, carbon dioxide, in the atmosphere. Indiscriminate and extensive deforestation and the burning of fossil fuels are the two major sources of increasing CO_2 levels (1). Currently, CO_2 concentration is approximately 350 ppm which is said to be twenty five per cent above the preindustrial level (2) and it is predicted

that it will double itself by the year 2025 (3). There is, therefore, considerable interest among scientists to assess the impact of the increased level of CO_2 on various plant processes.

In as much as CO_2 is the substrate for photosynthesis, several researches naturally have been conducted to study the effect of elevated CO_2 levels on photosynthesis and related processes. However, there are inherent limitations with their experimentation and also with their extrapolations and these aspects have been discussed subsequently in this article.

The present review is aimed to incorporate the following aspects:

1. The patterns of the influence of the elevated CO_2 levels on various aspects of photosynthesis process *per se* including the CO_2 diffusion and the biochemistry of carbon fixation and reduction.

2. To assess the differential responses of the C_3 and C_4 plants, woody plants, plants of determinate and indeterminate growth pattern and floricultural crops to en hanced CO_2 levels in terms of growth and yield.

3. The changes in water use efficiency (WUE) due to higher than ambient CO_2 levels.

4. Possible alterations in the rates and mode of partitioning of photoassimilates under elevated CO_2 levels between the sources and sinks as also the biochemical partitioning between starch and sucrose.

ELEVATED CO_2 LEVELS AND THE PHOTOSYNTHETIC PROCESS:

<u>CO_2 Diffusion</u>:

The diffusion of CO_2 from atmosphere to carboxylation site is under the regulation by stomata. Present evidences show that stomata respond to intercellular CO_2 concentration (Ci). They are relatively insensitive to the CO_2 concentration at the surface of the leaf and in stomatal pore (4).

The manner in which stomata respond to CO_2 is still not clear. Nonetheless, it has been suggested that stomatal response to CO_2 might depend on the response of photosynthesis to CO_2 in stomatal guard cells (5). Stomatal conductance has been observed to be decreased at elevated CO_2 levels irrespective of the fact whether the plants are exposed for short periods (hours) or for rather longer periods (days to weeks). As a consequence of decrease in stomatal conductance a lowering of intercellular CO_2 concentration would be noticed which in turn leads to lower photosynthesis rate (6) and lower transpiration rates. Such effects have been noticed irrespective of the type of plants, whether C_3, C_4, woody, determinate or indeterminate plants. Usually it takes several days for the stomata to exhibit normal conductance rates when transferred back to the lower CO_2 levels after longer exposure to elevated CO_2 levels (7) which indicates a fairly tight stomatal regulation at various levels.

Carbon Reduction Process (Carbon fixation process):

There are conflicting reports on the effect of elevated CO_2 levels on the rates of photosynthesis. The photosynthesis response could be categorized into short term and long term responses with respect to C_3, C_4 and woody plants.

C_3 PLANTS

Short term responses:

A linear increase in the rate of photosynthesis with increasing CO_2 concentration upto 3500 ppm has been shown in Alfa alfa, sugar beet and tomato in full sunlight with exposure to high CO_2 lasting for a few hours (8). Green & Wright (9) have reported an average increase in photosynthesis by 84% when branches of conifers were enclosed in cuvettes and subjected to concentration of 450 to 500 ppm of CO_2. Similarly, with short term CO_2 enrichment (1000 ppm) during photosynthesis measurements in tomato, light saturated rate of photosynthesis was more than double in mature leaves whereas young leaves did not show any response (10).

The mechanism of enhancement in the rates of photosynthesis during short term exposure to raised CO_2 in C_3 plants is largely the result of increasing RuBP carboxylation (11). Since Rubisco fixes CO_2 and O_2 competitively, increasing ambient CO_2 results in a steeper gradient of CO_2 between the atmosphere and the fixation site. This results in increasing gross photosynthesis due to an obvious advantage of CO_2 over O_2 in their competitive uptake (12). Oxygen concentration below and above 21% results in increased or decreased net assimilation rate, respectively. However, enrichment of the ambient CO_2 from 300 to 1000 ppm was found to increase the net assimilation rate of three tobacco cultivars at 0, 21 and 100% O_2 by a similar increment (13).

The quantitative relationship between the ambient CO_2 concentration and intensity of photosynthesis have not yet been fully understood because of limitations in the methodology for direct determination of photorespiratory flux (12).

Long term response:

Measurements made during relatively short periods of time apparently do not provide reliable information concerning what occurs when plants are grown at high concentrations of CO_2 for several weeks (14). In a large number of plants including cucumber, tobacco, *Pharbitis nil* showed a decline in photosynthesis rate after acclimation to high levels of CO_2 (15, 16, 17). In all these cases the photosynthesis rate at elevated CO_2 level was lower than the control values. Two species of tomato grown at high CO_2 concentration (900 ppm) for 10 weeks exhibited carbon exchange rates significantly higher in CO_2 enriched plants for the first few weeks of treatment but thereafter decreased rates were noticed as tomato plants acclimated to higher CO_2 levels (18). The minimum carbon exchange rates at elevated CO_2 levels were still higher than the control value.

Chenopodium album, Solanum tuberosum and *Phaseolus vulgaris* grown at high CO_2 levels (900-1000 ppm) possessed little higher CO_2 assimilation rates while *Beta oleracea* and *Solanum melongena* showed a reduction in CO_2 assimilation rate at the above elevated CO_2 levels (19).

It is suggested that the decrease in rate of photosynthesis following prolonged exposures could be caused by accumulation of starch in leaves (20,21, 22, 23) which could function as a feedback inhibitor of photosynthesis. An accumulation of starch within the chloroplast is believed to have damaged the thylakoid & grana (24, 25).

In contrast, an analysis of the fifth leaf of tomato plant grown at elevated CO_2 levels (900 ppm) after four weeks showed starch accumulation but without visible modification of the chloroplast ultrastructure. Even interspecific differences are noticed in this response. Chloroplast ultrastructure was more affected for instance in *Lycopersicon esculentum* than in *Lycopersicon chmielewskii* under high CO_2 (900 ppm) but *L. chmielewskii* showed much faster decline in photosynthesis. Thus, it could be concluded that starch accumulation is not the only cause for decline in photosynthesis rate at high CO_2 level. It is also to be noted that at high CO_2 the stomatal conductance is reduced but the Ci value does not decrease to the same extent to explain the reduction of photosynthesis rate (18, 19).

It has been shown that there is a loss in rubisco activity in a number of crops (26, 27, 28). Besford et al. (10) have observed a loss of rubisco protein to one-half in fully expanded leaves of tomato plants grown at higher CO_2 levels. It was further shown that rubisco activation state, which is believed to reflect the carbamylation state of the enzyme was significantly lower in leaves of five C_3 species grown at high CO_2 levels. Number of activated rubisco sites decline under high CO_2 conditions (27, 29).

Therefore, the reduced rates of photosynthesis on longer exposure to elevated CO_2 levels could be the result of effects at various levels of organization as discussed above including the nutrient limitations. However, there are very few reports indicating that plants reduce the allocation of nitrogen to rubisco in favor of light harvesting complexes and the enzymes of carbon reduction cycle (5).

C_4 PLANTS

Rubisco is located in the bundle sheath cells of C_4 plants where the CO_2 concentration largely saturates carboxylation and inhibits oxygenation (5). The CO_2 enrichment above 350 to 400 ppm has a diminished effect on net CO_2 fixation in a number of crops including maize and itchgrass (*Rottboellia exaltata*) (30, 31). In spite of very little effect of CO_2 enrichment in C_4 plants several benefits could occur as regards the dry matter production. Reasons for the same are examined later.

WOODY PLANTS

Woody plants (trees) are regarded to constitute a very strong sink for CO_2 since large fraction of their photosynthates are partitioned in favor of wood production. Therefore, the role of trees could be evaluated in effectively removing the excess atmospheric CO_2 (1).

Short term response:

All woody plants with the exception of some species of *Euphorbia* have the C_3 photosynthetic pathway that responds more to increased concentration of CO_2 when compared to the C_4 plants (32). Most of the data available for woody plants are only a short term measurements made on plants exposed to high CO_2 concentration for only a few hours. Such measurements are always not reliable indicators of the effects of prolonged exposure to high concentration of CO_2.

Short term experiments with ponderosa pine, *Quercus coccinea, G. alba, Pinus radiata, Populus deltoides, Encelia farinosa* and *Ceanothus megacarpus* have shown a definite increase in photosynthesis rate upto 40-80% under high CO_2 levels (600 ppm) (9, 32).

Long term response:

Long term experiments (2 years) with *Pinus radiata* have revealed higher photosynthesis rate at elevated CO_2 levels (660 ppm). The photosynthetic response to high CO_2 increased as the trees aged. the increased photosynthesis response could possibly due to higher size of the sink relative to the sources. However, elevated photosynthetic rates at high CO_2 did not sustain faster growth rates after week 39, presumably due to the limitation imposed by other factors (33). In mangosteen (*Garcinia mangostana* L.) also elevated CO_2 levels (800 ppm) resulted in an increase in CO_2 fixation rate by 40-60% (34).

Photochemical Processes:

These processes include capturing of light quanta and production of reducing power for CO_2 fixation. CO_2 concentration did not seem to affect the photochemical process in the ranges of normal & CO_2 enriched environment (35).

GROWTH, DRY MATTER PRODUCTION AND YIELD

Higher yield under greenhouse conditions is quite a common observation. Growth and yield depends upon many factors in addition to the rate of photosynthesis, such as the magnitude of dark respiration, nitrogen metabolism, translocation and partitioning of photosynthates to various organs. Environmental factors like, water, temperature, nutrition and irradiance are very important (14).

A comprehensive analysis of 770 observations on 38 agricultural crops and 18 other species resulted in a generalization that CO_2 enrichment could increase the economic yield of mature agricultural crops by 26%. Excluding flower crops, the mean yield increment of C_3 crops with CO_2 enrichment was 36% (35, 36). Cure and Acock (37) calculated a 28% increase in biomass under twice the normal CO_2 concentration, but increase of 100% or more in biomass has been reported for some C_3 plants (7, 22, 26). Results obtained for rice (C_3 plant) grown at elevated CO_2 level (500 ppm) were quite interesting. Net assimilation rate (NAR) increased with increasing the CO_2 levels upto 500 ppm in rice grown under controlled - Environment plant chambers with natural sunlight. The increment in NAR was observed only during earlier part of the growing season. NAR decreased during the growing season (38). Baker et al. (38) also claimed a 32% increase of grain yield in rice by doubling the CO_2 concentration from 370 to 660 ppm. The concentration of carbohydrates in the grain was similar in all CO_2 treatments. However, grain yield was higher at higher CO_2 concentration because of increase in number of panicles per plant.

On the contrary C_4 plants generally show very little increase in dry weight with CO_2 enrichment (30, 39).

Woody plants invariably show higher yields at elevated CO_2 levels. Stem dry weight, stem diameter, wood density, leaf number, leaf area and leaf dry weight have increased in most experiments. The increase in dry weight was 30-90% (33, 34, 41). Rogers et al. (42, 43) reported an increase in dry weight upto 157% in *Pinus taeda* grown in open top chambers in the field for 3 months at 910 ppm of CO_2 as compared with that grown at 350 ppm of CO_2.

While the differential responses of C_3 or C_4 annuals and of woody plants it is also necessary to assess if the determinate or indeterminate growth patterns would differ in the dry matter production under elevated CO_2 concentration. The best examples are that of soybean & cotton (indeterminate) and for sorghum and sunflower (determinate). At elevated CO_2 levels (660 ppm) soybean and cotton showed an increase in dry weight of 382% & 110% respectively. Sunflower and sorghum on the contrary showed an increase of 60 and 18%, respectively (22). This fact showed very clearly that sink strength is very important to control dry matter, production.

In flower crops, where yield is measured by number of blooms, CO_2 enrichment resulted in either positive or negative results depending upon climate conditions and irradiance available to the crop (44, 45). However, if factors like temperature, irradiance and water status are provided optimally CO_2 enrichment (1000-1200 ppm) could result into enhanced flower yields by 90% of the control values (46, 47).

Tissue cultured *Gerbera jamesonii* and *Rosa* showed much greener leaves and lesser number of senesced or abscised leaves when 5% CO_2 was supplied (48).

POSSIBLE REASONS FOR HIGHER YIELD AT ELEVATED CO_2 LEVELS:

Higher yields at elevated CO_2 levels could be the result of a favorable balance between carbon assimilation and respiration. Partitioning of photoassimilate would be important in determining the crop yield. Better carbon assimilation would be reflected in terms of better WUE. Unfortunately, there is absolutely no data on harvest index in the literature. Therefore, it is difficult to assess the role of partitioning in affecting the crop yield.

Water Use Efficiency (WUE) at Elevated CO_2 Levels:

In very simple terms WUE could be the ratio of CO_2 uptake to water lost by transpiration. Elevated CO_2 levels result in a partial closure of stomata which would reduce Ci as well as transpiration rates. No quantitative data are available to show the relative reduction in Ci and the transpiration rates to make further general conclusions.

When WUE is expressed as mg CO_2 uptake/g water transpired, an almost linear relationship is obtained both for C_3 and C_4 plants (5, 49). In case of woody plants also a doubling of CO_2 has been shown to result in doubling of WUE (28, 49, 50). The increase in WUE was greater in water stressed plants than in non-stressed plants (41). Such increments in WUE would depend upon the duration and intensity of water stress, as increasing the CO_2 concentration will have little effect if the stomata are already closed.

Partitioning of Photosynthates:

The partitioning of photosynthetic assimilates could be considered in two aspects:

a) Partitioning of photosynthates to different organs

b) Biochemical partitioning between starch and sucrose

Partitioning of photosynthates to various organs within the plant:

CO_2 enrichment has been shown to increase root to shoot ratio for barley, Kale, sugar beet, rice wheat & radish under greater higher light flux and plant age. On the other hand chrysanthemum and maize did not show any response (30, 51, 52, 53). It is also to be noted that inconsistent data were obtained for maize and soybean (50, 54, 55).

During vegetative development in rice, increasing the CO_2 concentration led to increase in leaf and culm total non-structural carbohydrate, leaf starch and leaf sucrose concentration.

At maturity the concentration of carbohydrates in the grain was similar in all CO_2 treatments. even the grain size, number of filled grain per panicle were similar in all the CO_2 treatment. However, there were more panicles per plant which resulted into higher yield (38). Leaf carbon budgets for the light period have revealed that plants grown under subambient CO_2 conditions, and thus reduced photosynthesis, gave priority to the storage of carbohydrates over the export of sucrose early in the season. With increasing CO_2 concentrations the priority has shifted more towards export than carbohydrate storage (38, 40).

The partitioning behavior is observed to be highly species specific. It was reported that family 20010 of *Pinus radiata* D. Don partitioned larger amounts of dry matter to trunks than family 20062 which favored roots and branches. Wood density has also increased at elevated CO_2 levels in family 20010 because of the thickening of tracheid walls (33). Mangosteen partitioned most of the photosynthate towards roots and trunk at elevated CO_2 levels (34).

Chemical Partitioning:

Analysis of leaf samples revealed that sucrose was the main soluble sugars and starch was the main non-soluble carbohydrate in rice plant. Leaf starch and sucrose concentration increased with increasing CO_2 at 19 days after planting wherein starch showed the maximum response. There was linear relationship between leaf starch/sucrose ratio and CO_2 enhancement. In soybean leaves nearly all of the extra carbon fixed as a result of CO_2 enrichment was partitioned into leaf starch (56). Tomato plants grown at 900 ppm CO_2 concentration contained more sugar and starch than the control. However, no significant accumulation of starch and sugar was observed in young leaves. The chloroplast of tomato plants exposed to higher CO_2 levels exhibited marked accumulation of starch (18).

In general, when exposed to elevated CO_2 levels, plants tend to export its photosynthate to other organs so as to check the feed back inhibition imposed by starch accumulation on photosynthesis.

GLOBAL CO_2 DOUBLING AND FUTURE SCENARIO

It is likely that, as predicted by Kimball (36, 57) and Cure and Acock (37), a doubling of CO_2 concentration would have beneficial effects on agricultural yield. However, all the experiments suffer from the limitations of being conducted under greenhouse conditions where only CO_2 concentration is changed and all the other environmental variables including temperature, humidity, irradiance are maintained constant. If CO_2 level is doubled by 2025 A.D., it would accompany changes in temperature and rainfall pattern too (58). If temperature and humidity are changed along with CO_2 concentration then the results of the experiments on photosynthesis are to be re-examined. Under most natural environment plants do not receive water and nutrients optimally for their growth which further complicates the whole scenario.

Planting of perennial woody species would certainly be useful in the context of global CO_2 change since they are regarded to function as strong sinks for atmospheric CO_2. Thus, acting as a buffer (1) to reduce the elevated atmospheric CO_2 levels.

Species specific response to elevated CO_2 levels may also lead to changes in vegetation patterns and the composition of some ecosystem (14). The higher than ambient levels of CO_2 in the atmosphere if maintained over long periods might lead to improved competitive ability for C_3 plants in relation to C_4 plants. In the long range it might also happen that the natural occurrence of the C_3/C_4 intermediate is diminished.

Much greater work is required to predict the effects of superambient CO_2 concentrations on overall plant growth and development, in relation to other environmental variables. The observed effects should also be analysed to assess whether they are the result of the influence of high CO_2 on photosynthesis per se or indirect influence through investment in new growth or development.

ACKNOWLEDGEMENTS

Preparation of this article was partially supported by USDA Grant FG-IN-680 (IN-ARS-403) to Prof.V.S.Rama Das. The help rendered by Dr.L.M.S. Palni and Dr.S.K. Nandi is also acknowledged.

REFERENCES

1. Gifford, R.M. (1989). Trees in the global greenhouse. Their role and potential. *Trees Natural Resour.* 31: 9-11.

2. Pearman, G.I.(1981). Greenhouse gases: evidence for atmospheric changes and anthropogenic causes. In: Greenhouse: Planning for Climatic Change, G.I.Pearman ed., E.J. Brill, The Netherlands, pp.3-21.

3. Gribbin, J. (1981). The politics of carbon dioxide. *New Sci.* 90: 82-84.

4. Mott, K.A. (1988). Do stomata respond to CO_2 concentrations other than intercellular? *Plant Physiol.* 86: 200-203.

5. Mott, K.A. (1990). Sensing of atmospheric CO_2 by plants. *Plant Cell Environ.* 13: 731-737.

6. Spencer, W. and Bowes, G. (1986). Photosynthesis and growth of water hyacinth under CO_2 enrichment. *Plant Physiol.* 82: 528-533.

7. Delucia, E.H., Sasek, T.W. and Strain, B.R. (1985). Photosynthetic inhibition after long term exposure to

elevated levels of atmospheric carbon dioxide. *Photosynth. Res.* 7: 175-184.

8. Thomas, J. and Hill, G.R. (1949). Photosynthesis under field conditions: In: Photosynthesis in Plants. J.Franck and W.E. Loomis, eds., Iowa State College Press, pp. 19-52.

9. Green, K. and Wright, R. (1977). Field response of photosynthesis to CO_2 enhancement in ponderosa pine. *Ecology* 58: 687-692.

10. Besford, R.T., Ludwig, L.J. and Withers, A.C. (1990). The greenhouse effect: acclimation of tomato plants growing in high CO_2, photosynthesis and ribulose-1, 5-bisphosphate carboxylase protein. *J. Exp. Bot.* 41: 925-931.

11. Besford, R.T., Withers, A.C. and Ludwig, L.J. (1985). Ribulose bisphosphate carboxylase activity and photosynthesis during leaf development in the tomato. *J. Exp. Bot.* 35: 495-504.

12. Bravdo, B.S. (1986). Effects of CO_2 enrichment on photosynthesis on C_3 plants. In: *Carbon dioxide enrichment of greenhouse crops.* Vol.II. H.Z. Enoch and B.A.Kimball ed., CRC Press, pp. 13-27.

13. Paz, N. (1983). Factors involved in photosynthetic efficiency of plants. Ph.D. Thesis, Herbew University of Jerusalem, 98 p.

14. Kramer, P.K. (1981). Carbon dioxide concentration, photosynthesis, and dry matter production. *Bio Science.* 31: 29-33.

15. Aoki, M. and Yabuki, K. (1977). Studies on the carbon dioxide enrichment for plant growth VII. Changes in dry matter production and photosynthetic rate of cucumber during carbon dioxide enrichment. *Agr. Mete-*

orol. 18: 475-485.

16. Raper, C.D., Jr., and Peedin, G.R. (1978). Photosynthetic rate during steady - state growth as influenced by carbon dioxide concentration. *Bot. Gaz.* 139: 147-149.

17. Hickelnton, P.R. and Joliffe, P.A. (1980). Carbon dioxide and flowering in *Pharbitis nil* Choisy. *Plant Physiol.* 66: 13-17.

18. Yelle, S., Beeson, RC. Jr., Trudel, M.J. and Gosselin, A. (1989). Acclimation of two tomato species to high atmospheric CO_2. 1. Sugar and starch concentrations. *Plant Physiol.* 90: 1465-1472.

19. Sage, R.F., Sharkey, T.D., Seeman, J.R. (1989). Acclimation of photosynthesis to elevated CO_2 in five C_3 species. *Plant Physiol.* 89: 590-596.

20. Madsen, E. (1975). Effects of CO_2 enrichment on growth, development, fruit production and fruit quality of tomato plant from a physiological viewpoint. *Phytotronics* 111: 318-330.

21. Thomas, J.F., Raper, C.D., Jr., Anderson, C.E. and Downs, R.J. (1975). Growth of young tobacco plants as affected by carbon dioxide and nutrient variables. *Agron. J.* 67: 685-689.

22. Mauney, J.R., Fry, K.E. and Guinn, G. (1978). Relationship of photosynthetic rate to growth and fruiting of cotton soybean, sorghum and sunflower. *Crop Sci.* 18: 259-263.

23. Sasek, T.W., Delucia, E.V. and Strains, B.R. (1985). Reversibility of photosynthetic inhibition in cotton after long term exposure to elevated CO_2 concentrations. *Plant Physiol.* 78: 619-622.

24. Carmi, A. and Shomer, I. (1979). Starch accumulation and photosynthetic activity in primary leaves of bean (*Phaseolus vulgaris* L.). *Ann. Bot.* 44: 479-484.

25. Wildman, S.G., Charlene, A., Jope, A. and Atchinson, B.A. (1980). Light microscopic analysis of the three-dimensional structure of higher plant chloroplasts, position of starch grains and probable spiral arrangement of stroma lamellae and grana. *Bot. Gaz.* 14: 24-36.

26. Wong, S.C. (1979). Elevated atmospheric partial pressure of CO_2 and plant growth. I. Interaction of nitrogen nutrition and photosynthetic capacity in C_3 and C_4 plants. *Oecologia* 44: 68-74.

27. Von Caemmerer, S., Farquhar, G.D. (1984). Effects of partial defoliation, changes of irradiance during growth, short term water stress and growth at enhanced p (CO_2) on the photosynthetic capacity of leaves of *Phaseolus vulgaris* L. *Planta* 160: 320-329.

28. Downton, W.J.S., Bjorkman, O. and Pike, C.S. (1980). Consequences of increased CO_2 atmospheric concentrations of carbon dioxide for growth and photosynthesis of higher plants. In: *CO_2 and climate, Australian Research,* G.I.Pearman, Ed. Australian Academy of Science, Canberra, Australia, pp. 143-151.

29. Sage, R.F., Sharkey, T.D. and Seemann, J.R. (1988). The in vivo response of the ribulose -1, 5-bisphasphate carboxylase activation state and the pool sizes of photosynthetic metabolites to elevated CO_2 in *Phaseolus vulgaris* L. *Planta.* 174: 407-416.

30. Moss, D.N., Musgrave, R.B. and Lemon, E.R. (1961). Photosynthesis under field conditions. III. Some effects of light, carbon dioxide, temperature and soil moisture on photosynthesis, respiration, and transpiration of corn. *Crop. Sci.* 1: 83-87.

31. Patterson, D.T. and Flint, E.P. (1980). Potential effects of global atmospheric CO_2 enrichment on the growth and competitiveness of C_3 and C_4 weed and crop plants. *Weed Sci.* 28: 71-75.

32. Sionit, N. and Kramer, P.J. (1986). Woody plants reactions to CO_2 enrichment. In: *Carbon dioxide enrichment of greenhouse crops.* Vol.II. H.Z.Enoch and B.A.Kimball eds. CRC Press, pp. 69-85.

33. Conroy, J.P., Milham, P.J., Mazur, M. and Barlow, E.W.R. (1990). Growth, dry weight partitioning and wood properties of *Pinus radiata* D. Don after 2 years of CO_2 concentration. *Plant Cell Environ.* 13: 329-337.

34. Downton, W.J.S., Grant, W.J.R. and Chacko, E.K. (1990). Effect of elevated carbon dioxide on the photosynthesis and early growth of mangosteen (*Garcinia mangostana* L.) *Sci. Hortic.* 44: 215-225.

35. Armond, P.A., Schreiba, U. and Bjorkman, O. (1978). Photosynthetic acclimation to temperature in desert shrub, *Larrea divaricata* II. Light harvesting efficiency and electron transport. *Plant Physiol.* 61: 411.

36. Kimball, B.A. (1986). Influence of elevated CO_2 on crop yield. In: *Carbon dioxide enrichment of greenhouse crops.* Vol.II. H.Z.Enoch and B.A.Kimball, eds. CRC Press, pp. 105-115.

37. Cure, J.D. and Acock, B. (1986). Crop responses to carbon dioxide doubling: a literature Survey. *Agric. Forest Meteorol.* 38: 127-145.

38. Baker, J.T., Allen, L.H., Jr. and Boote, K.J. (1990). Growth and yield responses of rice to carbon dioxide concentrations. *J. Agri. Sci.* 115: 313-320.

39. Curtis, P.S., Drake, B.G., Leadley, P.S., Arp. W.J. and Whigham, D.F. (1989). Growth and senescence in plant communities exposed to elevated CO_2 concentra-

tions on an estuarine marsh. *Oecologia* 78: 20-26.

40. Rowland-Bamford, A.J., Allen, L.H., Jr., Baker, J.T. and Boote, K.J. (1990). Carbon dioxide effects on carbohydrate status and partitioning in rice. *J. Exp. Bot.* 41: 1601-1608.

41. Tolley, L.C. (1982). The effects of atmospheric carbon dioxide enrichment, irradiance and water stress on seedling growth and physiology of *Liquidamber styeaciflua* and *Pinus taeda*. Ph.D. Dissertation, Duke Univ. Durham, N.C.

42. Rogers, H.H., Bingham, G.E., Cure, J.D., Smith, J.M. and Surano, K.A. (1983). Responses of selected plant species to elevated carbon dioxide in the field. *J. Environ. Qual.* 12: 569

43. Rogers, H.H., Thomas, J.F. and Bingham, G.E. (1983). Response of agronomic and forest species to elevated atmospheric carbon dioxide. *Science* 22: 428.

44. Nelson, P.V. and Larson, R.A. (1969). The effect of increased CO_2 concentrations on chrysanthemum (*C. morifolium*) and Snap dragon (*Antirrhinum majus*). N. Carorila. *Agric. Exp. Sta. Tech. Bull.* 194: 15.

45. Goldsberry, K.L. (1982). Greenhouse heat conservation: Covers vs. atmospheric conditions during five year period. *Colo Greenhouse Growers Assoc. Res. Bull.* 381: 1

46. Holley, W.D., Goldsberry, K.L. and Juengling, C. (1964). Effects of CO_2 concentration and temperature on carnations. *Colo. Flower Growers Assoc. Bull.* 174: 1-5.

47. Pettibone, C.A., Matson, W.R., Pfeiffer, C.L. and Ackley, W.B. (1970). The control and effect of supplemental CO_2 in air-supported plastic greenhouses. *Trans. ASAE.* 13: 259-262.

48. Woltering, E.J. (1990). Beneficial effects of carbon dioxide on development of gerbera and rose plantlets grown in vitro. *Sci. Hortic.*. 441: 341-345.

49. Carlson, R.W. and Bazzaz, F.A. (1980). The effects of elevated CO_2 concentrations on growth, photosynthesis, transpiration and water use efficiency of plants. In: *Proc. Symp. on environmental and climatic impact of coal utilization*, J.Singhjag and A.Deepek, eds., Academic Press pp. 609-612.

50. Rogers, H.H., Bingham, G.E., Cure, J.D., Heck, W.W., Heagle, A.S., Israel, D.W., Smith, J.M. Surano, K.A. and Thomas, J.F. (1980). Response of vegetation to carbon dioxide: Field studies of plant responses to elevated carbon dioxide levels. Report 001, U.S. Dep. of Energy, Carbon Dioxide Research Division and the U.S.Dep. of Agriculture, Agricultural Research Service, Washington, D.C.

51. Sionit, N., Strain, B.R., Hellmers, H. and Kramer, P.J. (1981). Effects of atmospheric CO_2 concentration and water stress on water relations of wheat. *Bot. Gaz.* 142: 191-196.

52. Ford, M.A. and Thorne, G.N. (1967). Effect of CO_2 concentration on growth of sugar beet, barley, kale and maize. *Ann. Bot.* 31: 629-644.

53. Sionit, N.H., Strain, B.R., and Hellmers, H. (1981). Effects of different concentrations of atmopsheric CO_2 on growth and yield components of wheat. *J. Agric. Sci.* Camb. 79: 335-339.

54. Acock, B., Baker, D.N., Reddy, Y.R., Mckinion, J.M., Wisler, F.D., Del Castillo, D. and Hodges, H.F. (1982). Soybean responses to carbon dioxide: measurement and simulation 1981. Report 004 in series, reponse of vegetation to carbon dioxide, U.S. Dep. of Energy and Agriculture, Washington, D.C.

55. Allen, L.H., Jr., Boote, K.J. Jones, J.W., Mishore, J.W., Jones, P.H., Vu, C.V., Valle, R. and Campbell, W.J. (1982). Effects of increased carbon dioxide on photosynthesis and agricultural productivity of soybeans, Report ØØ3 in series response of vegetation to carbon dioxide, U.S. Dep. of Energy and Agric. Washington, D.C.

56. Allen, L.H., Jr., Vu, J.C.V., Valle, R.R., Boote, K.J. and Jones, P.H. (1988). Non-structural carbohydrate and nitrogen of soybean grown under carbon dioxide enrichment. *Crop Sci.* 28: 84-94.

57. Kimball, B.A. (1983). Carbon dioxide and agricultural yield. An assemblage and analysis of 43Ø prior observations. *Agron. J.* 75: 779-788.

58. Adams, R.M., Rosenzweig, C., Peart, R.M., Ritchie, J.T., McCarl, B.A., Glyer, J.D., Curry, R.B., Jones, J.W., Boote, K.J. and Allen, L.H., Jr. Global climate changes and US agriculture. *Nature* 345: 219-224.

III

PHOTOSYNTHESIS AND ENVIRONMENTAL STRESS

FUNCTIONAL MODELS OF WATER OXIDATION COMPLEX IN PHOTOSYSTEM II

S.Padhye[1], R.Yerande[1], U.Hegde[1], A.Kumbhar[1], V. Klimov[2], G.Ananyev[2], S.Allakhverdiev[2], and S. Zharmukhamedov[2]; [1]Department of Chemistry, University of Poona, Poona -411007, INDIA; [2]Institute of Soil Science and Photosynthesis, USSR Academy of Science, Puschino, 142292 USSR.

A tetranuclear manganese cluster in Photosystem II (PS II) in green plants has been shown to play an important role in catalytic oxidation of water to molecular oxygen. Based on known properties and spectral profiles of this cluster attempts have been made in the past towards synthesis and characterization of spectral and structural analogs of the water oxidation complex (WOC). In no case the synthesized compounds were able to evolve molecular oxygen from water. A tetrameric manganese cluster incorporating carboxylate and catechol ligand bridges is capable of reconstituting the photoinduced oxygen evolution and variable fluorescence indicating its close similarity to structure and function of the native manganese compound. The compound thus constitutes one of the few examples of 'functional models' of the photosynthetic water oxidation complex.

Introduction

Natural photosynthesis converts large quantities of solar radiation into chemical energy which is used as fuel, food and fiber. The efficiency of natural system is apparently low (seldom exceeds 1%) for most of the cultivated plants. Attempts are underway presently to improve this efficiency by genetic and agronomic techniques. Another approach is to develop biomimetic synthetic systems which can avoid the unwanted physiological and agronomic side effects of natural photosynthesis which may have greater potential for solar energy bioconversions. Such artificial systems can also contribute to our understanding of the structure and functions of the natural assemblies [1].

In the present communication we report the synthesis and characterisation of a novel, tetrameric manganese cluster containing carboxylate and catechol bridges which is capable of photoreactivating variable fluorescence (ΔF) as well as O_2 evolution in Mn-depleted pea subchloroplast particles upon reconstitution. This synthetic cluster is modelled after the water oxidation complex (WOC) containing manganese ions present in photosystem II (PS II) of chloroplasts. There are a good number of reports available on the probable structural organization and function of WOC although none of them provide a complete picture. This has infact led to many controversial interpretations. There is, however, a general agreement on the view that the cluster consists of four manganese ions [2] and that it is capable of cycling between five discrete oxidation levels

(S_0 to S_4) during catalytic water oxidation cycle [3]. The manganese ions of the cluster seem to be inequivalent as they are released and reincorporated differently [2]. Speculations regarding the organization of manganese ions have largely come from EPR [4,5],EXAFS and XNES studies [6]. Based on the spectroscopic results several putative manganese compounds have been synthesised as the structural models of WOC [7] but none have actually been shown to evolve oxygen except a dimer described by McAuliffe [8].

Against this background we have chosen to employ "anchored catechol" ligands as close analogues of the peptide residues around manganese ions in the natural system and have prepared a tetrameric cluster which can be substituted for the native WOC in the Mn-depleted pea subchloroplast particles.

It has recently been reported that such Mn-depleted pea subchloroplast particles when re-constituted with monomeric Mn complexes with different oxidation levels are able to restore variable fluorescence (ΔF) to some extent but fail to reactivate O_2 evolution [9]. The work described in the present communication reveals that reactivation of O_2 evolution is possible only with the tetrameric Mn clusters which confirms earlier hypothesis that 4 Mn atoms act in unison during O_2 evolution. Our results also suggest that the present synthetic tetramer has a close analogy with the native WOC which opens up the possibility of achieving an important task in construction of artificial photosynthetic systems.

Experimental

The manganese tetramer used in the reactivation experiments was synthesized by the interaction of manganese (II) acetate with " anchored catechol " ligand (L1) in ethanol under N_2 atmosphere as described in case of quinonoidal complexes [10]. Calculated for $Mn_4 C_{34} H_{34} N_4 O_{13}$; Cal.: C,44.08; H,3.70; N,6.05; Mn,23.72%; Found : C,44.72; H,3.46; N,6.13; and Mn,23.06%.

Pea subchloroplast particles enriched in PS II (DT-20) were isolated by centrifugation of choloroplast suspension at 20 000*g pre-treated with 0.4% digitonin and 0.1% Triton X-100 as described by Klimov et al.[11]. The details of the complete extraction (more than 98%) of Mn and water soluble proteins with molecular masses 17, 23 and 33 kDa from DT-20 particles and reconstitution of these particles with synthetic model complexes alongwith the method of measurements of restoration of variable fluorescence (ΔF) and O_2 evolution have been described previously [12-15].

Results and Discussion

1. Description of the tetrameric cluster:

The interaction of manganese acetate with the "anchored catechol" ligand in ethyl alcohol results in a yellow green compound which on a slow evaporation yields crystals suitable for x-ray crystallography. The crystalline compound gives satisfactory fit for the formulation $[Mn_4 (L)_2 (CH_3 COO)_4].H_2O$ while cryoscopic measurements in acetonitrile solvent provide a molecular weight of 921.103 for it.

The structural features as revealed from our preliminary X-ray crystallographic studies indicate a linear arrangement of the manganese atoms with anchored catechol ligands providing the bridges for the inner manganese atoms while two carboxylate moieties furnishing additional bridges for the outer manganese atoms. With this kind of an arrangement the internuclear distances between the inner manganese pair turns out to be 3.05 A° while it is found to be 3.30 A° between the outermost manganese pair. These values are resonably close to the values provided by EXAFS measurements on chloroplast preparations which are found to be 2.7 A° and 3.3 A° for the inner and outer manganese pairs respectively [6].

The infrared spectra of the tetramer exhibits strong absorptions characteristic of acetate and catecholimine coordination. The splitted absorption for the acetate groups at 1580 and 1420 cm^{-1} probably originates from the symmetric vibrations of the acetate chromophore [16] and the span of these modes (Δ=160 cm^{-1}) is within the range generally observed for the symmetrically bridged carboxylate groups [17]. Other prominent IR spectral features of the compound include the (C=N) imine vibration which is shifted to lower wavenumbers upon coordination. Two strong absorptions at 1430 and 1270 cm^{-1} respectively are obviously diagnostic of catechol coordination [18].

The electronic spectra of the tetramer shows weak intensity spin forbidden d-d transitions characteristic of Mn(II) complexes flanked by intense charge transfer bands at 24210 cm^{-1} associated with the metal-quinone complexes . The solid state magnetic moment of the complex varies from 11.33 B.M. at 300 K to 3.24 B.M. at 5 K per tetranuclear manganese (II) cluster. The plot of the magnetic parameters with temparature (Fig. 1) shows different slope characteristics giving support to the observation that manganese pools in this tetramer are hetrogeneous in the sense that one pair of the ions are strongly coupled while in the other pair metal ions are weakly coupled. A detailed analysis of the magnetic data and EPR spectral features of this tetramer shall be reported elsewhere.

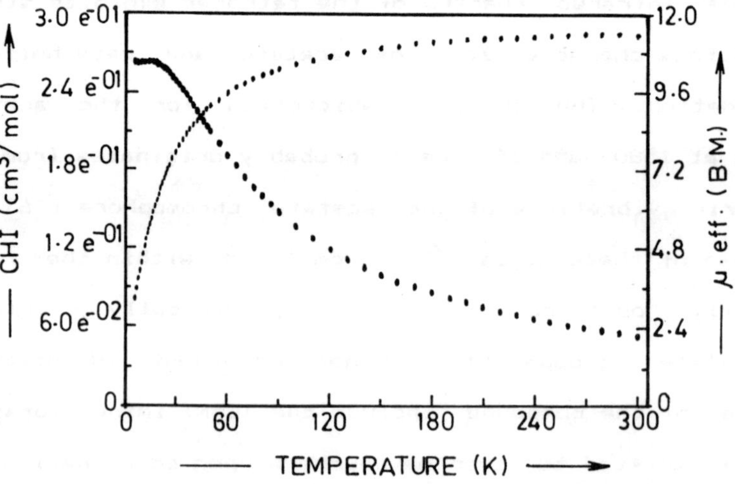

Fig.1. Temperature dependence of magnetic susceptibilities and effective magnetic moments ($\mu eff/Mn_4$) of the Manganese Tetramer.

2. Reconstitution Experiments:

Complete removal of Mn and water soluble proteins with moleculer weights 17, 23 and 33 kDa from pea subchloroplast particles (DT-20) shows a complete inhibition of oxygen evolution and photoinduced electron transfer [12,13]. Only a joint addition of $MnCl_2$ and $CaCl_2$ results in a gradual restoration of the O_2 evolution under continous illumination conditions [13]. Ca^{+2} is highly specific in the reactivation and the restoration requires a very high concentration of Ca^{+2} (Fig.2).

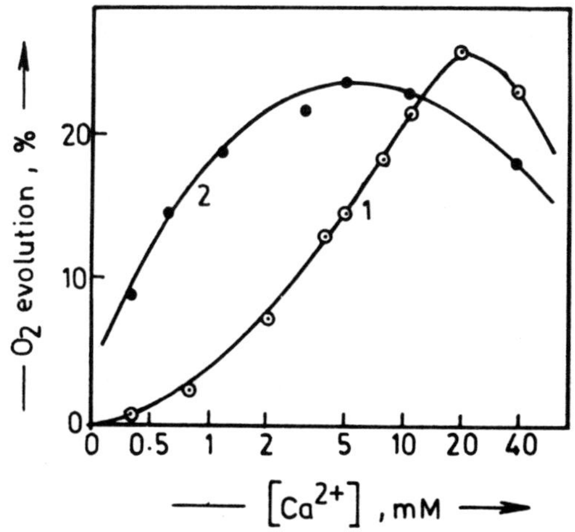

Fig.2. Dependence of the rate of oxygen evolution in DT-20 particles (lacking Mn & water soluble proteins) on concentration of $CaCl_2$ before (1) and after (2) readdition of the sum of the proteins with moleculer weights 17,23 & 33 KD in ratio 1:1 with P680. The medium contains (mM) : $MnCl_2$ (0.005), ferricyanide (0.25), MES-NaOH pH 6.5 (50), sucrose (400), NaCl (35) Actinic light > 600 nm, 55 W/cm.Chl.concentration 200 µg/ml.

Since a funtionally active oxygen evolving complex can be reconstituted after a complete removal of Mn from subchloroplast particles in the absence of water soluble proteins (17.23 and 33 kDa) it is clear that they are not the binding sites for Mn involved in oxygen evolution. When the sum of these proteins is added to the depleted particles, the reactivation is observed at a much lower concentration of Ca^{+2} although the maximal rate of oxygen evolution is not increased (Fig.2).

Illumination of the untreated DT-20 particles by actinic light induces an increase in ΔF by a factor of 2.5-3.0 and a complete extraction of Mn is accompanied by 15-20 fold decrease of the photoinduced ΔF [11].Addition of Mn^{+2} (as MnCl2) leads to an almost complete restoration of the ΔF(upto 85% of the original ΔF value observed for the untreated DT-20 particles) which on treatment with EDTA is found to be dramatically quenched indicating that in such a restoration manganese ions are essentially acting as free ions donating electrons to Z^+ or $P680^+$ species.

A very effective reactivation of the photoinduced ΔF is observed when $MnCl_2$ is replaced with the present synthetic tetrameric Mn complex and that also at a very low concentration i.e. 2 molecules (8 Mn atoms) per reaction center. Further, the restored ΔF is not inhibited by an addition of EDTA which indicates that the intact complex and not the free Mn ions (which could be released due to decomposition) are responsible for the reactivation (Fig.3).

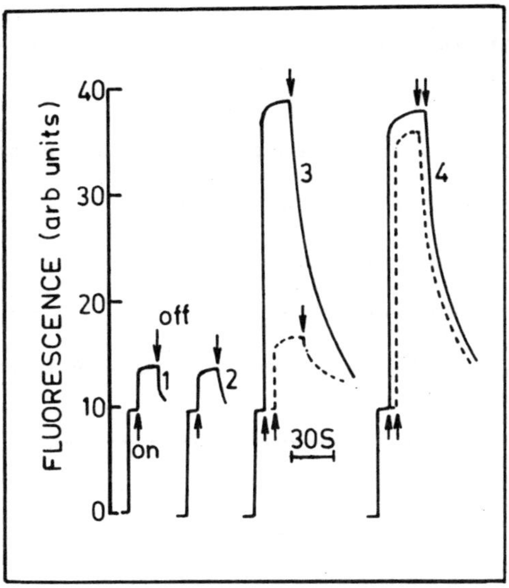

Fig.3. Reactivation of ΔF in DT-20 (lacking Mn & water soluble proteins) by synthesized Mn complexes. (1) No addition (2) After additions of monomeric Mn complex (0.4 μM, 8 molecules/RC), (3) $MnCl_2$ (0.4 uM, 8 Mn/RC), (4) tetrameric Mn complex (0.1 μM, 2 molecules , 8 Mn atoms/RC). Dashed line of (3) & (4) are same as (3) & (4) in presence of 50 μM EDTA. Chl. concentration 10 μg/ml.

Monomeric Mn complexes of quinone ligands had earlier been shown by us to restore Δ F to some extent but at very high concentrations [9]. It is , therefore, significant that the present manganese tetramer is able to reactivate ΔF at such a low concentration. Finally it is interesting to observe that the reactivation of oxygen evolution can be very effectively achieved by this tetramer (Fig.4). The rate of oxygen evolution in presence of 2 molecules of tetramer (8 Mn atoms/RC) is found to be 5-6 times higher than that observed in case of activation by $MnCl_2$. What is most peculiar about this photoreactivation is the fact that monomeric Mn complexes do not show any reactivation of oxygen evolution [9]. The reactivation of oxygen evolution

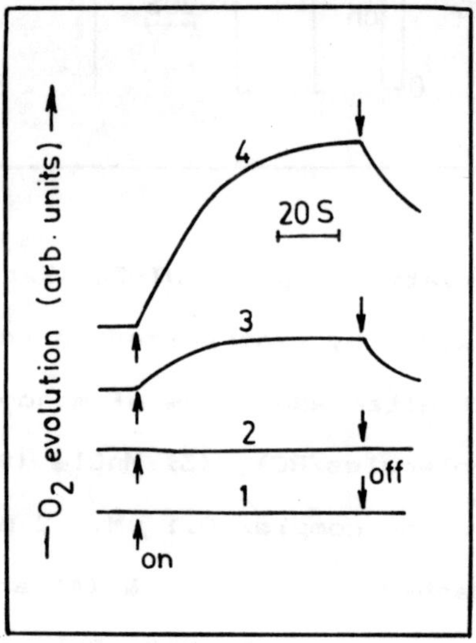

Fig.4. Reactivation of oxygen evolution in DT-20 (lacking Mn & water soluble proteins) by synthesized Mn complexes.

1. DT-20 (lacking Mn & water soluble proteins),
2. After additions of monomeric Mn complex,
3. $MnCl_2$,
4. Tetrameric Mn complex.

Concentrations same as in Fig.3. and conditions same as in Fig.2.

by the present tetramer requires an addition of high concentration of Ca^{+2} ions (20–30 mM) as well as photoreactivation procedures indicating that some additional interactions between the tetramer and PS II components are present which are being investigated further. However, the restoration of oxygen evolution and reactivation of photoinduced ΔF suggest that the structural conformation of our synthetic tetramer after incorporation into the subchloroplast particles must be very similar to that of the native Mn-cluster of PS II.

<u>References</u>

1. Hall,D.O., & Krishna Rao,K., (1989). Immobilized photosynthetic membranes and cells for production of fuels and chemicals. Chimicaoggi 41–47.
2. Babcock,G.T.,(1987).The photosynthetic oxygen evolving process.In Amesz,J.(ed)New Comprehensive Biochemistry: Photosynthesis, Chapter 6, Elsevier Science Publishers Amsterdam, pp. 125 – 158.
3. Kok, B., Forbush,B., & McGloin,M., (1970). Cooperation of charges in photosynthetic oxygen evolution. I. A linear four step mechanism. Photochem. Photobiol. 11: 457–475.
4. Dismukes,G.G. & Siderer,Y. (1981). Intermediates of a polynuclear manganese center involved in photosynthetic oxidation of water. Proc. Natl. Acad. Sci. USA. 78(1) : 274–278.
5. Zimmermann,J.L. & Rutherford,A.W. (1984). EPR studies of the oxygen evolving enzyme of Photosystem II. Biochem. Biophys. Acta. 767: 160–167.

6. For example:

(a) George,G.N., Prince,R.C. & Cramer,S.P. (1989). The manganes site of the photosynthetic water-splitting enzyme. Science 243: 789-791.

(b) Yachandra,V.K., Guiles,R.D., McDermott,A.E., Cole, J.L., Britt,R.D., Dexheimer,S.L., Sauer,K. & Klein, M.P.(1987). Comparison of the structure of the manganese complex in the S_1 and S_2 states of the photosynthetic oxygen evolving complex : An X-ray absorption spectroscopy study. Biochemistry . 26: 5974-5981.

(c) Yachandra,V.K., Guiles,R.D., McDermott,A.E., Britt, R.D., Dexheimer,S.L., Sauer,K. & Klein,M.P. (1986). The state of manganese in photosynthetic apparatus. 4.Structure of the manganese complex in photosystem II studied using EXAFS spectroscopy. The S_1 state of the O_2 evolving photosystem II complex from spinach. Biochim.Biophys.Acta 850 : 324-332.

(d) Penner-Hahn,J.E., Valdo,G.S., Riggs,P.J., Yu,S., Yocum,C.F. & Mei,R. (1990). Characterization of the manganese sites in manganese redox enzymes: Studies of Mn Catalase and the Photosynthetic oxygen evolving complex , Workshop on Bioinorganic Chemistry, Dec. 9-15, 1990. Indian Institute of Technology, Madras India .pp.L8.

7. For example :

(a) Mulay,M. & Padhye,S. (1987). Use and Practice of Biomimetic Models for water oxidation complex (WOC) in plant photosynthesis. Proc.Indian Natn.Sci.Acad. B53 No.5 & 6: 471-473.

(b) Wieghardt,K. (1989) The active sites in Manganese containing Metalloproteins and Inorganic Model complexes. Angew. Chem.Int.Ed.Engl. 28 : 1153-1172.

(c) Christou,G., (1989) Manganese Carboxylate chemistry and its biological relevance. Acc.Chem.Res.,22:329-335.

(d) Christou,G & Vincent,J.B.,(1989)Structural types in oxide-bridged manganese chemistry in'Metal clusters in Proteins', ACS symposium, Lawrence Que. Jr.ed., American Chemical Society,Washington DC.,pp 239-255.

(e) Pecoraro, V. L. (1988) Structural proposals for the manganese centers of the oxygen evolving complex.An inorganic chemist's perspective.Photochem.Photobiol. 48 : 249-246 and references therein.

8. Ashmawy, F.M., McAuliffe, C.A.,Parish,R.V., & Tames,V. (1985) Water Photolysis part I. The photolysis of coordinated water in [(MnL(H_2O)$_2$] [ClO_4] $_2$ (L=dianion of tetradentate O_2N_2 donor schiff base.A model for the manganese site in photosystem II of green plant Photosynthesis.J.Chem.Soc.Dalton Trans. 1319-1397.

9. Yerande,R., Hegde,U., Padhye,S., Klimov V., Ananyev,B. Allkhverdiev, S.I., & Zharmukhamedov, S.K., (1990). Artificial photosynthesis: monomeric quinone complexes of Iron (II) and Manganese (II),(III) and (IV) ions in Photoreactivation of Pea Subchloroplast Particles. In : ' Recent Advances in Bioenergetic Processes ', Singhal G.S ed., DBT, New Delhi.

10. Padhye,S.B.,Rane,S.Y. and Gupta,S.G. (1978). Manganese chelates of some naturally occuring vitamin K artefacts. Inorg. Nucl. Chem. Lett. 14: 83-86.

11. Klimov, V. V., Allakhverdiev, S. I., Shutilova, N. I., Krasnovsky, A. A. (1980).Study of pheophytic photoreduction and chlorophyll P680 photooxidation on photosystem II preparations from pea chloroplast and Chlamydomonas reinhardii. Fiziol.Rast.27: 315-326.

12. Klimov, V. V., Allakhverdiev, S. I., Shuvalov, V.A and Krasnovsky,A. A. (1982). Effect of extraction and readdition of manganese on light reactions of photosystem II preparations. FEBS.Lett. 148: 307-312.

13. Klimov, V. V., Allakhverdiev, S. I., Shafiev,M. A. and Demeter, S. (1985). Effect of complete extraction and readdition of manganese on thermoluminescence of pea photosystem II preparations. Biochem.Biophys.Acta 809: 414-420.

14. Ananyev,G.M.,Shafiev,M.A.,Isaenko,T.V. and Klimov,V.V. (1988). Photoreactivation of O_2 evolution function after complete removal of manganese from photosystem II particles. Biofizika (USSR), 33: 265-269.

15. Klimov,V.V., Ganago,I.B., Allakvberdiev,S.I., Shafiev, M.A. and Ananyev,G.M. (1987).Structural and functional aspects of electron transfer in photosystem II of O_2 evolving organisms in ' Progress in Photosynthesis', Biggins J.,Ed.,Martinus Nijhoff Publishers, Dordrecht, pp 1.5.581-1.5.584.

16. Suzuki, M.,Murata, S., Uehara, A, and Kida S., (1987) Synthesis and characterization of dinuclear manganese (II,III) mixed valence complex with dinulceating ligands. L= py 2,6-Bis [bis (2-pyridyl methyl) amino methyl] 4-methyl phenol. Chem.Lett.281-284.

17. Mikuria,M., Kida,S., and Murase,I. (1988) Preparation and structural characterization of binuclear manganese (III) complexes with 1, 5 -Bis(salicylideneamino)-3-phenol and its derivatives. Chem.Lett. 35-38.

18. Brown , D.G.,Johnson , W.L., (1979), Preparation and properties of two tris o-semiquinone complexes. Z.Naturforsch. Chem, 34b : 712-715.

CHEMICAL PROBES FOR WATER OXIDATION COMPLEX IN PHOTOSYSTEM II. REVERSIBLE CHLORIDE-ANION PAIRING BY K (PICRATE) (18-CROWN-6)

PRASANNA MOHANTY[1], S.C. SABAT[1] AND S. PADHYE[2]

[1]School of Life Sciences,
Jawaharlal Nehru University, New Delhi-110067.

[2]Chemistry Department, University of Poona,
Pune

INTRODUCTION

A tetranuclear manganese cluster is essential for oxygen evolving activity of photosystem II (PS II) along with Ca^2 and Cl^- ions (1). O_2 evolution from water is carried out by the transiently oxidized manganese cluster in a cycle of 4 flashes commonly referred as the S-Cycle (2). Ca^{2+} ions have been shown to be necessary for the maintenance of structural integrity of the polypeptides around manganese (3-5). On the other hand, chloride ions are required for the steady state O_2 evolution (6,7) as well as for the advancement of a particular S-state (8). Inspite of the rapid advances made in the past decade on the structural organization of water oxidation complex of PS II, our understanding of the molecular mechanism of water oxidation is far from complete. Several mechanistic models have been suggested to account for spectral and magnetic properties of the native manganese cluster. Amongst these the 'cubane - adamentane' model of Brudvig (9), the 'butterfly' model of Vincent and Christou (10) and the 'trimer-monomer' combination of Dismukes (11) figure prominently. Some of these models have suggested that chloride ion can be ligated to Mn (6,12), or may be required to stabilize a positive charge on the Mn-containing oxygen evolving enzyme (13,14). Kinetic studies suggest that the depletion of Cl^- ions inhibits the advancement of the Mn-oxygen evolving complex beyond the S_2-state (15,16), while other studies have implicated a role for Cl^- in the S_3 to S_o transition (17). Although the requirement for Cl^- ion on the donor side of PS II and its involvement with the oxygen evolving complex have been established, a selective structural probe to unravel its site and mode of action is very much needed.

Crown ethers are well suited for probing the role of Cl^- ions in the photosynthetic electron transport in the isolated thylakoid membranes due to their excellent complexing abilities towards alkali and alkaline earth cations as well as for the anions and neutral low molecular weight species (18,19). The crown ethers have been shown to discriminate between and within the group of alkali and alkaline earth cations, and also they exhibit preferential binding with the associated anionic species (20,21). We are thus interested in examining the effect of various crown ethers and their derivatives on the photosynthetic electron transport processes in the isolated thylakoid membranes. The work is thus a continuation of our recent studies on the use of developing chemical probes for the higher oxidation states of Mn ions in the transient S-states (22) and for revealing the amino acid residues involved in the binding of Mn ions (23).

In the present communication, we have extended our earlier studies on the picrate derivative of K (18-Crown-6) compound, (I), which is a versatile discriminating macrocyclic compound for examining the

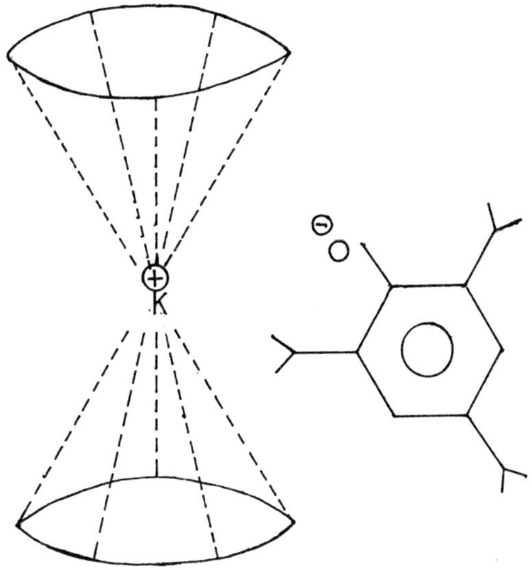

Fig.1 I, K(Picrate)(18-Crown-6)

reversibility of the chloride ion inhibition and corresponding effects on the photosynthetic electron transport in thylakoids of beet spinach (*Beta palanga*)

(24). The results suggest that the site of the action of crown compound is on the Cl^- ion pool in the thylakoid membranes. The reversibility and selectivity of Cl^- ion depletion by (I), (see Fig.1), lead us to assume that Cl^- ions are probably involved not in the direct bonding with the Mn-ions but as stabilizing anions towards the positively charged manganese clusters (25).

Experimental Methods and Materials

Thylakoid preparation

Broken chloroplasts (thylakoid membranes) were isolated from freshly harvested spinach (*Beta palanga*) leaves. Well washed, deribbed leaves were homogenized in ice cold isolation medium containing 300 mM NaCl, 5 mM $MgCl_2$ and 50 mM Hepes - KOH (pH 7.5). The homogenate was filtered through 4 layers of Mira-cloth and the filtrate was initially centrifuged at 300xg for 3 min to remove the cell debris. Supernatant from this centrifugation was recentrifuged at 6000xg for 5 min to pellet the thylakoid membranes. The pellet was once washed with the isolation buffer and finally suspended in the same medium, so as to obtain chlorophyll (Chl) concentration of 1 mg ml^{-1}.

Electron Transport Assay

Photosystem II (PS II) dependent (H_2O -->pBQ) photo-electron transport rates of isolated thylakoids were measured polarographically using a Clark - type O_2 electrode (24). Unless otherwise mentioned, the reaction mixture for the measurement of PS II catalyzed electron transport in a total volume of 1 ml contained 100 mM sucrose, 10 mM NaCl, 5 mM $MgCl_2$ and thylakoid equivalent to 15 ug Chl. pBQ at a concentration of 0.5 mM was used. The O_2 evolution rates were measured at rate saturating light intensity (470 Wm^{-2}) at $25 \pm 1^{o}C$. Chloride dependent stimulation of PS II activity in chloride depleted and crown compound, (I), treated thylakoids (see below) were also measured polarographically by addition of varying concentrations of NaCl. Other anions like Br^- (as NaBr) and SO_4^- (as Na_2SO_4) or cations like Ca^{+2} (as $CaCl_2$ or $CaSO_4$) or Mg^{+2} (as $MgCl_2$) were also used when required.

Preparation of chloride depleted thylakoids

Thylakoids depleted of chloride ions were prepared following the methods of Hind, Nakatani and Izawa (1969) and by Izawa, Heath and Hind (1969). In brief, thylakoids were isolated in a medium containing 0.1 M Na_2SO_4, 0.03 M TES-NaOH (pH 7.4). The thylakoids were pelleted at 6000xg for 5 min. Sedimented thylakoids were washed three times with a medium containing 0.1 M sucrose, 2 mM Tricine - NaOH (pH 7.8) and 0.5 mM Na_2-EDTA. The thylakoids were finally taken up in a medium of 0.1 M sucrose, 0.6 mM TES - NaOH (pH 7.4) and 5.0 mM $MgSO_4$. A brief exposure of thylakoid suspension to room temperature (25°C) during washing promoted the release of chloride from the lamellae.

Crown compound treatment of thylakoids

Isolated thylakoids, suspended in the isolation buffer containing 1 mg Chl ml^{-1} was incubated with 5, 20 and 50 uM crown compound (K-Pic-18-Crown-6) for 20 min at 4°C in dark. The crown compound (I), incubated thylakoids were finally spun down at 6000xg for 5 min and suspended in chloride free medium containing 100 mM sucrose, 5 mM $MgSO_4$ and 0.6 mM Hepes buffer adjusted to the pH 7.5 with KOH. Similarly crown untreated thylakoids were spun, washed down and suspended in the chloride free medium which served as the control.

Chemicals

Aqueous solution of K-(Picrate)-(18-Crown-6) was used in the experiment (see reference 24). pBQ was recrystalized through sublimation and used for assay as aqueous solution.

Results and Discussion

It had been shown earlier that K (Picrate) (18-Crown-6) compound, (I), inhibits the electron transport in the beet spinach (*Beta palanga*) thylakoid membranes to the extent of 75% at an optimum concentration of 10 uM (24). It was also observed that PSI catalyzed electron transport activity remained insensitive to the treatment with this crown compound even at higher concentrations (20 uM). The experiments with external electron donors like diphenyl carbazide (DPC) and NH_2OH had indicated that the site of inhibition of electron transport is probably close to the donation site of these exogenous electron donors. In view of the known discriminating properties of a charge-separated crown compound such as K (Picrate)(18-Crown-6) towards cations and anions (28) the most likely species to be affected by this compound is Cl^- ions and this hypothesis can be examined by reactivation of the crown-treated samples by the additions of exogenous Cl^- ions.

Treatment with the crown compound, (I), (see Fig.1) results in a substantial lowering of the O_2 evolving ability of the thylakoids which is very similar to the effect produced by Cl^- ion depletion (see Fig.3) following the procedures described by Izawa *et al*. (27). Additions of different concentrations of Cl^- ions (as NaCl) restored the electron transport activity (Fig.2(B), (C) and (D) nearly to the same extent as in case of the untreated samples (Fig.2A). The maximal restoration of electron transport was obtained at about 20uM and the restoration remained unaffected thereafter. The Cl^- ions depletion by repeated washings protocols developed by us (Fig.2(A) and by the method of Izawa's (Fig.2(D) yielded similar results on the restoration of

O_2 evolution indicating that both the procedures are effective in Cl^- ion removal and are compatible. Our present results thus lend support to the earlier suggestion of Cl^- ions being the target site of K-(Picrate) (18-Crown-6) treatment. Considering that Cl^- ions have been implicated as the stabilizing counter anions for higher S-states of the oxygen evolving complex (8,25) it is reasonable to suggest that they may be affected most by the present crown compound, (I), treatment. The ease with which the inhibitory effects of crown compound, (I), treatment are reversed by the addition of exogenous Cl^- ions make us also suggest that Cl^- ions are essentially acting as the stabilizing counter anions in the Mn-oxygen evolving complex (25).

The capabilities of various alkali and alkaline earths salt in restoring oxygen evolution in the chloride depleted beet spinach thylakoids is shown in the histograms in Fig.3. It can be seen that the most effective restoration of O_2 evolution is obtained with Cl^- ions followed by Br^- and SO_4^{2-} ions. The much lesser activation observed in case of $CaCl_2$ may probably be due to the formation of charge-separated and yet picrate-chelated compounds described in case of Cs-(Picrate)(Benzo-18-Crown-6) crown compound (28).

Figure Legends

Fig.2 Effect of addition of varying concentrations of NaCl on the photosystem II catalyzed electron transport activity in crown compound treated and control thylakoids [A], control-washed thylakoids to free Cl^- ions (Crown compound untreated), [B], [C] and [D] 5, 20 and 50 uM crown compound treated and then washed thylakoids respectively [E], electron transport activity of chloride depleted thylakoids prepared according to Izawa *et al.* (27). The insets (in Fig. A and E) show the effect of addition of NaCl on untreated thylakoids. All other details as in experimental methods. The data represent the mean value of three independent batches of thylakoid preparations. The variations in rates are shown as standard deviations.

Fig.3 Histogram showing the effect of addition of different anions and cations (indicated in the figure) on the photoelectron transport activity of [A], Chloride depleted and [B], Crown washed thylakoids. The added concentrations are as indicated in the figure. Chloride depletion was done as in Fig.2 [E].

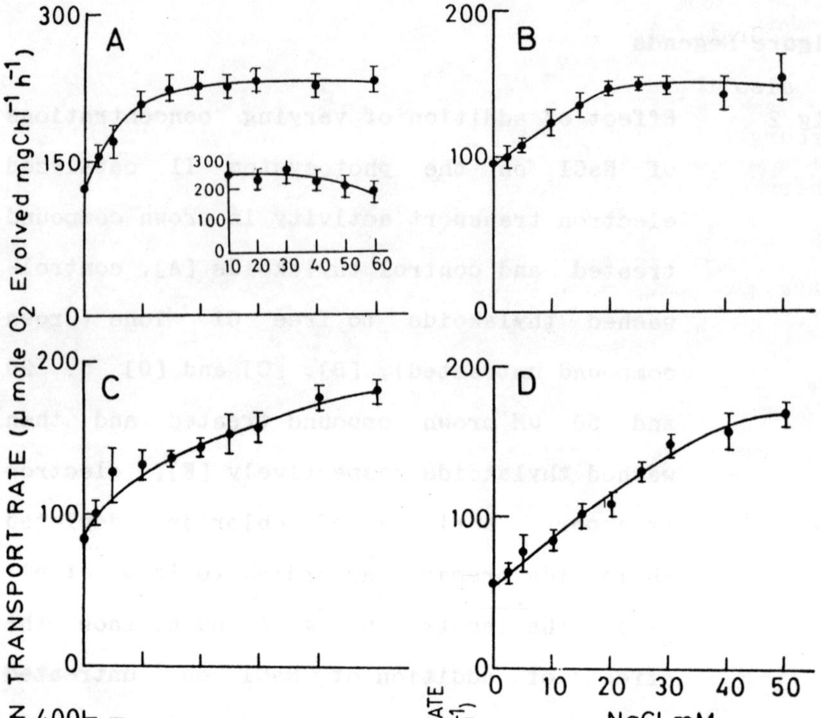
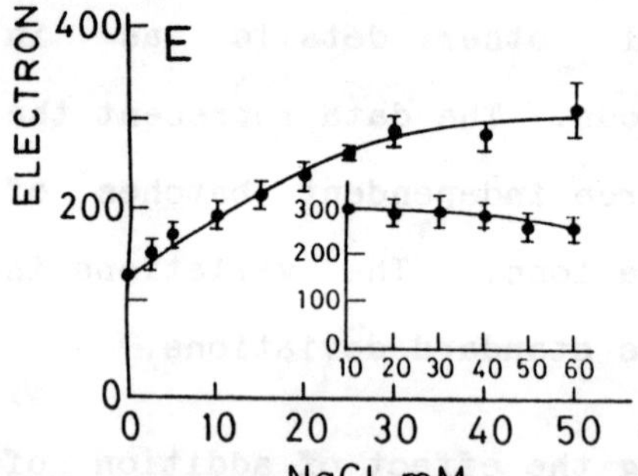
ELECTRON TRANSPORT RATE (μ mole O2 Evolved mgChl-1 h-1)
A
B
C
D
E
NaCl, mM
NaCl, mM
Fig. 2

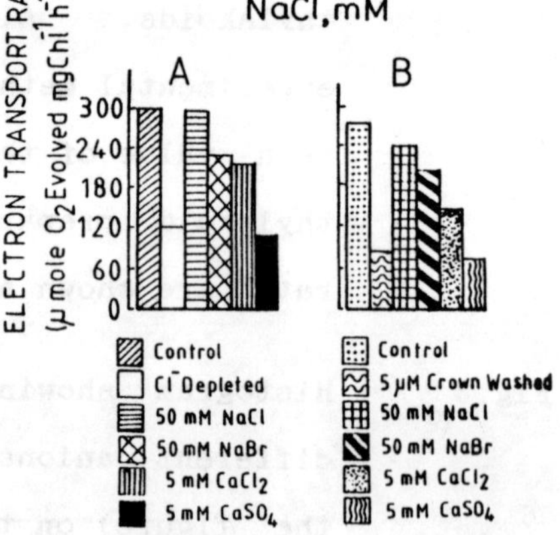
ELECTRON TRANSPORT RATE (μ mole O2 Evolved mgChl-1 h-1)
A
B
Control
Cl⁻ Depleted
50 mM NaCl
50 mM NaBr
5 mM CaCl2
5 mM CaSO4
Control
5 μM Crown Washed
50 mM NaCl
50 mM NaBr
5 mM CaCl2
5 mM CaSO4
Fig. 3

Acknowledgement

Grateful thanks to Dr. V. Vergiya for her help. We also thank Prof.(Dr.) Uli Schreiber, University of Wuzburg, Germany, for suggesting the use of crown ethers. Financial support of ICAR-USDA Grant No.FG-IN-679 (IN-ARS-402) and Department of Non-conventional Energy Sources (DNES) is gratefully acknowledged.

References

01 R. Radmer and G.M. Cheniae, Top. Photosynth. 2, 303-348 (1977)

02 B. Kok, B. Forbush and M. McGloin, Photochem. Photobiol., **11**, 457-475 (1970)

03 P.H. Homann, Biochim. Biophys. Acta, **934**, 1-13 (1988)

04 D.F. Ghanotakis and C.F. Yocum in 'Ann. Rev. Plant Physiol. and Molecular Biol., **43**, 255-276 (1990)

05 J. Brand, P. Mohanty and D.C. Fork, FEBS Letts. **155**, 120-123

06 P.M. Kelly and S. Izawa, Biochim. Biophys. Acta, **506**, 198-210 (1978)

07 S.M. Theg, P.A. Jursinic and P.H. Homann, Biochim. Biophys. Acta, **766**, 636-646 (1984)

08 Govindjee and P.H. Homann in 'Highlights of Modern Biochemistry', A. Kotyk, J. Skoda, V. Paces and V. Kostka, eds., VSP Press, Netherlands, Vol.I, P.933-961 (1989)

09 Brudvig, G.W. and R.H. Carbtree. Proceed. Natl. Acad. Sci. (USA) **83**, 4586-4588 (1986)

10 J. Vincent and G. Christou, Inorg. Chem. Acta **136**, 141-143, (1987)

11 G.C. Dismukes, Photochem. Photobiol. **43**, 99-115 (1986)

12 P.O. Sandusky and C.F. Youcum, FEBS Letts., **162**, 339-343 (1983)

13 C. Critchley, I.C., Baianu, Govindjee and H.S. Gutowsky, Biochim. Biophys. Acta, **682**, 436-445 (1982)

14 I.C. Baianu, C. Critchley, Govindjee and H.S. Gutowsky. Proc. Natl. Acad. Sci. U.S.A., **81**, 3713-3717 (1984)

15 S. Itoh, C.T. Yerkes, H. Koike, H.H. Robinson and A.R. Crofts, Biochim. Biophys. Acta **766**, 612-622 (1984)

16 S.M. Theg, P.A. Jursinic and P.H. Homann, Biochim. Biophys. Acta, **766**, 636-646 (1984)

17 J. Sinclair, Biochim. Biophys. Acta., **764**, 247-252 (1984)

18 J.M. Lehn, Structure and Binding, **16**, 1 (1973)

19 R.M. Izatt and J.J. Christensen, in., Synthetic Multidentate macrocyclic Compounds, Academic Press, New York (1978)

20 N.S. Poonia, J. Am. Chem. Soc., **96**, 1012 (1974)

21 N.S. Poonia in "Progress in Macrocyclic Chemistry". R.M. Izatt and J.J. Christensen, eds., Vol.1, Wiley-Interscience, New York, p.116 (1979)

22 U. Hegde, S. Padhye, M. Banerjee and P.B. Vidyasagar. Ind. Jour. Biochem. Biophys. **27**, 5-8 (1990)

23 M. Banerjee, P. Vidyasagar, U. Hegde and S. Padhye, Ind. J. Biochem. Biophys. **27**, 248-256 (1990)

24 S.C. Sabat, V. Vergiya, B.C. Tripathy and P. Mohanty Z.Naturforsch **46**C (in press) (1991)

25 G.C. Papageorgiou, Y. Fujimura and N. Murata. Biochim. Biophys. Acta (1991) in press

26 G. Hind, H.Y. Nakatani and S. Izawa. Biochim. Biophys. Acta. **172**, 277-289 (1969)

27 S. Izawa, R.L. Heath and G. Hind. Biochim. Biophys. Acta. **180**, 388-398 (1969)

28 N.S. Poonia, P. Bagde and K.S. Sindhu, J. Inclision Phenomena **4**, 167 (1986).

A UNIQUE ROLE OF CARBON DIOXIDE IN PHOTOSYSTEM II

Govindjee
Departments of Physiology & Biophysics, and Plant Biology,
University of Illinois at Urbana-Champaign, 505 South
Goodwin Avenue--289 Morrill Hall, Urbana, IL 61801-3793, USA

1. ABSTRACT

An evaluation of the effect of predicted global increases in CO_2 on photosynthesis should include not only an exmination of the effects on CO_2 fixation, but also on the so-called bicarbonate effect in photosystem II. This bicarbonate effect involves a unique stimulatory role of CO_2 (or HCO_3^-) in the conversion of plastoquinone to plastoquinol at the reaction center of photosystem II. After providing an introduction to the reaction centers of photosynthetic bacteria and photosystem II and to the bicarbonate effect, we will review here recent results that establish the following: (1) the bicarbonate effect is absent from the reaction centers of the green photosynthetic bacterium *Chloroflexus aurantiacus* and mutants of purple photosynthetic bacterium *Rhodobacter sphaeroides*; (2) inhibitors, such as formate, indeed release, and, thus, deplete CO_2 from thylakoid membranes; and (3) both the reaction center proteins D1 and D2 are involved in the bicarbonate effect since certain D1 and D2 mutants, altered in single amino acids, show differential sensitivity to bicarbonate-reversible formate effects. We will then discuss a current hypothesis, based on an iron-(bi)carbonate containing protein; it envisions CO_2/HCO_3^- binding on the Fe and certain specific amino acids, including arginines, in the PSII reaction center proteins D1 and D2. This binding is suggested to provide stability to the reaction center and stimulate electron flow and protonation required for plastoquinol formation. Finally, a possible relation of low affinity HCO_3^- binding sites to the phenomenon of photoinhibition will also be discussed.

2. INTRODUCTION

All life on earth has been divided into three domains: archea; bacteria; and eucarya [1]. True photosynthesis involving redox components occurs only in the latter two. Photosynthetic bacteria, such as the well-known purple bacteria *Rhodobacter sphaeroides* and *Rhodopesudomonas viridis*, the green bacteria *Chlorobium thiosulfatophilum* and *Chloroflexus aurantiacus*, and the bacteriochlorophyll-g-containing bacterium *Helibacterium chlorum*, are all anoxygenic, i.e., they do not evolve O_2. On the other hand, all cyanobacteria (example, *Synechocystis sp.*), algae (example, *Chlamydomonas reinhardtii*) and other photosynthesizing plants are all oxygenic, i.e., they evolve O_2.

Cyanobacteria and plants contain two types of reaction centers (I and II), that operate in series, to transfer electrons from H_2O to $NADP^+$ [2-4]. Purple photosynthetic bacteria and *Chloroflexus* type green bacteria contain reaction centers that are similar to reaction center II [5] although a major difference is that they do not oxidize H_2O and do not contain Mn (cf. [6]). A cyclic reaction produces high-energy intermediate that is used to reduce NAD^+ by reversed electron flow using external hydrogen donors. On the other hand, *Chlorobium* type green bacteria and *Heliobacterium chlorum* contain reaction centers that appear to be similar to reaction center I: NAD is reduced by externally added hydrogen donors. A possible scenario for the evolution of photosynthetic reaction centers has been discussed by Olson and Pierson [7].

2A. Reaction Centers

A major breakthrough in photosynthesis research has been the crystallization and X-ray structure of reaction centers of photosynthetic bacteria *Rhodopseudomonas viridis* and *Rhodobacter sphaeroides* [8-10]. The reaction center proteins L and M, that are similar in many respects to the reaction center proteins D1 and D2 of photosystem II of plants and cyanobacteria, harbor 4 bacteriochlorophyll, a or b, 2 bacteriopheophytin a or b, two bound ubi- or mena-quinones Q_A and Q_B and an iron atom. These reaction centers transfer electrons from their primary electron donor special pair bacteriocholorophyll P870 (*R. sphaeroides*) or P960 (*R. viridis*) to their, respective, electron acceptor bacteriopheophytin a or b within about 3 ps [11]. Recently, Holzapfel et al. [12] have established that a bacteriochlorophyll monomer is an intermediate between the primary electron donor and the bacteriopheophytin. However, the other bacteriochlorophyll monomer and the other bacteriopheophytin molecule, situated on the "M" side of the reaction center, do not participate in this photochemistry. The photosystem II reaction center has been shown to contain 6 chlorophyll a molecules and 2 bacteriopheophytin molecules [13]; the Q_A and Q_B are bound plastoquinone molecules. Wasielewski et al. [14] have shown that, like in photosynthetic bacteria, the charge separation, i.e., creation of oxidized reaction center chlorophyll a molecule, $P680^+$, and the reduced pheophytin molecule, $Pheo^-$, occurs in 3 ps. Furthermore, this charge separation, like that in photosynthetic bacteria [15], becomes faster (1.4 ps) at low temperatures (such as 15K) as compared to that at the ambient temperature [16]. As in photosynthetic bacteria, the primary electron donor, P680, also appears to be a dimer [17]. Whether chlorophyll a monomer is an intermediate between P680 and pheophytin is not yet known. By analogy to photosynthetic bacteria, we speculate that photosystem II primary photochemistry may occur as follows:

$$P680\ Chl\ Pheo + h\nu \xrightarrow{2fs} P680^{*}\ Chl\ Pheo \quad (1)$$

$$P680^{*}Chl\ Pheo \xrightarrow{3ps} P680^{+}\ Chl^{-}Pheo \quad (2)$$

$$P680^{+}\ Chl^{-}\ Pheo \xrightarrow{<1ps} P680^{+}\ Chl\ Pheo^{-} \quad (3)$$

The primary reactants return to their original state after $Pheo^-$ transfers its electron to the primary quinone Q_A within 200 ps [18,19], and $P680^+$ transfers its positive charge (hole) to the electron donor Z within 20-200 ns [20,21] depending upon the flash number that determines the state(s) of the charge accumulator (M) on the water oxidation side. Recently, it has been shown that Z is a tyrosine residue (Tyrosine-161) in the D1 protein [22]. (Since chlorophyll a flourescence yeild is extensively used to monitor photosystem II chemistry, we remind the readers that, in addition to the well-known quencher Q_A [23], $P680^+$ and $Pheo^-$ also act as quenchers of antenna chlorophyll a fluorescence.)

The major difference between photosynthetic bacteria and photosystem II lies in the inability of the former to oxidize water to O_2. $P680/P680^+$ has a very high ($E_{m,7}$ = +1.1V) redox potential [24] and is, thus, capable of water oxidation ($E_{m,7}$ for H_2O/O_2 = + 0.9V) whereas $P870/P870^+$ has too low a redox potential ($E_{m,7}$ = +0.4V). In addition, the amino acid sequences of D1 and D2 on the lumen side of photosystem II are such that they can easily bind Mn atoms [6,25,26] needed for O_2 evolution, whereas L and M bind, instead, cytochrome of the c type [10].

In addition to the similarity in the arrangement and the chemistry of the primary reactants of photochemistry, discussed earlier, the reduction of the secondary quinone (Q_B) to quinol (Q_BH_2) also appears quite similar [27,28]. A possible common scheme is:

$$Q_AQ_B + {}^1h\nu \rightarrow Q_A^-Q_B \quad (4)$$

$$Q_A^-Q_B \leftrightarrow Q_AQ_B^- \quad (5)$$

$$Q_AQ_B^- + H^+ \leftrightarrow Q_AQ_B^-(H^+) \quad (6)$$

$$Q_AQ_B^-(H^+) + {}^2h\nu \rightarrow Q_A^-Q_B^-(H^+) \quad (7)$$

$$Q_A^-Q_B^-(H^+) \leftrightarrow Q_AQ_B^{2-}(H^+) \quad (8)$$

$$Q_AQ_B^{2-}(H^+) + H^+ \leftrightarrow Q_AQ_B^{2-}(2H^+) \quad (9)$$

$$Q_AQ_B^{2-}(2H^+) \leftrightarrow Q_AQ_BH_2 \quad (10)$$

$$Q_AQ_BH_2 \leftrightarrow Q_A + Q_BH_2 \quad (11)$$

$$Q_A + \text{Quinone} \leftrightarrow Q_AQ_B \quad (12)$$

The above scheme includes the concept of (a) the "two electron gate" (i.e., Q_B has to be doubly reduced before the quinol, Q_BH_2, will be released from the reaction center) [29,30]; and (b) protonation involves initial binding of H^+ to an amino acid, rather than directly to Q_B^- [31]. The involvement of aspartic acid-213 on the L-subunit of *R. sphaeroides* in protonation steps has been elegantly shown by site-directed mutagenesis studies of Takahashi and Wraight [32]. Inspite of the suggested similarity of photosystem II and photosynthetic bacteria in reducing quinone to quinol, a remarkable difference has been observed. Neither the reaction centers of the purple photosynthetic bacteria [33,34,35], nor that of the green bacterium *Chloroflexus aurantiacus* [36] show inhibition of electron flow by formate [33-35] or nitric oxide

[35,36]. In contrast, both formate and nitric oxide cause remarkable bicarbonate-reversible inhibitory effects in plants [37,38] and cyanobacteria [39,40]. We refer the readers to reviews of Govindjee and Wasielewski [41] and of Hansson and Wydrzynski [42] for further details on photosystem II.

2B. <u>The Bicarbonate Effect</u>

Warburg and Krippahl [43] discovered that removal of CO_2 from photosynthetic samples inhibits Hill reaction. Readdition of CO_2/HCO_3^- restores the Hill reaction. This is called the bicarbonate effect. Warburg [44] argued that CO_2 must be the source of O_2 in photosynthesis contrary to the generally accepted idea that H_2O is the source of O_2 (see review [37]). Heavy oxygen-18 studies of Guy et al.[45] (J. Berry, personal communication, 1990) clearly show that H_2O is the source of O_2. Although a bicarbonate effect can be observed by removing CO_2 without the use of molecules that replace/displace CO_2 [46-49], a large reproducible effect is observed when formate [50,51] or nitric oxide [35,52] is used to replace/displace CO_2 or HCO_3^-. Addition of formate to thylakoids at pH 6.5 indeed releases CO_2 [53] although Stemler [54] did not observe any CO_2 release upon formate addition to osmotically shocked maize thylakoids at pH 6.0.

The bicarbonate effect has been monitored, over the years, by measuring saturating rates of electron flow (O_2 evolution or reduction of electron acceptors); O_2 evolution, proton release or thermoluminescence per flash in a sequence of bright, brief light flashes; chlorophyll <u>a</u> fluorescence transients; decays of chlorophyll <u>a</u> fluorescence yield, delayed light emission, and absorbance changes due to the formation of Q_A^- or Q_B^- after light flashes; Mossbauer and ESR signals due to the non-haem iron; changes in photosystem II herbicide binding characteristics, etc. The phenomenon of the bicarbonate effect has been reviewed at regular intervals since 1978 [37,38,55-60].

No bicarbonate effect is observed in photosystem I [51,61]. No significant effect has been observed in O_2 evolution kinetics <u>per se</u> after a flash [58,62], electron flow between the intermediate "M" and "Z" [63] and between "Z" and P680 [63,64] on the electron donor side of PSII. A major site of the bicarbonate effect has been located between Q_A and the plastoquinone pool through measurements on partial electron flow segments of the Hill reaction [51,61,65,66] and through kinetic meaurements of Q_A decay in repetitive [63,67] and after single flashes of light [35,68-73]. A much larger slowing down of electron flow occurs after second (or third) and subsequent flashes than after the first flash upon depletion of CO_2; this effect is fully restored upon the readdition of bicarbonate. A thorough depletion of CO_2 can easily lead to a 10-20 - fold inhibition of electron flow that can be fully restored to control values upon bicarbonate addition. In control membranes, the bottleneck reaction is the (transport and) reoxidation of plastoquinol: approximately 4 ms/e^- at room temperature. In CO_2-depleted thylakoids, the decay of [Q_A^-] after the 3rd and subsequent flashes is of the order of 150 ms [68]; a similar value was observed when the rereduction kinetics of $P700^+$ was measured [67]. In the first approximation, these slowed reactions, interpreted to be either in the production of quinol (Q_BH_2) and/or its release, seem sufficient to explain the inhibition of the Hill reaction. However, quantitation and precise explanation requires parallel measurements on the decay of various intermediates and the overall electron flow in a sequence of flashes with appropriate dark times between flashes in identically treated samples.

The location of the major effect between Q_A and the plastoquinone pool is supported by the following observations:

(1) Photosystem II herbicides, that are known to inhibit electron flow by displacing Q_B [74,75], affect the binding of herbicides [65,76] and *vice versa* [66,77].

(2) Nitric oxide, that binds to non-haem Fe[52] between Q_A and Q_B, shows bicarbonate-reversible effect in electron flow from Q_A to the plastoquinone pool [35]. Disulfiram, an iron chelator, is shown to cause effects similar to that of CO_2-depletion [78] although this effect cannot be reversed by bicarbonate addition. Furthermore formate, used to deplete CO_2, was shown to affect Q_A^-Fe ESR signal [40,79] and the Fe signal, as measured by Mossbauer spectroscopy [80,81].

(3) Herbicide-resistant D1 mutants, altered in single amino acids in the Q_B-binding region [82], show differential sensitivity to bicarbonate-reversible formate effects, both in *Synechocystis sp.* PCC 6803 [83] and in *Chlamydomonas reinhardtii* [84]. Furthermore, D2 mutants, that had been constructed in which arginine 233 was changed to glutamine (R233Q) or argine 251 was changed to serine (R251S), were found to be ten times more sensitive to formate than the wild type *Synechocystis sp.* PCC 6803 [85].

Consistent with the effect on D1 and D2, but on a reaction prior to Q_A, Mende and Wiessner [87] have observed a parallel decrease in variable chlorophyll a fluorescence and O_2 evolution when CO_2 was removed from *Chlamydobotrys stellata* cells. This phenomenon was confirmed in our laboratory in spinach leaf discs [49] and in *Chlamydomonas reinhardtii* cells [88]. We have however established that the same phenomenon persists in mildly heated (that lose water oxidation activity) and hydroxylamine treated samples. Thus, this effect appears to be located somewhere between Z and Q_A. C. Xu (personal communication, 1990) has observed that in spinach thylakoids, at pH 6.0, an effect prior to Q_A is observed at short times after mixing of thylakoids with formate; at later mixing times the effect beyond Q_A predominates. It is necessary to test if this effect is between Z and P680, and/or between pheophytin and Q_A.

We shall now summarize our recent results on the absence of formate and NO effects on the reaction centers of the green bacterium *Chloroflexus aurantiacus* [36]; CO_2 release from thylakoid membranes upon formate addition [53]; differential sensitivity of formate in D1 mutants of *Chlamydomonas reinhardtii [84]*, and D1 [83] and D2 [85] mutants of *Synechocystis* PCC 6803. Finally, we will discuss a current hypothesis for the function of bicarbonate in Photosystem II.

3. ABSENCE OF FORMATE AND NITRIC OXIDE EFFECTS IN THE REACTION CENTERS FROM THE GREEN BACTERIUM *CHLOROFLEXUS AURANTIACUS* (COLLABORATION WITH J. TROST AND R. BLANKENSHIP) [36]

In spite of the known similarity of the D1 and D2 proteins of photosystem II and the L and M subunits of the purple photosynthetic bacteria [5], the latter do not show the (bicarbonate-reversible) formate effect (see e.g. [34]), as mentioned earlier. Nitric oxide, which binds to Fe [52] between Q_A and Q_B, produces effects similar to formate in photosystem II, but not in purple photosynthetic bacteria [35]. These differences must be due to differences in the amino acid sequence and subsequently the architecture of the D1/D2 proteins versus L/M subunits. The photosynthetic reaction center from the green bacterium *Chloroflexus aurantiacus* has several differences with the purple photosynthetic bacteria: one of the monomer bacteriochlorophyll molecule is replaced by bacteriopheophytin; Fe between Q_A and Q_B is replaced by Mn (see e.g. [89]); and there are several significant differences in their amino acid sequences [90,91]. *Chloroflexus* reaction centers also appear to lack an H-subunit [89,92]. Thus, we used formate and NO to test if these differences are of any importance in producing the "bicarbonate effect". We concluded, from our observations presented below, that *Chloroflexus aurantiacus* reaction centers do not possess formate and NO binding sites that are responsible for the bicarbonate effect.

Using reaction centers containing Q_B, prepared as described elsewhere [89], we monitored [93,94] the decay of the oxidized reaction center ($P865^+$), either after the first flash or after the fifth flash (dark time between flashes, 3s), by back reactions with Q_A^- (t 1/2 = 60 ms) and with Q_B^- (t 1/2 = 0.5 - 3 s; 2 components). If formate and NO were to inhibit electron flow between Q_A and $Q_B^{(-)}$, as observed in plants [35,71,72] and cyanobacteria [35, 39], the amplitude of the fast component would increase and that of the slow component(s) would decrease. This is similar to what occurs [89] when the inhibitor O-phenanthroline is added; this inhibitor blocks electron flow between Q_A and Q_B by displacing Q_B. In contrast to PSII, but as observed in other photosynthetic bacteria [34], our results (see abstract [36]) showed no significant change between the control reaction centers and those treated with 100 mM formate. In addition, no difference was observed between the 300 μM NO treated (anaerobic) reaction centers and those treated with both NO and 10 mM bicarbonate. At pH 6.5, the lifetimes and (amplitudes) were 60 ms (50 + 5%) and 0.5 - 2 s (2 components; sum = 50 + 5%) for $P^+Q_A^-$ and $P^+Q_B^-$ recombinations, respectively. It thus appears that the differences in the amino acid sequence [90,91] in the L and M subunits of *Chloroflexus aurantiacus* from those of purple photosynthetic bacteria are not a sufficient condition to bring about the binding of formate or NO.

4. ABSENCE OF FORMATE EFFECTS IN GLU^{M234} MUTANTS OF *RHODOBACTER SPHAEROIDES* (COLLABORATION WITH X. WANG, J. CAO, P. MAROTI, H.V. STILZ, D. OESTERHELT AND C. WRAIGHT) [95]

Michel and Deisenhofer [5] suggested a possible reason for differences in the "bicarbonate effect" in photosystem II and photosynthetic bacteria: the iron atom in photosystem II lacks the fifth and sixth ligands provided by glutamate (M234) in reaction centers of *Rhodobacter sphaeroides*. Wang et al. [95] have, by site-directed mutagenesis, altered GLU^{M234} with valine (M234EV), glutamine (M234EQ), and glycine (M234EG). These mutants grew well under photoautotrophic conditions and showed no detectable effects of bicarbonate depletion (formate addition) on: (1) cytochrome b_{561} or cytochrome c_2 reduction; (2) the recombination of $P^+Q_A^-$ or $P^+Q_B^-$; (3) kinetics of electron transfer from $Q_A^-$to Q_B^-; and (4) flash dependent oscillation of semiquinone formation in the presence of exogenous donor to P^+ (Q_B turnover). Furthermore, addition of nitric oxide, instead of formate, also showed no effect on the electron acceptor complex of M234EV mutant (X. Wang and J. Cao, unpublished observations, 1990). The absence of a formate-bicarbonate effect in M234 mutants suggests that this amino acid may not be responsible for the absence of the phenomenon. These results are, however, interesting as they suggest that M234 may not be important for electron flow in the electron acceptor complex of *Rhodobacter sphaeroides*.

In conclusion, the existence of formate and NO effects in photosystem II lies strictly in the unique amino acid sequence of D1 and D2 proteins (see later sections). Now, I present data that show that formate indeed releases CO_2 from thylakoid membranes, and thus, depletes them of CO_2 when bound.

5. FORMATE RELEASES CARBON DIOXIDE/BICARBONATE FROM THYLAKOID MEMBRANES (COLLABORATION WITH H.G. WEGER, D.H. TURPIN, J.J.S. VAN RENSEN, O.J. DE VOS AND J.F.H. SNEL) [53]

There are two hypothesis for the action of formate: (1) Formate is an inhibitor of electron flow: it binds to empty sites, and bicarbonate displaces it and removes the inhibition. (2) Formate displaces the native bound bicarbonate and, thus, causes inhibition by the removal of the ion required for electron flow; readdition of bicarbonate displaces formate and restores electron flow. A third hypothesis may, however, invoke bicarbonate to be necessary for the most efficient electron flow, but in its absence a marginal or less efficient electron flow may proceed. Supportive evidence for hypothesis #2 was discussed earlier [37, 50]. However, absence of CO_2 release by formate addition to maize thylakoids at pH 6.0 [54] supported hypothesis #1. Using a sensitive membrane inlet mass spectrometer and a differential infra-red gas analyzer, we showed [53] that 100mM formate treatment released about 1 CO_2 per reaction center II at pH 6.5 from spinach and pea thylakoids within minutes of formate treatment; at pH 6.0, CO_2 release was reduced to a value of 0.4 CO_2 per reaction center (Fig. 1). Since pH of the stroma matrix is alkaline, it is reasonable to assume that bicarbonate is bound in native thylakoid membranes and functions to provide efficient electron flow [37]. However, it is possible that some inefficient electron flow may take place in the absence of bicarbonate in thylakoid membranes.

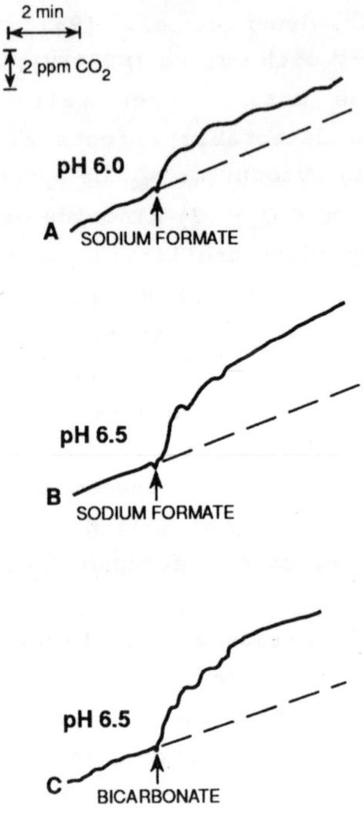

Figure 1: Formate-induced release of CO_2 from pea thylakoids as measured by a differential infra-red gas analyzer. Trace A is an illustration of CO_2 release at pH 6.0 upon addition of 100 mM formate to a thylakoid suspension containing 4 mg Chl/ml, recorded as the difference with the simultaneous addition of the same amount of formate to the second vessel containing the same medium without thylakoids. Trace B: Same as A, but pH 6.5. Trace C: recording of the difference of CO_2 in the gas phase upon addition of 10mM HCO_3^- to one of the vessels and water to the other, while both vessels contained 4 ml reaction medium without thylakoids at pH 6.5. At pH 6.5, 100 mM formate injection led to an estimated release of 1.2 HCO_3^-/CO_2 per PS II reaction center and about 0.4 HCO_3^- at pH 6.0 (After ref. [53]).

6. DIFFERENTIAL SENSITIVITY OF FORMATE IN D1 MUTANTS OF *CHLAMYDOMONAS*

REINHARDTII (COLLABORATION WITH R. STRASSER, B. SCHWARZ AND J-D. ROCHAIX) [84]

The secondary plastoquinone, Q_B, is in a pocket between the helices IV and V of the D1 protein; this pocket is also lined by the amino acids that are found to be altered in the herbicide-resistant mutants [96-98]. If bicarbonate is bound to Fe and/or any of the amino acids in the Q_B region, we expect to see differential sensitivity of the bicarbonate-reversible formate effects in the different mutants. The nature of the differences may ultimately provide clues as to the nature of the bicarbonate binding. Chlorophyll a fluorescence transients indeed revealed differential sensitivity to 25 mM formate treatment; these effects were fully reversed by 10 mM bicarbonate [84]. The most sensitive mutant was S264A (DCMU-4) and the most resistant mutant was L275F (Br-202). The order of resistance (highest to lowest) was L275F (Br-202) > A251V (Mz-2) >> wild type = F255Y (Ar-207) = V219I (Dr-18) >> S264A (DCMU-4) (Figure 2). These results clearly show the involvement of D1 protein in bicarbonate-reversible formate effects *in vivo*.

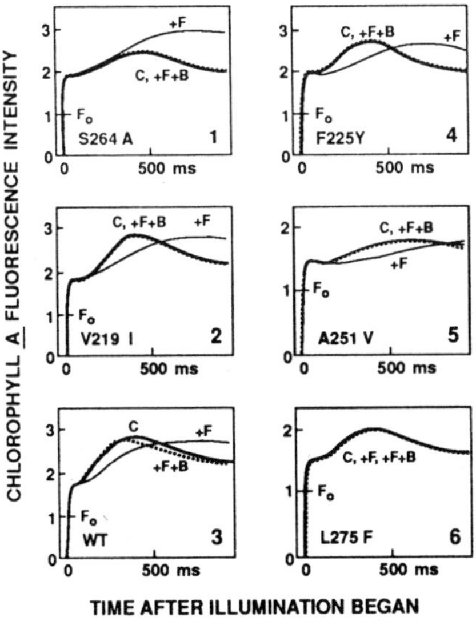

Figure 2: Chlorophyll a fluorescence transients of some *Clamydomonas reinhardtii* cells grown photoautotrophically and suspended in tris-phosphate medium at pH 6.5. C: control, +F for 25mM formate treated cells, and +F+B for 25mM formate followed by 10mM bicarbonate treated cells. Dark adaptation was 4 minutes. The ratio of variable fluorescence ($F_{1\ second}$ -Fo) after illumination with and without formate was 2.0 (for S264A [DCMU-4]), 1.48 (V291I [Dr-18]), 1.47 (for F255Y [Ar-207]), 1.4 (WT), 1.10 (A251V [MZ-2]) and 1.00 (L275F[Br-202]). Formate had no effect on $PSII^-$ (Fud-7) cells (data not shown) (After ref. [84]).

7. DIFFERENTIAL SENSITIVITY OF FORMATE IN D1 MUTANTS OF *SYNECHOCYSTIS* PCC 6714 (COLLABORATION WITH A-L. ETIENNE, C. VERNOTTE, B. PETERI AND C. ASTIER) [83]

The concept that bicarbonate binding may involve the region of helices IV and V and the loop connecting the two was supported further by measurements of formate sensitivity in D1 mutants of *Synechocystis* PCC 6714. Measurements, not only on chlorophyll a fluorescence transients, but on oxygen yield in a sequence of flashes and chlorophyll a fluorescence yield decay after a flash revealed differential sensivity of herbicide resistant mutant cells to formate treatment [83]. The order of resistance (highest to lowest) was : A251V/F211S (Az-V) > F211S (Az-I) = wild type > S264A (DCMU-IIA) (Figure 3). The extreme sensitivity of S264A mutant was confirmed. Bicarbonate addition restored all formate effects.

An interesting observation is that Az-V mutant, that was most resistant to formate, is the one most sensitive to exposure to excessive light [99]. Excess light, like formate treatment, not only produces photosystem II centers in which electron transfer is inhibited between Q_A and Q_B, but also centers in which Q_A^- formation is also inhibited [99]. We consider it a likely hypothesis that this mutant is resistant to formate because bicarbonate is poorly bound in it, and, the decreased bicarbonate binding makes it more suseptible to inactivation by excess light. Sundby et al. [100] and Sundby [101] have recently shown that bicarbonate protects against photoinhibition.

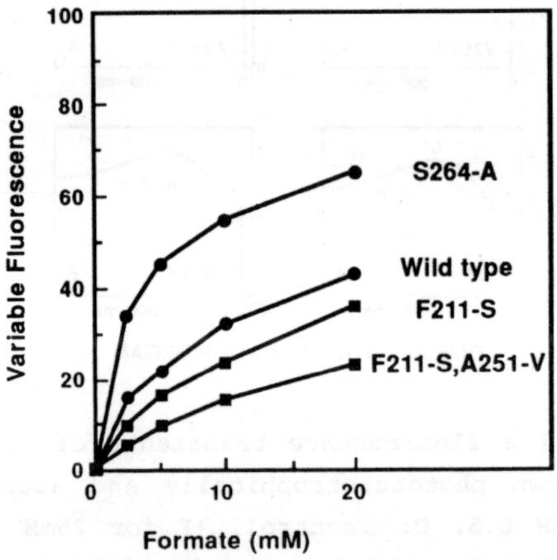

Figure 3. Chlorophyll a fluourescence as a function of formate concentration in cells of *Synechocystis* 6714 (wild type and mutants S264A (DCMUII-A) & F211S (AZ-I) and a double mutant F211S/A251V (AZ-V). Fifteen minutes darkness was given after the addition of formate, but before the measurements began. 100 indicates the value obtained with the

herbicide 10 μM diuron or 200 μM ioxynil (for DCMU II-A). (After ref. [83].)

8. DIFFERENTIAL SENSITIVITY OF FORMATE IN D2 MUTANTS OF *SYNECHOCYSTIS* PCC 6803 (COLLABORATION WITH J. CAO AND W. F. J. VERMAAS) [85]

Blubaugh and Govindjee [37] had discussed a role of specific arginine residues in the binding of bicarbonate on the D1 protein. Recently crystal structure of human serum lactoferrin at 2.8 A resolution has revealed that in this protein (bi)carbonate is bound to Fe^{III} as a bidentate ligand and the oxygens of (bi)carbonate form H-bonds with several amino acids including arginine. Thus, we considered it likely that one or more arginine residues in D2 may form H-bonds with (bi)carbonate. Cao et al. [102] constructed two arginine mutants in the D2 polypeptide of *Synechocystis* PCC 6803: R233Q and R251S. Measurements of steady-state rate of oxygen Hill reaction (H_2O to parabenzoquinone), as a function of increasing formate concentration showed that half maximal inhibition was at 50mM for the wild type, 4mM for R233Q and 3.5mM for R251S [85, 103] (Figure 4). Thus, the arginine mutants were several fold more sensitive to formate than the wild type.

Measurements of chlorophyll a fluorescence yield decay kinetics indicated that formate inhibited Q_A^- to Q_B^- reaction (flash 2) more than Q_A^- to Q_B reaction in the mutants as well as in the wild type as had been observed earlier in higher plant thylakoids [71, 72]. These results support the interpretation that CO_2 removal may interfere with the protonation at the reaction center II, that may not necessarily occur on D2. However, the halftime of the Q_A^- oxidation was increased by a factor of 2, 4 and 6 in the wild type, R251S and R233Q, respectively, confirming that the mutants were more sensitive to formate. Cao et al. [85] have suggested that arginine 233 and arginine 251 in the D2 polypeptide function to stabilize HCO_3^- binding in photosystem II of *Synechocystis* PCC 6803.

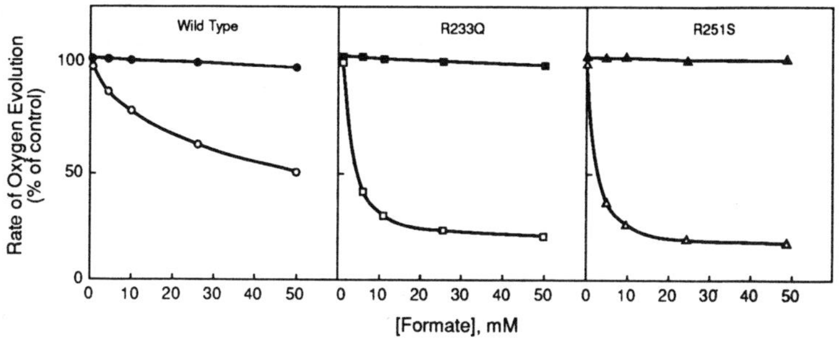

Figure 4. Steady-state oxygen evolution rate of Synechocystis 6803 cells of the wild type, and R233Q and R251S mutants. A combination of DMQ

(0.5 mM) and $K_3Fe(CN)_6$ (1 mM) was used as electron acceptor; measurements at pH 6.8. The samples were preincubated for one hour in the treatment medium (at pH 5.8) to which the indicated amount of sodium formate was added (open symbols). To restore the Hill reaction, 10 mM bicarbonate was added (solid symbols). (After [85]).

9. HYPOTHESIS

A working hypothesis for unique role(s) of HCO_3^-/CO_2 in photosystem II of plants and cyanobacteria is now emerging. It appears that, at physiological pH, HCO_3^-/CO_2 may H-bond to several amino acids on both D1 and D2, the reaction center proteins of photosystem II, and may form a ligand to iron that lies between Q_A (bound to D2) and Q_B (bound to D1). For a possible model, see binding of (bi)carbonate to Fe^{III} and to 4 amino acids including arginine in human serum lactoferrin [86] (Fig.5). HCO_3^-/CO_2 plays discrete and unique role(s) in photosystem II: its function may involve stabilization, by conformational means, of the reaction center protein that allows efficient electron flow and protonation of certain amino acids near Q_B^- [37]. Data on differential sensitivity of formate on the D1 mutants of *Chlamydomonas reinhardtii* (see section 6) and of *Synechocystis* 6714 (see section 7) clearly support binding domains of bicarbonate between helices IV and V and in the connecting loop of D1. Data on the extremely high sensitivity of R233Q and R251S mutants to formate sugggest that arginines may stabilize bicarbonate on the reaction center II. Absence of the "bicarbonate effect" in purple photosynthetic bacteria [34, 95] (see section 4) and in the green bacteria (see section 3) shows that we should concentrate on the unique differences between the D1 and D2 and the L and M proteins to unravel the binding niche of bicarbonate and, thus, its function.

Unpublished results of C. Xu (personal communication, 1990) suggest that the inhibitory species in formate solutions added to remove CO_2 may be the formic acid, not formate ions. If this is confirmed, we can begin to look for domains that would bind formic acid, not formate ions. Diner et al. [38] suggest the importance of Lysine-264 on D2 in the bicarbonate effect. It appears that we have to explore several possibilities before a final picture will emerge.

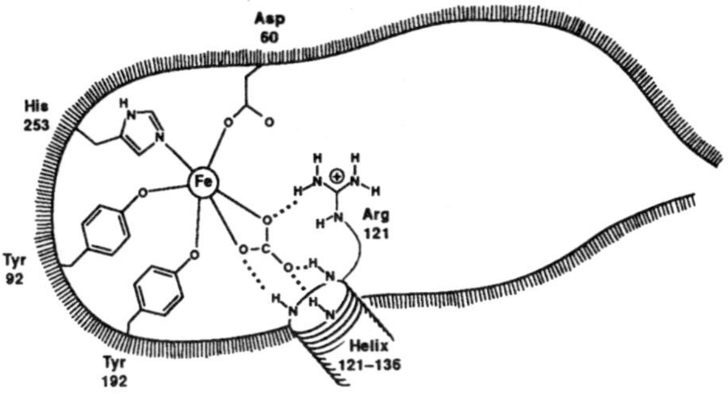

Figure 5. Schematic representation of the iron and anion-binding sites in lactoferrin. Numbering is for the N-lobe, but the same arrangement is found in the C-lobe. (After Anderson et al. [86].)

10. CONCLUDING REMARKS

It seems that binding of bicarbonate to both D1 and D2 affects the stability of the reaction center II and, thus, electron flow and protonation of the electron acceptor in a somewhat complicated manner, the nature of which still needs to be discovered. However, I suspect that specific roles of CO_2/HCO_3^- will emerge that may even include protection against photoinhibition. Further site-directed mutagenesis of the electron acceptor side of both D1 and D2 proteins is expected to unravel the molecular bases of the "bicarbonate effect".

No evidence has been found to support Warburg's concept that the CO_2 effect is related to O_2 evolution per se. We have, however, been able to explore, over the years, the original conclusion of Wydrzynski and Govindjee [104] that the CO_2 effect is on the electron acceptor side of photosystem II. Although tremendous amount of research is still needed, we feel that the end of the tunnel is in sight.

11. REFERENCES

1. Woese, C.R., Kandler, O. and Wheelis, M.L. (1990). Towards a natural system of organisms: proposal for the domains Archea, Bacteria and Eucarya. *Proc. Natl. Acad. Sci. USA* 87: 4576-4579.

2. Govindjee and Govindjee, R. (1974). Primary events in photosynthesis. *Sci. Am.* 231: 68-82.

3. Govindjee and Govindjee, R. (1975). Introduction to Photosynthesis. In: *Bioenergetics of Photosynthesis*, Govindjee ed., Academic Press, NY pp. 1-50.

4. Andreasson, L.E. and Vanngard, T. (1988). Electron transport in photosystems I and II. *Annu. Rev. Plant Physiol. Plant Mol. Biol.* 39: 379-411.

5. Michel, H. and Deisenhofer, J. (1988). Relevance of the photosynthetic reaction center from purple bacteria to structure of photosystem II. *Biochemistry* 27: 1-7.

6. Coleman, W. and Govindjee (1987). A model for the mechanism of chloride activation of oxygen evolution in photosystem II. *Photosynth. Res.* 13: 199-233.

7. Olson, J.M. and Pierson, B.K. (1987). Origin and evolution of photosynthetic reaction centers. *Origins of Life* 17: 419-430.

8. Deisenhofer, J. and Michel, H. (1989). The photosynthetic reaction centre from the purple bacterium *Rhodopsendomonas viridis*. *The EMBO J.* 8: 2149-2169.

9. Feher, G., Allen, J.P., Okamura, M.Y. and Rees, D.C. (1989). Structure and function of bacterial photosynthetic reaction centres. *Nature (London)* 339: 111-116.

10. Norris, J.R. and Schiffer, M. (1990). Photosynthetic reaction centers in bacteria. *Chem. & Eng. News* July 30, 1990: 22-37.

11. Woodbury, N.W., Becker, M., Middendorf, D. and Parson, W.W. (1985). Picosecond kinetics of the initial photochemical electron transfer reaction in bacterial photosynthetic reaction centers. *Biochemistry* 24: 7516-7521.

12. Holzapfel, W., Finkele, U., Kaiser, W., Oesterhelt, D., Scheer, H., Stilz, H.U. and Zinth, W. (1990). Intitial electron-transfer in the reaction center form *Rhodobacter sphaeroides*. *Proc. Natl. Acad. Sci. USA* 87: 5168-5172. Also see: Shuvalov, V.A. and Klevanik, A.V. (1987). The study of the state [$P870^+$ $B800^-$] in bacterial reaction centers by selective picosecond and low-temperature spectroscopies. *FEBS Lett.* 160: 51-56.

13. Kobayashi, M., Maeda, H., Watanabe, T., Nakane, H. and Satoh, K. (1990). Chlorophyll a and β-carotene content in the D1/D2/cytochrome b 559 reaction center complex from spinach. *FEBS Lett.* 260: 138-140.

14. Wasielewski, M.R., Johnson, D.G., Seibert, M. and Govindjee (1989). Determination of the primary charge separation rate in isolated photosystem II reaction centers with 500fs time resolution *Proc. Natl. Acad. Sci. USA* 86: 524-528.

15. Breton, J., Martin, J-L., Fleming, G.R. and Lambry, J-C. (1988). Low temperature femtosecond spectroscopy of the initial step of electron transfer in reaction centers from photosynthetic purple bacteria. *Biochemistry* 27: 8276-8284.

16. Wasielewski, M.R., Johnson, D.G., Govindjee, Preston, C. and Seibert, M. (1989). Determination of the primary charge separation rate in photosystem II reaction centers at 15K. *Photosynth. Res.* 22: 89-99.

17. van Kan, P.J.M., Otte, S.C.M., Kleinherenbrink, F.A.M., Nieveen, M.C., Aartsma, T.J. and van Gorkom, H.J. (1990). Time-resolved spectroscopy at 10K of the photosystem II reaction center; deconvolution of the red absorption band. *Biochim. Biophys. Acta* 1020: 146-152.

18. Eckert, H.J., Weise, N., Bernarding, J., Eichler, H.J. and Renger, G. (1988). Analysis of the electron transfer from $Pheo^-$ to Q_A in PSII membrane fragments from spinach by time resolved 325nm absorption

changes in the picosecond domain. *FEBS Lett.* 240: 153-158.

19. Nuijs, A.M., van Gorkon, H.J., Plijter, J.J. and Duysens, L.N.M. (1986). Primary charge separation and excitation of chlorphyll a in photosystem II particles from spinach as studied by picosecond absorption difference spectroscopy. *Biochim. Biophys. Acta* 848: 167-175.

20. Brettel, K., Schlodder, E. and Witt, H.T. (1984). Nanosecond reduction kinetics of photooxidized chlorophyll a II (P-680) in single flashes as a probe for the electron pathway, H^+ release and charge accumulation in the O_2-evolving complex. *Biochim. Biophys. Acta* 766: 403-428.

21. Sonneveld, A., Rademaker, H. and Duysens, L.N.M. (1979). Chlorophyll a fluorescence as a monitor of nanosecond reduction of the photooxidized primary donor P-680^+ of photosystem II. *Biochim. Biophys. Acta* 548: 536-551.

22. Metz, J.G., Nixon, P.J., Rogner, M., Brudvig, G.W. and Diner, B.A. (1989). Directed alteration of the D1 polypeptide of photosystem II: evidence that tyrosine is the redox component, Z, connecting the oxygen-evolving complex to the primary electron donor, P680. *Biochemistry* 28: 6960-6969.

23. Duysens, L.N.M. and Sweers, H.E. (1963). Mechanism of two photochemical reactions in algae as studied by means of fluorescence. In: *Studies on Microalgae and Photosynthetic Bacteria*, Jap. Soc. Plant Physiol. eds., The University of Tokyo Press, Tokyo, pp. 353-372.

24. Jursinic, P. and Govindjee (1977). Temperature dependence of delayed light emission in the 6 to 340 microsecond range after a single flash in chloroplasts. *Photochem. Photobiol.* 26: 617-628.

25. Babcock, G.T., Barry, B.A., Debus, R.J., Hoganson, C.W., Atamian, M., McIntosh, L., Sithole, I. and Yocum, C.F. (1989). Water oxidation in photosystem II: from radical chemistry to multielectron chemistry. *Biochemistry* 28: 9557-9565.

26. Rutherford, A.W. (1989). Photosystem II, the water-splitting enzyme. *TIBS* 14: 227-232.

27. Vermaas, W.F.J. and Govindjee (1981). The acceptor side of photosystem II in photosynthesis. *Photochem. Photobiol.* 34: 775-793.

28. Crofts, A.R. and Wraight, C. (1983). The electrochemical domain of photosynthesis. *Biochim. Biophys. Acta* 726: 149-185.

29. Velthuys, B.R. and Amesz, J. (1974). Charge accumulation at the reducing site of system II of photosynthesis. *Biochim. Biophys. Acta* 333: 85-94.

30. Wraight, C.A. (1977). Electron acceptors of photosynthetic bacterial reaction centers: direct observation of oscillatory behaviour suggesting two closely equivalent ubiquinones. *Biochim. Biophys. Acta* 459: 525-531.

31. Wraight, C.A. (1979). Electron acceptors of bacterial photosynthetic reaction centers: II. H^+ binding coupled to secondary electron transfer in the quinone acceptor complex. *Biochim. Biophys. Acta*

548: 309-327.

32. Takahashi, E. and Wraight C.A. (1990). A crucial role for Asp[L213] in the proton transfer pathway to the secondary quinone of reaction centers from *Rhodobacter sphaeroides*. *Biochim. Biophys. Acta* 1020: 107-111.

33. Beijer, C. and Rutherford, A.W. (1987). the iron-quinone acceptor complex in *Rhodospirillum rubrum* chromatophores studied by EPR. *Biochim. Biophys. Acta* 890: 169-178.

34. Shopes, R.J., Blubaugh, D., Wraight, C. and Govindjee (1989). Absence of a bicabonate-depletion effect in electron transfer between quinones in chromatophores and reaction centers of *Rhodobacter sphaeroides*. *Biochim. Biophys. Acta* 974: 114-118.

35. Diner, B.A. and Petrouleas, V. (1990). Formation by NO of nitrosyl adducts of redox components of the photosystem II reaction center. II. Evidence that bicarbonate/CO_2 binds to the acceptor side non-heme iron. *Biochim. Biophys. Acta* 1015: 141-149.

36. Govindjee, Trost, J.F. and Blankenship, R.E. (1991). Formate and nitric oxide fail to inhibit electron flow in reaction centers from green bacterium *Chloroflexus aurantiacus*. Book of Abstracts, *Am.Soc. Photobiol*. 19th Annual Meeting, San Antonio, Texas.

37. Blubaugh, D. and Govindjee (1988). The molecular mechanism of the bicarbonate effect at the plastoquinone reductase site of photosynthesis. *Photosynth. Res*. 19: 85-128.

38. Diner, B.A., Wendoloski, J.J. and Petrouleas, V. (1991). The iron-quinone electron-acceptor complex of photosystem II. *Physiologia Plantarum* 80: 502-511.

39. Cao, J. and Govindjee (1988). Bicarbonate effect on electron flow in a cyanobacterium *Synechocystis* PCC 6803. *Photosynth Res*. 19: 277-285.

40. Nugent, J.H.A., Corrie, A.R., Demetriou, C., Evans, M.C.W. and Lockett, C.J. (1988). Bicarbonate binding and properties of photosystem II electron acceptors. *FEBS Lett*. 235: 71-75.

41. Govindjee and Wasielewski, M.R. (1989). Photosystem II: from a femtosecond to a millisecond. In: *Photosynthesis*, Briggs, W. ed., Alan R. Liss, Inc., NY, pp. 71-103.

42. Hansson, O. and Wydrzynski, T. (1990). Current perceptions of photosystem II. *Photosynth. Res*. 23: 131-162.

43. Warburg, O. and Krippahl, G. (1960). Notwendigkeit der kohlensaure fur die chinon und ferricyanid-reaktionen in grunen grana. *Z. Naturforshg*. 15b: 367-369.

44. Warburg, O. (1964). Prefatory chapter. *Annu. Rev. Biochem*. 33: 1-18.

45. Guy, R.D., Fogel, M.F., Berry, J.A. and Hoering, T.C. (1987). Isotope fractionation during oxygen production and consumption by plants. In: *Progress in Photosynthesis Research*, Biggins, J. ed, Martinus Nijhoff, Dordrecht, vol III, pp. 597-600.

46. Eaton-Rye, J.J., Blubaugh, D. and Govindjee (1986). Action of bicarbonate on photosynthetic electron transport in the presence or absence of inhibitory anions. In: *Ion interactions in Energy Transfer Biomembranes*, Papageorgiou, G.C., Barber, J. and Papa, S. eds, Plenum Publishing Corporation, NY, pp. 263-278.

47. Ireland, C.R., Baker, N.R., and Long, S.P. (1987). Evidence for a physiological role of CO_2 in the regulation of photosynthetic electron transport in intact leaves. *Biochim. Biophys. Acta* 893: 434-443.

48. Garab, G., Rozsa, Zs. and Govindjee (1988) Carbon dioxide affects charge acccumulation in leaves. *Naturwissenschaften* 75: 517-519.

49. El-Shintinawy, F. and Govindjee (1990). Bicarbonate effects in leaf discs from spinach. *Photosynth. Res.* 24: 189-200.

50. Blubaugh, D. and Govindjee (1988). Kinetics of the bicarbonate effect and the number of bicarbonate binding sites in thylakoid membrane. *Biochim. Biophys. Acta* 936: 208-214.

51. Eaton-Rye, J.J. and Govindjee (1984). A study of the specific effect of bicarbonate on photosynthetic electron transport in the presence of methyl viologen. *Photobiochem. Photobiophys.* 8: 279-288.

52. Petrouleas, V. and Diner, B.A. (1990). Formation by NO of nitrosyl adducts of redox components of the photosystem II reaction center. I. NO binds to the acceptor-side non-heme iron. *Biochim. Biophys. Acta* 1015: 131-140.

53. Govindjee, Weger, H.G., Turpin, D.H., van Rensen, J.J.S., de Vos, O.J. and Snel, J.F.H. (1991). Formate releases carbon dioxide/bicarbonate from thylakoid membranes: measurements by mass spectroscopy and infra-red gas analyzer. *Naturwissenschaften,* submitted.

54. Stemler, A. (1989). Absence of a formate-induced release of bicarbonate from photosystem II. *Plant Physiol.* 91: 287-290.

55. Govindjee and van Rensen, J.J.S. (1978). Bicarbonate effects on the electron flow in isolated broken chloroplasts. *Biochim. Biophys. Acta* 505: 183-213.

56. Vermaas, W.F.J. and Govindjee (1981). Unique role(s) of carbon dioxide and bicarbonate in the photosynthetic electron transport system. *Proc. Indian Natn. Sci. Acad.* B47: 581-605.

57. Stemler, A. (1982). The functional role of bicarbonate in photosynthetic light reaction II. In: *Photosynthesis, Vol. II, Development Carbon Metabolism and Plant Productivity,* Govindjee ed, Academic Press, NY, pp. 513-539.

58. Vermaas, W.F.J. and Govindjee (1982). Bicarbonate or CO_2 as a requirement for efficient electron transport on the acceptor side of photosystem II. In: *Photosynthesis, Vol. II. Development, Carbon Metabolism and Plant Productivity,* Govindjee ed, Academic Press, NY, pp. 541-558.

59. Govindjee and Eaton-Rye, J.J. (1986). Electron transfer through photosystem II acceptors: interactions with anions. *Photosynth. Res.* 10: 365-379

60. van Rensen, J.J.S. and Snel, J.F.H. (1985). Regulation of photosynthetic electron transport by bicarbonate, formate and herbicides in isolated broken and intact chloroplasts. *Photosynth. Res.* 6: 231-246.

61. Khanna, R., Govindjee and Wydrzynski, T. (1977). Site of bicarbonate effect in Hill reaction: evidence from the use of artificial electron acceptors and donors. *Biochim. Biophys. Acta* 462: 208-214.

62. Jursinic, P.A. and Dennenberg, R.J. (1990). Oxygen release time in leaf discs and thylakoids of peas and photosystem II membrane fragments of spinach. *Biochim. Biophys. Acta* 1020: 195-206.

63. Jursinic, P., Warden, J. and Govindjee (1976). A major site of bicarbonate effect in system II reaction: evidence from ESR signal II vf, fast fluorescence yield changes and delayed light emission. *Biochim. Biophys. Acta* 440: 322-330.

64. Govindjee, Robinson, H., Crofts, A.R. and van Rensen, J.J.S. (1989). Bicarbonate does not influence electron transfer to the reaction center chlorophyll a fluorescence rise in microsecond. *Naturwissenschaften* 76: 719-721.

65. Khanna, R., Pfister, K., Keresztes, A., van Rensen, J.J.S. and Govindjee (1981). Evidence for a close spatial location for the binding sites for CO_2 and for photosystem II inhibitors. *Biochim. Biophys. Acta* 634: 105-116.

66. van Rensen, J.J.S. and Vermaas, W.F.J. (1981). Action of bicarbonate and photosystem II inhibiting herbicides on electron transport in pea grana and in thylakoids of blue-green algae. *Physiol. Plant.* 51: 106-110.

67. Siggel, U., Khanna, R., Renger, G. and Govindjee (1977). Investigation of the absorption change of the plstoquinone system in broken chloroplasts: the effect of bicarbonate depletion. *Biochim. Biophys. Acta* 462: 196-207.

68. Govindjee, Pulles, M.P.J., Govindjee, R., van Gorkom, H.J. and Duysens, L.N.M. (1976). Inhibition of the reoxydation of the secondary acceptor of photosystem II by bicarbonate depletion. *Biochim. Biophys. Acta* 449: 602-605.

69. Farineau, J. and Mathis, P. (1983). Effect of bicarbonate on electron transfer between plastoquinones in photosystem II. In: *The Oxygen Evolving System of Photosynthesis*, Inoue, Y., Crofts, A.R., Govindjee, Murata, N., Renger, G. and Satoh, K., eds, Academic Press Tokyo, pp. 317-325.

70. Robinson, H.H., Eaton-Rye, J.J., van Rensen, J.J.S. and Govindjee (1984). The effects of bicarbonate depletion and formate incubation on the kinetics of oxidation-reduction reactons of the photosystem II quinone acceptor complex. *Z Naturforschg.* 39C: 382-385.

71. Eaton-Rye, J.J. and Govindjee (1988). Electron transfer through the quinone acceptor complex of photosystem II in bicarbonate-depleted spinach thylakoid membranes as a function of actinic flash number and frequency. *Biochim. Biophys. Acta* 935: 237-247.

72. Eaton-Rye, J.J. and Govindjee (1988). Electron transfer through the quinone acceptor complex of photosystem II after one or two actinic flashes in bicarbonate-depleted spinach thylakoid membranes. *Biochim Biophys. Acta* 935: 248-257.

73. Govindjee, Nakatani, H.Y., Rutherford, A.W. and Inoue, Y. (1984). Evidence from thermoluminescence for bicarbonate action on there combination reactions involving the secondary quinone electron acceptor of photosystem II. *Biochim. Biophys. Acta* 766: 416-423.

74. Velthuys, B.R. (1981). Electron dependent competition between plastoquinone and inhibitors for binding to photosystem II. *FEBS Lett.* 126: 277-281.

75. Wraight, C. (1981). Oxidation-reduction physical chemistry of the acceptor quinone complex in bacterial photosynthetic reaction centers: evidence for a new model of herbicide activity. *Isr. J. Chem.* 21: 348-354.

76. Vermaas, W.F.J., van Rensen, J.J.S. and Govindjee (1982). The interaction between bicarbonate and the herbicide ioxynil in the thylakoid membrane and the effects of amino acid modification on bicarbonate action. *Biochim. Biophys. Acta* 681: 242-247.

77. van Rensen, J.J.S. (1991). Regulation of electron transport at the acceptor side of photosystem II by herbicides, bicarbonate and formate. In: *Photosynthesis and Plant Productivity*, Abrol, Y., Mohanty, P. and Govindjee eds, Oxford University Press, in press.

78. Blubaugh, D.J. and Govindjee (1988). Sites of inhibition by disulfiram in thylakoid membranes. *Plant Physiol.* 88: 1021-1025.

79. Vermaas, W.F.J. and Rutherford, A.W. (1984). EPR measurements on the effects of bicarbonate and triazine resistance on the acceptor side of photosystem II. *FEBS Lett.* 175: 243-248.

80. Diner, B.A. and Petrouleas, V. (1987). Q400, the non-heme iron of the photosystem II iron-quinone complex: a spectroscopic probe of quinone and inhibitor binding to the reaction center. *Biochim. Biophys. Acta* 895: 107-125.

81. Semin, B.K., Loviagina, E.R., Aleksandrov, A.Yu., Kaurov, Yu.N., and Novakova, A.A. (1990). Effect of formate on Mossbauer parameters of the non-heme iron of PS II particles of cyanobacteria. *FEBS Lett.* 270: 184-186.

82. Trebst, A. (1987). the three-dimensional structure of the herbicide binding niche on the reaction center polypeptides of photosystem II. *Z Naturforschg.* 42C: 742-750.

83. Govindjee, Vernotte, C., Peteri, B., Astier, C. and Etienne, A-L. (1990). Differential sensitivity of bicarbonate-reversible formate effects on herbicide-resistant mutants of *Synechocystis* 6714. *FEBS Lett.* 267: 273-276.

84. Govindjee, Schwarz, B., Rochaix-J.D. and Strasser, B. (1991). A herbicide-resistant mutant L275F of *Chlamydomonas reinhardtii* fails to show the bicarbonate-reversible formate effect in chlorophyll a fluorescence transients. *Photosynth. Res.* submitted for publication.

85. Cao, J., Vermaas, W.F.J. and Govindjee (1991). Arginine residues in the D2 polypeptide stabilizing bicarbonate binding in photosystem II of *Synechocystis PCC 6803*. Biochim. Biophys. Acta *Submitted for publication*.

86. Anderson, B.F., Baker, H.M., Norris, G.E., Rice, D.W. and Baker, E.N. (1989). Structure of human lactoferrin: crystallographic structure analysis and refinement at 2.8A resolution. *J. Mol. Biol*. 209: 711-734.

87. Mende, D. and Wiessner, W. (1985). Bicarbonate in vivo requirement of photosystem II in the green alga *Chlamydobotrys stellata*. *J. Plant Physiol*. 118: 259-266.

88. El-Shintinawy, F., Xu, C. and Govindjee (1990). A dual bicarbonate-reversible formate effect in *Chlamydomonas* cells. *J. Plant Physiol*. 136: 421-428.

89. Blankenship, R.E., Trost, J.T. and Manchino, L.E. (1987). Properties of reaction centers form the green photosynthetic bacterium *Chloroflexus aurantiacus*. In: *The Photosynthetic Bacterial Reaction Center: Structure and Dynamics*, Breton, J. and Vermeglio, A. eds, Plenum Press, NY, NATO ASI Series A: Life Sciences, Vol 149, pp. 119-127.

90. Ovchinikov, Yu., Abdulaev, N.G., Zolotarev, A.S., Shmuckler, B.E., Zargarov, A.A., Kutuzov, M.A., Telezhinskaya, I.N. and Levina, N.B.(1988). Photosynthetic reaction centre of *Chloroflexus aurantiacus* I. Primary structure of L-subunit. *FEBS Lett*. 231: 237-241.

91. Ovchinikov, Yu., Abdulaev, N.G., Shmuckler , B.E., Zargarov, A.A., Kutuzov, M.A., Telezhinskaya, I.N., Levina, N.B. and Zolotarev, A.S. (1988). Photosynthetic reaction centre of *Chloroflexus aurantiacus*: primary structure of M-subunit. *FEBS Lett*. 232: 364-368.

92. Shiozawa, J.A., Lottspeich, F. and Feick, R. (1987). The photochemical reaction center of *Chloroflexus aurantiacus* is composed of two structurally similiar polypeptides. *Eur. J. Biochem*. 167: 595-600.

93. Cho, H.M., Mancino, L.J. and Blankenship, R.E. (1984). Light saturation curves and quantum yields in reaction centers from photosynthetic bacteria. *Biophysic. J*. 45: 455-461.

94. Mancino, L.J., Dean, D.P. and Blankenship, R.E. (1984). Kinetics and thermodynamics of the $P870^{+}Q_{A}^{-}$ $P870^{+}Q_{B}^{-}$ reaction in isolated reaction centers from the photosynthetic bacterium *Rhodopseudomonas sphaeroides*. *Biochim. Biophys. Acta* 764: 46-54.

95. Wang, X., Cao, J., Maroti, Stilz, H.U., Oesterhelt, D., Govindjee and Wraight, C.A. (1991). Is bicarbonate in photosystem II the equivalent of the glutamate (M234) ligand of the iron atom in bacterial reaction centers? *Biophysical Society Abstracts*, Annual Meeting San Francisco, CA in press.

96. Erickson, J.M., Pfister, K., Rahire, M., Togasaki, R.K., Mets, L. and Rochaix, J-D. (1989). Molecular and biophysical analysis of herbicide-resistant mutants of *Chlamydomonas reinhardtii*: structure-function relationship of the photosystem II D1 polypeptide. *The Plant Cell* 1: 361-371.

97. Crofts, A., Robinson, H., Andrews, K., Van Doren, S. and Berry, E. (1987). Catalytic sites for reduction and oxidation of quinones. In: *Cytochrome Systems,* Papa, S., Chance, B. and Ernster, L. eds., Plenum Publishing Corporation, NY, pp. 617-624.

98. Trebst, A. (1991). The molecular basis of resistance of photosystem II herbicides. In: *Herbicide Resistance,* 11th Long Ashton International Synposium, Caseley, J.C. ed, Butterworths Publishers, London, pp. 52-61.

99. Kirilovsky, D. Ducruet, J-M. and Etienne, A-L. (1990). Primary events occurring in photoinhibition in *Synechocystis* 6714 wild type and atrazine-resistant mutant. *Biochim. Biophys. Acta* 1020: 87-93.

100. Sundby, C. Larsson, U.K. and Henrysson, T. (1989). Effects of bicarbonate on thylakoid protein phosphorylation. *Biochim. Biophys. Acta* 975: 277-282.

101. Sundby, C. (1990). Bicarbonate effects on photoinhibition-including an explanation for the sensitivity to photoinhibition under anaerobic conditions. *FEBS Lett.* 274: 77-81.

102. Cao, J., Vermaas, W.F.J. and Govindjee (1989). Construction of arginine 233 and 251 altered mutants in D2 Protein of PS II in cyanobacterium *Synechocystis* 6803. Abstract book, 15th Annual Midwest Photosynthesis conference, French Lick, In. (available from the author)

103. Cao, J. Vermaas, W.F.J. and Govindjee (1990). Arginine residues in D2 polypeptide may stabilize bicarbonate binding in photosystem II of the cyanobacterium *Synechocystis* 6803. Abstract book, 16th Annual Midwest Photosynthesis conference, Turkey Run, IN. (available from the author)

104. Wydrzynski, T. and Govindjee (1975). A new site of bicarbonate effect in photosystem II of photosynthesis: evidence from chlorophyll fluorescence transients in spinach chloroplasts. *Biochim. Biophys. Acta* 387: 403-408.

LIGHT-RESPONSIVE REGULATION OF PHOTOSYSTEM II GENES IN THE CYANOBACTERIUM *SYNECHOCOCCUS* SP. PCC 7942

S.S. Golden, S.A. Bustos, M.R. Schaefer, U.W. Mueller, and R.D. Kulkarni; Dept. of Biology, Texas A&M University, College Station, TX 77843 USA

The D1 and D2 proteins of the photosystem II reaction center are encoded by two small gene families in the cyanobacterium *Synechococcus* sp. strain PCC 7942. Each of these families contains members that are expressed at low levels under typical laboratory growth light intensities equal to or below 125 $\mu E \cdot m^{-2} \cdot s^{-1}$, but whose expression increases markedly upon transfer to higher light. The *psbAII*, *psbAIII*, and *psbDII* genes exhibit this mode of expression, which is regulated primarily at the transcriptional level. The *psbAI* and *psbDI* genes are expressed at higher constitutive levels than the other genes, and show little evidence of transcriptional regulation. However, the *psbAI* message appears to have a shorter half-life at high light than under low light conditions. Proteins from *Synechococcus* soluble extracts bind specifically to the upstream regions of members of both gene families, and may play a role in regulation of the genes. Current experiments are aimed at determining the primary sensor for light intensity, such as a photoreceptor or a change in redox state, and the signal transduction pathway that results in altered gene expression. A major goal is to understand whether the observed gene responses to changing conditions represent an adaptation to better utilize available light or to protect the photosynthetic apparatus from environmental stresses.

INTRODUCTION

Two polypeptides designated D1 and D2 are integral thylakoid proteins that constitute the reaction center of photosystem II, the site at which chlorophyll gives up an electron in response to excitation by light [1]. The genes *psbA* and *psbD*, which encode D1 and D2, respectively, are unique genes in the chloroplast genomes of most plants, but in cyanobacteria both genes are present as small multigene families [2,3]. The genome of the cyanobacterium *Synechococcus* sp. strain PCC 7942 (hereafter referred to as *Synechococcus*) contains a family of three distinct *psbA* genes which encode two different forms of D1 [2]. The three genes predict that the product of *psbAI* (D1 form I) differs from the product of *psbAII* and *psbAIII* (form II) at 25 out of 360 amino acids [2]. This organism also has two *psbD* genes, designated *psbDI* and *psbDII*. The *psbD* genes predict an identical D2 polypeptide [3], but differ in their organization in the chromosome. The *psbDI* gene, like chloroplast *psbD* genes, overlaps the open reading frame of *psbC*, the gene encoding a photosystem II protein called CP43 [3]; the *psbDII* gene does not overlap another photosystem II gene, and produces a monocistronic RNA.

We are studying the regulation of the *psbA* and *psbD* gene families and expression of their protein products to identify the regulatory signals that modulate transcription and translation in response to changes in the growth environment. We have shown previously that the three *Synechococcus psbA* genes are differentially regulated and respond to changes in light intensity [4]. Translational gene fusions between each of the *psbA* genes and the *Escherichia coli lacZ* gene have been recombined into the chromosome of *Synechococcus* cells to serve as reporters of *psbA* expression [4]. These experiments show that expression from the *psbAI-lacZ* gene fusion is highest at low light intensity, whereas expression from both the *psbAII-lacZ* and *psbAIII-lacZ* reporters is highest at high light intensity [4]. Western blot analysis of thylakoid membrane proteins using D1 form-specific antisera showed that in wild-type cells harvested at different light intensities, the ratio of the forms changed as predicted by the activities of the reporter genes [5]. Changes in RNA levels of each of the *psbA* genes are evident within minutes of transferring medium light-adapted *Synechococcus* cells to high or low light [6]. Northern blot analysis with gene-specific probes has revealed three different response patterns of the *psbA* transcripts: transfer to high light significantly increases the levels of the 1.2 kb *psbAII* and *psbAIII* transcripts and brings about a rapid loss of *psbAI* transcript. A longer transcript from *psbAII* that contains a second open reading frame of unknown function is unaltered by the light shift [6].

Recent experiments have focused on understanding the molecular events that mediate the light-responsive expression. Here we report that the increase in *psbAII* and *psbAIII* expression following a shift to higher light intensity is a transcriptional response, whereas *psbAI* may be regulated post-transcriptionally. One member of the *psbD* family, *psbDII*, also exhibits increased expression following a shift to higher light, whereas *psbDI* does not. All three members of the *psbA* gene family and one member of the *psbD* gene family bind proteins upstream of their open reading frames which may be involved in regulation of gene expression.

MATERIALS AND METHODS

Synechococcus Culture Conditions

Wild-type *Synechococcus* sp. strain PCC 7942 (previously referred to as *Anacystis nidulans* R2, Pasteur Culture Collection # 7942) was grown in liquid BG-11 medium [7] at 30°C. For controlled light studies, wild-type *Synechococcus* was grown in continuous culture using a turbidostat apparatus to maintain constant cell density (manuscript in preparation). An ethanol-sterilized immersible probe (Biospherical Instruments) was used to measure the photosynthetic photon flux density (PPFD, measured in $\mu E \cdot m^{-2} \cdot s^{-1}$) at the center of the cultures. In light-shift experiments for *psbA* and *psbD* transcripts, cultures were grown turbidostatically at PPFD of 125 $\mu E \cdot m^{-2} \cdot s^{-1}$, and a portion of the culture was harvested for RNA isolation. The light intensity was then adjusted to PPFD values indicated in the text for each experiment. Times of exposure to different light intensities prior to harvesting cells for RNA isolation are given in the text for each experiment.

DNA and RNA Methods

Most restriction and modifying enzymes were purchased from Bethesda Research Laboratories (Gaithersburg, Md.) or Boehringer Mannheim Biochemicals (Indianapolis, IN) and used as directed by the manufacturers. Isolation of restriction fragments from gels was by electroelution as described by Ausubel *et al.* [8].

Total RNA was isolated from *Synechococcus* as described by Reddy *et al.* [9]. During time-course experiments, samples at each time point were pelleted and frozen at -90°C until all samples were available for RNA extraction. For Northern blot analysis, RNA was denatured with formaldehyde, separated by electrophoresis on 1.2% agarose gels, and transferred to a charged nylon membrane as described by Ausubel *et al.* [8]. Radioactive anti-sense RNA probes for transcripts from each of the *psbA* and *psbD* genes were synthesized from restriction fragments subcloned into either pT3T7 (BRL) or pBluescript (Stratagene) vectors using T3 or T7 RNA polymerase and α-[^{32}P] UTP as recommended by the enzyme supplier.

Northern blots were hybridized to RNA probes at 50°C in a mixture of 50% (v/v) formamide, 5X SSPE [10], 1% (w/v) sodium dodecyl sulfate (SDS), and denatured salmon sperm DNA (100 μg/ml). Unhybridized probe was removed from the filters by washing in a mixture of 0.5 X SSPE and 0.1 % SDS at 65°C. To quantitate the relative transcript levels, Northern blots from duplicate RNA isolations were performed, and the autoradiographic bands were read in a BIO-RAD Video Densitometer (Model 620).

Fractionation of *Synechococcus* soluble protein extracts by heparin-sepharose chromatography and DNA gel mobility shift assays were carried out as described by Chastain *et al.* [11]. DNA footprinting by the copper ion method was performed according to Kuwabara and Sigman [12].

RESULTS

Transcriptional and Post-transcriptional Regulation of the *psbA* Genes

Transfer of *Synechococcus* cells from medium light intensity (125 $\mu E \cdot m^{-2} \cdot s^{-1}$) to a higher light intensity (500 $\mu E \cdot m^{-2} \cdot s^{-1}$) results in a rapid change in the levels of messages from the three *psbA* genes [6]. After 15 minutes of exposure to the new light intensity, transcripts from *psbAII* and *psbAIII* increase 4- to 6-fold [6]. However, the *psbAI* message level drops in response to the increase in light [6].

The increase in *psbAII* and *psbAIII* messages at high light was blocked by addition of the transcription inhibitor rifampicin, indicating that the increase in message level is a result of increased transcription of these genes (Fig. 1).

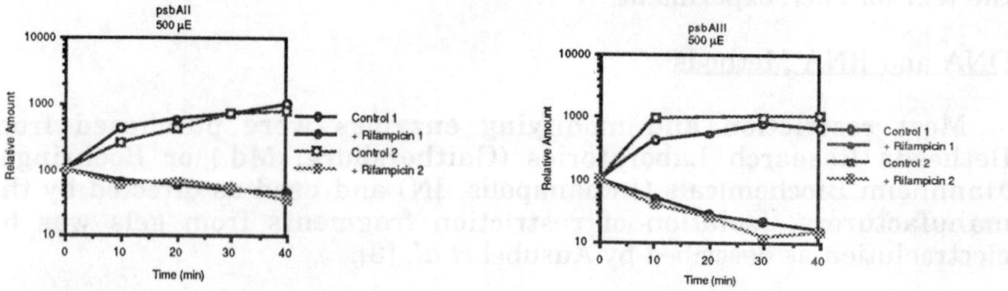

Fig. 1. Inhibition of the *psbAII* and *psbAIII* high light response by the transcription inhibitor rifampicin. *Synechococcus* cells were grown at 125 $\mu E \cdot m^{-2} \cdot s^{-1}$ and shifted to 500 $\mu E \cdot m^{-2} \cdot s^{-1}$ in the presence or absence of 50 µg/ml rifampicin (curves marked "+ rifampicin" or "control," respectively). Samples were harvested for RNA isolation before the light shift (time=0) and at the times indicated after the shift. Northern blots were probed with antisense RNA specific for the transcript indicated, and densitometry was used to compare RNA signal levels from different time points. The level of a given message at each time point was compared to the level in the unshifted control, which was designated as 100%. Results are graphed for two independent experiments.

Rifampicin had a surprising effect on the level of *psbAI* message, which was stabilized by the inhibitor (Fig. 2). This finding complicates determination of a native half-life for the *psbAI* message at high light, and suggests that *de novo* synthesis of a factor is required for rapid degradation of the *psbAI* transcript at high light. Addition of chloramphenicol prior to a light shift also blocked rapid degradation of the message (data not shown), although transcription is unaffected under these conditions.

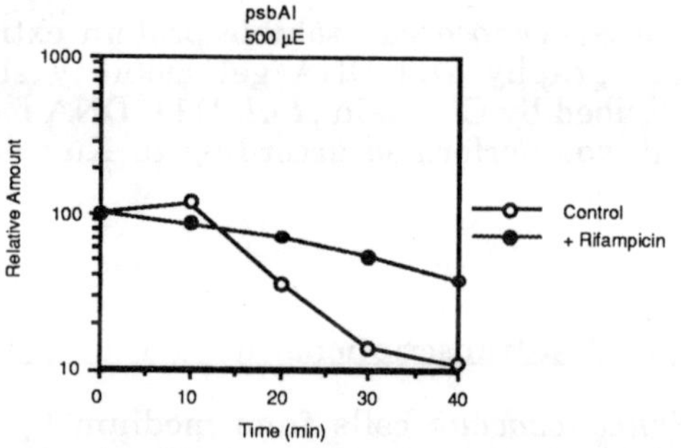

Fig. 2. Stabilization of the *psbAI* message at high light by the addition of rifampicin. Experimental design and quantitation of relative amounts of *psbAI* message was as described in the legend for Fig. 1. Data are shown from a single experiment.

Responses of the *psbD* Genes to an Increase in Light Intensity

The two *psbD* genes of *Synechococcus* encode an identical protein, but are arranged differently [3]. Northern analysis suggests that the monocistronic *psbDII* gene is expressed at a much lower level under normal laboratory growth conditions than the *psbDI/psbC* operon [3]. The response of each of these messages was analyzed following a shift in light intensity from 125 $\mu E \cdot m^{-2} \cdot s^{-1}$ to 360 $\mu E \cdot m^{-2} \cdot s^{-1}$ (Fig. 3). The dicistronic *psbDI* transcript was unaffected by the shift, but the level of *psbDII* transcript increased, reaching a peak increase of 5-fold over the starting level after 30 minutes. Experiments using gene fusions between regulatory regions of the *psbD* genes and the *E. coli lacZ* gene yielded similar results: the *psbDI-lacZ* fusion and a *psbC-lacZ* fusion did not respond to an increase in light intensity, whereas β-galactosidase production from a *psbDII-lacZ* fusion increased 80%-100% within 120 minutes after the light shift (data not shown).

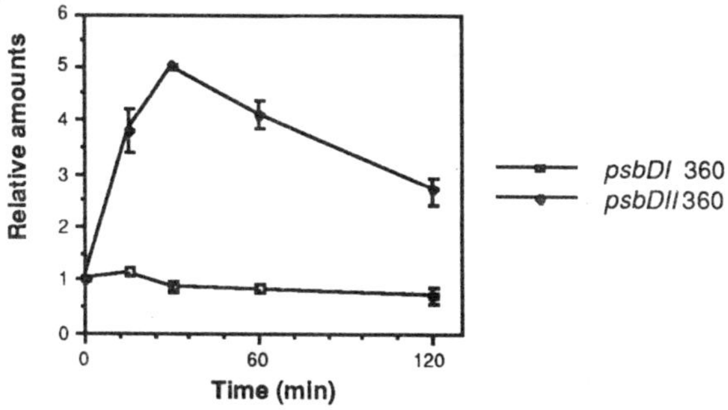

Fig. 3. Induction of the *psbAII* transcript by shifting cells to high light. Graphic representation of changes in abundance of transcripts from the *psbD* genes after cells were transferred from 125 $\mu E \cdot m^{-2} \cdot s^{-1}$ (time=0) to 360 $\mu E \cdot m^{-2} \cdot s^{-1}$ and sampled at the times indicated for each point. Northern blots of RNA from each time point were probed with antisense RNAs specific for the *psbDI/C* or *psbDII* transcript and autoradiographs were scanned to quantitate changes in RNA levels. Results from two experiments were averaged and the range of values is shown at each point. Relative amounts indicate the signal at a given time point divided by the time=0 value for that transcript.

Binding of Proteins from *Synechococcus* Cells to Upstream Regions of the *psbA* Genes and *psbDII*

Regulation of many genes in both prokaryotic and eukaryotic organisms has been shown to be mediated by the binding of protein factors to regulatory regions upstream of the genes. The *psbA* and *psbD* genes were tested for protein binding sites by DNA mobility shift assays [11]. A soluble protein extract from cells adapted to 125 $\mu E \cdot m^{-2} \cdot s^{-1}$ and shifted to 320 $\mu E \cdot m^{-2} \cdot s^{-1}$ for 30 minutes was enriched for DNA binding proteins by heparin-sepharose column chromatography [11]. Probes spanning the promoter regions of each of the five genes in this study were mixed with individual column fractions and subjected to polyacrylamide gel electrophoresis. The *psbAI*, *psbAII*, and *psbDII* probes showed peak binding with fractions 6 and 7, whereas *psbAIII* showed no binding with those fractions, and reacted with proteins in fractions 9-15. The *psbDI* probe showed no binding with any of the column fractions (data not shown).

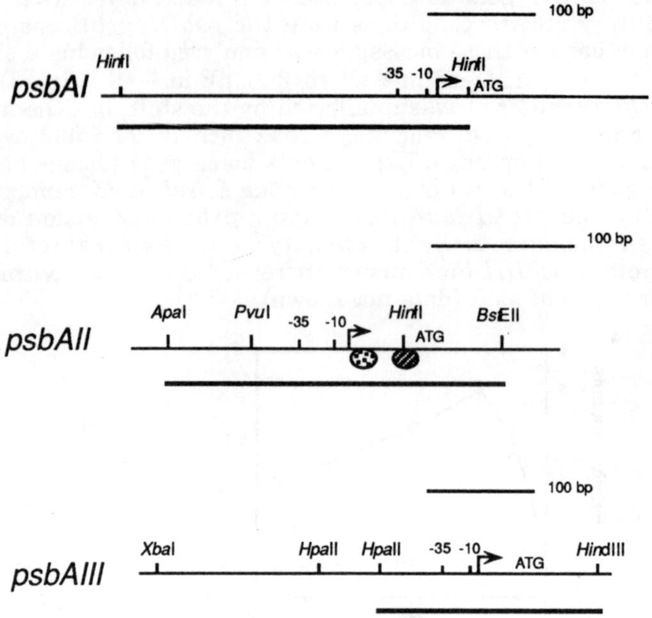

Fig. 4. Schematic representation of *psbA* gene upstream regions used in DNA-protein binding experiments. Each gene fragment is represented by a line, and sites for restriction enzymes are indicated. The letters "ATG" mark the start codon of each open reading frame. Sequences that have characteristics of *E. coli* promoter elements are indicated by "-35" and "-10". The transcription start site is shown as a rightward arrow. A dark bar under each gene diagram indicates the restriction fragment used in most DNA mobility shift assays. Two ovals are placed on the *psbAII* diagram to indicate the binding regions of proteins that protect the DNA in footprinting experiments.

The binding sites linked to *psbAII* and *psbDII* have been further delineated by using smaller fragments of each of the upstream regions and by DNA footprinting methods [12]. These results indicated that both genes bind polypeptides between the start site for transcription and the start codon of each gene. In the case of *psbDII*, DNA mobility shift assays suggest that there may be three binding sites in this region. The *psbAII* gene may also bind a factor that spans the start codon for D1. DNA mobility shift assays have localized protein binding to a 210 bp fragment containing the upstream region of the *psbAIII* gene. Deletion mutations upstream of a *psbDII-lacZ* reporter gene support the conclusion that light-responsive regulatory elements lie downstream from the promoter, since light regulation is retained until deletions reach the position which should correspond to a "-35" element (data not shown).

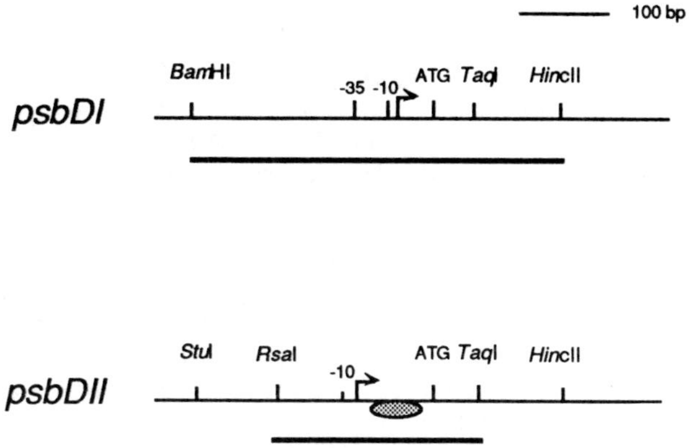

Fig. 5. Schematic representation of *psbD* gene upstream regions used in DNA-protein binding experiments. Symbols are as described in the legend to Fig. 4. The *psbDII* gene lacks a distinguishable "-35" element. An oval is placed on the *psbDII* diagram to indicate the binding region of a protein(s) that protects the DNA in footprinting experiments.

DNA mobility shift assays of the *psbDII* probe incubated with extract fraction 7 have been competed with unlabeled probe fragments. A *psbDII* probe effectively competed its own binding, whereas *psbDI* showed very little competition. The upstream fragment from *psbAII* effectively competed binding to two of the three *psbDII* complexes (data not shown), suggesting a shared binding factor.

DISCUSSION

The cyanobacterium *Synechococcus* has a complex mechanism for altering the expression of individual members of the gene families that encode the reaction center proteins of photosystem II in response to environmental changes. Current evidence indicates that transcriptional and post-transcriptional components are involved, and that at least two polypeptides interact with the members of the two gene families. Additional regulation at the level of D1 and D2 protein turnover has not yet been addressed. We presume that the existence of the regulatory mechanism is indicative of a physiological adaptation to new environmental conditions. This may be a protective mechanism to avoid photodamage or a modulation to improve photosynthetic capacity. At this time we have no direct evidence to indicate a change in photosynthetic function. Current efforts are aimed at uncovering such an adaptation and better defining the components of the regulatory network at the molecular level, including identification of the initial sensor that signals a change in light intensity, and the proteins that are involved in the sensory transduction pathway.

REFERENCES

1. Nanba, O. and Satoh, K. (1987). Isolation of a photosystem II reaction center consisting of D-1 and D-2 polypeptides and cytochrome *b*-559. *Proc. Natl. Acad. Sci. USA* 84: 109-112.
2. Golden, S.S., Brusslan, J. and Haselkorn, R. (1986). Expression of a family of *psbA* genes encoding a photosystem II polypeptide in the cyanobacterium *Anacystis nidulans* R2. *EMBO J.* 5: 2789-2798.
3. Golden, S.S. and Stearns, G.W. (1988). Nucleotide sequence and transcript analysis of three photosystem II genes from the cyanobacterium *Synechococcus* sp. PCC 7942. *Gene* 67: 85-96.
4. Schaefer, M.R. and Golden, S.S. (1989). Differential expression of members of a cyanobacterial *psbA* gene family in response to light. *J. Bacteriol.* 171: 3973-3981.
5. Schaefer, M.R. and Golden, S.S. (1989). Light availability influences the ratio of two forms of D1 in cyanobacterial thylakoids. *J. Biol. Chem.* 264: 7412-7417.
6. Bustos, S.A., Schaefer, M.R. and Golden, S.S. (1990). Different and rapid responses of four cyanobacterial *psbA* transcripts to changes in light intensity. *J. Bacteriol.* 172: 1998-2004.
7. Allen, M.M. (1968). Simple conditions for growth of unicellular blue-green algae on plates. *J. Phycol.* 4: 1-4.
8. Ausubel, F.M., Brent, R., Kingston, R.E., Moore, D.D., Seidman, J.G., Smith, J.A. and Struhl, K. (1987). *Current protocols in molecular biology*. Greene Publ. Assoc. and Wiley-Interscience, New York.
9. Reddy, K.J., Masamoto, K., Sherman, D.M. and Sherman, L.A. (1989). DNA sequence and regulation of the gene (*cbpA*) encoding the 42-kilodalton cytoplasmic membrane carotenoprotein of the cyanobacterium *Synechococcus* sp. strain PCC 7942. *J. Bacteriol.* 171: 3486-3493.
10. Maniatis, T., Fritsch, E.F. and Sambrook, J. (1982). *Molecular Cloning: a laboratory manual.* Cold Spring Harbor Laboratory, Cold Spring Harbor, New York.
11. Chastain, C.J., Brusca, J.S., Ramasubramanian, T.S., Wei, T-F. and Golden, J.W. (1990). A sequence-specific DNA-binding factor (VF1) from *Anabaena* sp. strain PCC 7120 vegetative cells binds to three adjacent sites in the *xisA* upstream region. *J. Bacteriol.* 172: 5044-5051.
12. Kuwabara, M.D. and Sigman, D.S. (1987). Footprinting DNA-protein complexes *in situ* following gel retardation assays using 1,10-phenanthroline-copper ion: *Escherichia coli* RNA polymerase-*lac* promoter complexes. *Biochemistry* 26: 7234-7238.

THE RELATIONSHIP BETWEEN PHOTOSYSTEM 2 ACTIVITY AND CO_2 ASSIMILATION IN LEAVES

Neil R. Baker

Department of Biology, University of Essex, Colchester, Essex, CO4 3SQ, U.K.

Abstract

Photosystem 2 plays an important role in the response of photosynthesis in higher plants to environmental perturbations and stresses. The relationship between photosystem 2 and photosynthetic CO_2 assimilation in leaves operating under physiological conditions is examined. It is demonstrated that light-dependent, non-photochemical quenching of excitation energy is an important factor determining the photochemical activity of photosystem 2, and consequently CO_2 assimilation, in leaves.

When light is limiting for leaf photosynthesis a linear relationship is observed between carbon assimilation and absorbed light, the slope of which estimates the maximum quantum efficiency of carbon assimilation (maximum ϕ_{CO2}). This parameter can be influenced by factors determining the conversion of absorbed light to ATP and reductants by the thylakoids (1). As the light intensity incident upon the leaf is increased the relationship between absorbed light and ϕ_{CO2} becomes non-linear, ϕ_{CO2} decreases progressively with increasing light absorption and the factors that determine the maximum ϕ_{CO2} will have a diminishing effect on leaf photosynthesis. In the field it would appear that the majority of leaves rarely receive light levels that are saturating for photosynthesis, thus a very considerable proportion of canopy photosynthesis is not light-saturated (2-5) and can potentially be influenced by changes in the quantum efficiency of thylakoid photochemistry (1).

The quantum efficiency of photosystem 2 (PS2) electron transport will be determined by (i) the efficiency with which an absorbed photon can reach a reaction centre and (ii) the proportion of reaction centres that are capable of transferring an electron to an acceptor at that point in time, i.e. which are 'open' (6). Since the efficiency of excitation energy capture by 'open' PS2 reaction centres is defined by the ratio of variable to maximal fluorescence, Fv/Fm (7,8), and the photochemical quenching coefficient, qP, is a measure of the proportion of oxidised or 'open' PS2 reaction centres, it has been argued (6) that the quantum efficiency of PS2 photochemistry (ϕ_{PS2}) in a leaf is defined by:

$$\phi_{PS2} = Fv/Fm \cdot qP \qquad (1)$$

In practice determination of ϕ_{PS2} for leaves under any given environmental condition can be made simply by monitoring the fluorescence in the given state (Fs), exposing the leaves to a saturating flash and measuring the maximal fluorescence level attained (Fm); the parameter (Fm - Fs)/Fm equates to (Fv/Fm).qP for the leaf under the given conditions and thus is a measure of ϕ_{PS2} (6). In maize leaves, which have minimal photorespiratory activity, modificaton of ϕ_{CO2} in response to changes in light intensity and atmospheric CO_2 concentration or during the induction of photosynthesis in a dark adapted leaf resulted in a direct proportionality between ϕ_{CO2} and ϕ_{PS2} being observed (6). Linear relationships between ϕ_{CO2} and ϕ_{PS2} were also observed in barley and ivy leaves exposed to a range of light intensities and in which photorespiration had been minimised by reducing the atmospheric O_2 concentration to 1% (6,9), although it should be noted that some deviation from linearity was found at very low light intensities. Clearly, under a wide range of physiological conditions, provided that the allocation of ATP and reductants produced by photosynthetic electron transport to processes other than CO_2 assimilation is minimal, a direct proportionality exists between ϕ_{CO2} and ϕ_{PS2} which has been defined as:

$$\phi_{CO2} = \phi_{PS2} \cdot I_a/I_i \cdot 1/k \qquad (2)$$

where I_a and I_i are the photon flux densities absorbed by PS2 antennae and the leaf respectively, and k is the number of electron equivalents required to reduce 1 mol of CO_2 (6). As would be expected for leaves of C3 plants under conditions favourable for photorespiratory activity where the products of electron transport are apportioned between carboxylation and oxygenation reactions, the strict linear relationship between ϕ_{CO2} and ϕ_{PS2} breaks down (9).

Recently non-invasive absorbance spectroscopic techniques have been developed to estimate the relative quantum efficiency of PS1 photochemistry (ϕ_{PS1}) in leaves (10-12). Simultaneous measurements of ϕ_{PS1} and ϕ_{PS2} made over a wide range of light intensities on leaves under both photorespiratory and non-photorespiratory conditions have shown that a good correlation exists between these parameters (9,13). This observation, together with the linear relationship found between ϕ_{PS2} and ϕ_{CO2}, suggests that ϕ_{PS2} is a good measure of the quantum efficiency of non-cyclic electron transport. The conserved linearity between ϕ_{PS1} and ϕ_{PS2} under a wide range of physiological conditions implies that the rate of PS1-mediated cyclic electron transport, if it occurs in vivo, is always proportional to the rate of non-cyclic electron transport. Consequently it would not appear that modulation of PS1-mediated cyclic electron transport is occurring to meet any changing metabolic demands under these experimental conditions.

If the allocation of the products of electron transport to sinks other than CO_2 assimilation is negligible or constant, it is apparent that ϕ_{PS2} provides a good relative measure of ϕ_{CO2}. Although it cannot be stated unequivocally from such studies whether ϕ_{PS2} determines ϕ_{CO2} or vice versa, it is unlikely that large changes in ϕ_{PS2} will not be reflected in parallel changes in ϕ_{CO2}.

Analyses of the fluorescence characteristics of leaves have shown that photochemical quenching of fluorescence (qP), which results from electron transfer from P680 to Q_A, is not necessarily related proportionally to PS2 photochemistry (6,9,11,13). This phenomenon has been attributed to non-photochemical quenching processes which compete with PS2 photochemistry for excitation energy (14-16). It has been suggested that increases in the degree of energisation of the thylakoids will increase the rate of dissipation of excitation energy by non-radiative processes and consequently decrease ϕ_{PS2} (14,15). Although increasing the proton electrochemical gradient across thylakoid membranes will increase non-photochemical quenching and decrease the rate of PS2 photochemistry (17-19), there has been no evidence from leaf tissue at steady state photosynthesis to implicate changes in thylakoid energisation as a determinant of non-photochemical quenching processes and the rate of PS2 photochemistry in vivo. Thylakoid energisation may well be a prerequisite for such non-photochemical quenching to occur but other factors, possibly carotenoid changes (20), are almost certainly involved.

Light-induced increases in non-photochemical quenching of excitation energy in a leaf and the consequences for ϕ_{PS2} and ϕ_{CO2} can be seen from Table 1. Fv/Fm, qP, ϕ_{PS2} and ϕ_{CO2} were determined for dark- and light-adapted (30 min at a PPFD of 1500 μmol m^{-2} s^{-1}) leaves when the leaves were exposed to a low PPFD of 34 μmol m^{-2} s^{-1}. Pretreatment of the leaves in high light induced a 26 % decrease in Fv/Fm, indicative of a large increase in non-photochemical quenching, with concomitant decreases of 12 and 14 % in ϕ_{PS2} and ϕ_{CO2} respectively. The decrease in ϕ_{PS2} must be attributed to an increase in non-photochemical quenching, since high light pretreatment results in an increase in qP, thus demonstrating that in high light-treated leaves a larger proportion of PS2 reaction centres are oxidised than in the dark-adapted leaves. Such an increase in qP would

Table 1. The relationships between the quantum yield of CO_2 assimilation (ϕ_{CO2}), the relative quantum yield of PS2 photochemistry (ϕ_{PS2}) and fluorescence quenching parameters in leaves of Silene dioica pretreated in the dark and in high light.

Leaves were kept in the dark or in a PPFD of 1500 μmol m^{-2} s^{-1} prior to exposure to a PPFD of 34 μmol m^{-2} s^{-1} to measure ϕ_{CO2}, ϕ_{PS2}, Fv/Fm and the photochemical quenching coefficient, qP. The leaves were exposed to 400 ppm CO_2 and 1 % O_2 in nitrogen to minimise photorespiration. The decreases in ϕ_{CO2}, ϕ_{PS2} and Fv/Fm after the high light pretreatment were fully reversible within 1 h of the leaves being placed in the dark. Data from Genty et al.,1989 (6).

Parameter	Dark	High Light	% Change
ϕ_{CO2}	0.051	0.044	- 13.8
ϕ_{PS2}	0.492	0.432	- 12.2
Fv/Fm	0.712	0.526	- 26.1
qP	0.690	0.824	+ 19.4

increase ϕ_{PS2} (see equation 1). Clearly the greater decrease in Fv/Fm, rather than the increase in qP, is the dominating factor in determining the change in ϕ_{PS2}, and consequently ϕ_{CO2}. Such light-induced non-photochemical quenching is reversible within 1 h of the leaves being placed in the dark and thus is unlikely to be associated with photoinhibitory damage to PS2 reaction centre proteins which recover considerably more slowly.

Increases in non-photochemical quenching of excitation energy with increasing light intensity is commonly referred to as a down regulation of photosynthesis with the implication that the increase in the dissipation of excitation energy by non-radiative decay processes is a regulatory mechanism that minimises damage to the D1 protein of the PS2 reaction centre as light levels increase. However, it may be more appropriate to consider the phenomenon as a protective adjustment rather than a regulation. It is evident that the phenomenon occurs in algae, lower and higher vascular plants and would appear to be a ubiquitous feature of the photosynthetic apparatus of thylakoids. However, it is not clear whether the quantitative relationship between the rates of dissipation of excitation energy by PS2 photochemistry and non-photochemical quenching processes is the same in all plants. It is possible that this relationship is highly conserved within the plant kingdom, thus suggesting that the phenomenon is an essential mechanism to reduce photodamage to PS2 and optimise the transduction of light energy by the thylakoid membranes, without which the observed photosynthetic efficiencies of photosynthetic organisms could not be achieved.

The molecular mechanism and the site of the light-induced non-photochemical quenching of PS2 excitation energy have not been definitively resolved. The effects of this non-photochemical quenching on the rate of PS2 photochemistry are attributable to a direct effect on the efficiency of photon capture by PS2 reaction centres (21). Determination of single turnover flash energy saturation curves for Q_A reduction in leaves exposed to low and high actinic light levels have demonstrated that the effective absorption cross section for PS2 photochemistry, which is the product of the effective antenna size and ϕ_{PS2} (22), is reduced under conditions producing light-induced, non-photochemical quenching (21). This would be consistent with observed decreases in Fv/Fm and Fo associated with increases in non-photochemical quenching (Table 1) and suggest that non-photochemical quenching decreases the rate of PS2 photochemistry by producing a decrease in the rate of excitation energy transfer from the antennae pigment matrices to PS2 reaction centres (21). Presumably this is indicative of a non-radiative energy dissipation process located in the antennae of PS2 which competes effectively with excitation energy transfer to PS2 reaction centres. It has also been suggested that such non-photochemical quenching may be associated with the creation of low photochemically efficient, highly quenching PS2 reaction centres (15) or electron cycling around PS2 reaction centres (19). However, there is no unequivocal demonstration that either of these mechanisms operate in leaves.

Light-induced, non-photochemical quenching does appear to occur in leaves in the field (23-25). Fv/Fm in leaves of a winter wheat crop, which were dark adapted for 15 min prior to measurement, decreased throughout the morning in parallel with the increasing light intensity received by the plants and then increased during the afternoon as light intensity decreased. This close antiparallel relationship between Fv/Fm and light intensity for leaves in the field during the day has also been observed in willow (24).

In conclusion, it is evident that ϕ_{PS2} reflects the quantum efficiency of non-cyclic electron transport, and under conditions where the distribution of the products of electron transport to sinks other than CO_2 assimilation is negligible, or constant, it can be used as an accurate, relative measure of ϕ_{CO2}. Since ϕ_{PS2} can be rapidly determined from two simple fluorescence parameters, it can be used as an effective and practical probe of photosynthetic performance of leaves operating under natural light conditions in the field. Consequently it is expected that measurements of ϕ_{PS2} will play an important role in studies upon which predictions of how climate change will affect the photosynthetic productivity of crops and natural vegetation.

Acknowledgements

I am most grateful to B. Genty, J. Harbinson and D. R. Ort for many helpful discussions. Particular acknowledgement must be given to Bernard Genty for his important contributions in elucidating the relationship between ϕ_{PS2} and ϕ_{CO2}. This paper constitutes part of a larger review on PS2 and environmental perturbations (26) currently in press.

References

1. Baker, N.R., Long, S.P. and Ort, D.R. (1988).Photosynthesis and temperature, with particular reference to effects on quantum yield. In: Plants and Temperature, S.P. Long and F.I. Woodward eds., The Company of Biologists Ltd., Cambridge, pp. 347-375.
2. Inoue, E., Uchijima, Z., Udagawa, T., Morie, T. and Kobayashi, K. (1968). Studies on energy and gas exchange within crop canopies. (2) CO_2 flux within and above a corn canopy. J. Agric. Met. 23: 165-176.
3. Beuerlein, J.E. and Pendleton, J.W. (1971). Photosynthetic rates and light saturation curves of individual soyabean leaves under field conditions. Crop Sci. 11: 217-219.
4. Hatfield, J.L. and Carlson, R.E. (1978). Photosynthetically active radiation, CO_2 uptake and stomatal diffusive resistance profiles within soyabean canopies. Agron J. 70: 592-596.
5. Ort, D.R. and Baker, N.R. (1988). Consideration of photosynthetic efficiency at low light as a major determinant of crop photosynthetic performance. Plant. Physiol. Biochem. 26: 555-565.
6. Genty, B., Briantais, J-M. and Baker, N.R. (1989). The relationship between the quantum yield of photosynthetic electron transport and quenching of chlorophyll fluorescence. Biochim. Biophys. Acta 990: 87-92.
7. Kitajima, M. and Butler, W.L. (1975). Quenching of fluorescence and primary photochemistry in

chloroplasts by dibromothymoquinone. Biochim. Biophys. Acta 376: 105-115.

8. Butler, W.L. and Kitajima, M. (1975). Fluorescence quenching in photosystem II of chloroplasts. Biochim. Biophys. Acta 376: 116-125.

9. Genty, B., Harbinson, J. and Baker, N.R. (1990). Relative quantum efficiencies of the two photosystems of leaves in photorespiratory and non-photorespiratory conditions. Plant Physiol. Biochem. 28: 1-10.

10. Harbinson, J. and Woodward, F.I. (1987). The use of light-induced absorbance changes at 820 nm to monitor the oxidation state of P-700 in leaves. Plant Cell Environ. 10: 131-140.

11. Weis, E., Ball, J.T. and Berry, J. (1987). Photosynthetic control of electron transport in leaves of Phaseolus vulgaris : evidence for regulation of photosystem 2 by the proton gradient. In: Progress in Photosynthesis Research, J. Biggins ed., Vol II, Martinus Nijhoff Publishers, Dordrecht, pp. 553-556.

12. Schreiber, U., Klughammer, C. and Neubauer, C. (1988). Measuring P700 absorbance changes around 830 nm with a new type of pulse modulation system. Z. Naturforsch. 43C: 686-698.

13. Harbinson, J., Genty, B. and Baker, N.R. (1989). Relationship between the quantum efficiencies of photosystems I and II in pea leaves. Plant Physiol. 90: 1029-1034.

14. Krause, G.H. and Laasch, H. (1987). Photoinhibition of photosynthesis. Studies on mechanisms of damage and protection in chloroplasts. In: Progress in Photosynthesis Research, J. Biggins ed., Vol. IV, Martinus Nighoff Publishers, Dordrecht, pp. 19-25.

15. Weis, E. and Berry, J. (1987). Quantum efficiency of photosystem 2 in relation to "energy" dependent quenching of chlorophyll fluorescence. Biochim. Biophys. Acta 894: 198-208.

16. Horton, P. and Hague, A. (1988). Studies on the inductin of chlorophyll fluorescence in isolated barley protoplasts. IV. Resolution of non-

photochemical quenching. Biochim. Biophys. Acta 932: 107-115.

17. Krause, G., Vernotte, C. and Briantais, J-M. (1982). Photoinduced quenching of chlorophyll fluorescence in intact chloroplasts and algae. Biochim. Biophys. Acta 679: 116-124.

18. Oxborough, K. and Horton, P. (1988). A study of the regulation and function of energy-dependent quenching in pea chloroplasts. Biochim. Biophys. Acta 934: 135-143.

19. Rees, D. and Horton, P. (1990). The mechanisms of changes in photosystem efficiency in spinach thylakoids. Biochim. Biophys. Acta 1016: 219-227.

20. Demmig-Adams, B. (1990). Zeaxanthin-associated energy dissipation and the susceptibility of various organisms to light stress. In: Current Research in Photosynthesis, M. Baltscheffsky ed., Vol II, Kluwer Academic Publishers, Dordrecht, pp. 357-364.

21. Genty, B. Harbinson, J., Briantais, J-M and Baker, N.R. (1990). The relationship between non-photochemical quenching of chlorophyll fluorescence and the rate of photosystem 2 photochemistry in leaves. Photosynthesis Res. 25: 249-257.

22. Chylla, R.A. and Whitmarsh, J. (1990). Light saturation response of inactive photosystem II reaction centres in spinach. Photosynthesis Res. 25: 39-48.

23. Adams, W.W., Terashima, I., Brugnoli, E. and Demmig, B. (1988). Comparisons of photosynthesis and photoinhibition in the CAM vine Moya australis and several C3 vines on the coast of eastern Australia. Plant Cell Environ. 11: 173-181.

24. Ogren, E. (1988). Photoinhibition of photosynthesis in willow leaves under field conditions. Planta 175: 229-236.

25. Groom, Q., Long, S.P. and Baker, N.R. (1990). Photoinhibition of photosynthesis in a winter wheat crop. In: Current Research in Photosynthesis, M. Baltscheffsky ed., Vol. II, Kluwer Academic Publishers, Dordrecht, pp. 463-466.

26. Baker, N. R. (1991) A possible role for photosystem II in environmental perturbations of photosynthesis. Physiol. Plant. in press.

MOLECULAR GENETIC ANALYSIS OF PHOTOSYSTEM I

R. Mannar Mannan[1], Salil Bose[1,2] and Himadri B. Pakrasi[1]
[1]Biology Department, Box 1137, Washington University, St. Louis, MO 63130-4899, USA. and [2]Center for Biotechnology, Jawaharlal Nehru University, New Delhi 110067, INDIA.

ABSTRACT

Biochemical and biophysical analysis have helped in identifying the redox components involved in the electron transfer reactions in the photosystem I (PS I) complex. Isolation of PS I particles through detergent solubilization of thylakoid membranes and subsequent SDS-polyacrylamide gel electrophoresis followed by N-terminal amino acid sequencing of the component polypeptides have resulted in the identification of number and nature of the polypeptides associated with PS I. Molecular cloning and sequencing of the genes for PS I polypeptides paved the way for deducing their amino acid sequence and in turn their primary structure. Further details regarding the potential function of individual polypeptides and the amino acid residues in the component polypeptides in PS I could be achieved through molecular genetic analysis. Here we discuss the present status of such an analysis and its scope in elucidating the structure, function and assembly of PS I. We have also made an attempt to describe the potential of using *Anabaena variabilis* ATCC 29413 as an organism of choice for the molecular genetic analysis of PS I.

INTRODUCTION

In oxygenic photosynthetic organisms, the transfer of electrons from water to $NADP^+$ is mediated by two different photosystems, namely Photosystem II (PS II) and Photosystem I (PS I). Each of these photosystems occurs as a supra-molecular membrane protein complex. Associated with these complexes are photosynthetic pigments and a variety of redox components directly involved in the process of electron transfer. PS II oxidizes water, leading to the liberation of molecular oxygen on one hand and the reduction of plastoquinone pool on the other hand. PS I mediates electron transfer from plastocyanin to NADP. The cytochrome b_6/f complex is responsible for the electron transfer from reduced plastoquinone pool to plastocyanin.

PS II reaction center shows many functional and structural similarities with the photosynthetic reaction center of purple photosynthetic bacterium. The determination of the structure of the reaction center in the photosynthetic purple bacterium from crystallographic analysis (1) and derived amino acid sequence of the component polypeptides have greatly stimulated work on PS II in oxygenic photosynthetic organisms. Moreover, during the last six years, the availability of an easily transformable unicellular cyanobacterium, *Synechocystis* sp. PCC 6803, capable of photoheterotrophic growth, has opened up the possibility of molecular genetic analysis of PS II associated polypeptides. This has been of great help in elucidating the contribution of each of the component polypeptides as well as the contribution of distinct domains or particular amino acid residues in the individual polypeptides of PS II reaction center towards the functional organization of PS II (2).

When compared to the development of our understanding of the PS II, the progress in our knowledge about PS I has occurred at a slower pace. PS I reaction center appears as a rather unique system without any well known bacterial counterpart. Its equivalent can perhaps be found in rather poorly-studied green sulfur bacteria (3). There is much to be learned about the structure and function of PS I and definitely there is vast scope for applying the techniques of molecular genetics in elucidating the structural and functional organization of this complex.

As we have learned from our experience with PS II, the application of molecular genetics comes into effective use only after adequate understanding of the structural and functional organization of the protein complex at the biochemical and biophysical levels. Therefore the question one is tempted to ask before venturing into any molecular genetic analysis of PS I is whether our current understanding about PS I is strong enough to allow us to do any molecular genetic analysis of this complex. The answer to this question is that our present knowledge about the primary structural organization of polypeptides in the PS I core and the nature of the redox components involved in PS I electron transfer is fairly complete (4-8) and the time is ripe for beginning the molecular genetic analysis of PS I. Here we have made an attempt to give an account of the current status of research in molecular genetics of PS I. We begin with a brief account of our current understanding about the structure and functional organization of PS I.

FUNCTIONAL ORGANIZATION OF PS I

The PS I reaction center in plants and cyanobacteria has a primary electron donor, P700 and a series of five electron acceptors: A_0, a chlorophyll monomer; A_1, probably a quinone, and F_X, F_A and F_B, all [4Fe-4S] clusters. The function of these acceptors is to prevent rapid charge recombination by delocalizing the electron from the initial site of charge separation between $P700^+$ and A_0^-.

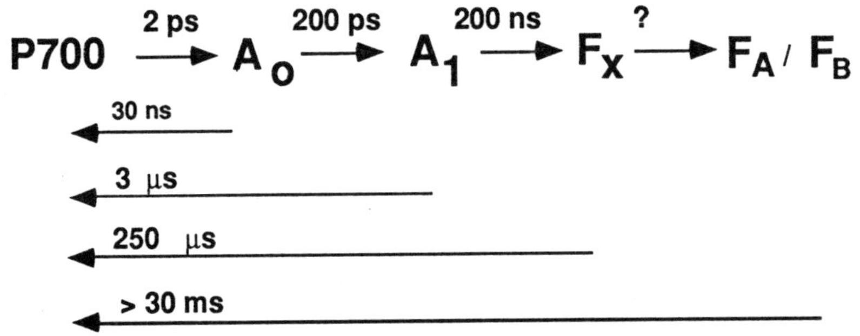

Fig. 1. Room temperature kinetics of electron flow in PS I. Estimates of forward electron transfer rates and back reaction rates are given.

The primary electron donor P700 is a chlorophyll *a* dimer and it derives its name from the wavelength maximum in the oxidized-minus-reduced difference spectrum. The optical difference spectrum of P700, obtained by photoxidation or chemical oxidation, shows an absorbtion decrease with peaks at 430 and 700 nm and an absorption increase with broad bands at 450 and 820 nm (9). Characterization of A_o, the primary electron acceptor by a variety of physical techniques indicates that it is a monomeric chlorophyll *a* and the strength of the EPR signal which characterizes it suggests that P700 and A_0 exists in an 1:1 ratio (10). Electron transfer from P700 to A_0 occurs in about 20 ps and the oxidation of A_o^- occurs in about 200 ps whereas the reduction of F_X, the next electron transfer intermediate occurs in 20-200 ns (Fig. 1). This observation suggests that there may be an electron transfer intermediate between A_0 and F_X and this electron transfer intermediate is designated as A_1. Under illuminated condition, this species shows as anisotropic ESR signal 10.3 G wide, centered at g= 2.0040(11). A_1 does not have any characterstic optical difference signal in the 350-750 nm range. However, an optical absorption change could be detected with a maximum at 287 nm, a minimum at 250 nm and an isobestic point at 269 nm, the appearence of which correlated with the amplitude of 10.3 G wide g= 2.0040 signal attributable to A_1 (12). This optical difference spectrum has similarities to the optical difference spectrum of certain quinone species that undergo redox changes between the quinone and semiquinone forms. The rates of appearence of the ESR and optical signal were also identical, which further suggests that the same chemical species is responsible for both signals. Thus A_1 appears to be a quinone in its chemical identity. However, a series of experiments involving organic extraction (13) or destruction of phylloquinone (14) as well as chemical modification of phylloquinone through deuteration during growth of the organism (15) have produced contradictory conclusions. Moreover, there seems to be a low temperature bypass around this acceptor (16).The possibility of an amino acid residue in the reaction center polypeptide functioning as A_1 has also been suggested (4). Todate, the chemical identity of electron acceptor A_1 and its function in the forward electron flow present an especially challenging problem in PS I research..

The electron acceptor acting next to A_1 is termed F_x , an iron-sulfur [4Fe-4S] center (17). F_A and F_B, the final electron aceptors in the PS I complex were the first to be discovered as two redox components acting on the reducing side of PS I. In the early 70's, optical spectroscopy and ESR spectroscopy helped in discovering these two iron-sulfur centers F_A and F_B with redox potentials -530 and -580 mv, respectively (18,19). Although the midpoint potentials suggest that F_B will precede F_A in electron transfer from F_x to ferredoxin, direct evidence for this view has not been obtained. Alternatively, it has been suggested that these centers could function in parallel.

STRUCTURAL ORGANIZATION OF PS I

In higher plants, PS I consists of a peripheral light-harvesting pigment-protein complex (LHC I) and a core complex carrying the electron transfer components and additional antenna pigments. LHC I is found to have 4-5 different polypeptides (20,21). The redox components involved in the PS I electron transfer are located in the core complex. In cyanobacteria, PS I contains only the core complex and LHC I is absent.

During the last 3-4 years, mild solubilization procedures accompanied by highly resolving electrophoretic procedures and N-terminal sequencing of the polypeptides have helped in identifying different polypeptides in the PS I core complex of higher plants and cyanobacteria (22-24). The genes for most of the PS I polypeptides have been cloned and sequenced and they have been denoted *psa*A, *psa*B, *psa*C etc. with the alphabetic order referring to the sequential order of their identification from a number of different species. The corresponding polypeptides are denoted PSI-A, PSI-B, PSI-C etc. (25). Figure 2 depicts a current model on the structural organization of PS I.

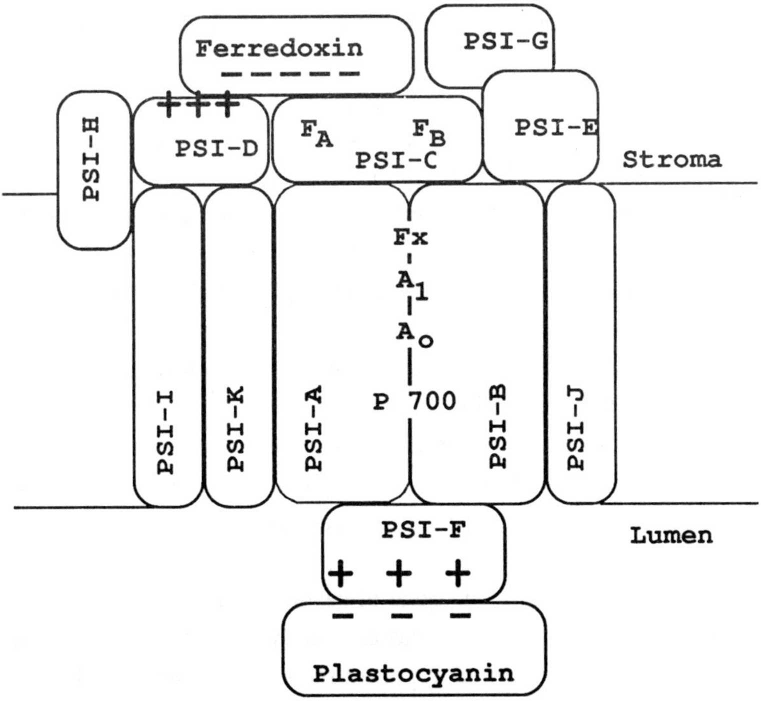

Fig. 2. Schematic model of the structural organization of photosystem I

The PS I electron transfer components, P700, A_0, A_1 and F_x are located in the heterodimer of PSI-A and PSI-B polypeptides, the products of *psa*A and *psa*B genes located in the chloroplast genome. From a number of species, these two genes have been cloned and sequenced and the deduced amino acid sequence of these gene products has helped in modeling the transmembrane organization of these two polypeptides as well as identifying the amino acid residues which act as the potential ligands for the iron-sulfur center F_X and the chlorophyll molecules (26). Hydropathy plots of these polypeptides predict that each of these polypeptides contain eleven membrane spanning helices. In all the species examined, both *psa*A and *psa*B genes specify a completely conserved stretch of 12 amino acid residues: Phe-Pro-Cys-Asp-Gly-Pro-Gly-Arg-Gly-Gly-Thr-Cys. The cysteine residues in these conserved regions are almost certainly the ligands of the center $F_{x,}$ since the PSI-B polypeptide contains no additional conserved cysteine residues and only one addtional cysteine residue is conserved in the PSI-A polypeptide. Iron-sulfur center F_X is therefore, concluded to be a [4Fe-4S] cluster bridged between the PSI-A and PSI-B heterodimer.

The PSI-A and PSI-B polypeptides contain 38 and 36 conserved histidine residues, respectively. A heterodimer of the two polypeptides thus contains 74 conserved histidine residues of which 52 are predicted to be positioned in the membrane spanning regions. Histidine residues have been implicated in the coordination of chlorophyll and heme groups in various polypeptides of photosynthetic apparatus. Thus the conserved histidine residues may provide binding sites for the majority of the 90 chlorophyll molecules associated with the PS I core complex. Very few histidine residues are found in the remaining PS I polypeptides.

The iron-sulfur centers F_A and F_B are located on a single 9 kDa polypeptide (PSI-C), the product of *psa*C gene located in the chloroplast genome. The *psa*C gene has been cloned and sequenced from a number of organisms and there are nine conserved cysteine residues, eight of which act as the ligands for the two iron-sulfur centers (27).

The nuclear-encoded *psa*D gene of several higher plants and several cyanobacteria have been isolated and completly sequenced (28-30). This gene codes for a polypeptide with an apparent molecular weight of 18-22 kDa and a deduced molecular weight of 17.5-18 kDa. PSI-D polypeptide is stromal-exposed (31) and is suggested to be involved in the docking of ferredoxin (32). This polypeptide carries a net positive charge which would greatly assist the binding of negatively charged ferredoxin. Cross linking studies have indicated that PSI-C and PSI-D polypeptides are situated close to one another (33).

*psa*E gene, located in the nuclear genome codes for a polypeptide with the apparent molecular weight of about 16 kDa. However, its deduced molecular weight is 9-10 kDa (29). This polypeptide is located on the stormal exposed portion of PS I.

PSI-F protein is a nuclear encoded hydrophilic protein and is located on the lumenal side of the thylakoid membrane. PSI-F protein appears to play a role in the transfer of electrons from plastocyanin to P700 and is considered to be the plastocyanin docking protein (34).

PSI-G polypeptide has a molecular weight of 10 kDa and it is coded by the nuclear gene *psa*G. The transit sequence of the precursor form of this gene product is typical of those directing proteins to the stroma (34). The functional significance of this polypeptide is not yet known.

*psa*H gene is nuclear-encoded and has been isolated as complete cDNAs which have been sequenced for barley (36), spinach (35), and *Chlamydomonas reinhardtii* (37). Although the higher plant sequences are highly conserved, the *C. reinhardtii* protein is only 40% identical in its primary sequence to the higher plant homologs. PSI-H polypeptide is rather polar overall and is probably located on the stromal side of the membrane. It is absent in cyanobacteria. In higher plants, it is believed to be involved in the docking of LHC I complex to the PS I core.

PSI-I polypeptide is chloroplast encoded and is very hydrophobic (38). This polypeptide is predicted to form a single transmembrane helix. The PSI-I polypeptide is weakly homologous to helix E of the D2 protein of PS II (*psb*D gene product). It has been speculated that the *psa*I gene product could be involved in cofactor (probably quinone) binding.

*psa*J gene is located in the chloroplast genome and it encodes a hydrophobic polypeptide expected to form a single transmembrane helix and believed to be involved in some cofactor binding (23).

*psa*K is located in the nuclear genome. The PSI-K polypeptide is highly hydrophobic and is expected to form a single transmembane helix (23). This polypeptide is found to be tightly bound to the reaction center polypeptides, PSI-A and PSI-B, and could be released from those two polypeptides by thiol treatment (39).

MOLECULAR GENETICS OF PS I

As described above, from the deduced amino acid sequence information, the primary structure of most of the PS I associated polypeptides has been determined (Fig. 2). The future challange will be to elucidate the organization of the various subunits within the PS I complex. Creation of well diffracting crystals of PS I would facilitate this analysis. Cross linking studies would also help to a certain extent in this connection. However, to elucidate the mechanism of PSI function, structure and assembly, molecular genetic techniques should be followed to introduce specific mutations in PS I. Analysis of the PS I complex from such mutants can be conveniently used to examine the role of the selected protein and the residues thererof in PS I.

Mutations in PS I may be created either through random mutagenesis or site-directed mutagenesis. In the random mutagenesis approach, the organism is treated with chemical mutagens, X-ray, or UV radiation and a certain phenotype is selected for. This type of classical mutagenesis in higher plants and *C. reinhardtii* has yielded a variety of PS I deficient strains. The genetic loci of these mutations are in both the nuclear and chloroplast genomes. In *C. reinhardtii*, 25 nuclear mutations have been mapped into 13 complementation groups. None of these mutations has been mapped to any of the known structural genes (40). The higher plant PS I mutations are generally seedling lethal (41). Even though a significant number of important contributions could be made in the study of PS I utilizing these mutants, this approach has its limitation because the determination of the site of the mutation is complicated by pleiotropic effects observed in the mutant.

During the last few years, with the development of molecular genetic techniques, the creation of mutations at any specific site in any particular gene has become feasible. Two different species of cyanobacteria can now be used to create specific mutations in PS II polypeptides: *Synechocystis* 6803 and *Synechococcus* sp. PCC 7002 (*Agmenellum quadruplicatum*). Both these organisms are photoheterotrophic and therefore they can be propagated in the absence of PS II activity when the appropriate carbon source is provided. These two organisms are naturally transformable with foreign DNA which can be incorporated into cyanobacterial genome by homologous recombination (2).

Three different specific mutagenic techniques are now available and all of them have been successfully applied in the analysis of PS II (2). In the first step of molecular genetic analysis, gene insertion/deletion method is followed. In this technique, a particular gene is selectively inactivated either by substituting the gene of interest with a DNA fragment containing a specific antibiotic resistance marker or inserting such a DNA fragment within the coding region. Both procedures allow for simple selection of the mutant due to acquisition of a gene coding for a protein that metabolizes a certain antibiotic that the organism is normally sensitive to. In the second stage of genetic manipulation, the homologous gene from another organism is substituted for the native gene in the organism of study. In the third level of mutagenesis, a single base in the gene could be mutated leading to a change in a particular amino acid. All of these three different mutagenic techniques have been effectively utilized in the molecular genetic analysis of PS II polypeptides (2).

As mentioned at the beginning, the molecular genetic analysis of PS I is at its early stages and currently the efforts are directed towards the creation of insertion/deletion mutations in PS I polypeptides. Chitnis *et al* ., have recently reported the insertional inactivation of *psa*D (42) and *psa*E (43) genes in *Synechocystis* 6803. The *psa*E mutant grows photoautotrophically at rates which are virtually identical to those of the wild type cells indicating that the organism can do away with the PSI-E polypeptide. *psa*D mutant grows very slowly under photoautotrophic conditions, but grows at rates equivalent to that of wildtype cells under photoheterotrophic condition in the presence of 5mM glucose. This mutant contains a PS I complex consisting of the PSI-A, PSI-B, PSI-C and PSI-F polypeptides. The abnormal PSI complex is unable to photoreduce ferredoxin but is, however, still be able to carry out electron transport from plastocyanin to methylviologen. These results are consistent with the conclusion that the PSI-D polypeptide functions as a docking protein for ferredoxin.

So far there is no information about the creation of mutant in *Synechocystis* 6803 which is totally devoid of PS I activity. In other words, there is no mutant yet available for the *psa*A, *psa*B and *psa* C genes in *Synechocystis* 6803. It is generally believed that during photoheterotrophic growth of *Synechocystis* 6803, PS I may be required to generate ATP through cyclic electron transport and therefore it may be difficult to obtain mutants in this organism for the *psa*A, *psa*B and *psa*C genes . Keeping this in mind, we have started developing another system for the molecular genetic analysis of PS I. Our system is based on the filamentous çyanobacterium *Anabaena variabilis* ATCC 29413. This organism is capable of chemotrophic growth and it grows well in complete darkness in the presence of 5 mM fructose (44). DNA can be introduced into this organism through conjugal transfer of plasmid DNA from *E. coli* (45). Once inside the organism, the introduced DNA get integrated into the chromosome either by single or double recombination (46). Double recombination leads to the replacement of wild type gene in the chromosome by the mutated copy on the plasmid. Elhai and Wolk (47) have developed a plasmid vector (pRL 277) for the introduction of DNA into *Anabaena* by conjugation and subsequent selection for double recombinants. In our laboratory, we are making use of this plasmid vector for the specific inactivation of *psa*A, *psa*B and *psa*C genes in the chromosomal DNA of *Anabaena* 29413 with the goal of creating mutations in these genes. As a way of describing our system, we will now provide an account of our experimental approach in the creation of a *psa*C mutant through the insertional inactivation of this gene .

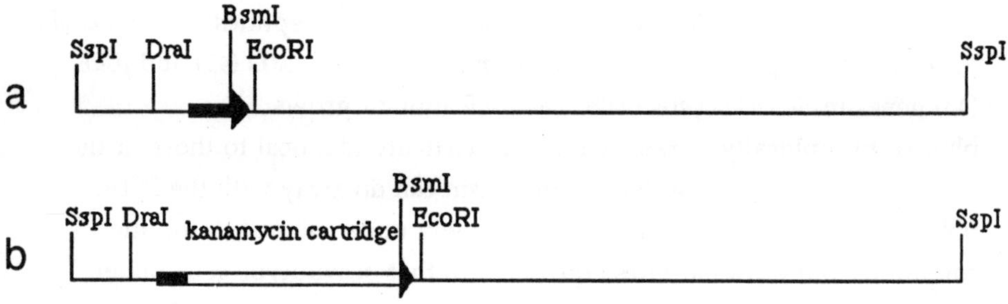

Fig. 3. (a) Restriction map of chromosomal DNA of *Anabaena* 29413 in the region flanking the *psa*C gene. (b)Insertion of a kanamycin resistance conferring cassette in the middle of the *psa*C gene at a Bsm I site.

In the first step, a 3.5 kbp SspI fragment containing the complete *psa*C and its flanking regions was cloned from a lambda EMBL3 library of *Anabaena* 29413 genomic library (48). In the second step, a kanamycin resistance cartridge was introduced in the middle of the gene at the Bsm I site (Fig. 3). This kanamycin cartridge is composed of the kanamycin resistence coding sequence from transposon Tn5 behind the promoter sequence of the *psb*A gene from the chloroplast genome of *Amaranthus hybridus* (49). In the third step, the construction made in the second step is cloned in the polylinker region of the transfer plasmid pRL277(47), resulting in the creation of pSL398 (Fig.4). pRL277 contains the streptomycin/

spectinomycin resistance coding sequence and a stretch of DNA that functions as the origin of conjugal transfer (*Bom* or *OriT*). In the fourth step, pSL398 created in the third step is transferred to a *E. coli* strain harboring another helper plasmid pRL528. pRL528 encodes two methylases that modify the recognition site of the restriction endonucleases Ava I and Ava II in *Anabaena* cells (45). It also carries the *mob* gene that encodes a DNA-nicking protein that recognizes the *bom* site on the conjugation vector (pSL398). In other words, the DNA-nicking protein essential for the mobilization of pSL398 into *Anabaena* cells during conjugation is provided in trans by pRL528.

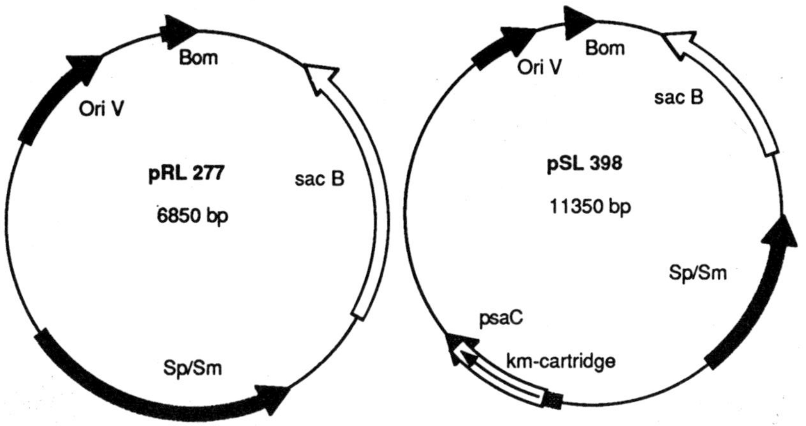

Fig. 4. Structure of the conjugal vector pRL277 and the transfer plasmid pSL398

The conjugation involves triparental mating between two *E. coli* strains and *Anabaena* cells. The first *E. coli* strain contains the two different plasmids, pRL528 and pSL398. The second strain of *E.coli* contains the plasmid pRL443. In the first stage of conjugation, pRL443 transfers itself into the first *E. coli* strain and then promotes the transfer of pSL398 into *Anabaena* cells (Fig. 5). After mating, the cells are plated on medium without selective antibiotics for 48 hours to allow recombination and gene expression. Then the cells are transferred to plates with antibiotics for selection of exconjugants and with fructose for the maintenance of photosynthetically deficient cells. Approximately one week is required before single antibiotic resistant colonies are visible. The colonies are then grown on liquid medium supplemented with kanamycin and fructose, sonicated to fragment the filament into single cells and replated. Several rounds of this purification are required because multiple copies of the chromosome are present in cyanobacterial cells. If there is a single recombination, the entire pSL398 would have got incorporated into the chromosome and the resulting exconjugant would be resistant to both streptomycin / spectinomycin and kanamycin. On the other hand, if there is a double recombination event , the *Anabaena* sequence present in pSL398 along with the interposed kanamycin cartridge would replace the homologous sequence in the chromosome. This double recombinants would be resistant to only kanamycin but sensitive to streptomycin / spectinomycin. In the double recombinants, the wild type *psa*C gene would have been replaced by the mutated copy of the *psa*C gene. In the same way efforts are being made to create *psa*A and *psa*B mutants.

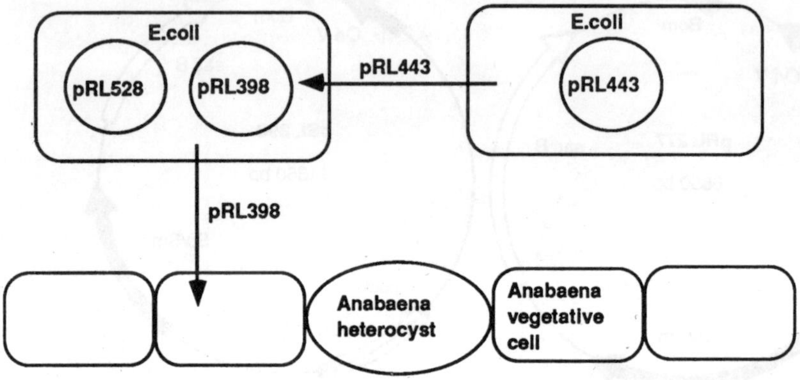

Fig. 5. Triparental mating for the transfer of DNA from *E. coli* to *Anabaena* 29413.

In the next step of our investigation we would like to make use of a shuttle vector which could replicate both in *E. coli* and *Anabaena* to complement the mutations created as described above. This would allow us to carry out all the manipulations in the gene of our interest in *E.coli* and introduce the manipulated copy of this gene into *Anabaena* 29413 to study the specific functions of the gene product.

ACKNOWLEDGEMENTS

This work was supported by funds from U.S. National Institutes of Health, National Science Foundation and Lucille Markey Charitable Trust to H.B.P. S.B. was supported by a short term Senior Research Fellowship from the Department of Biotechnology of Govt. of India.

REFERENCES

1. Deisenhofer, J., Epp, O., Miki, K., Huber, R. and Michel, H. (1985) Structure of protein subunits of the photosynthetic reaction center of *Rhodopseudomonas viridis* at 3 Å resolution. *Nature* 318 : 618-624.

2. Carpenter, S.D. and Vermaas, W.F.J. (1989) Site-directed mutagenesis to probe the structure and function of photosystem II. *Physiol. Plant.* 77: 436-443.

3. Nitschke, W., Feiler, U. and Rutherford, A.W. (1990) Photosynthetic reaction center of green sulfur bacteria studied by EPR. *Biochemistry* 29: 3834-3842.

4. Evans, M.C.W. and Bredenkamp, G. (1990) The structure and function of the Photosystem I reaction center. *Physiol. Plant.* 79: 415-420.

5. Scheller, H.V. and Mϕller, B.L. (1990) Photosystem I polypeptides. *Physiol. Plant.* 78: 484-494.

6. Margulis, M.M. (1989) Photosystem I core. *Plant Science* . 64: 1-13.

7. Lagoutte, B. and Mathis, P. (1989) The photosystem I reaction center: Structure and photochemistry. *Photochem. Photobio.* 49: 833-844.

8. Golbeck, J.H.(1987) Structure, function and organization of the photosystem I reaction center complex. *Biochim. Biophys. Acta* 895: 167-204.

9. Kok, B. (1961) Partial purification and determination of oxidation-reduction potential of the photosynthetic chlorophyll complex absorbing at 700 nm. *Biochim. Biophys. Acta* 48: 527 - 533.

10. Gast, P., Swarthoff, T., Ebskamp, F.C.R. and Hoff, A.J. (1983) Evidence for a new early acceptor in photosystem I of plants. An ESR

investigation of reaction center triplet yield and of the reduced intermediate acceptors. *Biochim. Biophys. Acta* 722:163-175.

11. Bonnerjea, J. and Evans, M.C.W. (1982) Identification of multiple components in the intermediary electron carrier complex of photosystem I. *FEBS Lett* . 148: 313-316.

12. Mansfield, R.W. and Evans, M.C.W. (1986) UV optical difference Spectrum associated with the reduction of electron acceptor A_1 in photosystem I in higher plants. *FEBS Lett*. 203: 225-229.

13. Itoh, S., Iwaki, M. and Ikegami, I. (1987) Extraction of vitamin K-1 from photosystem I particles by treatment with diethyl ether and its effect on the A_1-EPR signal and system I photochemistry. *Biochim. Biophys. Acta* 893: 508-516.

14. Biggins, J. and Mathis, P. (1988) The functional role of vitamin K-1 in the photosystem I reaction center of the cyanobacterium *Synechocystis* 6803. *Biochemistry* 27: 1494-1500.

15. Barry, B.A., Bender, C.J., McIntosh, L., Ferguson-Miller, S. and Babcock, G.T.(1988) Photoaccumulation in Photosystem I does not produce a phyloquinone radical. *Israel J. Chem.* 28:129-132.

16. Setif, P., Ikegami, I. and Biggins, J. (1987) Light-induced charge separation in photosystem I at low temperature is not influenced by vitamin K-1. *Biochem. Biophys. Acta* . 894:146- 156.

17. Scheller, H.V. and Moller, B.L. (1989) Subunit composition of photosystem I and identification of center X as [4Fe-4S] iron-sulfur cluster. *J. Biol. Chem.* 264: 6929-6934.

18. Ke, B. (1972) One-way electron discharge subsequent to the photochemical charge separation in photosystem I. *Biochim. Biophys. Acta* 267: 595-599.

19. Bearden, A.J. and Malkin, R. (1972) Quantitative EPR studies of the primary reaction of photosystem I in chloroplasts. *Biochim. Biophys. Acta* 283: 456-468.

20. Haworth, P., Watson, J.L. and Arntzen, C.J. (1983) The detection, isolation and characterization of a light-harvesting complex which is specifically associated with photosystem I. *Biochim. Biophys. Acta.* 724: 151-158.

21. Lam, E., Ortiz, W.,Mayfield, S. and Malkin, R. (1984) isolation and characterization of light harvesting chlorophyll a/b protein complex associated with PS I. *Plant Physiol.* 74: 650- 655.

22. Dunn, P.P.J., Packman, L.C., Pappin, D. and Gray, J.C. (1988) N-terminal amino acid sequence analysis of the subunits of pea photosystem I. *FEBS Lett.* 228: 157-161.

23. Ikeuchi, M., Hirano, A., Hiyama, T. and Inoue, Y. (1990) Polypeptide composition of higher plant photosystem I complex. Identification of *psa*I, *psa*J and *psa*K gene products. *FEBS Lett.* 263: 274-278.

24. Koike, H., Ikeuchi, M., Hiyama, T. and Inoue, Y.(1989) Identification of photosystem I components from the cyanobacterium, *Synechococcus vulcanus* by N-terminal sequencing. *FEBS Lett.* 253: 257-263.

25. Schantz, R. and Bogorad, L. (1988) Maize chloroplast genes *ndh*D, *ndh*E and *psa*C. Sequences,transcripts and transcript pools. *Plant Mol. Biol.* 11: 239-247.

26. Cantrall, A. and Bryant, D.A. (1987) Molecular cloning and nucleotide sequence of the *psa*A and *psa*B genes of the cyanobacterium *Synechococcus* sp. PCC 7002. *Plant Mol. Biol.* 9: 453-468.

27. Dunn, P.P.J. and Gray, J.C. (1988) Localization and nucleotide sequence of the gene for the 8 k Da subunit of photosystem I in pea and wheat chloroplast. *Plant Mol. Biol.* 11: 311-319.

28. Lagoutte, B. (1988) Cloning and sequencing of spinach cDNA clones encoding the 20 k Da PS I polypeptides. *FEBS Lett.* 232: 275-280.

29. Munch, S., Ljungberg, U., Steppuhn, J., Schneiderbauer, A., Nechushtai, R., Beyreuther, K. and Herrmann, R.G. (1988) Nucleotide sequences of cDNAs encoding the entire precursor polypeptides for subunits II and III of the photosystem I reaction center from spinach. *Curr.Genet*. 14: 511-518.

30. Reilly, P., Hulmes, J.D., Pan, Y.-C.E. and Nelson, N. (1988) Molecular cloning and sequencing of the *psa*D gene encoding subunit II of photosystem I of cyanobacterium *Synechocystis* sp. PCC 6803. *J. Biol. Chem*. 265: 17658-17662.

31. Ortiz, W., Lam, E., Chollar, S., Munt, D. and Malkin, R. (1985) Topography of the protein complexes of the chloroplast thylakoid membrane. Studies of photosystem I using a chemical probe and proteolytic digestion. *Plant Physiol*. 77: 389-397.

32. Zanetti, G. and Merati, G. (1987) Interaction between Photosystem I and ferredoxin: Identification by chemical cross-linking of the polypeptide which binds ferredoxin. *Eur. J. Biochem*. 169: 143-146.

33. Oh-oka, H., Takahashi, Y. and Matsubara, H. (1989) Topological considerations of the 9-kDa polypeptide which contains centers A and B, associated with the 14- and 19-kDa polypeptides in the photosystem I complex of spinach. *Plant Cell Physiol*. 30: 869-875.

34. Steppuhn, J., Hermans, J., Nechushtai, R., Ljungberg, U., Thummler, F., Lottspeich, F. and Herrmann, R.G. (1988) Nucleotide sequences of cDNA clones encoding the entire precursor polypeptides for subunits IV and V of photosystem I reaction center from spinach. *FEBS Lett*. 237: 218-224.

35. Steppuhn, J., Hermans, J., Nechushtai, R., Herrmann, G.S. and Herrmann, R.G. (1989) Nucleotide sequences of cDNA clones encoding the entire precursor polypeptide for subunit VI and of the plastome encoded gene for subunit VII of the photosystem I reaction center from spinach. *Curr. Genet*. 16: 99-108.

36. Okkels, J.S., Scheller, H.V., Jensen, B. and Moller, B.L. (1989) A cDNA clone encoding the precursor for a 10.2 kDa photosystem I polypeptide of barley. *FEBS Lett.* 250:575- 579.

37. Franzen, L.G., Frank, G., Zuber, H. and Rochaix, J.D. (1989) Isolation and characterization of cDNA clones encoding the 17.9 and 8.1 kDa subunits of photosystem I from *Chlamydomonas reinhardtii* . *Plant Mol. Biol.* 12: 463-474.

38. Scheller, H.V., Okkels, J.S., Hoj, P.B., Svendsen, I., Roepstorff, P. and Moller, B.L. (1989) The primary structure of a 4.0 kDa photosystem I polypeptide encoded by the chloroplast *psa*I gene. *J. Biol. Chem.* 264: 18402-18406.

39. Wynn, R.M. and Malkin, R. (1990) The photosystem I 5.5 kDa subunit (the *psa*K gene product). An intrinsic subunit of the PS I reaction center complex. *FEBS Lett.* 262: 45-48.

40. Girard, J., Chua, N.H., Bennoun, P., Schmidt, G. and Delosme, M. (1980) Studies on mutants deficient in the photosystem I reaction centers in *Chlamydomonas reinhardtii. Curr. Genet.* 2: 215-221.

41. Cook, W.B. and Miles, D. (1990) Anamalous electron transport activity in a photosystem I-deficient maize mutant. *Photosyn. Res.* 24: 81-88.

42. Chitnis, P.R., Reilly, P.A. and Nelson, N. (1989) Insertional inactivation of the gene encoding subunit II of photosystem I from the cyanobacterium *Synechocystis* sp. PCC 6803. *J. Biol. Chem.* 264: 18381-18385.

43. Chitnis, P.R., Reilly, P.A., Miedel, M.C. and Nelson, N. (1989) Structure and targeted mutagenesis of the gene encoding 8-kda subunit of photosystem I from cyanobacterium *Synechocystis* sp. PCC 6803. *J. Biol. Chem.* 264: 18374-18380.

44. Haury, J.F. and Spiller, H. (1981) Fructose uptake and influence on growth and nitrogen fixation by *Anabaena variabilis* . *J. Bacteriol.* 147: 227-235

45. Maldener, I., Lockau, W., Cai, Y. and Wolk, C.P. (1990) The calcium-dependent protease of the cyanobacterium *Anabaena* : Molecular cloning and expression of the gene in *Escherichia coli* and site-directed mutagenesis. *Mol. Gen. Genet.* (in press).

46. Elhai, J. and Wolk, C.P. (1988) Conjugal transfer of DNA to cyanobacteria. *Meth. Enzymol.* 167: 747-754.

47. Elhai, J., Thiel, T. and Pakrasi, H.B. (1990) DNA transfer into Cyanobacteria. In: *Plant Molecular Biology Manual*, Gelvin, S and Schilperroort, R.A. eds., Martinus Nijhoff, The Netherlands.

48. Mannan, R.M. and Pakrasi, H.B. (1991) Molecular analysis of the *psa*C gene encoding the F_A /F_B apoprotein of photosystem I from the filamentous cyanobacterium *Anabaena variabilis* ATCC 29413. *Plant Mol. Biol.* (Submitted)

49. Elhai, J. and Wolk, C.P. (1988) A versatile class of positive selection vectors based on the nonviability of palindrome-containing plasmids that allows cloning into long polylinkers. *Gene* 68: 119-138.

THE REGULATION OF PHOTOSYNTHETIC ELECTRON TRANSFER AND ITS POTENTIAL IMPACT ON PHOTOSYNTHETIC PERFORMANCE AT ELEVATED ATMOSPHERIC CO_2

Donald R. Ort *Photosynthesis Research Unit, USDA, Agricultural Research Service,*

& Department of Plant Biology , University of Illinois, Urbana, IL 61801 USA

The intent of this brief discussion is to consider how photosynthetic electron transfer is regulated and, in turn, how it may be expected to impact on the rate and efficiency of net photosynthesis under agricultural or natural habitats as atmospheric CO_2 levels increase. My focus will be on outlining the concepts and general mechanisms that govern photosynthetic electron transfer and this will unavoidably be at the expense of tracing the origin of ideas and development of the crucial supporting data. Traditionally, consideration of the control of photosynthetic electron transfer in higher plant chloroplasts has been dominated by discussions concerning the oxidation of plastoquinol by the cytochrome b_6f complex. This quinol-metalloprotein oxidoreductase is structurally and functionally analogous to complexes found in the photosynthetic and respiratory membranes from both prokaryotes and eukaryotes that oxidize a low potential quinol and reduce a high potential metalloprotein (1). In higher plants this complex oxidizes plastoquinol generated by photosystem II and reduces plastocyanin oxidized by photosystem I. The mechanism of both plastoquinone reduction and plastoquinol oxidation, in which intermediate reductive states are maintained bound to specific catalytic sites, ensures absolute asymmetry in proton uptake in keeping with the role of transmembrane proton asymmetries in energy coupling (2). This role that plastoquinol oxidation plays in energy coupling is especially important in considering the special role that it also plays in controlling the rate and efficiency of electron transfer.

The basis for underlying mechanism for this control is that plastoquinol oxidation by the cytochrome b_6f complex is the only appreciably reversible step in the sequence of photosynthetic electron transfer reactions.

$$PQH_2 \Leftrightarrow PQ + 2H^+ + 2e^-$$

Thus, as hydrogen ions are accumulated within the thylakoid vesicle as a result of light-driven electron transfer, the equilibrium increasingly favors the reverse reaction, that is, the formation of plastoquinol (PQH_2) by the reduction of plastoquinone (PQ) on the cytochrome b_6f complex. Thus, as the electrochemical potential of accumulated protons becomes large, the rate of net oxidation of plastoquinol by the cytochrome b_6f complex and therefore the rate of electron transfer becomes slower and, on an absorbed photon basis, less efficient. On thermodynamic grounds, the rate of the net reaction would be zero should the electrochemical potential of accumulated protons become equivalent to the redox free energy available in the electron transport couple involved with the proton uptake. In this case the net reaction would be zero if the electrochemical potential was equal to the free energy difference between PQ/PQH_2 and the one electron redox components of the cytochrome b_6f complex.

Up to this point only the role of the chemical potential of the accumulated protons has been considered in the thermodynamic control of photosynthetic electron transfer. However, in order to be complete, mention should be made of the control that can be exerted by the transmembrane electric field on the net rate of plastoquinol oxidation. Above, the reaction resulting in the oxidation of plastoquinol was viewed as an electrically neutral process in which no separation of charge across the thylakoid membrane took place. If this were actually the case, a transmembrane electrical field would not be expected to influence the rate of plastoquinol oxidation. However, the existence of an electrogenic reaction associated with the oxidation of plastoquinol by the cytochrome b_6f complex is now experimentally well established (1). A variety of models have been proposed to account for this electrogenic event and, while the alternative models differ in fundamental aspects of their mechanism, they share both the general name Q-cycle and the feature that one of the electrons released from plastoquinol during its oxidation is shuttled back across the membrane establishing a separation of charges. As predicted by the existence of an electrogenic reaction in the rate-limiting step of the chloroplast electron transfer sequence, control of the rate of electron transfer by the electrical component of the accumulated protons has been clearly demonstrated (3).

My purpose in reviewing what might be termed the thermodynamic control of photosynthetic electron transfer is to illustrate that this occurs as an inevitable energetic consequence of plastoquinol oxidation by the cytochrome b_6f complex. As we have seen the extent of this control is proportional to size of the transmembrane electrochemical potential. During continuous illumination the ΔpH component dominates the thermodynamic control. Recent evidence suggests that *in situ* the size of the ΔpH is invariant over a wide range of net photosynthesis rates (4). As such it is unlikely that control of electron transfer at the level of plastoquinol oxidation will play any significant role in determining how net photosynthesis responds to elevated CO_2 levels.

Only a few years ago this discussion would now be ended, however, as a result of the increasingly more sophisticated investigations that have been done in recent years on intact plants and on plants in natural habitats, evidence for a previously unrecognized regulation of light energy utilization has emerged. Recent data from several laboratories provide strong experimental evidence for the existence of an intricate mechanism(s) able to 'down regulate' photosynthetic efficiency as illumination levels exceed photosynthetic capacity (e.g. 5,6,7). While we were by no means the first to begin thinking along these lines, I would like to illustrate the phenomenon of 'down regulation' of photosynthetic efficiency with recent results that we obtained studying the response of photosynthesis to the daily cycling of leaf water potential in field-grown sunflower (*Helianthus annuus*).

On clear days mid-afternoon decreases in leaf water potential of about 15 bars and in net photosynthesis of up to 50% were typical for irrigated sunflower during seed filling. For instance, Figure 1 shows the typical diurnal pattern of photosynthesis that we observed. The highest rate of photosynthesis for the day was reached in the mid-morning before irradiance had reached half the maximal level for the day. Thereafter the rate of net photosynthesis declined even though irradiance increased during mid-morning and into the early afternoon. At 9 h the irradiance was near 900 mmol quanta$\cdot m^{-2} \cdot s^{-1}$ (Fig 1) and the leaf water potential was about -6 bars. By late afternoon (17 h), irradiance was once again near 900 mmol quanta $\cdot m^{-2} \cdot s^{-1}$ but the leaf water potential had declined to -15 bars and the rate of photosynthesis was less than 70% of the 9 h value. During the decline of net photosynthesis over the course of a cloudless day (Fig 1), a nearly linear relationship was maintained between stomatal conductance and photosynthesis (8). This relationship between stomatal conductance and the rate of net photosynthesis might seem consistent with strong stomatal control exerted on photosynthesis thereby accounting for declining assimilation rates as leaf water potential dropped throughout the day. Since direct control by stomatal aperture over the rate of photosynthesis is exerted through the control that stomata have on intercellular CO_2 concentration (C_i), lower C_i values are anticipated in the afternoon when the stomatal conductance was lowest. However, Figure 2 shows that the relationship between photosynthesis and C_i over the course of the day showed an unexpected inverse behavior. At 9 h when leaf conductance and photosynthesis were at their highest level of the day, the concentration of CO_2 within the leaf was at its lowest value. As the leaf water potential declined and the stomata began to close, C_i increased indicating that CO_2 utilization by the chloroplast, rather than the availability of CO_2, was limiting the rate of net photosynthesis .

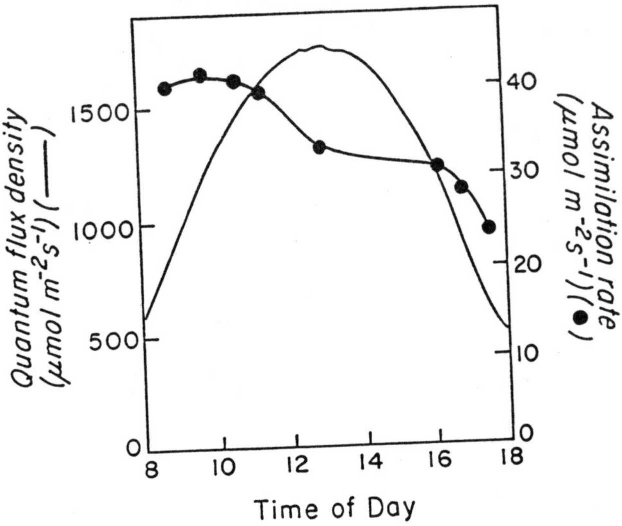

Figure 1. Profile of irradiance and net photosynthesis of attached sunflower leaves as a function of the time of day. These data are typical of those observed for irrigated sunflower on sunny days during the seed-filling stage. This figure is modified from Wise et al. (8).

To begin to understand the basis for the association between declining rates of photosynthesis and increasing C_i values observed over the course of the day in field-grown sunflower, we looked at what was happening to the intrinsic efficiency of photosynthesis by measuring the light-limited quantum yield of CO_2 assimilation. Figure 3 shows a declining quantum efficiency that qualitatively mimics the decline in the rate of net photosynthesis over the course of the day. In this case we cannot say that these are typical data as several replicates of this experiment gave somewhat different results. It is likely these differences reflect procedural aspects of the measurement because, as others have reported (*e.g.* 4), this depression of the quantum yield induced by high light reverses quite rapidly at the low light intensities used to make quantum yield determinations. Furthermore the half time of this reversal, which may be as short as several minutes, is influenced to an unknown degree by factors such as the intensity and duration of light (including therefore time of day) and possibly by leaf water status and conductance. In fact the rapidly reversing character of this effect of high light intensity on the quantum efficiency of CO_2 assimilation is a very pertinent point to this discussion. Seemingly comparable experiments in which growth chamber-grown plants are exposed to higher than growth chamber light intensities are often observed to respond with marked depression in quantum yield that can be traced to a light-dependent damage to photosystem II. Indeed, light-dependent damage to the photosynthetic apparatus is the best understood mechanism of photoinhibition. While the light-dependent decline in photosynthetic efficiency seen in Figure 3 might also be termed photoinhibition, its rapid reversal suggests, and a broad range of experiments done in other laboratories demonstrates (4), that this inhibition does not involve damage but is instead regulatory in nature.

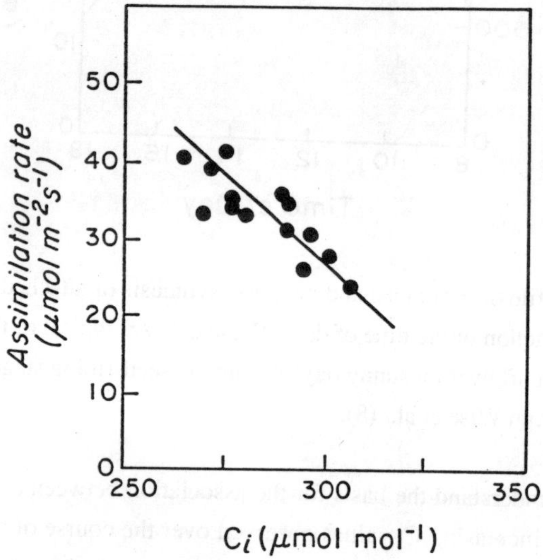

Figure 2. The relationship between photosynthesis and the intercellular CO_2 concentration. These data were obtained by measuring the rate of net photosynthesis and C_i at various times between 9 and 18 h on cloudless days. This figure is modified from Wise et al. (8).

Currently there is a good deal of activity and a similar level of controversy concerning the mechanism of this 'down regulatory' effect on photosynthetic efficiency. Most of the ideas presently under discussion involve radiationless deexcitation of antennae chlorophyll as a central feature of the mechanism. That is, fewer of the photons that are absorbed into the antennae chlorophyll arrays are transferred to the reaction centers where the photochemistry that leads to electron transfer occurs. It is postulated that large transmembrane pH differences that are associated with high irradiance levels (which would be particularly large if the rate of

ATP utilization in carbon dioxide assimilation were less than the capacity of the thylakoid membranes to generate ATP) induces a special, but wholly undefined, state in the antennae complexes with an enhanced probability of radiationless deactivation competing with excitation energy transfer to the reaction center.

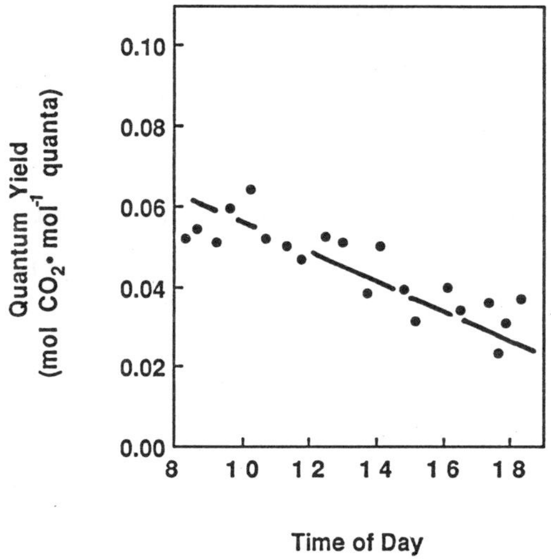

Figure 3. The light-limited quantum yield of photosynthesis measured on leaf discs punched from field-grown sunflower at different times of day. These are unpublished data of Dr. Robert Wise and Denise Sparrow.

Although at this time vanishing little is known about the mechanistic basis, it seems quite clear that chloroplasts the have capability to 'down-regulate' the rate and efficiency of net photosynthesis. It seems equally clear that significantly lower rates and efficiencies associated with 'down regulation' are commonplace under agricultural conditions (8 and references therein). More difficult is the question of how this 'down regulatory' mechanism would effect the response of net photosynthesis to elevated atmospheric CO_2. At the present state of knowledge we are convinced by our intuition that 'down regulation' is an important protective mechanism when the photon flux exceeds photosynthetic capacity as it may have in our field sunflowers in the afternoon when irradiances were high and stomata were partially closed to avoid further water loss. If this general line of thinking is correct then it may be that fewer photons will be diverted under away from photochemistry at elevated CO_2 under otherwise comparable conditions because of an improved water use efficiency. Nevertheless, my predilection is that sustaining the higher net rates of photosynthesis that are often transiently seen when CO_2 is increased will, in agricultural and natural habitats, be more profoundly influenced by mineral and water resources and by the influence that a high CO_2 environment has chloroplast development than by small gains in electron transport capacity that may be possible within known regulatory mechanisms.

REFERENCES

1. Cramer, W.A., Black, M.T., Widger, W.R. and Girvin, M.E. (1987) Comparative structure and function of the b cytochromes of the b-c_1 and b_6-f complexes. In: *The Light Reactions* , Barber, J. ed., Elsevier Sci Publ, Amsterdam, pp. 446-493.
2. Ort, D.R. (1986) Energy transduction in oxygenic photosynthesis: An overview of structure and mechanism. In: *Photosynthetic Membranes and Light Harvesting Systems*, Staehelin, L.A. and Arntzen, C.J. eds, , Springer-Verlag, New York, pp. 143-196.
3. Graan, T. and Ort, D.R. (1983) Initial events in the regulation of electron transfer in chloroplasts: The role of the membrane potential. *J. Biol. Chem.* 258: 2831-2836.
4. Foyer, C., Furbank, R., Harbinson, J. and Horton, P. (1990) The mechanisms contributing to photosynthetic control of electron transport by carbon assimilation in leaves. *Photosyn. Res.* 25: 83-100.
5. Demmig, B., Winter, K., Kruger, A. and Czygan, F-C. (1987) Photoinhibition and zeaxanthin formation in intact leaves. *Plant Physiol.* 84: 218-224.
6. Harbinson, J., Genty, B. and Baker, N.R. (1989) Relationship between the quantum efficiencies of photosystems I and II in pea leaves. *Plant Physiol.* 90: 1029-1034.
7. Weis, E. and Berry, J. (1987) Quantum efficiency of photosystem II in relation to 'energy'-dependent quenching of chlorophyll fluorescence. *Biochim. Biophys. Acta* 894: 198-208.
8. Wise, R.R., Frederick, J.R., Alm, D.M., Kramer, D.M., Hesketh, J.D., Crofts, A.R.and Ort, D.R. (1990) Investigation of the limitations to photosynthesis induced by leaf water deficit in field-grown sunflower (*Helianthus annuus* L.) *Plant, Cell & Environ* . in press

ROLE OF PHOTOSYNTHESIS AND PHOTORESPIRATION IN REGULATING ATMOSPHERIC CO_2 AND O_2 CONCENTRATION

N.E. Tolbert, Professor of Biochemistry
Department of Biochemistry
Michigan State University
East Lansing, MI 48824, USA

ABSTRACT

Because Rubisco activity determines the rate of CO_2 fixation by the C_3 reductive photosynthetic carbon cycle and the amount of CO_2 and O_2 exchange by the C_2 oxidative photosynthetic carbon cycle, photosynthetic carbon metabolism regulates CO_2 and O_2 compensation points. These levels are directly involved in establishing the atmospheric CO_2 and O_2 equilibrium levels. The C_2 cycle wastes excess photosynthetic energy. CO_2 equilibria can fluctuate between the CO_2 compensation point (60 ppm CO_2) and 1000 ppm CO_2 with present environmental conditions and plants. Concentrating mechanism utilizes part of the excess energy wasted by photorespiration to concentrate CO_2 in the chloroplasts by the C_4 cycle or by algal mechanisms for concentrating dissolved inorganic carbon (DIC pumps). Algal DIC pump activity is accompanied by accumulation of hydroxide ions in the medium, which in turn can be used for lowering atmospheric CO_2 and forming carbonate deposits. More C_4 plants is not an apparent solution to lower the atmospheric CO_2, because their photosynthetic CO_2 efficiency decreases rapidly above 350 ppm CO_2. More research is needed on photosynthesis by algae which form long-life deposits of reduced carbon and carbonates.

PHOTOSYNTHETIC CARBON METABOLISM

There are two large subject areas about the atmospheric CO_2 which need to be discussed with the same biochemical data on photosynthetic carbon metabolism. One is how the increasing CO_2 will effect the growth of plants and alter their distribution and composition (1). The other problem is how plants in turn regulate the atmospheric CO_2 and O_2. I shall only discuss the latter in this paper. From analyses of past atmospheric gases in ice cores going back 150,000 years, we know that the atmospheric CO_2 level had equilibrated between 180 ppm CO_2 in glacial periods to 280 ppm CO_2 during interglacial periods (2). How were these low CO_2 levels maintained? Now in a geological and ecological very brief period of only 100 years the atmospheric CO_2 has increased to 360 ppm. The photosynthetic CO_2 balance has been overwhelmed by conversion of much fossil photosynthate to CO_2.

In plants and algae the bifunctional riboluses bisphosphate carboxylase/oxygenase (Rubisco) initiates the C_3 reductive photosynthetic carbon cycle for CO_2 fixation and the C_2 oxidative photosynthetic carbon cycle for photorespiration (Figure 1) (3-8). The two cycles co-exist as competitive reactions for the two substrates CO_2 and O_2. Together they are photosynthetic carbon metabolism and together they determine the rate of net photosynthesis. Based on the rate of CO_2 exchange, the C_3 cycle fixes 2 or 3 times as much CO_2 as the C_2 cycle releases. Increasing CO_2 or lowering O_2 increases net photosynthesis, and lowering CO_2 or raising

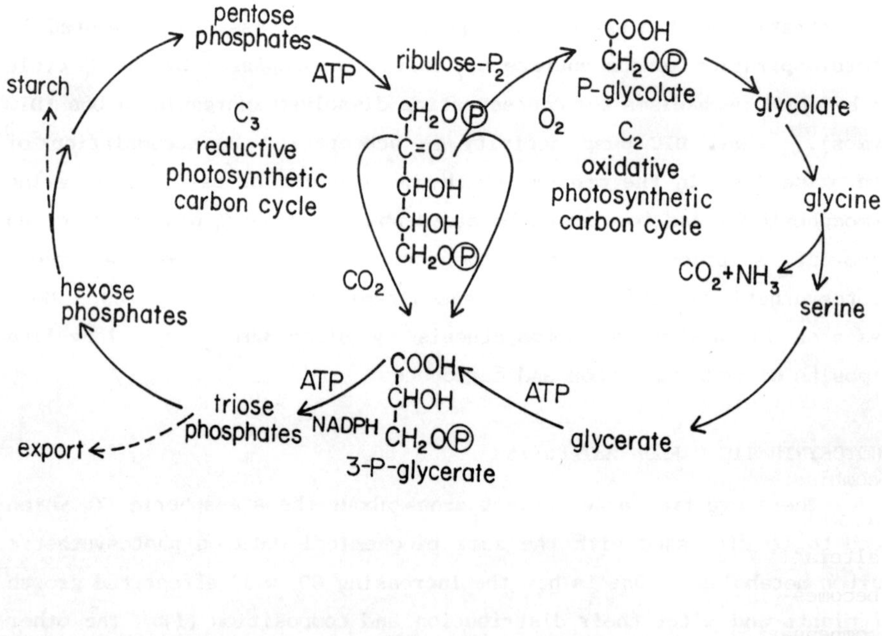

Figure 1. The C_3 Reductive And C_2 Oxidative Photosynthetic Carbon Cycles Initiated By Ribulose Bisphosphate Carboxylase/Oxygenase.

O_2 decreases net photosynthesis. The CO_2 compensation point is when net CO_2 exchange from both cycles is equal. Then net photosynthesis is zero and no growth occurs. For many C_3 crop plants in sun light the CO_2 compensation point is between 45 and 65 ppm CO_2 in 21% O_2 at 20°C, rising to over 100 ppm CO_2 at 30°C and to near atmospheric CO_2 levels around 37°C. One reason for this is that O_2 is relatively more soluble in water with increasing temperature than CO_2. The photosynthetic carbon system sets lower limits on CO_2 removal from air, and acts as a governing and limiting component of the global CO_2 equilibrium.

Reactions of the C_3 reductive cycle for fixing CO_2 are well known, and reactions of the C_2 oxidative cycle consists of the glycolate pathway for CO_2 evolution plus the reactions of the C_3 cycle for regeneration of ribulose bisphosphate (3-8). Reduction of 6 CO_2 to one hexose-P by the C_3 cycle uses 12 NADPH and 18 ATP of photosynthetic assimilatory capacity. Adding up all the energy utilizing steps during the release and refixation of 6 CO_2 by the C_2 cycle shows a burn up 36 NADPH and 60 ATP (5). This is approximately 3 times as much photosynthetic energy consumption per 6 CO_2 exchange as used to reduce 6 CO_2 to a carbohydrate. However CO_2 fixation rates exceeds photorespiratory CO_2 release by about 3 fold. Consequently energy waste by photorespiration has been about the same as energy used for CO_2 fixation during past environmental conditions.

When photorespiration wastes excess NADPH and ATP it is serving as a "Hill" process, an alternate photosynthetic electron acceptor rather than CO_2 reduction. Thus the C_2 cycle consumes excess photosynthetic energy when the CO_2 level is low. A consequence of photorespiration is to severely limit the rate and amount of CO_2 which plants can remove from the air. An increase in atmospheric CO_2 up to 1000 ppm CO_2 is necessary to decrease photorespiration by about 90% (9, more data is needed). The combination of the C_2 and C_3 cycle explains how plants can flourish over a wide range of CO_2 levels and light intensities without other alterations in photosynthesis. Above 1000 ppm CO_2 light intensity becomes limiting for C_3 plants. With increasing atmospheric CO_2 from the compensation point to 1000 ppm, the energy wasted by photorespiration at low CO_2 is used to reduce CO_2 at a faster rate. Thus photosynthesis is self regulated by its carbon metabolism.

DISSOLVED INORGANIC CARBON CONCENTRATING PROCESSES

Investigators of Rubisco seek to alter its specificity factor or the ratio of its carboxylase to oxygenase activity. Because Rubisco does not have a catalytic binding site for CO_2 or O_2 with a co-factor, alteration for a higher specificity factor has so far not been done by molecular methods. The Km(CO_2) for Rubisco is over 26 μM in air. In fact the lowest Rubisco specificity nature developed is in C_3 plants. Beyond that nature has used additional reactions to concentrate CO_2 in the chloroplast, in order to increase the carboxylase rate and decrease photorespiration.

There are two CO_2 concentrating processes: the C_4 cycle (10) and the algal dissolved inorganic carbon (DIC) concentrating mechnanisms (pumps) (11,12). Both are driven by extra photosynthetic energy, which would otherwise be lost by photorespiration. The phosphoenolpyruvate carboxylase of the C_4 cycle has a Km(HCO_3^-) of a few μM, and HCO_3^- is the main DIC substrate in the chloroplast. With a HCO_3^- binding site the enzyme has a rapid Vmax and it is not inhibited by O_2. Increasing atmospheric CO_2 around a plant is a better way to increase net photosynthesis as there is no energy drain for a CO_2 pump, but higher CO_2 levels will induce environmental changes.

LIMITATION TO C_4 PLANTS

Increasing atmospheric CO_2 has already reached an limiting ecological threshold for C_4 plants. In the past with only an average of 250 ppm CO_2, growth of C_4 plants was photosynthetically favored over C_3 plants, because the C_4 plant used some of its excess energy to concentrate the limiting supply of CO_2. Today the atmospheric CO_2 level at 360 ppm has reached a point above which the growth of C_4 plants becomes limited by light intensity and not by CO_2 concentration. At 350 to 400 ppm CO_2 the net rate of photosynthesis by the C_3 and C_4 plants are about equal. With further increase in CO_2 concentration up to 1000 ppm net photosynthesis by the C_3 plant continues to increase, but the C_4 plant, because it must continue to use energy on its C_4 cycle in the mesophyll cells, becomes less photosynthetically efficient than the C_3 plant. The C_4 plant ecologically was favored by low CO_2, but from now on with over 350 ppm CO_2 the C_4 plant is at a photosynthetic disadvantage.

Now we may need to look for C_3 varieties of C_4 plants, such as corn and sugar cane.

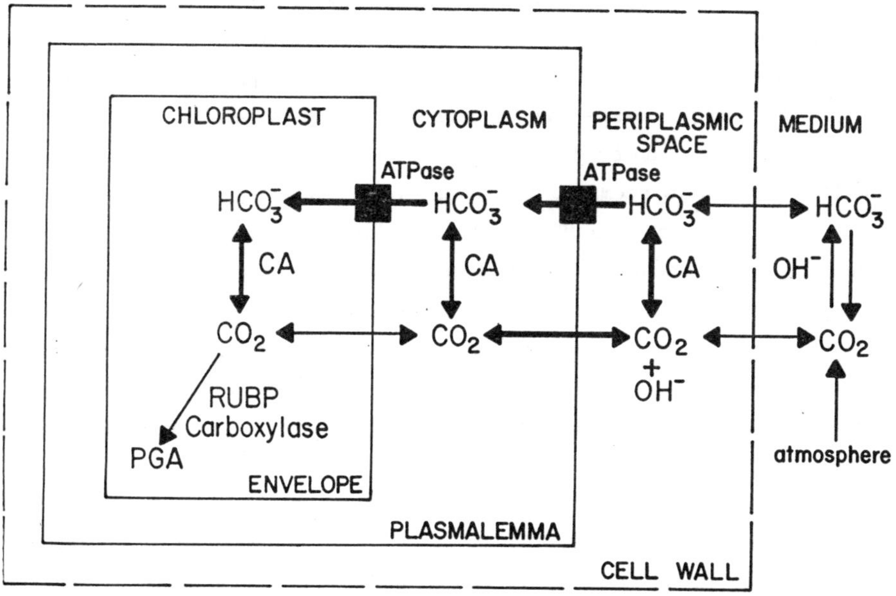

Figure 2. A working model for inorganic carbon accumulation by unicellular green algae. DIC may enter and accumulate in an algal cell by (a) passive CO_2 diffusion and conversion to HCO_3^- by pH gradients, (b) an active CO_2 pump involving at least an ATP driven transporter (adapted from 3) at the chloroplast envelope, and (c) an active HCO_3^- pump involving an additional ATPase exchange process at the plasmalemma.

COMPARISON OF THE CO_2 AND BICARBONATE CONCENTRATING PROCESSES IN UNICELLULAR GREEN ALGAE.

By the algal CO_2 pump, CO_2 is the DIC component which enters the cell and an external carbonic anhydrase (CA) is often present to facilitate rapid bicarbonate conversion to CO_2 (11,12). Most if not all algae from cyanobacteria to multicellular green algae form a CO_2 pump when grown on low CO_2. Some algae, which have a CO_2 pump, such as cyanobacteria or *Chlorella* 11h and *Chlamydomonas moeusii*, do not have measurable external CA, and it is not clear how the exterior HCO_3^- is used or converted to CO_2. The $K_{0.5}(CO_2)$ for O_2 evolution and DIC accumulation may be less than 1 μM, where as the $Km(CO_2)$ for Rubisco in these algae is from 26 to 60 μM CO_2. *Scenedesmus* cells have been used to study the HCO_3^- pump because when adapted to air levels of CO_2 for 4 to 6 hours in the light, the cells form two concentrating processes for dissolved inorganic carbon: one for utilizing CO_2 from medium of pH 5 to 8 and one for bicarbonate uptake and accumulation from medium of pH 7 to 11 (Figure 2) (13). The two processes can be differentiated by testing at different pH, by inhibitors and by differences in adaptation conditions to low CO_2. The CO_2 pump always seems to be present in air-adapted or air-grown green micro algae, where as a HCO_3^- pump in addition, as in *Scenedesmus*, is generally not present in fresh water unicellular algae that do not grow at a pH of over 8.5. Insufficient work has been done to date to know how prevalent is the HCO_3^- pump in marine algae. Salt tolerant *Dunaliella* (14) and the symbiotic *Zoxanthellae* (15, more data is needed) have CO_2 pumps, but HCO_3^- pumps for these marine algae have not been investigated extensively.

Isolated intact chloroplasts from air-adapted (but not high CO_2 grown) *Chlamydomonas* (16) or *Dunaliella* (14) can concentrate DIC in the light and this uptake is inhibited by vanadate or by inhibitors of photophosphorylation, so that during active CO_2 uptake, an ATPase, which uses photosynthetic energy, may be involved at the chloroplast envelope to concentrate DIC. External vanadate does not inhibit the CO_2 pump at the plasmalemma. It is doubtful that the chloroplast envelope ATPase is the only point of energy input for the CO_2 pump. Biochemical characterization of the steps and enzymes of the DIC pumps is an urgent research area.

For air-adapted *Scenedesmus*, the HCO_3^- concentrating process between pH 7 to 11 lowered the $K_{0.5}$ for O_2 evolution from 7300 μM HCO_3^- in high CO_2 grown *Scenedesmus* to 10 μM in air-adapted cells (13). To illustrate the efficiency of the HCO_3^- pump, a $Km(O_2)$ for photosynthesis with 10 μM HCO_3^- at pH 8.3 would be in the neighborhood of 0.1 μM CO_2. The HCO_3^- pump was inhibited by 150 μM vanadate in the medium, as if a perplasmic membrane ATPase linked HCO_3^- transporter were involved. Whereas, the CO_2 pump was formed on low CO_2 by high-CO_2 grown cells in growth medium within 4 to 6 hours in the light, the alkaline HCO_3^- pump was partially activated on low

CO_2 within 1 to 2 hours in the light or after 2-8 hours in the dark. A working hypothesis is that the HCO_3^- pump had one step (i.e. the plasmalemma HCO_3^- exchanger) that was activated in the dark on low CO_2, but that the maximum rate of HCO_3^- uptake was dependent on other steps, such as a chloroplast envelope ATPase that required light activation and low CO_2.

ALKALIZATION OF THE MEDIUM

During DIC accumulation, alkalization of the *Chara* cell surface has been extensively studied (17) and alkalization of the medium by micro blue-green and green algae has been recognized (12). Investigators of the DIC pump neutralize this phenomenon when they add mM amounts of buffer to their experimental algal medium. Because hydroxide production in the medium is in μmolar amounts during μmolar DIC removal, it is relatively low compared to mM buffer or even mM amounts of added $NaHCO_3$ which buffers the medium. Current investigations (Shiraiwa, Y., Goyal, A., and Tolbert, N.E., submitted) have been done with algae in water, and control pH titration curves were run for each experiment to convert the observed pH change in the buffered medium of the cells and HCO_3^- to equivalents of OH^- ions. Twelve different algae including *Chlorophyceae* strains of *Chlorella, Scenedesmus, Dunaliella,* and *Chlamydomonas*, and *Charaphyceae* and *Ulvaphyceae* were used and over the short experimental times of sec to 2 hrs the rates of O_2 evolution were not damaged by the unbuffered systems (in manuscript). Rates of alkalization, measured as pH increase and calculated as OH^- equivalents, were dependent on the presence of the CO_2 concentrating process in air-adapted cells, light, pH, and the external DIC concentration. Alkalization by high-CO_2 grown cells was very low. All the micro algae with a DIC pump rapidly (within 2 to 5 min) raised the pH from 7 to 9 or 10, until alkaline production and photosynthetic O_2 evolution abruptly stopped. Rates of OH^- formation were linear with time up to pH 9.5. If the pH were lowered to 6 or 7 with HCl, O_2 evolution and alkalization resumed, and this could be repeated numerous times over hours with the same algal suspension (in manuscript).

Alkalization of the medium was dependent upon external conversion of HCO_3^- to CO_2 + OH^- (Figure 2). Thus, alkalization did not occur if cells were using internal DIC or dissolved external CO_2. Alkali production was inhibited by CA inhibitors, AZ, EZ and COS, if external CA were present as part of the DIC pump. Our working hypothesis is that alkalization of the medium was due to conversion of extracellular HCO_3^- to CO_2 plus OH^-, as the CO_2 was actively and rapidly concentrated in the cells by the CO_2 pump. Thus OH^- was left outside the cell, and there has been no evidence that the process involved OH^- excretion or a plasmalemma ATPase for H^+ production to neutralize the base. Alkalization of the medium is a result of the green micro algal CO_2 pump. One role for the external base would be to titrated carbonic acid from newly dissolved CO_2. Thus alkali production could be neutralized by absorption of more atmospheric CO_2 in a global carbon cycle (equations 1 and 2).

Air-adapted, but not CO_2 grown, *Scenedesmus* alkalized the medium with DIC to pH 10 more rapidly than other green algae in the light. Even in the dark or without external DIC in the light or with DCMU, *Scenedesmus* cells alkalized the medium at a rate of about 50 to 90% of the rate in the light by other algae. This light and CO_2 independent alkalization of the medium by *Scenedesmus* would continue if the pH were lowered every few minutes with HCl. Presumably the dark alkalization was based on respiration, but low CO_2 adaptation was necessary to activate this dark alkalization process.

CALCIUM CARBONATE DEPOSITATION

Calcium carbonate deposition and biotic reef formation in antiquity and today occur because of the photosynthetic activity of calcareous algae. Although calcification has long been described by marine biologists and geologists, its association with photosynthesis has not been sufficiently investigated nor has its role as a CO_2 sink today been considered for the increasing atmospheric CO_2. $CaCO_3$ deposition by algae varies widely from amorphous or crystalline deposits outside the cell, in or on the cell surface, or inside the cell. Reviews (18) cite a photosynthetic carbon balance of two bicarbonates taken from the medium with one CO_2 reduced to organic carbon by the algae (equation 3 and 4) and one carbonate by-product (equation 5) that is deposited as a calcium or magnesium salt. When many algae, both marine and fresh water, accumulate DIC, they may alkalize the media (equation 3) and the hydroxy ion may be neutralized by the conversion of a second HCO_3^- to $CO_3^{=}$

(equation 5), which in turn is deposited as a salt (equation 6).

1. $CO_2 + H_2O \longrightarrow H_2CO_3$
2. $H_2CO_3 \longrightarrow HCO_3^- + H^+$
3. HCO_3^- (outside) $\longrightarrow CO_2$ (inside) + OH^- (outside) (photosynthetic DIC pumps)
4. $CO_2 \longrightarrow (CHO)_n$ (inside) (photosynthetic fixation)
5. $HCO_3^- + OH^- \longrightarrow CO_3^= + H_2O$ (outside)
6. $CO_3^= + Ca^{++} \longrightarrow CaCO_3\downarrow$ (outside)

Thus investigators of the DIC concentrating processes by algae are also investigating part of the process for calcium carbonate deposition by algae. The two areas need to be examined as a whole when considering global problems in marine biology and sequestering excess atmospheric CO_2. Every HCO_3^- converted to organic carbon during photosynthesis by many algae is potentially accompanied by deposition of a second HCO_3^- as $CaCO_3$. As a consequence of marine photosynthesis reduced organic carbon and carbonates may be deposited into pools with long half lives. Part of the excess CO_2 currently being put into the atmosphere is not accounted for by photosynthetic refixation by plants and algae. These unknown sinks may be in part algal organic and $CaCO_3$ deposits. Controlling or lowering atmospheric CO_2 by growing more plants for the global carbon cycle is only temporary, because the photosynthate would be continually recycled. To lower atmospheric CO_2, the excess carbon needs to be reburied, like the coal, gas and oil from which it comes, into a long life or inaccessible form or location. Increased formation of $CaCO_3$ deposits in the oceans by algae is a possible CO_2 depository today, as it has been in the past. Many natural limitations to $CaCO_3$ deposition and marine photosynthesis exist, such as algal nutrition, temperature, pH, and light. If photosynthetic options are to be investigated for controlling atmospheric CO_2, marine organic and carbonate deposition should receive highest consideration.

An additional speculation for future research would be to develop transgenic plants with the algal DIC pump. Because the DIC pump probably consists of only a few (4 to 6) enzymic components, such as CA and ATPases, its transfer into higher plants would be more feasible than conversion of C_3 plants into C_4 plants. However as a scientific community we are not yet prepared or financed to make these changes. In the mean time a survey for parts of the algal DIC pump in high plants could be undertaken.

EVOLUTIONARY CHANGES IN DIC CONCENTRATING PROCESSES

Peroxisomes evolved or became abundant and active with increasing atmospheric O_2 only in advanced algae of the *Charaphyceae* (19,20), which is the only line of algae that evolved into higher plants (21). Some of the biochemical and photosynthetic changes that occurred during evolution of this line are cited in Table I. Ancestral, unicellular, green and blue green algae all concentrate DIC for efficient photosynthesis.

Table 1

	Unicellular Green Algae, *Charaphyceae*	Multicellular Macro Algae and Plants
Peroxisomes	None	Present
Glycolate Metabolism	Dehydrogenase in mitochondria	Oxidase in peroxisomes
Pyruvate Reductase and D-lactate Dehydrogenase	Present	Absent
DIC Concentrating Mechanism	CO_2 or HCO_3^- pump	None or C_4 Cycle
Glycolate Excretion	Occurs	None

The internal high CO_2 competitively reduces the oxygenase activity of Rubisco for P-glycolate synthesis and photorespiration. Such algae have a low level of enzymatic activities for glycolate metabolism by a glycolate or D-lactate dehydrogenase in their mitochondria; they do not have leaf type peroxisomes. The developmental changes that occurred in the more advanced or complex algal forms resulted in leaf type peroxisomes with active photorespiration. Early forms of higher plants do not contain the algal DIC pumps, as far as we know. Some macro algal and plants may contain the C_4 cycle for concentrating CO_2. To evaluate these evolutionary changes we need much more data, of course, and we need to know why the changes occurred. Any algae that lost their DIC pumps would have become photosynthetically less CO_2 efficient, and they would waste more photosynthetic assimilatory capacity (ATP and NADPH) on photorespiration associated with leaf-type peroxisomes. As we currently understand this evolutionary change, it appears to have been unfavorable for the evolving algae. In fact in only one line of algae did this unfavorable change presist, but that line of *Charaphyaceae* did develop into higher plants. A global hypothesis may be that this change occurred so that advanced algae and plants could not remove all of the atmospheric CO_2, which is needed for the greenhouse regulation of the environment. However, these algal changes probably occurred long before the atmospheric CO_2 decreased to current levels. Thus other causes for these changes need to be probed. A review of the properties of Rubisco in ancestral algae to plants indicate little changes. Perhaps the development of more complex algal systems has been associated with an increase in atmospheric O_2. At present we have little evidence for the presence of the algal DIC concentrating processes in macro algae or plants or the accumulation of $CaCO_3$ deposits by higher plants. There are examples of $CaCO_3$ deposition in some plants (22), but these plants have not been investigated for CO_2 concentrating processes.

REGULATION OF ATMOSPHERIC O_2

The oxygenase activity of Rubisco should not have been limiting while algae and plants evolved, because atmospheric O_2 was low, CO_2 was high, and the algae had CO_2 concentrating mechanisms. The Rubisco oxygenase activity only became limiting on photosynthesis when oxygen increased to 21%. It is unlikely that the excess CO_2 was all removed and buried as fossil photosynthate at just the point when the atmospheric O_2 reached 21%. Rather an O_2 compensation point for plants may regulate the atmospheric O_2 level, just as the CO_2 compensation point influences the level of atmospheric CO_2. With an atmosphere of 250 to 300 ppm CO_2, net photosynthesis approaches zero between 40 to 99% O_2 due to increased photorespiration at near an O_2 compensation point. Such higher O_2 levels would exist in the chloroplast, were it not for O_2 diffusion outwards to given atmospheric O_2 levels. Just as limited CO_2 diffusion into the chloroplast from a low exterior CO_2 level sets up a CO_2 compensation point, the diffusion of O_2 out of the chloroplast is dependent on the atmospheric O_2 level. These limiting atmospheric gaseous exchange rates betwen chloroplasts and air for both O_2 and CO_2 regulate the activity of Rubisco, and in turn regulate the atmospheric compostion. At 21% O_2 in air the O_2 equilibrium between the chloroplast and air is sufficient to allow enough O_2 to escape so that photosynthetic CO_2 fixation occurs. A steady state lower O_2 level would have resulted in a lower CO_2 compensation point and a lower atmospheric CO_2 level, while a higher O_2 level would require a higher CO_2 level to sustain the same net photosynthesis. This is a difficult concept for which there is little supporting whole plant data. Quantitative control of O_2 and CO_2 levels is needed during growth of plants in chambers for measuring the effect of photosynthesis on the O_2 balance in the atmosphere.

I am proposing that the CO_2 and O_2 levels of the atmosphere are equilibrium levels controlled in part by photosynthesis, kinetics of Rubisco, and the ratio of carbon flow between the C_3 and C_2 photosynthetic carbon cycles. Both CO_2 and O_2 in the atmosphere are being turned over by photosynthesis. Photosynthesis by plants can deal with CO_2 levels between 80 to 1000 ppm CO_2 without significantly perturbing the O_2 level. Some present day plants may not do well in 1000 ppm CO_2, because they have not had time to evolve processes for utilizing the excess photosynthate. But photosynthesis itself will not be adversely effected by higher CO_2. More C_4 plants is not an apparent solution to lower the atmospheric CO_2 level. Instead we need to further examine photosynthesis by algae where the potential for removal of excess CO_2 as $CaCO_3$ deposits has long been in effect.

REFERENCES

1. Lemon, E.R. ed. (1983). CO_2 and Plants. The response of plants to rising levels of atmospheric carbon dioxide. Westview Press, Inc.

2. Branola, J.M., Raynaud, D., Korotkevich, Y.S., and Lorius, C. (1987). Vostok ice core provides 160,000-year record of atmospheric CO_2. *Nature.* 329:408-414.

3. Tolbert, N.E. (1980). Photorespiration. In: The Biochemistry of Plants, eds., Stumpf, P., and Conn, E., Vol. 2, Metabolism and Respiration, ed., Davies, D.D. Academic Press, pp. 488-525.

4. Husic, D.W., Husic, H.D. and Tolbert, N.E. (1987). The Oxidative Photosynthetic Carbon Cycle or C_2 Cycle. *Critical Reviews In Plant Sciences*, 5:45-100.

5. Tolbert, N.E. (1983). The oxidative photosynthetic carbon cycle. In: Current Topics in Plant Biochemistry and Physiology, ed., Randall, D.D., *et al.* Univ. Missouri. Vol. I, pp. 63-77.

6. Lorimer, G.H. (1981). The carboxylation and oxygenation of ribulose 1,5-bisphosphate. The primary event in photosynthesis and photorespiration. *Ann. Rev. Plant Physiol.* 32:349-383.

7. Ogren, W.L. (1984). Photorespiration: pathways, regulation, and modification. *Ann. Rev. Plant Physiol.* 35:415-447.

8. Osmond, C.B. (1981). Photorespiration and photoinhibition, some implications for the energetics of photosynthesis. *Biochem. Biophysica Acta.* 638:77-98.

9. Ku, S.B, Edwards, G.E., and Tanner, C.B. (1977). Effects of inhibition of photosynthesis and transpiration in *Solanum tuberosum. Plant Physiol.* 59:868-872.

10. Edwards, G.E., and Huber, S.C. (1981). The C_4 pathway. In: The Biochemistry of Plants, eds., Stumpf, P. and Conn, E., Vol. 8., eds., Hatch, M.D., and Boardman, N.K. Academic Press, pp. 237-281.

11. Lucas, W.J., and Berry, J.A., eds. (1985). Inorganic carbon uptake by aquatic photosynthetic organisms. Waverly Press, Baltimore, MD, USA.

12. Badger, M.R. (1987). The CO_2-concentrating mechanism in aquatic phototrophs. In: The Biochemistry of Plants, eds., Hatch, M.D., and Boadman, N.K., Vol. 10, *Photosynthesis*, Acad. Press, San Diego, CA, pp. 219-274.

13. Thielmann, J., Tolbert, N.E., Goyal, A., and Senger, H. (1990). Two systems for concentrating CO_2 and bicarbonate during photosynthesis by *Scenedesmus. Plant Physiol.* 92:622-629.

14. Goyal, A. and Tolbert, N.E. (1989). Uptake of inorganic carbon by isolated chloroplasts from air-adapted *Dunaliella*. *Plant Physiol*. 89:1264-1269.

15. Muscatine, L., Porter, J.W., and Kaplan, I.R. (1989). Resource partitioning by reef corals as determined from stable isotope composition. *Marine Biology*. 100:185-193.

16. Moroney, J.V., Kitayama, M., Togosaki, R.K., and Tolbert, N.E. (1987). Evidence for inorganic carbon transport by intact chloroplasts of *Chlamydomonas reinhardtii*. *Plant Physiol*. 83:460-463.

17. Lucas, W.J. (1985) Bicarbonate utilization by *Chara*: a re-analysis. In: Inorganic Carbon Uptake by Aquatic Photosynthetic Organisms, eds. Lucas, W.J., and Berry, J.A. Waverly Press, Baltimore, MD, pp. 229-254.

18. Borowitzka, M.A. 1982 Mechanisms in algal calcification. In: Progress in Phycological Research, Vol. 1, eds., Round, F.E., and Chapmann, D.J. Elsevier Biomedical Press, Amsterdam, pp. 137-177.

19. Stabenau, H. (1984). Microbodies in different algae. In: Compartments in algal cells and their interaction, eds, Wiessner, W., Robinson, D., Starr, R.C. Springer-Verlog, Berlin, Heidelberg, pp. 183-190.

20. Tolbert, N.E. (1974). Photorespiration by Algae. In: Algal Physiology and Biochemistry, Stewart, W.D.P., ed. Blackwell Scientific Publications Ltd., Oxford, England, pp. 474-504.

21. Bold, H.C. and Wynne, M.J. (1978). Introduction to the algae: structure and function. Prentice-Hall, Inc., Englewood Cliffs, New Jersey.

22. Okazaki, M. (1989) Calcium carbonate deposition by algal and land plants. In: Current topics in Marine Biotechnology, eds., Miyachi, S., Karube, I., and Ishida, Y. Fuji Technology Press Ltd., Tokyo, pp. 25-28.

RESPONSE OF PLANTS TO ELEVATED CARBON DIOXIDE: THE ROLE OF PHOTOSYNTHESIS, SINK DEMAND AND ENVIRONMENTAL STRESSES

D.W. LAWLOR

Institute of Arable Crops Research
Rothamsted Experimental Station
Harpenden, Herts., AL5 2JQ, U.K.

ABSTRACT

Increasing atmospheric carbon dioxide concentration will affect global climate and stimulate photosynthesis and dry matter production of plants, particularly those with C3 photosynthetic metabolism. Doubling the CO_2 increases dry matter by circa 30%, through increased carboxylation and a larger maximum rate of photosynthesis. These effects are mediated by the characteristics of ribulose bisphosphate carboxylase (RuBPc-o) and the competition between oxygen and CO_2 for ribulose bisphosphate from the photosynthetic carbon reduction cycle (PCRC). Accumulation of carbohydrates is characteristic of the effects of elevated CO_2 and regulation of CO_2 assimilation by feed-back control at the level of PCRC enzymes is considered. Decreased amount and activity of RuBPc-o may decreased assimilation as often observed with long-term exposure to elevated CO_2. The role of growth to provide sinks for assimilates, both in formation of new cellular components, in respiration and for storage, is emphasised as is the balance between the source and the sinks. The effects of environmental stresses are discussed as determining the capacity to form the CO_2 assimilation system and the sinks and thereby alter the ability of the plant to respond to elevated CO_2. Stresses of nutrition, water and temperature will decrease growth and also the rate of photosynthesis but the effect on the sinks may be larger, causing accumulation of carbohydrates. The nitrogen carbon ratio generally decreases in plants grown with elevated CO_2; it is discussed in relation to uptake of nitrate or other forms of N, to competition for products of the thylakoid reactions, to increased demand for growth and to senescence. The ability of plants to respond to elevated CO_2 is complex; better understanding of the mechanisms is required if models of the responses of vegetation to elevated CO_2 and climate change are to be produced.

INTRODUCTION

Carbon dioxide in the atmosphere has risen from a pre-industrial concentration of c. 280 $cm^3\ m^{-3}$ to over 350 $cm^3\ m^{-3}$ currently and the rate is accelerating (1,2,3,4). The change is very rapid on geological time scale, and is due to the increase in world human population and its increasing economic activity (4). This leads to consumption of fossil fuels with production of CO_2 and also to destruction of forests and other natural vegetation, releasing stored carbon by oxidative processes and reducing carbon storage in the standing vegetation. Consumption of fossil fuels will continue to rise for the foreseeable future (1) so that values of CO_2 double the current, i.e. 700 cm^3 m^{-3} or larger, must be expected early in the 21st century (1,5,6).

Carbon dioxide is the primary substrate for photosynthesis and carbohydrate synthesis (7,8). Therefore, it will have a direct, stimulating effect on the productivity of photosynthetic organisms, including terrestrial vegetation (8,9,10,11,12,13) and therefore elevated CO_2 will greatly affect the total amount of carbon stored (10,11,14). This may, to some extent, offset the rise in CO_2 by sequestering carbon in the biosphere (15). However, the increase in plant growth depends not only on the CO_2 supply; factors which regulate the development and growth of plants such as nutrition, water supply and temperature will influence the response of the vegetation to elevated CO_2 (16,17). The discrepancy between the conditions such as temperature or supply of energy or nutrients required for a plant to function effectively and those occurring in the environment may constitute a 'stress'. Stresses will affect the response of plants to elevated CO_2 (18).

There are very large differences in the response of plants of different species to CO_2 (7,8,19,20); the primary distinction is between C3 and C4 plants (8,9,13). Photosynthesis of those with the C3 mechanism is substantially stimulated by an increased supply of substrate CO_2, whereas that of plants with C4 metabolism is much less so (21,22). The reasons for this are considered in this review together with the likely effects of elevated CO_2 on productivity and the way other environmental factors may modify the responses (16,23,24,25).

The concerns about increasing CO_2 are not confined to the vegetation. A major aspect is the potential effect that CO_2 (and other gases such as methane and chlorofluorocarbons) will have on the global temperature (3,26,27,28). Carbon dioxide, like water vapour, methane and the CFC's, is a 'green house gas' which absorbs radiation in the infrared wavelengths, thereby warming the atmosphere (1,16,26) land and oceans. Indeed it may have already done so by some 0.5 °C over the last 100 years (1,29). The consequences for global climate patterns are potentially far-reaching. Global climate models, based on the physics of the atmosphere, predict that the global temperatures will increase, but the magnitude and rate of change are much disputed (26,27,28, 29,30). Models suggest that changes will not be uniform across the globe. Higher latitudes are expected to experience a larger proportional temperature increase than those near the equator. Areas of the continents are predicted to become drier although the global precipitation may increase by 15% (27,29,30). Many of these projected changes to climate are subject to great uncertainty and the scale of the modelling precludes specific predictions (30).

Clearly, modifications to climate on a large scale would substantially affect natural and managed vegetation (31,32,33) and there may be considerable interaction with the increase in CO_2. In order to understand and thereby to predict the effects of elevated CO_2 in combination with climate change on the productivity of vegetation, the response of plants to conditions relevant to the future environment must be addressed (34, 35). There is a large literature on aspects of plant response to CO_2, examining the effects on many species, although with a strong bias towards crops of temperate latitudes (3,11,12,14), under very different conditions. Relatively less work has been done on natural vegetation (9,36,37,38) or tropical species (39,40). Following the expectation that the CO_2 in the atmosphere will double in the near future many experiments contrast approximately 300 to 350 cm^3 CO_2 m^{-3} with 700 cm^3 CO_2 m^{-3}. Yet considerable uncertainty exists about the mechanisms of processes in the photosynthetic system and those using the assimilates it produces (40,41). Also the interactions between CO_2 and other environmental factors are very uncertain despite many experimental studies (42,43). Some work addresses the problem of modelling the responses to enable quantitative assessment of the plant responses to environmental change (34,43). However, uncertainty over the mechanisms hampers this approach (44,45). Currently the problem is to identify the features of plants which determine the responses, to understand the mechanisms and to incorporate these features into models (33,34,43). Only then can it be said that the problem is understood and the potential responses defined. Quantitative analysis is important for it will allow the contribution of the biosphere to amelioration of anthropomorphic CO_2 release to be assessed on a global scale (5,16) and will permit more effective response to climate change in terms of modifying human activities, for plant breeding and improving agricultural practice (31,33,34). Poor models lead to speculation without a solid foundation; many of the problems to be tackled in the arena of global climate change need large scale and therefore governmental action which can only be based on objective scientific evaluation.

In this paper I outline the evidence for the response of plants to elevated CO_2, how this is mediated through the effects on photosynthesis (the 'source') in different plant types and how demand for assimilates (the 'sink') may affect the responses. Also, the effects of nutrition, water and temperature, which may potentially affect the response either by direct influence on photosynthesis or via 'source-sink' interactions are considered. The many potential stresses that plants are exposed to and the quantitative nature both of the stress and of the plants response (7,43, 46,47,48) complicate understanding and assessment of the interactions with elevated CO_2 and will be analysed.

PLANT RESPONSE TO ELEVATED CARBON DIOXIDE

Production of Dry Matter and Organ Growth

The experimental evidence for the effects of doubling CO_2 concentration on C3 plants is overwhelmingly that dry matter production will increase by between 20 % and 40 %, with a mean of 33% The results of many experiments have been amply analyzed by Kimball (14) and by Cure (12) and it is unnecessary to repeat the evaluations. Productivity of C4 plants will be much less stimulated, rising by ca. 10 % to 20 %. Such increases have been observed for determinate, annual plants, for perennial and indeterminate plants, herbs and woody vegetation (40,49). The magnitude depends, however, on species and environmental conditions. The increases have been observed for plants grown in controlled environments, often with low irradiance (50) and under field conditions (51), where water supply and temperatures may be less suited for high levels of production. In general dry matter production is greatest, in absolute terms, for plants grown under conditions of ample water, nutrition etc. but the largest increases as a proportion to the growth rate are found in plants grown under less favourable conditions (e.g. in dim light or under water stress) with elevated CO_2 (18,24,52). The changes in plant parts are complex, as might be expected given the great range of species, ages and conditions observed. Generalising, the growth of shoots and roots is stimulated in approximately the same proportion, although storage organs may be considerably enhanced (23,53,54). In woody species wood production is increased (49).

There is much uncertainty about the effects of elevated CO_2 on the morphological characteristics of plants. In many, the production of organs is enhanced, e.g. grasses including cereals form more tillers (55, 56), although the apical development of wheat, for example, may not be stimulated (57). Woody species may branch more and produce more leaf area (17). Although some species produce larger leaves this is not always the case; more commonly leaves have greater mass per unit area (58). Changes in the morphology include increased stomatal numbers per unit area of leaf (59) or no change (58), although it has been suggested that a long-term effect of elevated CO_2 has been to cause the selection of plants with decreased stomatal frequency (60). The rates of production of leaves and other organs are generally little affected by elevated CO_2 (52,61) but are very variable depending on the species and conditions (52); it is possible that the accelerated development is more closely related to temperature than to the CO_2 as an effect of elevated CO_2 is to decrease the stomatal conductance (50,51,62,63), thus reducing water loss which may increase plant temperature, hence increasing the rate of development.

Productivity of the Photosynthetic System.

The C3 and C4 types of CO_2 assimilatory mechanism are well attested, and their broad response to CO_2 is clear (8). In the C3 type, ribulose bisphosphate (RuBP) produced by the photosynthetic carbon reduction (Calvin) cycle in the chloroplast combines with CO_2 in a reaction catalyses by the enzyme RuBP carboxylase-oxygenase (RuBPc-o). The products of the reaction (3-phosphoglyceric acid and triosephosphates) are metabolized to sugars, starch, cellulose and other carbohydrates. They also contribute to the formation of amino acids and proteins. The RuBPc-o is a bifunctional enzyme which also catalyses the reaction of RuBP with oxygen, producing phosphoglycolate in addition to 3-phosphoglyceric acid (64). Phosphoglycolate is metabolised to glycine, which is decarboxylated in the mitochondria, and the CO_2 is released as photorespiration. In the current atmosphere the CO_2 partial pressure is below that required for the saturation of RuBPc-o and there is competition between CO_2 and O_2 for RuBP resulting in synthesis of phosphoglycollate and in photorespiration. The latter may constitute 20 % to 30 % of net photosynthesis in C3 plants. In plants with the C4 form of photosynthesis the CO_2 partial pressure at the active sites of RuBPc-o is maintained at a concentration above atmospheric by the action of a 'pump' mechanism driven by the enzyme phosphoenol pyruvic acid carboxylase. This enzyme has a greater affinity for CO_2 than does RuBPc-o, consequently C4 plant photosynthesis is saturated with CO_2 at partial pressures substantially below those of the current atmosphere. Consequently C4 plants exhibit little photorespiration and are relatively insensitive to ambient CO_2.

The stomata of plants play an important role in regulating gas exchange between the atmosphere and the internal spaces of the leaf. They restrict not only the diffusion of water vapour out of the leaf but also the entry of CO_2 into it. This limits the assimilation of C3 plants more than that of C4's. In C3's a balance between the stomatal conductance and rate of photosynthesis is achieved, by as yet poorly understood mechanisms (45,62). Increasing the ambient CO_2 partial pressure increases the rate of photosynthesis whilst decreasing the stomatal conductance; thus increasing water use efficiency. In C4 plants the stomatal conductance is generally substantially smaller than in C3 species (8) yet faster rates of assimilation can be maintained at smaller ambient CO_2. Such physiological and biochemical distinctions determine the response of the plants to elevated CO_2; assimilation of C3 types may be stimulated by up to 50 % if the CO_2 is increased from 350 to 700 cm^3 m^{-3}, both because of increased assimilation and decreased photorespiration, despite the decrease in stomatal conductance. Thus CO_2 greatly increases the water use efficiency and slows the loss of water from the plant (65,66). However, there are complex interactions in the coupling of heat and energy exchange between plants and the atmosphere. The slower evaporation rate may increase leaf temperatures and also decrease the humidity of the air around the leaf, both of which may offset the advantage of smaller stomatal conductance (43). Such interactions must be addressed in a range of environments.

Regulation of Photosynthesis

In many experiments the exposure of plants to elevated CO_2 has been made shortly before measurements. The rates of photosynthesis of several species are high for a period after the initial exposure to elevated CO_2 but decrease substantially thereafter (67). This appears to happen in a range of species and correlates with the accumulation of products (68). In some cases the reduced rate of assimilation is larger than the rates in normal CO_2, so that a larger rate of accumulation of dry matter is still observed. The causes of the changes in assimilation are still controversial. There is evidence that loss of the enzyme proteins is responsible in addition to the loss of activity of the pre-formed proteins (69). This may be related to the decrease in the ratio of nitrogen to carbon which is frequently observed in plants grown in elevated CO_2 (36,47). Possibly the high concentrations of carbohydrates or, more probably, the decreased availability of nutrients (particularly nitrogenous compounds) trigger a senescence response of older leaves, similar to that observed in plants grown under normal atmospheric CO_2. The crucial difference in the effect of elevated CO_2 and the events causing normal senescence is not clear but have not been adequately addressed for plants in relation to any of the environmental factors interacting with CO_2.

The analysis of the response of plants to elevated CO_2 has concentrated on the responses at the leaf level and in the relatively short-term, often on plants grown at normal ambient concentrations, and the effects are rather well characterized. The increased production of carbohydrates by the chloroplasts leads in many experiments and for a range of different species to accumulation of carbohydrates (17,41,68,70). Interaction of production with demand for assimilates by growing and storage organs will be considered later. However, the accumulation of products may lead to feed-back inhibition of the photosynthetic cycle mechanisms. There are many potential sites of regulation within the photosynthetic system. Thus accumulation of sucrose in the cytosol of the cell slows the flux of triose phosphate out of the chloroplast, decreasing the inorganic phosphate and increasing formation of starch in that organelle (71). Accumulation of the intermediates of the cycle and the lack of phosphate decrease the rate of CO_2 assimilation (72). Thus, sequestration of phosphate in intermediates will inhibit reactions such as the conversion of ribulose-5-phosphate to RuBP by ribulose-5-phosphate kinase. This will reduce the rate of cycle turn-over despite the stimulation of other enzymes e.g. fructose bisphosphatase, by low concentration of inorganic phosphate. Also, accumulation of diphosphoglycerate inhibits phosphoglycerate kinase (8). The complex regulation of the PCRC is discussed in several publications (8,72,73,74). Increased carbohydrates (72,73,74,75) may result in synthesis of inhibitors of the photosynthetic carbon reduction cycle; the formation of carboxyarabinitol-1-phosphate (74) may be such a case. These regulatory processes lead to reduction in the rate of assimilation, called acclimation or 'down regulation' (19,69,72,75, 76,77). Acclimation may be related to loss of RuBPc-o active sites (78,79) and to decreased amount of enzyme protein (69,74,75,76,80) although the effect is not always as expected (79) if the protein is partially redundant in elevated CO_2. Other proteins of the PCRC do not appear to decrease, e.g. 3-phosphoglycerate phosphokinase or NADP-dependent glyceraldehyde 3-phosphate dehydrogenase (81); clearly more detailed analysis of the regulation of the composition of leaves under elevated CO_2 is needed.

Increased carbon assimilation will increase the use of ATP and NADPH, both products of the thylakoid reactions (8) but as elevated CO_2 decreases the flux of carbon through the glycolate pathway the consumption of ATP and NADPH diminishes, thus possibly decreasing competition for energy and reductant between photosynthesis and photorespiration and with nitrogen assimilation which requires reduced ferredoxin, NAD(P)H and ATP in the conversion of nitrate ions into glutamate via the nitrite-nitrate reductases and the GS/GOGAT cycle. Decreasing photorespiration by decreasing the oxygen content of the atmosphere whilst holding photosynthesis constant, which should increase growth if reductant etc were more freely available than in normal air, adversely affected many aspects of the growth and development of wheat (82). Although the comparison with elevated CO_2 is not precise, the changes do raise many questions about the role of photorespiration in the normal functions of C3 plants. Amino acid content decreases under elevated CO_2 and this may be caused by competition or it is possible that the demand for amino acids increases as the growth of plants is stimulated by elevated CO_2 hence decreasing the pools (83). The role of sink demand in this will be considered later.

Other areas of metabolism potentially modified by elevated CO_2 are the synthesis of pigments such as chlorophylls and of proteins. Decreased amino acid supply might affect the production of aminolaevulinic acid, the precursor for tetrapyrrole synthesis and would clearly decreased protein synthesis; this may provide an explanation of the observed reductions in the nitrogen to carbon ratios and decreased protein contents of plants exposed to elevated CO_2 (84); understanding the details of metabolism is of central importance for predicting whole plant responses to elevated CO_2.

Gene Expression and Regulation

The changes observed in plants may be by direct influence of elevated CO_2 on metabolism, e.g. altering rates of turn-over of components or by stimulating the synthesis of particular enzymes or by inducing senescence (which may require the production of proteolytic enzymes). It is unclear if these effects involve gene expression and regulation at the level of gene transcription and protein synthesis. There is little known about the interaction between nitrogen metabolism and tissue composition or of the role of carbohydrates in regulation. In the long-term there seems to be relatively little effect of CO_2 on gene expression, but changes such as the decrease in carbonic anhydrase and RuBPc-o (77) may reflect changes in the underlying regulation. There is inadequate information on the genetic control of the response to elevated CO_2. Comparison of the responses of cultivars show substantial differences in many aspects(75,76,85) of physiology, the greatest being in the harvest index rather than assimilation, with little evidence of specific genetic changes linked to metabolic alterations. Uncertainty exists about the ability of plants to undergo genetic selection and modification sufficiently fast to permit adjustment of the composition and function of photosynthetic systems or of organ and sink development in order to achieve optimum productivity in elevated CO_2 and changed climate. The evolutionary change is likely to be slow in comparison to the rapid environmental change. Such conditions may favour species with short generation times compared to slow growing species, e.g. ruderals compared to forest trees (49,86). If adjustment is related to rapid recombination of the gene pool and selection then such weed species may be favoured (86). The poor understanding of gene regulation by metabolism does not permit any sound projections of future plant responses to be made.

Sink Capacity and Control by Environmental Conditions

Sink capacity is the rate of demand for assimilates caused by the requirements for growth of new organs (stems, roots, leaves and reproductive structures), for storage (e.g. in stems and fruiting bodies) and for energy generation by respiration. In addition there may be consumption by futile cycles of metabolism (e.g. alternative pathways of respiration) and by loss of materials to the environment (e.g root exudation). Growth of organs constitutes a major sink which is very variable in magnitude over time, and in response to conditions such as temperature, water supply and nutrition (25,37). If source processes are to function at the maximum rate, sink capacity of the system must be able to utilize the assimilates both in the short- (hours to days) and long-term (days to weeks). Otherwise feed-back regulation may influence the source (72). The balance between sources and sinks may be disturbed by long-term exposure to elevated CO_2; increased content of carbohydrates indicates when assimilate production exceeds source demand although there is no evidence of the levels at which feed-back regulation is triggered (72,79). It is uncertain if sink demand will be adequate as global CO_2 concentrations increase. However, there is considerable unused capacity in current genotypes as shown by their response. Evidence from plant breeding suggests that sufficient genetic variation is available for selection in crops to modify the source-sink relations rapidly. So, despite the uncertainty, there seems little evidence that inadequate sink demand will greatly limit plant response to elevated CO_2 (17). Much depends on the way that the complete plant system responds, i.e. in long-term development and growth and functions of the photosynthetic system.

The development and growth of organs is genetically determined, often operating via environmental triggers such as day length and temperature. The absolute growth rate of organs depends on temperature and differs with species. For example the rate of leaf expansion in sunflower is much slower than that of oil seed rape at low temperature but the photosynthetic rate of sunflower is relatively less affected by cold so that soluble carbohydrates increase at low temperature more in sunflower than in rape. Dry matter accumulation is therefore very dependent on temperature. Root systems also respond similarly to the shoots thus decreasing the sink demand. If the rate of photosynthesis is increased by elevated CO_2, even if only for a few hours, soluble carbohydrates , starch and fructans accumulate; this is also observed with long-term exposure of plants to elevated CO_2 particularly at low temperature. However there is a limit to the storage capacity so that photosynthesis decreases (Lawlor and Paul, unpublished). Possibly the capacity of woody species to respond to additional assimilates is greater as the ability to branch and produce more wood is considerable (49).

Growth of the sinks depends not only on temperture but on the availability of nutrients,e.g. N to form proteins, P for intermediary metabolism and K for ionic balance (36,47,48). Growth decreases when the supply of any one falls below the level necessary for full expression of the genetically determined growth potential. Thus a shortage of N limits protein synthesis and slows growth of roots or leaves etc.and these form a smaller sink for CO_2 assimilates; a frequently observed effect of decreased N is starch and sugar accumulation (83). Each nutrient affects metabolic systems differently, deficient P appears to decrease the response to CO_2 substantially more than deficient N, but there is inadequate quantitation of complex and poorly described processes. If the nutrient supply permits leaf growth, photosynthetic capacity is generally less affected than the decrease in sink capacity, thus ensuring high levels of tissue carbohydrates. Nutrient deficiency, a common stress in natural conditions, decreases growth of organs, altering their composition, e.g. N deficiency decreases leaf area and assimilation of CO_2, thus decreasing productivity. Elevated CO_2 increases the capacity to form carbohydrates but the shortage of N as a substrate for protein synthesis still decreases growth, thus decreasing storage capacity. Nutrient shortage may limit the response of vegetation to elevated CO_2 depending on both the species and on the environment; these questions must be considered in assessing the effect of elevated CO_2.

Water stress is a recurrent limitation to the growth of plants in most parts of the world (18,23,51). Mild deficits decreased cellular turgor, reducing expansion of leaves and other organs and stomatal conductance for water vapour and CO_2. The effects of elevated CO_2 on the stomata have been mentioned; plants will benefit from slower rates of water loss thus delaying the onset of stress. This may permit greater growth of organs than in normal CO_2 and enhance assimilation under stress; greater leaf area will tend, despite smaller stomatal conductance, to increase the water loss (31,40). Despite considerable understanding of the physics there is only limited analysis of the effects of elevated CO_2 on plants. However, increased CO_2 alleviates the effects of stress on dry matter production and on grain growth in wheat (18,51). Stress decreases the expansion of organs more than it slows photosynthesis, consequently carbohydrates accumulation may increase although quality (N concentration) may suffer. With adequate water, CO_2 increased grain number and size but under stress it had less effect on size; the effect on yield is relatively much larger than without stress but total production is still decreased (18,51) Quantitative analysis is essential to establish how climate change will affect plant growth in elevated CO_2.

Extreme stress which damages the photosynthetic system will generally have severely affected sink development and thereby plant productivity. If metabolic damage occurs e.g. to the synthesis of ATP in photophosphorylation and the production of RuBP (74, 78), then the ability of the photosynthesising system to respond to increased CO_2 may be much impaired, as the short-term responses of stressed plants to elevated CO_2 indicates. Thus, the stimulation of photosynthesis by CO_2 become smaller as the degree of stress increases. However, the combined effect of improved water balance and increased substrate will probably increase the productivity of C3 plants under water stress by maintaining source and sink development and function, thus partially offsetting the damage. It is now necessary to quantify the relationships and to model the system for particular crops or types of vegetation.

Respiration consumes a substantial part of the assimilates produced by plants and the rate increases with temperature, suggesting that in the future climate the losses due to respiration may increase, particularly if wasteful respiration were stimulated by increased carbohydrates(72). This would offset accumulation of carbon. However, this scenario is not supported by the evidence (16,87): dark respiration of plants grown in elevated CO_2 is generally depressed compared to those from normal conditions and the absolute difference is bigger at high temperatures. The causes of depressed respiration are not known, but are not related simply to accumulation of non-metabolic dry matter which might alter the basis of expression. It may involve direct effects of CO_2 on mitochondrial electron transport. The contribution of respiration to dry matter accumulation and the effects of environment on the process has yet to be fully assessed.

The analysis based on the source-sink concept provides a way of relating the observations of the effects of high CO_2 and environment on the plant. Some of the very variable effects of long-term exposure of plants to high CO_2 may be explained by alterations to the source-sink balance. If growth is not limited by the environment and photosynthetic processes are sufficient to match the demands of growth, storage and respiration, then assimilates in the tissues will be maintained at a level governed by the characteristics of the metabolic systems e.g. for energy metabolism, ion transport etc. Under stress, the rate of growth falls and sink demand decreases. The balance between assimilate production and utilization will alter, depending on the relative effects on source and sink. Plants grown with the roots in very small volume of soil may, for example, suffer shortage of nutrients or water even if well supplied with both if transport is too slow to meet demand. Other factors may also be important; restricted growth may lead to production of abscisic acid which induces senescence. The effect will be to decrease the response to CO_2 compared to that expected if rooting is not restricted. One feature of plants which may play a large role in their response to elevated CO_2 is annual versus perennial. Perennials generally have vegetative organs of indeterminate growth which store assimilates, frequently carbohydrates. Woody plants also accumulate much carbohydrate as cellulose (17,49), lignin etc. and often respond greatly to elevated CO_2. In contrast, determinate annual plants often have specific growth periods and reduced vegetative growth. Their seeds may be of relatively fixed size, and the total sink for carbon depends on the seed number, often determined during early growth. The amount of carbon sequestered may be small in oily or fatty seeds; the limits to production may be more the energy supply rather than carbon.

In conclusion, it is established that C3 plants produce more dry matter with elevated CO_2, and that the decreased stomatal conductance will benefit the water balance to a limited extent but there is much less certainty about the longer-term adaptations. The ways that short-term photosynthetic metabolism is affected are also not well established. The role of sink capacity is crucial to the ability to respond to CO_2 and the effects of stresses of nutrition, temperature and water on the response to elevated CO_2 may explain variations in response observed in the literature. Quantitation of source-sink processes is required to assess their importance in regulating the long-term interactions between elevated CO_2 and environmental conditions for a range of vegetation.

References

1. Boden, T.A., Kancuruk, P. & Farrell, M.P. (1990). Trends '90: a compendium of data on global change. Carbon dioxide information analysis center, Oak Lodge, Tennessee.
2. Brewer, P.G. (1983). Past and future atmospheric concentration of carbon dioxide. In: *Carbon Dioxide Assessment Committee, U.S. National Research Council*, Changing Climate. National Academy Press, Washington, DC, pp. 186-215.
3. Allen, L.H. (1989). Global climate change and its impact on plant growth and development. In: *Proceedings of the Plant Growth Regulator Society of America, 16th Annual Meeting Aug. 6-10, 1989*, pp. 1-13.
4. Keeling, C.D. (1983). The global carbon cycle: What we know and could know from atmospheric, biosperic and oceanic observation. Proc. CO_2 Res. Conf., Berkeley Springer, WV, (DOE-CONF 820970), available from NTIS, Springfield, VA 22161.
5. Bacastow, R.B., Keeling, C.D. & Whorf, T.P., (1985). Seasonal amplitude increase in atmospheric CO_2 concentration at Mauna Loa, Hawai, 1959-

1982. *J. Geophysic. Res.* **90**: 10529-10540.

6. MacCracken, M.C. & Luther, F.M. (eds.) (1985b). *Projecting the climatic effects of increasing carbon dioxide*. U.S. Department of Energy, Carbon Dioxide Research Division, DQE/ER-0237, Washington, D.C.
7. Acock, B. & Allen, L.H. (1985). Crop responses to elevated carbon dioxide concentrations. In: *Direct effects of increasing carbon dioxide on vegetation.* Eds. Strain, B.R. & Cure, J.D. United States Department of Energy, Carbon Dioxide Research Division, DOE/ER-0238, Office of Energy Research, Washington, D.C. pp. 53-98.
8. Lawlor, D.W. (1987) *Photosynthesis: metabolism, control, physiology.* Longman Scientific and Technical, Harlow, Essex, pp. 266.
9. Lemon, E.R. (ed.) (1983). CO_2 and plants: the response of plants to rising levels of atmospheric carbon dioxide. *AAAS Selected Symposium 84*, Westview Press, Boulder, CO.
10. Bazzaz, F.A., Garbutt, K. & Williams, W.E. (1985). Effect of increased atmospheric carbon dioxide concentration on plant communities. In: *Direct effects of increasing carbon dioxide on vegetation.* Eds. Strain, B.R. & Cure, J.D. United States Department of Energy, Carbon Dioxide Research Division, DOE/ER-0238, Office of Energy Research, Washington, D.C., pp. 155-170.
11. Cure, J.D. & Acock, B. (1986). Crop responses to carbon dioxide doubling: a literature survey. *Agricultural and Forest Meteorology* **38**, 127-145.
12. Cure, J.D. (1985). Carbon dioxide doubling responses: a crop survey. In: *Direct effects of increasing carbon dioxide on vegetation.* Eds. Strain, B.R. & Cure, J.D. U.S. Department of Energy, Carbon Dioxide Research Division, DOE/ER-0238, Washington, D.C.
13. Gates, D.M. (1985). Global biospheric response to increasing atmospheric carbon dioxide concentration. In: *Direct effects of increasing carbon dioxide on vegetation*, B.R. Strain & J.D. Cure eds., Duke University, Durham, USA.
14. Kimball, B.A. (1983). Carbon dioxide and agricultural yield: an assemblage and analysis of 430 prior observations. *Agron. J.* **75**, 779-788.
15. Oechel, W.C. & Strain, B.R. (1985). Native species responses to increased atmospheric carbon dioxide concentrations. In: *Direct effects of increasing carbon dioxide on vegetation*. Eds. Strain, B.R. & Cure, J.D. pp. 117-154. U.S. Department of Energy, Carbon Dioxide Research Division, DOE/ER-0238, Washington, D.C.
16. Idso, S.B., Kimball, B.A., Anderson, M.G. & Mauney, J.R. (1987). Effects of atmospheric CO_2 enrichment on plant growth: the interactive role of air temperature. *Agri. Ecosystems Environ.* **20**, 1-10.
17. Krizek, D.T. (1986). Photosynthesis, dry matter production and growth in CO_2 - enriched atmospheres. In: *Cotton Physiology*, Mauney, J.R. & Stewart, J. McD eds., The Cotton Foundation, Publisher Memphis, Tennessee, USA.
18. Gifford, R.M. (1979). Growth and yield of CO_2-enriched wheat under water-limited conditions. *Aust. J. Plant Physiol.* **6**:367-78.
19. Sage, R.F., Sharkey, T.D. & Seemann, J.R. (1989). Acclimation of photosynthesis to elevated CO_2 in five C_3 species. *Plant Physiology* **89**, 590-596.

20. Strain, B.R. & Cure, J.D. (eds.) (1985). *Direct effects of increasing carbon dioxide on vegetation.* U.S. Department of Energy, Carbon Dioxide Research Division, DOE?ER-0238. Washington, D.C.
21. Downton, W.J.S., Björkman, O. & Pike, C.S. (1980). Consequences of increased atmospheric concentrations of carbon dioxide for growth and photosynthesis of higher plants. In: *Carbon dioxide and climate: Australian research.* Ed. Pearman, G.I. Australian Academy of Science, Canberra, pp. 143-151.
22. Drake, B.G. (1989). Photosynthesis of salt marsh species. *Aquatic Botany* **34**, 167-180.
23. Huber, S.C., Rogers, H.H. & Mowrey, F.L. (1984a). Effects of water stress on photosynthesis and carbon partitioning in soybean (*Glycine max* L. Merr.) plants grown in the field at different CO_2 levels. *Plant Physiol.* **76**, 244-249.
24. Paez, A., Hellmers, H. & Strain, B.R. (1984). Carbon dioxide enrichment and water stress interaction on growth of two tomato cultivars. *J. Agric. Sci. (Camb.)* **102**, 687-693.
25. Shaver, G.R., Chapin, F.S. III & Gartner, B.C. (1986). Factors limiting growth and biomass on vegetation. United States Department of Energy, Carbon Dioxide Research Division, DOE/ER-0238, Office of Energy Research, Washington D.C.
26. Plass, G.N. (1959). The carbon dioxide theory of climate change. *Tellus.* **8**:140-154.
27. MacCracken, M.C. & Luther, F.M. (1985a). *Detecting the climatic effects of increasing carbon dioxide.* U.S. Department of Energy, Carbon Dioxide Research Division, DOE/ER-0235, Washington, D.C.
28. Mitchell, J.F.B. (1989). The "Greenhouse" effect and climate change. *Reviews of Geophysics* **27**, 115-139.
29. Jones, P.D., Wigley, T.M.L., Folland, C.K., Parker, D.E., Angell, J.F., Lebedef, S. & Hansen, J.E. (1988). Evidence for global warming in the past decade. *Nature* **338**, 790.
30. Hansen, J., Lacis, A., Rind, D., Russell, G., Stone, P., Fung, I., Ruedy, R. & Lerner, J. (1984). Climate sensitivity: analysis of feedback mechanisms. In: *Climate Processes and Climate Sensitivity*, Eds. Hansen, J.E. & Takahashi, T., Geophys. Monogr. Ser., vol. 29, AGU, Washington, D.C.
31. Waggoner, P.E. (1984). Agriculture and carbon dioxide. *American Scientist* **72:** 179-189.
32. Strain, B.R. (1985). Background on the response of vegetation to atmospheric carbon dioxide enrichment. In: *Direct effects of increasing carbon dioxide on vegetation*, B.R. Strain & J.D. Cure eds., Duke University, Durham, USA.
33. Goudriaan, J., van Keulen H. & van Laar H.H. (editors). (1990). The greenhouse effect and primary productivity in European agro-ecosystems. PUDOC, Wageningen.
34. Reynolds, J.F. & Acock, B. (1985). Modeling approaches for evaluating vegetation responses to carbon dioxide concentration. In: *Direct effects of increasing carbon dioxide on vegetation,* B.R. Strain & J.D. Cure eds., Duke University, Durham, USA.
35. Kimball, B.A. (1985). Adaptation of vegetation and management practices to a higher carbon dioxide world. In: *Direct effects of increasing carbon*

dioxide on vegetation, B.R. Strain & J.D. Cure eds., Duke University, Durham, USA.
36. Curtis, P.S., Drake, B.G. & Whigham, D.F. (1989). Nitrogen and carbon dynamics in C_3 and C_4 estuarine marsh plants grown under elevated CO_2 in situ. *Oecologia* **78**, 297-301.
37. Curtis, P.S., Balduman, L.M., Drake, B.G. & Whigham, D.F. (1990). The effect of elevated atmospheric CO_2 on below ground processes in C_3 and C_4 estuarine marsh communities. *Ecology* (in press).
38. Tissue, D.L. & Oechel, W.C. (1987). Physiological response of *Eriophorum vaginatum* to field elevated CO_2 and temperature in the Alaskan tussock tundra. *Ecology* **68**, 401-410.
39. Oberbauer, S.F., Strain, B.R., Fetcher, N. (1985). Effect of CO_2 enrichment on seedling physiology and growth of two tropical tree species. *Physiol. Plant.* **65**, 352-356.
40. Dahlman, R.C., Strain, B.R. & Rogers, H.H. (1985). Research on the response of vegetation to elevated atmospheric carbon dioxide. *J. Environ. Qual.* **14**:1-8.
41. Ho, L.C. (1977). Effects of CO_2 enrichment on the rates of photosynthesis and translocation of tomato leaves. *Ann. Appl. Biol.* **87**, 191-200.
42. Goudriaan, J. & de Ruiter, H.E. (1983). Plant growth in response to CO_2 enrichment, at two levels of nitrogen and phosphorus supply. 1. Dry matter, leaf area and development. *Neth. J. Agric. Sci.* **31**:157-169.
43. Goudriaan, J., van Laar, H.H. van Keulen, H. & Louwerse, W. (1984). Simulation of the effect of increased atmospheric CO_2 on assimilation and transpiration of a closed crop canopy. *Wissenschaftliche Zeitschrift der Humboldt-Universität zu Berlin. Math/-Nat. R.* **XXXIII**:352-356.
44. von Caemmerer, S. & Farquhar, G.D. (1984). Effects of partial defoliation, changes of irradiance during growth, short-term water stress and growth at enhanced p(CO_2) on the photosynthetic capacity of leaves of *Phaseolus vulgaris* L. *Planta* **160**, 320-329.
45. Farquhar, G.D., von Caemmerer, S. (1982). Modelling of photosynthetic response to environmental conditions. In: *Encyclopaedia of Plant Physiology (NS)*. Ed. Lange, O.L., Nobel, P.S., Osmond, C.B. and Ziegler, H. 12B, 549-587. Springer-Verlag, new York.
46. Sionit, N. (1983). Response of soybean to two levels of mineral nutrition in CO_2-enriched atmosphere. *Crop Sci.* **23**, 329-333.
47. Cure, J.D., Israel, D.W. & Rufty, T.W. (1988). Nitrogen stress effects on growth and seed yield of non-nodulated soybean exposed to elevated carbon dioxide. *Crop Science* **28**, 4, 671-677.
48. Cure, J.D., Rufty, T.W. & Israel, D.W. (1988). Phosphorus stress effects on growth and seed yield responses of non-nodulated soybean to elevated carbon dioxide. *Agronomy Journal* **80**, 6, 897-902.
49. Conroy, J.P., Milham, P.J., Mazur, P. & Barlow, E.W.R. (1990). Growth, dry weight partitioning and wood properties of *Pinus radiata* D. Don after two years of CO_2 enrichment. *Plant, Cell and Environment*, **13**: 329-337.
50. Gifford, R.M. (1977). Growth pattern, carbon dioxide exchange and dry weight distribution in wheat growing under differing photosynthetic environments. *Aust. J. Plant Physiol.* **4**, 99-110.
51. Louwerse, W. (1980). Effect of CO_2 concentration and irradiance on stomatal behavior of maize, barley and sunflower plants in the field. *Plant,*

Cell Environ. **3**: 391-398.
52. Schönfeld, M., Johnson, R.C. & Ferris, D.M. (1989). Development of winter wheat under increased atmospheric CO_2 and water limitations at tillering. *Crop Sci.* **29**: 1083-1086.
53. Cure, J.D., Rufty, T.W. & Israel, D.W. (1989). Alterations in soybean leaf development and photosynthesis in a CO_2-enriched atmosphere. *Bot. Gaz.* **150** (4), 337-345.
54. Winzeler, M., Monteil, Ph. & Nösberger. (1989). Grain growth and tall and short spring wheat genotypes at different assimilate supplies. *Crop Sci.* **29**: 1487-1491.
55. Gifford, R.M., Evans, L.T. (1981). Photosynthesis, carbon partitioning and yield. *Ann. Rev. Plant Physiol.* **32**, 485-509.
56. Sionit, N., Hellmers, H. & Strain, B.R. (1980). Growth and yield of wheat under CO_2 enrichment and water stress. *Crop Sci.* **20**: 687-690.
57. Mohapatra, P.K. 1990. CO_2 enrichment and physiology of inflorescence development in wheat. *Photosynthetica*, **24**(1):9-15.
58. Thomas, J.F. & Harvey C.N. (1983). Leaf anatomy of four species frown under continuous CO_2 enrichment. *Bot. Gaz.* **144**(3):303-309.
59. Apel, P. (1989). Influence of CO_2 on stomatal numbers. *Biologia Plantarum*. **31**(1): 72-74.
60. Wong, S.C. (1979). Elevated atmospheric partial pressure of CO_2 and plant growth. I. Interactions of nitrogen nutrition and photosynthetic capacity in C_3 and C_4 plants. *Oecologia* **44**, 68-74.
61. Marc, J. & Gifford, R.M. (1983). Floral initiation in wheat, sunflower, and sorghum under carbon dioxide enrichment. *Can. J. Bot.* **62**: 9-14.
62. Farquhar, G.D., Sharkey, T.D. (1982). Stomatal conductance and photosynthesis. *Ann. Rev. Plant Physiol.* **33**, 317-345.
63. Jones, P., Allen, L.H., Jones, J.W. & Valle, R.R. (1985). Photosynthesis and transpiration responses of soybean canopies to short- and long-term CO_2 treatments. *Agron. J.* **77**, 119-126.
64. Laing, W.A., Ogren, W.L. & Hageman, R.H. (1974). Regulation of soybean net photosynthetic CO_2 fixation by the interaction of CO_2, O_2 and ribulose-1,5-bisphosphate carboxylase. *Plant Physiol.* **54**, 678-685.
65. Smith, S.D., Strain, B.R. & Sharkey, T.D. (1987). Effect of CO_2 enrichment on growth, CO_2 assimilation and water use efficiency in four great basin grasses. *Functional Ecol.* **1**, 139-143.
66. Valle, R., Mishoe, J.W., Jones, J.W. and Allen, L.H. (1985). Transpiration rate and water use efficiency of soybean leaves adapted to different CO_2 environments. *Crop Sci.* **25**, 477-482.
67. DeLucia, E.H., Sasek, T.W. & Strain, B.R. (1985). Photosynthetic inhibition after long-term exposure to elevated levels of atmospheric carbon dioxide. *Photosynthesis Res.* **7**, 175-184.
68. Mauney, J.R., Guinn, G., Fry, K.E. & Hesketh, J.D. (1979). Correlation of photosynthetic carbon dioxide uptake and carbohydrate accumulation in cotton, soybean, sunflower and sorghum. *Photosynthetica* **13**, 260-266.
69. Besford, R.T., Ludwig, L.J. & Withers, A.C. (1990). Acclimation of tomato plants growing in high CO_2, photosynthesis and ribulose-1,5-bisphosphate carboxylase protein. *J. Exp. Botany* **41**,
70. Vu, J.C.V., Allen, L.H. & Bowes, G. (1989). Leaf ultrastructure, carbohydrates and protein of soybeans grown under CO_2 enrichment.

Environ. Exp. Bot. **29**, 141-147.

71. Herold, A. (1980). Regulation of photosynthesis by sink activity: the missing link. *New Phytol.* **58**: 2181-2189.
72. Azcon-Bieto, J. (1983). Inhibition of photosynthesis by carbohydrates in wheat leaves. *Plant Physiol.* **73**:681-686.
73. Dietz, K.J. & Heber, U. (1986). Light and CO_2 limitation of photosynthesis and states of the reactions regenerating ribulose 1,5-bisphosphate or reducing 3-phosphoglycerate. *Biochim. Biophys. Acta* **848**, 392-401.
74. Woodrow, I.E. & Berry, J.A. (1988). Enzymatic regulation of photosynthetic CO_2 fixation in C_3 plants. *Ann. Rev. Plant Physiol. Plant Mol. Biol.* **39**, 533-594.
75. Yelle, S., Beeson, R.C., Trudel, M.J. & Gosselin, A. (1989a). Acclimation of two tomato species to high atmospheric CO_2. I. Sugar and starch concentrations. *Plant Physiology* **90**, 1465-1472.
76. Yelle, S., Beeson, R.C., Trudel, M.J. & Gosselin, A. (1989b). Acclimation of two tomato species to high atmospheric CO_2. II. Ribulose, 1,5-bisphosphate carboxylase/oxygenase and phosphoenolpyruvate carboxylase. *Plant Physiology* **90**, 1473-1477.
77. Porter, M.A. & Grodzinski, B. (1984). Acclimation to high CO_2 in bean. Carbonic anhydrase and ribulose bisphosphate carboxylase. *Plant Physiology*, **74**, 413-416.
78. Mott, K.A., Jensen, R.G. & Berry, J.A. (1986). Limitation of photosynthesis by RuBP regeneration rate. In: *Biological control of photosynthesis*, eds. Marcelle, R., Clijsters, H. & Van Pouche, M., pp. 33-43. Martinus Nijhoff, Dordrecht.
79. Mott, K.A. (1990). Sensing of atmospheric CO_2 by plants. *Plant, Cell and Environment*. **13**: 731-737.
80. Spencer, W. & Bowes, G. (1986). Photosynthesis and growth of water hyacinth under CO_2 enrichment. *Plant Physiol.* **82**, 528-533.
81. Besford, R.T. (1990). The greenhouse effect: acclimation of tomato plants growing in high CO_2, relative changes in Calvin cycle enzymes. *J. Plant Physiol.* **136**, 458-463.
82. Gerbaud, A., André, M., Gaudillère, J.-P. & Daguenet, A. (1988). The influence of reduced photorespiration on long-term growth and development of wheat. *Physiologia Plantarum,* **73**: 479-485.
83. Lawlor, D.W., Kendall, A.C., Keys, A.J. & Boyle, F.A. (1988). Nitrate nutrition and temperature effects on wheat : synthesis of plant growth and nitrogen uptake in relation to metabolic and physiological processes. *J. Expt. Botany,* **39**, 329-343.
84. Seemann, J.R., Sharkey, T.D. (1986). Salinity and nitrogen effects on photosynthesis, ribulose 1,5-bisphosphate carboxylase and metabolic pool size in *Phaseolus vulgaris* L. *Plant Physiol.* **82**, 555-560.
85. Gent, M.P.N. & Kiyomoto, R.K. (1985). Comparison of canopy and flag leaf net carbon dioxide exchange of 1920 and 1977 New York Winter wheats. *Crop Science,* **25**: 81-86.
86. Patterson, D.T., Flint, P. & Beyers, J.L. (1984). Effects of CO_2 enrichment on competition between a C_4 weed and a C_3 crop. *Weed Sci.* **32**, 101-105.
87. Gifford, R.M., Lambers, H. & Morison, I.I.L. (1985). Respiration of crop species under CO_2 enrichment. *Physiol. Plant.* **63**, 351-356.

WATER STRESS, STOMATA, AND ABSCISIC ACID

William H. Outlaw Jr. and Michael J. Harris
Department of Biological Science (B-157), Florida State University, Tallahassee, FL 32306-3050, USA

INTRODUCTION

Agricultural productivity is expected to be sensitive to [CO_2] increases and accompanying climatic changes [1, 2]. This and ancillary problems, such as decrease in precipitation, are currently a matter of broad public concern (*e.g.* [3, 4]), if not action. Many uncertainties (*e.g.*, the relationship between CO_2 assimilation and stomatal conductance [5], gross morphological changes in response to CO_2 enrichment [6], change in stomatal density as a function of growth-CO_2 concentration [7]) make predictions of plant responses to higher [CO_2] problematical. Stated another way, our failure to understand basic plant physiological responses to individual environmental perturbations preclude a quantitative synthesis with which most scientists will agree. However, during the past five years, progress on several fronts has converged to answer, or at least to redefine, the question of how water stress controls stomatal aperture. We shall present this new understanding with an emphasis on our own work after a basic review of stomatal mechanics.

---

[1]This contribution is taken, in part, from a grant proposal submitted to the U.S. Department of Energy by the first author in June 1990. Part of the progress report of that proposal has been published [8].

BASIC REVIEW OF STOMATAL MECHANICS

CO_2 concentration in the leaf intercellular spaces is lower than that of the atmosphere; CO_2 diffuses into the leaf through stomata, which are embedded in the leaf epidermis. In contrast, the water vapor pressure in the leaf intercellular spaces is almost at saturation and, thus, is much higher than that surrounding the leaf. Consequently, water vapor is lost from the leaf through the nonselective stomata. The control of gas exchange is crucial to the plant's survival and growth: although CO_2 is required for photosynthesis, water loss must be strictly limited to prevent desiccation. Here, "control of gas exchange" will be simplified to "control of stomatal aperture size," because stomata open and close, permitting more or less gas exchange. Before 1983 [9-11], the basic outline of stomatal aperture-size regulation was understood (Fig. 1). Speaking broadly, a stoma opens when the two guard cells flanking it accumulate K^+ salts. These salts lower the guard-cell ψ_s; as a result, water flows into guard cells, which increases their volume and ψ_p. The guard-cell walls distend asymmetrically as internal volume and pressure increase, the aperture between the cells widens, and diffusion of gases through it is enhanced. These and other general aspects of stomatal physiology are introduced in comprehensive books [12-15]. Recent and excellent reviews are offered by Raschke *et al*. [16], by MacRobbie [17], and by Mansfield *et al*. [18]. Finally, a special edition of *Physiol. Biochem. Pflanzen*. (H. Göring, H. Meidner, and T. Mansfield, eds.), which contains several

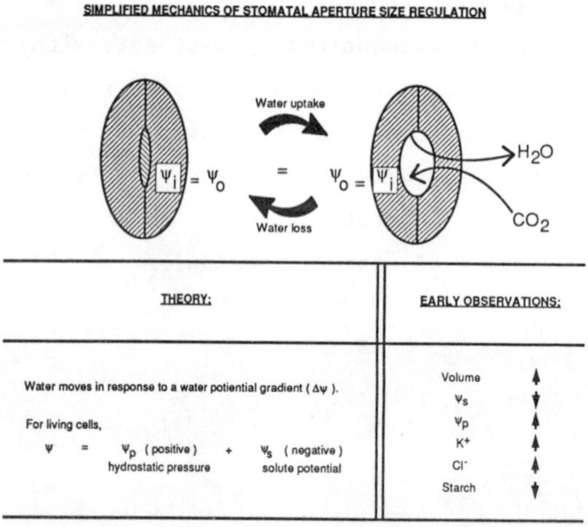

Fig. 1. Simplified outline of stomatal aperture size regulation. The biochemical bases for "Early Observations" are put forth in more detail in Fig. 2.

reviews of specific aspects of stomatal physiology, will be available soon.

CURRENT IDEAS ON GUARD-CELL K^+ MOVEMENT AND CONTROL BY ABSCISIC ACID

Giant strides have been made in our general understanding of the biochemical and biophysical bases for ion transport in plants [19-21], although much remains to be learned and, as the reader will note, there are cases of conflict. We summarize conclusions and conjectures from a large number of studies (Fig. 2), some documentation for which is provided (Table 1). For the present purposes, we will speak broadly and include current ideas even if they are disputed.

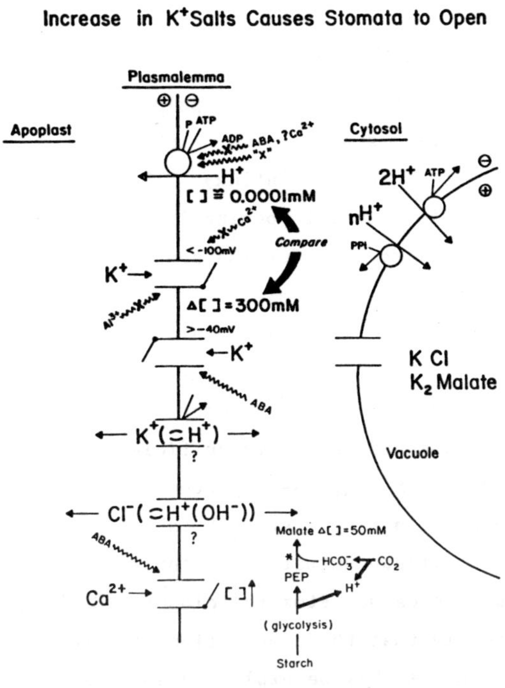

Fig. 2. A synthesis of the fundamental biochemical and biophysical bases of stomatal opening. Some components of this figure are conjectural. No implications concerning macromolecular relationships (configurations, size, proximity, number) are intended. Effectors may act indirectly; effector site should be interpreted as generalized. < indicates "more negative than"; the squiggly arrow indicates activation except when X'd, when it means inhibition.

Guard-cell plasmalemma contains a strongly electrogenic ATP-powered proton pump. For physiologically required activity, the pump is stimulated by "cytoplasmic factors" (≠ MgATP). This statement is based on higher currents in the slow-whole-cell configuration [22] and especially from microelectrode recordings [23], compared with currents measured with recording configurations that wash out cytoplasmic factors. Although it is unproven for guard-cell H^+-ATPase, other work indicates that the H^+-ATPase is reversibly phosphorylated [24], which is intriguing because blue light causes

Table 1. Examples of observations specifically on guard cells for construction of Figure 2.

Absence of functional plasmodesmata	Wille and Lucas [79]
Presence of ion channels in guard cells	Schroeder et al. [30]
Voltage dependence of K^+ channel(s)	Schroeder et al. [80]
ABA effects on K^+ channels	Schauf and Wilson [81]
Al^{3+} inhibition of K^+ in-channels	Schroeder [22]
Al^{3+} inhibition of stomatal opening (not closing)	Schnabl and Ziegler [82]
TEA inhibition of K^+ channels	Schauf and Wilson [81]
TEA inhibition of stomatal opening and closing	Karlsson and Schwartz [83]
FC stimulation of TEA-insensitive K^+ uptake	Blatt and Clint [84]
ABA effects on K^+	MacRobbie [85]
Importance of exogenous Ca^{2+} on K^+ efflux	Schwartz [86]
Cytosolic Ca^{2+} inhibition of inward K^+ in-channel	Schroeder and Hagiwara [87]
ABA-evoked mobilization of K^+ out-channel	Blatt [88]
Absence of long-term effect of ABA on channel	MacRobbie [89]
Ca^{2+} effects on outward-rectifying K^+ channels	Hosoi et al. [90]
Ca^{2+} dependence of ABA inhibition of stomatal opening	DeSilva et al. [91]
(disputed)	Smith and Willmer [92]
Ca^{2+}-channel blockers' effect on ABA efficacy	DeSilva et al. [93]
ABA elevation of cytosolic Ca^{2+}	McAinsh et al. [94]

stomatal opening, activates the pump [25], and inhibits specific protein phosphorylation in another plant system [26]. The proton pump is reported to be partially blocked by ABA, (e.g. [27]), and it is suggested that μM Ca^{2+} results in a decrease in its activity too [28] (see next section).

To borrow from MacRobbie [29], the paper by Schroeder et al. [30] is a landmark in the study of stomatal physiology. These authors identified (for the first time in plants) K^+ channels in membrane patches of Vicia faba guard cells. Hyperpolarization caused by the H^+ATPase opens a voltage-sensitive Ca^{2+}-inhibited "inward" K^+ channel[2] (which results in net K^+ influx). The inward K^+ channel is blocked by Al^{3+}, as is stomatal opening (but not stomatal closing); this correspondence--like that for TEA sensitivity--implicates the pump-channel "pair" as a biochemical mechanism underlying the accumulation of K^+ during stomatal opening. However, this explanation may be insufficient, as could be inferred from data published much earlier. E.g., even very negative membrane potentials (as reported by Gunar et al. [31]) could not drive sufficient K^+ uptake for stomatal opening from μM $[K^+]_{out}$ (as reported by Fischer [32]). Blatt [33] has elaborated in a single study by showing that, at low $[K^+]_{out}$, K^+ would tend to diffuse outward, not inward, during opening. In addition, K^+ ionophore causes stomatal closure [34-36]. (It should be noted parenthetically that these objections to the sufficiency of K^+ channels alone can possibly be explained away as

[2]Here and elsewhere, "inward" should be taken as an abbreviation for experimental demonstration of inward currents (and not as an intrinsic property of the channel itself).

physiologically irrelevant if the recently determined 3-70 mM values [37-39] for apoplastic [K^+] are accepted.)

Finally, ion channels are in the tonoplast too, where they may transport malate and K^+ [40], and, for completeness, we note that occasional papers [41-43] indicate that the concentrations of materials other than K salts must fluctuate during stomatal movements.

RELATIONSHIP OF (EXOGENOUS) ABA TO STOMATAL MOVEMENTS

ABA is the well-known endogenous antitranspirant that accumulates in water-deprived plants (review: [44-46]). Among many physiological effects, ABA causes stomatal closure. In this paragraph, we will only touch on ABA effects on guard cells in the context of Figure 2 (Table 1). Apparently, exogenous ABA causes the transient opening of Ca^{2+} channels, resulting in the elevation of cytosolic [Ca^{2+}]. Possibly, this elevation has two effects, viz., it blocks inward K^+ channels (see, e.g., [47]) and it may decrease H^+ extrusion by inhibiting the H^+-ATPase (but see [48]). It is unclear whether ABA itself directly affects the H^+-ATPase [49-51], but a direct effect of ABA on the outward K^+ channel has been demonstrated (Table 1). This effect (which alone would not require energy) is only a partial explanation, as metabolic inhibitors block ABA-induced stomatal closure [52]. It has been long known that ABA affects protein synthesis [53]. It is likely that some long-term ABA effects on guard-cells are due to an altered protein complement; in this regard, it is interesting to note that µM ABA alters the calmodulin complement of embryonic axis of *Cicer arietinum* [54]. If so, it would be interesting to learn whether this long-term effect in guard cells is calcium dependent (i.e., analogous to the effect observed by Napier *et al.* [55]). In brief summary, ABA appears to influence the plant's physiology through diverse mechanisms; the view that its effects on guard cells are external is too simple an explanation to accommodate such observations as differential effects of endogenous and exogenous ABA [56].

A ROLE FOR ABA IN THE WATER-STRESS EFFECTS ON STOMATA

Environmental factors--light, CO_2 concentration, and, perhaps most powerfully, water status--are integrated to achieve the requisite balance in gas exchange. Water stress is accompanied by a dramatic increase in leaf ABA concentration [57]. Spraying ABA onto leaves [58], supplying it to leaves through the transpiration stream [59], or floating detached epidermal peels on an ABA-containing solution [60] causes stomatal closure. Importantly, Hoglund and Klockare [61] showed that norflurazon-treated plants (which do not accumulate ABA) do not close stomata in response to rapidly imposed water stress. As pointed out in the previous paragraph, others recently have learned that ABA affects solute transport in guard cells and, perhaps separately, elevates cytosolic Ca^{2+}. Such correlative findings, as well as the work on mutants, which began early [62], would seem to imply that ABA mediates the water-stress effect on stomata, except that two incongruous phenomena have long been raised as objections (see [63, 64]). First, following imposition of water stress, stomatal closure occurs before leaf [ABA] increases detectibly (many reports, e.g. [65]). Second, leaf [ABA] returns to prestress values following relief of water stress before stomatal aperture size does (again, many reports, e.g. [66]). To a lesser extent, the noncoincidence of diurnal changes in [ABA] and stomatal conductance have also been difficult to explain straightforwardly [67]. The simplest way to reconcile these "hitherto unexplained discrepancies" [68] is to invoke compartmentation (e.g. [69, 70]), but attempts to demonstrate compartmentation have not yielded unequivocal results. (E.g., Dörffling and Tietz [71] could not detect a rise in epidermal [ABA] consistent with the increase in diffusive resistance following water-stress imposition in Commelina communis.) In an attempt to determine whether guard-cell ABA could be involved in the initial stomatal response to water stress, Vicia faba leaflets were dehydrated to 90% of fresh weight (~15 min). At various times, leaflets were frozen and freeze-dried. Samples were also taken after relief from water stress (relief by rehydration, 5 min in H_2O). [ABA] was assayed in individually excised cells by means of an ultrasensitive assay (= 200 x 10^{-18} mol [72, 73]). An important result (Fig. 3, from Harris and Outlaw [8]) that emerged from these analyses is that ABA accumulates very rapidly in guard cells (~9x increase after 15 min, before any other cell type had altered [ABA]). These findings close the 15-year argument against the possibility of ABA's involvement in initial responses and substantiate the assertion of Zeevaart and Creelman [45] that "knowledge of the distribution of ABA within cells and organs is essential to understand its mode of action." As mentioned, we also determined [ABA] in various cells after relief from water stress. (Recall that stomata do not open normally for up to several days following water-stress relief; for data on *Vicia*, our experimental plant, see Stålfelt [74] and Fischer *et al.* [75] and subsequent literature.) The kinetics of [ABA] decline in guard cells were different from those in other cells, but the more important result is that [ABA] in guard cells had declined to near control values within six hours after stress was relieved. Thus, it appears that continued elevated guard-cell [ABA] itself is not the cause of the after-stress effect (one of the three hypotheses presented by Raschke [44]).

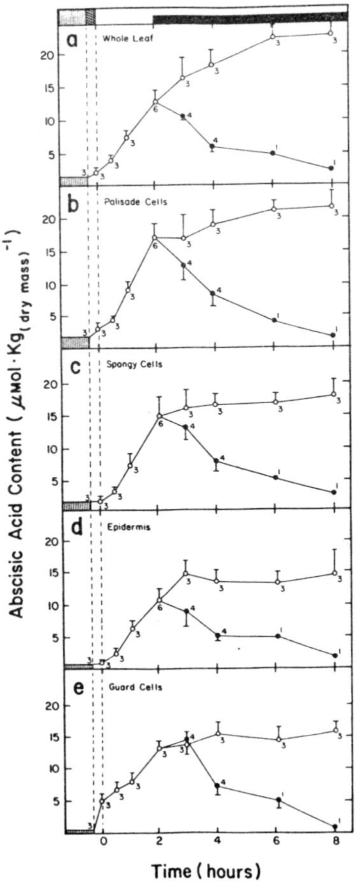

Fig. 3 ABA concentration kinetics in various cell types of *Vicia faba* L. leaves following stress imposition and relief from stress. Control values--from well watered plants--are given by the bars at the left of each panel. During stress imposition (a 15-min period during which detached leaves were dehydrated to 90% fresh weight) guard-cell ABA concentration increased 9x while in other cells [ABA] did not change detectably. Six h after stress relief (submersion in H_2O, 5 min), guard-cell [ABA] had declined like that in other samples.

However, the stress-relief results do not permit a clear-cut conclusion, as did the stress-imposition studies. (For example, it is possible that guard-cell [ABA] would remain high in attached leaves that could continuously receive an ABA supply from the roots.) Many hypotheses are consistent with extant data for aftereffects of water stress; finally, a resolution will depend on further information about the [ABA] in guard cells and the bathing apoplastic solution. At the moment, we would favor an alteration in cellular protein as the first hypothesis to test. (Reduction in turgor stimulates ABA accumulation [76, 77] and rapid changes in leaf translatable RNA [78].

REFERENCES

1. Adams, R.M., Rosenzweig, C., Peart, R.M., Ritchie, J.T., McCart, B.A., Glyer, J.D., Curry, R.B., Jones, J.W., Boote, K.J. and Allen, L.H., Jr. (1990). Global climate change and U.S. agriculture. Nature 345:219-224.
2. Waggoner, P.E. (ed.) (1990). Climate Change and U.S. Water Resources. Wiley.
3. Linden, E. (1990). The last drops. Time 136:58-61.
4. Matthews, S.W. (1990). Under the sun. National Geographic 178:66-99.
5. Ball, J.T., Woodrow, I.E. and Berry, J.A. (1987). A model predicting stomatal conductance and its contribution to the control of photosynthesis under different environmental conditions. Prog. Photo. Res. IV.5.221-224.
6. Mousseau, M. and Enoch, H.Z. (1989). Carbon dioxide enrichment reduces shoot growth in sweet chestnut seedlings (Castanea sativa Mill.). Plant Cell Environ. 12:927-934.
7. Woodward, F.I. (1987). Stomatal numbers are sensitive to increases in CO_2 from preindustrial levels. Nature 327:617-618.
8. Harris, M.J. and Outlaw, W.H., Jr. (1991). Rapid adjustment of guard-cell abscisic acid levels to current leaf water status. Plant Physiol. 95, in press.
9. Outlaw, W.H., Jr. (1982). Carbon metabolism in guard cells. Recent Adv. Phytochem. 16:185-222.
10. Outlaw, W.H., Jr. (1983). Current concepts on the role of potassium in guard cells. Physiol. Plant. 59:302-311.
11. Zeiger, E. (1983). The biology of stomatal guard cells. Annu. Rev. Plant Physiol. 34:441-475.
12. Sen, D.N., Chawan, D.D. and Bansal, R.P. (eds.) (1979). Structure, Function and Ecology of Stomata. Bishen Singh Mahendra Pal Singh.
13. Jarvis, P.G. and Mansfield, T.A. (1981). Stomatal Physiology. Cambridge University Press, Cambridge.
14. Willmer, C.M. (1983). Stomata. Longman.
15. Zeiger, E., Farquhar, G.D. and Cowan, I.R. (eds.) (1987). Stomatal Function. Stanford University Press.

16. Raschke, K., Hedrich, R., Reckmann, U. and Schroeder, J.I. (1988). Exploring biophysical and biochemical components of the osmotic motor that drives stomatal movement. Bot. Acta 101:283-294.

17. MacRobbie, E.A.C. (1988). Control of ion fluxes in stomatal guard cells. Bot. Acta 101:140-148.

18. Mansfield, T.A., Hetherington, A.M. and Atkinson, C.J. (1990). Some current aspects of stomatal phyiology. Annu. Rev. Plant Physiol. Plant Mol. Biol. 41:55-75.

19. Hedrich, R. and Schroeder, J.I. (1989). The physiology of ion channels and electrogenic pumps in higher plants. Annu. Rev. Plant Physiol. Plant Mol. Biol. 40:539-569.

20. Serrano, B. (1989). Structure and function of plasma membrane ATPase. Annu. Rev. Plant Physiol. Plant Mol. Biol. 40:61-94.

21. Tester, M. (1990). Tansley Review No. 21, Plant ion channels: whole cell and single channel studies. New Phytol. 114:305-340.

22. Schroeder, J.I. (1988). K^+ transport properties of K^+ channels in the plasma membrane of Vicia faba guard cells. J. Gen. Physiol. 92:667-683.

23. Blatt, M.R. (1987a). Electrical characteristics of stomatal guard cells: the ionic basis of the membrane potential and the consequence of potassium chloride leakage from microelectrodes. Planta 170:272-287.

24. Schaller, G.E. and Sussman, M.R. (1988). Phosphorylation of the plasma-membrane H^+-ATPase of oat roots by a calcium-stimulated protein kinase. Planta 173:509-518.

25. Assmann, S.M., Simoncini, L. and Schroeder, J.I. (1985). Blue light activates electrogenic ion pumping in guard cell protoplasts of Vicia faba. Nature 318:285-287.

26. Short, T.W. and Briggs, W.R. (1990). Characterization of a rapid, blue light-mediated change in detectable phosphorylation of a plasma membrane protein from etiolated pea (Pisum sativum L.) seedlings. Plant Physiol. 92:179-185.

27. Nejidat, A., Itai, C. and Bejerano, N.R. (1989). Regulation of Mg, K-ATPase activity in guard cells of Commelina communis by phytochrome and ABA. Plant Cell Physiol. 30:945-949.

28. Inoue, H. and Katoh, Y. (1987). Calcium inhibits ion-stimulated stomatal opening in epidermal strips of Commelina communis L. J. Exp. Bot. 38:142-149.

29. MacRobbie, E.A.C. (1985). Ion channels in plant cells. Nature 313:529.
30. Schroeder, J.I., Hedrich, R. and Fernandez, J.M. (1984). Potassium selective single channels in guard cell protoplasts of Vicia faba. Nature 312:361-362.
31. Gunar, I.I., Zlotnikova, I.F. and Panichkin, L.A. (1975). Electrophysiological investigation of cells of the stomate complex in spiderwort. Fiziol. Rast. 22:810-813.
32. Fischer, R.A. (1968). Stomatal opening: role of potassium uptake by guard cells. Science 160:784-785.
33. Blatt, M.R. (1987b). Electrical characteristics of stomatal guard cells: the contribution of ATP-dependent, "electrogenic" transport revealed by current-voltage and difference-current-voltage analysis. J. Memb. Biol. 98:257-274.
34. Richardson, C.H., Truter, M.R., Wingfield, J.N., Travis, A.J., Mansfield, T.A. and Jarvis, R.J. (1979). The effect of benzo-18-crown-6, a synthetic ionophore, on stomatal opening and its interaction with abscisic acid. Plant Cell Environ. 2:325-327.
35. Gepstein, S., Jacobs, M. and Taiz, L. (1982). Inhibition of stomatal opening in Vicia faba epidermal tissue by vanadate and abscisic acid. Plant Sci. Lett. 28:63-72.
36. Dahse, I., Schnabl, H., Hampp, R., Ziegler, H., Müller, E. and Liebermann, B. (1988). Inhibition of light-induced stomatal opening and of guard-cell-protoplast swelling in Vicia faba L. by tentoxin, an inhibitor of photophosphorylation. Planta 173:391-396.
37. Bowling, D.J.F. (1987) Measurements of the apoplastic activity of K^+ and Cl^- in the leaf epidermis of Commelina communis in relation to stomatal activity. J. Exp. Bot. 38:1351-1355.
38. Lowen, C.Z. and Satter, R.L. (1989). Light-promoted changes in apoplastic K^+ activity in the Samanea saman pulvinus, monitored with liquid membrane microelectrodes. Planta 179:421-427.
39. Starrach, N. and Mayer, W.-E. (1989). Changes of the apoplastic pH and K^+ concentration of the Phaseolus pulvinus in situ in relation to rhythmic leaf movements. J. Exp. Bot. 40:865-873.
40. Hedrich, R., Flügge, U.I. and Fernandez, J.M. (1986). Patch-clamp studies of ion transport in isolated plant vacuoles. FEBS Lett. 204:228-232.

41. MacRobbie, E.A.C. and Lettau, J. (1980). Potassium content and aperture in "intact" stomatal and epidermal cells of Commelina communis L. J. Memb. Biol. 56:249-256.
42. Jinno, N. and Kuraishi, S. (1982). Acid-induced stomatal opening in Commelina communis and Vicia faba. Plant Cell Physiol. 23:1169-1174.
43. Maier-Maercker, U. (1983). A critical assessment of the role of potassium and osmolarity in stomatal opening. J. Exp. Bot. 34:811-824.
44. Raschke, K. (1987). Action of abscisic acid on guard cells. In: Stomatal Function, Zeiger, E., Farquhar, G.D. and Cowan, I.R., eds., Stanford University Press, pp. 253-280.
45. Zeevaart, J.A.D. and Creelman, R.A. (1988). Metabolism and physiology of abscisic acid. Annu. Rev. Plant Physiol. Plant Mol. Biol. 39:439-473.
46. Creelman, R.A. (1989). Abscisic acid physiology and biosynthesis in higher plants. Physiol. Plant. 75:131-136.
47. Schroeder, J.I. and Hagiwara, S. (1990). Voltage-dependent activation of Ca^{2+}-regulated anion channels and K^+ uptake channels in Vicia faba guard cells. In: Plant Growth and Development, Leonard, R.T. and Hepler, P.K., eds., American Society of Plant Physiologists Symposium Series, Vol 4.
48. Miller, A.J. and Sanders, D. (1987). Depletion of cytosolic free calcium induced by photosynthesis. Nature 326:397-400.
49. Raghavendra, A.S., Rao, L.M. and Das, V.S.R. (1976). Adenosine triphosphatase in epidermal tissue of Commelina beghalensis: possible involvement of isoenzymes in stomatal movement. Plant Sci. Lett. 7:391-396.
50. Kasamo, K. (1979). Characterization of membrane-bound Mg^{++}-activated ATPase isolated from the lower epidermis of tobacco leaves. Plant Cell Physiol. 20:281-292.
51. Blum, W., Key, G., Weiler, E.W. (1988). ATPase activity in plasmalemma-rich vesicles isolated by aqueous two-phase partitioning from Vicia faba mesophyll and epidermis: characterization and influence of abscisic acid and fusicoccin. Physiol. Plant. 72:279-287.
52. Weyers, J.D.B., Paterson, N.W., Fitzsimons, P.J. and Dudley, J.M. (1982). Metabolic inhibitors block ABA-induced stomatal closure. J. Exp. Bot. 33:1270-1278.

53. Skriver, K. and Mundy, J. (1990). Gene expression in response to abscisic acid and osmotic stress. Plant Cell 2:503-512.

54. Hernández-Nistal, J., Rodriguez, D., Nicolas, G. and Aldasoro, J.J. (1989). Abscisic acid and temperature modify the levels of calmodulin in embryonic axes of Cicer arietinum. Physiol. Plant. 75:255-260.

55. Napier, J.A., Chapman, J.M. and Black, M. (1989). Calcium-dependent induction of novel proteins by abscisic acid in wheat aleurone tissue of different developmental stages. Planta 179:156-164.

56. Wardle, K., Dobbs, E.B. and Short, K.C. (1982). Effect of abscisic acid pretreatments on stomatal reopening in Vicia faba. Physiol. Plant. 56:312-316.

57. Wright, S.T.C. and Hiron, R.W.P. (1969). (+)-Abscisic acid, the growth inhibitor induced in detached wheat leaves by a period of wilting. Nature 224:719-720.

58. Jones, R.J. and Mansfield, T.A. (1972). Effects of abscisic acid and its esters on stomatal aperture and the transpiration ratio. Physiol. Plant. 26:321-327.

59. Mittelheuser, C.J. and van Steveninck, R.F.M. (1969). Stomatal closure and inhibition of transpiration induced by (RS)-abscisic acid. Nature 221:281-282

60. Horton, R.F. (1971). Stomatal opening: the role of abscisic acid. Can. J. Bot. 49:583-585.

61. Höglund, H.-O. and Klockare, R. (1987). Stomatal responses to rapidly imposed water stress and light/dark transition in norflurazon-treated wheat plants. Physiol. Plant. 69:477-480.

62. Tal, M. and Imber, D. (1970). Abnormal stomatal behavior and hormone imbalance in flacca, a wilty mutant of tomato. II. Auxin- and abscisic acid-like activity. Plant Physiol. 46:373-376.

63. Raschke, K. (1982). Involvement of abscisic acid in the regulation of gas exchange: evidence and inconsistencies. In: Plant Growth Substances, Wareing, P.F., ed., Academic Press, pp. 581-590.

64. Radin, J.W. and Ackerson, R.C. (1982). Does abscisic acid control stomatal closure during water stress? What's New in Plant Physiology 12:9-12.

65. Beardsell M.F. and Cohen, D. (1975). Relationships between leaf water status, abscisic acid levels, and stomatal resistance in maize and sorghum. Plant Physiol. 56:207-212.

66. Henson, I.E. (1981). Abscisic acid and after-effects of water stress in pearl millet (*Pennisetum americanum* (L.) Leeke). Plant Sci. Lett. 21:129-135.

67. Henson, I.E., Alagarswamy, G., Mahalakshmi, V. and Bidinger, F.R. (1982). Diurnal changes in endogenous abscisic acid in leaves of pearl millet (*Pennisetum americanum* (L.) Leeke) under field conditions. J. Exp. Bot. 33:416-425.

68. Quarrie, S.A. (1984). Abscisic acid and drought resistance in crop plants. News Bull. Brit. Plant Growth Reg. Group 7:1-15.

69. Raschke, K. Zeevaart, J.A.D. (1976). Abscisic acid content, transpiration, and stomatal conductance as related to leaf age in plants of *Xanthium strumarium* L. Plant Physiol. 58:169-174.

70. McMichael, B.L. and Hanny, B.W. (1977). Endogenous levels of abscisic acid in water-stressed cotton leaves. Agron. J. 69:979-982

71. Dörffling, K. and Tietz, D. (1985). Abscisic acid in leaf epidermis of *Commelina communis* L.: distribution and correlation with stomatal closure. J. Plant Physiol. 117:297-305.

72. Harris, M.J. and Outlaw, W.H., Jr. (1990). Histochemical technique: a low-volume, enzyme-amplified immunoassay with sub-fmol sensitivity. Application to measurement of abscisic acid in stomatal guard cells. Physiol. Plant. 78:495-500.

73. Harris, M.J., Outlaw, W.H., Jr., Mertens, R. and Weiler, E.W. (1988). Water-stress induced change in the abscisic acid content of guard cells and other cells of *Vicia faba* L. leaves as determined by enzyme-amplified immunoassay. Proc. Natl. Acad. Sci. USA 85:2584-2588.

74. Stålfelt, M.G. (1963). Diurnal dark reactions in the stomatal movements. Physiol. Plant. 16:756-766.

75. Fischer, R.A., Hsiao, T.C. and Hagan, R.M. (1970). After-effect of water stress on stomatal opening potential. I. Techniques and magnitudes. J. Exp. Bot. 21:371-385.

76. Pierce, M. and Raschke, K. (1980). Correlation between loss of turgor and accumulation of abscisic acid in detached leaves. Planta 148:174-184.

77. Creelman, R.A., Zeevaart, J.A.D. (1985). Abscisic acid accumulation in spinach leaf slices in the presence of penetrating and nonpenetrating solutes. Plant Physiol. 77:25-28.
78. Guerrero, F.D. and Mullet, J.E. (1988). Reduction of turgor induces rapid changes in leaf translatable RNA. Plant Physiol. 88:401-408.
79. Wille, A. and Lucas, W. (1984). Ultrastructural and histochemical studies on guard cells. Planta 160:129-142.
80. Schroeder, J.I., Raschke, K. and Neher, E. (1987). Voltage dependence of K^+ channels in guard cell protoplasts. Proc. Natl. Acad. Sci. USA 84:4108-4112.
81. Schauf, C.L. and Wilson, K.J. (1987). Effects of abscisic acid on K^+ channels in Vicia faba guard cell protoplasts. Biochem. Biophys. Res. Comm. 145:284-290.
82. Schnabl, H. and Ziegler, H. (1975). The influence of aluminum ions on the movement of the stomata in Vicia faba epidermis strips. Z. Pflanzen Physiol. 74:394-404.
83. Karlsson, P.E. and Schwartz, A. (1988). Characterization of the effects of metabolic inhibitors, ATPase inhibitors and a potassium-channel blocker on stomatal opening and closing in isolated epidermis of Commelina communis L. Plant Cell Environ. 11:165-172.
84. Blatt, M.R. and Clint, G.M. (1989). Mechanisms of fusicoccin action: kinetic modification and inactivation of K^+ channels in guard cells. Planta 178:509-523.
85. MacRobbie, E.A.C. (1981). Effects of ABA on isolated guard cells of Commelina communis L. J. Exp. Bot. 32:563-572.
86. Schwartz, A. (1985). Role of Ca^{2+} and EGTA on stomatal movements in Commelina communis L. Plant Physiol. 79:1003-1005.
87. Schroeder, J.I. and Hagiwara, S. (1989). Cytosolic calcium regulates ion channels in the plasma membrane of Vicia faba guard cells. Nature 338:427-430.
88. Blatt, M.R. (1990). Potassium channel currents in intact stomatal guard cells: rapid enhancement by abscisic acid. Planta 180:445-455.
89. MacRobbie, E.A.C. (1989). Calcium influx at the plasmalemma of isolated guard cells of Commelina communis. Effects of abscisic acid. Planta 178:231-241.

90. Hosoi, S., Iino, M. and Shimazaki, K.-I. (1988). Outward-rectifying K^+ channels in stomatal guard cell protoplasts. Plant Cell Physiol. 29:907-911.

91. DeSilva, D.L.R., Cox, R.C., Hetherington, A.M. and Mansfield, T.A. (1985). Synergism between calcium ions and abscisic acid in preventing stomatal opening. New Phytol. 100:473-482.

92. Smith, G.N. and Willmer, C.M. (1988). Effects of calcium and abscisic acid on volume changes in guard cell protoplasts of Commelina. J. Exp. Bot. 39:1529-1539.

93. DeSilva, D.L.R., Cox, R.C., Hetherington, A.M. and Mansfield, T.A. (1985). Suggested involvement of calcium and calmodulin in the responses of stomata to abscisic acid. New Phytol. 101:555-563.

94. McAinsh, M.R., Brownlee, C. and Hetherington, A.M. (1990). Abscisic acid-induced elevation of guard cell cytosolic Ca^{2+} precedes stomatal closure. Nature 343:186-188.

ASSOCIATIONS OF PHOTOSYNTHESIS AND WATER RELATIONS WITH PAR AND RH IN BLACK GRAM AND CLUSTER BEAN UNDER RAINFED CONDITIONS

V. Balasubramanian, S. Venkateswarlu, M. Maheshwari and G.R.M. Sankar

Division of Crop Sciences
Central Research Institute for Dryland Agriculture
Hyderabad, India

ABSTRACT

During rainy season, crop growth is subjected to slow or rapid changes in soil water availability, air temperature, solar radiation intensity and atmospheric humidity caused by variations in amount and frequency of rainfall and cloud intensities and in duration of rainless period. For better understanding of crop productivity under such agroclimatic conditions, black gram and cluster bean were raised in drylands during rainy season in 1986 and 87, and the correlations of leaf water potential, stomatal conductance, photosynthesis, transpiration, leaf temperature and canopy temperature with photosynthetically active radiation (PAR) and relative humidity (RH) during reproductive phase were studied. PAR and RH were inversely associated. The linear inverse associations of leaf water potential and stomatal conductance with PAR revealed that the two crops experienced water stress at high PAR. Yet the transpiration rate increased with increase in PAR. The polynomial relationship of photosynthesis rate with PAR and RH showed that the productivity of both the crops was limited by low PAR, when leaf water potential and stomatal conductance were high, and by low leaf water potential, when PAR was high. However, when leaf water potential, stomatal conductance and transpiration rates were lower than their maximum, photosynthesis rates were high at PAR of about 1400 μ mol $m^{-2}s^{-1}$and RH of about 50 per cent.

INTRODUCTION

In drylands of India, crops grow under continuous variations in soil moisture, atmospheric humidity, air temperature and solar radiation intensity during rainy season. The variations in the environmental components, caused by variation in amount and frequency of rainfall and cloudy weather, influence crop growth and productivity. For example, the responses of crops to rainfall in terms of dry matter production during the first half of monsoon season in India and West Africa is ascribed not only to soil moisture availability but also to increase in atmospheric humidity (1).

Dry matter production is related to transpiration and ratio of assimilation to transpiration (2). Both assimilation and transpiration are influenced by environmental factors, which are more critical to crop yield in drylands than in irrigated agriculture. Information on influence of variations in environmental components on crop physiological processes will enrich our knowledge base and help in predicting effects of slow but steady change in climate on crop productivity and in crop modelling. This paper reports the results of a field study on correlations of photosynthesis and water relations with solar radiation and relative humidity (as a measure of rainfall) during reproductive phase in two legumes (black gram and cluster bean) for better understanding of crop growth during rainy season in drylands.

MATERIALS AND METHODS

Black gram (<u>Vigna mungo</u> (L.) Hepper) cv T-9 and cluster bean (<u>Cyamopsis tetragonoloba</u> (L.) taub.) cv Pusa Navbahar were raised under dryland conditions during rainy season (June to mid-October) in 1986 and 1987 at the Research Farm of the Institute as per agronomic recommenddations on plant density, fertilizer dose and plant protection for the crops. Soil (Alfisol) was sandy loam with a mean depth of about 30 cm and bulk density of 1.6

g/cc. Soil moisture (by weight) was 14 per cent at field capacity and 4 percent at permanent wilting point. Total rainfall from seeding to harvest was 275 mm in 1986 and 406 mm in 1987 for black gram and 281 mm in 1986 and 513 mm in 1987 for cluster bean. During reproductive phase, rainfall amount was 203 mm in 1986 and 278 mm in 1987 for black gram and 210 mm in 1986 and 357 mm in 1987 for cluster bean.

Both the crops started flowering about a month after seeding. From flowering to near maturity, leaf water potential was determined using PMS Pressure Chamber, and stomatal conductance, photosynthesis, transpiration and leaf temperature were measured using LI-6000 Portable Photosynthesis System, which also recorded photosynthetically active radiation (PAR) and relative humidity (RH) at the time of physiological observations. Data were collected between 1000 and 1100 h in fully expanded young leaves. One plant per plot, selected at random for observations, constituted a replication and there were six to eight replications on each occassion. Canopy temperature was also measured using Barnes Infra-red Thermometer in 1987. Observations were made three times in both the crops at 20-day interval in 1986 and six times in black gram and eight times in cluster bean at 10-day interval in 1987.

Linear or polynomial (where best fit) correlation coefficients of physiological variables with PAR and RH were computed.

RESULTS

Photosynthetically active radiation (PAR), when the physiological observations were recorded, ranged from 627 to 2251 μ mol m^{-2} s^{-1} for black gram and from 767 to 2554 μ mol m^{-2} s^{-1} for cluster bean in 1986. The range of PAR in the 1987 season was wider (363-2084 μ mol $m^{-2} s^{-1}$ for black gram and 186-2179 μ mol m^{-2} s^{-1} for cluster bean). The associations of water relations and photosynthesis with PAR during the reproductive growth phase are presented in Fig.1 (black gram) and Fig.2 (cluster bean). The correlation co-efficients are given in Table 1.

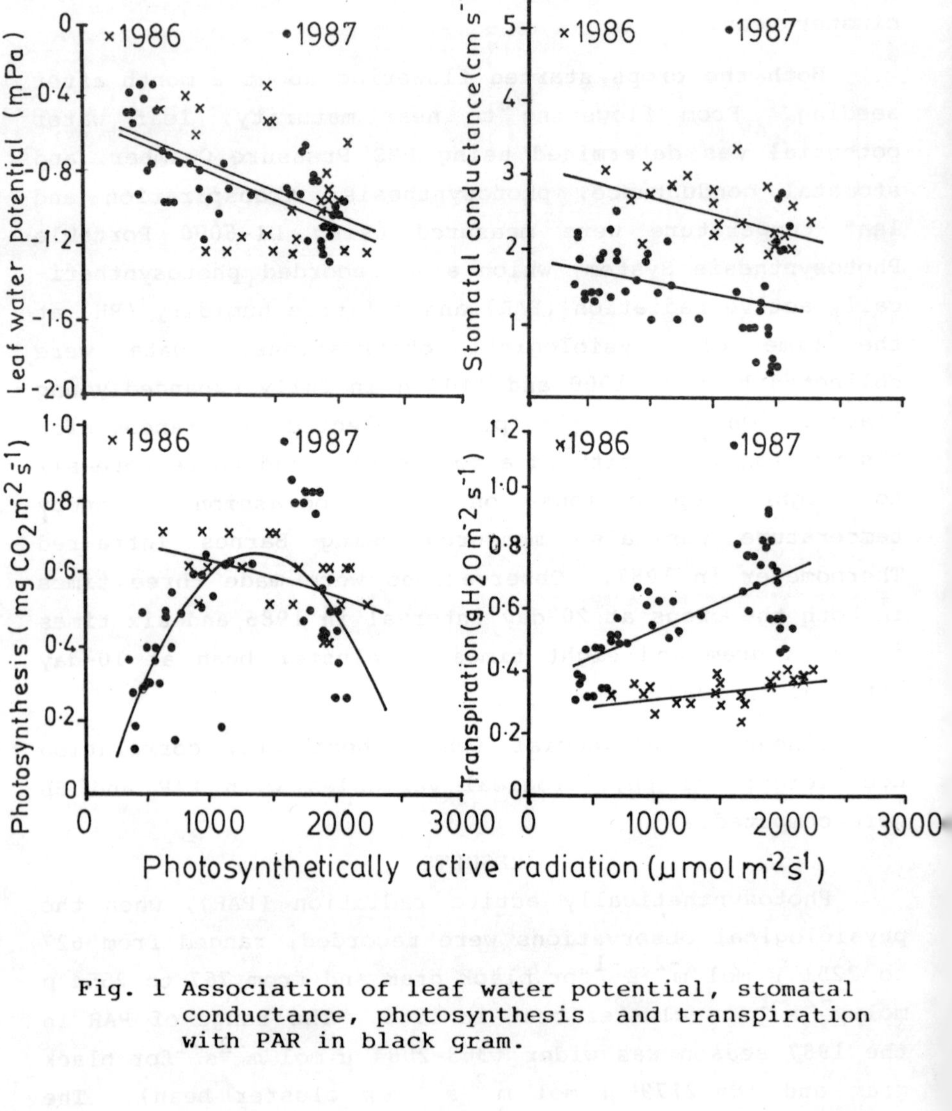

Fig. 1 Association of leaf water potential, stomatal conductance, photosynthesis and transpiration with PAR in black gram.

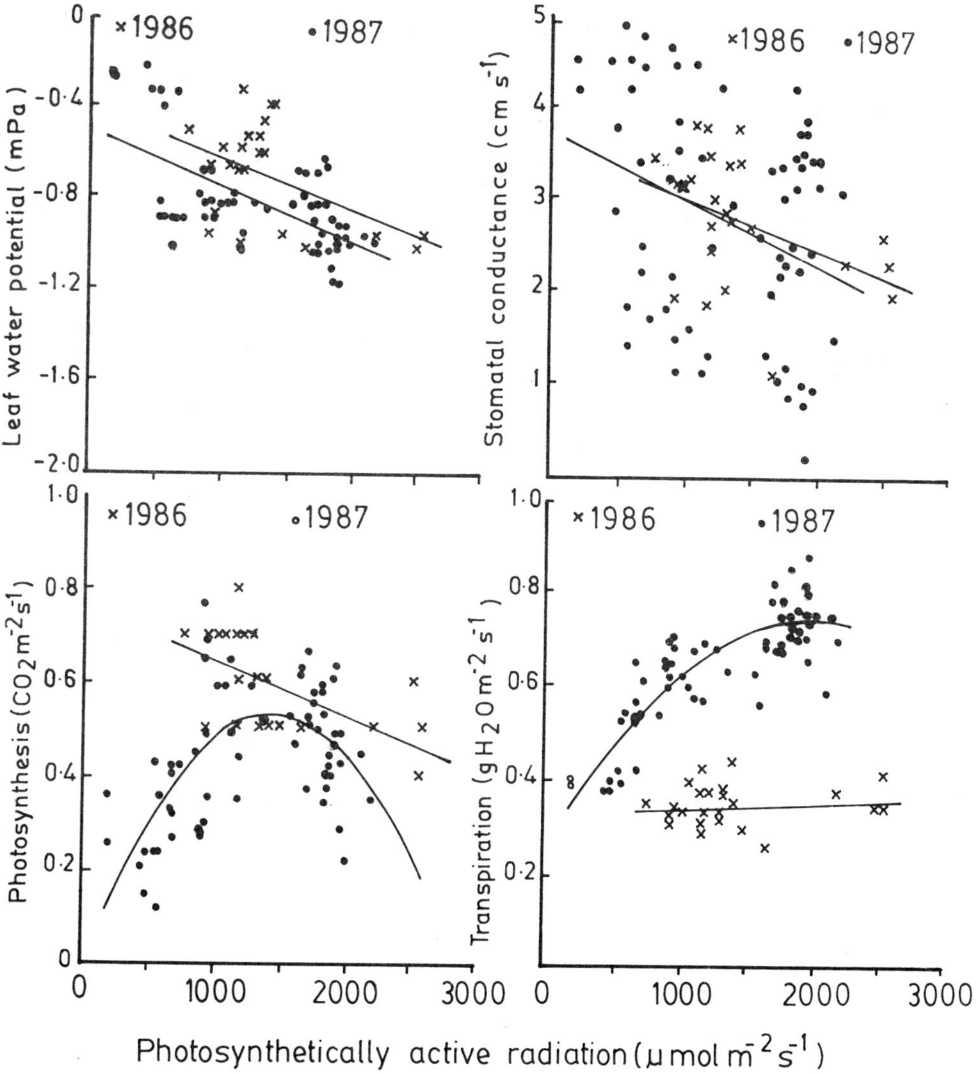

Fig. 2. Association of leaf water potential, stomatal conductance, photosynthesis and transpiration with PAR in cluster bean.

Table 1: Correlation coefficients (r) between PAR and physiological parameters in black gram and cluster bean under rainfed conditions

Physiological parameter	Black gram		Cluster bean	
	1986	1987	1986	1987
Leaf water potential	-0.40*	-0.69**	-0.51**	-0.67**
Stomatal conductance	-0.51**	-0.79**	-0.41*	-0.38**
Photosynthesis	-0.40*	0.42**@	-0.57**	0.36**@
Transpiration	0.47*	0.87**	0.18NS	0.71**@
Leaf temperature	0.64**	0.96**	0.20NS	0.90**
Canopy temperature	--	0.76**	--	0.53**

@ Coefficient of determination $(R)^2$ of the polynomial associations

* Significant at P = 0.05; ** Significant at P = 0.01;
NS Nonsignificant

Leaf water potential and PAR were inversely associated. The maximum of leaf water potential in black gram and cluster bean was about -0.35 MPa in both the seasons. The two crops differed in the minimum of the leaf water potential, which decreased to a mean value of -1.26 MPa in black gram and -1.10 MPa in cluster bean.

Stomatal conductance decreased with increase in PAR. The mean stomatal conductance was higher in cluster bean (about 3 cm s^{-1}) than in black gram (about 2 cm s^{-1}). There was noticeable seasonal difference in the magnitude of stomatal conductance in black gram; it was higher in 1986 than in 1987.

Black gram and cluster bean had more or less similar rates of photosynthesis and transpiration in a season. Association between PAR and photosynthesis rate was polynomial in both the crops in 1987. The rate of photosynthesis initially increased rapidly with increase in PAR but decreased steeply with further increase in PAR above 1500 μ mol m^{-2} s^{-1}. In 1986, however, only the negative relationship between photosynthesis rate and PAR was apparent.

Transpiration rate and PAR had high positive association in 1987 in both the crops. In 1986, however, the correlation was low in black gram and was nonsignificant in cluster bean. The transpiration rates were much higher in 1987 (0.35-0.97 g H_2O m^{-2} s^{-1}) than in 1986 (0.26-0.42 g H_2O $m^{-2}s^{-1}$).

Leaf temperature had high positive associations with PAR in black gram in both the seasons and in cluster bean in 1987 only. Canopy temperature was also positively associated with PAR but the correlation was higher in black gram than in cluster bean.

The associations of water relations and photosynthesis with relative humidity (RH) are presented in Fig.3 (black gram) and Fig.4 (cluster bean). The correlation co-efficients are given in Table 2. The range of RH within which the physiological observations were made was

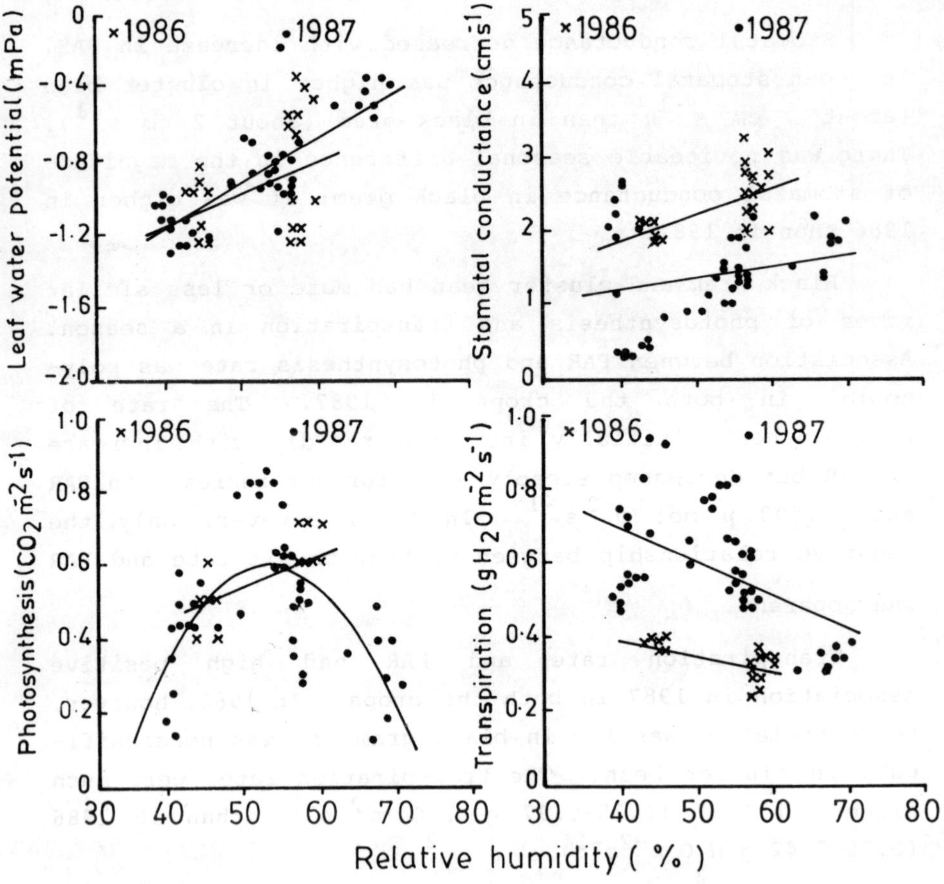

Fig. 3. Association of leaf water potential, stomatal conductance, photosynthesis and transpiration with RH in black gram.

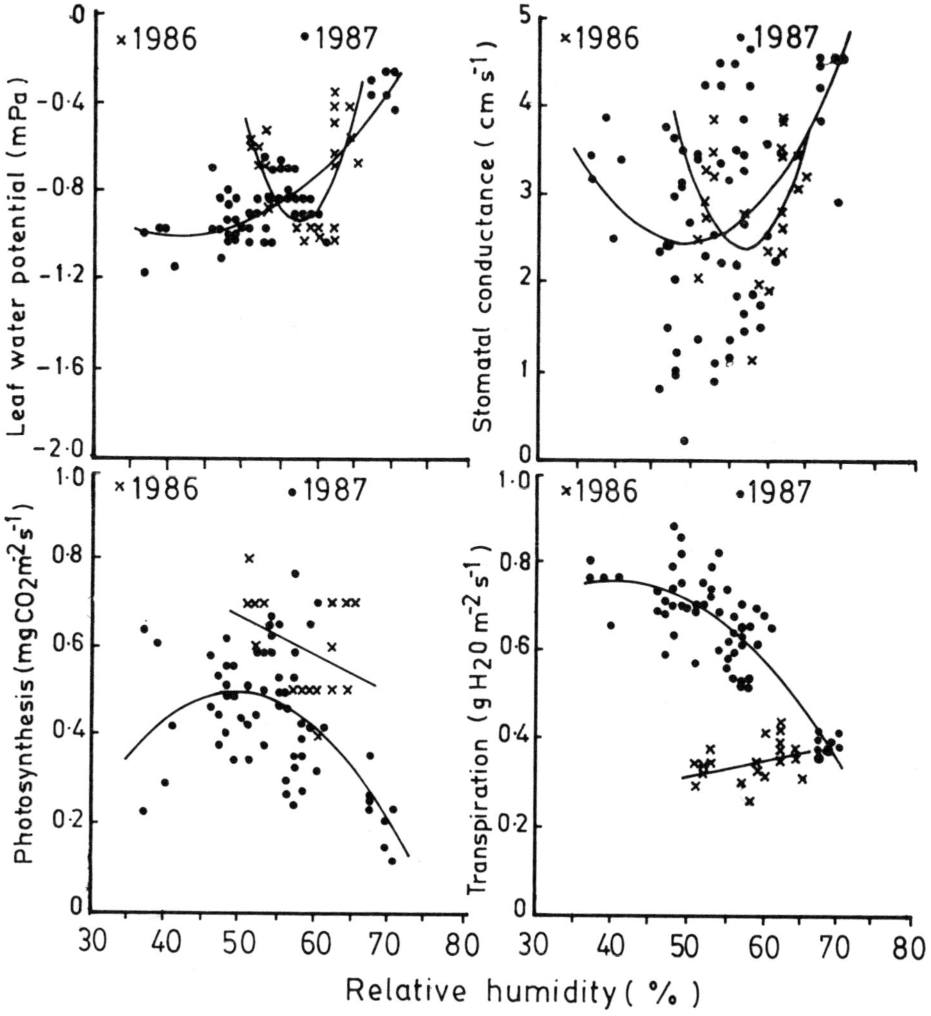

Fig. 4. Association of leaf water potential, stomatal conductance, photosynthesis and transpiration with RH in cluster bean.

Table 2: Correlation coefficients (r) between RH and physiological parameters in black gram and cluster bean under rainfed conditions

Physiological parameter	Black gram		Cluster bean	
	1986	1987	1986	1987
Leaf water potential	0.35NS	0.89**	0.39**@	0.70**@
Stomatal conductance	0.68**	0.85**	0.17NS	0.23**@
Photosynthesis	0.59**	0.43**@	-0.39*	0.32**@
Transpiration	-0.63**	-0.78**	0.41*	0.67**@
Leaf temperature	-0.83**	-0.94**	0.41*	-0.93**
Canopy temperature	--	-0.74**	--	-0.54**

@ Coefficient of determination (R^2) of the polynomial associations

* Significant at P = 0.05; ** Significant at P = 0.01; NS Non-significant

43 to 60 per cent for black gram and 50 to 65 per cent for cluster bean in 1986. RH ranged from 37 to 70 per cent in 1987 for both the crops.

Leaf water potential and RH were positively associated. However, this relationship was not very distinct in the 1986 season, as leaf water potentials were higher or lower at similar RH or were similar at different RH.

Stomatal conductance increased with increase in RH in black gram. But the association in cluster bean was nonsignificant in 1986 and was polynomial in 1987. Stomatal conductance decreased with increase in RH at lower levels but it increased steeply with increase in RH above 50-60 per cent.

The association between photosynthesis rate and RH was polynomial in both the crops in 1987. The photosynthesis rate increased with increase in RH upto 50-55 per cent and with further increase in RH, photosynthesis rate decreased fast. In the 1986 season, the linear association between photosynthesis rate and RH was positive in black gram and negative in cluster bean.

Transpiration rate and RH were inversely related in black gram in both the seasons and in cluster bean in 1987 (above 45 per cent RH). There was a weak but positive association, on the contrary, in cluster bean in 1986. The correlation between leaf temperature and RH paralleled that of transpiration rate and RH.

Canopy temperature and RH were inversely related in both the crops, the correlation co-efficient being higher in black gram than in cluster bean.

DISCUSSION

Inverse relations of leaf water potential and stomatal conductance with PAR suggest that both the crops experienced increasing water stress at higher PAR. In contrast, inverse relation was found between stomatal resistance and changes in radiation intensity in a day in cotton grown under different irrigation levels, and it was inferred that radiation intensity was the main factor affecting stomatal resistance in well watered plants (3). Stomatal conductance is likely to be more decisively influenced by plant water status, when water is limiting as in drylands, and higher radiation intensity is likely to accentuate plant water stress.

Cluster bean had relatively higher minimum leaf water potential when soil moisture was limiting. Visual field observations revealed deeper root system in cluster bean than in black gram. Probably for this reason, cluster bean had a higher mean stomatal conductance, and variation in it within a season was larger. However, the presence of seasonal difference in stomatal conductance of black gram in spite of similar leaf water potential and higher degree of association of stomatal conductance with PAR and RH suggest that stomatal opening in black gram was more sensitive to changes in environment than in cluster bean.

In cotton, stomatal resistance was more sensitive under lower solar radiation and high humidity conditions (4). Such differential sensitivity of stomatal conductance to radiation was not observed in black gram and cluster bean within the range of PAR recorded in both the seasons.

The polynomial associations between leaf water potential and RH and between stomatal conductance and RH were most likely the results of light rains which would have caused some increase in atmospheric humidity but would not have been sufficient to increase soil moisture and to relieve plant water stress. Under such circumstances, stomatal conductance could only decrease in spite of some increase in RH. It is not surprising that these anomalous relationships of leaf water potential and stomatal conductance with RH were observed at lower levels of RH. Although stomatal conductance is known to respond to atmospheric humidity, it is also influenced by soil water potential (5,6).

Notwithstanding some differences between the two crops in the minimum of leaf water potential, stomatal conductance rates and their responses to changes in RH, both black gram and cluster bean had similar rates of photosynthesis and transpiration.

The polynomial correlations of photosynthesis to increasing PAR and RH were similar in 1987, because of significant inverse association between PAR and RH (r:-0.73 for black gram and -0.60 for cluster bean, when data of both the seasons were pooled). These correlations suggest that a) at lower PAR levels, photosynthesis rate was limited by PAR although leaf water potential and stomatal conductance were higher; b) at higher PAR levels, photosynthesis rate was limited by leaf water potential and stomatal conductance, and c) the photosynthesis rate was higher when leaf water potential and stomatal conductance were lower than maximum and when PAR was around 1400 μ mol m^{-2} s^{-1} and RH was around 50 per cent. In juvenile Douglas - fir stand, higher CO_2 exchange was observed on cloudy days with low air temperature, and vapour pressure deficit, suggesting maximum productivity under these conditions (7).

The inverse association between photosynthesis rate and PAR in 1986 was probably due to the fact that observations were fewer and hence the positive association of photosynthesis with lower levels of PAR as seen in 1987 was not brought out. Similarly, there was only an inverse relationship between photosynthesis rate and RH in cluster bean in 1986, because the photosynthesis rates were within the limited RH range of about 50-65 per cent. The response in this range of RH was similar to that observed in 1987.

The progressive fall in leaf water potential and steep fall in photosynthesis rate with increasing PAR, at higher PAR levels, support the conclusion that photosynthesis of water stressed plants is more strongly inhibited at high than at low radiation intensities (8). The progressive fall in stomatal conductance and steep rise in photosynthesis rate with increasing PAR at lower levels of PAR suggest that the decreasing stomatal conductance was still high enough for photosynthesis. Photosynthesis rate was limited by stomata under wellwatered or mild-stress conditions (3, 9). The results of the present study show that stomatal limitation of photosynthesis rate occurred only when PAR became non-limiting. At high photosynthetic photon flux densities, both photosynthesis rate and leaf conductance decreased linearly in water-stressed cotton (10). In black gram and cluster bean, stomatal conductance continued to decrease progressively but photosynthesis rate declined faster at higher PAR. Under severe moisture stress conditions, non-stomatal factors start limiting photosynthesis (3, 9, 11).

Transpiration rate continued to increase despite decreases in leaf water potential and stomatal conductance, with increase in PAR. Radiation seemed to overwhelm the plant factors in controlling transpiration. Positive linear regression between transpiration and solar radiation was found in cucumber, grown under green house conditions (12).

The increasing transpirational loss with increasing water stress suggest that both the leguminous crops are water spenders for postponing dehydration under water stress conditions.

Transpiration is closely correlated with crop yield (2, 13). But, the rapid fall in photosynthesis rate at higher PAR (lower RH), despite increase in transpiration rate, would only result in decrease in plant productivity. So, transpiration is functionally more related to plant survival rather than plant productivity under severe water stress.

The relatively higher air temperature (and hence higher vapour pressure deficit) that prevailed during reproductive growth in 1987 (29.0 to 34.5^{o}C) than in 1986 (28.1 to 32.1^{o}C) was most likely the reason for higher transpiration rate in the 1987 season in the two crops.

The correlation between transpiration rate and PAR was not significant in 1986 in cluster bean. Crops may differ in the sensitivity of their responses in plant water relations to variations in PAR, but it cannot be inferred with certainity because of limited observations, in 1986. If it is true, then this may also be the basis for difference observed between the two crops in the degree of association of canopy temperature with PAR and RH. Canopy temperature (infrared thermometry) is related to crop water status (14).

In conclusion, productivity of black gram and cluster bean during reproductive phase was limited by low solar radiation, when water was available, and on the other hand, by low water availability, when solar radiation was high during rainy season under dryland conditions.

REFERENCES

1. Monteith, J.L. (1986). Significance of the coupling between saturation vapour pressure deficit and rainfall in monsoon climates. Exptl. Agric. 22: 329-338.
2. Fischer, R.A. and Turner, N.C. (1978). Plant productivity in the arid and semi-arid zones. Annu. Rev. Plant Physiol. 29: 277-317.
3. Ephrath, J.E., Marani, A. and Bravdo, B.A. (1990). Effects of moisture stress on stomatal resistance and photosynthesis rate in cotton (Gossypium hirsutum). 1.Controlled levels of stress. Field Crops Res. 23: 117-131.
4. Inoue, Y., Kimball, B.A., Jackson, R.D., Pinter Jr, P.J. and Reginato, R.J. (1990). Remote estimation of leaf transpiration rate and stomatal resistance based on infrared thermometry. Agric. For. Meteorol. 51: 21-33.
5. Schulze, E.D. (1986). Carbondioxide and water vapour exchange in response to drought in the atmosphere and in the soil. Annu. Rev. Plant Physiol. 37: 247-274.
6. Gollan, T., Passioura, J.B. and Munns, R. (1986). Soil water status affects the stomatal conductance of fully turgid wheat and sunflower leaves. Aust. J. Plant Physiol. 13: 459-464.
7. Price, D.T. and Black, T.A. (1990). Effects of short

term variation in weather on diurnal canopy CO_2 flux and evapotranspiration of a juvenile Douglas-fir stand. Agric. For. Meteorol. 50: 139-158.

8. Bjorkman, O. and Powels, S.B. (1984). Inhibition of photosynthetic reactions under water stress: interaction with light levels. Planta 161: 490-504.

9. Ogren, E. and Oquist, G. (1985). Effects of drought on photosynthesis, chlorophyll fluorescence and photoinhibition susceptibility in intact willow leaves. Planta 166: 360-388.

10. Turner, N.C., Hearn, A.B., Begg, J.E. and Constable, G.A. (1986). Cotton (Gossypium hirsutum L.) physiological and morphological responses to water deficits and their relationship to yield. Field Crops Res. 14: 153-170.

11. Farquhar, G.D. and Sharkey, T.D. (1982). Stomatal conductance and photosynthesis. Annu. Rev. Plant Physiol. 33: 317-345.

12. Yang, X., Short, T.H., Fox, R.D. and Bauerle, W.L. (1990). Transpiration, leaf temperature and stomatal resistance of a green house cucumber Crop. Agric. For. Meteorol. 51: 197-209.

13. Stanhill, G. (1986). Water Use efficiency. Adv. Agron. 39: 53-85.

14. Jackson, R.D., Idso, S.B., Reginato, R.J. and Pinter, Jr. P.J. (1981). Canopy temperature as a crop water stress indicator. Water Resour. Res. 17: 1133-1138.

INDUCTION OF PROTEIN SYNTHESIS IN RESPONSE TO HEAT SHOCK AND VARIOUS HEAT SHOCK MIMICKING AGENTS IN *VIGNA SINENSIS*

S. Krishnasamy** and A. Gnanam$
School of Biological Sciences, Madurai Kamaraj University
Madurai 625 021, India

Abstract

Leaves of V. sinensis synthesized an array of new temperature-dependent proteins called the heat shock proteins (HSPs) when the growth temperature was raised from 25°C to 40°C for a brief period of 2 h. This was accompanied by the suppression of the synthesis of normal cellular proteins. Previously, we have shown that the low molecular weight HSPs are synthesized on cytoplasmic ribosomes and transported into chloroplasts and localized in the thylakoid membranes. In the present study, we have examined the specific induction of HSPs by a number of compounds that mimic the heat shock response in other organisms. Sodium arsenite induced HSPs synthesis in V. sinensis, whereas compounds that generate free radicals and those which cause DNA damage did not induce HSPs. On the basis of these observations, we suggest that the synthesis of HSPs is not mediated either by free radical formation or by damage caused to DNA.

---

* This work was supported by a research grant PL-480 USDA (FG-In-681) to A.G.

** Present address: S. Krishnasamy, Department of Botany, The University of Iowa, IOWA City, IA 52240.

$ Present address: A. Gnanam, Vice-Chancellor, Pondicherry University, R.V. Nagar, Kalapet, PONDICHERRY 605 014.

Introduction

A wide range of organisms ranging from bacteria to humans, including plants have been shown to respond to thermal stress by synthesizing a specific set of proteins, the heat shock proteins (HSPs) and repressing the synthesis of most of the normal proteins (for reviews see Kimpel and Key 1985, Lindquist 1986, Nover and Scharf 1984, Neumann et al 1989). At the elevated temperature, HSP mRNAs are selectively translated relative to the mRNAs of the normal proteins which seem to be sequestered rather than degraded. The exact role (or function) of these HSPs is not known, although a strong positive correlation between the accumulation of HSPs and the acquisition of thermal-tolerance for the organism to otherwise lethal temperatures are reported (Marimiroli et al. 1986, McAlister and Finkelstein 1980). Thermotolerance has also been demonstrated where HSPs are induced by stress other than heat shock (Lin et al. 1984, Plesset et al. 1982).

In Drosophila and E. coli, numerous agents other than heat shock have been shown to induce the heat shock response (Neidhardt et al. 1984, Nover and Scharf 1984). But in plants, an array of stress conditions, both biotic and abiotic, induce the synthesis of specific proteins called 'stress proteins', which totally differ from HSPs induced by heat shock (Matters and Scandalias 1986). However, there are also reports that abiotic stress agents like arsenite, cadmium (Czarnecka et al. 1984), and wounds (Heikilla et al. 1984) induce HSP-like proteins in plants. Although different stress agents induce heat shock-like responses, the molecular mechanism by which they induce HSPs in plants is not understood, though there have been some suggestions. One proposed mechanism suggests that the reduction product of oxygen could initiate events leading to the synthesis of HSPs (Morgan et al. 1986, Ropp et al. 1983). Alternatively in E. coli (Kruger and Walker 1984) and in cyanobacteria (Nicholson et al. 1987), agents which cause damage to DNA induced the synthesis of HSPs. In this report we present the results of our studies on the effect of oxidative and DNA damaging agents in inducing heat shock-like proteins in V. sinensis.

Materials and Methods

Plant material and growth condition

Seeds of V. sinensis were allowed to germinate in moist vermiculite for 7 days at 25°C in a controlled growth temperature provided with illumination at a photon flux density of 45 uE/m^2/sec. A 12 h dark/light cycle was maintained.

In vivo protein synthesis with leaf discs

About 1 g of leaves were cut into 1 mm^2 segments and transferred to a 25 ml conical flask containing 10 ml of sterile water. These flasks were kept in a water bath shaker at a specified temperature. Illumination was provided at a photon flux density of 37.2 uE/m^2/sec. The leaf discs were preincubated for 30 min at specified temperature and then 150 uCi of ^{14}C-Chlorella protein hydrolysate (specific activity : 26 mCi/m atom of C) was added. Labelling was allowed to continue for another 90 min. While studying the effect of various agents that mimic heat shock response, the leaf segments were treated with these agents at 25°C for 30 min before the addition of the labelled amino acid mixture. Labelling was continued for another 90 min at the same temperature.

Sample preparation

After labelling, leaf segments were thoroughly rinsed in deionized water and homogenized by using a buffer containing 50 mM Tris-HCl (pH 8.3), 1 mM $MgCl_2$ and 2 mM EDTA. The homogenate was filtered through four layers of cheese cloth and the proteins contained in the filterate were precipitated with 10% (w/v) tricholoroacetic acid in acetone for 30 min at 4°C. The precipitate was washed twice with ice-cold acetone and thrice with ice-cold diethyl ether. The final pellet was dissolved in a small volume of 10% SDS and subsequently used for SDS-polyacrylamide gel electrophoresis (SDS-PAGE).

Polyacrylamide gel electrophoresis and fluorography

SDS-PAGE analysis was carried out according to the procedures described by Laemmli (1970) with minor modifications. The gel contained a linear 7.5% to 15% acrylamide concentration gradient with 5% to 10% glycerol co-gradient. The stacking gel contained 4% acrylamide. Electrophoresis was carried out at 25°C for 10 h at 17 mA (initially at 5 mA for 1 h). Gels were stained, destained and processed for fluorography as described by Lasky and Mills (1975). A screen type X-ray film was exposed over the dried gel at -80°C for pre-determioned time. The gel was calibrated using the following molecular weight standards : phosphorylase A, Mr 97000; bovine serum albumin, Mr 66000; egg albumin, Mr 44000; carbonic anhydrase, Mr 29000; B lactoglobulin, Mr 18400; lysozyme, Mr 14300.

Results

Effects of heat shock and arsenite on the protein synthesis pattern in V. sinensis leaves

Leaf discs from 7 day-old seedlings of V. sinensis were incubated at 25°C or 40°C for 2 h with radiolabelled (^{14}C)- amino acids mixture. After 2 h, the total leaf proteins were extracted and subjected to SDS-PAGE followed by fluorography. Heat shock at 40°C resulted in the reduction of protein synthesis and caused several changes in the pattern of protein synthesis. The synthesis of most of the normal cellular proteins was inhibited at 40°C but in lieu a new set of proteins, the HSPs, were synthesized (Fig.1). The total number of low molecular weight HSPs species (between 15 and 30 KDa) synthesized were greater in number than the high molecular weight HSPs (60 KDa and above).

The effect of arsenite was tested by incubating the leaves in four different concentrations (25, 50, 75 and 100 uM) of sodium arsenite at a normal growth temperature of 25°C, in the presence of (^{14}C) amino acids. Then the pattern of protein synthesis was analysed by SDS-PAGE and fluorography as in heat shock studies. Unlike heat shock, arsenite did not affect the total protein synthetic activity of the leaves at all the four concentrations used. But arsenite did cause synthesis of a set of new proteins. Since arsenite-induced proteins and the heat shock-induced proteins (HSPs) had a similar molecular weight, as determined by SDS-PAGE, the arsenite induced proteins are referred to in this paper as HSPs. Arsenite induced a spectrum of proteins similar to heat shock (Fig. 2). A notable difference was that the high molecular weight HSPs were induced to a higher extent on arsenite treatment than the HSPs which were induced by heat shock. The small molecular weight HSPs were made at low levels, even at the highest concentration of arsenite used.

Effect of oxidative stress on in vivo protein synthesis

Earlier reports have surmised based on certain observations that all chemicals inducing free radicals mimic the heat shock like response (Ropp et al. 1983, Morgan et al. 1986). In order to test this hypothesis in plant systems, in vivo studies on protein synthesis in the leaves was carried out both in the presence and absence of chemicals which are known to induce free radicals, like methyl viologen and juglone.

Methyl viologen accepts electron on the reducing side of photosystem I in a manner very similar to that of pseudocyclic electron transport and produces free radicals after autooxidation. When the fluorographic analysis was carried out with the samples prepared from leaves treated with methyl viologen (Fig. 3) it was seen that, unlike heat shock, methyl viologen did not induce any specific response leading to the synthesis of new proteins. On the other hand, methyl viologen decreased the net protein synthesis and consequently very few proteins were made. The only polypeptide which was detected in the leaves treated with 100 uM of methyl viologen was the large subunit of ribulose-1, 5-bisphosphate carboxylase. The synthesis of most of the normal cellular proteins also gradually declined as the methyl viologen concentration was increased from 25uM to 100 uM.

Juglone is known to induce free radicals without involving any oxidative-reductive processes (Matters and Scandalias 1986). In order to further confirm the effect of free radical on HSPs synthesis, the leaves were pretreated at four different concentrations (100, 200, 300 and 500 uM) of juglone and the in vivo protein synthetic pattern was studied. Induction of new proteins like HSPs was not observed by juglone treatment (Fig. 4). These results are comparable to the effect of methyl viologen on

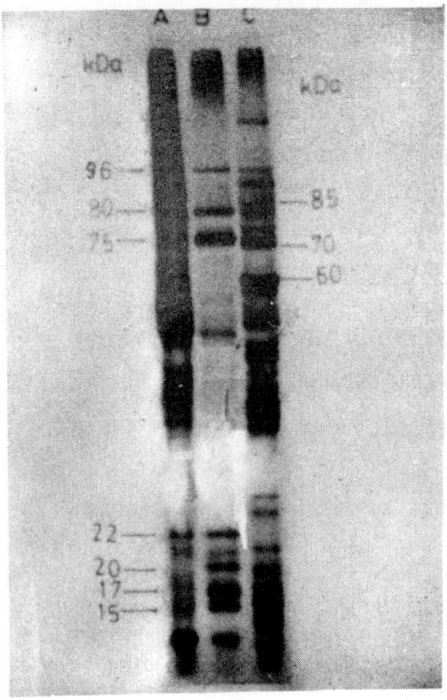
A B C
kDa
96
80
75
22
20
17
15
kDa
85
70
60

1

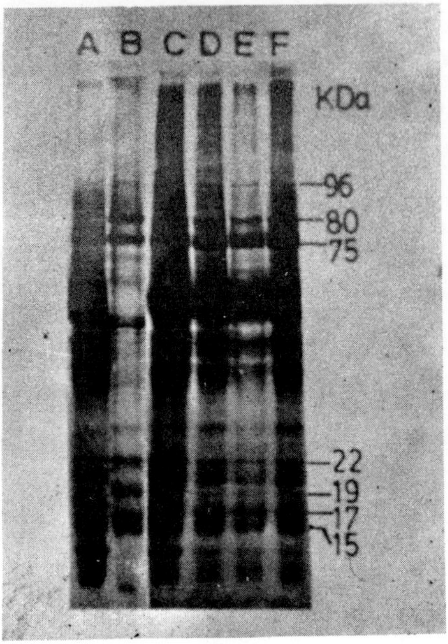
A B C D E F
KDa
96
80
75
22
19
17
15

2

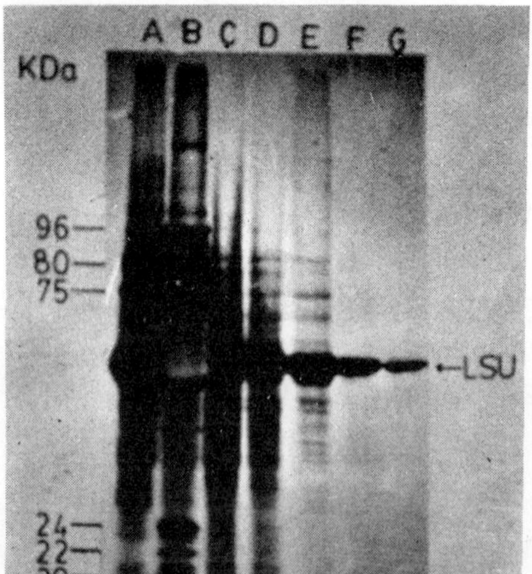
A B C D E F G
KDa
96
80
75
LSU
24
22
20
17
15

3

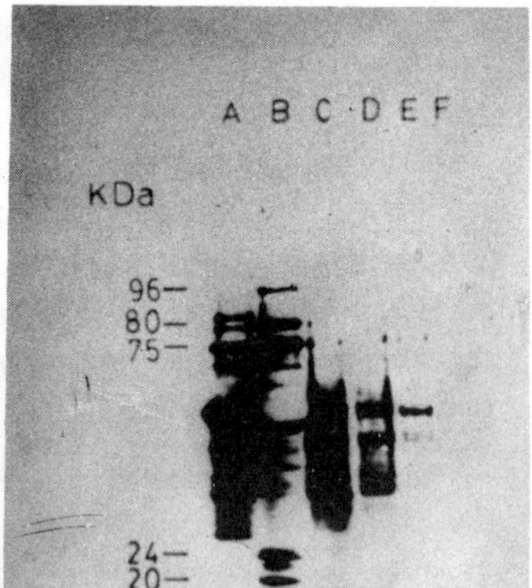
A B C D E F
KDa
96
80
75
24
20
18
17
16
15

4

Fig. 1. Effects of temperature and arsenite on the pattern of protein synthesis in V. sinensis leaves. Leaf discs were preincubated for 30 min either at 25°C (Lane A) or at 40°C (Lane B) and labelled at the same temperature for 90 min with (^{14}C)-labelled amino acids mixture. At 25°C, leaves were treated with sodium arsenite at concentrations of 25 uM (Lane C), 50 uM (Lane D), 75 uM (Lane E) and 100 uM (Lane F). After labelling, total leaf proteins were extracted and analysed by SDS-PAGE followed by fluorography as described in detail in "Materials and Methods". An equal amount of protein (150 ug) was loaded in each lane.

Fig. 2. Fluorographic profile of proteins synthesized by the leaves of V. sinensis incubated in the presence of sodium arsenite at various concentrations. Lane A, total homogenate of leaves incubated at 25°C for 30 min and labelled at the same temperature for 90 min; lane B, total homogenate from leaves incubated at 40°C for 30 min followed by labelling for 90 min; lanes C, D, E and F, total homogenate from leaves incubated at 25°C for 30 min in the presence of sodium arsenite at 25, 50, 75 and 100 uM concentrations, respectively, followed by labelling for 90 min at the same temperature.

Fig. 3. Fluorographic profile of the proteins synthesized in V. sinensis leaves incubated with methyl viologen. Leaves were incubated at 25°C (Lane A), at 40°C (Lane B) and at 25°C with methyl viologen at 0.1 mM (Lane C), 1 mM (Lane D), 5 mM (Lane E), 25 mM (Lane F) and 50 mM (Lane G) concentrations. The detailed procedure of methyl viologen treatment, protein labelling, extraction and visualization are as described in Fig. 1. An equal amount of protein (150 ug) was loaded in all the lanes.

Fig. 4. Fluorographic profile of the proteins synthesized in V. sinensis leaves treated with juglone. Leaf discs were incubated at 25°C (Lane A), at 40°C (Lane B), at 25°C with juglone at 100 uM (Lane C), 200 uM (Lane D), 300 uM (Lane E) and 200 uM (Lane F). The conditions of juglone treatment and other details are as given in "Materials and Methods" section. An equal amount of protein (150 ug) was loaded in all the lanes.

protein synthesis. As in the case of methyl viologen, an inhibition of total cellular protein synthesis was also observed by increasing the concentration of juglone. The results clearly reveal that neither of these chemicals that induce free radical generation mimic the heat shock-like response in V. sinensis leaves.

DNA damaging agents and their effect on protein synthesis in V. sinensis leaf segments

It has been reported previously that E. coli cells pretreated with agents that cause damage to DNA, such as irradiation with UV light, mitomycin C and nalidixic acid, induced the synthesis of HSPs. The same kind of results were also obtained recently in a cyanobacterium by Nicholson et al. (1987). To investigate the effect of DNA damaging agents on plants, leaves of V. sinensis were incubated with mitomycin C and nalidixic acid and their effect on in vivo protein synthesis was studied. The synthesis of HSP-like proteins was not induced by either of the two DNA damaging agents tested (Fig. 5 and 6). The overall protein synthetic activity was inhibited by both of these agents but the effect was more pronounced in nalidixic acid-treated leaves. But, these DNA damaging agents induced a set of new low molecular weight polypeptides (15 to 17 KDa), which are entirely different from the low molecular weight forms of HSPs induced by heat shock or arsenite.

Discussion

Alterations in protein synthesis readily occurs in many organisms exposed to a variety of stresses. Heat shock is known to be the best stimuli to induce alteration of gene expression and also serves as a good model system to study the gene regulation. Among the various plant species, soybean has been used extensively to study the heat shock as well as the various other stresses (Czarnecka et al. 1984, Kimpel and Key 1985). The leaves of V. sinensis synthesized a set of HSPs when subjected to heat shock. In our earlier reports we have shown that the maximum heat shock response was observed when the leaves were treated at 40°C for 2 h.

Fig. 5. Fluorographic profile of the proteins synthesized in V. sinensis leaves treated with mitomycin C. Leaves were incubated at $25^{o}C$ (Lane A), $40^{o}C$ (Lane B), and at $25^{o}C$ with mitomycin C at 25 ug/ml (Lane C), 50 ug/ml (Lane D), 100 ug/ml (Lane E), 200 ug/ml (Lane F) and 400 ug/ml (Lane G) concentrations. The details of mitomycin treatment, protein labelling, extraction and separation was as described in "Materials and Methods" section. An equal amount of protein was loaded in all the lanes.

Fig. 6 Fluorographic profile of the proteins synthesized by green leaves of V. sinensis treated with different concentrations of nalidixic acid. The procedure of nalidixic acid treatment and other details are given in "Materials and Methods" section. An equal amount of protein (150 ug) was loaded in all the lanes. Lanes: (A) leaves incubated at $25^{o}C$, (B) at $40^{o}C$, at $25^{o}C$ with nalidixic acid at concentrations of 50 ug/ml (C), 200 ug/ml (D), 300 ug/ml (E) and 500 ug/ml (F).

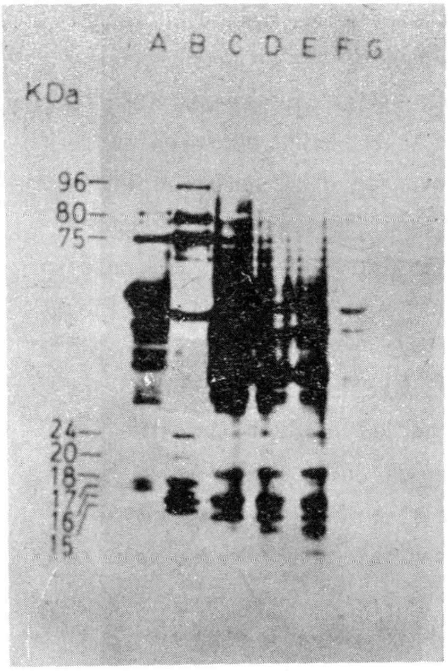

5

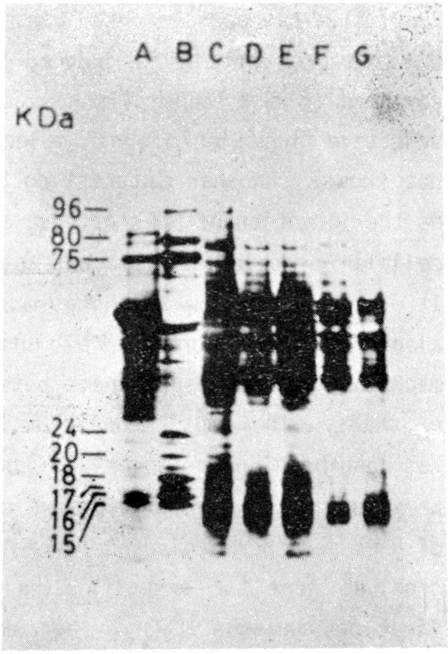

6

Besides, a set of low molecular weight HSPs were transported into chloroplasts and localized on thylakoid membranes (Krishnasamy et al. 1987, 1988). In this paper, we report, the effect of agents that mimic heat shock on the synthesis of HSPs in V. sinensis.

Arsenite has been shown to mimic the heat shock response in a number of eukaryotic organisms (Vincent and Tanguay et al. 1983), including plants (Czarnecka et al. 1984, Edelman et al. 1988, Lin et al. 1984). In accordance with these observations, leaves of V. sinensis treated with sodium arsenite induced the heat shock response by inducing the full spectrum of HSPs. In the presence of arsenite, the high molecular weight HSPs induced are as much and sometimes even more than the levels seen with heat shock. On the other hand, the low molecular weight HSPs are induced to a lesser extent by arsenite as compared to heat shock. Similar effect of arsenite on HSPs synthesis has been reported in soybean by Lin et al. (1984). In Drosophila also, arsenite induces predominantly the high molecular weight HSPs (Vincent and Tanguay 1982). It has been reported earlier that the low molecular weight H_2B histone protein is induced by arsenite to a lesser extent compared to hear shock (Tanguay et al. 1983). The reason for the selective induction of high molecular weight HSPs by arsenite is not known. Another interesting observation with sodium arsenite is the induction of HSPs without affecting the synthesis of normal cellular proteins. This contrasts with the heat shock response. It is not clear how arsenite maintains the HSP mRNA translation along with the normal mRNA and allows the protein synthetic machinery not to discriminate between these two different species of mRNAs. The impairment of mitochondrial activity by arsenite has long been known (Nover and Scharf 1984). Recently, there have been many reports of a positive correlation between the synthesis of HSPs and thermotolerance of mitochondria. It is reasonable to conclude that the arsenite paralyzes the mitochondrial activity first causing the HSPs to be induced to overcome. this stress. This conclusion is supported well by the findings of Lin et al. (1984) that soybean seedlings treated with arsenite provided thermotolerance to heat shock. They also observed that HSPs induced by arsenite are transported into their respective organelle in a temperature-dependent manner. This effect has yet to be elucidated in V. sinensis leaves.

The reduction products of oxygen such as superoxide and hydrogen peroxide were shown earlier to initiate events leading to HSPs synthesis (Ropp et al. 1983, Morgan et al. 1986). Methyl viologen and juglone are the two compounds tested in this study to generate free radicals to induce the HSPs synthesis. In maize leaves, Matters and Scandalios (1986) have successfully used these compounds to generate free radicals. Further, heat shock as well as cells returned to air from anoxia induced the synthesis of HSPs through the production of free radicals (Lee et al. 1983, Ropp et al. 1983). Instead of subjecting the plants to anoxia, we directly exposed the plants to a free radical to see the effect. In contrast to the above observation, V. sinensis leaves treated with free radical generating chemicals did not synthesize HSPs. Our results are in close agreement with the recent report that the induction of HSPs in rice embryo is not mediated by free radical formation (Mocquot et al. 1987). The methyl viologen-mediated inhibition of protein synthesis is greater than that of juglone. This effect may be due to juglone being less efficient in generating free radicals in comparison to methyl viologen, as suggested by Matters and Scandalias (1986). The observed decrease in number of polypeptides in methyl viologen-treated leaves may be due to either a free radical-induced proteolysis as noted by Ropp et al. (1983) or the direct interference of methyl viologen on protein synthesis. The possibility of the former explanation was not found to be true, because we have not observed any sign of proteolysis either in Coomassie blue stained gel (our unpublished observation) or in fluorographic picture (Fig. 2). The possible reason for the failure of the free radicals to induce HSPs synthesis in plants may be that plants in general may have an efficient free radical scavenging (quenching) system.

Finally, when the DNA damaging agents (mitomycin C and nalidixic acid) were tested to study the heat shock response in V. sinensis leaves, once again a negative correlation with heat shock was noticed. But in E. coli (Neidhardt et al. 1984) and recently in a cyanobacterium, Phormidium laminosum (Nicholson et al. 1987), the agents that cause damage to DNA induced proteins similar to HSPs and these proteins are believed to be involved in DNA repair process. At this stage we do not know whether these two DNA damaging agents can induce damage to plant DNA or if their mechanism of DNA repair is different from that of the bacteria. As we observed in this study, Czarnecka et al. (1984) have found that when the soybean seedlings were treated with a number of various stress inducing agents, only arsenite and to some extent cadmium induced the heat shock response.

In conclusion, a comparison of heat shock response with the response elicited by a variety of stress-inducing agents in V. sinensis indicate that while several parallels exist between the heat shock response and other stress responses, there are also major differences in the mechanism by which the heat shock responses and other stress responses are evoked.

Acknowledgement

We are indebted to Dr. K. Veluthambi, Madurai Kamaraj University, for critically reading the manuscript and Mary Andrews for preparing the manuscript.

References

Czarnecka E, Edelman L, Schoffl F and Key JL (1984) Plant Mol. Biol. 3: 45.

Edelman L, Czarnecka E and Key JL (1988) Plant Physiol. 86: 1048.

Heikkila JJ, Papp JET, Schultz GA and Bewely JD (1984) Plant Physiol. 76: 270.

Kimpel JA and Key JL (1985) Plant Physiol. 79: 672.

Krishnasamy S, Krishnan M, Mannar Mannan R and Gnanam A (1987) In: Current Trends in Life Sciences, XIII Biomembranes:

Structure, Biogenesis and Transport, pp. 233-238, Today & Tomorrow's Printers and Publishers, New Delhi.

Krishnasamy S, Mannar Mannan R, Krishnan M and Gnanam A (1988) J. Biol. Chem. 263: 5104

Kruger JH and Walker G (1984) Proc. Natl. Acad. Sci., USA 81: 1499.

Laemmli UK (1980) Nature 227: 680-685

Lasky RA and Mills AD (1975) Eur. J. Biochem. 56: 335-341.

Lee Ph C, Bochner BR and Ames AN (1983) Proc. Natl. Acad. Sci., USA 80: 7496

Lin CY, Roberts JK and Key JL (1984) Plant Physiol. 74: 152.

Lindquist S (1986) Ann. Rev. Biochem. 103: 321.

Marimiroli N. Restivo FM, Odoardi-Stanca M, Terzi V, Giovanelli B, Tassi E and Lorenzoni C (1986) Genet. Agri. 40: 9.

Matters GL and Scandalios JG (1986) Dev. Gen. 7: 167.

McAlister L and Finkelstein DB (1980) Biochem. Biophys. Res. Commun. 93: 819.

Mocquot B, Richard B and Pradet A (1987) Biochimie 69: 677.

Morgan RW, Christman MF, Jacobson FS, Storz G and Ames BN (1986) Proc. Natl. Acad. Sci., USA 83: 8059.

Neidhardt FC, Van Bogelen RA and Vaughan V (1984) Ann. Rev. Genet. 18: 295.

Neumann D, Nover L, Parthier B, Scharf K, Wollgrehn R and Nieden U (1989) Heat shock and other response systems of plants. Biologisches Zentralhlatt (Biol., Zent. bl. Leipzig 108 (6): 1-156.

Nicholson P, Osborn RW and Howe CJ (1987)
FEBS Lett 221: 110

Nover L and Scharf KD (1984)
Eur. J. Biochem. 139: 303.

Plesset J. Palm C and McLaughlin CS (1982)
Biochem. Biophys. Res. Commun. 108: 1340.

Ropp M, Courgeon AM, Calvayrac R and Best-Belpomme M (1983)
Can. J. Biochem. Cell Biol. 61: 456.

Tanguay RM, Camato R, Lettre F and Vincent M (1983)
Can. J. Biochem. Cell Biol. 61: 414.

Vincent M and Tanguay RM (1982)
J. Mol. Biol. 162: 365.

IV

CROP MODELLING

MODELLING THE EFFECTS OF CO_2-INDUCED CLIMATIC CHANGE ON CEREAL CROPS

J.R. PORTER Department of Agricultural Sciences, University of Bristol, AFRC Institute of Arable Crops Research, Long Ashton Research Station, Bristol BS18 9AF, UK.

F. MIGLIETTA Institute of Environmental Analysis and Remote Sensing for Agriculture, Consiglio Nazionale delle Ricerche, Piazzale delle Cascine 18, 50144 Firenze, Italy.

ABSTRACT

Climatic change has implications for agricultural primary production which are starting to be appreciated. Using a well-tested model of the growth and development of wheat, the change in dry matter productivity for coincidental increases in CO_2 concentration and temperature are explored. The main prediction is for harvest indices to fall. The model also suggests that there will also be an increase in nitrate-loss from crops under warmer, wetter conditions. In attempting to develop strategies which will ameliorate the combined effects of increased temperature and changes in water supply, use is made of the model to define crop ideotypes that will be better adapted to altered conditions. The paper will conclude with discussion of the areas of investigation necessary for further progress to be made.

INTRODUCTION

Ambient carbon dioxide levels have risen in the past 150 years from ca. 280 ppm to about 350 ppm and are predicted to increase to about 600 ppm within the next 50 years or so. It is estimated that the combustion of fossil fuels and other emissions of CO_2 and other gases are contributing ca. 6×10^9 tonnes of C per annum to the atmosphere and this represents three times the amount emitted in the 1950s.

Mean global temperatures are higher now than at any time since records began. In the 1980s, five of the years have been, successively, the warmest ever recorded. The consensus predicted rise in global mean temperature associated with the build-up of 'greenhouse' gases is 0.3°C per decade over the next century with the amount of warming increasing with latitude. There is uncertainty attached to this forecast with a lower limit for temperature change of 0.2°C and an upper limit of 0.5°C per decade (1). Predictions of changes in other climatic features, such as precipitation, are even less certain. This is especially true for the edges of continents (2, 3).

The purpose of this paper is threefold. First, to illustrate how simulation of the growth and development of crops in a series of 'what-if' scenarios about the greenhouse effect provides us with indications of possible outcomes in terms of whole crop productivity, grain yield and consequent nitrogen balance. Secondly, the model can be used to identify the main features of wheat varieties that are better suited to a changed climate. Lastly, several experimental avenues which are open for consideration in concert with simulation modelling of the processes will be suggested.

MODELLING

Potential Productivity

From sowing to harvest, the growth and development of crops, and other plants, is influenced by a number of climatic and soil factors (Fig. 1) which interact in very complex ways.

Figure 1. Environmental factors controlling the growth and development of a wheat crop

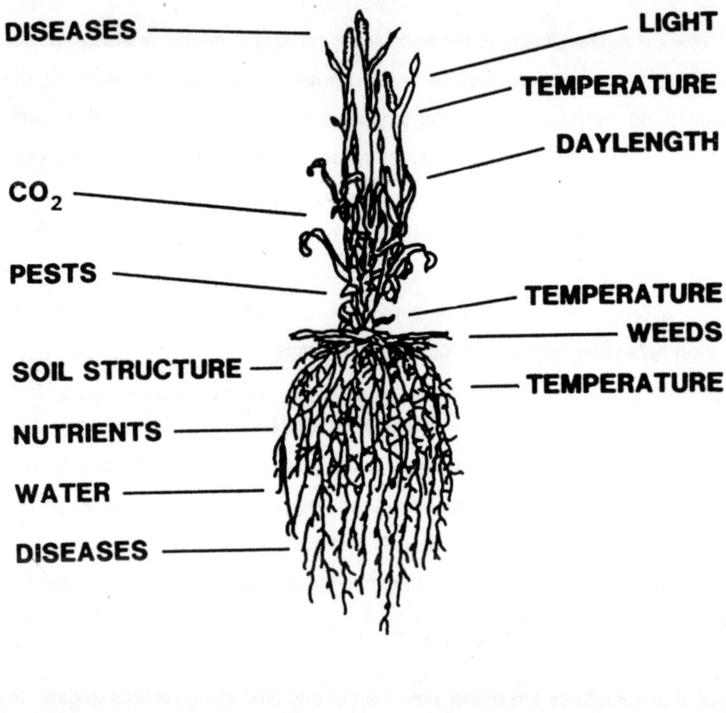

It is the goal of plant physiologists to disentangle, by experiment and observation, the cause and effect relationships within such intricacy. However, the complexity is not such as to inhibit the possibility of developing models of growth and development which serve both to evaluate the relative importance of various factors in elucidating variation in yield and to predict the future succession of growth under different conditions. A flow diagram of such a model [AFRCWHEAT2, (4)] is shown in Fig. 2. AFRCWHEAT2 is an updated version of a previous crop simulation model [AFRCWHEAT1, (5)] which includes modifications to growth as a result of water or nitrogen.

The guiding principle of such a model is that we can make estimates of the growth and development of a crop for two levels of environmental input. We estimate the production of a crop assuming there are no limitations to growth other than those imposed by the amount of radiation entering the system and the temperatures experienced by the crop. Given that dry matter growth and plant development are broadly separate processes, it is possible to generalise that temperature exerts overriding control on crop phenology, whereas radiation influences the amount of carbon fixed. This is not to omit influences such as photoperiod which are also important in controlling morphic changes, often in collaboration with temperature (6).

Most crops are often limited in potential by, in the first instance, sub-optimal levels of water and nutrients which may affect different physiological processes and components of the crop. Furthermore, when limiting, they can have differential severity. The latest AFRC model (AFRCWHEAT2) includes only such abiotic modifiers to potential growth (Fig. 2).

AFRCWHEAT2 was run to explore responses to simultaneous increases in CO_2 concentration and mean daily temperature and the interaction between climatic change and the nitrogen economy of a crop. Wheat is a determinate C_3 crop and would, thus, be expected to show an increase in dry matter productivity for a doubling of the ambient CO_2 concentration (7). The effect of a concomitant increase in temperature is harder to predict. Increasing temperatures would decrease the number and range of sites where coldness was a limiting factor controlling crop distribution. On the other hand since the speed of development of crops is, within limits, positively correlated with temperature (8), higher temperatures should shorten life-cycles and, thus, give less time for fixing carbon. The rates of other plant processes such as respiration also increase at higher temperatures which may lead to increased loss of carbon, although the relationship is not thought to be linear (9). Water saturation deficits would also increase following a rise in mean daily temperature and the associated increase in the potential evaporation rate. Usually, crops produce less dry matter per unit of transpired water when growing in a drier atmosphere (10).

In order to see the whole-system effects of these environmental changes AFRCWHEAT2 was run for a single site in the east of the UK under conditions in which a doubling of CO_2 was coincidental with rises in mean daily temperature up to a maximum of 4°C [Fig. 3, (11)]. Doubling the level of CO_2 led to an increase in total above-ground crop and grain dry matter of 25-30% (Fig. 3).

Such a modelled increase is in good agreement with published results for controlled environments (12). Raising the temperature affected both the vegetative and reproductive phases, although it should be remembered that only a single sowing date was examined. Changing the date of sowing can affect the rate of development, particularly for early phenological stages (13). The main effect during the vegetative phase was on the leaf area which increased slightly with a consequent small rise in total productivity up to a 2°C rise in temperature. Following a 3°C rise in temperature, leaf senescence became more important, and less dry matter was produced.

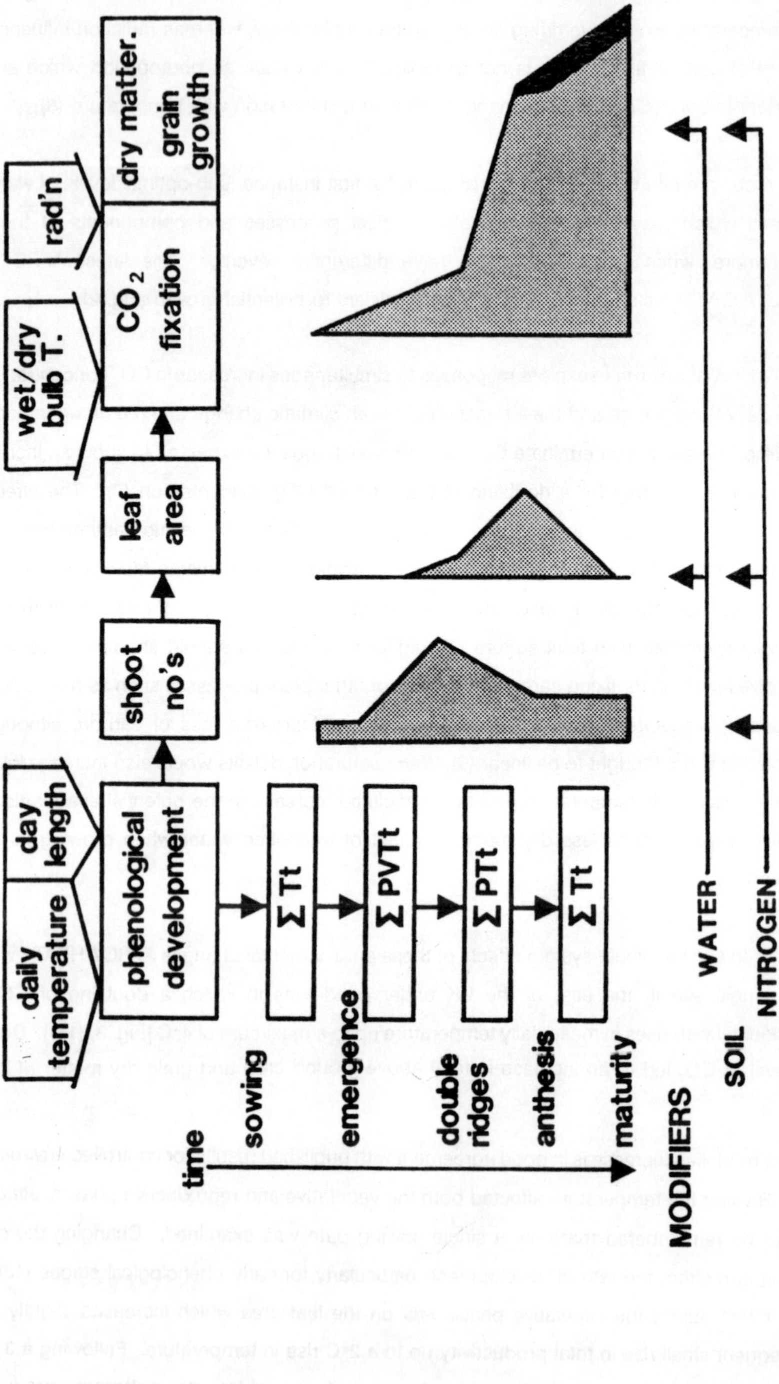

daily temperature
day length
phenological development
wet/dry bulb T.
rad'n
shoot no's
leaf area
CO_2 fixation
dry matter. grain growth
Σ Tt
Σ PVTt
Σ PTt
Σ Tt
time
sowing
emergence
double ridges
anthesis
maturity
MODIFIERS
SOIL
WATER
NITROGEN

Figure 2. Diagrammatic representation of the computer simulation model for winter wheat, ARCWHEAT2. This shows the interaction between the submodels: phenological development, tiller and leaf growth, light interception and photosynthesis, dry matter partitioning and grain growth (shown in black). The arrowed boxes show the driving variables. Tt, thermal time; PTt, thermal time modified by photoperiod; PVTt, thermal time modified by photoperiod and vernalisation. Thermal time is calculated as the sum of mean daily temperatures above a base temperature.

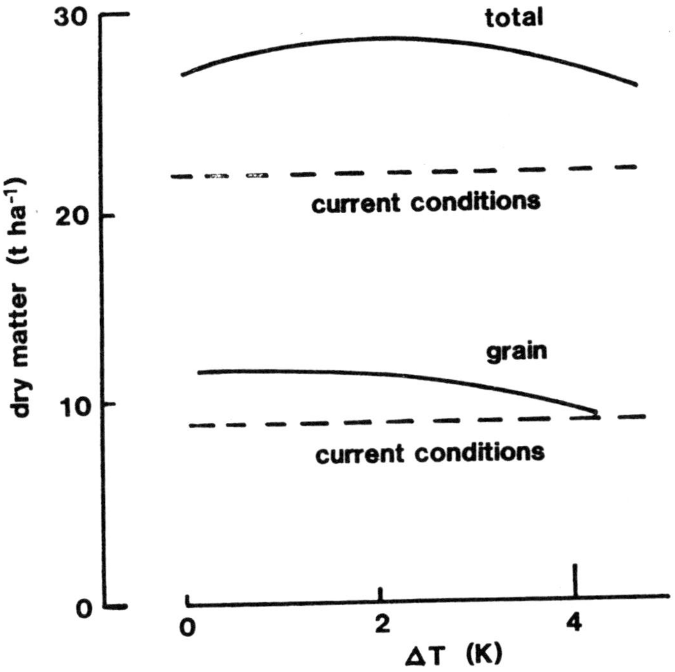

Figure 3. Modelled effects of a rise in mean temperature on the total dry matter production and grain yield of wheat. The temperature at T = 0 (K) corresponds to the 1981 temperature at Brooms Barn, Bury St Edmunds, UK. The dashed lines show dry matter production and grain yield at the current CO_2 level and for T = 0 (K).

The effect of increased temperature on post-flowering growth was solely to reduce the period for grain-filling. The final grain weight for a 4.5°C increase in mean daily temperature and a doubling of CO_2 was predicted to be the same as that for no change in CO_2 concentration. Thus, the 'greenhouse climate' is predicted to result in lower harvest indices for annual crops.

Nitrogen Economy

A current matter of deep concern for all sectors of the arable crop community is the pollution of drinking water with nitrates leached from agricultural land (14). Beside alterations to the thermal and gaseous environment of plants, it is postulated that there will be alterations in both the amount and pattern of precipitation as a result of climate change. Squire and Unsworth (11) reported their findings on the basis of a maximum 40% increase in precipitation for the UK during winter months and a -40% to +40% change during the summer.

Using these guidelines, the AFRCWHEAT2 model was run for four scenarios of climate change;

(1) Baseline: Simulation of a winter wheat crop sown in October 1985 at Rothamsted Experimental Station. The crop was well irrigated and adequately fertilised with nitrogen (*ca* 150 kg ha-1). The simulation used present-day CO_2 and temperature levels.

(2) Baseline with a doubled CO_2 level and a 4•C increase in mean daily temperature throughout the season, but without any change to the pattern or amount of rainfall.

(3) As (2) but with a 40% increase (above observed) in rainfall for October 1985 to the end of February 1986 and a 40% decrease (below observed) in rainfall for March 1986 until August 1986, thus representing a wet winter and a dry summer.

(4) As (2) but with a 40% increase (above observed) in rainfall throughout the season, thus representing a wet winter and summer.

These simulations show that, where there was a simple increase in temperature and CO_2 [scenario (2)], the results generally agree with previous simulations. That is, the total crop dry matter was little affected but grain yield was noticeably decreased (Table 1).

Table 1. The simulated effects on growth and the nitrogen balance of an autumn sown winter wheat crop grown at Rothamsted in 1985 for changes in climate. Scenario 1, simulation with present-day conditions; Scenario 2, simulation with double the CO_2 concentration and a 4°C increase in mean daily temperature; Scenario 3, as 2 but with a 40% increase in daily precipitation from September to February and a 40% decrease in precipitation for the rest of crop growth; Scenario 4, as 3 but with a 40% increase in precipitation throughout the growing season.

	SCENARIO			
	1	2	3	4
CROP GROWTH				
Total crop dry weight at maturity (t ha-1)	19.4	23.5	20.6	22.5
Grain yield (t ha-1)	9.4	7.0	7.1	7.1
Harvest index (%)	48	30	34	31
NITROGEN/NITRATE BALANCE (kg ha-1)				
Input:				
mineralised N	51		51	58
fertiliser N	154		54	154
soil residual N	27		27	26
TOTAL N	232		232	238
Output:				
plant uptake N	226		216	223
drainage N	6		16	15
drainage NO_3	26		70	66
TOTAL N	232		232	238

In scenario (3), over three times as much N per hectare was leached when compared with the simulation for present-day conditions. From deeper analysis of the output, it is clear that this was because the warmer, wetter winter allowed more mineralisation to occur in the soil. Also, since the crop was in the soil for a shorter period because of the higher temperatures and, thus faster development, more N could be leached through the profile.

The combination of a wet winter with a wet summer [scenario (4)] also resulted in more N leaching but no more than in (3). Either the amount leached in scenario (3) represented a maximum for the profile or the increased precipitation in the winter months was mostly responsible for increased nitrogen losses.

The implications from these runs of AFRCWHEAT2 point to possible serious agricultural and environmental problems stemming from the interaction between nitrate leaching and climate change. These simulations reinforce the experimentally derived conclusions that the major source of leached nitrates cannot be attributed to applied fertiliser but arise from the enhanced mineralisation of organic matter in the soil. Thus, under projected conditions of global warming and the increase of precipitation over evaporation, nitrate from mineralisation is expected to increase in amount in the soil. Work involving models is presently in hand to explore, over a wider geographical scale, the nature of the interactions between nutrition and climate change. Further experimental studies in this area are also required. They must be performed in conjunction with modelling work, with particular emphasis on the process of demineralisation (which is also temperature and moisture-sensitive) and the role of soil microbes in providing an absorbing sink for nitrogenous compounds.

Crop Ideotypes for a Greenhouse Climate

As stated above, major uncertainties exist with respect to changes in precipitation following any change in climate. However, crop models can be useful in defining the type of varieties of wheat and other crops that could be expected to cope with a warmer climate.

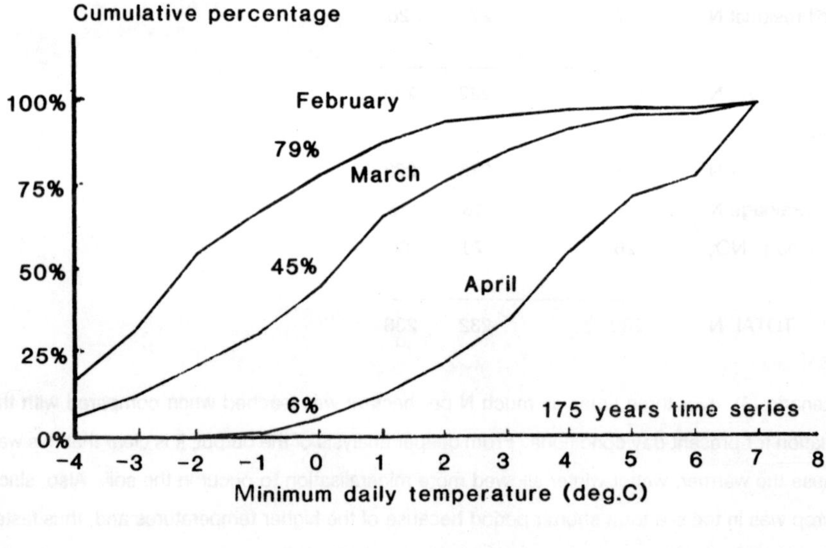

Figure 4. The percentage cumulative frequency distribution of the minimum daily temperatures for three months from the 1813-1988 meteorological record of Florence, Italy.

In order to come up with ideotypes for a warmer climate it is necessary to examine long-term meteorological data sets from which both reliable average and extreme temperature conditions can be estimated. For southern Europe, an example is the 175-year long record of maximum and minimum daily temperatures from Florence, Italy. From such a record it is possible to calculate the cumulative percentage frequency distribution of minimum temperatures for any month (Fig. 4). Thus, in about 75% of the years between 1813 and 1988, the last freezing frost occurred before the end of February (day number 60); in 45% of the years before the end of March. Late spring frosts can damage cereal crops if they occur following the double ridge stage of the apical meristem (15, 16). Similarly, drought and high temperatures at and following flowering can cause reductions in grain yield (17). Both the number of fertile grains and the rate of grain filling are reduced. Thus, a well adapted variety for these conditions is one that, for a variety of sowing dates, will reach double ridges after the last frost but flower before the onset of the drought period. From the above analysis day 60 would be a suitable target date after which double ridges would occur in non-freezing conditions and similar calculations have shown that day 140 (May 20th) is the earliest date on which the dry season started in 75% of the 175 years (Maracchi, personal communication).

Current general circulation models (GCMs) do not provide information on the effect of changing CO_2 concentration upon diurnal variation in temperature. Thus, at present, it is not possible to predict the effect that changing the climate will have on the date of the last annual frost. It can only be assumed that, given a general warming and without a change in the diurnal variation in temperature, that this date will be later for a warmer climate. Similarly, since the GCMs predict for the 'Business-as-Usual' scenario an average decrease in rainfall over southern Europe of 5-15% and a decrease in soil moisture of 15-25% by the year 2030 (1) it is likely that the dry season will start earlier than at present.

Model simulations were performed using the AFRC Wheat Model taking a particular year, 1965-66, from the Florence time-series as an example of a growing season where the average temperatures for all months were above the mean of those for the whole set. The observed temperatures for 1965-66 were not as high as those predicted by GCMs for the region as a result of a CO_2 doubling but give an indication of the conditions that crops may experience as the climate moves to a warmer equilibrium state. We were interested in the responses of two broad types of varietal group; those that develop relatively slowly with a well pronounced vernalisation requirement and faster developing varieties without a vernalisation requirement. The model was run for 10-day intervals in date of sowing from day 290 (October 17th) to day 360 (December 26th). We acknowledge that the latter dates are unrealistic from the viewpoint of arable crop practice but they were included in order to test the behaviour of the model at the extremes.

The slow variety consistently reached double ridges later than day 40 for all sowing dates, indicating that they would escape late frosts and primordial damage (Fig. 5). Fast varieties sown prior to day 310 (November 6th) did not come to double ridges before day 40 and thus could be considered at risk from late frosts.

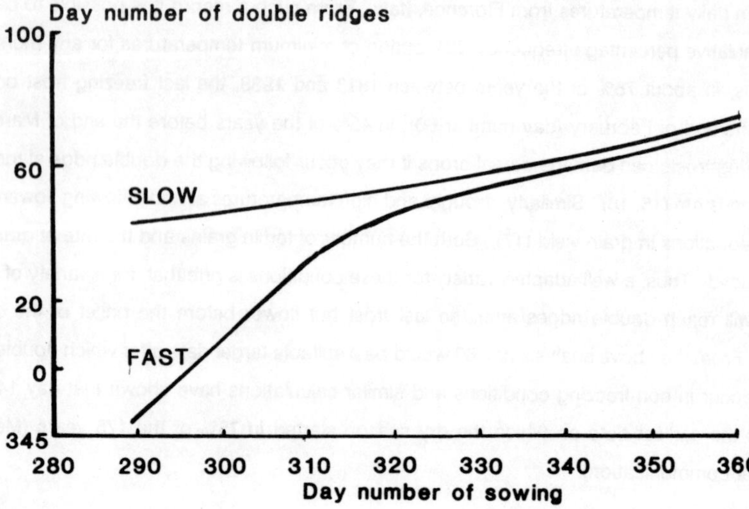

Figure 5. The predicted date of double ridges for fast and slowly developing varieties of wheat for a range of sowing dates using the AFRC Wheat Model. The temperature data used to drive the model are from the season 1965-66 at Florence, Italy. Mean monthly temperatures for this year were consistently higher than those for the 175-year data set.

Except for unrealistically late sowing dates both fast and slow developing varieties headed by day 130 (May 10th) which is 10 days before the onset of the dry period in the Florence time series (Fig. 6). Whether this is sufficient acceleration in order to cope with a warmer and drier climate is open to further question and experiment.

The conclusion from this exercise is that slower, rather than faster, developing varieties would be better adapted to a warmer climate for both the avoidance of late frosts and arrival at heading before the onset of severe drought. It is important that these conclusions now be tested by experiment. Since durum wheat is economically more important in southern Europe than bread wheat it is important to develop slower developing varieties of this type of wheat. Finally, since slower developing varieties tend to have higher final leaf numbers and thus more internodes and thus are taller than faster cultivars it is important to investigate the control of internode extension in cereals and its genetic basis in order to develop varieties that will resist lodging.

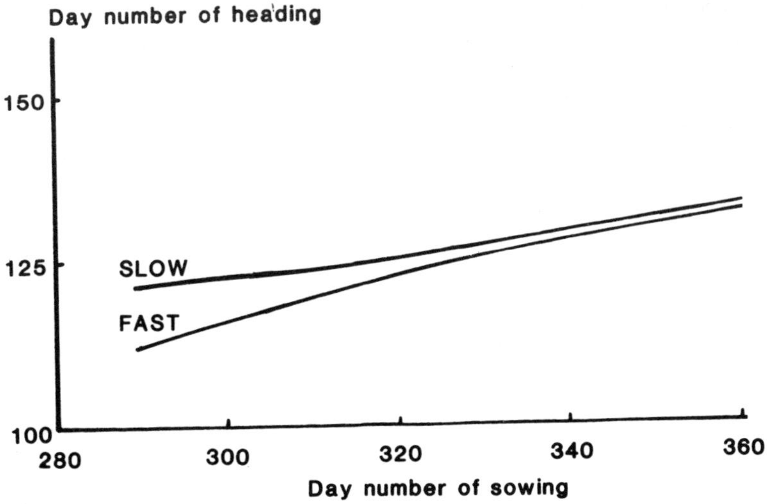

Figure 6. The predicted date of heading for fast and slowly developing varieties of wheat for a range of sowing dates using the AFRC Wheat Model. The temperature data used to drive the model are from the season 1965-66 at Florence, Italy. Mean monthly temperatures for this year were consistently higher than those for the 175-year data set.

FUTURE WORK

The concluding aim of this paper is to identify those areas of science where research on the impact of climate change on plants needs to be concentrated in the future, particularly with respect to collaborative research.

Such a list is bound to be subjective but we would offer the following topics as; (i) receiving rather low levels of interest at present but which are central to producing better predictive models of the effects of climate change; (ii) being areas which are supported at present but where significant effort still needs to be made. In no particular order of priority, we feel that we need work that looks at;

1. The responses of plants and crops to long-term exposure to changed CO_2 environments. Most predictions of changes in productivity rely on short-term experiments. The effects of warming and CO_2 change on aspects of physiology such as partitioning and crop morphology also need to be assessed in a rigourous manner. Whether such experiments are best done under semi-controlled conditions or by using field chambers is open to debate.
2. The available and desired characteristics of crop varieties for use in a greenhouse climate need urgent consideration in view of the projected speed of climate change vis-a-vis the conventional breeding cycle.

3. The simulations which looked at prospective nitrogen balances for crops grown in a greenhouse climate illustrated the importance of synergisms in assessing the overall effect of climatic change on features of the agricultural and/or natural environment. Work in this area needs to commence using congruent experimental and modelling approaches.
4. Work is in progress to connect global climate models to regional weather models and, thence, to crop models. Efforts in this area are crucial and need to continue. The interface between regional agricultural models and more detailed simulation models of crop growth needs to be established.

CONCLUSIONS

In the short-term, farming has always had to deal with problems caused by variation in weather. On a longer timescale changes in climate have been one of the factors that has determined the development of different types of agriculture. Political and socio-economic pressures have been, and continue to be, other hugely important agents of agricultural change.

It is not possible to predict which of these forces will influence arable farming more in the future. However, given the best available guesses for climatic change over the world, we can say that climate-induced changes have the potential to be profound and the time-scale is short (18). The lack of precision in present appraisals should not deter action either to ameliorate the effects of, or to estimate the risks associated with, the establishment of greenhouse conditions. For policy makers the greater failing would be to do too little rather than too much. It would be better to be approximately correct rather than to be precisely wrong (19).

ACKNOWLEDGEMENT

We thank Gareth Bewes and Rodney Parfitt for help in the production of the Figures.

REFERENCES

1. IPCC (1990). Scientific Assessment of Climate Change. The Policymakers' Summary of the Report of Working Group I to the Intergovernmental Panel on Climate Change. World Meteorological Organisation/United Nations Environment Programme.
2. Houghton, R.A. and Woodwell, G.M. (1989). Global Climate Change. Sci. Amer., 260: 18-26.
3. Natural Environment Research Council (1989). Biogeochemical Ocean Flux Study (BOFS) Report. Oceans and the Global Carbon Cycle. An Introduction to the Biochemical Ocean Flux Study of NERC Marine Sciences Directorate. Swindon, UK.
4. Porter, J.R., Weir, A.H., Rayner, J.H. and Parry, L. (1989). AFRCWHEAT2. Unpublished computer program.
5. Weir, A.H., Bragg, P.L., Porter, J.R. & Rayner, J.H. (1984). A winter wheat crop

simulation model without water or nutrient limitations. J. Agric. Sci., Camb., 102: 371-382.

6. Masle, J., Doussinault, G., and Sun, B. (1989). Responses of wheat genotypes to temperature and photoperiod in natural conditions. Crop Sci., 29: 712-721.

7. Kimball, B.A. (1983). Carbon dioxide and agricultural yield: an assemblage and analysis of 770 prior observations. WCL Report 14, Water Conservation Laboratory, Agricultural Research Service USDA, Phoenix, Arizona, USA.

8. Vincent, C.D. (1989). Recent advances in modelling crop responses to temperature. Outlook Agric., 18, 54-57.

9. McCree, K.J. (1974). Equation for the rate of dark respiration of white clover and grain sorghum, as functions of dry weight, photosynthesis rate and temperature. Crop Sci., 14: 509-514.

10. Monteith, J.L. (1981). Climatic variation and the growth of crops. Quart. J. Roy. Met. Soc., 107: 749-774.

11. Squire, G.R. and Unsworth, M.H. (1988). Effects of CO_2 and Climatic Change on Agriculture. Contract Report to the Department of the Environment. University of Nottingham, Sutton Bonington, UK.

12. Morison, J.J.L. (1988). Effect of increasing atmospheric CO_2 on plants and their responses to other pollutants, climatic and soil factors. In Environmental Aspects of Applied Biology, Aspects of Applied Biology, 17: 113-122.

13. Porter, J.R., Kirby, E.J.M., Day, W.. with 17 other authors. (1987). An analysis of morphological development stages in Avalon winter wheat crops with different sowing dates and at ten sites in England and Scotland. J. Agric. Sci., Camb., 109: 107-121.

14. Agricultural and Food Research Council (1989). Annual Report for 1987/88. London, UK.

15. Kirby, E.J.M. (1988). Significant stages of crop development. In: Models in Agriculture and Forest Research, F. Miglietta ed., IPRA-CNR, pp. 79-95.

16. Morison, J.J.L. and Butterfield, R.E. (1990). Cereal crop damage by frost. Weather, 45: 308-312.

17. Lawlor, D.W., Day, W., Johnston, A.E., Legg, B.J. and Parkinson, K.J. (1981). Growth of spring barley under drought: crop development, photosynthesis, dry-matter accumulation and nutrient content. J. Agric. Sci., Camb., 96: 167-186.

18. Parry, M.L. (1990). Climate Change and Agriculture, Earthscan, 192 pp.

19. Myers, N. (1988). Tropical deforestation and climatic change. Envt Conserv., 15: 293-298.

ENVIRONMENTAL STRESS, PHOTOSYNTHATE SUPPLY AND DEMAND, AND PLANT BEHAVIOR

Hesketh, J. D., Ephrath, J. E., and Alm, D. M.

Photosynthesis Research Unit, ARS, USDA, 1102 S. Goodwin, Agronomy Department, University of Illinois, Urbana, IL 61801, U.S.A.

ABSTRACT

We discuss literature sources for plant morphology and phenology which we found useful for developing generic plant models. One use for such models is to integrate physiological detail for predicting how plants might respond to stresses associated with climate change. Materials budgets should be used to determine when the timing, intensity and duration of stress result in more photosynthate supply than demand. Before the supply of scientific generalists or ecologists dwindles, literature syntheses describing complex ecological systems need to be written for an audience of scientific specialists. Our sources for phenology, morphology and plant geography should be helpful in developing such syntheses, but this is only a beginning.

THE MATERIALS BALANCE APPROACH

Much has been written recently about potential climate change and its effects on plant behavior. Recent emphasis has been on responses at the whole plant level, or the results of many CO_2:yield dosage response experiments, stress interaction studies, and related research, as well as what research needs to be done (2,5,6,16,33,36-38,53). There have been many significant discoveries recently about the physiology and biochemistry involved (29,56), which are being discussed at this symposium. The problem, of course, is an ecological one, at the geographic level. While many discoveries at the specialized level frequently override any system hierarchical barriers, we need to discuss the hierarchy problem.

Location x genotype x year effects and interactions limit the value of field plot, dosage response results. The systems modeling approach, supported liberally with laboratory techniques for analysis of a system, offers a way of overcoming some of these experimental limitations, provided emphasis is put on the biological logic involved and the systems approach is rigorously adhered to. Central to such an approach is a materials (C, N, H_2O, etc.) budget for any system under study.

Computers have made it easier to monitor the carbon economy of plants. Related research and modeling activities have indicated many real-world situations where plant stress prevents or greatly reduces any potential benefit from CO_2 enrichment. The quantification of the many possible kinds of stress involved (39) required a careful materials balance and systems analysis. There were differential effects of stress on supply and demand processes, which had to be quantified carefully, taking into account acclimation to stress, as well as recovery dynamics after stress relief.

In the cotton (Gossypium hirsutum L.) model (42), growth rates were considered to be linear between 12.5 and 30° C; whereas canopy photosynthesis was linear between 12.5 and 20° C, with no further change up to 30° C. In the short term, supply can exceed demand under these conditions at the lower temperatures; however, in the long run, more vegetative or reproductive units can start to grow or more fruit can be set at the lower temperatures, balancing demand with supply. Leaf starch levels offer a probe for how well supply and demand are balanced; high leaf starch levels have been reported frequently for cold-, N- or water-stressed plants.

If one has a reasonable model for all these processes and their interactions, if one defines carefully experimental conditions, and if one asks the critical questions and makes the critical experimental tests, the many location x growing period effects and interactions that limit field research can be partially overcome. Obviously such an approach is both intellectually and financially demanding; most scientific research is. For the test crops used, considerable morphological and phenological detail had to be generated from the literature and controlled environment experiments. Materials balances were needed for different kinds of plant organs growing under different, well defined conditions. Supply thresholds for new growth, curtailed growth, abscission (death) or induced dormancy were also needed for the same organs. Field experiments at different latitudes, elevations, seasonal growth periods, fertility and water levels, different plant densities or shade levels, and artificially extended daylengths or controlled night temperatures, with careful materials budgets, soil and weather data, and morphological descriptions have provided an information base for quantifying stress effects. Controlled environment experiments have added to this information base.

One, of course, needed a photosynthesis model, which took into account the makeup of the crop leaf canopy and how it intercepted light. Again stress effects and interactions had to be established, based upon considerable physiological detail. Logic used for differential effects of stresses on photosynthate supply and growth demand frequently controlled the model prediction.
The next question to be addressed was how to extend this physiology, morphology and phenology (PMP, 21) information

base to all plants. The crop or forest test plants already modeled included many PMP differences which could serve as examples. In order to predict morphology and phenology for other species, or to search for further major differences in both, we felt we needed to find literature sources for a wide range of plants, similar to what we had available for the few species already modeled. Progress was slow, but after three years we have found much of what we felt we needed in introductory and advanced textbooks (1,9,24,26-28,49,51,54) and books dedicated to one or many different kinds of species (3,30). These sources, found in the botanical, horticultural, grass, crops, ecological, and taxonomic literature, have been discussed elsewhere (15,30). We did not include the forestry literature; models for wood growth and tree branching behavior have been discussed elsewhere (30,49). Every research group studying the effect of environmental change on plant behavior should have access to these sources.

Obviously, even with all the background information available, one still needs coefficients for quantifying PMP characteristics and materials balances for the life cycle of various kinds of plant parts, including roots and N-fixing root nodules, for each species under a wide range of conditions. The botanical intricacies, however, determine the functional relationships, and how these coefficients interact. Phenological logic, of course, accounts for temperature, photoperiod and other environmental effects on growth and development.

Next one needs a classification system for the many strategies plants invoke to cope with a stress, including those stresses associated with plant competition and pest damage. Numerous authors have summarized such information; timing, intensity, and duration effects greatly complicate such a classification system. It is fascinating to watch what plants do under stress in a specific situation, even when the strategies deployed are anticipated.

Finally one needs to be aware of the principles involved in classifying topography, soils and the climate, or soil-plant-climate systems. These can be found in the soils, ecology and vegetation books (7,11,13-14,17,20,22,31,43, 47,51,57,59,60).

We have summarized above, with some greatly simplified phrases, the many complex ways global vegetation might be expected to respond to climate change. We are not the first to address this problem. The message, if anyone is missing it at this point, is that it is very difficult to generalize for a single plant species at a particular climate under a particular set of conditions, to say nothing of all plants under all prevailing global conditions. At many locations, year-to-year differences in growing season can be considerable. This, of course, is the nature of ecology, as is evident in introductory ecological textbooks. Also this is the rationale for systems modeling; or more appropriately the rationale for everyone making some effort to become a systems generalist. Systems modelers advocating this synthesis approach frequently have exhibited all the characteristics of narrow specialists, including the avoidance of multidisciplinary research. Fortunately others, totally unaware of the systems approach, have inadvertently shown the way.

Here we emphasize morphology, one of the PMP trio. While the references we found were adequate for our purposes, other scientists will require more detail for studying other problems. We need to identify what information is needed for each species, to attract better textbooks and literature syntheses and to design experiments for obtaining mechanisms and coefficients with the minimum amount of effort. For example, a large-scale study of many species in field plots at different elevations at several latitudes, with facilities for visiting scientists, might provide what is needed for the least cost. One must be very careful, however, when planning a large scale, expensive experiment; all too often such experiments only result in the depletion of much needed research resources.

GROWTH ANALYSIS MODELS

The original growth analysis model based upon measurements of dry weight (W) and leaf weight (L) over time still plays a role in a materials balance analysis. Tilman (58) used it to indicate how partitioning among leaves (L), stems (S) and roots (R) affects the RGR or (1/W)*(dW/dt). Perennials storing assimilate in vegetative organs or woody species have low RGR because of high amounts of assimilate diverted to storage or woody organs.

Tilman (58) assumed that dW/dt = (L*Pm)-(r*W), where Pm is the specific leaf photosynthetic rate (rate of accumulation of mass per unit leaf mass L) and r the respiration rate (all parameters are in dry weight mass units). With L = W-(S+R), dW/dt = (W-S-R)*(Pm-rW), and RGR or (1/W)*(dW/dt) = [1-(R+S)*(1/W)]*Pm-r, showing the effects of photosynthesis, respiration and partitioning on RGR. He also added a term (-dH/dt) for losses due to herbivoire activity or dying tissue.

Tilman (58) discussed how different species might survive or compete using such a model; annuals grow faster presumably because they allocate less to stems and roots. Woody stems and roots are costly, but like storage organs in non-annuals, represent an investment in the future. Annuals, of course, eventually allocate photosynthate to seed and the future. Under water of N stress, more photosynthate is allocated to roots, decreasing RGR.

For a more detailed balance, one needs a factor combining the effects of growth respiration and of losses in converting photosynthate to protein, cellulose, fats, other organic plant components, and ash or 'g'. One needs to subtract some leaf respiration associated with photosynthesis or 'p'.

One needs to convert L to leaf area A with a value for the specific leaf area 'SLA' (area/leaf mass L). Then one needs a factor for how well this area intercepts light ('q', depending upon phytomer placement, row effects, etc., 3,4, see below) as well as the effect of leaf age on Pm and p ('z') (55). Now our equation for RGR = [1-(R+S)*(1/W)]*[1/(1+g)]*q*[(z*Pm)-(z*p]-r-(1/W)*(dH/dt), showing the effects of many more physiological or biophysical processes on RGR. A recent review of the various components of this equation indicated further detail that can be quantified (23,55).

NAR = (1/A)*(dW/dt) can be obtained from (W/A)*RGR, keeping track of the spatial coordinants and age of A at the leaf or leaf fraction level (3,4,23). One needs a similar set of equations for a N and H_2O balance, with interactions between these three components of the materials balance. One can do other balances such as energy to predict plant tissue temperatures.

In most balance studies, materials usually balance, which proves that many of the techniques used provided good numbers. Of course, one hopes to gain a better understanding of how plants might respond to climate change in using this approach. The equations suggest how the biochemistry associated with Pm, g, r, and p can be incorporated into a whole plant synthesis. Fundamental differences among plants such as the light or dark pathways for photosynthesis can impact whole plant systems in ways that don't require a complex synthesis for analysis.

CROP-WEED COMPETITION AS AN EXAMPLE

An important factor controlling the resource economy of a plant is the expansion, light interception characteristics or spatial location, photosynthetic activity and senescence of its leaves (4,19,52,55,62). It logically follows that these same factors control the ability of a plant to establish itself from a propagule or to maintain itself among a group of competing plants. Leaf area expansion had been quantified for the modeled crop species; once we attempted to quantify crop-weed competition we had to extend the crop modeling methods to weeds. Shade tolerance during seedling growth, leaf appearance and expansion rates, mature internode lengths and extension rates, and the redistribution of N from old to young leaves played an important role in how well the crop or weed competed; these were quantified for several important weed species in a Midwest USA corn (Zea mays L.)-soybean [Glycine max (L.) Merr.] production system (unpublished). The addition of N fertilizer for corn but not for soybean had to be accounted for.

We built a generic model based upon the growth and development of each vegetative phytomer (leaf and leaf parts, internode, axillary buds) unit and its vertical placement, taking into account the distance between rows, plant population density, leaf angles, and branching (3,4). Logistic equations were used for leaf expansion and internode extension, with the phyllochron or total leaves that had appeared on the main stem as the time unit. A temperature function for leaf appearance rate related the phyllochron to real time.

The supply and demand for N and C was established for phytomer components, based upon results from plant population and N dosage studies. Branching depended upon this N and C economy. Such an economy can be used to predict fruit set, growth rate and duration of growth, or ultimately yield. We go through all this here to indicate the complexity of whole plant systems. The success of a crop or weed depended upon how much light each could intercept, along with the supply of N and water.

Besides predicting canopy light interception and photosynthetic behavior, another objective of our phytomer approach was to add some materials detail to the 'stem' and 'leaf' components of any allocation model. Tilman (see above, 58) described the effect of stem and root allocation of RGR. Without competition, the relative rate of dry matter production depended upon the amount of photosynthate allocated to leaves and the specific leaf photosynthetic rate. Stem growth, while causing RGR to decrease, was important for maintaining RGR in plants competing for light. In our studies (3), some weed plants overtopped the crop, after switching from a shade-tolerant strategy (long internodes, thin leaves) to a sun-tolerant one. Some overtopped only enough to smother the adjacent crop plants; others waited in the shade until the crop canopy senesced.

Many plants, such as the grasses, allocate photosynthate to stems after the leaf area is built up to a threshold level, either for a seedling or within a phytomer (30). Photosynthate supply also is phytomer dependent; the specific leaf photosynthetic rate changes with leaf age and position on the plant (23) and leaves higher on the plant sometimes have higher peak photosynthetic rates and mass densities. A detailed phytomer model is needed to account for such effects; prior to our work such models were mostly conceptual or dry weights and leaf areas were missing (30).

THE CORN:SUNFLOWER C_4:C_3 COMPARISON

Both corn and sunflower (Helianthus annuus L.) have large seeds, expand leaves rapidly, and have high leaf photosynthetic rates. Both corn and sunflower share a common strategy for producing large plants quickly; corn differs more from a great many other C_4 plants than it does from sunflower in this respect. The sunflower leaf photosynthetic rate was usually less than that for corn, but its transpiration rate was much higher. The dry matter production rate per unit ground area was occasionally reported to be similar for the two species; we found that for sunflower to be about 80% of the corn crop growth rate (27 vs 33.5 g $m^{-2}d^{-1}$, full ground cover) at Urbana IL. Both rates are lower than the reported maximum of 50 for C_4

grasses in the tropics. Sunflower plants recently were reported to extract more water from the lower part of the soil profile (12,25), than corn or other crop species, and had less than 1/2 the dry matter:water use efficiency than C_4 species. Consequently, given a deep profile available for root growth and water extraction, sunflower plants can withstand a drought longer (25). We now seem to have an explanation for the rapid growth rates seen for corn and sunflower: they both have large seeds and rapid leaf area expansion rates. It takes more transpiration, however, for sunflower to show rapid dry matter production rates. Other C_3 plants exhibited similar leaf photosynthetic and transpiration rates as sunflower for a short time at the early seed growth stage -- under humid conditions. Differences in the length of the growing season and cold tolerance were confounded with differences or similarities in total water use or dry matter produced.

CO_2 ENRICHMENT STUDIES

Krizek and coworkers at Beltsville (5,35-37) established some time ago that the best response of vegetative growth in seedlings to CO_2-enriched atmospheres was under warm temperatures, intense light, and an increased supply of mineral nutrients. They were concerned with seedling production for the horticultural trade. Their results have been frequently confirmed since.

The effect of photosynthate supply on growth or yield of a single plant in a plant stand can be researched and modeled, without actually doing a CO_2 enrichment experiment, for many stand densities and environmental conditions. There are many other ways for manipulating photosynthate supply than varying CO_2 in the air. A CO_2 enrichment experiment, where the appropriate conditions are imposed and the appropriate measurements are made, however, can be used to verify or improve a such a model.

One of us (JDH) was involved in CO_2 enrichment research with the ARS in Durham NC and Phoenix AZ (41,42) before the USA DOE-ARS (2,6,33,53) effort began. In those earlier studies, we established that cotton plants could grow and mature twice as many bolls in an enriched CO_2 environment, depending upon temperature and mineral nutrition. These results were used with photosynthetic information to develop a computer model for the C and N economy of the cotton plant, which predicted that yield responses to CO_2 depended upon the temperature and how the plant flowered and set fruit. Given adequate photosynthate and nitrogen the cotton plant can continuously set and mature bolls; in this case the length of the growing season determines yield. A period of cloudy weather or a drought can upset this process. This work established in the early 1970's the importance of a whole system, materials balance approach for CO_2 enrichment research.

We (41) also followed the accumulation of starch in plants under CO_2 enrichment, obtaining storage values equal to those for the non-starch portion of the leaf. Such leaves were brittle when handled and appeared chlorotic, although they had the same amount of chlorophyll as leaves grown at ambient CO_2. Rates of starch accumulation and decline were obtained for leaves switched back and forth between ambient and enriched CO_2 atmospheres. Sugar levels did not increase much, when starch increased. Sunflower plants did not store large amounts of starch; the extra photosynthate was quickly converted into stem tissue. Soybean plants became stunted under enriched conditions in the greenhouses at Phoenix; in growth cabinets at high temperatures at Durham, soybean plants in the vegetative phase grew all possible axillary branches, with numerous secondary branches. Squash (Cucurbita maxima L.) leaves died in enriched atmospheres. These results, obtained during the early 1970's, formed a base for much of the subsequent DOE effort.

There should be several effects of CO_2 enrichment on leaf photosynthetic activity, and these effects should be studied separately. Under optimum conditions, the rate will be enhanced if starch levels are kept within limits; above some level of starch the rate will be depressed because of feedback control (41,56). A biochemical model for feedback control has been proposed, with considerable supporting evidence (56). Finally, the supply of photosynthate during leaf development and expansion will determine the structure and makeup of its photosynthetic apparatus. Short-term stresses, or suboptimal conditions, will affect all these processes, depending upon the stress timing, duration and intensity.

At Urbana we (8,10,18,45) set out to determine how high a photosynthetic rate we could measure, after 10 to 20 min of exposure to enriched CO_2. We also supplied extra light to the bottom surface of exposed leaves to make sure the system was light saturated. While we only barely reached the maximum photosynthetic rates reported by Ludlow and Wilson (40), C_3 dicots with low rates under ambient conditions did better than sunflower. In the end we concluded that the only limiting factor was the ability of the leaf to handle the heat load involved; we were able to maintain maximum photosynthetic rates for 30 min before stopping the experiments. Our maximum rates of 120 μmol CO_2 $m^{-2}h^{-1}$ occurred with fully turgid leaves with open stomata when the treatment was imposed.

Enriched CO_2 has been used this way to overcome stomatal resistances to photosynthesis due to stress (18,54); we completely overcame such resistances under very severe water and heat stress conditions during a very severe drought and associated heat wave at Urbana IL (18). Others have almost overcome such resistances under similar conditions; it was apparent during our experiments that non-stomatal controls of photosynthesis were about to assert themselves.

HIERARCHY

We discussed the hierarchy problem a bit above; it needs more emphasis. There are various hierarchical barriers to be overcome. 'Getting from the laboratory to the field' is one; much lab work has been on light-stressed (shaded) and water stressed plants without knowing it. Yet research on 'field plants' has its location x year x cultural method limitation, or just what is a 'field-grown plant'? Getting from a monoculture to a complex ecosystem is as big a hurdle as any, although studies of interactions between pests and plants in monocultured systems are a beginning. Great strides have been made getting from a leaf to a canopy, although much remains to be done (32,55). Getting from a reaction site to a leaf is equally complex and it is not clear if progress here has been as great. Little has been done with light interception by different layers of reactions sites, although some of the leaf-canopy synthesis is applicable. 'Stems and roots' are still a big black box, but progress is being made in overcoming this problem. We are only beginning to quantify soil macropore dynamics and their role in an ecological synthesis.

The introduction of pests into a monoculture synthesis confronts the monumental spatial variability problems that plagues all ecological syntheses; these problems are being addressed (46). Soil scientists have dealt with these problems for years; as have crop science statisticians involved in experimental field plot design and analysis. Ecologists, or course, have always had to deal with these problems

IRREVERSIBLE STRESS EFFECTS

Stress, of course, may limit the response of plants to increased CO_2. Short-term severe stresses at critical stages of growth may reduce the potential size or growth rate of developing organs, or such organs may be abscised or rendered irreversibly dormant, limiting the future response of such plants to increased CO_2 once the stress is relieved. Field studies are subject to such stresses; cloudy weather at a critical stage of growth may cause irreversible damage. However, cloudy weather is usually associated with cool temperatures, which can decrease the demand for photosynthate for growth. Sometimes water stress for a short period at a critical time can greatly reduce yields, even though the stress is relieved soon after.

CLIMATE CHANGE IN THE PAST

Pollen deposits, vegetative relics, and comparisons of alpine and arctic-tundra plants with altitude and latitude suggest that it was much warmer in North America some 7000 years ago (The Hypersithermal Interval, 13). It was also drier in the U.S. Midwest, with the prairie shifting much farther east than it is now. The sea level was much higher. Evidence for changes in vegetation and animals from when the Wisconsin glacier peaked (16000-18000 years ago) through the Hypersithermal Interval and subsequent periods of warming and cooling suggests what climate change might bring (7,11,14).

RESEARCH PHILOSOPHY

Biophysical principles need to be used to quantify ecosystem processes. Such an approach should lead to research on critical processes that hasn't yet been done. Scientific discoveries should occur more frequently, with less resources. There needs to be a productive balance between research and theory; most researchers may be able to synthesize to the ecosystem level, knowing and applying the principles involved.

Morphological and phenological studies are urgently needed but are labor intensive requiring little equipment other than a standard weather station and a dissecting microscope. One does have to be clever in designing experiments and analyzing results, to separate variables. It is possible to invest heavily in computer equipment, but one can avoid it by being clever in developing mathematical models and simple inexpensive experimental tests. It does get a little frustrating when the products of such clever scientists are used to generate support for less clever ones. India is not alone in lacking support for equipment intensive research; the recent books on whole plant syntheses or materials balances reveal how extensive this problem is.

To get a good feel for science, one should consult Popovsky (48), not for the names, places and details, but for the clash between the basic and applied research approach in the agricultural sciences and how scientists interact. What he reported knows no national boundary; today the clash is more between scientific disciplines, deciding how to allocate resources. Peters and coworkers (44) have critically analyzed how research is planned, how discoveries are made, and, as a result, how science should be managed, or how scientists should collaborate. Their analysis, which is somewhat novel, is important in an era of limited resources. The common pitfalls of a modeling or mathematical component of a systems analysis effort involving team research has also been critically analyzed (29,50), but the approach has contributed to progress in some cases, particularly how leaf canopies behave (55).

NEEDED LITERATURE SYNTHESIS AND RESEARCH

Textbooks should cover the literature syntheses that all specialists, including those working on the biochemistry and biophysics of the photosynthetic light and dark reactions, need to become generalists or ecologists, overcoming somehow the horrifying nomenclature problem. Such syntheses should include quantitative analyses of test systems, with user friendly computer aids for solving problems. Videotape aids are needed for descriptions of global vegetation, to overcome some of the nomenclature problem. Pictures help (7); plant ecology could take better advantage of video tapes and laser disks. Communication with the uninitiated depends upon actual pictures of plant systems, short of actually visiting representative field sites. Certainly this is one way of converting specialists into generalists, and focusing research from the many plant disciplines into an ecological synthesis. Until more progress is made, one research group cannot do all this alone. Much of what is needed is already available somewhere in some form; it just needs to be made available to more people.

Complex plant strategies for coping with the timing, duration and intensities of stresses, and associated interactions among stresses, need to be quantified and classified. While the generalists approach may well predict much of how plants respond to climate change, one should be on the alert for unusual, unexpected complex responses. It also goes without saying that we should define and focus on the critical physiological processes involved (2,33,34,36,38,56,61). All this seems trite, but the systems analysis-computer simulation approach has been formally stated, over and over, for some 20 years, while frequently ignored by its proponents in their related research. Others, innocent of such formal logic, utilized the approach in their research, with remarkable success. Somehow we need to foster the use of the approach, without undue emphasis on the religious philosophy associated with it.

REFERENCES

1. Aitken, Y. (1974) Flowering Time, Climate and Genotype. Melbourne Univ. Press, Melbourne. 193 pp.

2. Allen, L. H. (1990) Plant responses to rising carbon dioxide and potential interactions with air pollutants. J. Environ. Qual. 19: 15-34.

3. Alm, D.M., McGiffen, M.E., and Hesketh, J.D. (1990) Weed phenology. In: Predicting Crop Phenology. Hodges, T., ed. CRC Press, Boca Raton FL. pp.

4. Alm, D.M. and Hesketh, J.D. (1990) A generic model for phytomer ontogeny and vertical placement in competitive weed-crop communities. In: Proc. Workshop on Modeling Pest-Crop Interactions. Teng, P. and Yuen, J., eds. Res. Ext. Series, Hawaii Inst. Tropical Agric. Human Res., Honolulu. pp. 8-10.

5. Bailey, W.A., Klueter, H.H., Krizek, D.T., and Stuart, N.W. (1970) CO_2 systems for growing plants. Trans. A.S.A.E. 13: 263-268.

6. Baker, J.T., Allen, L.H., Boote, K. J., Jones, P.H. and Jones, J.W. (1989) Response of soybean to air temperature and carbon dioxide concentration. Crop Sci. 2f9: 98-105.

7. Barbour, M. G. and Billings, D.W., eds. (1988) North American Terrestrial Vegetation. Cambridge Univ. Press, Cambridge. 434 pp.

8. Begonia, G.B., Hesketh, J.D., Woolley, J.T., and Peters, D.B. (1987) Variability in leaf photosynthetic CO_2 exchange rates near saturating light and CO_2. Photosynthetica 21: 584-587.

9. Benson, L. (1979) Plant Classification. D.C. Heath, Lexington, Massachusetts. 901 pp.

10. Breen, P.J., Hesketh, J.D. and Peters, D.B. (1986) Further field measurements of leaf photosynthesis in enriched light and CO_2. Photosynthetica 20: 281-285.

11. Collinson, A.S. (1989) Introduction to World Vegetation. Unwin Hyman, London. 325 pp.

12. Cox, W. J. and Jollif, G.D. (1986) Growth and yield of sunflower and soybean under soil water deficit. Agron. J. 787: 226-230.

13. Crawford, R.M.M. (1989) Studies of Plant Survival, Ecological Case Histories of Plant Adaptation to Adversity. Blackwell Scientific, Oxford. 296 pp.

14. Daubenmire, R. (1978) Plant Geography, with Special Reference to North America. Acad. Press, New York. 338 pp.

15. Elmore, C.D., Ephrath, J.E., Alm, D.M., and Hesketh, J.D. (1991) Physiological, morphological, and phenological aspects of Crop-Weed Competitive Strategies. submitted.

16. Enoch, H.Z. and Kimball, B.A. eds. (1986) Carbon Dioxide Enrichment of Greenhouse Crops. Vol. II, Physiology, Yield and Economics. CRC Press, Boca Raton FL.

17. Fitter, A.H., and Hay, R.K.M. (1987) Environmental Physiology of Plants. Acad. Press, London. 423 pp.

18. Frederick, J.R., Alm, D.M., Hesketh, J.D., and Below, F.E. (1990) Overcoming drought-induced decreases in soybean leaf photosynthesis by measuring with CO_2-enriched air. Photosynthesis Research 25: 49-57.

19. Grant, R.F., Frederick, J.R., Hesketh, J.D. and Huck, M.G. (1989) Stimulation of growth and morphological development of maize under contrasting water regimes. Can. J. Plant Sci. 69: 401-418.

20. Grime, J.P. (1979) Plant Strategies and Vegetation Processes. John Wiley, Chichester. 222 pp.

21. Gutschick, V.P. (1987) A Functional Biology of Crop Plants. Timber Press, Portland OR. 230 pp.

22. Harlan, J. R. (1975) Crops and Man. American Soc. Agron, Madison. 295 pp.

23. Harper, J.L. (1989) Canopies as populations. In: Plant Canopies: Their Growth, Form and Function. Russell, G., Marshall, B. and Jarvis, P.G. eds. Cambridge Univ. Press, Cambridge. pp. 105-128.

24. Hartmann, H.T. and Kester, A.W. (1983) Plant Propagation. Prentice Hall, Englewood Cliffs, N.J. 727 pp.

25. Hattendorf, M.J., Redelfs, M.S., Amos, B., Stone, L.R., and Gwin, R.E. (1988) Comparative water use characteristics of six row crops. Agron. J. 80: 80-85.

26. Hay, R.K.M. and Walker, A.J. (1989) An Introduction to the Physiology of Crop Yield. John Wiley, New York. 292 pp.

27. Hayward, H.E. (1938) The Structure of Economic Plants. Macmillan, New York. 674 pp.

28. Hector, J.M. (1936) Introduction to the Botany of Field Crops. Vol. I. -- Cereals; Vol.II.-- Non-Cereals. Central News Agency, Johannesburg. 478 and 1127 pp.

29. Hesketh, J.D. and Alm, D.M. (1990) Synthesis: Modelling Field-Crop Ecosystems and Issues for the Future. In: Field Crop Ecosystems of the World. Pearson, C.J., ed. Elsevier, New York. in press.

30. Hesketh, J.D., Wisiol, K., Pettigrew, W.T., and Alm, D.M. 1990. Stems: Patterns and duration of development. In: Predicting Crop Phenology. Hodges, T., ed. CRC Press, Boca Raton FL. pp.

31. Howard, J.A. and Mitchell, C.W. (1985) Phytogeomorphology. John Wiley, New York. 222 pp.

32. Hughes, G. (1990) Inpact of pests on crops: Spatial issues. In: Workshop on Modeling Pest-Crop Interactions. Teng, P. and Yuen, J., eds. Research Extension Series, Hawaii Inst. Tropical Agric. Human Res., Honolulu, Hawaii. pp. 23-25.

33. Kimball, B.A. (1990) Impact of Carbon Dioxide, Trace Gases, and Climate Change on Global Agriculture. Amer. Soc. Agron., Madison. 133 pp.

34. Kramer, D.M., Wise, R.R., Frederick, J.R., Alm, D.M., Hesketh, J.D., Ort, D.R. and Crofts, A.R. (1991) Regulation of coupling factor in field-grown sunflower: A redox model relating coupling factor activity to the activities of other thioredoxin-dependend chloroplast enzymes. Photosynthesis Res., in press.

35. Krizek, D.T. (1979) Carbon dioxide: Guidelines. In: Controlled Environment Guidelines for Plant Research. Tibbitts, T.W. and Kozlowski, T.T. (eds.) Acad. Press, New York. pp. 241-269.

36. Krizek, D.T. (1986) Photosynthesis, dry matter production and growth in CO_2 enriched atmospheres. In: Cotton Physiology. Mauney, J.R. and McD.Stewart, J., eds. The Cotton Foundation, Memphis TN. pp. 193-225.

37. Krizek, D.T., Bailey, W.A., Klueter, H. and Liu, R.C. (1974) Maximizing growth of vegetable seedlings in controlled environments at elevated temperature, light and CO_2. Acta Hort. 39: 89-102.

38. Lemon, E. R. ed. (1983) CO_2 and Plants, The Response of Plants to Rising Levels of Atmospheric Carbon Dioxide. Westview Press, Boulder CO.

39. Levitt, J. (1980) Responses of Plants to Environmental Stresses, Vol. I. Chilling, Freezing and High Temperature Stresses; Vol. II. Water, Radiation, Salt, and Other Stresses. Acad. Press, New York. xx and xx pp.

40. Ludlow, M.M. and Wilson, G. L. (1971) Photosynthesis of tropical pasture plants. I. Illuminance, carbon dioxide concentration, leaf temperature, and leaf-air vapour pressure difference. Aust. J. Biol. Sci. 24: 449-470.

41. Mauney, J.R., Guinn, G., Fry, K.E. and Hesketh, J.D. (1979) Correlation of photosynthetic carbon dioxide uptake and carbohydrate accumulation in four crop species. Photosynthetica 13: 260-266.

42. McArthur, J.L., Hesketh, J.D. and Baker, D.N. (1975) Cotton. In: Some Case Histories of Crops Physiology. Evans, L.T., ed. Cambridge Univ. Press. pp. 297-325.

43. Pears, N. (1985) Basic Biogeography. Longman, London. 358 pp.

44. Peters, T.J. (1987) Thriving on Chaos: A Revolutionary Agenda for Today's Manager. Knopf, New York. pp.

45. Pettigrew, W.T., Hesketh, J.D., and Peters, D.B. (1990) CO_2-saturated photosynthesis as affected by a rapdily induced phosphate stress. Photosynthetica 24: 209-216.

46. Pinnschmidt, H., Teng, P.S. and Yuen, J. (1990) Pest effects on crop growth and yield. In: Proc. Workshop on Modeling Pest-Crop Interactions. Teng, P.S. and Yuen, J., eds. Res. Ext. Series, Hawaii Inst. Tropical Agric. Human Resources, Honolulu, HA. pp 26-29.

47. Polunin, N. (1960) Introduction to Plant Geography and Some Related Sciences. McGraw-Hill, New York. 640 pp.

48. Popovsky, M. (1984) The Vavilov Affair. Archon Books, Shoestring Press, Hamden CN. 216 pp.

49. Porter, J.R. (1989) Modules, models and meristems in plant architecture. In: Plant Canopies: Their Growth, Form and Function. Russell, G., Marshall, B. and Jarvis, P.G., eds. Cambridge Univ. Press, Cambridge. pp. 143-159.

50. Quade, E.S. (1988) Quantitative methods: uses and limitations. In: Handbook of Systems Analysis, Craft Issues and Procedural Choices. Miser, J.J. and Quade, E.S., eds. Elsevier, New York. pp. 283-324.

51. Radford, A.E. (1986) Fundamentals of Plant Systemics. Harper and Row, New York. 498 pp.

52. Reid, J.F., Zur, B. and Hesketh, J.D. (1990) The dynamics of a maize canopy development. 2. Leaf area growth. Biotronics 19: in press.

53. Rogers, H.H., Cure, J.D., Thomas, J.F. and Smith, J.M. (1984) Influence of elevated CO_2 on growth of soybean plants. Crop Sci. 24: 361-366.

54. Roshevits, R. Yu. (1937 - 1980) Grasses, An Introduction to the Study of Fodder and Cereal Grasses. Indian Nat'l. Sci. Doc. Ctr., New Delhi (English translation). 635 pp.

55. Russell, G., Marshall, B. and Jarvis, P.G. (1989) Plant Canopies: Their Growth, Form and Function. Cambridge Univ. Press, Cambridge. 178 pp.

56. Stitt, M. (1990) Fructose-2,6-Bisphosphate as a regulatory molecule in plants. Annu. Rev. Plant Physiol. Plant Mol. Biol. 41: 153-185.

57. Svafer, W. (1975) General Plant Geography. Foreign Sci. Pub. Dept., Warsaw. (English Translation) 430 pp.

58. Tilman, D. (1988) Plant Strategies and the Dynamics and Structure of Plant Communities. Princeton Univ. Press, Princeton NJ. 360 pp.

59. Walter, H. (1985) Vegetation of the Earth and Ecological Systems of the Geo-biosphere. Springer-Verlag, Berlin (English translation). 318 pp.

60. Wilsie, C.P. (1962) Crop Adaptation and Distribution. W.H. Freeman, San Francisco. 448 pp.

61. Wise, R.R., Frederick, J.R., Alm, D.M., Kramer, D.M., Hesketh, J.D., Crofts, A.R. and Ort, D.R. (1990) Investigation of the limitations to photosynthesis induced by leaf water deficit in field-grown sunflower (Helianthus annuus L.). Plant, Cell and Environment. 13: in press.

62. Zur, B., Reid, J.F. and Hesketh, J.D. (1989) The dynamics of a maize canopy development. 1. Leaf ontogeny. Biotronics 18:55-66.

ADDENDUM

The following references should have been included in our list. References (63,65,66) obviously covered primary productivity at the global level, which is a place to start for global modeling. Gribbin (64) is an excellent source for background information on climate change; he listed other similar sources. The last one (67) discussed taxonomy, similar to Radford (51), but from a different perspective.

63. Cooper, J. P., ed. (1975) Photosynthesis and Productivity in Different Environments. Cambridge Univ. Press, Cambridge. 1975. 715 pp.

64. Gribbin, J. (1990) Hothouse Earth - The Greenhouse Effect and Gaia. Grove-Weidenfield, New York. 273 pp.

65. Hall, D. O., Combs, J. and J. M. O. Scurlock. (1985) Biomass production and data. In: Techniques in Bioproductivity and Photosynthesis. Coombs, J., Hall, D. O., Long, S. P. and Scurlock, J. M. O., eds. Pergamon Press, New York. pp. 274-287.

66. Leith, H. F. H., ed. (1978) Patterns of Primary Production in the Biosphere. Dowden, Hutchinson, and Ross, Stroudsburg, PA. 342 pp.

67. Soule, J. (1985) Glossary for Horticultural Crops. John Wiley, New York. 898 pp.

DYNAMIC LIMITATIONS TO PHOTOSYNTHETIC GAS EXCHANGE IN A SOYBEAN CANOPY

R. W. Pearcy[1], T. L. Pons[2] and G. F. Sassenrath[1]; [1] Department of Botany, University of California, Davis, CA 95616, USA; [2]Department of Plant Ecology, University of Utrecht, 3512 PN Utrecht, The Netherlands.

ABSTRACT

Leaves at mid level in a soybean canopy received as much as 50 to 90% of the available PAR in the form of sunflecks. These sunflecks were found to be typically very brief, with most lasting less than 5 seconds, but also very frequent. Steady-state gas-exchange measurements do not provide an adequate description of the limitations to photosynthesis under these conditions. Field and laboratory studies with gas-exchange systems designed for measurements of rapid transients showed that the induction state of a soybean leaf, which determined the capacity to respond to a sudden increase in irradiance, was a major limitation to the use of sunflecks. Leaves that were initially shaded for long periods but then exposed to either constant high irradiance or to sequences of lightflecks (simulated sunflecks) exhibited an increase in induction state that resulted from an apparent increase in RuBP regeneration capacity, light activation of ribulose-1,5-bisphosphate carboxylase and an increase in stomatal conductance. Leaves that were initially in high light but then shaded exhibited a rapid decay in induction state. Estimates of the limitation to respond to a rapid increase in PFD were obtained by measuring the photosynthetic rate 5 and 60 s after a light increase. An induction state (IS_5 and IS_{60}, respectively) was calculated as the rate expressed relative to the steady-state light-saturated assimilation rate. After only 5 min in shade, IS_5 had decayed to only 0.4 and IS_{60} had decayed to 0.8. By contrast, leaves shaded for 1 min or less maintained IS_5 and IS_{60} at values above 0.9. The equilibrium induction states depended on the mean PFD but differed significantly if the PFD consisted of a series of lightflecks as compared to constant irradiance.

INTRODUCTION

Leaves within crop canopies experience highly dynamic and spatially variable light environments resulting from the frequent occurrence of sunflecks. These sunflecks are transmitted through gaps in the canopy to leaves at lower levels where they may contribute 40 to 90 percent of the daily PFD (photon flux density) available for photosynthesis (14). The rotation of the Earth and canopy movement due to wind cause high and low frequency changes in the PFD (11). In addition, intermittent cloudiness can cause rapid changes in the PFD available for photosynthesis at all levels in the canopy.

Despite the highly dynamic nature of the light environment, there has been little research on the photosynthetic utilization of transient light by crop species and its consequences for photosynthesis by canopies. Many studies have examined the steady-state responses of photosynthesis which result when the environment is held constant for a sufficient period so that a constant response is obtained. Leaves in the canopy, however, may rarely experience conditions that approximate a steady-state. The environmental and physiological controls on photosynthesis are quite different under transient as compared to steady state conditions. Studies with forest understory plants, which depend on sunflecks for a large part of the light available for photosynthesis, have revealed that when sunflecks are infrequent the induction requirement of the photosynthetic apparatus limits their utilization (1, 7). However, frequent short sunflecks cause induction to occur. Moreover, when leaves are fully induced short sunflecks can be used with high efficiency because of the contribution of post-lightfleck CO_2 fixation (2, 9).

The objective of this study was to examine the responses of soybean leaves to increases in PFD, or to sequences of lightflecks (artificial sunflecks). Particular attention was paid to the limitation imposed by the induction requirement of photosynthesis. The results show that the induction requirement develops rapidly in soybean leaves following shading and poses a significant limitation to the use of subsequent sunflecks.

MATERIALS AND METHODS

Field studies

Soybeans (Glycine max (L.) Merr. cv. 'Williams') were planted in rows with 75 cm spacing. Williams is a tall, indeterminate type that allowed access to the middle and lower portions of the canopy with minimal disturbance. After establishment, the plants were thinned to a 10 to 20 cm spacing within rows. The plots were fertilized with a commercial NPK fertilizer at the rate of 300 kg ha^{-1} and irrigated 2-3 times weekly. Measurements were started after the canopy had closed and had reached a height of 1-1.2 m. The plants were in a flowering or early pod-filling stage at this time.

Measurements of PFD were made within the canopy with gallium arsenide phosphide (GaAsP) photocells (Hanamatsu model G1118) that were calibrated under sunlight against a LI-COR model 190s Quantum Sensor. These sensors respond to wavelengths from 400 to 680 nm and under canopy shade exhibit a spectral error relative to a quantum sensor of about 3% (12). They are therefore a reasonable substitute for much more expensive Quantum Sensors when more than a few sensors are required. The output from the photocells was converted to a millivolt signal by a 40 ohm shunt resistor. A total of 8 sensors were mounted on wooden stakes at different heights within the canopy and the PFD values from each were logged at either 1-s (for full-day measurements) or 0.05-s (for 1-hr periods) intervals with a Campbell Scientific model CR 21X Micrologger data acquisition system. The readings were recorded on magnetic tape and later analyzed on a microcomputer with a program, HISTO, that constructed histograms of sunfleck durations, the maximum PFD of a sunfleck, and the length of the low-light interval between sunflecks. A sunfleck was defined as a continuous excursion of the PFD above a threshold of 200 μmol photons m^{-2} s^{-1}. The sensors were moved daily to increase the number of positions within the canopy that were sampled. The leaf area index (LAI) above each sensor was estimated with a LI-COR model LAI-2000 plant canopy analyzer (20).

Measurements of leaf gas exchange were made with a transportable, open-system, gas-exchange apparatus that has been described elsewhere (15). A leaf in its natural position within the canopy was enclosed in the leaf chamber, which was constructed of nickel-plated brass with a glass-windowed top. Fans within the chamber kept the air well mixed and served to maximize the boundary layer conductance. Chamber temperature was controlled by thermoelectric heat-exchange modules attached to a finned heat sink in the bottom of the chamber. In some experiments, natural sunlight was utilized as the light source and was suddenly increased by parting the canopy so that the time course of photosynthetic induction could be followed. In other experiments, artificial light from a 150-W quartz-iodide projection lamp was used. The beam from the lamp was focused through an electronic shutter onto the leaf via a dichroic reflector. The shutter was then opened and closed to create sequences of lightflecks (artificial sunflecks). Data from the gas exchange apparatus were logged a 5-s intervals via a Hewlett Packard model 3421a data acquisition system connected to a portable microcomputer. Calculations of gas exchange parameters were done using the equations of von Caemmerer and Farquhar (19).

Laboratory measurements

Plants were grown in 4-l pots of a soil/sand/peat mixture (1:1:1) in a growth chamber with a 26° C day, 18° C night temperature regime and a fluorescent/incandescent lamp mixture that gave 400 μmol photons m^{-2} s^{-1} at plant height. The daylength was 14 hours. Photosynthetic gas-exchange rates were measured on the third or fourth trifoliate leaf in an apparatus specially designed for rapid transient measurements. A 22 cm^2 portion of a leaflet was enclosed in a chamber that was only 2 mm deep, giving a total chamber volume, including the inlets, of 4.5 cm^3. The adaxial surface of the leaf was appressed directly to the glass window of the water jacket in the chamber lid with silicon stopcock grease. Thus, gas exchange occurred only through the abaxial surface. With the flow rates used, the chamber completely flushed within 0.5 s. The response time of the system was then limited by the time to flush the cells of the LI-COR model 6252 CO_2 analyzer (volume= 12 cm^3) and was 1.5 s for a full-scale response to a step change in CO_2 within the chamber (9). Light was supplied from a quartz-iodide lamp and shutter system as described above. Data from the gas exchange apparatus were logged at 0.25-s intervals via an Hewlett Packard model 3497A data acquisition system connected to an IBM PC/AT microcomputer.

Activities of ribulose-1,5-bisphosphate carboxylase (rubisco) were assayed according to procedures given in ref 15. The initial activity measured immediately after extraction and before any activation by mg^{2+} or CO_2 could occur was taken as being equal to the *in vivo* CO_2 and RuBP-saturated rate of the enzyme.

RESULTS

Field Studies

The light environment at nearly all levels within the soybean canopy was characterized by frequent sunfleck activity. An example showing the highly transient nature of the PFD measured by a single sensor is given in figure 1. This sensor had a LAI above it, as viewed by the LI-COR plant canopy analyzer, of 2.3 and received 9.3 mol photons m^{-2} day^{-1} on this day. This daily total was 21% of the PFD received at the top of the canopy. The 1042 sunflecks contributed 78% of the total PFD and had a mean length of 5.5 s. Overall, the light environment was characterized by periods of very frequent sunflecks, which were created by movement of the leaves in the wind, separated by periods of low PFD when the sensor was completely in the shade of leaves. Even leaves close to the top of the canopy (cumulative LAI of 0.5 to 1) experienced highly dynamic light environments consisting of rapidly alternating periods of sun and shade.

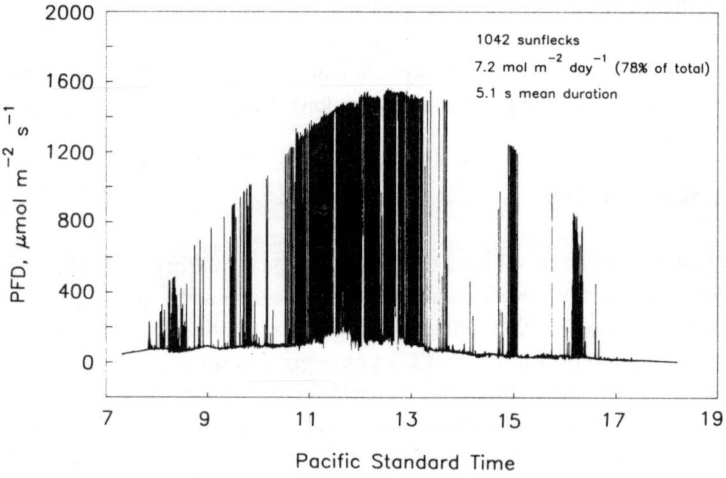

Figure 1. Representative daily course of photon flux density on a horizontal sensor at mid level in a soybean canopy. The estimated leaf area index above this sensor was 2.32. Measured on Sept. 6, 1988.

Analysis of all the records showed that the most frequent sunflecks were 0.4 to 0.8 s in duration but reached a low maximum PFD of only 400-500 μmol photons m^{-2} s^{-1} (see ref. 14). These short sunflecks were created when the shadow of a leaf edge was close to the sensor and rapid vibrations of the leaf created intermittent shading (11). The low PFD during the sunflecks was caused by the penumbra of the leaf shadows. Although much less frequent, longer sunflecks contributed most of the PFD available for photosynthesis. 79% of the sunflecks were $<$1.6 s in duration but these contributed only 6.7% of the PFD. By contrast, sunflecks longer than 12.6 s contributed 77% of the sunfleck PFD. Sunflecks longer than 3 s generally had PFDs that were within 90% of those measured above the canopy at the same time. Thus, penumbral effects were minimal for sunflecks of this or longer duration.

It was not possible to study the response of CO_2 assimilation to natural sunflecks since often only part of the leaf in the chamber was illuminated with a sunfleck. We therefore examined the response of assimilation to either 1.) step increases in PFD created by parting the canopy to allow the direct solar beam in, or 2.) sequences of lightflecks created with a lamp and shutter system. When a leaf that had been in the shade was suddenly exposed to full sunlight, the assimilation rate jumped almost immediately to a value usually no greater than 10 to 20 percent of the light-saturated CO_2 assimilation rate (Fig. 2A). Assimilation then typically exhibited a sigmoidal increase lasting 20 to 60 minutes until the light-saturated, steady-state rate was reached. This is the classic induction response of photosynthesis (3), which was very evident in most leaves sampled in the canopy. It was not possible to separate the response of the plant from the time lags due to the chamber during the fast increase (initial 30 to 60 s of induction), but by 60 s a plateau was usually reached, and thereafter the change was slow enough that the lags in the gas-exchange system had little effect. After a short lag, stomatal conductance (g_s) increased and did not reach its maximum until considerably after assimilation had already approached its maximum. Intercellular CO_2 pressure (C_i) initially decreased from greater than 300 μbar to about 200 μbar within 1 min but then gradually increased to the steady-state level of 260 μbar at full induction.

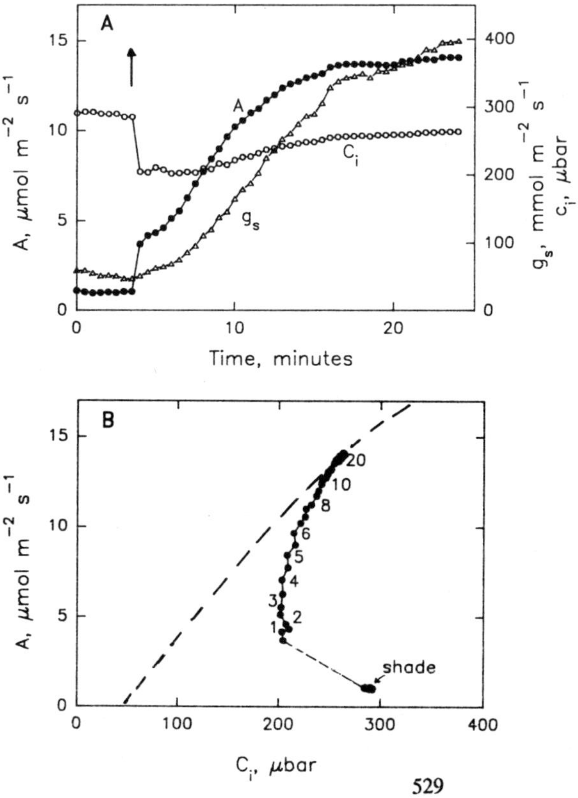

Figure 2A. Time-course of CO_2 assimilation (A), stomatal conductance (g_s) and intercellular CO_2 pressure (C_i) during induction. At the arrow, the PFD was increased from 50 to 1550 μmol m^{-2} s^{-1}. Fig. 2B shows the assimilation rate plotted against intercellular CO_2 pressure. The time in minutes from the light increase is shown next to selected data points.

It is evident that the increase in CO_2 assimilation during induction (1 to 25 min) occurs with only a rather small change in C_i. Fig 2B shows the trajectory of assimilation versus C_i during induction. The dashed line gives the steady-state dependence of assimilation on C_i, which is a function of the biochemical (primarily carboxylation) capacity of the leaf for CO_2 assimilation (19). If the increase in CO_2 assimilation was solely due to the increase in g_s, then the data points during induction should fall along this steady-state curve. Instead, the data points start well below the curve but approach it by 10 minutes after the light increase. This trajectory is consistent with the increase in CO_2 assimilation being due primarily to an increase in biochemical capacity (7). In effect, the data points lie on a series of individual curves that converge at the CO_2 compensation point and increase in slope as induction proceeds. In studies reported elsewhere (15), it was found that the activity of ribulose-1,5-bisphosphate carboxylase (rubisco) increased 5-fold by 20 minutes after shaded leaves within the soybean canopy were exposed to high light.

The decay of induction was followed by measuring induction state in leaves following reillumination after increasing periods of low light exposure. We used the value of CO_2 assimilation achieved 60 s after a light increase, expressed relative to the final, light-saturated value of assimilation, as a measure of the induction-state (IS_{60}). When fully-induced leaves were shaded, the IS_{60} declined (Figure 3). After only 5 minutes in low light, a significant induction limitation to a subsequent light increase developed.

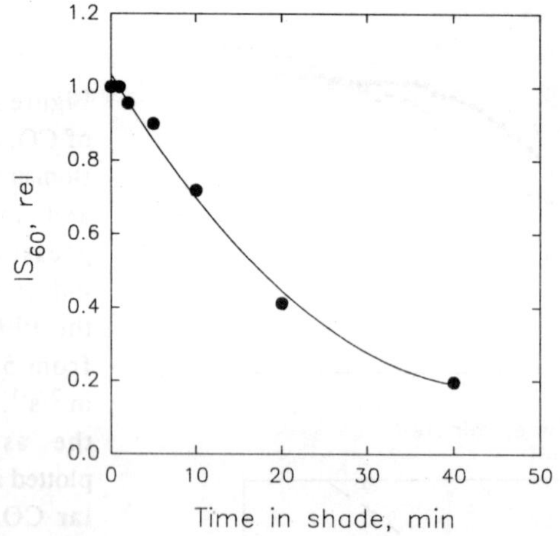

Figure 3. Decay in induction state (IS_{60}) following a decrease in PFD from saturating levels (1500 to 1700 μmol m^{-2} s^{-1}) to approximately 50 μmol m^{-2} s^{-1}.

Leaves within the canopy are rarely exposed to periods of high PFD that would result in full induction of the photosynthetic apparatus. Instead most light is received during periods of rapid light fluctuations. We investigated the response to light fluctuations similar to those experienced by leaves by giving different frequencies of lightflecks. In most cases, the lightflecks were 5 s in duration, but in some they were as short as 1 s. The leaf was exposed to a given frequency of lightflecks for 20-25 min so that a steady response was achieved. The PFD was then increased and held constant so that the resulting induction state could be measured. For some measurements, the sequence of lightflecks was started after the leaf had been fully induced whereas in others the leaf was initially in the shade and therefore at a low induction state.

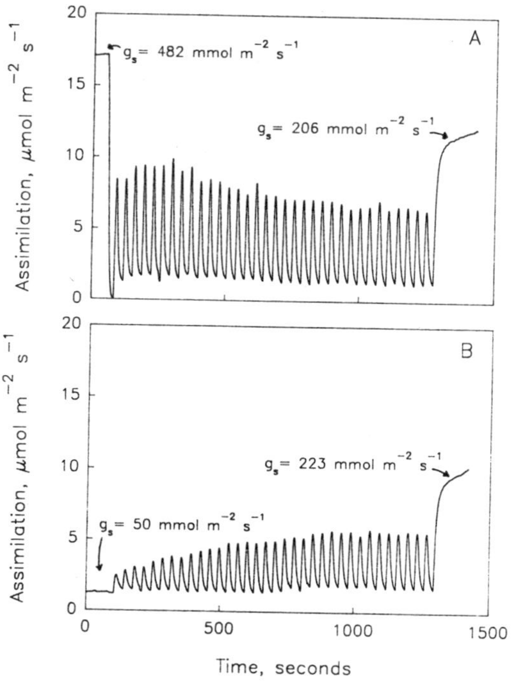

Figure 4. Response of assimilation to a series of 5-s sunflecks (1750 μmol photons m^{-2} s^{-1}) separated by 30-s of 50 μmol photons m^{-2} s^{-1}. At the end of the sequence the light was increased and held constant. A. leaves started from full induction. B. leaves started from a low induction state.

Figure 4 shows the response of assimilation to a series of 5-s lightflecks separated by 30 s of shade. Leaves were initially at a high (A) or a low (B) induction state prior to beginning the sequence of lightflecks. When started from a low induction state, CO_2 assimilation increased during each successive lightfleck until a steady response was obtained. g_s also increased over this period. Conversely, when started from an IS_{60} equal to one (i.e. fully induced), the response to a lightfleck decreased during successive lightflecks. Similarly, g_s decreased over this period. The values of CO_2 assimilation achieved during the lightflecks were much higher than those shown because the field gas-exchange system had a relatively slow response time (30s for a full-scale response). Nevertheless, it is evident that the capacity to respond to a lightfleck changes as induction state increases or decreases.

We compared the IS_{60} of leaves achieved after exposure to different lightfleck frequencies and durations to those achieved under constant PFD. The IS_{60} under constant PFD was measured by first exposing the leaf for at least 45 min to the constant PFD and then increasing the PFD up to saturating values for the measurement of IS_{60}. At low PFDs, IS_{60} was always high enough to allow the leaf to immediately increase its photosynthetic rate, typically by 2-3 μmol m^{-2} s^{-1}. Thus, induction was regulated in a way that allowed an immediate increase in assimilation but only to a level well below the steady-state light saturated assimilation rate. The immediate increase was then followed by the slower induction response that brought the assimilation rate to its steady-state value. The resulting light dependence of IS_{60} (Fig. 5) was similar to the light response of CO_2 assimilation except that the relative values were higher, especially at low PFD.

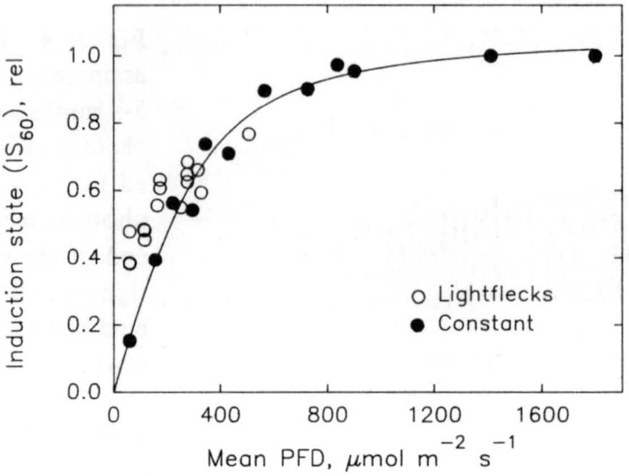

Figure 5. Dependence of induction state (IS_{60}) on the mean PFD during constant irradiance (solid symbols) or sequences of lightflecks (open symbols).

Different mean PFDs were obtained in the flashing light and it was therefore possible to compare the IS_{60} achieved to that observed under constant PFD. IS_{60} under flashing light also decreased as mean PFD decreased. At low mean PFDs, IS_{60} was higher under flashing as compared to constant light but as PFD increased the difference disappeared. There was no discernible difference in the final induction state achieved under flashing light for leaves started from a high or from a low induction state. Therefore the higher values at low mean PFD in flashing light were not the result of a failure to achieve an equilibrium.

Laboratory Studies

The faster gas exchange apparatus available in the laboratory was used to examine the rapid transient responses that may regulate the capacity to utilize short lightflecks. These experiments revealed complex transients during the first minute that were not evident in the field because they were hidden by the slower response time of the gas-exchange apparatus. The increase in CO_2 assimilation during the first 60 s consisted of 2 phases: 1.) a rapid jump that was complete within 5 s, and 2.) a second increase that lasted for 60 to 90 s. These initial phases were followed by the slow, often sigmoidal, increase to a steady-state as was observed in the field. The second increase was absent in fully-induced leaves that had been shaded for only 30 s, but was well developed in leaves shaded for 5 minutes (Fig. 6). Following shading of fully induced leaves, this limitation, as measured by the induction state at 5 s (IS_5), developed with a half time of 3.5 min. By contrast IS_{60} and stomatal conductance decreased with half times of approximately 9 and 30 min, respectively.

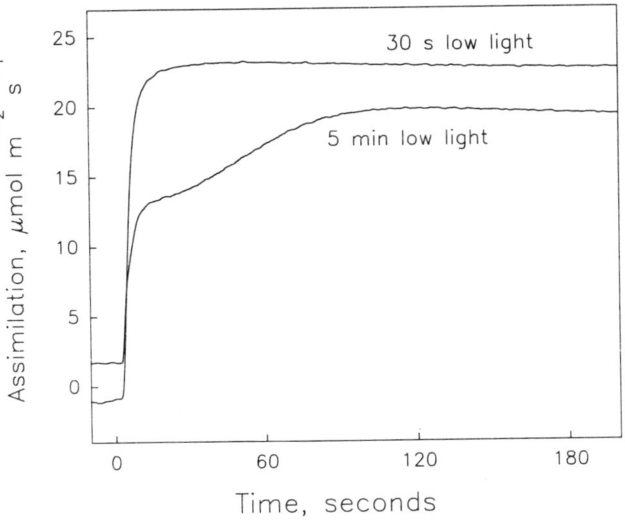

Figure 6. Increase in assimilation rate following an increase in PFD from 35 to 1500 $\mu mol\ m^{-2}\ s^{-1}$ for leaves that had been in low light for 30 seconds or 5 minutes. Prior to the low light treatment, the leaves were brought to full induction in saturating light.

The response of IS_5 and rubisco activity to flashing light with different frequencies and durations of lightflecks was investigated. Lightfleck frequency (1 second lightflecks, 1450 $\mu mol\ m^{-2}\ s^{-1}$) was varied from 0.5 to 0.025 Hz (mean PFD of 730 to 46 $\mu mol\ m^{-2}\ s^{-1}$). The background PFD was kept at 10 or 50 $\mu mol\ m^{-2}\ s^{-1}$, depending on the experiment. The IS_5 achieved after long exposure to the flashing light regime increased as lightfleck frequency and the resulting mean PFD increased. Similarly, under constant light, IS_5 increased with increasing mean PFD. However, when comparisons of IS_5 in flashing and constant light at the same mean PFD were made, IS_5 was lower in flashing light. This difference was largest for shade-grown leaves, which had a 60% lower photosynthetic capacity and 70 % lower light saturation point (data not shown) than the high-light grown leaves. The initial rubisco activity in the leaf also increased with increasing lightfleck frequency and was lower in flashing as compared to constant light of the same mean PFD.

In contrast to the large change in IS_5 with lightfleck frequency, changes in lightfleck duration from 0.2 to 25 s (with a concomitant change in the low light duration so that the mean PFD was 370 $\mu mol\ m^{-2}\ s^{-1}$), caused IS_5 to increase by only 0.1. Initial rubisco activity, however, was influenced by lightfleck duration and was 40% higher with long as compared to short lightflecks (data not shown).

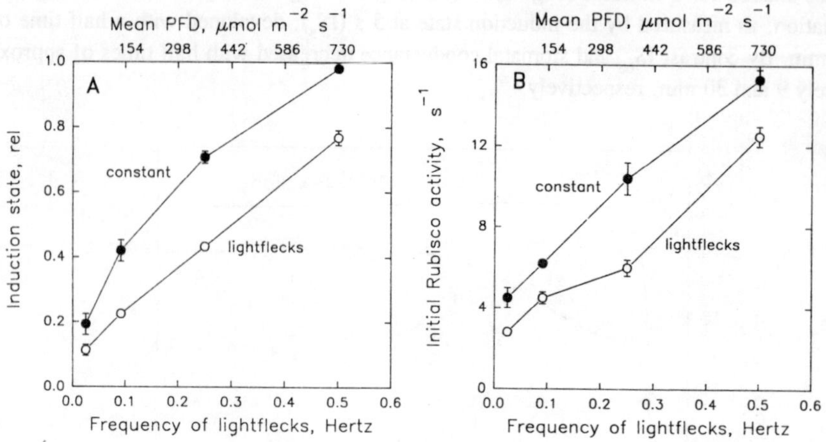

Figure 7. Dependence of induction state (IS_5) and initial rubisco activities on the mean PFD and the lightfleck frequency. Values for constant light are plotted against the corresponding mean PFD.

DISCUSSION

The results of this study show that taking into account the dynamic responses of photosynthesis to variable light is likely to be of considerable importance in efforts to scale up from leaf-level gas exchange characteristics to canopy photosynthesis. Measurements of the light environments within the soybean canopy reveal that perhaps only the uppermost leaves and a few at the very bottom of the canopy experience light environments where most of their photosynthesis occurs under conditions that would give rise to a quasi steady-state (14). For the vast majority of leaves, 40 to 90 percent of the PFD available for photosynthesis is in the form of sunflecks. Since these sunflecks are generally much shorter than the time required for induction and stomatal opening, a knowledge of the dynamics of the photosynthetic response to light is necessary in order to adequately describe their photosynthetic behavior.

The induction state is a measure of the readiness of a leaf to respond to a sudden increase in PFD (13). It is a function of the light modulation of rubisco and stomatal conductance, which determine the slow component of induction, and of limitations in RuBP regeneration, which determine the fast inducing component. The dynamics of the stomatal response to transient light and of the light regulation of Rubisco are therefore of great importance in determining how a leaf will respond to a sunfleck or a series of sunflecks. Under natural canopy conditions, many leaves are at induction states that are considerably less than 1. Pearcy and Seemann (15) reported that leaves randomly sampled at mid-canopy height had an IS_{60} of only 0.52. IS_{60} was correlated with both stomatal conductance and Rubisco activity and varied widely between leaves, presumably because they had experienced different patterns of sunfleck and shade. An IS_{60} of 0.52 means that if a 1-min sunfleck is received, the leaf will only be able to achieve 52 percent its steady-state light-saturated photosynthetic rate by the end of the sunfleck. Even this low value may overestimate the potential for utilization of sunflecks since IS_{60} does not include the limitations imposed by the fast component of the induction response. This is measured by IS_5, which decays much faster than either stomatal conductance or rubisco activity when a leaf is shaded (8, Pons, Pearcy and Seemann, unpublished results).

The limitations on sunfleck utilization imposed by the light modulations of stomatal conductance, rubisco activity and RuBP regeneration will depend in part on the time constants for each response and the initial starting values. In soybeans, the increase in stomatal conductance has a time constant (time for 0.63 of the total change) of 5 to 8 min and is the slowest factor in the induction response. This long time constant does not necessarily mean that stomatal conductance will be limiting since the limitation also depends on the value of g that the increase starts from. In soybean, this value is usually sufficiently high that, overall, stomatal conductance poses little limitation for the first 5 to 10 min after the light increase (see Fig. 2B). Consequently, the limitations are shifted to rubisco and RuBP regeneration. Similar results have been reported for forest understory species (7, 17). Although there have been few detailed studies of the response to light transients, stomatal closure appears to be slower than stomatal opening in many species (4, 6, 22). There is also substantial variation between species in the rates of stomatal opening and closure (4, 5, 22) so that in other species, stomatal conductance may be a more important factor (10). Moreover, environmental stresses could shift more of the limitation to stomatal conductance. Rubisco activity increases with a time constant of about 5 min and decreases with a time constant of 25 to 30 min (18, 21). By contrast, the apparent limitation in RuBP regeneration that causes the fast component of induction is removed within 60 to 90 s after a light increase and develops with a time constant of 5 to 6 min after a light decrease.

The differences in time constants for the different components of the induction response result in each being potentially limiting under different conditions. Leaves near the top of the canopy that experience intermittent shade may be most limited by the fast induction component. This is because the decline in stomatal conductance and rubisco activity is much slower than the decline in the induction state of the fast component. The fast component then limits the increase when full light returns. On the other hand, near the bottom of the canopy, rubisco activity and, to a lesser extent, stomatal conductance may be so limiting to CO_2 assimilation during a light increase that any additional limitation imposed by the fast-inducing component may be small. Moreover, each component may respond to flashing light differently. The longer time constants for the decrease than the increase in rubisco activity and the fast inducing component should result in higher induction states under flashing light than constant PFD of the same mean value. However, this did not occur. IS_{60} tended to be higher whereas IS_5 and rubisco activity tended to be lower under flashing as compared to constant PFD (Figs. 5 and 7). The reasons for this difference are unclear. Possible explanations are differences between the field and laboratory grown plants used to measure IS_{60} and IS_5, respectively, or differences in the regulatory mechanisms for each. It may also be that the longer lightflecks used in the field measurements influenced the response. Support for this possibility comes from the higher rubisco activities under long versus short lightfleck regimes of the same mean PFD (Fig. 7).

Limitations on sunfleck use imposed by the induction requirement will be partially offset by post-lightfleck CO_2 assimilation, which for short sunflecks may contribute over 50 percent of the carbon fixed (2, 9). Since short sunflecks are abundant in soybean canopies, post-lightfleck CO_2 fixation potentially could make a substantial contribution to total carbon gain. The capacity for post-lightfleck CO_2 fixation is at least as great in soybeans as in understory plants (Pons and Pearcy, unpublished results). However, two factors indicate that the overall contribution of post-lightfleck CO_2 fixation to the assimilation of soybean leaves may be relatively small. First, the efficiency of lightfleck utilization is dependent on induction state and will be relatively low for most leaves in the canopy. Second, sunflecks less than 5 s in duration, for which the post-lightfleck CO_2 fixation is proportionally high, contribute only a small fraction of the total sunfleck PFD in a soybean canopy. Longer sunflecks, for which induction state is the primary determinant of their utilization, contribute far more of the available PFD.

Scaling up from leaf to canopy photosynthesis in crops typically has been done by combining steady-state models of CO_2 assimilation with canopy models of light penetration. Models of this type ignore the highly dynamic nature of the light environment within a crop canopy as well as the dynamic responses of CO_2 assimilation itself. The models contain the assumption that CO_2 assimilation can be adequately described by a series of "steady states" that are dependent only on the current environmental conditions. The results of this study suggest that significant overestimates of the photosynthesis by sunlit leaf area within the canopy may result from ignoring the induction requirement. Moreover, prediction of the limitation by induction state from some measure such as the mean PFD will clearly include some inaccuracies that need to be better understood. Comparison of steady-state predictions with measurements in an understory plant showed that the steady state model overestimated daily carbon gain by about 20% (16). A similar estimate for crop canopies is not yet available. Obtaining one for a crop such as soybean is a more difficult proposition since an evaluation would need to be made at several canopy levels with the results integrated to obtain an estimate of the consequences for the whole canopy.

ACKNOWLEDGEMENTS

This research was supported by USDA-CRGO Grant 89 37231-4469 and by fellowships from OECD and the Netherlands Organization for Scientific Research to TLP.

REFERENCES

1. Chazdon, R. L. and Pearcy, R. W. (1986a). Photosynthetic responses to light variation in rainforest species. I. Induction under constant and fluctuating light conditions. *Oecologia*. 69:517-523.

2. Chazdon, R. L. and Pearcy, R. W. (1986b). Photosynthetic responses to light variation in rainforest species. II. Carbon gain and photosynthetic efficiency during lightflecks. *Oecologia.* 69:524-531.

3. Edwards, G. and Walker, D. A. (1983). *C_3, C_4: Mechanisms, and Cellular and Environmental Regulation of Photosynthesis.* University of California Press, Berkeley. 542 pp.

4. Grantz, D. A. and Zeiger, E. (1986). Stomatal responses to light and leaf-air water vapor pressure difference show similar kinetics in sugarcane and soybean. *Plant Physiol.* 81:865-868.

5. Johnsson, M., Issaias, S., Brogardh, T. and Johnsson, A. (1976). Rapid, blue-light-induced transpiration response restricted to plants with grass-like stomata. *Physiol. Plant.* 36:229-232.

6. Kirschbaum, M. U. F., Gross, L. J. and Pearcy, R. W. (1988). Observed and modelled stomatal responses to dynamic light environments in the shade plant Alocasia macrorrhiza. *Plant, Cell Environ.* 11:111-121.

7. Kirschbaum, M. U. F. and Pearcy, R. W. (1988a). Gas exchange analysis of the relative importance of stomatal and biochemical factors in photosynthetic induction in Alocasia macrorrhiza. *Plant Physiol.* 86:782-785.

8. Kirschbaum, M. U. F. and Pearcy, R. W. (1988b). Gas exchange analysis of the fast phase of photosynthetic induction in Alocasia macrorrhiza. *Plant Physiol.* 87:818-821.

9. Kirschbaum, M. U. F. and Pearcy, R. W. (1988c). Concurrent measurements of oxygen- and carbon-dioxide exchange during lightflecks in Alocasia macrorrhiza (L.) G. Don. *Planta.* 174:527-533.

10. Knapp, A. K. and Smith, W. K. (1990). Stomatal and photosynthetic responses to variable sunlight. *Physiol. Plant.* 78:160-165.

11. Norman, J. M. and Tanner, C. B. (1969). Transient light measurements in plant canopies. *Agron. J.* 61:847-849.

12. Pearcy, R. W. (1989). Radiation and light measurements. In: *Plant Physiological Ecology: Field Methods and Instrumentation*, Pearcy, R. W., Ehleringer, J. R., Mooney, H. A. and Rundel, P. W., eds., Chapman and Hall, New York, pp. 97-116.

13. Pearcy, R. W. (1990). Sunflecks and photosynthesis in plant canopies. *Annual Review of Plant Physiology and Plant Molecular Biology.* 41:421-453.

14. Pearcy, R. W., Roden, J. S. and Gamon, J. A. (1990). Sunfleck dynamics in relation to canopy structure in a soybean canopy. *Agric and Forest Meteorol.* 52:359-372.

15. Pearcy, R. W. and Seemann, J. R. (1990). Photosynthetic induction state of leaves in a soybean canopy in relation to light regulation of ribulose-1,5-bisphosphate carboxylase and stomatal conductance. *Plant Physiol.* 94:628-633.

16. Pfitsch, W. A. and Pearcy, R. W. (1989a). Daily carbon gain by Adenocaulon bicolor (Asteraceae), a redwood forest understory herb, in relation to its light environment. *Oecologia.* 80:465-470.

17. Pfitsch, W. A. and Pearcy, R. W. (1989b). Steady-state and dynamic photosynthetic response of Adenocaulon bicolor (Asteraceae) in its redwood forest habitat. *Oecologia.* 80:471-476.

18. Seemann, J. R., Kirschbaum, M. U. F., Sharkey, T. D. and Pearcy, R.W. (1988). Regulation of ribulose 1,5-bisphosphate carboxylase activity in Alocasia macrorrhiza in response to step changes in irradiance. *Plant Physiol.* 88:148-152.

19. von Caemmerer, S. and Farquhar, G. D. (1981). Some relationships between the biochemistry of photosynthesis and the gas exchange of leaves. *Planta.* 153:376-387.

20. Welles, J.M. (1990). Some indirect methods of estimating canopy structure. In: *Remote Sensing Reviews*, Goel, N.S. and Norman, J.M., eds., Harwood Academic Publishers, pp. 31-43.

21. Woodrow, I. E. and Mott, K. A. (1989). Rate limitation of non-steady-state photosynthesis by ribulose 1,5-bisphosphate carboxylase in spinach. *Aust. J. Plant Physiol.* 16:487-500.

22. Woods, D. B. and Turner, N. C. (1972). Stomatal response to changing light by four species of varying shade tolerance. *New Phytologist.* 70:77-84.

V

AGRO AND NATURAL ECO-SYSTEMS

AGRO-ECO CLIMATES OF INDIA

A.R.SUBRAMANIAM
Dept.of Meteorology and Oceanography
Andhra University
VISAKHAPATNAM, INDIA.

ABSTRACT

Earlier workers used rainfall and soils alone for identification of agroecoclimates (Panabokke, 1979, Murty and Pandey, 1978 etc.). However these approaches have not very clearly brought out the ecoclimatic features as mere rainfall alone is taken along with soils. Hence in this paper a index of moisture adequacy which combines both rainfall and temperature and is a ratio of actual water need to potential water need is utilised. The use of this index in combination with the soils of a given area reveals the agro ecoclimates on the lines done by Panabokke for FAO using rainfall alone and soils. The agroecoclimates so obtained also provide useful information of the potential crops and vegetation supported in the area. Such a delineation of agroclimates for the whole of Indian subcontinent and also for Andhra Pradesh and Maharashtra states is presented. The crop potential in different agroclimates are given for optimising the cropping pattern.

INTRODUCTION

For better crop yields, a knowledge of environment in relation to crops will enable not only to suggest a suitable cropping pattern but also to modify various agronomic practices. Based on the climatic zonation the identification of cropping potential often leads to irrational situations. It is therefore desirable to identify areas based on ecological and agricultural factors which take a comprehensive view of suitable crop potentials and the basis for a cropping pattern which can exploit the resources of the area to the maximum. For the development of crop production, particularly rainfed crop production the delineation of Agro-ecological zones is hence important. Since better rainfed crop production is dependent on the selection of suitable cropping pattern at a region, delineation of agro-ecological zones forms a useful basis.

On the basis of Index of moisture adequacy, a derived water balance element, instead of mere rainfall, the agro-ecological zones of India area classified. The index of moisture adequacy is taken into account because rainfall never fully reflects the moisture status of a region, and also plants do not fully depend on rainfall alone, besides the significant role played by soil as the storage of water cannot be overlooked. The concept of water balance utilises the principle of potential evapotranspiration (PE) and the Actual evapotranspiration (AE). The ratio of AE to PE expressed as percentage and known as index of moisture adequacy, varies with the available moisture and indicates the moisture status of the soil, and as such is to be considered in studying the suitability of any region for agricultural development.

Data regarding climatic normals is obtained from information published by India Meteorological Department (1966) for 160 stations in the country. Monthly PE, AE, soil moisture storage and other water balance elements are calculated following the modified scheme of Thornthwaite (1955). Based on the moisture adequacy values six moisture adequacy regions are classified and shown below.

Symbol	Description
MA_1	Sufficient moisture adequacy throughout the year.
MA_2	Sufficient moisture adequacy throughout the year with a weekly expressed intervening period of moderately low moisture adequacy.

MA_3 Sufficient moisture adequacy during the main cropping season followed by a well expressed dry season of at least two consecutive months.

MA_4 Moderate degree of variability of moisture adequacy between years but usually sufficient moisture adequacy with in the main cropping season followed by a well expressed dry season of at least three consecutive dry months.

MA_5 Moderate to high degree of variability of moisture adequacy between years as well as within the main cropping season followed by a well expressed dry season of at least three consecutive dry months.

MA_6 Inadequate moisture adequacy throughout the year.

Utilising the available data, the moisture adequacy regimes are first delineated and they are super imposed on the soil map of India. According to F.A.O. (UNESCO, 1975), Scheme (S_1) Soils of India are classified and there are six out of seven soil types according to the above scheme, making it possible to show a variety of crops. After some generalisation and smoothening of boundaries, the resultant agro-ecological zones are identified and presented in Fig.1. (Subramaniam 1983).

From the Fig.1 it can be seen that a major portion of northern India falls under the agro-ecological zones MA_5S_6 and MA_1S_6 portions of MA_5S_1, MA_6S_3, MA_1S_3, MA_5S_5, MA_5S_3, MA_1S_1, MA_3S_1, MA_2S_3 and MA_2S_1 zones are also found in the northern region. Gangetic plans come under MA_3S_3, MA_3S_1 and MA_2S_3. The western Indian arid region comes under MA_6S_2 and MA_5S_2. The central India is dominated by MA_5S_4, MA_3S_4 and MA_3S_5. Bihar, West Bengal and Orissa regions come under MA_2S_5, MA_2S_6, MA_3S_1, MA_3S_5 and MA_4S_1. The west coast of the peninsular India is dominated by MA_3S_6 and MA_3S_1, whereas the east coast by MA_4S_1. The western ghat

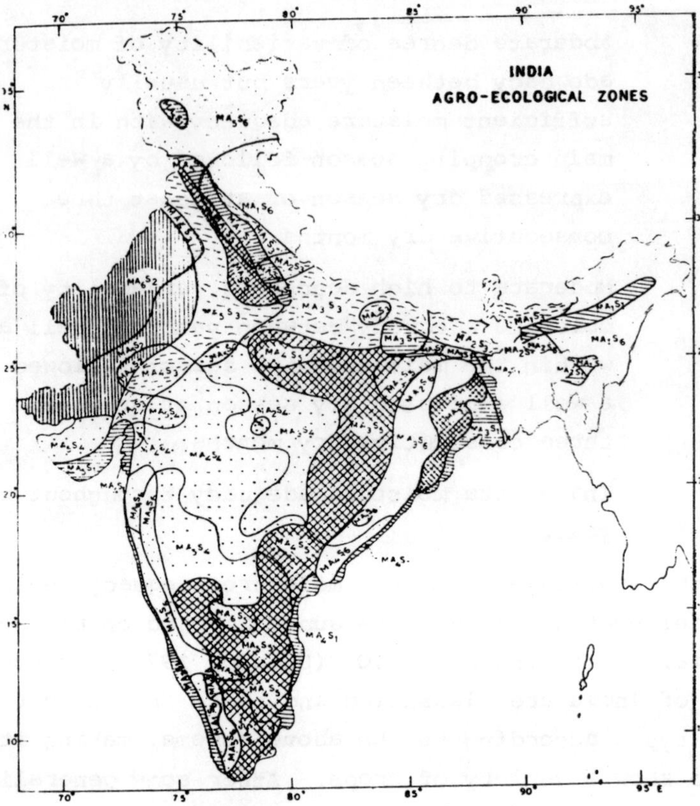

Fig. 1. India agro-ecological zones

region with high rainfall and relatively low temperatures comes under MA_1S_6 and MA_3S_6. The Deccan plateau is dominated by MA_5S_6 and MA_5S_4. The arid region around Bellary in South India comes under MA_6S_5. A major portion of Tamilnadu state comes under MA_4S_5 and MA_5S_5.

A table of agro-ecological zones of India is presented in the Table 1 and it can be seen from the Table that according to the above classification there are 28 different agro-ecological zones of the possible 36 combinations in India, indicating the potential for a variety of crops that can be grown. According to the dominant cropping systems that occur within each of the agro-ecological zones, the following agronomic groupings can be made.

Group 1 : Mainly rainfed or irrigated rice.
MA_1S_1, MA_1S_6, MA_2S_1, MA_2S_5, MA_2S_6, MA_3S_1, MA_3S_5 & MA_3S_6.

Group 2 : Mainly irrigated rice during wet season followed by other arable crops during dry cool season (rice, pulses, ragi, groundnut etc.)
MA_2S_1, MA_2S_3, MA_2S_4, MA_2S_5, MA_3S_1, MA_3S_3, MA_3S_4, MA_3S_5, MA_3S_6, MA_4S_1 and MA_4S_6.

Group 3 : Mainly moisture demanding perennial trees (banana coconut, cashew, arecanut, jack fruit, sapota, pine apple etc.)
MA_1S_6, MA_3S_1, MA_3S_3, MA_3S_4, MA_3S_6 and MA_4S_1.

Group 4 : Mainly rainfed dry crops with moderate drought stress (bajra, jowar, ragi, cotton, etc.).
MA_2S_6, MA_3S_5, MA_4S_4, MA_4S_5, MA_4S_6, MA_5S_2, MA_5S_4, MA_5S_5, MA_5S_6 and MA_6S_2.

Group 5 : Mainly rainfed dry crops with moderate to high drought stress (jowar, sunflower, coston, etc.)
MA_4S_4, MA_4S_5, MA_5S_4, MA_5S_5 and MA_6S_5.

Group 6 : Assortment of green and other horticultural crops and some hydromorphic rice (rice, cardamom, citrus, coffee, rubber, tea, vanilla, cloves, pepper etc.)
MA_1S_1, MA_1S_3, MA_1S_6, MA_3S_1, MA_3S_5 and MA_3S_6.

TABLE 1

Moisture adequacy regimes	Soil groupings	S_1	S_2	S_3	S_4	S_5	S_6
MA_1		MA_1S_1		MA_1S_3		MA_1S_5	MA_1S_6
MA_2		MA_2S_1		MA_2S_3	MA_2S_4	MA_2S_5	MA_2S_6
MA_3		MA_3S_1		MA_3S_3	MA_3S_4	MA_3S_5	MA_3S_6
MA_4		MA_4S_1	MA_4S_2	MA_4S_3	MA_4S_4	MA_4S_5	MA_4S_6
MA_5		MA_5S_1	MA_5S_2	MA_5S_3	MA_5S_4	MA_5S_5	MA_5S_6
MA_6			MA_6S_2	MA_6S_3		MA_6S_5	

Based on the above classification the agro-ecological zones of Andhra Pradesh and Maharashtra are presented.

1. Andhra Pradesh: In Andhra Pradesh there are six out of seven soil types according to the above scheme, making it possible to grow a variety of crops. The resultant Agro-ecological zones are identified and shown in Fig.2 and Table 2 (after Subramaniam and T.S.Mani, 1987).

From the Fig.2 it is observed that the northern region of Andhra Pradesh is covered by MA_3S_3 except Nizamabad and Adilabad where MA_3S_4 is found. Irrigated rice, ragi, pulses and groundnut can be grown here. The MA_3S_3 covers Karimnagar, Warangal, Medak, Khammam, East Godavari, Visakhapatnam, Vizianagaram and parts of Krishna, West Godavari and Srikakulam districts where irrigated crops like rice, pulses, ragi and groundnut, rainfed crops like bajra, jowar, cotton etc., horticultural crops; plantation crops like cardamom, citrus, coffee, rubber, tea, vanilla, cloves, pepper etc., can be grown. MA_3S_3 is found in parts of Srikakulam, East Godavari, West Godavari and Krishna areas where perennials like banana, coconut, cashew, jack fruit, sapota, pine apple , rice, pulses, ragi and groundnut can be grown. A small portion in west Godavari is

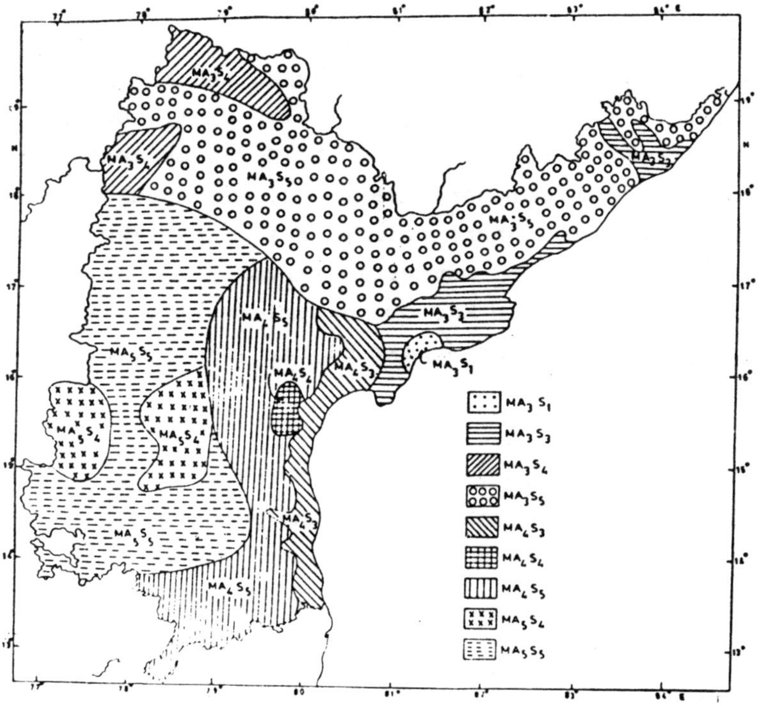

Fig.2. Agro-ecological zones of Andhra Pradesh.
(After Subramaniam and T.S.Mani, 1987).

occupied with MA_3S_1 where irrigated or rainfed rice can be grown.

The eastern parts of Krishna, Guntur, Prakasham and Nellore districts are covered by MA_4S_3 where mainly rainfed dry cr crops like bajra, jowar, ragi, cotton etc. are grown. Except a small patch in Guntur, parts of Nalgonda, Guntur, Nellore, Kurnool, Prakasham, Cuddapah and entire Chittoor occupied by MA_4S_5 where bajra, jowar, ragi, cotton, sunflower, castor etc. are grown. MA_5S_5 is present in Medak, Hyderabad, Mahaboobnagar, Anantapur, Cuddapah and a small portion in Kurnool where rainfed dry crops like bajra, jowar ragi, cotton etc. and commercial crops like sunflower, castor are grown. In Kurnool districts MA_5S_4 is present. Mainly rainfed dry crops like bajra, jowar, ragi, cotton etc. are grown.

According to the above classification there are nine different Agro-ecological zones of the possible 36 combinations in Andhra Pradesh, indicating potential for a variety of crops that can be grown.

TABLE 2

Moisture adequacy regimes	Soil Grouping	S_1	S_2	S_3	S_4	S_5	S_6
MA_1							
MA_2							
MA_3		MA_3S_1		MA_3S_3	MA_3S_4	MA_3S_5	
MA_4				MA_4S_3	MA_4S_4	MA_5S_5	
MA_5					MA_5S_4	MA_5S_5	
MA_6							

Based on the groupings of zones, 9 agro-ecological zones with 5 groups, viz., Group 1, Group 2, Group 3, Group 4 and Group 5. Under Group 1 MA_3S_1 is present where rainfed or irrigated rice could be grown; under group 2, MA_3S_3, MA_3S_4 and MA_3S_5 are present where irrigated rice during wet season followed by other crops during dry or cool season are grown; under group 3, MA_3S_3 is present where mainly moisture demanding perennial trees are grown; under Group 4, MA_4S_3, MA_4S_4 and MA_4S_5 are present where mainly rainfed dry crops are grown; and in Group 5 MA_5S_4 and MA_5S_5 are present where mainly rainfed dry crops can be grown.

Maharashtra:- The major soil types of the state can be classified into vertisols, alfisols, ultisols and oxisols. The map delineating the moisture regimes according to the Subramaniam classification (1983) was super imposed on the soil map and the resultant Agro-ecological zones were identified (Sambasivarao 1983). According to the classification there are six major ecological zones (Fig.3) in the state. These are MA_3S_4, MA_3S_6, MA_4S_4, MA_4S_5, MA_5S_4 and MA_5S_6; where S_4S_5 and S_6 refers to the soil zones of types vertisols, alfisols, ultisols and oxisols. According to the classification, the following agronomic grouping could be made.

Group 3 : Mainly moisture demanding perennial tree crops (Banana, Coconut, Cashew, Arecanut, Jack fruit, sapota, Pineapple, palm). MA_1S_6, MA_3S_4, MA_3S_6, MA_4S_1 (S_1 refers to Entisols).

Group 4 : Mainly rainfed dry crops with moderate drought stress (Bajra, Jowar, Ragi, Cotton).
MA_2S_6, MA_4S_4, MA_4S_5, MA_5S_4, MA_5S_5, MA_5S_6.
stress (Jowar, Sunflower, Safflower, Castor).
MA_6S_4, MA_6S_5.

This approach utilising the index of moisture adequacy in the delineation of agro-ecoclimates, if felt to be more rational and widely supported by the information of cropping pattern in India and also in the state of India. In conclusion, it is felt that this method can be a useful tool in the delineation of agro-ecoclimates of tropical regions.

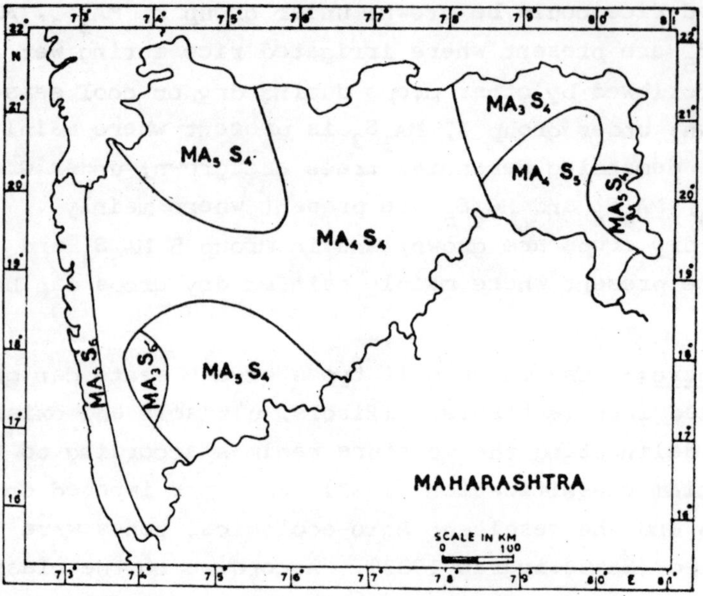

Fig.3. Agro-ecological zones of Maharashtra.
(After A.Sambasiva Rao, 1983).

References

1. Thornthwaite, C.W., Mather, R.W. (1955). The water balance. Publ. in Clim. Drexl Inst. Tech. Note No.8.

2. Murthy, R.S., Pandey, S. (1978). Delineation of Agro-ecological regions of India. National Bureau of Soil survey and land use planning (ICAR), New Delhi.

3. Panabokke, C.R. (1979). Agro-ecological zones of south and south-east Asia. FAO Regional Office for Asia and the Far East Bangkok, Thailand.

4. Subramaniam, A.R. (1983). Agro-ecological zones of India. Arch. for Met. Geophy. Biocl., Ser.B., 32, 329-333, Verlag.

5. Sambasiva Rao, A. (1983). Some Agro-climatic studies of Maharashtra. Thesis submitted to Andhra University, Waltair, India.

6. Subramaniam, A.R., Mani, T.S. (1987). Biometeorological studies of Andhra Pradesh in Peninsular India. Presented at XI International Biometeorological Congress, held at Purdue University, Indianapolis.

GLOBAL CLIMATIC CHANGE: REGIONAL SCENARIO OVER INDIA

D.R.Sikka and G.B.Pant
Indian Institute of Tropical Meteorology
Dr. Homi Bhabha Road, Pune-411 008, India

ABSTRACT

An overview of the global climatic change is presented in relation to contemporary regional changes over India. Documented evidences and observations on climatic change scenario over India are presented using proxy data, historical accounts and available instrumental records. The major global warm and cold periods/epochs are examined in relation to the performance of the monsoon rainfall over India. The possible contemporary changes in the climate of the earth caused by the activities of man, particularly by the release of CO_2 into the atmosphere, are also considered. The results of GFDL, GISS and NCAR Atmospheric General Circulation Model simulations for a doubling of CO_2 scenario are examined. The projections of these models regarding the changes in precipitation, temperature and soil moisture distribution for a doubling of the CO_2 have been closely examined by zooming in on the monsoon region, and interpreted in terms of the climatic change and consequent agricultural production over India.

1. INTRODUCTION

The climate of the earth has been changing on different time scales. The regional manifestations of these global changes have affected the societies to different extents on many occasions. Over the past few decades, the possible contemporary changes in the climate of the earth due to the activities of man has become the focus of scientific and social attention. The most imminent of the climatic change of the earth is the significant increase in the atmospheric temperatures due to the increased levels of CO_2 and other green house gases. Estimation of radiative effects of CO_2 and other trace gases for the future is uncertain, because of the uncertainties in the projected concentration of these gases. Many of these gases are generated by human activities such as, burning of fossil fuels and release of CO_2 from biotic sources and many other industrial activities which are on continuous increase during the last 4 to 5 decades.

The observed and projected increase in CO_2 concentration alone may not fully account for the climate changes. Palaeoclimatic evidences have demonstrated that the climate has varied substantially in the past and natural causes of climatic changes have always been operative on various time scales. The changes in the eccentricity of the earth's orbit, the variation of the perihelion and the tilt of its axis (Milankovitch effect) are believed to be important as they might have induced small variations in the solar flux reaching the top of the atmosphere in a quasiperiodic manner. Similarly the solar constant may have small perturbations due to the activities on the surface of the Sun such as sunspots, solar flares etc.

The normal climate is an ensemble of interactions between the atmosphere, ocean, sea and land ice and the land surface including the biosphere. Because of the complex couplings (feedback relations) between physical, chemical, radiative and dynamical processes in the climate system, it is not possible to derive simple equations relating the changing atmospheric composition to climatic parameters. The relations are nonlinear and need numerical solutions and simulations for future projections. Interpretation of climatic patterns become difficult due to a variety of time and space scales involved.

In a critical review of the climatic trends as seen by the recorded data, Ellsaesser et al. [1] remark that all the trend analyses based on instrumental records start from a minimum point of the curve corresponding to the last phase of the Little Ice Age. The observed increasing trend in Northern Hemispheric (NH) temperature may perhaps be a return to the normal levels. All observed patterns of surface temperature change display cellular areas of both warming and cooling, whereas the models generate warmings with latitudinal gradients. These latitudinal variations may be more of an inbuilt nature of the model characteristics rather than real detectable warming signals.

Bradley et al. [2] have studied the precipitation fluctuations over NH land areas since the mid-19th century and found significant increase in mid-latitude precipitation and concurrent decrease in low-latitude precipitation over the last 30 to 40 years. Vinnikov et al. [3] conclusively established global warming in both the hemispheres, with an average rate of 0.5°C/100 years and found an increase of 6%/100 years in the mean annual precipitation on the NH continents in the 35°-70°N zone.

The Indian subcontinent is a highly populated region having a critical dependence on the monsoon rains concentrated in just four months of the year. Arid and semiarid regions occupy large areas of the subcontinent for which any significant change in climate is of far reaching consequences. The dependence of agriculture on monsoon rains and river and groundwater for irrigation makes it a vital issue to examine these sources of water in relation to the changing climatic scenario.

In the context of the global warming scenario, the regional climatic change over India could be in the form of change in north-south temperature gradients, changes in seasonal temperature contrasts, changes in sea surface temperature over the monsoon region etc., which may induce net change in monsoon rainfall, changes in its interannual to decadal scale variability, and even shift in the monsoon belt. Other equally important effects may be changes in the frequency and intensity of tropical cyclones and other disturbances, jet streams and changes in the boundaries of arid and semiarid zones in the region. Enhanced convective activity in the northern latitudes due to a shift of the monsoon belt to higher northern latitudes may result in more extensive floods affecting the agricultural production and causing soil erosion in the northern parts of the country.

To examine these aspects, in this paper, the instrumental data on Indian rainfall, temperature, and an index of area under arid conditions are examined in relation to NH temperature anomalies. Warm and cold epochs as well as some selected extreme warm and cold years over the NH and over India are examined in relation to the performance of monsoon rains over India. The data of the last 4 decades (1950-1989) on All-India rainfall, temperature and agricultural production are also examined to assess the regional impacts of recent global temperature changes.

The results of three atmospheric general circulation models, namely GFDL, GISS and NCAR are examined for monsoon (JJA) and winter (DJF) months. The changes projected for a doubling CO_2 scenario in temperature, precipitation and soil moisture are compared with the trends in observed instrumental data and their future tendencies are discussed.

2. SOURCES OF DATA

The All-India summer monsoon rainfall series is obtained from Parthasarathy et al. [4]. The NH temperature anomalies are updated from Jones [5]. The all-India mean surface air temperature series has been updated from Hingane et al. [6]. The information on area under arid conditions over north India is obtained from Singh et al. [7]. The model output results of three GCM experiments are taken from Schlesinger and Mitchell [8].

3. EVIDENCES OF CLIMATIC CHANGE FROM PROXY AND INSTRUMENTAL CLIMATIC DATA

Over the Indian region there are evidences of appreciable climatic change in the past for which reliable data are available from the regions of northwest India. Pollen sequences from lake beds of northwest India suggest that periods with subdued monsoon activity existed during the recent glacial maximum [9] but the epoch 10,000-4,500 BP experienced warm-humid climate with frequent floods. The conditions turned more hospitable with equable climate during the following 1000 year period. Archaeological evidences suggest that during this period, 4500 to 3700 BP, the Harappan civilization flourished in the region [10]. Fluctuations of lake levels and change of salinity of lake sediments also support these century-scale changes in the region [11,12]. The arid conditions with shifting sand dunes are reported to have dominated the region since the last 3000 years or so.

An examination of the historical records of droughts over the Indian subcontinent for the last 200 years have shown that extreme droughts have been a part of Indian rainfall variability. More frequent droughts with more number of two consecutive drought years is observed in the 19th century than in the 20th century, which may perhaps suggest an improvement in rainfall amounts with the transition from colder to warmer periods [13]. Analysis of a representative rainfall series over the past 176 years for India as a whole does not suggest any significant trend [14]. The long term statistical characteristics of the series are practically stable and the extremes appear to be a part of its natural variability. An analysis of the mean annual surface air temperature over India using data for the period 1901-88 based on 73 stations shows a significant warming of about 0.4°/100 years [6]. This warming trend in general is comparable to the global mean trend of 0.5°C/100 years [3].

To examine the manifestations of the global climate change in the Indian region, a detailed study of the data on NH surface air temperature in relation to the Indian summer monsoon rainfall and surface temperature is done. Fig. 1 shows the smoothed curve of mean annual surface air temperature of the NH in association with the curves of All-India mean annual surface air temperature, All-India summer monsoon rainfall and the arid region area of northwest India. Though the Indian temperatures show the general increasing trend similar to the NH temperatures, the post-1940 cooling of the NH is not conspicuous over India. The correlation between the two series is significant during 1901-81 and

1901-40, but is not significant during 1941-81 [6]. The Indian monsoon rainfall does not show any continuous trend; only some high and low rainfall epochs are seen. Thus, there is no consistent long-term association between the Indian monsoon rainfall and NH annual surface air temperature series. There is a slight decrease in the arid area of northwest India, which could be a pointer towards the possible effect of NH warming on the climatic patterns of the country [7].

Earlier studies have indicated that the winter temperatures of the NH have an important bearing on the Indian monsoon performance [15]. The departures of decadal means of all-India monsoon rainfall, NH and all-India winter temperatures from their long-term mean are presented in Table 1. During the period 1900-19, the low rainfall is associa-

Table 1. Departures of decadal means from the long-term average (1890-1989).

Decade	All-India Summer Monsoon Rainfall (cm)	NH Surface Air Temp. Anomaly (Jan+Feb mean), °C	All-India Temp. Anomaly (Jan+Feb Mean), °C
1890-99	+1.68	-0.56	-0.03
1900-09*	-3.23	-0.24	-0.33
1910-19*	-0.64	-0.12	-0.07
1920-29	-2.17	+0.01	-0.12
1930-39	+1.68	+0.15	-0.13
1940-49**	+3.65	+0.19	+0.02
1950-59**	+2.44	+0.02	+0.23
1960-69	-1.61	+0.03	+0.15
1970-79	-0.34	+0.01	-0.12
1980-89	-1.44	+0.49	+0.41

* decades of consistent negative anomalies

** decades of consistent positive anomalies

ted with low temperatures on NH as well as all-India scale and during 1940-59, the high rainfall is associated with high temperatures. During other periods the relationships are not consistent. During the period of warming trend in the NH (1901-40), the mean all-India monsoon rainfall was 83.6 cm while during the period of cooling trend in the NH (1941-70), it was 84.4 cm. The extremes of the winter temperature anomalies over NH (Table 2) and India (Table 3) are also studied with reference to the anomalies in the monsoon rainfall during the period 1890-1989. In view of the inconsistency in the relationships over the full data period as indicated above, the anomalies are also considered for the most recent 4 decades, 1950-1989. When considered for the whole data period (Table 2a), out of ten highest positive winter anomalies of NH, there were two heavy (1983 and 1988)

Table 2. All-India summer monsoon rainfall for ten/five years of highest positive/negative anomalies of NH surface air temperature (Jan+Feb mean).

Year	Positive anomalies (°C)		Rainfall (cm)	Year	Negative anomalies (°C)		Rainfall (cm)
			(a) Data Period: 1890-1989				
1981	0.94	(+0.01)	85.2	1893	-2.09	(-0.02)	95.3
1983	0.82	(+0.14)	95.5	1895	-1.48	(+0.01)	82.5
1989	0.76	-	86.6	1929	-1.08	(-0.04)	81.9
1944	0.76	(0.00)	92.1	1891	-0.96	(-0.01)	78.9
1926	0.72	(+0.11)	90.1	1972	-0.81	(-0.02)	65.3
1988	0.58	(+0.03)	99.1	1905	-0.81	(+0.07)	71.5
1958	0.58	(+0.05)	88.6	1969	-0.70	(+0.02)	82.9
1987	0.57	(+0.08)	68.8	1917	-0.81	(-0.13)	100.3
1960	0.49	(-0.01)	83.9	1907	-0.69	(-0.04)	77.6
1914	0.49	(+0.04)	89.9	1911	-0.65	(-0.04)	73.3
Mean	+0.67		88.0		-1.01		81.0
			(b) Data Period: 1950-1989				
1981	0.94		85.2	1972	-0.81		65.3
1983	0.82		95.5	1969	-0.70		82.9
1989	0.76		86.6	1950	-0.63		87.5
1988	0.58		99.1	1951	-0.55		73.7
1958	0.58		88.6	1974	-0.44		74.7
Mean	0.74		91.0		-0.63		76.8

Note: (1) Values in bracket are the temperature anomalies attributed to ENSO effect.
(2) Mean All-India Rainfall (1890-1989) is 84.8 cm and standard deviation (s.d.) is 8.3 cm.

and one poor (1987) monsoon years. Similarly, out of ten highest negative anomalies, there were two heavy (1893 and 1917) and three poor (1905, 1911 and 1972) monsoon years . The others were within the normal limits (±1 s.d.) The difference in the mean rainfall for the ten highest positive and negative winter temperature anomalies is 7 cm, which is less than the s.d. of the monsoon rainfall. Thus, during the full data period, the NH winter temperature does not seem to be a crucial parameter influencing the monsoon. This fact was also brought out by Pant et al. [16] while studying the long-term variability of the monsoon and related variables. However, during the last 4 decades (Table 2b), the signal slightly improves, the difference in mean rainfall corresponding to the five highest positive and negative NH winter temperature anomalies being 14.2 cm. A majority of the extreme years of NH temperature anomalies, positive as well as negative, are also El Niño years.

Jones [17] found significant high frequency association between El Niño and hemispheric and global temperatures and subtracted the El Niño influence from the corresponding annual temperatures. Based on his data, the authors have quantified the El Niño signal in the NH annual temperatures (Raw temperature minus El Niño subtracted temperatures) for the extreme years of winter temperatures (see Table 2a). Interestingly, the El Niño signal tends to be negative in the case of negative extremes of winter temperatures. It is well-known that the Indian drought is frequently associated with El Niño, however, based on the above data, it is difficult to arrive at a definitive conclusion as to how the El Niño modulates the relationship between NH surface temperature and Indian monsoon.

On a regional-scale (Table 3a) out of the ten highest positive anomalies of all-India winter temperatures, two were heavy monsoon years (1892 and 1988) and two poor monsoon years (1966 and 1987) and out of the ten highest negative anomalies, only one was heavy monsoon year (1895) and four were poor monsoon years (1899, 1918, 1968 and 1972). The mean monsoon rainfall for the ten warmest winter years was 86.3 cm and that for the ten coldest winter years was 78.4 cm, the difference being 7.9 cm. The signal is not very clear but it may be stated that the cold winters are more frequently followed by poor monsoons, even on a regional scale. However, this signal still weakened during the recent 4 decades (Table 3b), though generally showing the same picture as above. It may be noted here that the pre-monsoon temperatures rather than winter temperatures over India have been found to more significantly related to the monsoon rainfall [18]. However, since Hingane et al.

[6] found that the pre-monsoon temperatures do not show any significant trend but the winter temperatures do show warming, the above observations are relevant to anticipate the consequent changes in the monsoon rainfall.

Table 3. All-India summer monsoon rainfall for ten/five years of highest positive/negative anomalies of the all-India mean surface air temperature (Jan+Feb mean).

Year	Positive anomalies (°C)	Rainfall (cm)	Year	Negative anomalies (°C)	Rainfall (cm)
		(a) Data Period: 1890-1989			
1988	1.33	99.1	1905	-1.95	82.5
1966	1.07	73.5	1893	-1.68	95.3
1952	1.00	79.2	1925	-1.10	80.3
1958	0.92	88.6	1972	-0.78	65.3
1892	0.85	99.0	1945	-0.70	90.7
1897	0.80	89.0	1891	-0.69	78.9
1987	0.77	68.8	1918	-0.65	64.8
1973	0.73	91.2	1906	-0.63	88.3
1989	0.70	86.6	1968	-0.59	75.4
1980	0.66	88.3	1899	-0.57	62.8
Mean	+0.88	86.3		-0.93	78.4
		(b) Data Period: 1950-1989			
1988	1.33	99.1	1972	-0.78	65.3
1966	1.07	73.5	1968	-0.59	75.4
1952	1.00	79.2	1962	-0.43	80.7
1958	0.92	88.6	1975	-0.42	96.0
1987	0.77	68.8	1974	-0.41	74.7
Mean	+1.02	81.8		-0.53	78.4

4. MODELLING THE PATTERN OF CLIMATIC CHANGE

The most developed tools for prediction of future climate are the Atmospheric General Circulation Models (GCM). These models are based on the basic laws of physics and use the description of smaller scale processes in the atmosphere such as clouds in physically parameterised form. Atmospheric-GCMs are coupled with Ocean-GCMs to account for the role of oceans in the atmospheric circulation.

The measurements taken at Mauna Loa, Hawaii show that the CO_2 concentration has increased from 316 ppmv in 1959 to 342 ppmv in 1983 [19]. Reports [20] suggest that the average rate of global increase of CO_2 emission from fossil fuels in the last 30 years has been around 3.5% per year. To address this question, the GCMs have been used to simulate the change in the equilibrium climate resulting from a doubling of the CO_2 concentration, such as from 300 to 600 ppmv. The equilibrium climate change studies of doubling CO_2 scenario as used in three important GCMs namely GFDL, GISS, NCAR are critically examined in this study. The long term change in surface air temperature following doubling of CO_2 (referred to as climate sensitivity) is generally used as bench mark reference of a model. The global mean surface air temperature change given by the simulations of the three recent models mentioned above range from 3.5 to 4.2°C. There are large variations in simulated changes from region to region and model to model. Simulation of the climatic change induced by a doubling of the CO_2 concentration using these models exhibit on increase in the global mean precipitation rate with values that range from 7.1 to 11.0% of their respective $1xCO_2$ levels [8]. The observed variations of precipitation coincide with the model results by sign but contradict by scale [3].

In weather forecasts, the predictions can be compared with actual weather and then their skill is assessed. We cannot do that for climatic prediction. The models may contain different compositions of greenhouse gases and observed boundary conditions. Their simulation of the present climate is in general adequate, capturing major features such as ITCZ, monsoons, mid-latitude storm belts, as well as major contrasts between summer and winter circulations.

A different but potentially useful way of predicting the pattern of change is to search for periods in the past when the global mean temperature were similar to those we expect in future to find analogue for greenhouse change. This makes the palaeoclimatic studies useful in model validation.

5. RESULTS OF SIMULATED DOUBLING CO_2 SCENARIO FOR SOME IMPORTANT TROPICAL REGIONS USING GENERAL CIRCULATION MODELS

The IPCC report [21] has given the regional estimates of climatic effects for doubled CO_2 which may occur by 2030 with business as usual scenario. The results for the selected tropical regions are given below. (Note that the models considered here are not exactly the same as the three model results discussed above).

(a) South East Asia (5-30°N 70-105°E): The warming varies from 1 to 2°C throughout the year. Precipitation changes little in winter and generally increases throughout the region by 5-15% in summer. Summer moisture increases by 5 to 10%.

(b) Sahel (10°-20°N, 20°W-40°E): The warming ranges from 1 to 2°C. Area mean precipitation increases and area mean soil moisture decreases marginally in summer. However there are areas of both increase and decrease in both parameters throughout the region which differ from model to model.

There are problems relating to changes in climatic extremes such as, intense rainstorms, number of days with break monsoon (decadal average estimate show a decrease) and minimum and maximum temperatures. It is also important to know whether the number of rainy days/length of growing season/crop yield etc. also change. Tropical storm intensity is expected to increase theoretically but the data does not support that; perhaps the GCMs are not able to simulate the individual storms.

6. CLIMATIC SCENARIO FOR THE INDIAN REGION AS SEEN IN THE GCM SIMULATIONS FOR A DOUBLED CO_2

Ability of the models to simulate the regional climate anomalies in the monsoon region not only depends on the accuracy of the seasonal totals but also requires spatial and temporal resolutions. In general most atmospheric GCMs simulate the gross features of the summer monsoon precipitation though there are significant differences from model to model. Sensitivity studies of sea surface temperature change and changes of monsoon circulation has brought out the importance of ENSO in the monsoon variability [22, 23]. In the present study the equilibrium climatic change of doubling CO_2 scenario over the Indian region is examined using the GFDL [24], GISS [25] and NCAR [26] models. The details of the models and the global mean and regional climatic distributions of these models are described by Schlesinger and Mitchell [8]. The results are summarised in Table 4 and the simulated results for the Indian region are reproduced in Figure 2(a,b,c). A careful examination of these results indicates that the temperature increases by about 2 to 4°C in winter (DJF) and 1-2°C in summer (JJA, monsoon over India) . Even if CO_2 take the next 100 years to get doubled, the rate of warming is almost 5 to 10 times larger than that seen in the historical data. All the three models show warming in winter and summer (monsoon) with the

Table 4. Change in climatic parameters over the Indian region as demonstrated by three GCM simulations for doubling CO_2 scenario.

Climatic parameters	Models		
	GFDL	GISS	NCAR
	DJF (winter)		
Temperature	Increase 3-4°C	Increase 3-4°C	Increase ~2°C
Precipitation	Slight (< 1 mmd^{-1}) decrease in north and increase in south	Increase 1-2 mmd^{-1}	Increase of 1-2 mmd^{-1} in north and 1 mmd^{-1} decrease in south
Soil Water	Decrease < 1 cm	Decrease < 1 cm	Increase 1-2 cm
	JJA (monsoon)		
Temperature	Increase 2°C	Increase 1-2°C	Increase 2°C
Precipitation	3 mmd^{-1} increase mainly over East India	1 mmd^{-1} decrease	1-2 mmd^{-1} increase
Soil water	slight increase	slight decrease except in high northern latitudes	slight increase except a slight decrease in Himalayan belt

highest estimates given by the GFDL model. The historical data on rainfall over the Indian region does not show any appreciable trend thus the GFDL simulated increase in precipitation seems to be too high. The other models show a decrease/increase of 1 mmd^{-1} which can be estimated to be about 10 cm for the season, This is within one standard deviation of the rainfall series for the last 100 years [4].

Years of extreme warm anomaly and cold anomaly for the winter season on hemispheric scale show a weak signal on the all-India rainfall and also the recent global warm epoch is responding reasonably well with the rainfall amounts. Palaeoclimatic records of lake levels and pollen distribution from the climatologically most sensitive regions of northwest India show that the warmer periods have been the periods of established habitations whereas the colder periods have been the periods of enhanced aridity. Thus the projected warming of the doubling CO_2 scenario over India may alter the climatic regimes of the present but the extensions of monsoons to the north and increase in summer rainfall may have beneficial effects.

The agricultural economy is closely linked to the monsoon rainfall amount as well as distribution. Parthasarathy et al. [27] found a correlation of 0.82 between all-India monsoon rainfall and the total foodgrains production in the country, during the data period 1961-85. Parthasarathy et al. [28] found a correlation of 0.88 between the monsoon rainfall and kharif foodgrains production, during the data period 1964-88. In these studies, the technological trend has been accounted for. In view of this, the global warming scenario, through consequent increases in monsoon rainfall, may have a positive impact on the agricultural production in India.

7. ANTHROPOGENIC FACTORS OF CLIMATE CHANGE : EVIDENCES FROM THE INDIAN REGION

It is difficult to interpret the observed changes in the climatic parameters over India, in terms of cause and effect. It is likely that some of these changes may be part of global-scale variations. However, any human intervention in the natural environmental equilibrium has to be viewed with concern. It is in this context that the intense deforestation activities along the foothills of the Himalayas and Assam region cause much apprehension regarding their possible role in changing the climate. Changes in land-use pattern also have significant climatic effects, as seen by Pant and Hingane [29] for the northwest Indian region. They found significant surface cooling associated with significant rainfall increase and proposed the increased area under irrigation as one of the main causal factors. The warming observed over India cannot be fully attributed to industrialization and urbanization, as neither the warming trend is confined to the industrial stations nor the cooling trend confined to the non-industrial stations. For example, Rupa Kumar and Hingane [30] found significant differences in the nature of temperature trends over six major industrial cities of India. This suggests that no generalization can be made on the effect of industrialization and urbanization on the ambient temperature. Natural events like volcanic eruptions also have important climatic implications. Handler [31] observed that the Indian summer monsoon rainfall was deficient following major low-latitude eruptions and Hingane et al. [32] found abrupt decrease in Indian temperatures following NH low latitude eruptions. Mukherjee et al. [33] observed that the winter monsoon rainfall was deficient in Sri Lanka following major low-latitude eruptions. As the observed changes are a resultant of all such effects, it becomes necessary to improve our modelling capabilities to isolate the various causal factors and assess their impact.

To avert the situation where the regional mechanisms of global warming may start adversely affecting the climate over the region, it is necessary to take preventive measures like reduction in fossil fuel use, reduction in the rate of deforestation, accelerated afforestation along with a scientific planning and management of natural resources.

There is at present no obvious mechanism that will slow, stop or otherwise deflect the warming through stabilization of the composition of the atmosphere by human action. The warming is excepted to be continuous, progressive and rapid in comparison to human experience. The warming trend if not slowed down may result into rapid changes in climate zones of the earth which may cause rapid changes in zone of natural vegetation, distribution of arable lands and productivity of agriculture, availability of water and other resources. The magnitude of these changes is significant; however, details will depend on topography soil and other features.

8. CONCLUSIONS

An attempt has been made to bring out the regional features of climatic changes over India, in the context of global change and model projections. The following points are noteworthy :

(1) The Indian monsoon rainfall has shown stable long-term characteristics since the historical times with extremes being a part of its natural variability.

(2) Mean annual surface-air temperature shows a significant warming of 0.4°C/100 years, which is comparable to the global mean trend.

(3) There is no consistent long-term association of the NH surface air temperature trend and the Indian monsoon rainfall as well as temperatures.

(4) There is a weak signal of warm and cold epoch of NH winter temperatures associated with heavy and poor monsoon rainfall situations.

(5) On a regional scale, colder winters are more frequently followed by poorer monsoons.

(6) Many uncertainties are associated with the model projections for regional climatic changes.

(7) The projected warming of the doubled CO_2 scenario may lead to changes in the climatic regimes over India, which could be for the better, particularly over north India, due to increased summer rainfall.

(8) Changes in land-use pattern and deforestation over India have indicated important implications for the local climate.

References

1. Ellsaesser, H.W., MacCracken, M.C., Walton, J.J. and Grotch, S.H. (1986). Global climatic trends as revealed by the recorded data. Rev. Geophys., 24, 4 : 745-792.

2. Bradley, R.S., Diaz, H.F., Eischeid, J.K., Jones, P.D., Kelly, P.M. and Goodess, C.M. (1987). Precipitation fluctuations over Northern Hemisphere land areas since the mid-19th century. Science, 237 : 171-175.

3. Vinnikov, K.Ya, Groisman, P. Ya and Lugina, K.M. (1990). Empirical data on contemporary global climate changes (Temperature and precipitation). J. Climate, 3, 662-677.

4. Parthasarathy, B., Sontakke, N.A., Munot, A.A. and Kothawale, D.R. (1990). Vagaries of Indian Monsoon rainfall and its relationships with regional/global circulations. Mausam, 41, 2 : 301-308.

5. Jones, P.D. (1988). Hemispheric surface air temperature variations : Recent trends and an update to 1987. J. Climate, 1 : 654-660.

6. Hingane, L.S., Rupa Kumar, K. and Ramana Murty, Bh.V. (1985). Long-term trends of surface air temperatures in India. J. Climatol. 5 : 521-528.

7. Singh, N., Pant, G.B. and Mulye, S.S. (1990). Arid area of northwest India, its fluctuations and association with the general circulation of the Atmosphere. J. Climate (communicated).

8. Schlesinger, E. and Mitchell, John F.B. (1987). Climate Model Simulations of the equilibrium climate response to increased carbon dioxide. Rev. Geophys. 25, 4 : 760-798.

9. Singh, G., Joshi, R.D., Chopra, S.K. and Singh, A.B. (1974). Late quaternary history of vegetation and climate of the Rajasthan desert, India. Phil. Trans. Roy. Soc. 267(B) : 467-561.

10. Pant, G.B. and Maliekal, J.A. (1987). Holocene climatic changes over northwest India : An appraisal. Climatic Change, 10 : 183-194.

11. Singh, G. (1971). The Indus valley culture. Archaelogy & Physical Anthropology in Oceania. VI, 2 : 177-189.

12. Rajguru, S.N. (1983). Problems of late pleistocene aridity in India. Man and Environment, VII : 107-111.

13. Mooley, D.A. and Pant, G.B. (1981). Droughts in India over the last 200 years, their socio-economic impacts and remedial measures for them. In : Climate and History - Studies in Past Climates and their Impact on Man. TML Wigley et al. Eds. Cambridge Univ. Press, pp. 465-477.

14. Sontakke,N.A. (1990). Indian Summer Monsoon rainfall variability during the longest Instrumental period 1813-1988. M.Sc. Thesis, University of Poona.

15. Verma, R.K., Subramaniam, K. and Dugam, S.S. (1985). Interannual and long-term variability of the summer monsoon and its possible link with northern hemispheric surface air temperature. Proc. Indian Acad. Sci. (Earth and Planet. Sci.), 94 : 187-198.

16. Pant, G.B., Rupa Kumar, K., Parthasarathy, B. and Borgaonkar, H.P. (1988). Long-term variability of the Indian summer monsoon and related parameters. Advances Atmos. Sci., 5, 4 : 469-481.

17. Jones, P.D. (1988). The influence of ENSO on global temperatures. Climate Monitor, 17 : 80-89.

18. Parthasarathy, B., Rupa Kumar, K. and Sontakke, N.A. (1990). Surface upper air temperatures over India in relation to monsoon rainfall. Theor. Appl. Climatol., 42 : 93-110.

19. Elliott, W.P., Machta, L. and Keeling, C.D. (1985). An estimate of the biotic contribution to the atmospheric CO_2 increase based on direct measurements at Mauna Loa observatory. J. Geophys. Res., 90 : 3741-3746.

20. WMO (1983). Report of the WMO (CAS) meeting of experts on the CO_2 concentration from pre-industrial times to IGY. WCRP (WCP-55). World Meteorological Organization, Geneva, Switzerland.

21. WMO/UNEP (1990). Climate Change : The IPCC Scientific Assessment, J.T. Houghton et al., Eds. Cambridge Univ. Press, Cambridge, U.K. pp. 364.

22. Keshavamurty, R.N. (1982). Response of the atmosphere to sea surface temperataure anomalies over the equatorial pacific and the teleconnections of the southern oscillation. J. Atmos. Sci., 39 : 1241-1259.

23. Keshavamurty, R.N. (1983). Southern oscillation : Further studies with a GFDL general circulation model. Mon. Wea. Rev., 111 : 1988-1997.

24. Wetherald, R.T. and Manabe, S. (1986). An investigation of cloud cover change in response to thermal forcing. Climatic Change, 8 : 5-23.

25. Hansen, J. Lacis, A., Rind, D., Russel, G., Stone, P., Fund, I., Ruedy, R. and Lerner, J. (1984). Climate sensitivity : Analysis of feedback mechanism in climate processes and climate sensitivity. Geophys. Mono. Ser. Vol. 29, J.E. Hansen and T. Takahashi Eds. pp. 130-163.

26. Washington, W.M., and Meehl, G.A. (1984). Seasonal cycle experiment on the climate sensitivity due to a doubling of CO_2 with an atmospheric General Circulation model coupled to a simple mixed-layer ocean model. J. Geophys. Res., 89 : 9475-9503.

27. Parthasarathy, B., Munot, A.A. and Kothawale, D.R. (1988). Regression model for estimation of foodgrain production from summer monsoon rainfall. Agric. Forest. Meteorol., 42 : 167-182.

28. Parthasarathy, B., Rupa Kumar, K. and Munot, A.A. (1990). Long-range prediction of Indian monsoon rainfall for decision making in agricultural production. Tropical Meteorological Research Programme Report Series No. 36 (WMO/TD-No. 353) : 139-143.

29. Pant, G.B. and Hingane, L.S. (1988). Climatic changes in and around the Rajasthan desert during the 20th century. J. Climatol., 8 : 391-401.

30. Rupa Kumar, K. and Hingane, L.S. (1988). Long-term variations of surface air temperature at major industrial cities of India. Climatic Change, 13 : 287-307.

31. Handler, P., (1986). Stratospheric aerosols and the Indian monsoon. J. Geophys. Res., 91 : 14475-14490.

32. Hingane, L.S., Rupa Kumar, K. and Patil, S.D. (1990). Low latitude volcanic eruption signal in Indian surface air temperatures. Int. J. Climatol., 10 : 703-709.

33. Mukherjee, B.K., Indira, K. and Dani, K.K. (1987). Low-latitude volcanic eruptions and their effects on Sri Lanka rainfall during the northeast monsoon. J. Climatol., 7 : 145-155.

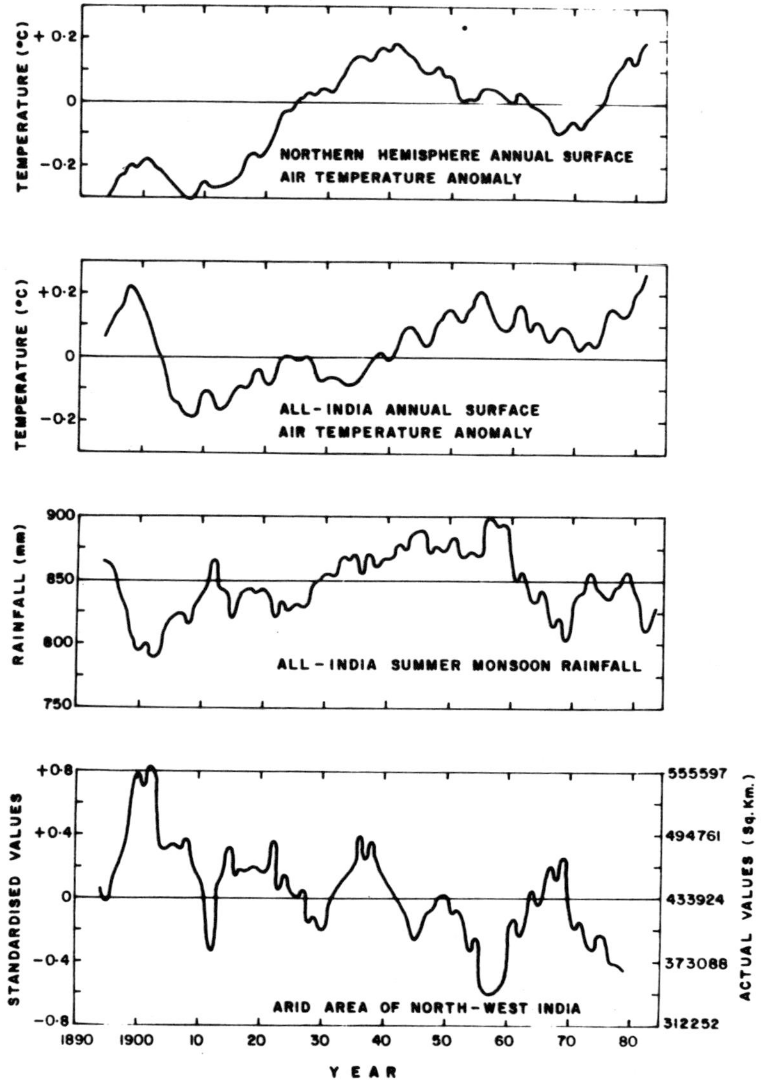

Fig. 1. Ten-year moving averages of selected hemispheric/ regional climatic parameters.

TEMPERATURE DIFFERENCES
2 X CO_2 - 1 X CO_2

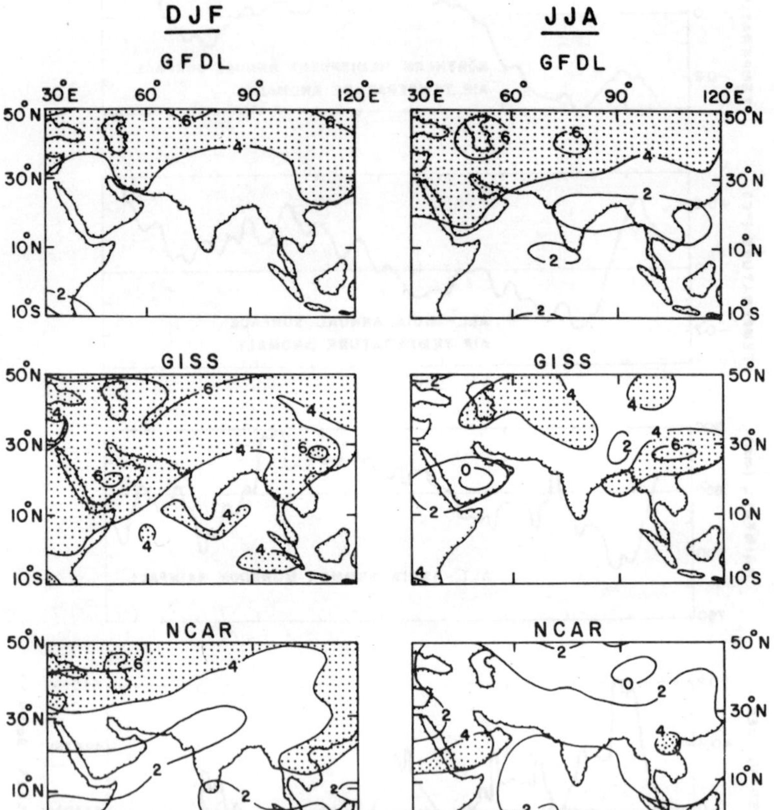

Fig. 2a. Surface air temperature change, $(2CO_2)$ - $(1CO_2)$, in °C for (30°E-120°E and 10°S-50°N) sector, from the global simulations of GFDL, GISS and NCAR models for DJF and JJA seasons (*adapted* from Schlesinger and Mitchell, 1987). Stippling indicates temperature increase larger than 4°C.

PRECIPITATION DIFFERENCES
$2 \times CO_2 - 1 \times CO_2$

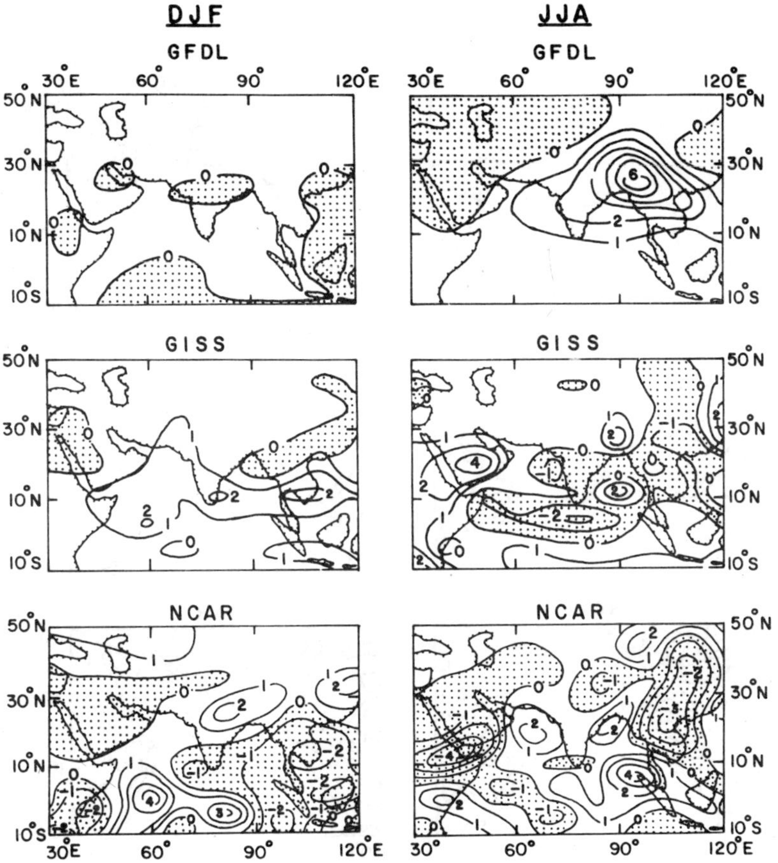

Fig. 2b. Precipitation rate change, $(2CO_2) - (1CO_2)$ in mm/day for (30°E-120°E and 10°S-50°N) sector, from global simulations of GFDL, GISS and NCAR models for DJF and JJA seasons (*adapted* from Schlesinger and Mitchell, 1987). Stippling indicates a decrease in precipitation rate.

SOIL WATER DIFFERENCES
2 X CO_2 - 1 X CO_2

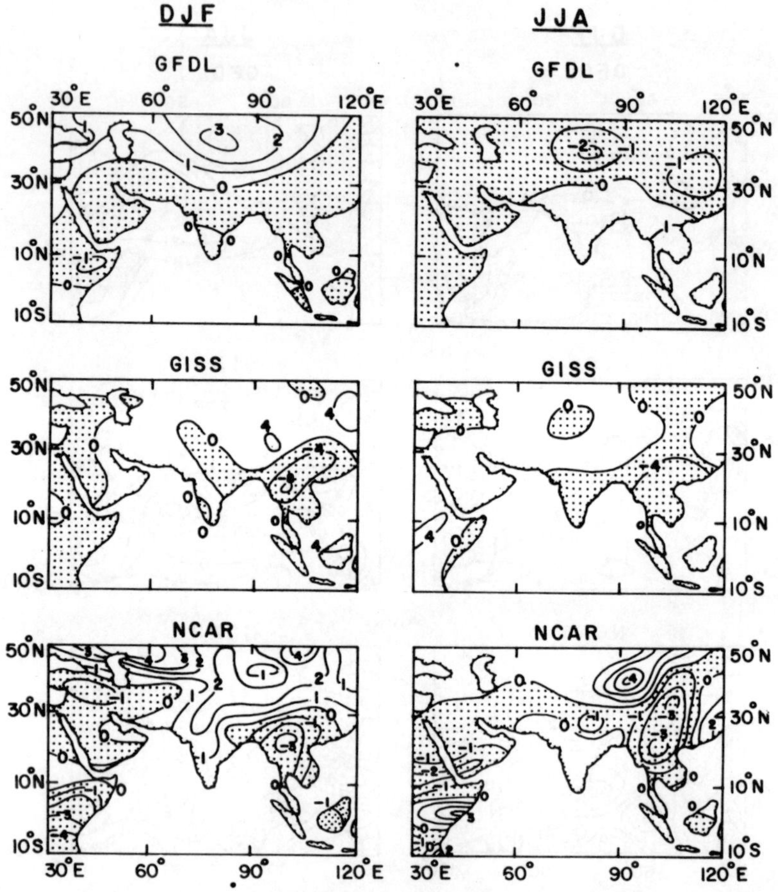

Fig. 2c. **Soil Water Change, ($2CO_2$) - ($1CO_2$) in cm for (30°E - 120°E and 10°S - 50°N) sector, from global simulations of GFDL, GISS and NCAR models for DJF and JJA seasons (*adapted* from Schlesinger and Mitchell, 1987). Stippling indicates a decrease in soil water.**

CLIMATIC CHANGE AND PLANT GROWTH

B.L.Deekshatulu and R.K.Gupta
National Remote Sensing Agency
Balanagar, Hyderabad 500 037

ABSTRACT

The contribution of green house effect on agriculture directly as well as thru hydrological responses has been discussed. Role of satellites in describing large spatial scale agricultural status and its relationship with air temperatures based parameters, description of satellite based sensing capabilities to monitor atmospheric ozone, temperature and humidity profiles, Earth Radiation Budget and the year 1998 onwards opportunities in agricultural remote sensing have been discussed.

INTRODUCTION

On a time scale of next hundred years the most important of interactions among Biosphere-Atmosphere-Cryosphere-Hydrosphere-Solid Earth would be those between the behaviour of living organisms and climatic change. Measurements from space have established that earth radiates energy at wavelength and intensity characteristics of a body at -18°C(1) yet the average temperature at the surface is some 33°C higher. This indicates that heat is trapped between surface and the level high in the atmosphere from where radiation escapes. Paper presents the discussion of climatic temperature rise and its possibility of effecting plant growth processes in a direct mode and thru hydrospheric modifications. The advantage of space based measurements with mention of existing and future (till 2010 A.D.) space based measurement techniques for surface and atmospheric temperature and humidity profiles, and Earth Radiation balance have been discussed.

Present status of remote sensing techniques in studying plant processes at agricultural and agrometeorological spatial scales has been discussed.

Threat to ozone concentration in stratosphere and consequential damage to crop biological processes thru modification in Ultra Violet (UV)-

B radiation at Earth's surface has been discussed. Existing space based observation systems for measuring total ozone in a vertical atmospheric column of unit cross-section and variation of its concentration with height in stratosphere and above have been discussed. Enhanced measurement capabilities for ozone and its vertical profile thru polar platforms in 1998-2010 time zone have also been discussed.

Existing remote sensing techniques to study plant growth processes at agricultural and agrometeorological spatial scales have been discussed. Capabilities of polar plaform based measurement systems to study crop growth processes and stress due to nutrients, water deficiency etc. have also been discussed.

GREENHOUSE GASES AND TEMPERATURE RISE

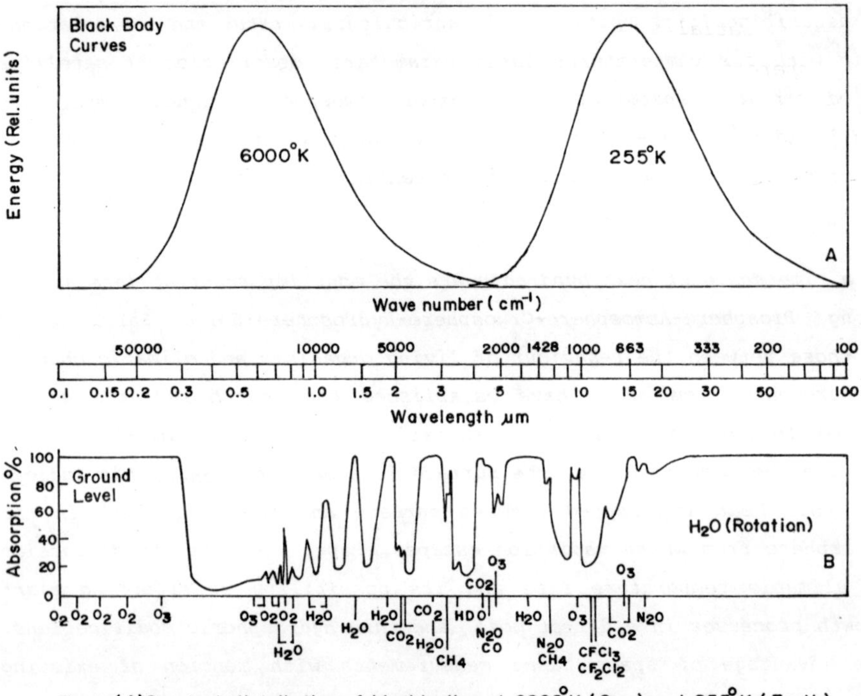

Fig. 1: (A) Spectral distribution of blackbodies at 6000°K (Sun) and 255°K (Earth), and (B) Atmospheric absorption characteristics of various gases (2).

Increasing concentrations of Carbon dioxide (CO_2), Methane (CH_4), Nitrous oxide (N_2O) and Chlorofluorocarbons in atmosphere are considered to be the primary cause of greenhouse effect. The water vapor and CO_2 absorb over a wide range of frequencies (figure 1) and absorption is weak in 8-12 μm spectral region except in 9.6 μm region due to ozone (2). Pressure and doppler broadening are the cause of increased bandwidth

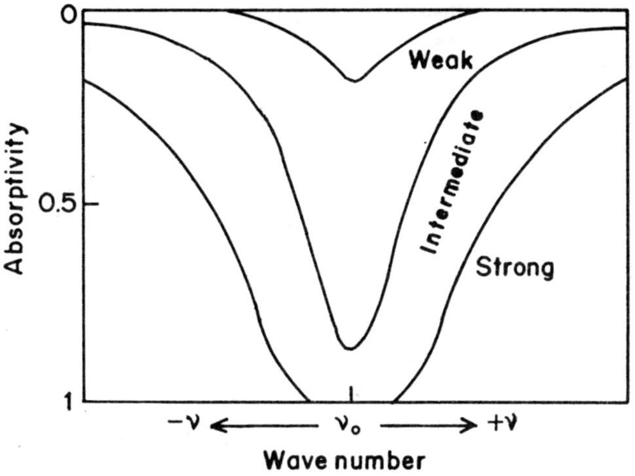

Figure.2 : Schematic illustration of weak,intermediate and strong line absorption(2).

of a spectral line. As CO_2 and H_2O vapour are available in relatively high concentration compared to CH_4 , N_2O and Chlorofloromethanes the absorption by CO_2 and H_2O vapour leads to strong absorption (figure 2) in the central portion of the spectral line. Any further increase in the concentration of CO_2 and H_2O vapour would limit the absorption to the wings of spectral lines (2). Thus the radiative heating due to increasing concentration of CO_2 and H_2O vapour would be a logarithmic function as depicted in figure 3. The O_3, CH_4 and N_2O have absorption bands either in or near the 8-12 μm range (3). There is partial overlap between CH_4 and N_2O absorption lines (figure 1) and their effect increases as square root of their concentration (figure 3). The $CFCl_3$ (CFC11) and CF_2Cl_2 (CFC12) have strong absorption lines in 8-12 μm window region. Their low concentrations are such that absorption in the middle of spectral lines is not saturated, so their effect increases linearly with concentration. It is seen from figure 3 that very low concentration of CF_2Cl_2 and $CFCl_3$, CH_4 and N_2O as compared to that of CO_2 are able to give rise to considerable heating. Radiative properties of methane make it 20 times more effective molecule for molecule than carbon dioxide in absorbing radiant heat. The band strength of absorption bands (a measure of probability of a molecule absorbing a photon at the band wavelength) for CH_4 (at 1306 Cm^{-1}), N_2O (at 1285 Cm^{-1}) and Ozone (at 1041 Cm^{-1}) are 185, 235 and 376 atm^{-1} Cm^{-1} STP respectively(3). The absorption band strengths for $CFCl_3$ at 846 and 1085 Cm^{-1} are 1965 and 765 atm^{-1} Cm^{-1} STP respectively while for CF_2Cl_2 these at 915,1095 and 1152 Cm^{-1} are 1568, 1239 and 836 atm^{-1} Cm^{-1} STP respectively.

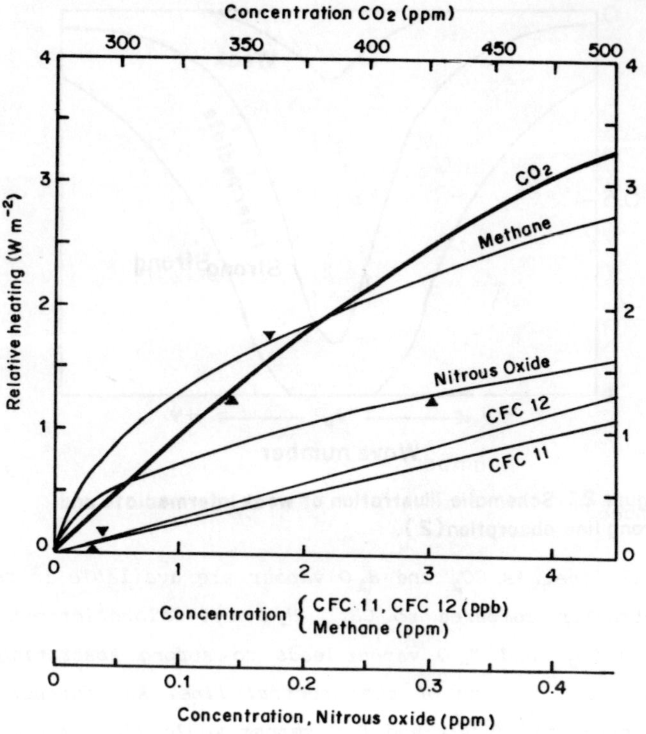

Figure.3: Greenhouse heating as a function of CO_2, CH_4, CFC11, CFC12 and Nitrous oxide concentration. The CO_2 concentration baseline is 275 ppm. Curve shows heating above that level. The traingle denotes 1985 concentration (3).

Methane in the atmosphere is contributed by paddy fields, ruminants and wet lands. Using ratios of $^{14}C/^{12}C$ and $^{13}C/^{12}C$, Lowe et al.(4) have suggested that 32% of atmospheric methane originates from fossil fuels. Concentration of methane was about 0.7 ppm 400 years ago (5). It has been estimated (6) that atmospheric CH_4 has been increasing at the rate of 1% per year over the last decade. The methane is removed from the atmosphere thru photochemical oxidation by hydroxyl radical. India is considered to be a dominant producer of methane.

Nitrous oxide is produced by microbial processes in soil and water, soil in tropical forests, nitrogen fertilizers, agricultural and fossils fuel burning and is destroyed in stratosphere by photolysis with O (^1D) state of atomic oxygen. Bolin et al.(7) and Weiss (8) has worked out a growth rate of 0.2% per year, which, if maintained, would produce a concentration of 0.332 ppm in 2035. The life time of N_2O is of 150 years order.

Chlorofluoromethanes are used by refrigeration industry and in aerosol sprays and have life times of 75 (for $CFCl_3$) to 111 (for CF_2Cl_2) years. For comparison, the life time of CH_4 is 11 years. With existing production rate, $CFCl_3$ and CF_2Cl_2 will rise to 0.5 and 1.0 ppb within 40 years (9). With accumulation due to increased life time the effect on heating could be considerable and Montreal Protocol implementation would control the use of these gases (10).

The annual carbon flux in units of billion metric tons getting into atmosphere is 5, 50, 2, 50 and 100 due to fossil fuels burning, respiration by plants, deforestation, soil respiration and physico-chemical diffusion (from ocean) respectively while the fixing of carbon flux from atmosphere is 100 units thru photosynthesis by plants and 104 units by physico-chemical diffusion to sea (11). This puts addition of 3 units of CO_2 to atmosphere every year. The CO_2 emission due to fossil fuels and deforestation could be related to population thru

Total CO_2 emission = (Engineering Effect) x (Standard of Living) x (demography)

= CO_2 emission/technology x technology/capita x Total population size

The size of carbon reservoir in world vegetation, world soils, atmosphere, Ocean and Fossil Fuel reserves is 560, 1500, 735, 36000 and 5000 to 10000 billion metric tons respectively.

As per the records of Mauna Loa Observatory the CO_2 concentration has increased from about 310 ppm to more than 350 ppm since 1958. The typical annual growth rate for CO_2 due to fossil fuels use is 0.5 to 2% implying that CO_2 concentrations will double (to 660 ppm) in 21st century (12-14). Study of the ratio of O^{18} and O^{16} as well as of Deutrium to Hydrogen in the trapped air bubblesin the deep ice region of Greenland (15) and Vostok station in Antarctica suggests that CO_2 levels have been reasonably close to the pre-industrial value of 280 ppm during past several thousand years of present interglacial cycle. At the maximum of last ice age 18,000 years ago, CO_2 levels were roughly 25% lower than the pre-industrial value (16). Sine 1850 the CO_2 has risen approximately by 25%. Study of 2,200 meters long ice core study at Vostok shows that there is a close resemblance among CH_4, CO_2 concentrations and Temperature curve over past 160,000 years (11).

Figure 4 (3) schematically depicts the effect increasing CO_2 on surface temperature rise. Process of surface temperature rise due to other

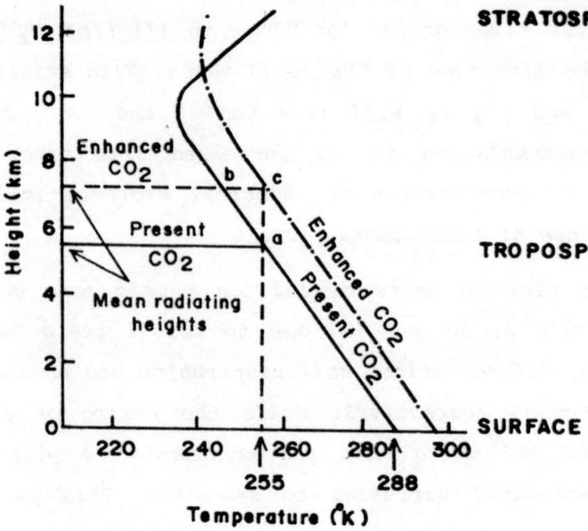

Figure.4: Schematic representation of the effect of increasing CO_2 on the vertical profile of temperature.

greenhouse gases would be similar. Due to longwave emission from atmosphere the effective temperature (Te) could be assigned to a layer raised above the surface. The Earth has a mean radiating height of 5.5 Km (point 'a' in figure 4) and a global mean lapse rate of 6°K per Km gives a current greenhouse temperature of 33°C. The increased absorption of radiation by greenhouse gases would raise the mean radiating height to an altitude where the temperature is less than the effective radiating temperature Te. Radiation balance of Earth-Atmosphere system would lead to warming of surface temperature till the new mean radiating height (point 'c' in figure 4) is equal to temperature Te.

The issue is how much would be the heating due to increased concentration of greenhouse gases. Higher CO_2 concentration could lead to increased photosynthesis. The anticipated 3 billion metric tonnes per year CO_2 enhancement in already photosynthesis based fixation of 100 billion metric tonnes per year of CO_2 is not a very large quantity. The soils have about twice the amount of carbon in the form of dead organic matter as in the atmosphere and this is slowly decomposed into CO_2 and other gases thru microbial decomposition. The degree of this biological positive feedback with increase in temperature could enhance CO_2 build up. Carbon dioxide is slowly removed thru the biological and chemical processes in the oceans. A large perturbation in atmospheric CO_2 concentration may need decades to centuries to attain equilibrium with oceanic sink processes. With all this uncertainty typical estimates put this so-called 'airborne' fraction at about 50%.

If the temperature lead to arctic warming then the clathrates in tundra and off continental shelves could release vast quantities of methane lying trapped below the frozen sediments (17). This melting would cause decrease in albedo and further rise in temperature (18). Even with all these uncertainties, many workers project the CO_2 concentration at 600 ppm level between 2030 and 2080 and faster rate of growth for other trace greenhouse gases. Projections of global temperature rise due to increase in CO_2 from 300 to 600 ppm level have ranged from 1.5 to 5°C. Global Circulation Models (GCMs) predict a rise of surface temperature by 3.5 to 5°C for the doubling of CO_2 or an equivalent increase in other trace gases. This would be 10 to 100 times faster as compared to 5°C warming since the peak of ice age 18,000 years ago. The feedback and sub-grid scale processes are not properly represented in GCMs which have grid size of 4 to 7.5° in latitude and 5 to 10° in longitude. It has to be kept in mind that the time series of data sets for past 100 years would have bias due to physical movement of observatories over the course of the century and skewing of observations by the heat released by machinery or stored by buildings and pavement. By applying multiple-window time-series methods, Kus et al. (19) have shown that from 1880 to 1988, the average global temperature increased by 0.0055 ± 0.00096°C per year. This gives an increase of 0.5°C since the beginning of century. Further, the six warmest years on the record were 1988, 1987, 1983, 1981, 1980 and 1986 in that order. To have a feel of substantiality of 0.5°C temperature change, it is worth noting that the year 1816, the "year without a summer" had less than 1°C drop in the mean global temperature. Figure 5 gives the estimate of the possible trace gases contributions to global warming and its comparison with respect to CO_2 warming (20). The uncertainties indicated by vertical bar are considerably large in case of trace gases.

GREEN HOUSE EFFECT, PLANT GROWTH AND SATELLITES MEASUREMENTS

Photosynthesis depends on Solar irradiance, carbon dioxide, temperature, oxyzen, water, plant age and nutrient factors and such a complex mechanism had been studied using variations in one contributing factor at a time and this way Law of Minimum was extended to the Principal of Limiting Factors by Blackman (21). The photochemical process is not affected by the temperature. Electron transport is neither accelerated by temperature rise nor inhibited by extremely low temperatures. However, the enzyme controlled biochemical portion of photosynthesis is strictly temperature dependent. Thus the light use efficiency is affected by

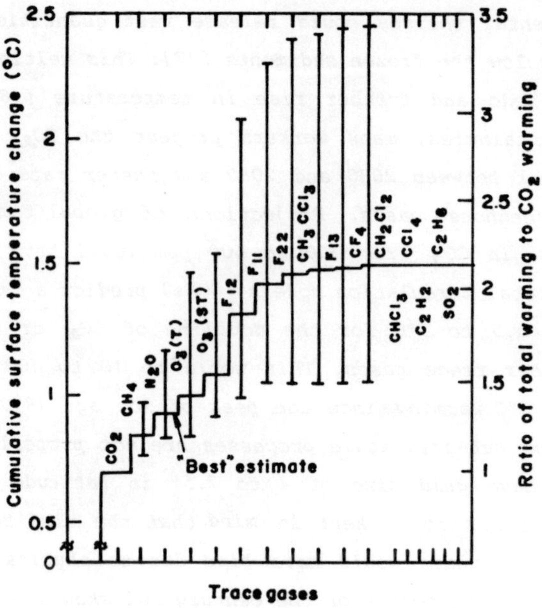

Fig.5 : Various trace "greenhouse gases" contribute about as much to equilibrium global warming (see right-hand scale) as CO_2 for the "best" estimate case, but uncertainties in the projected scenarios of these trace gases are large (see vertical bars). Additional uncertainties in equilibrium temperature response from climate model assumptions are not included in the figure (20)

temperature. The dark respiration rate is more affected by the temperature as compared to photosynthesis efficiency. The close interconnections between dark respiration and light reactions make the overall photosynthesis process quite thermally dependent and the curve between apparent photosynthesis would be of a type where the growth is reduced below some threshold of minimum temperature and above some threshold of maximum temperature, and is optimum within some temperature. The rate of respiration reaches maximum around 40-45°C and photosynthesis rate becomes zero around 50°C while temperature around 60°C kills the plant. Idso (22) considers that the surface temperature sensitivity employed by the models (23) was fully an order of magnitude too great than the actuals seen thru analysis of real-world observations. A variation of about 0.6°C has been observed by Parker and Folland (24) over Northern Hemisphere land region during 1860 to 1980. Kimball and Idso (25) have found that doubling of CO_2 concentration could reduce trans-

piration by 34%. Here comes the effect of feedbacks which have been missed while estimating evapotranspiration loss due to increased temperature. The atmospheric CO_2 increase would induce partial stomatal closure and thereby reduce the plant evaporative loss. The natural ground water recharge, a small difference between rainfall and evapotranspiration, would increase several times due to little decrease in evapotranspiration (26,27). Walker and Dickinson (28) visualize greater water demand by the crop due to warmer climate. A rise of 1-4°C in temperature is expected to induce a 5-15% change in potential evaporation. They anticipate greater amounts of rainfall in tropics with extreme events in Southern Hemisphere (20-40°S), longer sequence of rainy days and flood rainfalls of long duration. The evaporation losses from reservoirs would increase with increase in temperature. Further, the increased water temperatures of reservoirs are likely to promote algal growth possibly resulting in new levels of eutrophication. Increased rainfall kinetic energy is expected to give exponential rise to sedimentation with associated repercussions in reservoir siltation and estuarine system.

Reviewing 590 research papers, Strain (29) noted that increased level of CO_2 would, in general, lead to increase in photosynthesis, plant tolerance to atmospheric pollution, leaf area, leaf numbers, branch numbers, fruit numbers, fruit size and number of seeds produced per plant; decrease in plant water loss via transpiration; better germination of seeds and spores; earlier production of flowers; and accelerated maturity of crops. Rise in temperature would lead to melting of polar ice resulting in sea level rise and flooding of coastal low lands. Melting of polar ice would reduce albedo for the incoming solar radiation and would result in further warming of polar regions.

The earth surface temperature, temperature and humidity profiles are obtainable from NOAA (National Oceanic and Atmospheric Administration, U.S.A.) polar orbiting meteorological satellites thru Advanced Very High Resolution Radiometer (AVHRR) and Tiros Operational Vertical Sounder (TOVS) sensing systems (30). These and Earth Radiation Budget Experiment (ERBE) on board NOAA satellites (31) provide opportunity to measure earth warming. The measurements in red and near IR spectral bands in earth resources satellites such as LANDSAT, SPOT and IRS; and in visible (0.55-0.68 micron) and near IR (0.725-1.1 micron) bands of AVHRR provide opportunity to compute vegetation indices thru ratioing and normalized differencing (32) which are indicator of Leaf Area Index. These AVHRR vegetation indices have been found to be related with district level spatial scale Growing Degree Days Agrometeorological growth indicators at 98 to 99% confidence level (33).

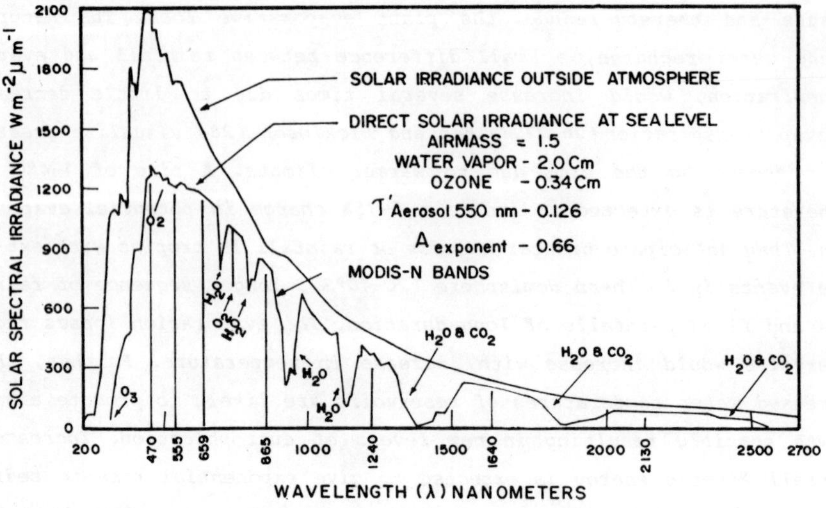

FIGURE.6: SOLAR REFLECTANCE BANDS FOR MODERATE RESOLUTION IMAGING SPECTRORADIOMETER IN NADIR VIEWING CONFIGURATION (MODIS-N) ON BOARD THE POLAR PLATFORM (34).

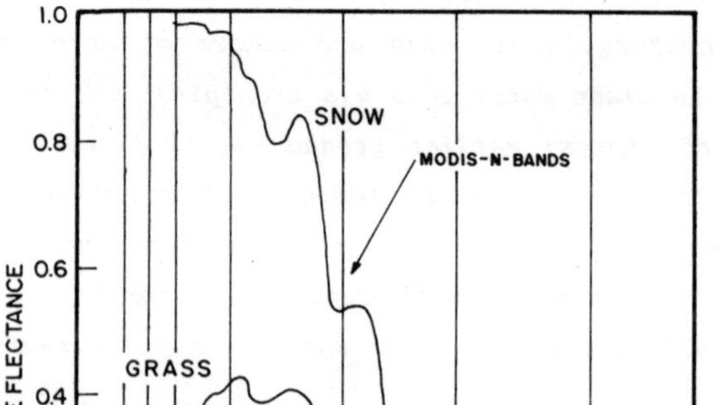

FIG.7. SPECTRAL REFLECTANCE CHARACTERISTICS OF SNOW, GRASS AND SOIL IN MODIS-N SPECTRAL BANDS(34).

During 1998-2010 A.D. the polar platforms would have advanced version of AVHRR as AMRIR (Advanced Medium Resolution Imaging Radiometer) which would improve the existing spatial resolution of 1.1 Km to 500 m and would provide enhanced 0.45-0.52, 0.52-0.60, 0.58-0.68, 1.58-1.64, 2.08-2.35 micron spectral channels enabling detailed study of biomass. The High Resolution Imaging Spectrometer (HIRIS) would enable detection of stress in crop due to water, minerals, nitrogen and lignin deficiencies through the flexibility of 200 spectral bands in 20 m spatial resolution. The Moderate Resolution Imaging Spectrometer (MODIS) providing 40 spectral bands in Nadir (N) mode in 0.47-2.3, 3.7-4.56 and 6.7-14.2 micron spectral region would enable study of agriculture leaf/canopy characteristics simultaneously at 250 m, 500m and 1000m spatial resolutions (34). Figure 6 gives the details of reflective spectral bands of MODIS-N while Figure 7 gives the typical reflection curves for grass, soil and snow in these MODIS-N bands. The all microwave bands measurements would be available for estimation of humidity (at 15 Km spatial resolution) and temperature (at 50 Km spatial resolution) profiles through Advanced Microwave Sounding Unit (AMSU). Improved version of ERBE as ERBI (Earth Radiation Budget Instrument) in scanning as well as non-scanning mode would be available on polar platforms (35). Altimeter to measure sea level rise would become available on European Remote Sensing Satellite (ERS-1) in 1991 and also on polar platforms. Short-term climatic fluctuations study would be possible through the use of data from Laser Atmospheric Wind Sounder (LAWS) on board polar platforms.

OZONE VARIATIONS, PLANT GROWTH AND SATELLITE MEASUREMENTS

Atmospheric ozone selectively absorbs the UV radiations in 0.2-0.4 μm spectral range. Variations of ozone in the atmosphere do not significantly change the radiations in UV-A (0.32-.4 μm : plays role in photodamage as well as photorepair in plants) and in UV-C (0.2-0.28 μm: extremely injurious to organisms) spectral regions. The ozone depletion would lead to significant variations in UV-B (0.28-0.32 μm) radiation which could cause extreme damage to the physiology and biochemistry of many forms of life (36, 37). Here, the weighting function or action spectra conform to the absorption spectra of either proteins or nucleic acid (DNA). Figure 8 depicts the relative biological efficiency curve for DNA (ϵ_{DNA}, DNA molecule is responsible for the transmission of genetic information) and for photoinhibition (ϵ_{PI}) of chloroplasts together with the downward spectral irradiance, Ed(λ ; for

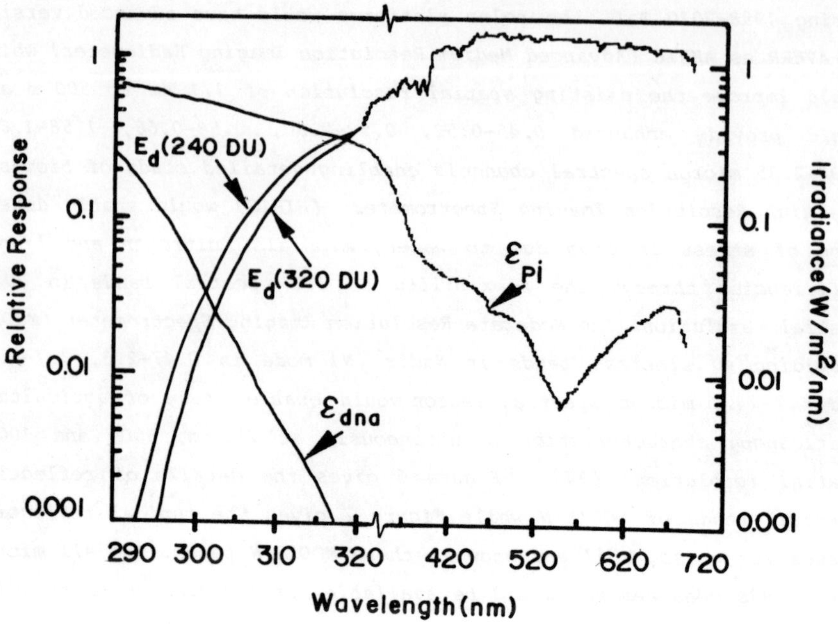

Fig 8 : Relative biological efficiencies, $\varepsilon_{PI}(\lambda)$ and $\varepsilon_{DNA}(\lambda)$, and downward spectral irradiance, $E(\lambda)$, for two stratospheric ozone concentrations (240 DU, 320 DU) plotted against wavelength (nm). Note that the wavelength scale is expanded below 320 nm in order to show greater detail in the UV-B region of the spectrum(36).

240 and 320 Dobson Unit (DU) ozone concentrations. The ϵ DNA is heavily weighted in UV-B spectral region while ϵPI shows significant absorption throughout the UV and visible region spectrum. The 320 DU is the normal mid-latitude stratospheric O_3 concentration and 240 DU refers to reduced ozone situation. The biologically weighted fluence rate for a particular photoeffect is the product of the appropriate biological efficiency and the spectral irradiance integrated over the wavelength.

The problem of ozone variations impact is two fold i.e. accurate estimation of biological weighting function as well as ozone concentration and its vertical profile especially in stratospheric region.

Existing NOAA polar orbiting meteorological satellites provide measurements of total ozone in a vertical unit cross section atmospheric column thru measurements in 9.67 micron band(TOVS), and its stratospheric concentration profiles using Solar Backscattered Ultra Violet (SBUV-2nd

Version) radiometer. SBUV/2 measures backscattered radiations in 11.3° FOV (Field Of View) at nadir in 12 discrete, 11nm bandwidth spectral bands in 0.252-0.3398 µm spectral region; and the direct Solar radiation (38). Besides this discrete mode it also has sweep mode wherein the Solar irradiance or the atmospheric backscattered radiance is measured continuously in nominal increments of 0.148 nm over the 0.160 to 0.400 µm spectral region. Sweep mode would provide information on atmospheric photochemical processes. The Total Ozone Mapping System (TOMS) on board Nimbus-7 satellite sequentially sampled the atmosphere at six wavelengths from 0.3125 to 0.380 µm in 3° steps in the ± 51° cross-track scan from nadir. Measurements for estimating stratospheric concentrations of H_2O, HNO_3, NO_2 and O_3 are available from Nimbus-7/Limb Infrared Monitor of Stratosphere (with 1.8 Km vertical and 28 Km horizontal spatial resolutions) while the Nimbus-7/Stratosphere and Mesosphere Sounder (SAMS) provides measurements for estimating CH_4, CO, H_2O, NO and N_2O concentration in stratospheric and mesospheric region. Ozone is produced in mesosphere and sinks to stratospheric levels with maximum concentration around 25 Km altitude. The advanced version of these instruments would be available on polar platforms as Global Ozone Monitoring Radiometer (GOMR). GOMR would provide Nadir as well as limb-to-limb measurements in 6 spectral channels in 0.305-0.340 µm, and 3 or more channels in 6 to 18 µm to measure O_3 (total amount as well as its vertical distribution), temperature and perhaps other species information. Its IFOV (Instantaneous FOV) would be 43 x 43 Km^2 and vertical resolution would be 2 Km (35).

CONCLUSIONS

The temperature of the atmosphere is increasing but the existing models do not represent the sub-scale processes and feedbacks in a proper manner. This leads to conflicting estimates for the magnitude of heating. Similar is the situation for the impact of temperature rise on plant growth directly and through hydrologic changes. Satellites provide data over the whole globe with good precision. These may be exploited for studying change detection and in estimating the rate of change. It is the rate of change which would become important in providing evolutionary adjustment for the natural processes. The regional variability due to climatic change would be much different from the global average and development of regional and meso scale models is the necessity. Biological weighting functions for various processes in UV-B as well as in the Photosynthetically Active Radiation (PAR) spectral regions need better estimation. The magnitude of greenhouse effect variability at regional and meso scales may cause geographical shifts in agriculture crops cultivation zones making one geographical area to loose while the other geographical area gains. This would raise the issue of redistributive justice.

ACKNOWLEDGEMENTS

The authors are thankful to their organisation for permitting the publication of this contribution and to Mr.V.V.S.B.Sastry for the secretarial care to this manuscript.

REFERENCES

1. *Schneider, S.H.(1989 'a'). The Changing Climate, Sci. Amer., 70-79.*
2. *MacCracken, M.C. and Luther, F.M.(Eds.), (1985). Detecting the climatic effects of increasing carbon dioxide. Rep DOE/ER-0235, U.S. Deptt. of Energy, Washington, D.C.*
3. *Mitchell, J.F.B. (1989). The "Greenhouse" effect and climate change. Revs. of Geophys., 115-139.*
4. *Lowe, D.C., Brenninkmeijer, C.A.M, Manning, M.R., Sparks, S. and Wallace, G. (1988). Radiocarbon determination of atmospheric methane at Baring Head, Newzealand. Nature, 522-525.*
5. *Craig, H. and Chou, C.C. (1982). Methane: The record in polar ice cores. Geophys. Res. Lett., 1221-1224.*
6. *Blake, D.R. and Rowland, F.S. (1988). Continuing worldwide increase in tropospheric methane; 1978 to 1987. Science, 1129-1131.*
7. *Bolin, B., Doss, B.R., Jaeger, J. and Warrick, R.A. (Eds.), (1986). The Greenhouse Effect, Climatic Change and Ecosystem. SCOPE Rep., 1-541.*
8. *Weiss, R.F. (1981). The temporal and spatial distribution of tropospheric nitrous oxide. J.Geophys. Res., 7185-7195.*
9. *Dickinson, R.E. and Cicerone, R.J. (1986). Future global warming from atmospheric trace.gases. Nature, 109-115.*
10. *Wigley, T.M.L. (1988). Future CFC concentrations under Montreal Protocol and their greenhouse-effect implications. Nature, 333-335.*

11. Houghton, R.A. and Woodwell, G.M. (1989). Global Climatic Change, Scientific American, 36-44.

12. Nordhaus, W. and Yohe, G. (1983). Report of Carbon Dioxide Committee. In Changing Climate, National Academy Press, Washington, D.C., U.S.A., 87-153.

13. Mintzer, I.N. (1987). A matter of degrees: The potential for controlling the greenhouse effect. World Resources Institute, Washington, D.C., U.S.A.

14. Chandler, W.U., Geller, H.S. and Ledbetter, N.R. (1988). Energy Efficiency, A New Agenda. The American Council for an Energy - Efficient Economy Washington, D.C., U.S.A.

15. Neftel, A., Moor,E., Oeschger,H. and Stauffer,B. (1985). Evidence from polar ice cores for the increase in atmospheric CO_2 in past two centuries. Nature, 45-47.

16. Schneider, S.H. (1989 'b'). The Greenhouse effect: science and policy. Science, 771-781.

17. Revelle, R. (1989). In Changing Climate. Report of the Carbon Dioxide Assessment Committee. National Academy Press, Washington, D.C., U.S.A, 252-261.

18. Schlesinger, M.E. and Mitchell, J.F. (1987). Climate model simulations of the equilibrium climatic response to increased carbon dioxide. Revs. Geophys, 760-798.

19. Kuo, C., Lindberg, C. and Thomson, D.J. (1990). Coherence established between atmospheric carbon dioxide and global temperature. Nature, 709-714.

20. Ramanathan, V., Cicerone, R.J., Singh, H.B., and Kiehl, J.T. (1985). Trace gases trends and their potential role in climate change. J. Geophys. Res., 5547-556.

21. Bluckman, F.F. (1905). Optimum and limiting factors. Ann. Bot., 281.

22. Idso, S.B. (1980). The climatological significance of a doubling of Earth's atmospheric carbon dioxide concentration. Science, 1462-1463.

23. National Research Council (U.S.), (1982). Carbon dioxide and climate: a second assessment. National Academy Press, Washington, D.C.

24. Parker, D.E. and Folland, C.K. (1988). The nature of climatic variability. The Met. Mag., 201-210.

25. Kimball, B.A. and Idso, S.B. (1983). Increasing atmospheric CO_2 : Effects on crop yield, water use and climate. Agric. Water. Manage., 55-72.

26. Bouwer, H. (1984). Water conservation in agricultural and natural systems. Procs. Conf. Water for the 21st century: Will it be there. Southern Methodist University, Dallax, TX, U.S.A.

27. Rogers, H.H., Thomas, J.F. and Bingham, G.E. (1983). Response of agronomic and forest species to elevated atmospheric carbon dioxide. Science, 428-429.

28. Walker, B.H. and Dickinson, R.E. (1988). Southern Hemisphere perspectives of global change: Scientific Issues, Research needs and proposed activities. IGBP Global Change, Report No.9, 55 pp.

29. Strain, B.R. (1978). Report of the workshop on anticipated plant responses to global carbon dioxide enrichment. Dept. of Botany, Duke University. Durham, N.C. 20776, U.S.A.

30. Gupta, R.K. and Gopalrao, U.V. (1983). Survey of satellite Meteorology, Procs. Indian Acad. Scis., 255-277.

31. Barkstrom, B.R. and Smith, G.L. (1986). The Earth radiation budget experiment: Science and implementation. Revs. Geophys., 379-390.

32. Gupta, R.K. (1989). Satellite Vegetation indices as District Level Agromet/Growing Degree Days Indicator. Ph.D. Thesis, Jawaharlal Nehru Technological University, Hyderabad, India.

33. Gupta, R.K. (1990). NOAA AVHRR vegetation indices and agriculture-meteorology processes. Accepted for publication in Advances in Space Research, U.K., 4 pp.

34. Salomonson, V.V. (1990). Execution phase (C/D) spectral band characteristics of the EOS Moderate Resolution Imaging Spectrometer - Nadir (MODIS-N) facility instrument. To appear in Advs. Space Res.

35. Gupta, R.K. (1989). Meteorological satellites: Sensing systems, application aspects and anticipated growth till 2010 A.D.. Lecture at 32nd IETE Annual Convention, Deptt. Elec. Engg., Ind. Inst. Techno., Madras, 23 pp.

36. *Jagger, J. (1985). Solar-UV actions on living cells. Praeger, New York, U.S.A.*

37. *Giese, A.C. (1976). Living with our Sun's ultraviolet rays. Plenum Press, New York, U.S.A., 185 pp.*

38. *Schwalb, A. (1982). Modified version of the TIROS-N/NOAA A-G Satellite Series (NOAA E-J)-- Advanced TIROS-N (ATN). US NOAA Tech. Memo. NESS 116, US. NOAA NESS, Washington, D.C., U.S.A, 23 pp.*

POTENTIAL IMPACT OF GLOBAL CLIMATIC CHANGE IN HIMALAYA

A. N. PUROHIT
G.B. Pant Institute of Himalayan Environment & Development
Kosi, Almora 263643, U.P., INDIA

SUMMARY

Based on the general pattern of climatic gradient along the altitude in Himalaya, it is predicted that the direct and indirect impacts of global climatic change will become apparent at a much faster rate in Himalaya and the effect will increase with altitude. The climatic warming may reduce the altitudinal contrasts in species diversity and population status leading to an upward shift in species colonization. It is also predicted that with global warming the climatic sensitivity of geomorphological processes, such as weathering of rocks and mass movement will increase. The permanent snow-line will be disturbed and also the hydrological cycle. A comparative study of CO_2 and temperature effects on low and high altitude Himalayan species would increase our predictive capacity for changes in ecosystem dynamics due to global climatic changes in the near future.

1. INTRODUCTION:

Global climatic change is a normal process on a geological time scale. However, recent and projected changes in the atmospheric chemistry resulting from human activities point to rapid changes by the beginning of the next century. Since 1861 there have been suggestions that the Carbon dioxide content of the atmosphere controls temperatures, because the CO_2 molecules in the air absorb infra-red radiation (Plass, 1956). Attention, therefore, has been directed at the artificial production of Carbon dioxide from the consumption of fossil fuels and other activities as a possible cause of climatic change in a very near future. It is a well established fact that Carbon dioxide and other radiatively active gases are being added to the atmosphere at a rapid rate (Crutzen and Graedel, 1986; Ramanathan, 1988) and CO_2 and temperature are highly correlated (Genthon et al., 1987). Therefore, the conclusion that a rise in the atmospheric CO_2 concentration might lead to rapid increase in the global temperature has been advocated in recent past. However, underestimation and overestimation in the magnitude of temperature rise are expected due to the fact in addition to the greenhouse gases there are other significant changes taking place in the atmospheric chemistry. Those who give less importance to greenhouse effect support their arguments on the basis of changes in global reflectivity, atmospheric turbidity and naturally produced aerosols along with the changes in greenhouse gases which counter act greenhouse effect by lowering the mean air temperature (Lovelock, 1971). However, the net increase in global temperature is unquestionable otherwise the earth would have been covered with ice even today.

The advocates of greenhouse effect propose that due to large scale combustion of fossil fuels, deforestation and other land use changes, the atmospheric CO_2 concentration has reached to a level of about 340 ppm which is about 20 to 25 per cent higher than that of a century ago (Boer et al.,1990). Depending upon the growth rate of earth energy use, the CO_2 levels may rise upto as much as 400 to 600 ppm by the first half of next century, which will lead to an increase in global mean temperature by 1.5 to 4.5°C (Gates et al., 1981). This change in the global temperature is certainly alarming and needs both protective as well as preventive measures. It is also necessary to analyse the sequential changes at local as well as global scale.

Despite changes in absolute values, climatic gradients would persist and the changes will vary between regions depending upon the landscape and their regional climatic scenarios. It will, therefore, be necessary to identify the most sensitive landscapes and geographical regions to this climatic change. This article attempts to focus on the possible sensitivity of Himalayan region to global warming. The conclusions are mainly drawn on the basis of known interaction of factors which generally operate in mountains along an altitudinal gradient.

2. CLIMATIC SCENARIO

In the Himalayan context neither we have long-term climatic records nor the exact records of past vegetational pattern. Therefore, the predictions made here are based purely on general pattern of climatic elements along the altitudinal gradient (as projected by Baumgartner, 1980-Table 1). The pattern of air pressure, air density, concentration of CO_2, water vapour and air temperature is almost identical in all mountain zones. However, the day and night fluctuation in the air temperature is more contrasting in Himalayan region as compared to those reported for other mountain zones. The characteristic feature of altitudinal climatic scenario in Himalaya further reveals that even today the CO_2 concentration in the mountains at about 3000 meters is that of pre-industrial period (270 ppm) where as in valleys at about 500 m it is around 330 ppm which is almost identical with present day CO_2 concentration at the global level. These differences in CO_2 concentration along the altitudinal gradient reflect the changes which have taken place during last one century at the global level. Even the air density shows the trends of past and present environment.

It is, therefore, reasonable to speculate the global warming due to greenhouse effect will be more severe at higher elevations than in the plains. The reason being thin atmosphere at relatively higher elevations. The possible modifications in climatic parameters are shown in Table 2. The pattern of climatic scenario along the altitudinal gradient therefore seems to be analogous to past and present climatic patterns. Since the analogue scenarios are considered to be important tools for predicting the future changes, a comparision of existing landscape and ecosystems along the altitudinal gradient can provide information about the future course of changes.

3. IMPACT ON NATURAL ECOSYSTEMS

Depending on the prevalence of other particulates in the environment, the global change in CO_2 level might lead to the following three combinations with temperature:

(a) High CO_2 and high temperature,

(b) High CO_2 and low temperature due to dominance of global reflectivity, atmospheric turbidity and other atmospheric chemical factors over atmospheric CO_2,

(c) High CO_2 and normal temperature due to counter action of CO_2 effect on temperature by potential effects of atmospheric turbidity.

The impact on ecosystem should be reviewed separately under these three categories. Without having much fluctuation in the atmospheric temperature the increasing atmospheric Carbon dioxide is known to increase the plant productivity through increasing photo-synthetic rates and inhibitory effect on photorespiration. The water use efficiency is also known to increase under high CO_2

concentrations due to the effect of CO_2 on stomatal conductance (Warrick et al., 1986). A comparative study of high and low altitude plants also reflect such behaviour, as most of the high altitude plants have relatively higher stomatal density as compared to that of low altitude plants. However, Korner (1988) has found no statistically significant differences in stomatal density in low as well as high altitude plants over the 7 to 12 decades. The physiological estimates on the basis of instantaneous plant responses also indicate that the present increase in global CO_2 will increase assimilnation in plants from higher elevations (Korner and Diemer (1987). In addition to these effects Carbon dioxide is also known to induce or inhibit flowering (Wittwer, 1967; Posner, 1971; Purolint and Tregunna, 1974), supress or stimulate bud break in plants (Purohit & Tregunna, 1976), supress the expression of symptoms of the viral disease but not the multiplication (Purohit et al., 1975). All such changes brought out by the Carbon dioxide on plants will lead to changes in primary producer of the ecosystem and thereby leading to a considerable change in the ecosystem in due course of time. However, one can not predict at present that increasing Carbon dioxide concentration in the atmosphere along with the increasing or decreasing temperatures will also have similar impact on the vegetation pattern. In fact, there is hardly any knowledge about the low temperature and high carbon dioxide interactions. Therefore, the predictions at present would lead to misleading conclusions. However, one can speculate that if Carbon dioxide accumulation is associated with increase in the temperature the entire vegetational belt in mountains will be shifted upwards due to warmer climate as well as favourable carbon dioxide concentration. However, this shift will be selective and species specific, because the different species will respond differently in their own way to changing environmental conditions. The species which are adopted to wider environmental gradient would spread at a faster rate and would take over those which have restricted distribution at present. The species which are adopted to a more restricted environment in sub- alpine and alpine zone will be main losers.

4. IMPACT ON CROPS & CROPPING PATTERNS

In mountains, the productivity of crops (agricultural as well as forestry) is primarily limited by the cool climate and correspondingly short growing season. These restrictions increase with altitude.

Since the length of the growing season is defined as the number of days during the year with a mean temperature of at least +6°C and in principals for each 100 meters increase in altitude, the length of growing season decreases by 5-6 days, the CO_2- induced climatic changes would certainly prolong the growing season and would result in potentially higher crop yields and forest production. In addition, there might be an increase in plant production due to the direct effect of CO_2 enrichment and possible higher nutrient availability caused by higher soil temperature and correspondingly higher rate of decomposition and mineralisation at higher altitudes. Recently, the estimates worked out by Korner and Diemer (1987) indicate that if the global CO_2 level increases by 100 ppm it would increase net photosynthesis by 9% at lower elevations where as at higher elevations the increase will be about 21 per cent. However, the positive perspectives for total biomass production might not be reflected in the higher production of economic yields, especially in plants which have chilling requirement for the onset of flowering. A counter-balance system might also start operating in relation to forests because if the climate becomes more favourable more land might be put under agricultural operations leading to further reduction in forest areas in future.

5. IMPACT ON GEOMORPHOLOGY

Many of the geomorphological process are affected

directly by climatic conditions or indirectly through the climatic influence on attributes like vegetation cover, soil properties and ground hydrology especially in younger mountain zones like Himalaya. With increasing altitude frost play an important role in sediment transport and holding the large blocks together. If due to global warming winters get shorten, the frost weathering will be affected to a considerable extent. The freeze-thaw cycles will change which will lead to release of finer sediment load at a faster rate and increase the sediment transport in the rivers. Similarly, the magnitude of mass movement, which is affected by soil moisture and temperature conditions, will also be affected to a considerable extent. Shorter and warmer winters will lead to higher mass movement events, like rock falls. The seasonal precipitation will increase at higher elevations. This will also add to increase in mass movement and landslides. The debris flow and soil slips will increase as a consequence of increase in soil moisture as well as temperature. Therefore, the climatic sensitivity of mountains at higher elevations will increase leading to slope failures at a faster rate. If all these events speed up by global warming it will add to the process of erosion and environmental perturbations. However, more emphasis can not be laid on these variables in absence of the data. The effects of global warming will be apparent on the permanent snow line and recession of glaciers, which has already been taking place at an alarming rate (Valdiya, 1988). This will again add to the geomorphological instability in the area and will cause serious problems at many places.

At higher altitudes, the CO_2 induced climatic change would lead to increasing precipitation and therefore more water will be available which will increase the runoff. In the early stages of greenhouse warming and melting of glaciers or permafrost would add an additional quantity of water to the rivers draining the high mountain areas. Therefore, the water resources would probably increase at higher altitude with CO_2 induced climatic change. However, in valleys there will be an adverse impact due to increase in the intensity of radient fog for longer period during the morning hours. The water vapours will get condensed closer to the soil surface. Infact for the last several years an increase in the morning fog density in valleys in majority of the Himalayan tracks is an indication of this process. These potential geomorphological changes are summarised in table 3.

The above account, although speculative at present, indicates that if global warming speeds up, like the biomass at higher latitudes, Himalayan ecosystem would be affected seriously like the biomes at higher latitudes.

6. THE APPROACH TO PREDICT THE FUTURE IMPACTS

Undisputed is the fact that the fast climatic change due to change in the environmental chemistry would bring about major impact on ecosystems. How quickly we can predict the possible impact on the basis of authentic data will depend on how soon we can identify the most sensitive systems to climatic change. In the arguments presented here attempts have been made to advocate that, in the Indian context, Himalaya is the most sensitive zone to changing global climate and its sensitivity increases with the altitude. Since Himalaya covers a large altitudinal range representing the climatic senario at higher altitudes of pre-industrialisation period and post industrialisation climate at lower elevations in valleys, the eco-physiology of flora and fauna of this region, therefore, offer interesting opportunities to develope predictive ecosystem models. Studies on the soil-vegetation-atmosphere analysis, water and energy fluxes, species spread and turn over, landscape analysis, agricultural & forest crop productivity under natural environmental conditions along the altitudinal gradient supported with data from laboratory experiments on CO_2-temperature-water interactions in plant developmental processes will help in formulating the desired eco-system model. Such short as well as long-term studies will also help in finding out proper bioindicators with high sensitivity to the climatic changes. The data generated from the continuous monitoring of geomorphological process along the altitudinal gradients will increase the reliability of simulated parameter values of the model. The modeling should aim at deriving the information on possible protecting and preventing majors if the climatic change takes place even at higher rates than that of today. To increase our predictive capacity for ecosystem dynamics, we also need to investigate the interaction of CO_2 and temperature in well defined systems like, nitrogen fixing plants, evergreen and deciduous plants, conifers and even the lower plants at community level. A comparative study of low and high altitude plants will be most rewarding to predict the effect on ecosystem dynamics.

BIBLIOGRAPHY

Baumgartner, A. (1980). Mountain climates from a perspective of forest growth. In: Mountain Environments and Subalpine Tree growth (U. Benecke and M.R. Davis eds.) New Zealand Forest Service, Wellington, pp. 27-39.

Boer, M.M., Koster, E.A. and Lunberg H. (1990). Greenhouse impact in Fennoscandia-Preliminary findings of a European workshop on the effects of climate change. Ambio 19 (1) 2-10.

Crutzen, P.J. and Graedel, T.E. (1986). The role of atmospheric chemistry in environment-development interactions. In: Sustainable Development of the Biosphere. (Clark W.C. and Munn, R.E. eds.) Cambridge Univ. Press, Cambridge. pp. 213-250.

Gates, W.L., Cook, K.H. and Schlesinger, M.E. (1981). Preliminary analysis of experiments on the climatic effects of increased CO_2 with an atmospheric general circulation model and a climatological ocean. J. Geophys. Sc. 86, 6385-6393.

Genthon, C.; Barnola, J.M.; Raynaud, D.; Lorius, C.; Jouzel, J.; Barkov, N.I.; Koratkevids, Y.S., and Kotlyakov, V.M. (1987). Vostok ice core: climatic response to CO_2 and orbital forcing changes over the last climatic cycle. Nature 3290 414-118.

Korner, Ch. (1988). Does global increase of CO_2 alter stomatal density? Flora 181: 253-257.

Lovelock, J.E. (1971). Air pollution and climatic change. Atmos. Envir. 5: 403-411.

Plass, G.N. (1956). The Carbon dioxide theory of climatic changes. Tellus, 8: 140-153.

Posner, H.B. (1971). Inhibitory effect of carbohydrate on flowering of Lamna perpusilla. III. Effects of respiratory intermediates, amino acids and CO_2, glucose-phosphate dehydro- genase. Plant Physiol. 48:361-365.

Purohit A.N. and Tregunna, E.B. (1974). Effects of Carbon dioxide on Pharbitis, Xanthium and Silene in short days. Can. J. Bot. 52: 1283-1291.

Purohit, A.N. and Tregunna, E.B. Ragethi, H.W.J. (1975). CO_2 effects on local-lesion production by Tobacco-mosaic virus and Turnip mosic virus. Virology 65: 558-564.

Purohit, A.N. and Tregunna, E.B. (1976). Effects of Carbon dioxide on the growth of Douglasfir seedlings. Indian J. Plant Physiol. 19: 164-170.

Ramanathan, V. (1988). The greenhouse theory of climate

change: a test by an inadvertent global experiment. Science 240: 293-299.

Valdiya, K.S. (1988). Dwindling water resources of India and their conservation. Proc. Nat. Acad. Sci. India, Annual Number, pp. 37-55.

Warrick, R.A., Shugart, H.H., Antonovsky, M.J., Tarrant, J.R. and Tucker, C.J. (1986). The effect of increased CO_2 and climatic change on terrestrial ecosystems. In: The Greenhouse Effect, Climatic Change and Ecosystems, (Bolin, B., Doos, B.R., Jager, J. and Warrick. R.A-eds) SCOPE 29: 363-382. John Wiley & Sons, Chichester.

Wittwer, S.H. (1967). Carbon dioxide and its role in plant growth Proc. XIII Int. Hort. Congr. 3: 311-322.

Table 1. **EFFECTS OF ALTITUDE ON AIR-PRESSURE, AIR-DENSITY, AND CONTENT OF OXYGEN AND CARBON DIOXIDE**

The figures are shown as percentage at sea level values

		Altitude (m a.s.l.)						
		0	500	1000	2000	3000	4000	5000
Air-pressure	(%)	100	94	89	78	69	60	51
Air-density	(%)	100	95	89	82	75	66	60
Content of oxygen	(%)	100	95	89	82	75	66	60
Content of carbon dioxide	(%)	100	95	89	82	75	66	60

CONTENT OF WATER VAPOUR WITHIN THE FREE ATMOSPHERE IN RELATION TO AIR TEMPERATURE ALTITUDE

(asuming a relative humidity of 80%, a summer temperature lapse rate of -5°C/1000m, and a winter lapse rate of -4°C/1000m. Calculations per a standard atmosphere)

		Altitude (m a.s.l.)						
		0	500	1000	2000	3000	4000	5000
					Summer			
Air-temperature	(°C)	20	17.5	15	10	5	0	-5
Water vapour	(g/m^3)	14	12	10	7	6	4	1.5
					Winter			
Air-temperature	(°C)	0	-2	-4	-8	-12	-16	-20
Water vapour	(g/m^3)	4	3.5	2.5	1.5	1.2	1	0.6

Table 2. POTENTIAL ASSOCIATED PHYSICAL CHANGES AT HIGHER ALTITUDES IN HIMALAYAN REGION DUE TO TWO-FOLD INCREASE IN THE ATMOSPHERIC CO_2 CONTENTS

	Parameter	Positive Change
1.	Annual temperature	+
2.	Winter temperature	+
3.	Summer temperature	+
4.	Annual precipitation	+

The intensity of change will increase with altitude.

Table 3. POTENTIAL ASSOCIATED GEOMORPHOLOGICAL CHANGES AT HIGHER ALTITUDE IN HIMALAYAN REGION DUE TO TWO-FOLD INCREASE IN THE ATMOSPHERIC CO_2 CONTENTS

Parameter	Change (+ Positive: - Negative)
1. Frost weathering	+
2. Mass movements	+
3. Erosion process	+
4. Permafrost	-
5. Cloud cover and cloud brust	+

The intensity of change will increase with altitude.

PRIMARY PRODUCTIVITY OF INDIAN WETLANDS AND THEIR FUTURE SCENARIO

R.S. Ambasht & N.K. Srivastava

Department of Botany, Banaras Hindu University, Varanasi-221 005, India.

ABSTRACT

Wetlands are wide ranging natural and man-made permanent, temporary or seasonal water bodies or water saturated lands, lentic or lotic, fresh, brackish or saline such as lakes, ponds, many paddy fields, rivers and flood plains, impoundments, streams, peats, bogs, marshes, swamps, marine tidal belts and mangroves, deltas, estuaries and brackish back waters. They have profound ecological and economic importance. Mostly they are highly productive but ecologically fragile, liable to degradation and degeneration under the prevailing anthropogenic pressures and climatic changes due to dumping of solid and liquid wastes, toxic material, runoff inputs of pesticides and fertilizers, silt from degrading watersheds, dam constructions and overexploitation. Standing crop biomass and net production rates in Indian wetlands have been measured by a number of University botanists. The studied wetlands range from cold to very cold Himalayan lakes like Dal, Anchar, Manasbal, Surinsar, Nilnag in Kashmir to strongly seasonal cold winter and hot summer North Indian plains around Varanasi (ponds and river corridors), Jaunpur (Gujar Lake), Ballia (Surha Lake), Gorakhpur (Ramgarh and other lakes), Bhagalpur (Ganga River), Udaipur, to warm belts of Madhya Pradesh (Doodhadhari Lake) and Andhra (Kolleru Lake). Maximum standing crop dry biomass and annual net production is in emergent and marsh zone vegetation of <u>Eleocharis</u>, <u>Cyperus</u>, wild rice and cultivated floating rice (upto 10-40 t ha^{-1} yr^{-1}) followed by prolific growth of the ubiquitous <u>Eichhornia crassipes</u>. On the banks of rivers (R. Gomati, R. Ganga and R. Rihand) and Rihand reservoir, we have estimated the quantities of soil erosion, water runoff, nutrient loss against simulated rainfall and it is found that destruction of vegetation increases the intensity of above forces tremendously. Some of the selected riparian plants were grown in culture to quantify the conservation value of component species.

Unless the watershed vegetation is protected, application and runoff of pesticides and fertilizers in catchment are regulated, effluents are adequately pre-treated and judicious management steps are taken, the future scenario of Indian wetlands are too bleak.

INTRODUCTION

Wetlands are wide ranging natural and man-made permanent, temporary or seasonal shallow waterbodies or water saturated lands. They are transitional areas between the terrestrial and deep aquatic ecosystems. They are stationary (lentic) or running (lotic), fresh, brackish or saline. Lakes, ponds, many paddy fields, rivers, flood plains, impoundments, streams, peats, bogs, marshes, mangroves, swamps, estuaries and brackish back waters are the common examples. There is considerable difference of opinion as to how deep or shallow aquatic bodies should be included in the wetlands. Ramsar convention agrees to a maximum of 6 metres depth during low-tides for coastal wetlands. They have defined wetlands as "Areas of marsh, fen, peatland or water whether natural or artificial, permanent or temporary with water that is static or flowing, fresh, brackish or salt including areas of marine water the depth of which at low-tides does not exceed 6 metres". Wetlands are characterized by an excess of water with undrained hydric soil covered with hydrophytes and even in temporary bodies the hydrophase is an essential feature. The World Conservation Union (IUCN-Dugan 1990) has made a comprehensive classification of the world wetlands. The saltwater wetlands are (1) marine, (2) estuarine, (3) lagoonar and (4) salt lakes. The marine and estuarine have characteristic subtidal and intertidal wetlands. The fresh water wetlands are essentially of three types (1) riverine which may be permanent rivers, streams, waterfalls and inland deltas, and temporary, seasonal and irregular streams and rivers, riverine flood plains, seasonally flooded grasslands and flat river basin. (2) Lacustrine are permanent freshwater lakes of usually more than 8 ha size with seasonal or irregular inundation of the shore area and the permanent freshwater ponds of even smaller sizes. (3) Palustrine wetlands are seasonal freshwater lakes including the flood plain lakes and permanent freshwater marshes and swamps, peatlands and seasonally flooded meadows. There are extensive forested shrub swamps in the river basin and low lying flat plains. Besides the above natural wetlands there are vast varieties of man-made wetlands for aquaculture, fish and shrimp culture, farm ponds, temple and village tanks, irrigation channels, canals, embanked rice fields, salt pans or salines and excavation pits around mining areas, villages, brick kilns, etc. There are water storage dams which have extensive marginal wetlands. The important functions of the wetlands are ground water recharge and discharge, flood control, shore line stabilization and erosion control, retention of nutrients, sediments and toxicants, export of biomass, water transport and stabilization of microclimate. Wetlands are important sources of biomass especially from the forested and shrubby wetlands including mangroves, production of agricultural resources like paddy, makhana (*Euryale ferox*) and water chestnuts (*Trapa bispinosa*), production and culture of fishes and production of forage and paper pulp material. Above all the wetlands are among the most important storages of readily available water supply. They have profound biological diversity and are associated with cultural activities.

Water is a raw material of photosynthesis and all plants and animals essentially need it. Through recharge water from the wetland filters down into the ground water and moves laterally where from it is drawn for human consumption. The reverse process of upward movement of ground water into a wetland is known as the ground water discharge. Wetlands store the precipitation water and regulate the runoff in such a way that the onslaught of flood in the downstream region is considerably reduced. The present trend of shrinkage and filling up of wetlands is bound to cause considerable flood problem in future. The marginal wetland vegetation is efficient stabilizer of the shore lines. Plants have high soil binding capacity and they retard water, nutrient and soil runoff. Wetlands serve as pools for settling of the toxicants and some wetland plants are adapted to absorb them in high concentration as a result of which the downstream rivers and other water storages remain relatively free from pollutants. Thus they act as an efficient harvesters of toxicants. In fact, in modern water treatment units there is an important use of plants in the filtration process. Nutrient retention is an important function of wetlands especially the vegetation accumulates nitrogen and phosphorus which prevents eutrophication. Wetlands also serve as sinks for wastes. The toxicants are slowly broken down. Excess nitrates are converted into gaseous nitrogen through denitrification and circulated to the atmosphere. Wastes dumped in the wetlands up to a certain extent act as nutrient for rapid and excessive production of plants that support rich growth of fish and shrimps. Thus wetlands while acting as sink act as exporter of valuable foods.

Primary Productivity

Global

The continents are spread over 149 x 10^6 km^2 of the total earth surface area of 510 x 10^6 km^2. Golley and

Lieth (1972) have given that the freshwater bodies occupy only about 4 x 10^6 km^2 of which the swamps and marshes occupy 2 x 10^6 km^2 and lakes and streams also 2 x 10^6 km^2 (Table 1). Coastal estuaries and reefs occupy 2 x 10^6, upwelling zone 0.4 x 10^6 and the continental shelf 26.6 x 10^6 km^2. Swamps and marshes are estimated to account for net productivity of 4.0 x 10^9 t yr^{-1} at an approx. average of 2000 g m^{-2} yr^{-1} in the range of 800-4000 g m^{-2} yr^{-1}. Lakes and streams have a much lower range of 100-1500 g m^{-2} yr^{-1} and a total of 1.0 x 10^9 tons yr^{-1} is produced at the average rate of 500 g m^{-2} yr^{-1}. Reefs and estuaries are estimated to have a net production of 4 x 10^9 tons yr^{-1} and the continental shelf upwelling coastal zones together for 9.5 x 10^9 tons yr^{-1}. It may be of interest to note that the total continental net primary production including agriculture is about 100.2 x 10^9 tons and oceanic as 55.0 x 10^9 tons yr^{-1} i.e. the global net primary production is approximately 155.2 x 10^9 tons yr^{-1}. All these production values are in terms of oven-dry weight.

Primary productivity refers to the amount of carbon fixation by the autotrophic plants. Part of the primary production is used in respiration. The difference between gross production and respiration is referred as net primary production. The net accumulation of biomass over a unit area and a unit time is referred as net primary productivity. It is expressed in terms of oven dry biomass or the energy content. In wetlands we come across a variety of life forms but principally they are divided into two categories - the phytoplankton and the macrophytes. In phytoplankton the primary production is expressed in terms mg C m^{-3} hr^{-1} and measured by light and dark bottle method or by using radioactive carbon (^{14}C). Macrophyte productivity is measured by harvest method where the positive difference of biomass in successive sampling periods over one year is added to obtain the net productivity in terms of g m^{-2} yr^{-1} or tons ha^{-1} yr^{-1}. The summation of the above two life forms gives the primary productivity of the ecosystems.

In such estimates usually the values provide information on the available net production and not the total photosynthesis. Not only the loss on account of respiration is there but a sizeable portion is transferred to the consumers i.e. animals through grazing.

As would be evident from following review of productivity studies, the phytoplankton productivity is usually for the assemblage of all categories of algal flora found suspended in water while for macrophytes the data are available for different vegetational units like the marsh zone, submerged zone, submerged with floating leaf zone or with free floating zones. They are also sometimes described for the individual dominant species.

Phytoplankton

Phytoplankton consists of a variety of algae and the main represented classes are Chlorophytes, Bacillariophytes and Cyanophytes. In the industrial belt of Obra-Renukoot region in India, the polluted wetlands has the dominance of Bacillariophytes and in the unpolluted of Chlorophytes. In a case study of upstream and downstream parts in respect to the effluent discharge point it is found that the net available primary production is decreased from 375 g C m^{-2} yr^{-1} in the upstream at Renukoot to 136 g C m^{-2} yr^{-1} in the downstream part (Table 2) (Srivastava and Ambasht 1990). In the Rihand river at Obra it decreased from 169 in upstream to 89 g C m^{-2} yr^{-1} in the downstream region. In the temperate conditions of Kashmir, Kaul (1971) has given the range of phytoplankton production from 195 mg C m^{-2} day^{-1} in winter season in Dal and Manasbal lakes to 600 mg C m^{-2} day^{-1} in spring in Manasbal. But the production is much lower in the shallow lake basins from 35 mg C m^{-2} day^{-1} in summer to 84 mg C m^{-2} day^{-1} in autumn.

Macrophytes

In contrast to the microscopic phytoplankton, the larger aquatic and marsh zone plants mostly angiosperms some aquatic pteridophytes and big sized algae are called as macrophytes. Depending upon their life form and position in the wetland ecosystem, they are broadly classified into: (i) marsh zone emergent plants, (ii) floating rooted, (iii) free floating, and (iv) submerged plants. Marsh zone vegetation is mostly dominated by emergent species in which the lower rhizomatous or cormous undermud part and a few centimetres of shoot are covered with water and the rest emerge out aerially. The common examples are Eleocharis spp., Typha sp., Polygonum amphibium, Cyperus spp., Scirpus spp., Oryza rufipogon. Among the floating rooted plants the common examples are Nelumbo nucifera, Nymphea stellata, Trapa bispinosa. The free floating examples are Pistia stratioitis, Lemna minor, Echhornia crassipes, Azolla sp. and Salvinia molesta. The examples of the submerged plants are Hydrilla verticillata, Najas graminea, Potamogeton sp., Aponogeton sp., etc.

Primary productivity in macrophytes is usually expressed for the net gain in weight over a period of time in a given area and the values are expressed on oven dry weight basis for the entire plant body, whereas the agricultural production is usually referred on air dry weight basis of the usable harvested part. In India, primary productivity in the wetland has been measured by different workers in different climatic belts. Kaul (1971), Kaul & Zutshi (1979), Zutshi et al. (1980) and Zutshi & Vass (1982) have worked on Kashmir lakes. They have reported an organic production of 27 ton ha^{-1} yr^{-1} for emergent zones, 8.8 tons ha^{-1} yr^{-1} in floating zone and 4.6 tons ha^{-1} yr^{-1} in the submerged zone of Dal lake. In the Gangetic plains there are primary production measurements in the wetlands around Varanasi, Jaunpur and Gorakhpur. There are some data available from Madhya Pradesh, Rajasthan (Table 3).

Srivastava (1989) has recorded 945 g m^{-2} yr^{-1} production in the upstream and 896 g m^{-2} yr^{-1} in the downstream segment of the submerged zone of river Ganga at Varanasi (Table 3). Verma (1979) has recorded 2021 g m^{-2} yr^{-1} production in the emergent zone, 874 g m^{-2} yr^{-1}

in the floating zone and 620 g m^{-2} yr^{-1} in the submerged zone of Gujar lake, Jaunpur. In Gorakhpur region, Ramgarh lake has been extensively worked by Sahai and his student Sinha (1969) who have reported the productivity of 930, 103 and 105 g m^{-2} yr^{-1} in the emergent, floating and submerged zones, respectively. Jain and Vyas (1980) have found the average productivity of 654 g m^{-2} yr^{-1} in the dry bank community of a pond in Udaipur. Prakash (1982) has found a poor production of only 228 g m^{-2} yr^{-1} in the natural vegetation of Yamuna River bank community at Agra. Singh (1984) has reported 476.9 g m^{-2} yr^{-1} production on the bank of Gomati river at Jaunpur. Ambasht (1988) has found that on Ganga river bank on the upstream side of Varanasi, the net production in winter season was 900 g m^{-2} yr^{-1} and in summer 100 to 200 g m^{-2} yr^{-1}. In the downstream region at Varanasi the winter value was 900 to 1100 and in summer 100-250 g m^{-2} yr^{-1}. In the Ganga river water the aquatic macrophyte productivity varied widely for different species. Potamogeton crispus and P. pectinatus together accounted for the highest value of 1080 g m^{-2} yr^{-1} (Ambasht 1988).

Ambasht et al. (1983) have briefly reviewed the work on aquatic productivity in India. It is found that invariably the marsh zone productivity is strikingly high as it combines the best of the aquatic and terrestrial conditions. This zone attracts large population of migratory birds during winter when the wetland zone is most extensive and productive.

Wetlands river corridors, flood plains and watershed lands are facing extreme anthropogenic pressures, particularly in respect to degradation of vegetation and increased soil erosion, water and nutrient runoff. In view of the important roles as described in the introductory part of this paper it is necessary to evaluate the role of native and planted vegetation in the flanking zones of wetlands and of the watersheds. We have evaluated the accelerated movement of water, soil particles and nutrients under the protection of some common watershed vegetation in relation to denuded condition.

On the banks of Rihand river and Pant Sagar reservoir at Renukoot, India ,we have estimated the quantities of erosion of soil, water runoff, nutrient loss against simulated rainfall and it is found that destruction of vegetation increases the intensity of above forces tremendously. Some of the selected riparian plants were grown in culture to quantify the conservation value of component species. The soil conservation value of five native species is found to vary widely from 33% in *Sida acuta*, 45% in *Parthenium hysterophorus* to 57% in *Ageratum conyzoides*, 69% in *Cassia tora* and 84% in *Leonotis nepaetifolia* (Table 4). The water conservation value varied in the same order among the species but it ranged between 19-60%. Among nutrients the overall nutrient conservation against soil erosion and water runoff ranged from 41 to 53% for total phosphorus and from 26% to 63% for total nitrogen. Ambasht (1988) has estimated that the soil and water conservation percentage values of *Xanthium strumarium* (a common wetland marginal plant) are respectively 58.4% and 42.7%. The corresponding values of a fast growing nitrogen fixing tree *Leucaena leucocephala* recommended for embankment plantation was 62.7% and 53.3%. The common native hardy and soil binder grass on river banks, *Saccharum benghalensis* has 95% soil conservation values (Ambasht 1963). Ambasht et al. (1984) have found that the common riparian weeds *Cynodon dactylon* has 96% soil Cv and 74% water Cv, *Phyla nodiflora* has 94% and 73% soil and water Cvs respectively. Other weeds have less conserving efficiencies (Table 4).

Future Climatic Changes Especially in Context of Wetlands

It is now well realized that the life sustaining environment of the earth is on the threshold of a change at global scale largely due to the anthropogenic activities, and unfortuntely the change in most cases is ecologically not desirable. Realizing this, the

International Council of Scientific Unions (ICSU) has started the International Geosphere-Biosphere Programme (IGBP) and World Meteorological Organization (WMO) has begun the World Climate Research Programme (WCRP) to monitor the kind and extent of climatic changes. The WCRP addresses more on the physical aspects of the climate system while IGBP on the biogeochemical aspects of the global change of the environment.

The changes taking place are in the atmospheric gaseous content particularly CO_2, CH_4, N_2O and volatile organic compounds (VOC) and they play key role in atmospheric chemistry particularly by affecting the tropospheric and stratospheric concentration of ozone and production of OH radicals. There is a definite evidence of rise in CO_2 concentration from 280 ppm to 350 ppm and of methane from 0.7 ppm to 1.7 ppm during the last 200 years or so. The current per year rate of increment of these gases is alarming. Among the industrially produced gases, the CFCs with a long residence time in the atmosphere are most important. These gases such as CO_2, CH_4, N_2O CFCs and water vapour are together called as greenhouse gases because of their ability to absorb infra-red heat waves causing a global warming of the atmosphere.

Thus the main kinds of global changes are rise in the concentration of greenhouse gases, a warming up of the atmosphere, depletion in the thickness of ozone layer and incidence of solar ultra violet radiations in the biosphere.

Atmosphere has an estimated 711 x 10^9 tons of carbon (calculated from an average 0.0335%) of which the annual exchange rate between atmosphere and terrestrial biosphere is 56 x 10^9 t yr^{-1}. Besides this an additional 5 x 10^9 t are added every year through combustion of fossil fuel and this rate is on increasing trend. The problem needs a two pronged attack, (i) of maximizing photosynthetic process by additional green cover and increased cultivation of C_4 plants, and (ii) a reduction in fossil fuel combustion by developing alternative less pollutive technology in power generation and energy source for conventional transports. Oceans are the largest store-house of carbon containing 580 x 10^9 t in the surface layer and above 38,400 x 10^9 t in the middle and lower depths. The estimated exchange rate between ocean surface water and atmosphere is 90 x 10^9 t ha^{-1} yr^{-1} which is rather fast and fairly balanced. The overall storage of carbon in fossil fuels and shales is about 12000 x 10^9 t, of which 7500 x 10^9 t is ultimately recoverable. In estuarine marsh vegetation, Curtis, Drake and Whigham (1989) and Curtis, Drake and Leadley (1989) have worked on the effect of ambient (350$\pm$22 µl l^{-1}) to elevated (686$\pm$30 µl l^{-1}) CO_2 concentration in open top chambers over C_3 Scirpus olenyi and C_4 Spartina patens. Elevated CO_2 increased shoot density, delayed senescence and increased biomass in Scirpus but there was no effect on Spartina. C:N ratio was also studied. While C percentage did not change in Scirpus leaves, the nitrogen content decreased by as much as 40%. Bazzaz (1990) has concluded that increased CO_2 content in the air surrounding plants enhance photosynthesis and growth, increase allocation to underground parts and the water use efficiency is increased. But at community level the response may not be as evident as in certain individual species.

Methane gas is produced from the anaerobic situations in wetlands, rice fields, fossil fuel and biomass combustion, enteric fermentation in cattle, sheep and other animals, etc. This gas slowly rises up in the atmosphere and increases stratospheric water vapour on oxidation. The rise in water vapour is really more important source of greenhouse effect than the direct effect of methane. About 400 to 765 x 10^9 kg ha^{-1} of methane is estimated to be the annual input and there is an increase in the input rate by 1% per year. Besides the greenhouse effects some of the discharged gases to the atmosphere such as SO_2 and oxides of nitrogen combine with the atmospheric nitrogen and precipitate in form of acid rains. Besides modifying the edaphic and aquatic environment by lowering the pH the rate of primary production is decidedly decreased. Sheehan (1984) has quoted the work of Niboer et al. (1976) in which the rate of carbon fixation by Steriocolon paschale initially increases in a sulphur dioxide mixed solution but then it decreases rapidly. In open cast coal mining and through factory effluents there is a common contamination of acidic material into the wetlands which reduces their primary productivity.

CFCs refer to chloro-fluoro-carbon compounds synthesized by man for use in refrigeration and airconditioning. CFC 11 is $CFCl_3$ and CFC 12 is CF_2Cl_2 or Freon. These are highly stable and non-destructive gases which move from the biosphere to the upper layers of atmosphere very slowly and there they are broken by the ultraviolet radiation. The released chlorine atom acts upon ozone and produces oxygen and ClO ($Cl+O_3 = ClO+O_2$). The ClO in turn acts upon another ozone molecule to produce oxygen and chlorine ($ClO+O_3 = 2O_2+Cl$). Thus one atom of chlorine released by the action of ultraviolet radiation of CFC breaks two molecules of ozone into 3 molecules of oxygen and again the same chlorine atom act afresh on other ozone molecule. So in essence the chlorine released from the CFC acts as a catalyst and in a chain breaks down millions of molecule of O_3 before it finally precipitates in form of hydrochloric acid.

Future climatic changes are bound to alter the biosphere in variety of ways for instance a greater evapotranspiration, an increase in rainfall, increase in melting of snow, rise in the ocean level, submergence of coastal lands, mass scale migration of flora due to alteration in the climatic belts, change in forest structure, agricultural operation, incidence of disease and pests, genetic changes in plant and animal life due to U.V. radiation, readjustment in the trophic structure of the ecosystem etc. There can be a number of unforeseen direct and indirect effects we would need to develop strains of crops adjusted to the modified drought and disease conditions.

REFERENCES

1. Ambasht, R.S. (1963). Ecological problems in the tropics - erosion in relation to vegetation cover in two areas of Varanasi. Proc. Nat. Acad. Sci. (India) 33: 158-162.

2. Ambasht, R.S. (1988). Community structure, biomass production and energy aspects of aquatic macrophytes and soil, water and nutrient conservation of the Ganga. In: The Ganga - A Scientific Study., Krishnamurty, C.R., Bilgrami, K.S., Das, T.M. and Mathur, R.P. eds.

3. Ambasht, R.S., Singh, M.P. and Sharma, E. (1983). Changes with plant age in fractionated phytomass, productivity and chlorophyll density of crops on a riparian agro-ecosystem. Proc. IX APWSS Conference, Manila, Phillippines, pp. 491-501.

4. Ambasht, R.S., Singh, M.P. and Sharma, E. (1984). Soil water and nutrient conservation by certain riparian herbs. J. Environ. Manage. 18: 99-104.

5. Bazzaz, F.A. (1990). The response of natural ecosystems to the rising global CO_2 levels. Ann. Rev. Ecol. Syst. 21: 167-196.

6. Curtis, P.D., Drake, B.G., Leadeley, P.W. and Whigham, D.F. (1989). Growth and senescence of plant communities exposed to elevated CO_2 concentration on an estuarine marsh. Oecologia 78: 20-26.

7. Curtis, P.D., Drake, B.G. and Whigham, D.F. (1989). Nitrogen and carbon dynamics in C_3 and C_4 estuarine marsh plants grown under elevated CO_2 in situ. Oecologia 78: 297-301.

8. Golley, F.B. and Lieth, H. (1972). The bases of tropical production. In: Tropical Ecology, Golley, F. B. & Golley, P., eds., p. 1-26.

9. Jain, S.L. and Vyas, L.N. (1980). Dry matter transfer relationship in dry bank community of Govardhan Vilas

Lake, Udaipur. Inst. Indian Conf. Ecol. Environ. Sci., Jaipur. Abstract: 32.

10. Kaul, V. (1971). Production and ecology of some macrophytes of Kashmir lakes. Hidrobiologia 12: 63-69.

11. Kumar, R., Ambasht, R.S. and Srivastava, N.K. (1990 communicated). Nitrogen conservation efficiency of five riparian weeds in runoff experiments on slopes.

12. Lieth, H. (1975). Productivity of major vegetation units of the world. In: Primary Productivity of Biosphere, Lieth, H. and Whittaker, R.H. eds., Springer Verlag. pp. 119-130.

13. Prakash, V. (1982). Ecological Studies on River Bank Vegetation at Agra. Ph.D. Thesis, Agra University.

14. Sahai, R. and Sinha, A.B. (1969). Investigations on bioecology of inland waters of Gorakhpur, India. I: Bioecology of Ramgarh Lake. Hydrobiologia 34: 433-447.

15. Sheehan, P.J. (ed.) (1984). The Effect of Pollutant at Ecosystem Level. SCOPE Publishing House, Burn.

16. Singh, M.P. (1984). Ecological Investigation of Gomati River Bank Vegetation (Jaunpur). Ph.D. Thesis, Banaras Hindu University.

17. Singh, S.P. (1979). Investigation of the Role of Distillary and Domestic Wastes on Productive Potential of Freshwater of Gorakhpur. Ph.D. Thesis, Gorakhpur University.

18. Sinha, A.B. (1969). Investigation on the Ecology of Ramgarh Lake, Gorakhpur. Ph.D. Thesis, Gorakhpur University.

19. Srivastava, A.K. (1989). Primary Productivity and Nutrient Study of Aquatic Macrophytes of River Ganga with Respect to Water Quality at Varanasi. Ph.D. Thesis, Banaras Hindu University.

20. Srivastava, N.K. (1990). Primary Productivity and Nutrient Studies of Receiving Waters in Industrially Polluted Obra-Renukoot Region. Ph.D. Thesis, Banaras Hindu University.

21. Srivastava, N.K. and R.S. Ambasht. (1990). Impact of industrial effluent on limnology of Pant Sagar and Rihand river. In: Environmental Degradation of Obra-Renukoot-Singrauli Area and its Impact on Natural and Derived Ecosystems, Singh, J.S., Singh, K.P. and Agrawal, eds., Final Tech. Report, DOEn. pp. 265-284.

22. Srivastava, V.C. (1983). The Limnology, Primary Productivity and Energetics of Chilwa Lake, Gorakhpur. Ph.D. Thesis, Gorakhpur University.

23. Verma, K.R. (1979). Phytosociology, Productivity and Energetics of Macrophytes of Gujar Lake (Kheta Sarai) Jaunpur. Ph.D. Thesis, Banaras Hindu University.

24. Zutshi, D.P., Subla, B.A., Khan, M.A. and Wangneo (1980). Comparative limnology of nine lakes of Jammu and Kashmir Himalaya. Hydrobiologia 72: 101-112.

25. Zutshi, D.P. and Vass, K.K. (1982). Limnological studies on Dal Lake, Srinagar III. Biological features. Proc. Ind. Nat. Sci. Acad. 48: 234-241.

Table 1. Global land area, major biomes, wetland area with average production rate.

	Area (km^2)	Average Annual Production
Global surface area	($510x10^6$)	$16.4x10^{10}$ t yr^{-1}
Continental	($149x10^6$)	$100.2x10^9$ t yr^{-1} (Av. 730 g m^{-2} yr^{-1})
Oceans	($361x10^6$)	$5.5x10^{10}$ t yr^{-1} (Av. 155 g m^{-2} yr^{-1})
Tropical forests	(13% land area)	$4x10^{10}$ t yr^{-1} (Range 1-5 kg ha^{-1} yr^{-1})
Temperate forests	(12% land area)	$2.34x10^{10}$ t yr^{-1} (R=0.6-2.5 t ha^{-1} yr^{-1})
Tropical grasslands	($15x10^6$ km^2)	$1.05x10^{10}$ t yr^{-1}
Temperate grasslands	($9x10^6$ km^2)	$0.45x10^6$ t yr^{-1}
Deserts (ice and dry sand)	($155x10^6$)	3.0 g m^{-2} yr^{-1} (negligible)
Cultivated agricultural lands	($140x10^6$)	$1.0x10^{10}$ t yr^{-1}
Wetlands		
Freshwater swamps and marshes	($2x10^6$)	$4x10^9$ t yr^{-1}
Freshwater lakes and streams	($2x10^6$)	$1x10^9$ t yr^{-1}
Coastal estuaries and Reefs	($2x10^6$)	$4x10^9$ t yr^{-1}
Continental shelf	($26.6x10^6$)	$9.5x10^9$ t yr^{-1}

Table 2. Comparison of primary production in Indian freshwater ecosystems.

Study site	Primary production ($g\ C\ m^{-2}\ yr^{-1}$)	References
Renukoot site (G.B. Pant Sagar reservoir)		
Upstream	375	Srivastava (1990)
Downstream	136	
Obra site (Rihand river)		
Upstream	169	
Downstream	89	
Elsewhere	130-400	
Baghela lake	374.1	Radhakrishnan (1980)
Pichhola lake	549.7	
Naranbagh lake	134	Khan and Zutshi (1979)
Nilnag lake	90-100	Zutshi et al. (1980)

Table 3. Macrophyte production.

Lake	Production ($g\ m^{-2}\ yr^{-1}$)	References
Ramgarh (Gorakhpur)		
Emergent zone	930	Sinha (1969)
Floating zone	103	
Submerged zone	105	
Chilwa lake (1970-72)		
Free floating zone	890	Srivastava (1983)
Attached floating zone	130	
Submerged zone	130	
Emergent zone	370	
Marshy zone	360	
Jalwania pond (1970-72)		
Free floating zone	410	Singh (1979)
Attached floating zone	1130	
Submerged zone	860	
Emergent zone	520	
Marshy zone	280	
Gujar lake (Jaunpur)		
Emergent zone	2021	Verma (1979)
Floating zone	874	
Submerged zone	628	
Pichhola lake (Udaipur)		
Emergent zone	46.3	Billore and Vyas (1982)
Floating	88.8	
Submerged	555.9	
Dal lake (Kashmir)		
Emergent zone	2800	Kaul (1971)
Floating zone	880	
Submerged zone	460	
Drybank community (Udaipur)	654	Jain and Vyas (1980)
Ganga river bank (Varanasi)		
Upstream Winter	900	
Summer	100-200	

Contd. ...

Table 3 contd.

Downstream	Winter	900-1100	
	Summer	100-250	
Yamuna River bank (Agra)		228.8	Prakash (1982)
Gomati River bank (Jaunpur)		476.9	Singh (1984)

Table 4. Soil and water conservation (%) of some selected riparian plants.

Name of the species	Conservation value(%)		Remarks
	Water	Soil	
1. *Xanthium strumarium*	42.7	58.4	Average of seven showering treatments (Ambasht 1988)
2. *Leucaena leucocephala*	53.3	62.7	
3. *Saccharum benghalensis*	77.2	83.4	
4. *Cynodon dactylon*	74.2	95.4	Average of three showering treatment (Ambasht et al. 1984)
5. *Phyla nodiflora*	72.9	93.8	
6. *Cyperus rotundus*	70.9	91.1	
7. *Crotalaria medicaginea*	27.2	52.8	
8. *Digitaria adscendens*	65.1	87.6	
9. *Leonotis nepetaefolia*	50.0	84.0	Average of three showering treatments (Kumar et al. 1990 communicated)
10. *Cassia tora*	34.0	84.0	
11. *Ageratum conyzoides*	44.0	57.0	
12. *Sida acuta*	19.0	33.0	
13. *Parthenium hysterophorus*	25.0	45.0	

MEASUREMENT OF GREENHOUSE GAS EMISSIONS IN INDIA

D.C.PARASHAR, PRABHAT K.GUPTA, J.RAI, R.C.SHARMA, N.SINGH and B.M.REDDY

National Physical Laboratory,
Dr. K.S. Krishnan Road, New Delhi-110 012.

Measurements of methane and other green house gases carried out in India to estimate Indian contribution to the global budget are reported. Methane emission from rice paddyfields grown under different geographical environmental, soil, pH parameters and having different paddy varieties has been studied at New Delhi, Karnal, Dehradun, Hyderabad and West Bengal. Effects of different parameters on methane emission is discussed. A first order estimate indicates a contribution of 3 to 9 tg yr^{-1} of methane from Indian rice paddy fields and is only 6% of the global methane budget due to rice cultivation. Methane emissions from sources other than paddy fields are also discussed. Results obtained on nitrous Oxide emissions show significant efflux from grass land and waste land in India.

INTRODUCTION

Like carbon dioxide, methane andnitrous oxide are important green house gases. High atmospheric concentration of carbon dioxide already prevailing leads to lower increase in green house warming due to additional CO_2 loading of the atmosphere. The situation with CH_4 & N_2O is different and each additional mole of CH_4 and N_2O added to atmosphere would have orders of magnitude higher green house effect (1) than a mole of carbon dioxide. Methane and nitrous oxide are also chemically active and lead to increase in water vapour and nitric oxide respectively which are very important in ozone layer chemistry. Recent discovery that methane and nitrous oxide are increasing in the atmosphere at a rate of 1% and 0.25% per year respectively has led to renewed interst in the global cycles of carbon and nitrogen and particularly their source strengths. The major sources of methane are rice paddy fields, natural wet lands and enteric fermentation and the minor ones include biomas burning, termites, gas drilling, land fills and a host of others whereas nitrous oxide is mainly emitted by soils of tropical forests and oceans and to some extent by fertilizers, combustion and biomass burning. Several efforts have been made in Europe to evaluate the contributions of different sources of methane budget (2 to 6) and nitrous oxide budget (7 to 10) in the atmosphere. Recently Chinese rice fields and biogas generators have been studied (11) for trace gas emissions. The authors have published (12-14) the measurements of methane and other trace gases emitted from agricultural practices including rice cultivation and Jhum cultivation in India. In this paper the authors have attempted to evaluate the effect of various parameters like the pH of the soil, temperature of the soil, rice variety and fertilizer used on methane emission from rice paddy fields. Some results obtained on nitrous oxide emissions from forest and waste lands are also reported here.

EXPERIMENTAL DETAILS

The experimental arrangement used in this work has been reported earlier (12 - 14). The soil pH was measured in the laboratory using 30 g of soil sample duly dried and powdered and mixed with 75 ml distilled water and kept for an hour with occasional stirring. The temperature of the soil was measured in the field 6 to 10 cm below the su rface using soil thermometer.

CH_4 concentration in field was determined using an AID (USA) model 511 portable gas chromatograph which was calibrated periodically using 2-100 ppm methane gas standards prepared from 0.5% methane in nitrogen standard obtained from Indian Oxygen Limited, New Delhi by the standard dilution technique. 10′x1/8" s.s. column filled with molecular sieve 5A maintained at 80°C with carrier nitrogen gas at flow rate of 20 c.c.min^{-1} and flame ionization detector were used for methane determination. Simultaneous air samples collected in tedlar air sampling bags (Aerovironment Inc.Monrovia Ca. USA)were brought to NPL New Delhi and analyzed for methane and nitrous oxide using a microprocessor controlled Perkin Elmer model Sigma 2000 gas chromatograph coupled with data station and printer. Nitrous oxide concentration was measured by using two 6′ x 1/8" s.s. columns connected together and packed with 80-100 mesh porapaq Q and porapaq R respectively and electron capture detector. The detector and columns were maintained at 400°C and 50°C respectively and 10% methane in Argon was used as carrier at 30 c.c.min^{-1} flow rate. Nitrous oxide standards were prepared by static dilution technique using 1.0% nitrous oxide in nitrogen standard obtained from EDT U.K. The maximum variation in nitrous oxide efflux value is less than ± 10%. 1 ppm methane in nitrogen standard obtained from Matheson Gas Products USA wasused for calibration as primary standard. The value of methane mixing ratio determined in the field using portable gas chromatograph and those determined in the laboratory agreed within 1.5%. The maximum variation in methane efflux values reported here are less than ± 4%.

RESULTS AND DISCUSSION

Table 1 shows the area distribution in terms of topo-sequence for Kharif crop (July to September) which is the main season for rice paddy in India (15). It is seen from the table that the major Kharif crop is produced from the low-lying ill-drained, irrigated and upland parts of the country. In addition about 5% of the

area is cultivated for rice in the Rabi (January to March) season also. The actual areas under various catagories (Table 1) are variable but the total area under rice cultivation is around 41 m.ha.

TABLE 1

Rice Area Distribution in terms of Topo-sequence

Situation	Area(m.ha)	Percentage of area
Irrigated area	11.0	27
Low lying ill-drained	16.4	40
Upland	8.6	21
Deep water	2.5	6
Arid high temperature	2.0	5
Long days of low temperature	0.41	1

Methane emission study from rice paddy has been carried out in four typical locations at New Delhi (22-28 June 1989` and 25-28 Sept. 1989), Dehradun (6-7 Oct.1989), Karnal (13-1 Oct.1989) and Hyderabad (20-24 March 1990). The environmental and soil parameters oftheselocations are given in Table 2. It is seen from Table 2 that the soil at these locations has pH 7 to 8.8, soil temperature (23-34°C) and ambient temperature (20-39°C). Other parameters covered in Table 2 are nature of fertilizer, sulpate content and geographical parameters.

Fig.1 shows the typical insitu methane efflux measurement set-up in rice paddy fields at Central Soil & Salinity Research Institute, Karnal during October 1989 with static box covering some plants.

Table - 2: Different Locations with their environmental and soil parameters.

Place	N. Delhi (N.P.L)	Dehradun (C.S.W.C.R.T.I)	Karnal (C.S.S.R.I)	Hyderabad (D.R.R)
Dates	22-28 June 89 25-28 Sept.89	6-7 Oct. 89	13-14 Oct.89	20-24 March 90
Geographical locations	28.38N 77.12E	30.19N 78.04E	29.42N 77.02E	17.6N 78.0E
Height from sea level	218m	682m	245m	545m
Annual total rainfall(mm)	430	2120	630	1020
Ambient temperature (^{o}C)	23 to 39.6	23 to 33	20 to 28	27.5 to 39
Soil temperature (^{o}C)	27 to 32	23 to 30	23.5 to 26	23 to 34
Soil group	Old Alluvium	Alluvium	Alluvium	Red loam (Chalka)
Texture	Sandy loam	Silty clay loam	Sandy loam	Sandy clay loam
Dominant clay minerals	Illite	-	Illite	Illite,Montmorillonite
Soil pH (IS:2720(Part 26)1973)	8.8	7.0	8.3	8.2
Sulphate (SO_4^{2-})(mg/kg)	230	110	265	309
Fertilizer applied (Kg/Ha)	Normal farm yard manure	NPKZn(100,60,60,10)	N(120)	NPKZn(150,90,90,50)
Methane efflux (mg $m^{-2}hr^{-1}$)	0.19 to 2.90	0.07 to 0.13	1.47 to 57.98	1.39 to 80.00

Fig.1 - Methane efflux study in Central Soil and Salinity Research Institute paddy farms at Karnal (Haryana)

Fig.2 shows methane efflux versus pH of the soil at different paddy fields. It is seen from the figure that methane emission varies by about two orders of magnitude at some location but is observed to be maximum at pH value around 8.2. Measurement at pH 7.0 and at pH 8.8 show significantly smaller CH_4 efflux values. These results indicate that the soil pH in the range 7.5 to 8.5 may be most favourable for methanogenesis. The dominant clay mineral and texture at Delhi and Karnal are similar but much higher flux was observed at Karnal which had puddled soil and is situated at low lying planes where water level is a few feet deep. Further the puddled soil becomes impervious to water and thus making water stagnate permanently. It seems, therefore, that the texture of the soil also plays significant role in the methane emission from paddy. Fig.3 shows the average value of methane efflux for different varieties of rice paddy grown under identical environmental and soil parameters. The fertilizer applied was normal NPKZn corresponding to 150, 90, 90, and 50 Kg ha^{-1} of nitrogen as urea, phosphorus

Table 3 - Methane efflux study for Rassi (IET-1444)paddy variety with different fertilizer treatment.

Fertilizer applied (Kg/ha)	Fertilizer applied as	Flux(mg $m^{-2}hr^{-1}$)
NPK (150, 90, 90)	Urea,DAP* MOP**	3.2 to 4.0
NPK-Zn (150, 90, 90, 50)	Urea, DAP, MOP $ZnSO_4$	15.5 to 19.3
NPK-Zn with sulphur (150, 90, 90, 50)	Urea, SSP# MOP, $ZnSO_4$	1.6 to 2.1
Control (No fertilizer)	-	6.6 to 8.1
Green leaf manure	Doze equivalent to normal nitrogen doze.	80

* DAP Diammonium phosphate, ** Muriate of potash (K_2O),

SSP Single super phosphate [$CaSO_4Ca_3(PO_4)_2$]

as single super phosphate, potassium as muriate of potash and zinc as zinc sulphate respectively. It is seen from the figure that the rate of methane emission is greatly influenced by paddy variety. The effect of different fertilizers for the same variety and same soil parameters has also been studied as given in table 3. It is seen from the table that the rate of methane emission is also influenced by the kind of fertilizer applied. It is further clear from table 3 that the presence of SO_4^{-2} bearing fertilizer reduces the methane emission and the application of green leaf manure enhances the methane emission by about two orders of magnitude. The effect of different

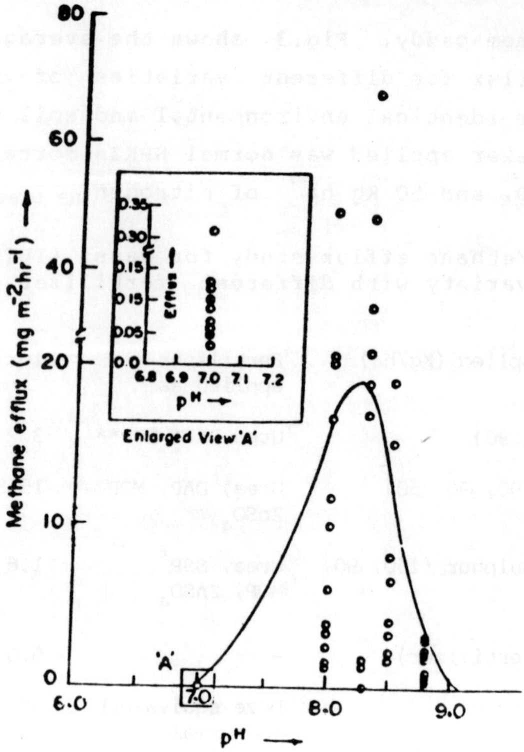

Fig 2- Variation of methane efflux with soil pH

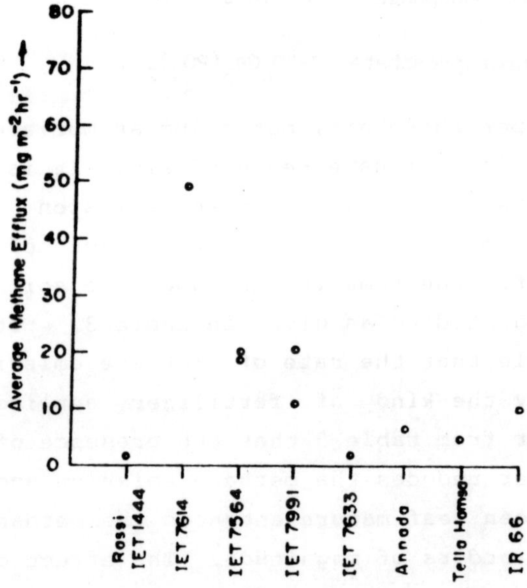

Fig 3- Variation of methane efflux with paddy variety

nitrogen fertilizers on methane emission has also been studied by Cicerone and Shetter (16). These authors have also observed low methane eflux with ammonium sulphate as fertilizer. These results are quite understandable as the presence of SO_4^{-2} is shown to inhibit methanogenesis (17).

Figs. 4a and 4b show the variation of methane efflux with soil temperature in the upland and lowland paddy fields respectively. It is seen from the figures that the rate of change of methane efflux with temperature varies between 0.01 to 14 mg $m^{-2}hr^{-1}$ from place to place. The percentage increase in methane efflux per degree rise in soil temperature ranges from 10 to 30.

METHANE BUDGET FROM INDIAN RICE PADDY

In order to estimate methane budget from India we have to consider the rice cultivation area distribution. Methane budget over India from rice cultivation would require, amongst other parameters, consideration of various soil characteristics which include texture, pH and soil mean monthly temperature. In addition to soil characteristics, the modes of cultivation in India viz. rain fed low-land (40% which includes coastal regions) and others (60%, which are mainly irrigated) (15) are also considered. Because of sea the coastal regions may be rich in salinity which is known to inhibit methane emissions (18). Total area of 41 m.ha. under Kharif cultivation and 2 m.ha. under Rabi cultivation has been broadly divided into upland (60%) and low-land (40%) rice cultivation. These areas are further subdivided into acidic and alkaline categories. The alkaline category under low-land cultivation includes coastal saline, saline and non saline regions (The measurements, at Dehradun represents acidic upland, at New Delhi represent upland alkaline, mean of New Delhi and Dehradun has been considered for lowland acidic, Hyderabad measurements represent coastal).

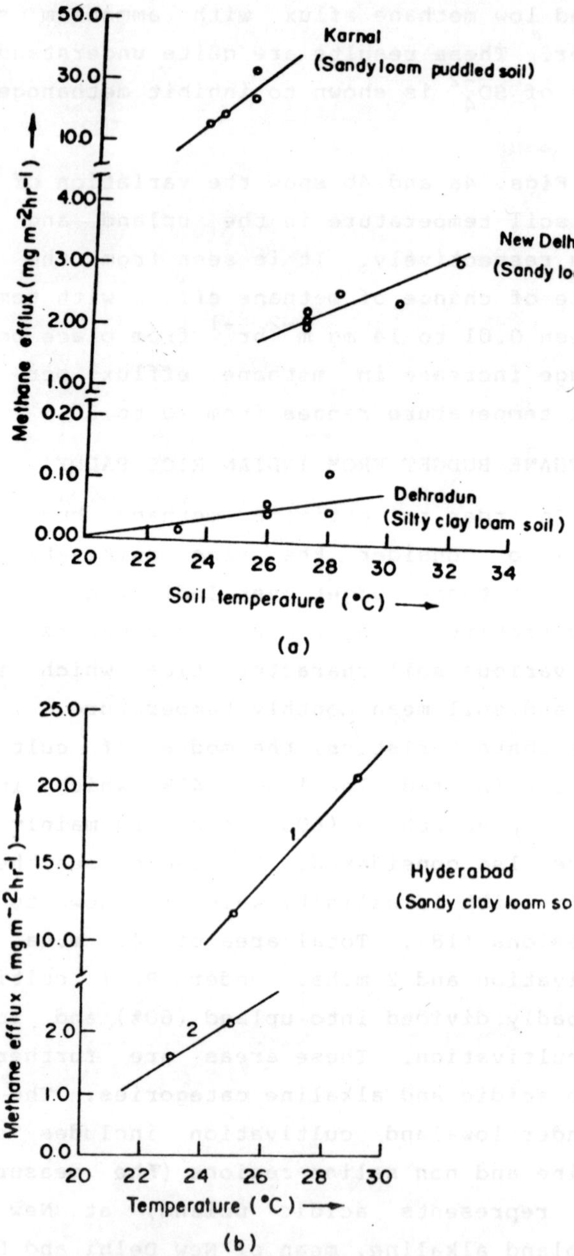

Fig. 4 – Variation of methane efflux with temperature
(a) Upland soils
(b) Lowland soil (1) Rice-rice cultivation and
(2) Rice-no rice cultivation.

The methane efflux measurements from four different locations in India have been used to calculate the methane emission from the various categories described above for rice cultivation.

Methane budget has been estimated by taking the highest emission rates as well as average emission rate. These considerations lead to methane budget estimate from Indian rice paddy to be 3 to 9 tg yr^{-1}. This is a preliminary estimate based on the data collected till now and show the Indian contribution to only 6% of the projected global methane budget due to rice cultivation. Further observations in other rice growing regions and natural wet-land in India are being done for more accurate budget estimate. The results presented in this paper suggest reconsideration of the global methane budget estimates from rice paddy fields with due consideration to parameters strongly affecting the methane efflux.

Nitrous oxide emission from soil require Aerobic conditions capable of oxidising nitrogen of the soil into nitrous oxide. Nitrous oxide emission rates observed in different plots viz. IA-containing weed (Trianthima) in soil having cowdung manure, IB-where weeds have been removed from plot IA, IC-containing weed in raw soil with no cowdung manu re, ID-where weeds have been removed from plot IC, IIA-paddy (PUSA-169) field with plant height of 120 cm, IIB-paddy (PUSA-106) field with plant height of 100 cm, IIC-paddy (PUSA-33) field with plant height of 90 cm, IID-paddy (Tai chu ng, PUSA2-21N) field with plant height of 80 cm, IIE-paddy (BASMATI-370) field with plant heights of 90 to 150 cm, plot III-grassy lawns and plot IV-forest dry land , are given in Table 4. It is seen from the Table that nitrous oxide emission varies from negative value to a maximum of 0.92 mg $m^{-2}hr^{-1}$. The negative values were observed in rice paddy fields in general and also in raw soil indicating anerobic processes resulting in nitrous oxide sinks on the ground. The variation in the strengths of sources and sinks in the

TABLE-4

Emission of Nitrous Oxide from different locations.

Plot No.	Date	Time (HRS)	N_2O Emission rate ($mg\ m^{-2}\ hr^{-1}$)
Plot IA	14.9.90	17.20	0.917
	15.9.90	12.30	0.920
	19.9.90	12.00	0.580
Plot IB	19.9.90	12.00	0.104
Plot IC	17.9.90	12.00	0.104
Plot ID	17.9.90	16.30	-0.129
	17.9.90	17.30	0.024
Plot IIA	13.9.90	12.15	0.142
Plot IIB	13.9.90	12.15	0.115
Plot IIC	13.9.90	13.00	0.125
Plot IID	13.9.90	13.00	0.061
Plot IIE	10.9.90	13.30	0.003
	12.9.90	12.00 to 17.00	-0.064 to 0.009
	14.9.90	17.00	0.019
	15.9.90	12.30	0.006
	17.9.90	10.45	0.003
Plot III	10.9.90	10.30	0.003
	14.9.90	17.20	0.015
Plot IV	17.9.90	16.45	0.019

paddy soil seem to be related with soil condition and paddy growth stage approaching the ripening of the crop. Further studies are in progress to correlate the various parameters of the paddy soil or paddy stage with the nitrous oxide source or sink strength. It is further seen from the Table that soil with Trianthima weeds (Fig. 5) show higher nitrous oxide emission. It may be

Fig.5. Nitrous Oxide flux study in soil with trianthima weeds in NPL campus, New Delhi.

mentioned that trianthima growth in alkaline soils is common in temparate-tropical climate regions and may be one of the stronger sources of nitrous oxide. Low nitrous oxide emission from grassy lawns and dry forest land indicate their low contribution to world's nitrous oxide budget. These observations are in general agreement with the results reported for dry land forests by Schmidt et al.(8). The results discussed above have been obtained in spring season and work is in progress to study the seasonal variation and other sources of nitrous oxide to be able to estimate nitrous oxidebudget from Indian sub-continent.

ACKNOWLEDGEMENTS

The authors are grateful to Dr. A.P. Mitra, DGSIR for his continued inspirations and guidance and to Dr.S.K. Joshi, Director NPL for his keen interest in this work. Discussion with Prof. S.K. Sinha, Professor, IARI, New Delhi were of great help in many ways. We would like to place on record the enthusiastic help given b y Dr.Gurmail Singh, Director, Central Soil & Water Conservation Research & Training Institute, Dehradun, Dr. R. Chhabra, Central Soil and Salinity Research Institute, Karnal and Dr. E.A. Siddiq and Dr. D.K. Kundu of D.R.R., Hyderabad during measurement in paddy fields of their respective institutes.

REFERENCES

1. Henning Rodhe (1990) A comparison of the contribution of various gases to the green house effect. Science 248: 1217-1219.

2. Whalen, S.C. and Reeburgh, W.S. (1988) A methane flux time series for tundra environments. Global Biogeochemical Cycles. 2: 399-410.

3. Cicrone, R.J. and Oremland, R. (1988) Biogeochemical aspects of atmospheric methane. Global Biogeochemical Cycles 2:299-327.

4. Bingmer, H.G. and Crutzen, P.J. (1987) The production of methane from solid waste. J.Geophys Res.92:2181-87.

5. Crutzen, P.J., Aselmann, I. and Seiler, W.S. (1986) Methane emission from domestic animals, wild ruminants, other herbivorous fauna and humans. Tellus 38B:271-284.

6. Schutz, H., Holzapfel-Pschorn, A., Conrad, R., Rennenberg, R., and Seiler, W. (1989) A three year continuos record on the influence of daytime, season and fertilizer treatment on methane emission rates from an Italian rice paddy. J.Geophys. Res. 94:16405-16416.

7. Cicerone, R.J. (1989) Analysis of sources and sinks of atmospheric nitrous oxide. J.Geophys.Res.94:18265-18271.

8. Schmidt, J., Seiler, W. and Conrad, R. (1988) Emission of nitrous oxide from temperate forest soils into the atmosphere. J. Atmospheric Chemistry 6:95-115.

9. Hao, W.M., Wofsy, S.C. McElroy, M.B., Beer, J.M. and Togan, M.A. (1987) Sources of atmospheric nitrous oxide from combustion. J.Geophys.Res. 92(D3):3098-3104.

10. Ronen, D., Mordeckai, M. and Almon, E. (1988) Contaminated aquifers are a forgotton component of the global nitrous oxide budget. Nature 335:57-59.

11. Khalil, M.A.K., Rasmussen, R.A., Wang, M.X. and Ren.L. (1990) Emisions of trace gases from Chinese rice fields and biogas generators CH_4, N_2O, CO, CO_2, chlorocarbons and hydrocarbons. Chemosphere 20:207-226.

12. Saha, A.K., Rai, J., Raman. V., Sharma R.C., Parashar D.C., Sen,S.P. and Sarkar B. (1989) Methane emission from inundated fields in a monsoon region. Ind. J. Radio & Space Physics 18:215-217.

13. Parashar, D.C., Rai, J., Gupta, Prabhat K., and Singh, N. (1990) Paramenters affecting methane emission from Indian Rice paddy fields. Ind.J. Radio and Space Physics (In Press).

14. Parashar,D.C., Rai. J., Gupta, Prabhat K., Sharma, R.C., Singh, N. and Reddy, B.M. (1990) Trace gas emissions from Jhum cultivation and rice paddy fields in India.
Proceedings of the Chapman Conference on Global Biomass Burning, Atmospheric, Climatic and Biospheric Implications. March 19-23, 1990, Williamsburg, Virginia, USA (In Press).

15. Sinha, S.K., Indian Agricultural Research Institute, New Delhi, India. Private discussions.

16. Cicerone, R.J. and Shetter, J.D. (1981) Sources of atmospheric methane: Measurement in Rice paddies and a discussion. J.Geophys.Res.86:7203-7209.

17. Martins, C.S., and Berner, R.A. (1974) Methane production in the interstitial waters of sulphate depleted marine sediments. Science 185: 1167-1169.

18. Koyama, T. (1963) Gaseous metabolism in lake sediments and paddy soils and the production of atmospheric methane and hydrogen. J. Geophys. Res. 68:3971-3973.

THE GLOBAL METHANE CYCLE

M.A.K. Khalil and R.A. Rasmussen
Center for Atmospheric Studies
Department of Environmental Science & Engineering
Oregon Graduate Institute
Beaverton, Oregon 97006 U.S.A.

Abstract

The global cycle of methane is driven by emissions around 550 tg/yr from both natural and sources related to anthropogenic activities, particularly the production of food and energy. Major sources are rice agriculture, domestic ruminants, and wetlands. Methane is removed from the atmosphere mostly by reacting with OH radicals. Some methane is removed by the soils. Over the past decade methane concentrations have been increasing at about 1%/yr or 16 ppbv/yr. A record of atmospheric methane extending back 150,000 years has been constructed from the analysis of polar ice cores. It shows that, during this time, methane concentrations have never been more than half of present levels. The recent increase of methane is probably caused by increasing emissions. Some of the increase may also be attributed to decreasing levels of OH.

1. INTRODUCTION

It has been nearly 10 years since it was conclusively established that methane concentrations are increasing in the earth's atmosphere (Rasmussen and Khalil, 1981a, 1981b). Since then, considerable progress has been made towards an understanding of the global methane cycle and the causes of the increasing trends. This paper is a review of the current understanding. We will first examine the recent global budgets and then the recent and long-term trends of atmospheric methane. Next we will review the trends of emissions and the role of human activities. Finally we will relate the trends of emissions and atmospheric concentrations to establish a coherent view of the global cycle of methane.

2. GLOBAL BUDGETS

2.1 Sources and Sinks

At least 11 complete budgets of methane have been proposed in recent years, most of them over the last 10 years (Ehhalt, 1974; Ehhalt and Schmidt, 1978; Donahue, 1979; Sheppard et al., 1982; Khalil and Rasmussen, 1983; Blake, 1984; Bolle et al., 1986; Bingemer and Crutzen, 1987; Cicerone et al., 1988; see also Warneck, 1988). A number of other studies have concentrated on specific sources and their global distributions. The published budgets are listed in Table 1 (Khalil and Rasmussen, 1990a). The main sources affected by human activities are rice fields and ruminants, mostly cattle. Other anthropogenic sources also include biomass burning, coal mining, oil and gas use, landfills, automobiles, and a variety of other even smaller sources. The main natural source is from the wetlands, with smaller

Table 1. *Estimates of methane emissions from various anthropogenic and natural sources: 1978–1988*

	Ehhalt and Schmitt (1978)		Donahue (1979)		Shep-pard et al. (1982)	Crutzen (1983)		Khalil and Ras-mussen (1983)	Blake (1984)		Seiler (1984)		Crutzen (1985)		Seiler (1986)		Bingemer and Crutzen (1987)		Cicerone and Oremland (1988)	
ruminants	100–220	160	100–220	160	90	60	60	120	70–160	115	70–100	85	60	60	70–100	85	70–80	75	65–100	80
rice paddy fields	280	280	140–280	210	39	30–60	45	95	149–189	165	30–75	53	120–200	160	70–170	120	18–91	54	60–170	110
biomass burning					60	30–110	70	25	25–110	68	50–100	75	20–70	45	55–100	78	30–100	65	50–100	55
landfills																	30–70	50	30–70	40
coal mining	8–28	18	16–50	33							30	30	34	34	35	35	35	35	25–45	35
natural gas flaring					50	20	20	40			20–30	25	33	33	30–40	35	0–35	18	25–50	45
automobiles	1	1													1–2	2				
other anthropogenic	7–21	14	110–210	160	100			40	62–100	81										
d*S*	0–50	25	0–50	25	18	0–25	13	20	0–25	13	0–30	15			0–10	5				
swamps and marshes	190–300	245	200–300	250	39	30–220	25	150	120–190	155	15–60	38	70–90	80	25–70	48	26–137	82	100–200	115
lakes	1–25	13			35			10	13	13	1–7	4			15–35	25			1–25	5
oceans	1–17	9			30			13	5–21	13									5–20	10
tropical forests					767				60–400	230										
tundra	0–3	2	3–50	27				12												
other natural						150	150	48			5–15	10					0–30	15	10–100	40
total anthropogenic	396–550	473	366–760	563	339	140–250	195	320	297–560	429	200–335	268	267–397	332	261–447	354	183–411	297	255–535	365
total natural	192–345	269	203–350	277	871	180–370	275	233	198–624	411	21–82	52	70–90	80	40–105	73	26–167	97	116–345	170
total	588–895	741	569–1110	840	1210	320–620	470	553	495–1184	840	221–417	319	337–487	412	301–552	427	209–578	394	371–880	535
F	67–61	64	64–68	67	28	44–40	41	58	60–47	51	90–80	84	79–82	81	87–81	83	87–71	75	69–61	68

Notes: The first column under each budget is the range of estimated emissions, the second column is the central value. *F* is the ratio of anthropogenic emission to total emissions.

contributions from a number of other sources, such as the tundra, lakes and rivers, the oceans, and termites.

The total emissions are about 550 tg/yr (1 tg = 10^{12} gm), of which some 60% of the emissions are from anthropogenic sources. Once methane gets into the atmosphere it has a lifetime of 8-10 years. It is removed primarily by reacting with OH radicals (~490 tg/yr), but smaller amounts (~20 tg/yr) are also removed by the soils, and even smaller amounts are removed by other chemical processes in the troposphere and stratosphere. The present imbalance between sources and sinks is about 40 tg/yr, which is observed as an increasing atmospheric trend of about 1% per year.

2.2 Present Global Distribution

The present global distribution is shown in Figure 1. Because most of the sources are land-based and many are related to human activities, concentrations of methane are higher over continents and over the northern hemisphere in general. The present global average concentration is about 1680 ppbv.

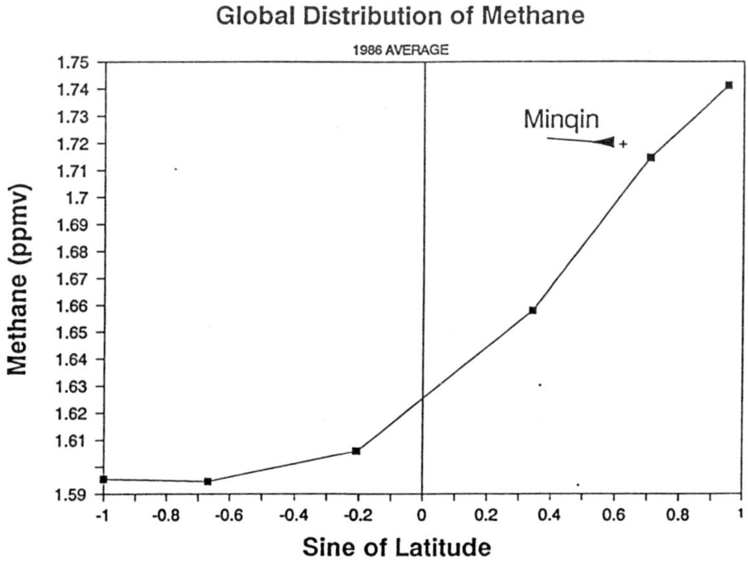

Figure 1. The latitudinal distribution of methane. Data are average values for 1986 at six long-term flask sampling stations from the arctic circle to the south pole. The annual average concentrations at Minqin in the Gansu Province of China are shown for comparison.

3. GLOBAL TRENDS

3.1 Trends of Concentrations

The global atmospheric trends are summarized in Figure 2 from Khalil and Rasmussen (1990b). It shows the results of three systematic global studies spanning the decade between 1978 and 1988 (Steele et al., 1987; Blake and Rowland, 1988; Khalil and Rasmussen, 1990b). There is remarkably good agreement among the studies even though the methods and strategies were quite different.

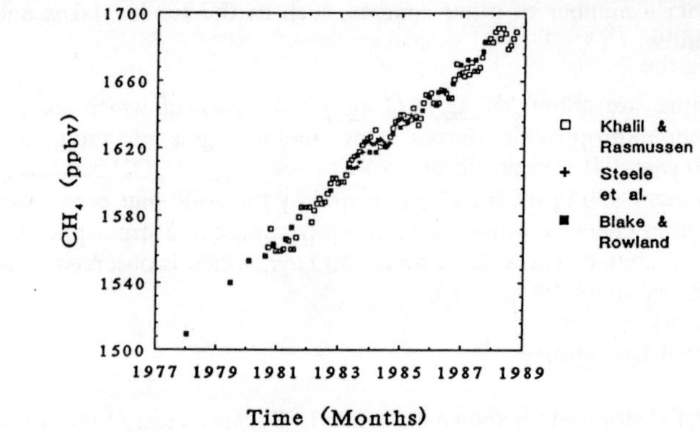

Figure 2. Comparison of methane trends from three systematic global long-term studies (Blake and Rowland, 1988; Steele et al., 1987; and Khalil and Rasmussen, 1990a).

From the systematic studies, it is apparent that the trends have not been constant even over the past decade. The early parts of the record in the late 1970s shows a faster rate of increase than the more recent measurements in the later part of the 1980s. This variation of the rate of increase has led to some confusion as to how fast methane is really increasing. Our early work showed rates of increase of somewhat less than 2%/yr; our later work showed increasing rates of 1.4%/yr; Blake and Rowland (1988) found rates of increase of 1%/yr and Steele et al. (1987) reported rates of increase of somewhat less than 0.8%/yr. All these findings are consistent when the variability in the rate of increase is taken into account. This is shown more clearly in Figure 3 from Khalil and Rasmussen (1990b). Over the decade the average rate of increase is 1%/yr, although within this decade there were 2-year periods when the rate of increase was as low as 0.7%/yr or as high as 2%/yr (in the earlier years).

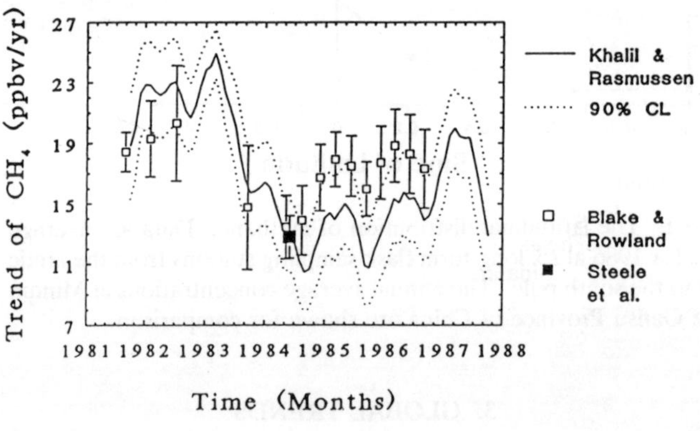

Figure 3. The trends over each successive two-year period and the 90% confidence limits plotted in the middle of the time period over which the trend is calculated. The figure shows that the rate of increase is not constant, and trends over short times may not reflect long-term tendencies. It also shows that the apparently different trends reported in previous studies agree once the variability of the trends is taken into account.

Systematic measurements were not taken before this period; however, there are published data taken during the 1960s and 1970s. An analysis of these data shows trends of similar magnitude during the 2 decades as shown in Figure 4 (Khalil et al., 1989a).

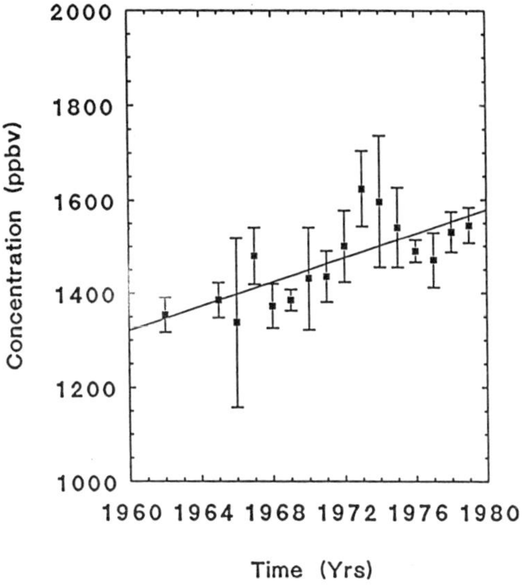

Figure 4. Annual average concentrations during the1960s and 1970 (Khalil et al., 1989).

Over still longer periods spanning the last 150,000 years, there is a remarkable record of atmospheric methane from the analysis of polar ice cores (Figure 5) (Khalil and Rasmussen, 1989b; Stauffer et al., 1985; Raynaud et al., 1988). This record shows that methane concentrations have varied naturally because of changing climatic conditions from ice ages to inter-glacial periods. The record also shows that the present increase started about only a hundred years ago with small trends going as far back as two hundred years (Figure 6). The pattern of the present rapid increase is therefore likely to be caused by human activities linked to the rapidly rising population. During the 150,000-year record concentrations have never been more than half of present levels.

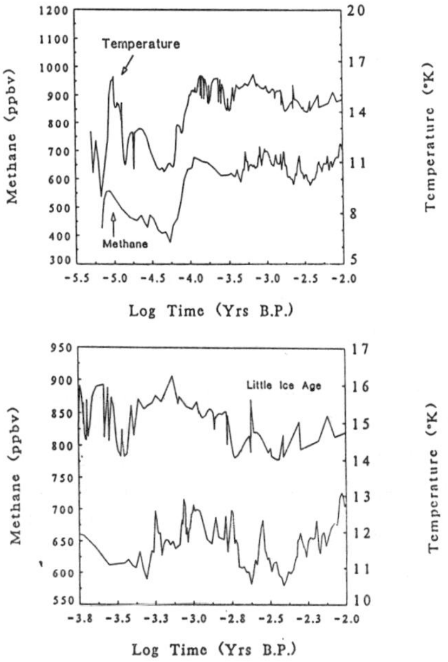

Figure 5. Long-term variation of atmospheric methane and global temperature (lower curves are for methane and upper curves are for temperature).

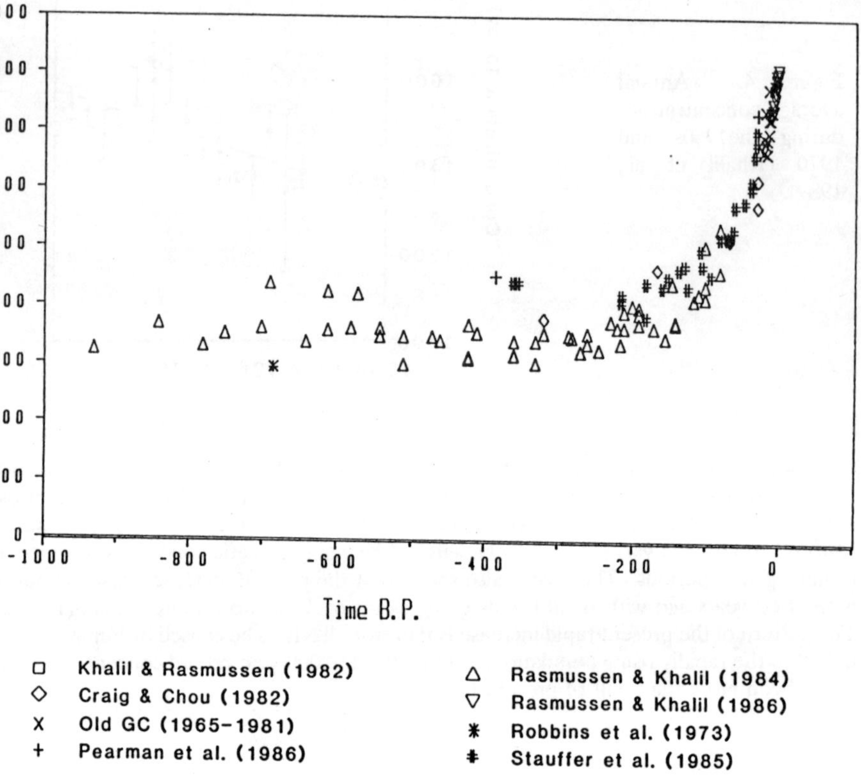

Figure 6. The atmospheric concentrations of CH_4 over the last 1000 years (in ppbv). The figure shows the complementary relationship between different studies, the time when CH_4 concentrations began to increase rapidly, and the doubling of CH_4 over the last several hundred years (Khalil & Rasmussen, 1987).

3.2 Trends of Sources

The emissions of methane from domestic ruminants and rice fields have increased steadily over the past 100 years. More recently, emissions from other human activities such as oil, gas and coal production, landfills, and other waste-processing have also been increasing. We have compiled data from a number of sources to estimate the increase of emissions from anthropogenic activities. The results are summarized in Figure 7. The increase of these sources follows closely the increase of population, but the sources are no longer increasing as fast as in earlier post-World War II decades.

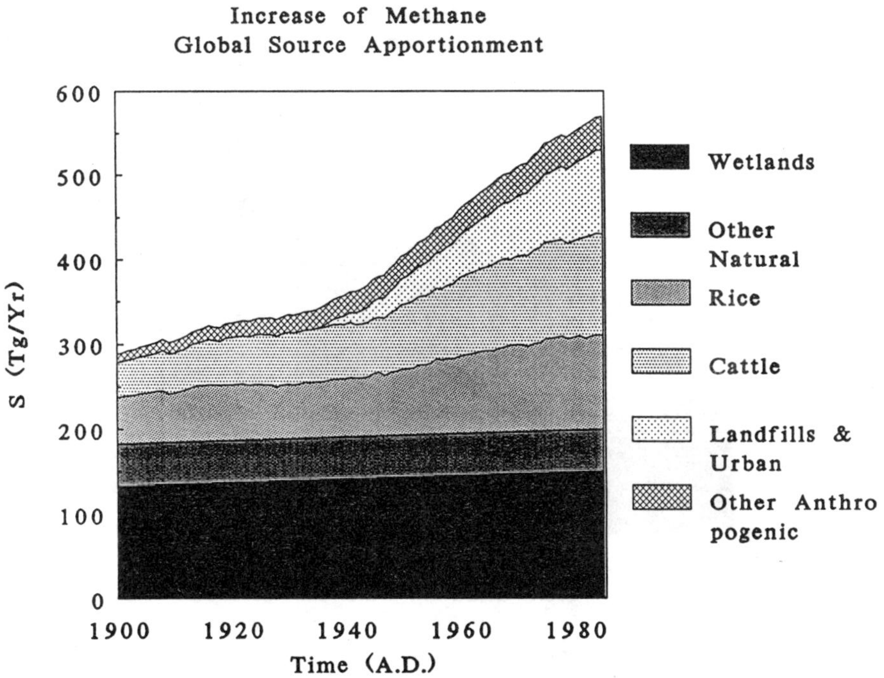

Figure 7. Estimated annual emissions of methane from various sources.

4. Causes of Increasing Methane

The mass balance of methane in the atmosphere can be written as:

$$\frac{dC(t)}{dt} = S(t) - \frac{C(t)}{\tau(t)} \tag{1}$$

where C is the concentration, S the emission rate, and τ is the atmospheric lifetime. The concentration can increase only if the sources increase or the sinks decrease. For methane, both possibilities are plausible. Since methane is removed mostly by reacting with OH, a reduction in OH could lead to an increase of methane. OH can be reduced because of the increasing emissions of CO and also of methane. Nonetheless, there is no clear evidence of large changes in OH that would explain the increase of methane to more than double the levels of a hundred years ago. On the other hand, there is very good evidence that the sources have increased, although uncertainties and issues still remain. A good case can be made that most of the increase of methane has occurred because of increasing emissions. This still leaves a margin for the contribution of declining OH (Khalil & Rasmussen, 1985, 1987).

When the sources are put into Eqn. 1 with a reduction in OH of about 20% over the last century, the resulting expected concentrations of methane are shown in Figure 8. The trend of the calculated concentrations is decreasing in recent times, which agrees with the observations mentioned earlier.

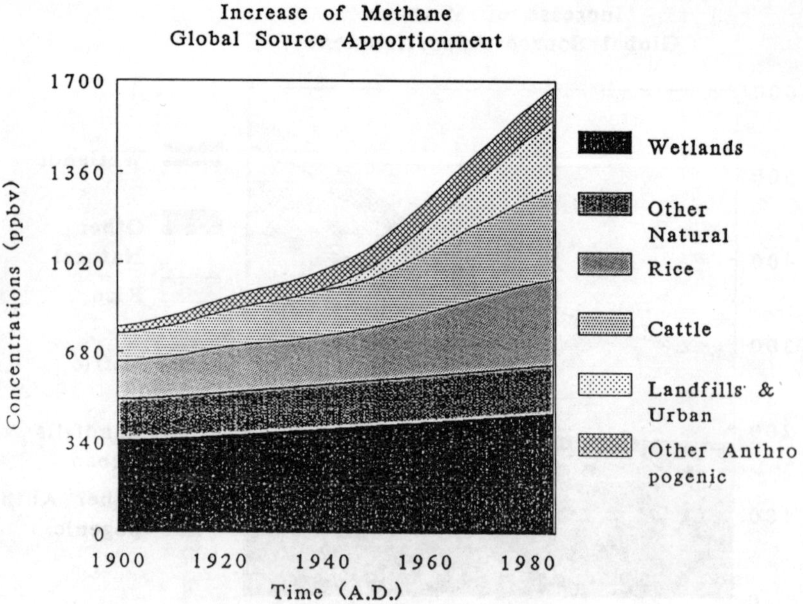

Figure 8. Estimated contribution of each source to the increase of methane and to the present concentration in the atmosphere. Increasing contributions from wetlands is due to decreasing OH and increasing global temperatures included in the model.

5. CONCLUSIONS

We have shown a coherent picture of the global methane cycle in which the trends from different experiments agree and the increase both at present and over the past 100 years is dominated by increasing emissions from human activities.

In spite of our present knowledge of the methane cycle, there are still too many unknowns to predict future levels. It seems, however, that there are natural and economic limits to the increase of the main present anthropogenic sources such as cattle populations and acreage of rice fields. The increases of these sources are already slowing and may severely limit the increase of methane in the future. The effect of draining wetlands over the past century and the continuing loss of wetlands at present is not fully understood but may lead to a reduction in natural emissions. On the other hand, there are new sources such as (sanitary) landfills that are proliferating, especially in western countries. Moreover, there are possibilities that the warming of the world will lead to an increase of methane emissions from natural sources or even destabilize large reservoirs of methane such as the permafrost, thus overcoming the diminishing role of the sources we know today.

Acknowledgements

We thank Bob Dalluge for field work in China, Don Stearns for laboratory work, and Martha Shearer for data management. Funding for this research was provided by grants from the National Science Foundation (#ATM84-8414020; DPP8717023) and the U.S. Department of Energy (#DE-FG06-85ER6031). Additional support was provided by the Biospherics Research Corporation and the Andarz Co. This paper is also a part of the Proceedings of the ICGREAC conference held in Beijing, PRC, in 1989.

References:

Bingemer, H.G., and Crutzen, P.G., 1987. The production of methane from solid wastes. *J. Geophys. Res. 92*: 2181-2187.

Blake, D.R., 1984. Increasing concentrations of atmospheric methane. Ph.D. dissertation, University of California at Irvine, U.S.A.

Blake, D.R., and Rowland, F.S., 1988. Continuing worldwide increase in tropospheric methane, 1978 to 1987. *Science 239*: 1129-1131.

Bolle, H.-J., Seiler, W., and Bolin, B., 1986. Other greenhouse gases and aerosols. IN: *The Greenhouse Effect, Climate Change and Ecosystems*, Ch. 4 (SCOPE 29) (J. Wiley & Sons, N.Y., 1986), pp. 157-198.

Cicerone, R.J., and Oremland, R.S., 1988. Biogeochemical aspects of atmospheric methane. *Global Biogeochemical Cycles 2*: 299-327.

Donahue, T.M., 1979. The atmospheric methane budget. In: *Proceedings of the NATO Advanced Study Institute on Atmospheric Ozone: Its Variation and Human Influences* (Ed. A.C. Aikin). (U.S. Department of Transportation, Washington, D.C.

Ehhalt, D.H., 1974. The Atmospheric cycle of methane. *Tellus 26*: 58-70.

Ehhalt, D.H., and Schmidt, U., 1978. Sources and sinks of atmospheric methane. *PAGEOPH 116*: 452-464.

Khalil, M.A.K., and Rasmussen, R.A., 1983. Sources, sinks and seasonal cycles of atmospheric methane. *J. Geophys. Res. 88*: 5131-5144.

Khalil, M.A.K., and Rasmussen, R.A., 1990a. Constraints on the global sources of methane and an analysis of recent budgets. *Tellus 42B*: 229-236.

Khalil, M.A.K., and Rasmussen, R.A., 1990b. Atmospheric methane: recent global trends. *Environ. Sci. Technol. 24*: 549-553.

Khalil, M.A.K., and Rasmussen, R.A., 1985. Causes of increasing atmospheric methane: depletion of hydroxyl radicals and the rise of emissions. *Atmos. Environ. 19*: 397-407.

Khalil, M.A.K., and Rasmussen, R.A., 1987. Atmospheric methane: trends over the last 10,000 years. *Atmos. Environ. 21*: 2445-2452.

Khalil, M.A.K., Rasmussen, R.A., and Shearer, M.J., 1989a. Trends of atmospheric methane during the 1960s and 1970. *J. Geophys. Res. 94*: 18279-18288.

Khalil, M.A.K., and Rasmussen, R.A., 1989b. Climate-induced feedbacks for the global cycles of methane and nitrous oxide. *Tellus 41B*: 554-559.

Rasmussen, R.A., and Khalil, M.A.K., 1981a. Atmospheric methane (CH_4): trends and seasonal cycles. *J. Geophys. Res. 86*: 9826-9832.

Rasmussen, R.A., and Khalil, M.A.K., 1981b. Increase in the concentration of atmospheric methane. *Atmos. Environ. 15*: 883-886.

Raynaud, D., Chappellaz, J., Barnola, J.M., Korotkevich, Y.S., and Lorius, C., 1988. Climatic and CH_4 cycle implications of glacial-interglacial CH_4 change in the Vostok ice core. *Nature 333*: 655-657.

Sheppard, J.C., Westberg, H., Hopper, J.F., Genesan, K., and Zimmerman, P., 1982. Inventory of global methane sources and their production rates. *J. Geophys. Res. 87*: 1982.

Stauffer, B., Fischer, G., Neftel, A., and Oeschger, H., 1985. Increase of atmospheric methane recorded in Antarctic ice core. *Science 229*: 1386-1388.

Steele, L.P., Fraser, P.J., Rasmussen, R.A., Khalil, M.A.K., Conway, T.J., Crawford, A.J., Gammon, R.H., Masarie, K.A., and Thoning, K.W., 1987. The global distribution of methane in the troposphere. *J. Atmos. Chem. 5*: 125-171.

Warneck, P., 1988. *Chemistry of the Natural Atmosphere*. Academic Press, N.Y.

The activities of the Center for Atmospheric Studies include measurements of methane and other environmentally important trace gases (chlorocarbons, hydrocarbons, CO, N_2O, and sulfur gases), statistical analyses, global budgets of trace gases, tracer circulation and mass balance modeling, 1-d and 2-d climate modeling, and global tropospheric chemistry, particularly OH. Publications on these subjects may be obtained by writing to Professor M.A.K. Khalil.

THERMAL RELATION OF LIGHT UTILIZATION AND BIOMASS PRODUCTION IN UPLAND COTTON (*GOSSYPIUM HIRSUTUM* L.)

S.N. Bhardwaj and Munshi Singh

Divisions of Plant Physiology and Genetics, Indian Agricultural Research Institute, New Delhi- 110 012 (India)

ABSTRACT

The excessive use of fossil fuel resources is likely to enhance global temperature owing to green house effect. An understanding of the possible impact of the rising temperature on crop photosynthesis and productivity would help in mitigating the above adverse effects. The present study indicates that increase in temperature (prevailing range under Delhi conditions)affected adversely the quantum of solar radiation absorption vis-a-vis biomass production. It appears that cotton crop followed a definite strategy in combating the adverse influence of increasing temperatures experienced during the cropping season. To begin with increase in leaf area ceases and crop growth rate reaches the compensation point. As the temperatures increase further, the interception and absorption of photosynthetically active radiation becomes almost negligible. Promising genotypes like '15-2-1' and '19-27' were identified for their performance under tropical summer (March-June).

INTRODUCTION

Investigations on morpho-physiologial basis of biomass production have highlighted numerous associations (positive and negative) between traits and characteristics contributing to the build up of this parameter. Important traits are foliage area, its duration and assimilatory rate (Asana 1975, Baker and Myhre 1969, Bhardwaj et al. 1971 and 1975, Brown 1973, Karami 1972, Singh et al. 1984) and these are not only inter-dependent but are also controlled by macro- and micro-climatic variations during the cropping season. Temperature, by and large, is the most important and any increase in temperature resulting from 'green house' effect is likely to depress biomass production. To counter such an effect its impact on components of biomass production and of solar energy interception and utilization merits investigation.

Biomass and its components

Growth potential of a crop is built up through a series of growth and developmentory processes. Photosynthesis is the only source of biomass production and is governed by foliage area, its duration and partitioning efficiency between photosynthetic (foliage) and non-photosynthetic (stem, root, etc.) components. It is suggested that improved adaptation of the photosynthetic process to sub- and supra-optimal conditions and more effective exploitation of leaf area index are more relevant for increased biomass production rather than any attempt to improve the efficiency of the process per se (Beedle et al. 1985).

Growth parameters controlling biomass production

Analysis of biomass production in upland cotton by Asana (1975) and Bhardwaj et al. (1975) highlighted the contribution of parameters like foliage area, its duration and assimilation rate. Importance of leaf area index and net assimilation has also been emphasized by Watson (1963). Muramoto et al. (1965) opined that difference in biomass production could be primarily traced to the difference in the area of the foliage rather than its assimilatory rate. Similar results were reported by Singh and Bhardwaj (1983) and have suggested that leaf area controlled dry matter (biomass)production.

Subsequent studies by Singh et al. (1984) highlighted the importance not only of leaf size as the determinant of biomass formation but its thickness (SLW) is equally important. Thicker leaf (high SLW) is relatively more thermostable and suited for arid tropical climate which is most suited for cotton cultivation. It was further shown that small to medium sized leaves were conducive for higher productivity owing to their higher leaf conductance and consequently, high photosynthetic rate (Bhardwaj et al., 1986). Difference in leaf area alone accounted for 90% variation in biomass, whereas SLW increased it further to the extent of 1.3% only (Bhardwaj et al. 1988).

As regards growth components, difference in biomass production is comparable with mean leaf area index (LAI), whereas the difference in net assimilation rate (NAR) or leaf weight ratio (LWR) are rather marginal and generally not significant, though some fluctations are seen in respect of specific leaf weight (SLW) which decreased at relatively higher leaf area indices. Thus a suggestion was advanced that biomass is built up through increase in leaf area, irrespective of individual leaf size or its SLW, though these parameters do have a relevance in determining seed-cotton productivity (Singh et al. 1987).

Influence of temperature on components of biomass

Studies by Jackson (1967) in Sudan (N.E. Africa) revealed that RGR and NAR increased with increasing temperature during August-mid October, but subsequent decrease in temperature lowered NAR. Rajan et al. (1973) studied the impact of temperature on growth components at seedling stage within the range of 10-35°C, NAR and LAI increased with increasing temperature. LAR was also enhanced.

Studies by Singh et al. (1987) showed that increasing temperature (under Delhi conditions) decreased crop growth rate (CGR) and mean LAI but improved NAR, SLW and LWR. LAR was, however, not influenced by change in temperature. It appears that temperature regime(s) experienced by the crop were supra optimal and any further increase in maximum and/ or minimum temperature (s) would lead to decrease in leaf area and biomass production.

Solar radiation and biomass production

Solar radiation provides the basic energy for all plant processes. Biomass production is an effective measure of radiation utilization for photosynthesis (Monteith 1958) and explained this relationship between the solar radiation and productivity in tropical ecosystems (Montieth 1972). Biscoe and Gallagher (1977) also opined that at least during vegetative growth, crops accumulate dry matter at rates proportional to intercepted radiation. Significance of leaf angle distribution (Dewit 1965, Duncan 1971, Kimes et al. 1980) and plant height, width, leaf area index and leaf angle (Marani and Ephrath 1985) in radiation interception and photosynthetic efficiency was also highlighted. It appears that erectophile canopies are more efficient than planophile ones, especially at medium to high leaf area indices. Thus it might be argued that factors limiting photosynthesis as well limited potential production.

Constable (1986) working on growth and light receipt by main stem leaves of cotton in relation to plant density highlighted the significance of light extinction coefficient (K) having a large effect on calculated photosynthesis of lower leaves in crop model based at the single leaf level.

Components of Solar energy interception and utilization

* Light utilization efficiency (LUE) depends primarily on availability of photosynthetically active radiation (PAR) at the crop canopy level and its absorption (Fp) is regulated by light extinction coefficient (K) and leaf area index (LAI). Temperature is an important factor modulating the interrelationship (s) in between the above parameters.

Computation of inter-correlations between the above parameters including temperature for 3 elite cotton genotypes (31-38, 595-BN-3 and H-777) grown at 3 plant densities (44, 67 and 133 thousand plants per hectare) reveals that CGR is positively correlated with PAR (absorbed), Fp (fraction of PAR absorbed) and $\bar{L}$ (mean leaf area index) but negatively with maximum and minimum temperatures (Table 1). On the other hand, PAR absorbed is

Table 1 : Intercorrelations (r values) between CGR and related parameters

(number of cases= 45)

	PAR (ab)	Mean leaf area	Fp	Temperature		SLW
				max.	min.	
CGR	0.72	0.83	0.78	-0.54	-0.58	-0.17
PAR (ab)	-	0.91	0.96	-0.89	-0.91	-0.21
Mean leaf area	-	-	0.88	-0.77	-0.78	-0.24
Fp	-	-	-	-0.75	-0.79	-0.29
Temperature max.	-	-	-	-	0.99	0.10
Temperature min.	-	-	-	-	-	0.15

correlated positively with Fp and $\bar{L}$ but negatively with maximum and minimum temperatures. $\bar{L}$ and Fp also had negative correlation with temperature (maximum and minimum), whereas Fp was correlated negatively with SLW. Further, breakup of these correlations into direct and indirect effects through path analysis (Table 2) indicates the

Table 2 : Break-up of correlations between CGR and related parameters into direct and indirect effects

(number of cases = 45)

	PAR (ab)	Mean leaf area	Fp	Max. temp.	Min. temp.	SLW	Correlation with CGR
PAR (ab)	-3.35	1.17	2.08	-0.42	1.29	-0.05	0.72
Mean leaf area	-3.05	1.29	1.90	-0.36	1.11	-0.05	0.83
Fp	-3.22	1.13	2.16	-0.35	1.12	-0.06	0.78
Temperature max.	2.98	-0.99	-1.62	0.47	-1.40	0.02	-0.54
Temperature min.	3.05	-1.00	-1.71	0.47	-1.42	0.03	-0.58
SLW	0.70	-0.31	-0.63	0.05	-0.21	0.22	-0.17

direct of PAR absorbed on CGR is of negative nature but appeared positive owing to the indirect positive effects of Fp, $\bar{L}$ and minimum temperature. Again, the positive relationship of CGR with $\bar{L}$ is a direct one but is also contributed indirectly by PAR absorbed, Fp and minimum temperature, whereas the relationship of Fp with CGR is contributed directly as well as indirectly through the effectws of PAR absorbed, $\bar{L}$ and minimum temperature. The effect of maximum temperature on CGR is indirectly contributed by minimum temperature, Fp, PAR absorbed and $\bar{L}$. The impact of SLW on CGR, though positive is not significant.

Since $\bar{L}$ and Fp have a direct bearing on CGR, impact of temperature and SLW on $\bar{L}$ has been analysed (Table 3). Both

Table 3 : Break-up of correlations between mean leaf area and related components into direct and indirect effects

(number of cases = 45)

	Temperature maximum	Temperature minimum	SLW	Correlation with mean leaf area
Temperature maximum	-0.22	-0.53	-0.01	-0.77
Temperature minimum	-0.22	-0.53	-0.02	-0.78
SLW	-0.02	-0.08	-0.13	-0.24

maximum and minimum temperatures affected $\bar{L}$ directly, whereas the effect of SLW was not found significant. Fp was directly influenced by minimum temperature as well as indirectly through maximum temperature.

CGR has been compared with LUE along with $\bar{L}$, mean temperature and SLW (Table 4). CGR possessed a positive

Table 4 : Inter-correlations (r values) between LUE and related parameters

(number of cases = 45)

	$\bar{L}$	Temperature	SLW	CGR
LUE	0.665	-0.191	-0.147	0.932
$\bar{L}$	-	-0.207	-0.231	0.835
Temperature	-	-	-0.100	-0.245
SLW	-	-	-	-0.171

correlation with LUE and $\bar{L}$, while the relationship with mean temperature and SLW was of negative nature but not significant.

Effect of temperature on light utilization components

Under the prevailing maximum and minimum temperatures during the cropping season (May-November) at Delhi, CGR, $\bar{L}$, PAR absorbed and Fp showed negative correlation with temperature (Table 5) and temperature maxima were calculated through computation of linear regression equations (Table 6).

Table 5 : Inter correlations between temperature and growth parameters

Between	Correlation (r value)			
Temperature	$\bar{L}$	CGR	PAR (ab)	Fp
Maximum	-0.77	-0.54	-0.89	-0.75
Minimum	-0.78	-0.58	-0.91	-0.79

Table 6 : Temperature Maxima for growth and light utilization

	Temperature maxima (°C)			
Temperature	$\bar{L}$	CGR	PAR(ab)	Fp
Maxima	40.9	42.0	43.0	45.5
Minimum	30.2	30.8	31.9	33.7

It appears that the crop follows the following strategy to combat higher temperatures. When temperatures approach 41°/30°C (maximum and minimum respectively) increase in leaf area ceases and is followed by no crop growth (i.e. Zero CGR) at 42°/31°C. Photosynthesis is evenly balanced by respiratory activity (i.e. compensation point). PAR was not absorbed by the crop at 43°/32°C, whereas light interception ceases at 46°/34°C. At this stage the crop would register negative growth owing to stoppage of photosynthesis but continuation of photo and dark respirations.

Breeding for temperature tolerance

Twenty-one high yielding hirsutum genotypes derived from intraspecific crosses between exotic american and indigenous types were grown at 3 distinct environments experienced during March-June, May-November and July-December at two consecutive cropping seasons of 1989 and 1990. 2 Strains '15-2-1' and '19-27' were identified which yield more than 10 q/h seed cotton both under arid summer (March-June) and humid cooler (July-December) seasons (Table 7).

Table 7 : Yield performance of three *hirsutum* strains at different sowing times

Sowing Time	Season	Seed-cotton yield (q/h)		
		15-2-1	19-18	19-27
March	1989	5.46	4.41	4.35
	1990	10.90	4.39	10.67
May	1989	15.50	14.60	11.40
	1990	16.19	13.62	11.89
July	1989	9.48	13.65	13.40
	1990	10.82	6.00	12.36

REFERENCES

Asana RD, (1975). Physiological research in cotton- past and present. Indian Farming (Advances in Cotton Research), **24** : 17-19.

Beadle CL, Long SP, Imbamba SK, Olembo R, Hall DO. (1985). Photosynthesis in relation to plant productivity Tycooly International, Oxford (UK), 182 pp.

Bhardwaj SN, Ruwali KN, Singh M. (1975). Physiological parameters for breeding for higher seed cotton productivity in American cottons. Indian J. Agric. Sci., **45** : 124-127.

Bhardwaj SN, Singh M, Singh VP, Singh KP (1986). Leaf size, its thickness and conductance in genotypes of upland cotton. Indian J. Agric. Sci., **56** : 840-843.

Bhardwaj SN, Sani AD, Singh M and Singh KP. (1988). Contribution of area, thickness and conductance of leaf in biomass production in upland cotton (Gossypium hirsutum L.). Indian J. Agric. Sci., **58**(2) : 100-103.

Biscoe PV, Gallagher JN. (1977). Weather, dry matter production and yield. In : JJ Landsberg and CV Cutting (eds.), Environmental Effects on Crop Physiology. Academic Press, London, pp. 75-100.

Constable GA. (1986). Growth and Light receipt by main stem cotton leaves in relation to plant density in the field. Agric. & Forest Met., **37** : 279-292.

De Wit CT. (1965). Photosynthesis of leaf canopies. Agric. Res. Rep. No. 663, Pudoc, Centre Agric. Publ.DOC., Wageningen, The Netherlands.

Duncan WG. (1971). Leaf angle, leaf area and canopy photosynthesis. Crop Sci., **11** : 482-485.

Jackson JE. (1967). Studies on the sowing dates of cotton in Sudan Gezira.I. The influence of date of sowing on plant growth development. J. Agric. Sci. Camb., **69** : 305-15.

Kimes DS, Ranson KJ, Smith JA. (1980). A Montecarlo calculation of the effects of canopy geometry on PAR absorption. Photosynthetica, **14** : 55-64.

Marani A, Ephrath J. (1985). Penetration of Radiation into cotton crop canopies. Crop Sci., **25** : 309-313.

Monteith JL. (1958). The reflections of short wave radiation by vegetation. Quart. J. Met. Soc., **85** : 386-92.

Monteith JL. (1972). Solar radiation and productivity in tropical ecosystems. J. appl. Ecol., **9** : 747-766.

Muramoto H, Hesketh JD, El-Sharkawy M. (1965). Relationship among rate of leaf area development, photosynthetic rate and rate of leaf area development, photosynthetic rate and rate of dry matter production among American cultivated cotton and other species. Crop Sci., **5** : 163-166.

Rajan AK, Betteridge B, Blackman GE. (1973). Differences in the interacting effects of light and temperature on the growth of 4 species in the vegetative phase. Ann. Bot., **37** : 287-313.

Singh D, Bhardwaj SN. (1983). Physiological analysis of yielding ability in hirsutum cottons. II. Parameters controlling dry matter accumulation, boll number and seed-cotton yield. Indian J. Pl. Physiol., **26** : 264-75.

Singh KP, Bhardwaj SN, Singh M. (1987). Parameters controlling crop growth rate and their relationship with photosynthetically active radiation and temperature in upland cotton. Indian J. Agric. Sci., **57** : 828-32.

Singh M, Singh VP, Bhardwaj SN, Paul K. (1984). Breeding for desirable leaf types in cotton. Indian J. Pl. Physiol., **27** : 247-254.

Watson DJ. (1963). Climate, weather and plant yield. In : Environmental control of plant growth (LT Evans, Ed.), pp. 337-350, Academic Press, New York, London.

ACQUISITION OF CYTOSOLUTES IN YOUNG WHEAT SEEDLINGS IN RESPONSE TO HIGH TEMPERATURE STRESS

Sanjeev K. Thind and C. P. Malik

Department of Botany
Punjab Agricultural University
Ludhiana

Stress from high temperature during wheat seedling development induces heat shock protein after 3 h of incubation. During initial period cellular adaptations are accomplished by the accumulation of cytosolutes e.g. sugars. The authors conclude that increase in sugar is correlated with enhanced activities of amylases. Decline in amylases with stress from high temperature for long duration, is attributed to the disruption of RER.

INTRODUCTION

In response to heat stress, 'heat-shock proteins'are synthesized ubiquitously and such a response to high temperature stress has also been observed in wheat seedlings (Krishnan *et al*.1980; Thind and Malik,1986). With the induction of heat-shock proteins (hsp's) there is concomitant supression of normal protein synthesis(Schleeinger *et al*., 1982; Sachs and Ho,1986). This alteration of gene expression is accompanied by sequestering of many preexisting mRNAs rather than being degraded. We have reported earlier(Thind and Malik,1986) that heat shock (3 h at 45°C) induces synthesis of hsp's in wheat seedlings. Till the time hsp's are synthesized, it is essential to maintain cytoplasmic ionic composition compatible with metabolic activity, and the maintenance of an adequate turgor pressure in cells by some cytosolutes. The reconciliation of these two criteria requires achievement of cytoplasmic osmotic adjustment by the accumulation of non-toxic organic solutes. The objective of this research was to determine if hydrolases and level of free sugars were altered when wheat seedlings were exposed to high temperature shock for variable periods.

MATERIAL AND METHODS

Fifteen healthy seeds of Sonalika cultivar of wheat (*Triticum aestivum* L.) were surface sterilized with 0.1% mercuric chloride and germinated in Petri-dishes lined with two layers of blotting sheets, saturated with distilled water. In all studies, the seeds were incubated at optimum temperature (25 ± 2°C) in the dark. Three-day old seedlings were given high temperature shock(45 ± 2°C) for 1,2,3 and 4 hours, were not allowed to recover at optimum temperature; and were immediately tested for the activities of amylases and free sugars.

α - and β amylases were extracted in phosphate buffer (0.1M) pH 7.0 by homogenising seedlings by freezing and thawing. The supernatant collected after centrifugation (10,000 xg) for 10-15 min was taken as crude enzyme extract. α- amylase activity was measured as suggested by Murata et al.(1968) and enzyme activity expressed as ΔA620 mg protein^{-1} h^{-1} . To analyse β -amylase activity 0.2 ml of enzyme extract was added to 0.1 ml of freshly prepared starch solution kept at 30°C for 1 hour and reaction was terminated by adding 1.0 ml of DNSA and diluted to 5 ml after boiling in water bath for 10 min. Absorbance was recorded at 560 nm (Shuster and Gifford,1962) and activity expressed as ug glucose and fructose released h^{-1} mg protein^{-1}. Acid invertase activity was measured in a reaction mixture consisting of 0.6 ml of 0.2M acetate buffer, pH 4.8, 0.3 ml of 0.4M sucrose and 0.1 ml of enzyme extract (Singh et al. 1978). After incubation at 30°C for 30 min. reducing sugars released were measured by dinitrosalicylic acid method and enzyme activity was expressed as µg glucose and fructose released h^{-1} mg protein^{-1}.

Sugars were extracted from heat shocked seedlings by repeated homogenisation in 80% ethanol. Water-alcohol column (Dowex AG 50W x8, 200-400 mesh H^+ exchange) and anion exchange column (Dowex AG 1x8, 200-400 mesh formate column) were employed. The two columns were neutralized by washing with 80% ethanol and a tandem column - the cation exchange column over the anion exchange column - was prepared by connecting them with a short piece of rubber tubling and 40 to 60 ml of neutral fraction of sugar below the anion exchange column collected. This extract was used for the analysis of total free sugars and individual sugars were separated by paper chromatography(Shallenberger and Moores, 1957).

RESULTS

Free sugars concentration increased with enhancement in the exposure time, the increase was more than two-fold over control with 1 h heat shock(Table 1). The increase was three-fold over the control in 3 and 4 h treatment. Glucose concentration decreased in 1 h shock but its content increased enormously with further increase in time of shock (Table 2). Fructose concentration increased linearly with increasing period of high temperature shock.

No oligosaccharides were sampled in control seedlings but their concentration was maximum in seedlings subjected to high temperature shock for 1 h.

The activity of α- amylase increased 3-fold(Table 3) following 1 h exposure to high temperature but decreased tremendously in seedlings exposed to high temperature for 2 and 3 h and only negligible activity was reported following exposure for 3 h. β - amylase activity increased 2-fold in seedlings exposed to high temperature for 1 h but decreased when exposure time was increased (Table 3). 4 hours heat shock decreased activity over control.

It was noted that invertase activity decreased linearly on exposure to heat shock for 1 to 4 hours(Table 1).

DISCUSSION

Higher plants experience myriads of environmental and biological stresses, and they adversely affect normal plant growth and development(Sachs and Ho,1986). Plants appear to respond to stresses by evolving developmental, biochemical and physiological adaptations to resist their assault. There is general alterations in the level of phytohormones, different macromolecules, especially the proteins. The role of proline and glycine betaine has been repeatedly elaborated and debated in imparting environmental resiliency. The acquisition of stress resistance has been linked to the synthesis of specific proteins such as heat shock proteins, dehydrins and osmotin, etc. The induction of hsp's allows the cells to establish thermotolerance in many organisms (Schlesinger *et al*.1982) although their exact function is still debated. The induction of hsp's is accompanied by the alteration of expression of other genes and consequently normal protein synthesis is suppressed.

However, because the rate of total protein synthesis is unchanged, there appears to be a suppression of synthesis of some of the normal proteins and to compensate hsp's synthesis increases. In wheat seedlings we have already reported that hsp's appear after 3 h of high temperature shock. In the present study we have shown that before the induction of hsp's, wheat seedlings try to adjust to heat stress by the accumulation of soluble carbohydrates. In fact soluble sugars are among the first substances accumulating as possible protective agents against heat injury. Several investigators have reported marked protection of seedlings with sugars. Sucrose, glucose, rhamnose and maltose prevent the coagulation of proteins due to heat injury in different species(Levitt, 1980). The activites of α - and β - amylase enzymes presently increased following 1 h heat shock though with enhanced period of heat shock, their activities decreased. After seedlings were incubated at 45°C for 3 h (the time for induction of hsp's in wheat), the activities of amylases and invertase decreased tremendously. Since all the secretory proteins are synthesized on RER, their disruption could be correlated to the decline to the synthesis of α - amylase, and the increased degradation of their mRNA following high temperature shock(sensu Sticher et al.,1990).

In summary, we are tempted to suggest that accumulation of free sugars in wheat seedlings acts as proteive agents against high temperature stress till the time there is synthesis of specific proteins.

REFERENCES

1. Krishnan, M., Nguyen, H.T. and Burke, J.J.(1980). Heat shock protein synthesis and thermal tolerance in wheat. Plant Physiol.90:140-145.

2. Levitt, J.(1980). Response of plants to environmental stresses. Vol.I. High Temperature Heat Stress. pp.347-370. Academic Press, New York, London, Toronto, Sydney, San Francisco.

3. Murata, T., Akazawa, T. and Fukuchi, S.(1968). Enzymatic mechanism of starch breakdown in germinating rice seeds.I. An analytical study. Plant Physiol.43:1899-1904.

4. Sachs, M.M. and Ho, T-H.D.(1986). Alteration of gene expression during environmental stress in plants. Annu. Rev. Plant Physiol.37:363-376.

5. Schlesinger, M.J., Ashburner, M., Tissieres, A.(1982). Heat Shock from Bacteria to Man. New York Cold Spring Harbor Lab. pp.440.

6. Shallenberger, R.S. and Moores, R.G.(1957). Quantitative determination of sugars and sucrose separated by paper chromatography. Analytic. Chem.29:27-29.

7. Shuster, L. and Gifford,R.H.(1962). Changes in nucleotidase during the germination of wheat embryos. Arch. Biochem. Biophys.96:532-540.

8. Singh, M.B., Malik, C.P. and Thapar, N.(1978). Changes in the activities of some enzymes of carbohydrate metabolism in Amaryllis vittata pollen suspension cultures. Plant Cell Physiol.19:677-684.

9. Sticher, L., Biswas, A.K., Bush, D.S. and Jones, R.L. (1990). Heat shock inhibits α - amylase synthesis in barley aleurone without inhibiting the activity of endoplasmic reticulum marker enzymes. Plant Physiol.92:506-513.

10. Thind, S.K. and Malik, C.P.(1986). Heat shock induced proteins in wheat seedlings. Jour. Pl. Sci. Res. 2:37-42.

Table 1: Effect of high temperature shocks(45°C) on total free sugar and invertase activity in 3 day old seedling of wheat

Duration of treatment at 45° C (h)	Free - sugars (mg g^{-1} FW)	Invertase (µg glucose released mg $protein^{-1}$ h^{-1})
Control (25 ± 2°C)	5.50	3.47
1	12.81	2.54
2	14.42	1.45
3	16..07	0.46
4	16.21	0.0

Table 2: Effect of high temperature (45°C) shocks for variable periods(0,1,2,3,4 h) on non-reducing sugars (mg g^{-1} FW) in three day old seedlings of wheat

Sugars	Duration of treatment (h)				
Reducing	0	1	2	3	4
Glucose	2.10	1.80	2.40	2.40	4.32
Fructose	0.90	1.80	2.70	5.10	1.80
Non-reducing					
Sucrose	1.40	4.35	4.05	3.00	2.22
Oligosaccharides	0.00	4.05	2.10	1.20	1.80

Table 3: α- ($\Delta A620$ mg protein^{-1} h^{-1}) and β -amylase ($\Delta A560$ mg protein^{-1} h^{-1}) activities in three day old seedlings of Sonalika cv. of wheat in control and after 1 to 4 h high temperature shock (45 °C)

Duration of treatment	Enzyme	
	α- amylase	β- amylase
Control (25 ± 2 °C)	9.0	18.5
1 h at 45°C	22.6	40.0
2 " "	17.0	18.9
3 " "	10.2	12.0
4 " "	4.3	8.2

IMPACT OF GLOBAL CLIMATIC CHANGES ON PHOTO-SYNTHESIS AND PRODUCTIVITY OF TROPICAL RICE

K.S. Murty
Central Rice Research Institute,Cuttack-753006, India

ABSTRACT

Impact of likely global weather and associated changes in carbondioxide, temperature, UV radiation, drought spells, flooding and salinity on tropical rice has been analysed. The favourable influence on photosynthesis by 1.5 to 2.0 fold increase in CO_2 and other gases may not out weigh the unfavourable affects of rising temperature. The increase in temperature by 2 to 4°C in tropical rice areas may reduce net photosynthesis due to high photorespiration and dark respiration. The average temperature during the wet season may exceed the optimum of 30°C for various growth processes. Increase in temperature especially night temperature during reproductive and ripening stages of rice is detrimental for spikelet production and sink potential. The dry season temperature may exceed the threshold value of 35°C during anthesis of rice resulting in acute spikelet sterility. These affects are further aggravated by low solar radiation during wet season. The stresses due to aberrant rains, water-logging and salinity may further increase warranting the need to tailor the rice varieties to the changing climate.

Currently an area of concern widely publicised is the proposed global climatic changes especially rising temperature that may soon be initiated by increasing concentrations of carbondioxide and other gases in the atmosphere. These gases liberated through mounting use of fossil fuels absorb heat from earth and cause global warming simulating green house effects. Reduction of ozone in the stratosphere through release of chlorofluro-carbons (CFC) in Industries and Agriculture will rise the level of UV radiation creating problems for crops and human and animal health. Temperature increase especially in polar region and mountain peaks causes extensive melting of ice eventually leading to rise in sea level (by 5-24 cm every decade), flooding and coastal salinity. Large scale deforestation and destruction of vegetation and the impending global warming may induce greater uncertainty in rainfall with sporadic droughts and risk in Agricultural production especially in tropical Asia. The disease and pest problems will be further exacerbated(1).

Rice though a warm season crop will still be largely affected by changing weather especially temperature. The possible implications of climatic and other changes on photosynthesis and productivity of tropical rice with specific relevance to rice crop during wet season in traditional rice areas of Eastern and North East India are analysed basing on the relevant literature.

Carbondioxide

CO_2 concentration in the atmosphere is estimated to be increasing at 1.5 ppm per year and the rate may further escalate with ever increasing use of fossil fuels (2). The level of 270 ppm during the pre-industrial period has now reached 350 ppm and may steeply enhance to 600 ppm CO_2 by 2050 AD.

The influence of CO_2 enrichment on field crops including rice has been studied (Table 1). CO_2 enrichment (900 ppm) for rice especially 30 days before flowering to flowering increased grain yield by 30% through increae in grain number (3). The root shoot ratio, water use efficiency and specific leaf weight also increased, enabling better crop photosynthesis, accumulation of soluble sugars and their transport to developing grains

(4). However, higher CO_2 may often induce partial closure of stomata and its effects on photosynthesis depend on the prevalent solar radiation which is low during the cloudy monsoon season in the tropical rice areas. Higher CO_2 is also reported to induce early leaf senescence in crop plants affecting the ultimate photosynthetic productivity.

Table 1. Effect of carbondioxide enrichment before and after heading on growth and grain yield of IR8*

Treatment	Yield (t/ha)	Sugar & starch in shoot (%)	Grain Wt. (mg)	Grains ($10^3/m^2$)	Filled spikelets (%)
Wet season					
Control	5.7	18	24.3	28.8	78
CO_2 before heading	7.7	25	26.2	34.1	81
CO_2 after heading	6.9	18	26.1	29.5	85
Dry season					
Control	9.0	22	23.1	45.7	74
CO_2 before heading	11.6	30	25.9	50.9	77
CO_2 after heading	10.9	22	25.1	44.6	86

* Adapted from Yoshida (5), CO_2 = 900 ppm

Temperature

Respiration in rice increases with rise in temperature from 15-40°C, the Q 10 for respiration being 2. The root respiration increases linearly up to 32°C, followed by slow rise up to 38°C and a decrease later while shoot respiration increases linearly up to 44°C. Dark respiration is about 10-30% of net photosynthesis depending on mean daily temperature. Panicle respiration is high and accounts for about 50-60% of total respiration. Maintenance respiration which is normally 40% of total respiration increases with temperature resulting in greater loss of photosynthetic CO_2 (5). Photorespiration accounts for 40-50% of photosynthesis in tropical rice and high temperatures further escalate the loss through greater activity of oxygenase over carboxylase in the carboxylating enzyme, Rubisco.

The optimum temperature for photosynthesis for indica rice is 25-35°C with a Q 10 of 1.1. Photosynthetic rate (Pn) decreases below 20°C and above 40°C with highest Pn at day night temperature of 30°/25°C (7). Light induced photorespiration may heavily influence temperature -Pn curve the optimum being below 30°C. Freshly fixed CO_2 in photosynthesis is mostly incorporated to ethanol soluble fractions like sugars and aminoacids with less conversion to storage proteins at high (32°C) than low (22°C) temperature in tropical rice, IR 8 (8). The apparent diffusion constant of sugars to conducting vessels rises with temperature enabling greater mobilisation of sugars. Evidently the high temperature during ripening stage of rice enhances translocation of assimilates and induce faster grain filling. However, such temperature hastens senescence of leaf and conducting tissues restricting substrate availability and its continued translocation for optimum grain filling.

Growth and yield

Temperature is a dominant climatic factor affecting growth processes and yield in rice, the critical low and high temperature being normally 20°C and 30°C respectively. The critical temperatures for various growth organs and processes are summarised in table 2.

Higher temperature reduces the viability of seed especially during humid monsoon season and modest estimates indicated a reduction in viability period by half for every 5°C raise in storage temperature. Germination and early growth and tillering are enhanced with temperature upto 35°C. During reproductive stage, within temperature range of 22-31°C, the number of spikelets per plant decreases as temperature increases. Spikelet production depends on nitrogen uptake and higher temperature decreases the efficiency of absorbed N to produce spikelets.

Table 2. Response of the rice plant to varying temperatures at different growth stages(5)*

Growth stages	Critical temperatures (°C)		
	Low	High	Optimum
Germination	10	45	20-35
Seedling emergence and establishment	12-13	35	25-30
Rooting	16	35	25-28
Leaf elongation	7-12	45	31
Tillering	9-16	33	25-31
Initiation of panicle primordia	15	-	-
Panicle differentiation	15-20	38	-
Anthesis	22	35	30-33
Ripening	12-18	30	20-25

*Daily mean temperature except for germination.

Evidently the optimum temperature of rice decreases as growth advances from vegetative to reproductive stage. Rice is most sensitive to high temperature at flowering, a day time temperature higher than 35°C at flowering time may increase spikelet sterility due to disturbed pollen shedding and impaired pollen germination (9,10).

Solar radiation

When solar radiation is not limiting, yield is correlated negatively with mean daily air temperature the relation being:

$$Y=S[1.20-0.021(t-21.5)^2]$$

where Y,S and t are yield (kg/0.1 ha), solar radiation (cal/cm^2 per day) and mean temperature (°C) respectively during August-September.

The temperature higher or lower than 21.5°C decreases yield in rice (5,11). Average night temperatures rather than mean temperature per day and diurnal variation in temperature are more critical for productivity in view of better photosynthate conservation. Prediction models computed from global weather yield nursery (IRWYN) trials indicated:

$$Y=4.8+0.5743(DNB-3.1)-0.1492(TDB-28.1)+0.06384(RSB-30.446)+0.1665(RSC-13.964)-0.1327(TNC-24.0),$$

in which

yield average=4.8 t/ha;

DNB = preflowering day-night temperature difference, average 3.1°C;

TDB = preflowering day temperature, average 28.1°C;

RSB = preflowering radiation sum, average 30.466 X 10^3 mWh/cm^2;

RSC = postflowering radiation sum, average 13.964 X 10^3 mWh/cm^2; and

TNC = postflowering night temperature, average 24°C.

Thus greater difference in day/night temperature prior to flowering and higher radiation sum during crop period have positive effect while higher preflowering day temperature and postflowering night temperature negatively influence the yield (12).

Higher temperature especially average minimum temperature reduces total duration of a variety the reduction being more apparent from average day temperature of 21°C to 24°C for photo-insensitive varieties like IR 8. Likewise the optimum and critical photoperiod increases with increase in temperature which also accelerates photoperiodic response and panicle development (13).

Incident solar radiation especially photosynthetic radiation is likely to decrease with increase in atmospheric gases, pollutants, water vapour and smog. Ozone and other gases near troposphere partially shield incoming radiation and may considerably affect photosynthesis due to reduced stomatal conductance. Ozone (1-2 ppm) has a tendency to damage palisade cells while SO_2 injures mesophyll cells and ozone and SO_2 along with NO_2 synergetically reduce stomatal conductance and apparent photosynthesis (14).

<u>Weather - Eastern India</u>

Cuttack represents typical weather conditions of

Eastern India which is the rice bowl in the country. The wet season (WS) crop (July-Oct, Nov) is subjected to a mean maximum temperature of about 31.5°C and minimum temperature of 26°C during reproductive and ripening stages (July-Sep) of early and medium rice varieties. During dry season (Jan-April) the crop is subjected to a maximum and minimum temperature of 34°C and 21.4°C respectively at reproductive stage and 36.5 and 24.7°C at ripening stage. The solar radiation is about 350 (WS) and 520 (DS) $mWh/cm^2/day$ during the above critical growth periods (Table 3).

Table 3. Weather variables at Central Rice Research Instt., Cuttack, India (1980-89)

Month*	Temperature(°C d^{-1})		Radiation	Sun shine
	Max.	Min.	(mWhr $cm^{-2}d^{-1}$)	(hr d^{-1})
January	27.8	14.3	420	8.5
February	31.0	17.5	475	8.7
March	34.1	21.4	480	8.2
April	36.5	24.7	520	8.3
May	39.2	26.3	515	8.8
June	35.5	26.2	420	5.8
July	31.9	25.2	350	4.5
August	31.4	25.5	360	4.7
September	31.6	25.3	370	5.6
October	31.0	23.6	395	8.2
November	29.2	18.4	390	8.8
December	27.6	14.1	395	8.6

*Mean for 10 years; Wet season (July to September/October), Dry season (February to April) for early rice varieties.

The dry season yields are about 1.5 times higher than of wet season for early varieties. The low yields during wet season are attributed to unfavourable weather conditions like low solar radiation, high night temperature and low diurnal variation in temperature which are detrimental for photosynthetic productivity (15). The impending global weather changes especially temperature may further aggravate the undesirable affects of enhanced temperature on yield of both season crops and on high temperature induced spikelet sterility in dry season crop as critical temperature at anthesis in April may exceed 35°C.

UV Radiation

UV-B radiation is likely to increase due to gradual destruction of stratosperic ozone layer by chlorofluro-carbons. The UV radiation has adverse influence on chlorophyll, cell division and elongation due to decrease in endogenous IAA levels. Plants are partly shielded from UV radiation by accumulation of pigments like flavanols through activity of key enzyme, phenylalanine aminolyase (PAL). Irradiation with UV radiation affects germinability of seed especially at high seed moisture (16).

Other stresses

About 60% of rice area is subjected to adverse situations like drought, waterlogging, coastal salinity and low soil fertility limiting the production of rice. Adverse climate of low light, high night temperature and RH further confounds the problems and impairs yield even in optimum soil and water management. Maintenance of high relative water content and leaf turgidity by osmotic adjustment under moisture stress; less uptake of toxic sodium salts, maintenance of high K/Na ratio with low salt accumulation in younger leaves with high sugar concentration in shoot for salinity; high oxidising power of roots and early sheath elongation through ethylene as growth stimulant at early stages and release of bound GA from nodes for stem elongation at later stages under water stagnation and flooding and better root activity and foraging effect of root to absorb nutrients from low fertile problem soils are considered to be desirable crop traits for varietal improvement to such situations. Certain varieties tolerant to specific stress have been identified (Table 4) and special emphasis is being given for intensive breeding to such location specific problems (17).

Table 4. . Sources of resistance (donors) for stresses and elite entries for physiological efficiency in rice

Stress/Physiological traits	Donors/elite cultures
Stresses	
Heat	N 22, Dular, Carreon, IET 4658, IET 5714
Low light	Ptb 10, Mahsuri, Swarnaprabha, Vijaya
Drought	Mtu 17, Lalnakanda 41, CR 143-2-2
Salinity	CSR 1, CSR 2, Nonabokra, Pokkali, CSR 6
Alkalinity	PVR 1
Water logging/flooding	Nagaribao, Jalamagna, Jaladhi (Deep water) Suresh, Tulasi, Panidhan (Water-logged, flooding)
Low nitrogen	Moti, IET 7590 (Salivahan)
Low phosphorus	H 4, IR 5
Iron toxicity	Mahsuri, IR 36
Aluminium toxicity	Karangiya gora, MW 10
Physiological traits	
Photosynthetic efficiency (Low light)	Ptb 10, Mahsuri, Swarnaprabha, Vijaya
Low photorespiration	Triveni, Kanchan, Pusa 33
Low maintenance respiration	Rasi, Kanchan, Swarnaprabha
High translocation	IR 50, Swarnaprabha
Nitrogen use efficiency	Pallavi, Swarnaprabha
Slow leaf and panicle senescence	Pallavi, Swarnaprabha
UV radiation	T(N) 1
General high physiological efficiency	Ptb 10(traditional), Swarnaprabha(HYV)

Future strategies

Environmental pollution and impending weather changes are now considered as a major global problem and Government efforts are being directed to reduce emission of CO_2, CFC and other gases by 20% by 2050 AD and to trap increasing CO_2 by vegetation through large scale forestation. Higher CO_2 in atmosphere may have favourable biological effects like enhancing photosynthesis and biomass production. However, the negative effects of high temperature, stomatal resistance and senescene and the prevelant low light in monsoon season may outweigh the favourable influence. Evidently there is need to investigate the overall effects of increase in CO_2 on food crop like rice.

Considerable variability in temperature tolerance among rice varieties exists. Such sources of tolerance to be further identified and introduced in breeding programmes. Emphasis needs to be given for heat tolerance at flowering. Heat tolerance combined with early maturity are desirable for heat prone areas. Heat stress variation is evident even in C_3 crops, groundnut being more tolerant than soyabean. Such tolerance is considered to be associated with stability of carboxylation over oxygenation of Rubisco under heat stress. Varietal variation in rice in stability of carboxylase to heat needs investigation.

UV radiation affects are partially nullified through absorption by pigments in epidermal portion of leaves and production of flavanols through the key enzyme PAL. Photo reactivation of UV induced pyrimidine dimer production in DNA increases plant efficiency to utilise short wave radiation. Application of growth regulators like IAA may compensate partial photoreduction of endogenous IAA by UV-B radiation. These aspects need to be examined in rice to identify UV-B tolerant varieties and devise methods to moderate its adverse affects.

Efforts are already in progress to intensify varietal improvement programmes for stresses like drought, water-logging, salinity and nutritional deficiency. New sources of tolerance including wild rice sources are to be identified. New biological tools and innovative breeding techniques like wide hybridization, tissue culture and hybrid rice need to be explored for stress areas to achieve greater success in the varietal improvement programmes.

Since coastal rice areas may partially go out of rice cultivation due to expanding sea and the yields in normal areas are likely to be depleted by rising atmospheric temperature and pollutants, concerted efforts are warranted to fix priorities and reorient the programmes now onwards keeping the impending climatic and associated changes during the 21st century.

REFERENCES

1. Sinha, S.K. (1988). How India can weather impact. In: Hindu Survey Indian Agri., pp.238-239.

2. Chakravarty, S. (1990). Threat of the green house effect. Curr. topics., Jan.90,pp.30-31.

3. Yoshida, S. (1976). Carbondioxide and yield of rice. In: Int. Rice Res. Inst., 1976. Climate and Rice, pp. 211-221.

4. Enoch, H. and Naftaly, Z. (1988). Growth and development of plants response to CO_2 concentrations. Appl. Agril. Res. 3: 248-256.

5. Yoshida, S. (1981). Fundamentals of Rice Crop Science, Int. Rice Res. Inst., Philippines, 269 pp.

6. Tanaka, I. (1976). Climatic influence on photosynthesis and respiration in rice. Int. Rice Res. Inst., 1976, Climate and Rice, pp.223-247.

7. Murty, K.S. (1989). Photosynthesis of rice plant. Agrophysics Monograph 1: 62-73.

8. Murty, K.S. (1971). Tracer studies on the effect of different temperatures on photosynthesis in rice. In: Int. Symp. on The use of Isotopes and Radiation in Agriculture and Animal Husbandry Research, Nov. 30-Dec.2, 1971, IARI, New Delhi,pp. 511-517.

9. Owen, P.C. (1971). The effects of temperature on the growth and development of rice. Field Crop Abstr. 24: 1-8.

10. Yoshida, S., Satake, T. and Mackill, D.S. (1981). High Temperature Stress in Rice. IRRI Research Paper Sr. 67, 15 pp.

11. Venkateswarlu, B. and Visperas, R.M. (1987). Solar radiation and rice productivity. IRRI Research Paper Sr. 129, 22 pp.

12. Oldeman, L.R., Seshu, D.V. and Cady, F.B. (1987). Response of rice to weather variables. In: Weather and Rice, Int. Rice Res. Inst., Philippines,pp.5-39.

13. Vergera, B.S. and Chang, T.T. (1976). The Flowering Response of the rice Plant to Photoperiod: A review of literature, Int. Rice Res. Inst., Tech. Bull. 8, 75 pp.

14. Sheoran, I.S. (1989). Response of plants to environmental pollution. In: Natl. Seminar on Plant Physiology, Kanpur, invited talks, pp. 137-138.

15. Murty, K.S. and Sahu, G. (1987). Impact of low light stress on growth and yield of rice. In: Weather and Rice, Int. Rice Res. Inst., Philippines, pp.93-101.

16. Murali, N.S. and Teramura, A.H.(1986). Intraspecific differences in *Cucumis sativus* sensitivity to ultra-violet-B radiation. Physiol. Plant. 68: 673-677.

17. Murty, K.S. (1989). Physiological research on rice and future perspectives. In: Natl. Seminar on Plant Physiology, Kanpur, Invited talks, pp. 1-9.

WHAT DRIVES GLOBAL CLIMATE AND ECOSYSTEMS?: AN OVERVIEW

PHOOL SINGH,
DIRECTORATE OF RESEARCH,
HARYANA AGRICULTURAL UNIVERSITY,
HISAR-125004, INDIA

ABSTRACT

The earth climate responds to astronomical events, solar variaion,internal atmospheric processes,southern oscillation and Cenozoic uplift of mountain range. Vesuvius/Avellino may be one source of senventeenth century BC climate disturbances. Production of greenhouse gases from human activities has begun to warm the globe, particularly at high latitude, cause less precipitation and less moisture in the soil at lower latitude. A minority contradicted the global warming. But an International Panel assessing greenhouse warming denied the validity of objection raised by them. Analysis indicates that the ozone layer depletion due to CFCs reduces stratospheric heating rate and delays te break down southern polar vortex. Increase in the flux of uv-B radiation at the earth surface alters its energy balance. Main culprits are the glass-house gases and ozone layer depletion, also driving the global ecosystems. Global warming may migrate prairie forest border to north, decrease crop yield by 3-17% and shift in geographical locations of potential crop regions by several hundred kilometers. Ozone hole could reduce the rate of production of phytoplankton harming the entire southern ocean ecosystems.

INTRODUCTION

Changes to the climate have various causes including oscillation in the heat output of the sun or in the orbit of the Earth around the sun, the primary cause of the ice ages. In addition volcanic eruptions casting clouds of ash high in the atmosphere cause global cooling over a period of a few years. Today, however, attention is being focussed on the 'greenhouse effect' - the term used to describe the way in which certain gases in the atmosphere act as a heat retainig 'blanket' over the Earth. The greenhous effect is of particularl importance in the context of changes to the climate and ecosystems because the atmosphere concentrations of these gases are being increased by the actions of mankind (1)

ASTRONOMICAL EVENTS

With in past million years, the earth climate equation changed eight times. Past 30 years experience shows that the glacial cycles are ultimately driven by astronomical factors. The two important astronomical factors are eccentricity of the earth's orbit and the tilt and orientation of its spin axis. The third factor is the interaction between accentricity and precession and tilt frequency of earth spin axis. Quoted by Broecker and Denton (2), the Yugoslav astronomer Milutin Milankovitch in 1920's and 1930's advocated that these are three components of astronomical pace maker, two that change the intensity of the seasons and a third that affects the interaction between the two driving factors. The first is the tilt of the earth's spin axis. Currently about 23.5 degrees from the vertical, it fluctuates from 21.5 degrees to 24.5 degrees and back every 41,000 years. The greater the tilt is, the more intense seasons in both hemisphere become; summers get hotter and winter colder. The second weaker factor controlling seasonality is the shape of the earth's orbit. Over a period of 100,000 years, the orbit stretches into a more eccentric ellipse and then grows more nearly circular again. As the orbital accentricity increases the difference in the earth's distance from the sun at the orbit's nearest and farthest points grows intensifying the seasons in one hemisphere and moderating them in the other (at present the earth reaches its farthest point during the Sourthern Hemispher winter: as a result, southern winters are little cooler - and summers a little warmer - than their northern counter parts.) A third astronomical fluctuation governs the interpaly between the tilt and accentricity effects. It is the precession of the earth's spin axis, which traces out a complete circle on the background of the stars about every 23, 000 years. The precession determines whether summer in a given hemisphere falls at a near or far point in the orbit - in other words, whether tilt seasonbility is enhanced or weakened by distance seasonality, whether these two contributors of seasonality reinforce each other in one hemisphere, they oppose each other in the different hemisphere.

Over the past 800,000 years, the global ice volume has packed every 100,000 years, matching the period of the eccentricity variation. In addition, wrinkles" superposed on each cycle - small decreases or surges in ice volume - have come at intervals of roughly 23,000 and 41,000 years in keeping with the precession and tilt frequencies. Imbrie (quoted by Broecker & Denton (2)) working with a group called SPECMAP, later strengthened the case for the astronomical theory even more when he showed that the amplitude of the shorter signals have varied exactly as one would expect if the signals were having modulated by distance seasonality.

Much of the Northern Hemisphere warming and increased rainfall in the Holocene and Eemian was due to changes in the Earth's orbit, rather than carbon dioxide, climate modellers say (3).

GEOLOGICAL EVENTS

Workers have sought answers to both puzzles in the physics of the ice-sheets and the underlying rock which sinks under the weight of the ice. For example William R. Peltier and William T. Hyde of the University of Toronto (quoted by Broecker and Denton (2)) have built a theoretical model. It takes nearly 100,000 years for an ice-sheet to reach a critical size, at which point the rock below the earth's crust begins to flow rapidily and allows the burdened crust to sink. The surface of the ice sheet drops, warmed by the lower elevation, the ice can melt rapidily when the shorter period cycles bring the next episode of strong northern summers.

Uplift of Mountain Range

The uplift of mountain range may be one of the causes of the global climatic change or vice-versa. Molnar and England (4) reported that the high altitude of most mountain range have commonly been ascribed to late Cenozoic uplift, without reference to when crustal thickening and other tectonic processes occurred. Deep incision and recent denudation of these mountain ranges, abundant late Cenozoic coarse sediment near them, and paleobotanical evidence for warmer climates, where high mountain climates today are relatively cold, have traditionally been interpreted as evidence for recent uplift. An alternative cause of these phenomena is late Cenozoic global climate change: towards lower tempertatures, increased alpine glaciation, a stormier climate, and purturbations to humidity, vegetation cover and precipitation.

Volcanic Eruptions

Volcanic eruptions cause the climatic perturbations by injecting large quantities of sulphur compounds into the atmosphere. A principal component is sulphur dioxide, subsequent oxidation of sulphur dioxide in the atmosphere to sulphuric acid aerosol may lead to climate perturbations. Laj et al., (5) presented evidence to suggest that SO2 oxidation may occur at high latitutde by reaction with H2O2. Variation in sulphate and H2O2 concentration in four sections of greenland ice-core (corresponding to four volcanic events) show that high sulphate concentrations, from volcano fallout are accompanied by depletion of H2O2. They suggested that some of the volcanic sulpher was still in the form of SO2 when it reached greenland, and then was oxidized by H2O2 in greenland precipitation. As SO2 oxidation in the stratosphere is very slow, volcanic events that inject SO2 into the stratosphere could produce large depletions of distant H2O2 reserviors when the material is latter reinjected into the troposphere.

Postdiction studies showed that one possible source of seventeenth century BC climatic disturbances was the eruptions of vesuvius in AD 79 (as described by Pliny) is the archetype of explosive 'Plinian' eruptions which can cause recurrent local destruction and which may affect global climate through aerosol emissions Studies of the past eruptions of such volcanoes require accurate age determinations, especially if the eruptions are to be correlated with distant events. Vogel et al., (6) used materials from two Plianian eruptions of vesuvius, the AD 79 and the preceding avellino event, to test the 14C methods that often provide these ages. 14c ages for charred samples and corrected soil reservoirs buried by the Avellino eruption averaged to 3,360 $\pm$ 40 BP (1617-1703 BC), 200-500 years later that the original 14c ages of soil humic carbon and total organic carbon. Furthermore, a revaluation of the 14c data for the Mount st. Helens Yn and Aniakchak II eruptions suggested that these major plinian eruptions may also have occurred in the seventeenth century BC. The 14c ages of these three eruptions are identical, within uncertainties, to that of the 'Minoan'eruption of Thera (Santorini), which has often been correlated with mild-seventeenth century BC disturbances recorded in tree-ring series. Additional information is still required to distinguish which, if any, of the eruptions relate to these records.

There are some reports which prove that some volcanic eruptions emit carbon-dioxide, which may enhance the global warming and harmful for the human population. Recent investigations on Mount Etna (sicily) have revealed that volcanoes may release abundant carbondioxide not only from their active craters, but also from their flanks, as diffuse soil emanations. Baubron et al., (7) presented analysis of soil gases and air in the water wells on Vulcano Island which provide further evidence of such lateral degassing. Nearly pure carbon dioxide, enriched in helium and radon escapes from the slopes of the fossa active cone, adding a total output of 30 tonnes per day to the fumarolic crater discharge (about 180 tonnes CO_2 per day). This emanation has similar He/CO_2 and $^{13}C/^{12}C$ ratios to those of the crater fumaroles (300-500°C) and therefore a similar volcanic origion. Gases rich in carbondioxide also escape at sea level along the isthmus between the fossa and Vulcanello volcanic cones, but their depletion in both He and 13 C suggests a distinct source. Diffuse volanic gas emanations, once their genetic link with central fumarole degassing has been demonstrated, can be used for continuous volcano monitoring, at safe distances from active craters. Such monitoring has been initiated at Vulcano, where soil and well emanations of nearly pure CO_2 themselves represent a threat to the local population.

These volcanic eruptions may influence the global warming caused by glasshouse gases. These may have positive or negative feedback on greenhouse global warming depending upon the nature of the volcanic eruptions. GCMs should also consider the volcanic component. Mass and Portman (8) claim that actinometric observations of total solar radiation following major volcanic eruptions show a higher percentage of depletion in the polar regions than at lower latitudes. Mass and Portman used their observations to suggest that a 2% solar reduction, with a half- life of 1 year, following major volcanic eruptions is hemispherically representative and when introduced into a climate model produces results that are consistent with observed global temperature variations. Comparison of available data show that future observations will need to be made to quantify the role of tropospheric volcanic aerosols (9).

SOLAR VARIATION

The changes of solar irradiance in the past decade are associated with changes in the area of sunspots, which are dark cool regions of reduced irradiance, and faculae, which are bright regions of increased irradiance (10,11). The faculae effect dominates, and thus the irradiance increases as period of maximum solar activity. If the empirical relationships between the area of sunspots and faculae and the solar irradiance deduced from the data of past decade was valid on all time scales, longer-term changes of solar irradiance would hardly exceed 0.1%. Nor are large fluctuations of energy flow from the solar interior expected, even if the rate of nuclear energy production varied, the long diffusion time for a photon to reach the solar surface about 10000 years, would smooth out the variation (12). But changes in the efficiency of convection in the outer convective layers of the sun, caused, for example, by fluctuations in magnetic field strength can alter the rates of energy storage and release. Given the magnitude of surface layer energy reservoirs, this implies that there may be important changes in the solar irradiance on time scales from decades to centuries (12).

Hansen and Lacis (9) noted that the climate systeme would not be expected to respond in exactly the same way to a change of solar forcing as it would to a change of greenhouse forcing on the same magnitude and timing. Solar heating, for example, is concentrated towards low latitudes, whereas greenhouse heating is more uniformly distributed.

Eddy (13) has argued that reduced solar irradiance during the Maunder minimum of solar activity (1640-1720) may have been responsible for the little ice age when global temperature is estimated to have been as much as 1°C colder than today (14). Other possible causes for that climate change exist, including fluctuations of ocean heat transport and the increase of greenhouse gases between the 1700 and 1800s, but solar variability is one of the plausible mechanisims. Is it likely that a future decrease in solar activity may cancel greenhouse warming? That tack is taken in a recent report (15) to the chief of staff of U.S. president Bush, which contends that solar irradiance can be expected to decline early in the twenty first century. Thus the authors agreed against measures to slow down the increse in greenhouse forcing because such efforts "could turnout to be unnecessary or even harmful if a substantial natural cooling occurs in the twenty first century".

The possibility of a decline of solar irradiance by 2% although 20 times larger than the changes measured in the past 11 years, has not been ruled out categorically. The irradiance of some solar type stars has been observed to change by several tenths of a percent (16). The Smithsonian Observatory monitored the sun from mountain tops for the period 1902-1962, initially reporting variations of more than 1%. But comprehensive reviews of these data (17) concluded that the variations were due primarily to fluctuations of atmospheric transparency, and that they established only an upper limit of about 0.3% for solar variations. Evidence of solar change over a longer period is provided by measurements of the solar diameter. The diameter of the sun is affected by thermal energy storage in its outer layers because it must maintain global hydrostatic equilibrium on timescales greater than an hour. The small changes of solar diameter measured over the past 275 years are near the limit of detectability (18). Although the exact relations between the solar diameter and luminosity is uncertain, there probably were secular luminosity variation during that period, but they did not exceed several tenths of a percent (14,19,20).

These results do not imply that solar varibility has been unimportant in past climate change or that the sun's effects will be negligible in the future. For example, it has been shown that an increase of solar irriadiance by only about 0.3 -0.4%, just within the range of possibility, is one conceivable explanations of the observed global warmth of the 1930 and 1940 (21). Also, Wigley and Kelly (14) have shown that solar variation of several tenths of a percent may explain many of the climate variations of the past 1000 years.

It may be concluded only that solar irradiance would need to decline by about 2% to counter greenhouse climate forcing by all anthropogenic gases that will have accumulated in the atmosphere by the middle of next century, assuming no reductions in greenhouse emission. Although such a solar decline is not impossible, it is much larger than existing indications of solar variability.

Furthermore, solar fluctuations can be positive as well as negative. Measurement of solar irradiance during the next decade or two are crucial for direction of any long term trends and to help sort out cause and effect of observed climate changes (9).

Kelly and Wigley(22) found that solar variablity is unlikely to have accounted for more than a small fraction of the observed warming. Solar variation compete with anthropogenic greenhouse gases as cause of global climate change. Comparisons of available data show that solar variability will not counteract greenhouse warming(9). An international panel assessing greenhouse warming pointedly denies the validity of objections raised by a prominent minority (23).

SOUTHERN OSCILLATION

The discovery of a 2-year ticking in the record of EL Nino is fueling a growing awareness of biennial climate variations (24). Fire scare and tree growth chronologies (1700 to 1905) and fire statistics (since 1905) from Arizona and New Maxico show that small areas burn after wet springs associated with the low phase of the Southern Oscillation (SO), whereas large areas burn after dry springs associated with the high phase of the SO. Through its synergistic influence on spring weather and fuel conditions, climatic variability in the tropical pacific significantly influences vegetations dynamics in the south western United States. Synchrony of fire-free and severe fire years across diverse southwestern forests implies that climatic forces fire regims on a subcontinental scale, it also underscores the importance of exogenous factors in ecosystems dynamics (25).

Southern oscillation also effects the rainfall. The relationship between the relative variability of annual rainfall, the longterm mean annual rainfall, the latitude and the correlation between annual rainfall and the Southern Oscillation Index is examind, using data from 974 stations. A non-liniar relationship between these variables accounts for 94% of the variance in annual rainfall variability. Relative variability typically increases as mean annual rainfall decreases, as latitude decreases, and as the effect of the Southern Oscillation on varibility weekens as latitude increases (26).

The interannual variability of tropical convection related to the Southern Oscillation (SO) and regional climate anomallies was studied (27) from satellite-derived estimates of highly reflective clouds (HRC) during 1971-87. The Novel HRC data bank provides a particularly useful measure of tropical convection for the purpose of climate diagnostics, because of its length and continuity of record. For the first time, maps were presented of the patterns of correlation between the SO as well as regional rainfall anomalies, and convection over the global tropics.

Throughout the year, the SO (high SO phase defined by anomalously high/low pressure at Tahiti/Darwin) exhibits a highly significant nagative correlation with HRC in the equatorial Pacific but a much weaker positive correlation with Indonesia. The SO is correlated positively with HRC in the Amazon basin in boreal winter but negatively with HRC over central Africa throughout most of the year. The three equatorial convection centers tend to vary in unison, in particular those over the Amazon basin and central Africa, while the positive correlations of any of these centres with the SO are much weaker. Copious precipitation during the March-April rainy season of northern Brazil is associated with a southward displaced low-pressure trough and embedded wind confluence, as well as a southward shift of the convection belt in the sector extending from South America across the Atlantic into equatorial Africa. During abundant Nordeste rainy seasons, as in the high SO phase, convective activity tends to be enhanced over Indonesia but reduced in the equatorial Pacific. Copious rainfall in Subsaharan West Africa (Sahel) tends to be associated with high SO phase and thus intense convection over Indonesia and reduced convective activity in the equatorial central Pacific. Another new finding was the strong inverse relationship of Sahel rainfall with the convection over central Africa. Abundant Indian summer monsoon rainfall is accompanied by enhanced convective activity over the Indian Ocean and Indonesia and reduced convection in the equatorial Central Pacific, characteristics of the high SO phase (27).

The western equatorial Pacific warm pool (sea-surface temperature > 29°C) was observed to migrate eastward across the date line during the 1986-1987 El Nino-Southern Oscillation event(28).It seems that all the global climatic model should consider the southern oscillation as one of the mechanisms.

GREEN HOUSE GASES

Since the dawn of the industrial era, the atmospheric concentrations of seveal radiatively active gases have been increasing as a result of human activities. The radiative heating from this inadvertent experiment has driven the climate system out of equilibrium with the incoming solar energy. According to the greenhouse theory of climate change, the climate system will be restored to equilibirium by a warmig of the surface-troposphere system and a cooling of the stratosphere. The provided changes, during the next few decades, could far exceed natural climate variations in historical times. Hence, the greenhouse theory of climate change has reached the crucial stage of verification. Surface warming as large as that predicted by models would be unprecedented during the interglacial period such as the present. The theory, its scope for verification, and the emerging complexities of the climate feedback mechanisms are the themes of most importance (29).

Advances in the observational area of in situ gas sampling, Rasmussen and Khalil (30) have shown that the concentrations of several atmospheric trace gasses have increased significantly all over the globe during the last decade. Chemical analysis of gases trapped in ice cores reveal that the increase began during the 19th century. Model studies of the chemical balance of the atmosphere, the ocean and biosphere provide compelling arguments for attributing the observed increase to human activities (31,32).

The trace gases absorb infrared (IR) radiation(also known as terrestrial and thermal radiations) emitted by the relatively warmer surface and emit radiation to space at the colder atmospheric temperatures, laeading to a net trapping of IR energy with in the atmosphere (the greenhouse effect). The long term climate is governed by a balance between the absorbed solar radiations and the emitted IR. When there is enhanced IR trapping, for example, as a result of an increase in the concentrations of gases, excess energy is suddenly available to drive the climate system. The result, according to the theory, is a 'Vigorous" climate system, that is, a warmer globe with a more active hydrological cycle. The earth warms until the excess IR energy trapped by the greenhouse gases is radiated to space as IR emission. Because of the nonlinear interactions between the atmosphere, the cryosphere (ice and snow), the oceans, and the land, the predicted warming is not uniform but varies significantly with latitude, longitude, altitude, and season. The temperature and gradients that result from the nonuniform warming patterns can alter the general circulation of the atmosphere and ice oceans (29).

The greenhouse effect of the atmosphere was pointed out perhaps for the first time, by Fourier(33), who also suggested that human activity can modify climate. The formal foundation for the greenhouse theory of climate change was laid in 1896 by Arrhenius (34) who dealt with the climatic effects of changes in atmospheric CO_2. On the basis of his finding that a doubling of atmospheric CO_2 would warm the globe by 5 K. Arrhenius concluded that past glacial epochs may have occurred largely because of reduction in atmospheric CO_2. This conclusion is rapidly gaining favour because analysis of air trapped in glacier ice have revealed that CO_2 concentrations in the past fluctuated from about 270 to 300 parts per million by volume (ppmv) during interglacial periods to about 180 ppmv during periods of glacial advance (35-37). Finally, Callender (38) concluded in the late 1930s that human activities were causing an increase in atmospheric CO_2, and that this increase could initiate a global warming. In 1967, Manabe and wetherald (39) provided quantitative results for the CO_2 induced global warming, on the basis of a model that treated the global energy balance and radiative convective equilibirum. Manabe and Wetherald suggested that the surface and the troposphere are so strongly coupled by convection, i.e., vertical, heat transfer processes that the surface temperature governed not only by the radiative heating at the ground but by the heating of the joint surface troposphere system. Furthermore, this development made the stratosphere and stratospheric O_3 an integral part of the climate system. For example, a reduction in stratospheric O_3 would reduce the solar heating, which in turn would cool the stratosphere.The stratosphere would then emit less IR to the troposphere. Because of conventive coupling of the entire troposphere with the surface, the reduction in downward IR emission from the stratosphere can cause a surface cooling. The theory of climate change inolving CO_2 variations is now viewed as one of the major mechanisms for understanding climate variations of the past, including the Archean (40,41), the cretaceous (40,41) and the ages of Pleistocene (35-37).

The importance of the other trace gases was not recognised until the last decade when it was found that the addition of one molecule of CFC-11 ($CFCl_3$) and CFC-12 (CF_2Cl_2) can have the same greenhouse effect as the addition of 10000 molecules of CO_2 to the present atmosphere (42). The important non-CO_2 trace gases that may govern climate change include CFCs, CH_4, N_2O, Ozone(O_3) and more than a dozen synthetic chemicals (43-49). These developments set the stage for a number of theoretical and model studies (31,32,43-49). The conclusions of these studies can be summarized as follows: (1) the observed increases the trace gases concentrations from the mid 19th century to the present have significantly increased the radiative heating of the surface and atmosphere, (ii) there will be a severe reduction of the upper stratosphere O_3 and a cooling of the mid to upper stratosphere and warming of the surface and troposphere; and (iii) if the observed decadal trace gas growth rate continues unabated, the cumulative surface warming (predicted theoretically) would, in the next two decades, become large enough to manifest itself unambiguously in the global temperature records (29).

The Natural Greenhouse Effects On Climate

The important radiative active gases in the present atmosphere are H_2O, CO_2 and O_3. Their ability to intercept IR radiation (6.2-25 μm) has been demonstrated by satellite measurements (50). The absorption features of CO_2 lies between 13-17 um, O_3 between 9 to 10 μm and water in the entire spectrum domain (6.2-25 μm). On a global annual average basis, the surface emits about 390 W m^{-2}, whereas the radiation emitted to space as measured by satellite, is only about 237 W m^{-2}. The balance of 153 W m^{-2} is the IR radiation effectively trapped by the atmosphere (including clouds) which should be compared with the absorbed solar radiation of 237 W m^{-2}. Without the greenhouse effect, the surface would cool to about 255 K (49) from its present day value of 288 K, and the planet would be covered with ice.

The Man-made Greenhouse Effects On Climate

The human is dirtying the atmospheric window. Nearly 80% of the radiation emitted by the surface in the region from 7 to 13 μm escapes to space because the atmosphere, atleast in preindiustrial atmosphere, is quite transparent in this region. Because of the transparency, this region is referred to as the "Window". Outside the window region, most of the surface radiation is absorbed by the atmosphere and then reemitted to space at much colder atmospheric temperature(29).

Remarkably, most synthetic gases absorb strongly in the window region (49) and this absorption by pollutants has literally an effect of "dirtying the atmospheric window". The synthetic gases are extremely effective compared with CO_2. The relative energy trapped in the surface-atmosphere system is estimated to be (31,32) radiative heating (CO_2; from 300 to 600 ppmv) about 4 W m^{-2}; and radiative heating (CFC-11 and CFC-12; from 0 to 2 parts per billion by volume) about 1 W m^{-2}. CFCs are so effective because (i) their absorption occurs in the window region; (ii) their absoprtion band strengths are significant stronger than that of CO_2 (49) and (iii) the concentration of CO_2 is so large (300 ppm), that this gas is optically thick, and has its greenhouse effect scales logrithmically with the concentration whereas the greenhouse effect of CFCs scales almost linear. (49,29).

Climate Response To An Increase In The Greenhouse Forcings

The response of the climate to an increse in the greenhouse forcing consists of restoring the energy balance at the top of the atmospheric (39,49).

The role of feedback

An amount of negative feedback is the simple case of that due only to the radiation of heat by the atmosphere. A doubling of CO_2 would lead to an increase in heating of 4 W m-2, however as the atmospheric temperature rose in response, the amount of heat radiated would also increase, offsetting the initial heating. A temperature rise of 1.1 K would enhance the loss of heat by radiation just sufficiently to balance the gain of heat due to CO_2 (1).

More complex feedbacks, which may be positive or negative, involve other aspects of the system - water vapor, snow, sea-ice, the amount of cloud and its radiative properties. Detailed studies of such effects are described by Mitchell (51), who characterizes their strengths by a feedback parameter which is the change in radiative flux needed to cause a temperature change of 1 K.

Water vapor

A warmer atmosphere will hold more water vapour and enhance the original change in radiation due to CO_2. This is easily the strongest positive feedback and there is good agreement between different models (1).

Recently, Khalil and Rasmussen(64) showed that the concentration of H_2 increased at an average rate of 3.2 parts per billion by volume per year(a relative increase of 0.6 per cent per year. These increases originate from anthropogenic sources. Higher level of hydrogen will add more water vapor to the stratosphere, where it can affect stratospheric ozone.

Snow and sea-ice

A warmer atmosphere will reduce the highly reflective snow and ice cover, rendering in increased absorption of solar radiation and further increased in temperature (1).

Cloud cover

In most models with increased CO_2, cloud amount decreases and leads to increased absorption of solar radiation and a warming effect. In addition there is an increase in mean cloud height, resulting in colder cloud tops, which reduces the rate of emission of long wave radiation and therefore produces a further effective warming. The expected atmospheriic warming will increase the amount of water cloud relative to ice cloud. There is some evidence that water cloud is more persistent, at least at lower level and that this would substantially reduce the positive feedback to due to cloud cover (51).

Cloud radiative properties

The representation of cloud and their radiative properties is a source of major uncertainty in modeling climate change. Estimates of the effect of range from strong negative feedback to a weak positive feedback (1).

Slingo (63) presented the results of a study that used a three dimensional GCM, which should give mean reliable estimates. The top of atmosphere rediative forcing by doubling carbondioxied concentrations can be balanced by modest relative increases of about

15-20% in the amount of low clouds and 20-30% in liquid water path, and decreases of 15-20% in the mean drop radius (depending on the version of the model). This indicates that a minimum relative accuracy of about 5% is needed, both to simulate these quantities to climate model and estimate climate response by monitoring them over extended periods from satellite plateforms.

Ocean-atmosphere interaction

The oceans influence the climate response in two fundamentally important ways. First, because of the importance of the H2O greenhouse feedback, the response of the tropospheric temperatures and even the land surface temperatures would be governed by warming of the ocean surface. If for some reasons, the ocean does not respond to the greenhouse heating, H2O feedback would be turned off since increased evaporation from the warming ocean is the primary source of increasing atmospheric water. Second, oceans can sequester the radiative heating into the deeper layers, which, because of their enormous heat capacity, can significantly delay the warming (52-54).

Ozone Layer Depletion

Stratospheric ozone

There are no known important sinks for CFCs in the troposphere(55). They are ultimately broken down in the atmosphere above 25 Km by photoloysis which frees the highly reflective chlorine. As a result, their tropospheric life time are long; about 65 years for CFC13 and 100 years for CF2C12. The chlorine competes with nitrogen oxides for odd oxygen species including 03, and the result is an efficient catalytic destruction of 03 by chlorine until it is removed from the system as HCl. Atmospheric chemistry models predict that, at present rates of emission of CFCs, severe 03 reduction ranging from 15% near 30 km to about 40% near 45 km (55) will result. The stratosphere derives its name because of the increase in the temperature with altitude, that is, a stably stratified region. This thermal inversion is maintained by 03 solar heating, and hence 03 destruction would cool the stratosphere. For example, a 50% reduction in 03 would cool the upper stratosphere by as much as 20 K (49). furthermore, 03 modulates the solar and IR flux incident on the troposphere. A reduction in stratospheric 03 would allow more sunlight to reach the surface and tend to warm it. This warming will be offset by reduced IR fluxes emitted by the cooler atmosphere and also by a reduced 03 green house effect. The net effect, i.e.,surface warmig or cooling, critically depends on the veritical distribution of the 03 change (39,49).

Tropospheric ozone

The 9.6 μm band of 03 is an important source of IR opacity for the troposphere (56-58). Photolysis of 03 in the troposphere produces oxygen which in turn reacts with water to release hydroxy (OH) redicals through H2O+O=2OH. The highly reactive OH is the primary sink for many tropospheric gases and pollutants including 03, CH4, CO, and NO (56,57). Hence, increases in CH4 such as those during the last century, could have caused a substantial (20 to 40%)

reduction in OH (56,57), which in turn, could cause an increase in tropospheric O_3 by as much as 20% (56-58). Since CH_4 oxidation leads to the formation of H_2O, an increase in CH_4, an important greenhouse gas, can lead to an increase in H_2O in the atmosphere. Likewise, an increase in CO concentration can tie up more OH in the oxidation of CO. Thus through chemical reactions, an increase in either a radiatively active gas such as CH_4 or even a radi-a-tively inactive gas such as CO can increase the concentration of several important greenhouse gases (56,57).

Integration

The fundamental effect of the time-dependent trace gas radiative heating is to push the climate system into a state of disequilibrium with the incoming and outgoing energy fluxes. The theory states that the earth tries to restore the equilibirium by warming and by emitting the excess energy to space, about 0.5 to 1 K of this warming should have already occurred. Observed records do reveal a warming of the order of about 0.5 K, but the temporal history of the warming is unlike the pattern anticipated by the theory. The warming occurred abruptly and the bursts. Either the observed warming is not related to the increase in the trace gases or current theories and models of ocean-atmosphere interactions are inadequate to captive the transient response of the climate system to a time-dependent variation in the external forcing.The theortical understanding of the climate system is by no means complete. Some issues need to be resolved (29).

If the trace gases continue to increase, the warming, according to the sensitivity of GCM, can reach 5 K during the next century. A warming of this magnitude would be unprecedented because the present climate is just coming out of the peak of the interglacial period. A warming of 5 K above this peaks would likely be beyond the range of validity of current models, particularly with respect to the response of the world's ice sheet and changes in the ocean circulation. On the other extremes, the Vostok ice core (Antarctica) data (35-37) reveal that CO_2 has decreased from about 280 ppm during interglacial to about 190 ppm during glacial time. Furthermore, it has been suggested that the CO_2 variations themselves may help explain the glacial interglacial cycles (35-37). The radiative cooling due to a CO_2 decrease from 280 to 190 ppm is only 2.5 $W\ m^{-2}$; for this cooling to institute a major ice age requires a climate sensitively even larger than those of current GCMs (35-37). Studies (59-62) have made compelling arguments to link cloud microphysics with climate change.

The increase in absolute humidity with surface temperature, could lead to an increase in cloud liquid water content (59). The resulting increase in cloud visible optical depth will enhance cloud reflection of solar radiation and provide negative feedback (59). Another mechanism of negative feedback (60) suggests that human activities have also increased tropospheric particulates, and this will lead to an increase in number of cloud droplets. A variant of this feedback, which involves biogenic processes, has been proposed by Charlson et al (61). They pointed out that the value of number of droplets over the oceans is predominantly sulfuric acid droplets derived from emission of dimethyl sulfide by marine organisms. They suggested that an increase in surface temperature may cause an increase in emission of dimethyl sulfide and hence number of droplets. These suggestions received indirect response from a satellite study (62) in which ship tracks that had been spotted amidst marine stratocumulus clouds were consistently brighter in visible wave length. This brightness was not due to particulates from the ship smoke (these would be darker) but was due to enhanced cloud condensation nuclei, from the sulfur in the smoke, which enhances cloud visible optical depth.

Slingo (63) presented the results of a study that used a three dimensional GCM, which should give mean reliable estimates. The top of atmosphere radiative forcing by doubling CO_2 concentrations can be balanced by modest relative increases of about 15-20%. in the amount of low clouds and 20-30% in liquid water path, and decreases of 15-20% in the mean drop radius (depending on the version of the model). Recently, Khalil and Rasmussen (64) showed that the concentration of H_2 increased at an average rate of 3.2 ppbv by volume (a relative increase of 0.6% per year). These increases originate from anthropogenic sources. Higher level of hydrogen will add more water vapor to the stratosphere, where it can affect stratospheric ozone.

The increase in CH_4 of nearly a factor of 2, along with the various ways it can influence the atmosphere, makes it one of the most intriguing greenhouse gases. An increase in CH_4, because of its coupling with OH, can cause an increase in tropospheric O_3. Also, CH_4, which is oxidized to H_2O in the atmosphere, is an important source for stratospheric H_2O (49,65). In principle, CH_4 can be oxidized to two H_2O molecules. The observed increase in CH_4 from 0.7 to 1.7 ppmv (32) can lead to a maximum increase in H_2O of 2 ppmv. The observed H_2O concentration in the stratosphere increases from about 2 to 5 ppmv near the tropical tropopause (10 to 16 km) to about 5 to 8 ppmv above 30 km (66). Most of this increase could be due to an increase in CH_4 from the preindustrial to the present time. Stratospheic H_2O enhances the greenhouse effects substantially, a doubling of stratospheric H_2O (entire column) from 5 ppmv can have nearly the same warming effect(39,43-48) as that due to the observed CO_2 increase during the last century. Because H_2O and other gases are also transported poleward and downward in the stratosphere (67), H_2O produced in the tropical middle and upper stratosphere by CH_4 oxidation can be transported to the polar lower stratosphere. This H_2O may contribute to the recently discovered polar stratospheric clouds(68) in two ways. First, H_2O may provide the required source for the liquid and solid H_2O in the clouds (69,70). Second, the local radiative cooling from the increased H_2O (71,72) may provide the temperature drop needed for cloud formation. Since the radiative response time of the lower stratosphere is of the order of 50 to 100 days, an increase in the cooling rate of 0.02 to 0.05 K day-1 may cool the region by 2 to 3 K (71,72). This cooling is sufficient for H_2O even at 5 ppmv , to condense in the antarctic lower straosphere during winter. This climatic effects of CH_4 may involve complex interactions among atmospheric chemistry, radiation, thermodynamics and dynamics.

GLOBAL CLIMATE, SEA LEVEL RISE AND ECOSYSTEMS

Global sea level could rise due to climatic changes by as little as 4 cm or as much as 40 cm by the year 2030. However, since this range represents the combination of extreme assumptions, these values should be considered highly unlikely, though still possible.

The best estimate is that by the year 2030 global sea level will be 14-24 cm higher than today. The implied rate of rise is about 4-6cm per annum, which is 2-6 times faster than that over the last one hundred years. Regional changes in sea level may differ from the global average. These regional changes may be due to land movements, subsidence due to sedimentation and ground water pumping, as well as lingering crustal movement following the last era of glaciation. Even if changes in green house forcing stopped abruptly in the year 2030, global sea level would continue to rise for many decades, and possibly for hundreds of years. This sea level rise is a result of the long response time of the polar ice sheets and the slow processes of heat transfer from the atmosphere to the ocean (73).

The Ganges-Brahmaputra delta, much of which is less than one meter above sea level and home to million of people, is vulnerable to flooding from storm surges and to change in ecosystems due to sea level rise. Encrochment of the sea in the Nile delta also threatens to rob Egyt of a substantial portion of agricultural land as well as making thousands of people homeless. Other deltas at risk include the Huang-Ho, Indus, Amazon, Mississipi and PO (74).

Most of the land area of the 1,190 small islands constituting the Republic of Maldives is less than two meters above mean sea level. Last November, the Govt. of Maldives hosted a small states conference on sea level rise, a ministerial level meeting designed to identify common problems associated with global climate changes. Geological data suggest that the growth rate of the coral can match the rate of sea level rise - provided the rise is not too rapid. How fast the coral can grow depends on the species and the environment; in some circumstances it can grow as fast as 10 mm per annum (75).

In the maldives and elsewhere coral is mined for building material. Maintaining the health of the coral may be critical for island survival. The mangrove forests, an important resource, are also a natural means of island protection and maintenance. These, too, could be threatened by fast rate of sea level rise, exacerbated by excessive cutting for fuel wood and other uses. In southern Kiribati and in Tuvalu island states some fresh water wells are already contaminated by salt water. Many pacific islands are dependent on root crops for a principal part of their diet and crops such as pulaka and taro are grown in shallow pits which risk salt contamination (73).

It is already a problem in polder areas in the Netherlands, which are below sea leve. Flushing and rinsing of the polders and canals with fresh water is necessary, but in dry summer crop damage is unavoidable due to insufficient supplies of fresh water (76). An example of a change in land use of the agricultural area is in Bangladesh, where deeper inundation in the interior flood plain may lead to a reduction in the area suitable for cultivating the dominant strain of paddy rice and necessitate a shift to a lower yielding variety (73).

for most small developing countries, the construction of sea walls, dykes (Levees) and storm surge barriers, such as the delta works in the Netherland and the Thames Barrier in UK, are not feasible options. One of the main lines of defence is the careful management of coral reefs, mangroves and other natural features, and this should form the key element in national precautions against sea level rise (74).

The impending problems of global warming are causing alarm

in countries like the Republic of the Maldives. Here the general public as wll as the Government is aware of the possibility of rising sea level and their vulnerability to the ecosystems. This is a sense of frustration, if not anger, that they could be the victims of a global change not of their making. Their massage is directed at the industrialized countries of the world and comes across loud and clear, STOP (73).

GLOBAL CLIMATE AND AGROECOSYSTEMS

There is increasing evidence that rising emission of carbon dioxide, methane, nitrous oxide and other radiatively active gases will lead to an increase in the average surface temperature of the Earth. In addition to changes in global temperature there will be changes in precipitation. The rate and magnitude of these climatic changes is such that they could have a significant impact on agricultural potential in many parts of the world. The nature and extent of such changes will vary regionally and are dependent upon the level of climatic change in each region, the agricultural and management practices used at present and the ability of agricultural systems to adapt to changes (77).

If greenhouse gas emission continues to increase at their present rates, the best estimates indicate a global warmig of approximately 0.5 ^{o}C by 1995-2005, 1.5^{o}C by 2015 to 2050 and 3°C by 2050-2100 with greater increases evidence at higher latitudes (78). Such increases appear to be small, yet the Earth's temperature has not varied by more than 1-2°C in the past 10,000 years. One area of uncertainty is how the climate may change en route to the new high CO2 climate. Although the broadscale effect is for a smooth global increase in temperature, there may be sudden and step like changes in some regions (77).

At high latitudes and high midlatitudes (i.e. over 45°N and S) rainfall may increase by 5% in summer and possibly upto 15% in winter (79). At lower midlatitude (30°-45°N and S) summer rainfall is likely to become very limited and winter rainfall may decrease by 5-10%. Global warming may migrate Prairie forest border to north, decrease crop yield by 3-17 per cent and shift in geographical locations of potential crop regions by several kilometers (80). At low latitudes (0-30°N and S) rainfall is expected to be enhanced by about 5-10%. Increases in potential evaporation can be expected to accompany the increase in temperature. Marked water stress may occur in areas where there is no accompanying increase in precipitation. There are important differences between GCMs in the magnitude of changes in temperature they project and more importantly for agriculture, there is very little agreement on the likely changes in regional patterns of precipitation. However, there is some agreement concerning what may prove to be the most important changes of climate for agriculture. These critical changes are warming in the high latitudes, northern advance of monsoon rainfall and reduced soil water availability. Warming in high latitude will reduce temperature constraints on high latitude agriculture, increase the competition for land here and result in the northward retreat of the southern margin of boreal forest. In a warmer world intertropical convergence zone would be likely to advance further northward into Africa and Asia. If this were to occur then the total rainfall in the Sahel and India could increase. Rainfall could also be more intense in its occurrence and propagate flooding and erosion (77).

Probably the most important consequences for agriculture would be stem from a reduction in soil moisture due to high rate of transpiration from plants and of evaporation from soil surfaces exposed to higher temperatures. As yet there are no certainties about the regional pattern of soil water changes that may occur. However, there are a number of regions where decreases in soil water are predicted by three GCMs (81-83). Areas which may suffer decreases of soil water between December and February include the Horn of Africa, Southern Africa, and Western West Africa, parts of southern Asia, and the Arabian Peninsula, eastern Australia, and the southern half of North America. Decreases in soil water between June and August may occur in West Africa, the Maghreb and Horn of Africa, Western Europe, China and Soviet Central Asia, S.W. United States, Mexico and Central America, Eastern Brazil and North eastern and western Australia.

Research suggests that for many species a doubling of CO_2 will lead to a 10 to 15% increase in dry matter production providing all other factors remain constant. For wheat and barley, yield increses of as much as 40% have been suggested (84). The typical aerable crops of central and Northern Europe and similar latitudes should benefit more than those of tropical areas, where maize, sorghum, sugarcane and millet are staples. At some point most plants will suffer from water stress when evaporative demands are greater than water supply. Under such conditions plants being to wilt causing the stomata to close. This restricts water loss but also limits CO_2 assimilation, and, thereby plant growth and yield. Higher level of atmospheric CO_2 will lead to increase efficiency of water use by reducing transpiration rates (77).

While agricultural potential at midlatitudes generally decreases towards the poles due to smaller thermal inputs, so the some increases of temperature will have greater relative effects on crop potential at higher latitudes than at lower latitudes. In addition because the size of CO_2 induced temperature increase is likely to be greater at higher latitudes, substantial effects on crop potential can be anticipated in the northern regions. For example, under one scenario of climate change for a doubling of CO_2, the mean annual temperature increase in Northern Finland is 4.7 °C related to present temperature, whereas the Southern finland it is 4.1 °C (85). For this reason, quite large increases in productive potential are likely at high midlatitude.

The growing period is taking as the number of months with average temperatures above 5 °C and rainfall exceeding half of its potential evapotransiration (86). In Northern Europe there would be an increase in the potential growing periods, where as around the Mediterranean the growing season could shorter significantly due to warming and drier conditions in spring and autumn. There is a shift of cropping potential from southern Europe countries to Northern Europe.

A number of studies have examined changes in the climatic limits for a range of crops to northen hemisphere countries under a variety of climatic scenarios (87). They suggest that a 1 °C increase in mean annual temperature would tend to advance the thermal limit of cereal cropping in midlatitude northern hemisphere regions by about 150-200 km, and to raise the altitudinal limit to arable agriculture by about 150-200 m. A 4 °C warming which is of the order projected for a doubling of CO_2 would probably raise climatic zones in the European Alps by 450-650 m, making them similar to the levels of those today in the Pyrences which lie 300 km south of the alps.

In cool temperature and cold regions yields of most crops can be expected to increase with increasing temperature, except where moisture is a limiting factor. Indeed it has been suggested that Fennoscandia could gain from global warming more than any other part of the world. Under a 2XCO2 climate, yield of barley and oat in Finland would be raised by 9-18%, depending on the region (88).In Iceland the carrying capacity of improved grassland for sheep is estimated to increse by about two and a half times, and on unimproved range land by more than a half (89). This increase of biomass potential in northern Europe is in contrast in the decreases in Sourthern Europe that are implied by current GCM projections of changes in temperature and rainfall. It has been suggested that there thus could be a singificant northward shift of the balance of agricultural resources in the Europe (90).

There are indications that global warming could lead to a decrease in cereal production in Northern America mainly due to the accompanying reduction in soil moisture (77). Under a 2XCO2 climate the yield of maize is likely to fall by 16-25%, even assuming the crop is irrigated (91).In other areas, such as Mexico the combination of decreases in soil moisture and increases in heat stress are likely to result in a lowering of wheat yields (92).

In the Leningrad region of the USSR the rises in temperature and precipitation anticipated under a 2 XCO2 climate could initially raise the yields of rye. However, the higher rainfall is likely to cause more leaching and erosion, eventually decreasing soil fetility and therefore, rye yields (93). In other areas increased rainfall could be beneficial to crop yields. In China there are indications that global warming could lead to increased rainfall during the summer monsoon. Under a 1°C warming with precipitation increases of 100 mm, national yield of rice, maize and wheat are estimated to increase by 10% (94). Yields of rice are also expected to increase in Japan (95), but decrease in other rgions of Southern Asia due to more rapid crop growth (cited by Parry et al. (77)). In Northern and East Africa any change in precipitation could substantially affect maize yields and grass growth and thus the carrying capacity of range lands. Any change in the frequency of dry years would significantly influence the average output of agriculture in these areas (77).In the U.K. the temperature limit for the successful ripening of grain maize, which at present lies in the extreme south of England, would gradually shift northwards as temperatures begin to rise (77).

Under a 2XCO2 scenario the climate of Iceland is similar to the climate of Northern Britain today. The present-day alarming types in analogue regions are a useful indicator of the adaptive stratgies likely to be required to retune agriculture to altered climatic circumstances in the study regions (96). It is estimated that barley, which is at present a highly marginal crop, may become cultivable throughout lowland iceland due to longer growing seasons but losses due to pests and diseases (which are at present minimal in Iceland) could increase to as much as 15% (89).

In midlatitude, midcontinental areas there are indications that reduced amounts of rainfall and snowfall could significantly curtail the rates of ground water re-charge and accelerate the rates of ground-water depletion. In areas where increases in the intensity of rainfall occur, there may be more surface runoff, less percolation of water through the soil and less available soil moisture. Such effects could occur in tropical regions,particularly with monsoon rainfall,and in some maritime mid-latitude regions where it has been suggested that more rain may fall as convective thunderstorms (97). In these areas any decrease in water availability will place an additional strain on crops already stressed by higher temperature. It has been estimated that the yield of quick maturing rice varieties currently grown in Northern Japan would probably increase by about 4% with the temperature increas estimated for a 2XCO2 atmosphere (95). In Saskatchewan (98) and the central region of USSR (93) winter wheat would give higher yields than spring cereals in a high CO2 environments. The potential for the production of sunflower in the UK could shift northwards by about 300 km for each 1 °C rise in mean annual temperature (90). Similarly, northward shifts of citrus, olives and vines have been projected for southern Europe (99) and southward shifts of land use have been suggested in the southern hemisphere (100).

Very substantial increases in the need for and the cost of irrigation,are likely to occur which will probably lead to significantly higher costs of production and possible shifts towards less water demanding uses (101). Conversely increases in the amount or intensity of rainfall particularly in regions characterized by monsoon rainfall, will require changes in management to prevent soil erosion. In addition, increases in fertilizer use may be required in aeas where greater rainfall is likely.

Present indications are that with a changing climate global food production could be sustained at levels sufficient to meet world demand. It is likely that overall levels of production will be maintained through a combination of shifts of agricultural zones and adjustments in agricultural technology and management. However, there are several inherently vulnerable regions,particularly in Africa and South America and elsewhere in the lower middle and lower latitudes, where changes in temperature and precipitation may further stress the already limited productive capacity. Threse regions have identified as the areas most at risk from the effect of climatic changes.If however, we are to improve our understanding of significance of changes in our climate and the consequences for mankind it is important that the increase substantially the range of current research into how agricultural can best adapt to counter or even profit from such changes.(77).

OZONE LAYER DEPLETION AND ECOSYSTEMS

If the ozone layer is depleted by a chlorine from CFCs, more UVB radiation reaches ground level. For a 10 per cent ozone reduction, the flux of the biologically damaging UVB radiation (with wave lengths between 290 and 320 nanometers) at the Earth's surface increases by 20 per cent (102). Such a change will increase the incidence of cataracts and skin cancer, both melanoma and nonmelanoma, particularly in the fair-skinnned portion of the population. Non-melanomic cancers are expected to increase by 40 per cent, for a 10 per cent ozone depletion.

An increased flux of UV radiation could also damage the eyes of seals basking on the ice and of male Emperor penguines- and their chicks- who stand for many weeks under the depleted ozone region. It can damage the DNA molecules in the cells of living organisms, that is, it causes mutations.The increased intensity of UVB radiation under the ozone hole could reduce the rate at which phytoplankton (or

Diatoms) are prodt ced at and near the surface of the southern ocean. Phytoplankton, by photosynthesis, use chlorophyll to convert visible sunlight into carbohydrates.Since the organisms from the basis of the southern ocean food web, on which the shrimps like krill feed, the krill would be under-nurished. In turn, this would reduce the food source for fish, squid, birds, seals and whales, all of which feed on krill. The entire southern ocean ecosystems could be harmed (103).

There can also be consquences of the ozone depletiun on atmospheric dynamics. Reduced ozone reduced stratospheric heating rates. There is evidence, that, in mid-October, the lower stratospher over antarctica is now 5 degree Celcius cooler than it was 15 years ago (104). Further, the break down of the southern polar vortex is delayed by a month or so. Could this delay be significantly increased, so that the vortex never broke down from one year to the next (103)?

The greatst changes between 1979 to 1980 and 1986 to 1987 are reduced ozone amounts, by 10 per cent or more over the Antarctica from september to November. Ozone reduction of 5 per cent over 8 years have occurred at 50 degree latitude south in most months of the year. This is uncomfortably 'close to home' for people living in Tasmania, New zealand and southern south America (105). There are also some indications of reduction in the arctic, especially in February and March and a slight increase in the tropical northern hemisphere in spring (103).

If the observed yearly average Antarctica ozone reduction over 8 years (which is about 10 per cent) were spread over the globe, the depletion would effectively be diluted to about half a per cent in the 8 years. But the global mean indicates a global depletion of about 2 per cent in the 8 years. So some other- as yet unknown- processes are important world wise (103).

FUTURE RESEARCH ACTIONS

There is a need to integrate GCMs models to pridict global warming due to 2xCO_2 situations with crop models, because there are major difficulties in applying these results to biological model. The problems associated with extrapolating estimates of changes in global climate to the probable regional changes in weather during the transition period are formidable. Many of the existing weather driven biological models contain empirical constants which rely on the present intercorelation of the various weather factors. Some of these relationships may well alter under the changing climate, eg., that of temperture with solar radiation.

There is a need for modellers of crop pest diseases systems to be aware of the uncertainty in the climate models. A strong case can be made for the promotions of multidisciplinary studies involving meteorologists and biologists so that the fullest possible benefit may be drived from the investment of brain and computer power in climate studies which will take place over the next few decades (106).

What is needed most is support for research using existing data, continuation and improvement of observations and, perhaps most of all, training of people to define and analyse future data. Future observations will need to be made to quantify the role of tropospheric aerosols. Policy makers may be advised for phasing out

CFCs, improving energy efficiency, increasing recycling, reducing deforestation and planting trees in appropriate places. They can also be advised for unprecedented international cooperation to achieve a turnaround of global greenhouse-gas emission. It is difficult to predict its prospects for success. Government should give much higher priority to research and development on energy source that produce little or no greenhouse gases. It must foster conditions leading to population stabilization, if we are to preserve the global climate and environment (9).

Photosynthesis on land removes about 100 billion tons of carbon from the atmosphere annually in the form of CO2. Plant and soil respiration each return about 50 billion tons. Fossil-fuel burning and deforestation release in to the atmosphere respectively about 5 and 2 billion tons. Physiochemical processes at the sea surface release about 100 billion tons into the atmosphere and absorb about 104. The net atmospheric gain is about 3 billion tons annually (107). There is a need to establish such research institute which may help in finding out the aquatic and terrestrial plants which could use the CO2 luxuriently with less respiration.

More research on the atmosphere is required, to understand the spring time Antarctica ozone depletion in the global context. Such rsearch will need more labortaroy studies, more Antarctica experiments, more theory, and more modelling, particularly involving supercomputers to study whether southern hemisphere dynamics of the global climate may be changed by the spring time Antarctica ozone depletion. More measturements of the spectrum and intensity of UVB radiation at different latitudes and longitudes are desirable. More research on the many possible atmospheric and other consequences of the ozone hole also needed.

The ozone hole has worsened drastically in only a decade. But the atmospheric life time of freons is so long, typically 75 years or 100 years for CFC-12, that the ozone problem will undoubtedely continue to worsen. The deleterious effects of most of the CFCs that man has produced have not yet occurred. So should we carry on manufacturing freons (103)?

In the west, the CFCs production rate is still about 100 thousand tonnes per year, for CFC-11 and for CFC-12 (UK Stratospheric ozone review group 1987). The worldwise production rate is of the order of a million tonnes per year. This gives rise to atmospheric concentration of CFC 11 of only about 0.2 parts per billion by volume, and CFC 12 of 0.3 parts per billion by volume, both concentrations rising by about six percent per year. This must be folly. Government action is required. The September 1987 Montreal convention to halve the chlorofluorocabons production in the world by the end of Century is to be welcomed, but the problem will still worsen with that halving. A much more drastic reduction in the production and use of chlorofluorocarbon is called for. That many nations are committed to banning the further production of CFCs, as reported (Times, 5 May 1989) at the united nations Environment programme Conference held in Helsinki, is excellent news. International regulations to eliminate the manufacture of CFCs must be framed and agreed soon. CFCs should only be used where use is essential. Non-essential uses should be banned. Surely there is no justification for releasing CFCs to the atmsophere by using CFCs as aersol can propellants - other propellants are already available. Their use should be encouraged by clearly labelling them as "ozone friendly" i.e. not harmful to the ozone layer. Alternative methods for the production of hard plastic foams must be found. Now non-toxic safely bio-degradable refrigerants for refrigerations, freezers and air conditioners, and new solvents for industry, need to be designed and produced. Such CFCs replacement should neither lead to the destruction of ozone layer nor contribute to the greenhouse effect. Already existing CFCs should not be allowed to escape in the atmosphere from refrigerators and air conditioners that has outlived its useful life. The CFCs should be liquidified and stored until the chemical industry can destroy them effectively into environmentally harmless products (103). Future research actions for Indian context have also been sugosted (108).

REFERENCES

1. Hume, C.J. and Cattle, H. (1990). The greenhouse effect - Meteorological mechanisms and models. Outlook on Agriculture 19: 17-23.
2 Broecker, W.S. and Denton, G.H. (1990). What drives glacial cycles? VIGYAN Scientific American January: 47-56.
3. Aldhous, P. (1990). World climate conference. Nature 348: 98
4. Molnar, P. and England, P. (1990). Late Cenozoic uplift of mountain ranges and global climate change: Chicken or Egg? Nature 346: 29-34.
5. Laj, P., Drummey, S.M., Spencer, M.J., Palais, J.M. and Sigurdsson, H. (1990). Depletion of H2O2 in a greenland ice core: Implications for oxidation of volcanic SO2. Nature 346: 45-48.
6. Vogel, J.S., Cornell, W., Nelson, D.E. and Southon, J.R. (1990). Vesuvius/Avellino, one possible source of seventeenth century BC climatic disturbances. Nature 344: 534-537.
7. Baubron, J.C., Allard, P. and Toutain, J.P (1990). Diffuse volcanic emissions of carbon dioxide from Vulcano island, Italy. Nature 344: 51-53.
8. Mass, C.F. and Portman, D.A. (1989). Major volcanic eruptions and climate. A critical review. J. Climate 2: 566-593.
9. Hansen, J.E. and Lacis, A.A. (1990). Sun and dust versus greenhouse gases: an assessment of their relative roles in global climate change. Nature 346: 713-719
10. Sofia, S., Demarque, P and Endal, A. (1985). Am. Sci. 73: 326-333.
11. Foukal, P. and Lean, J. (1990). Science 247: 556-558
12. Gough, D. (1977). The solar output and its variations : 451-474 (Colorado Univ. Press, Boulder).
13. Eddy, J.A. (1976). Science 192: 1189-1202.
14. Wigley, T.M.L. and Kelly, P.M. (1990). Phil. Trans. R. Soc. A 330: 547-560.
15. Sentz, F. Jastrow, R. and Nierenberg, W.A. (1982). Scientific perspectives on the greenhouse problem (Marshall Institute, Washington, DC.
16. Radick, R.R., Lockwood, G.W. and Baliunas, S.L. (1990). Science 247: 39-44.
17. Hoyt, D.V. (1979). Rev. Geophys. Space Phys 17: 429-458.
18. Gilliand, R.L. (1981) Astrophys. J 248: 114-1155.
19. Gilliand, R.L. (1982). Astrophys. J 253: 399-405.
20. Endal, A.S., Sofia, S. and Twigg, L.W. (1985). Astrophys. J 290; 748-757.
21. Hansen, J. et al. (1981). Science 217: 957-966.

22. Kelly, P.M. and Wigley, T.M.L. (1990). The influence of solar forcing trends on global mean temperature since 1861. Nature 347: 460-462.
23. Kerr, R.A. (1990). New green hosue report puts down dissenters. Science 249: 481-482.
24. Kerr, R.A. (1990). The climate system as a ticking clock. Science 249: 1246- 1247.
25. Swetnam, T.W. and Betancourt, J.L. (1990). Fire-Southern Oscillation relations in the Southwestern United States. Science 249: 1017-1023.
26. Nicholis, N. and Wong, K.K. (1990). Dependence of rainfall variability on mean rainfall, latitude, and the Southern Oscillation. Journal of climate 3: 163-170.
27. Hastenrath, S. (1990). The relationship of highly reflective clouds to tropical climate anomalies. Journal of climate 3: 353-365.
28. McPhaden, M,J. and Picaut, J. (1990). El Nino - Southern Oscillation displacement of the Western Equatorial Pacific warm pool. Science 250: 1385-1388.
29. Ramanathan, V. (1988). The greenhouse theory of climate change: A test by an inadvertent global experiment. Science 240: 293-299.
30. Rasmussen, R.A. and Khalil, M.A.K. (1986). Science 232: 1623.
31. Bolin, B., Doos, B.R., Jager, J., and Warrick, R.A. (1986). Eds., The Greenhouse Effect, Climate Change and Ecosystems. Wiley, New York.
32. Dickinson, R.E., and Cicerone, R.J. (1986). Nature 319: 109.
33. Fourier, J.B. (1827). Mem. Acad. Sci. Inst. Fr. 7: 569.
34. Arrhenius, S. (1896). Philos. Mag. 41: 237.
35. Jouzel, J. et al. (1987). Nature 329: 403.
36. Barnola, J.M., Raynaud, D., Korotkevich, Y.S., Lorius, C. (1987). Nature 329: 408.
37. Genthon, et al. (1987). Nature 329: 414.
38. Callendar, G.S. (1938). Q. J. R. Meteorol. Soc. 64: 223.
39. Manabe, S., and Wetherald, R.T.(1967). J. Atmos. Sci. 24: 241.
40. Kiehl, J.T. and Dickinson, R.E. (1986). J. Geophys. Res. 92: 2991.
41. Barron, E.J. and Washington, W.M. (1984). J. Geophys. Res. 90: 1267.
42. Ramanathan, V. (1975). Science 190:50.
43. Wang, W.C., Yung, Y.L., Lacis, A.A., Mo, T., Hansen, J.E. (1976). Science 194: 685.
44. Ramanathan, V. (1980). Energy climate Interactions , Reidel, Dordrecht, pp. 269-280.
45. Lacis, A. et al. (1981). Geophys. Res. Lett. 8: 1035.
46. Wang et al. (1986). Rev. Geophys. 24: 110.
47. Tricott, C. and Berger, A. (1987). Climate Dyn. 2: 39.
48. Bruhl, C. and Crutzen, P.J. (1988). Climate Dyn. 2: 173.
49. Ramanathan, V. et al. (1987). Rev. Geophys. 25: 1441.
50. Hanel, R.A. et al. (1972). J. Geophys. Res. 77: 2029.
51. Mitchell, J.F.B., Senior, C.A. and Ingram, W.J.(1989). CO2 and climate: a missing feedback? Nature 341: 132-134.
52. National Research Council (1982). Carbon Dioxide and climate. A second Assessment. National Academy Press, Washington DC.
53. Bryan, K. et al. (1984). In: Climate Process and Climate Sensitivity, Hansen, J.E., and Takahasi, T. eds., American

Geophysical Union, Washington, DC, Vol. 5, pp. 29-38.
54. Wigley, T.M.L. and Schlesinger, M.E.(1985). Nature 315:649.
55. Cicerone, R.J.(1987). Science 237: 35.
56. Thompson, A.M. and Cicerone, R.J. (1986). J. Geophys. Res. 91,10: 853.
57. Hameed, S. et al. (1980). J. Geophys. Res. 85: 7537.
58. Fishman, J., Ramanathan, V., Crutzen, P.J., Liu, S.C. (1979). Nature 282: 818.
59. Somerville, R.C.J. et al. (1984). J. Geophys. Res. 89: 9668.
60. Twomey, H.B. et al. (1984). Tellus 36B: 356.
61. Charlson, R.J., Lovelock, J.E., Andreae, M.O., Warren, S.O. (1987). Nature 326: 655.
62. Cookley, Jr. J.A., Bernstein, R.L., Durkee, P.A. (1987). Science. 237-1020.
63. Slingo, A. (1990). Sensitivity of the Earth's radiation budget to changes in low clouds. Nature 343: 49-51.
64. Khalil, M.A.K., and Rasmussen, R.A. (1990). Global increase of atmospheric molecular hydrogen. Nature 347: 743-745.
65. Hansen, J. et al. (1984). In: Climate Processes and climate sensitivity , Hansen, J.E. and Takahashi, T. eds., American Geophysical Union, Washington DC, Vol. 5, pp 130-163.
66. Ellsaesser, H.W. et al. (1980). Planet Space Sci. 28: 827.
67. Mahlman, J.D. et al. (1986). J. Geophys. Res. 91: 2687.
68. McCormick, M.P., and Trepte, C.R. (1987). J. Geophys. Res. 92: 4297.
69. Ramanathan, V. (1987). Paper presented at the Dahlem Workshop, Berlin, 1 to 6 November,1987.
70. Blake, D.R., and Rowland, F.S. (1988). Science 239: 1129.
71. Ramanathan, V. et al. (1983). J. Atmos. Sci. 40: 605.
72. Fels, S.B. et al. (1980). J. Atmos. Sci. 37: 1077.
73. Warrick, R.A., and Barrow, E.M. (1990). Climate and sea level change: A perspective. Outlook on Agriculture. 19:(1): 5-8.
74. Holdgate, M.W. (1989). Climate change: Meeting the challenge. Report by a Commonwealth Group of Experts. Commonwealth Secretariat,131 pp.
75. Wells, S. and Edwards, A. (1989). Gone with the waves. New Scientist 1690: 47-51.
76. De Ronde, J. (1990). What will happen to the Netherlands if sea level rise accelerates? In: Climate and Sea level Change, Observations, Projections, and Implications, Warrick, R.A. and Wigley, T.M.L., eds., Cambridge University Press, Cambridge (In Press).
77. Parry, M.L., Porter, J.H. and Carter, T.R. (1990). Climatic change and its implications for Agriculture. Outlook on Agriculture 19 (1): 9-15.
78. Wigley, T.M.L. (1989). The greenhouse Effect: Scientific Assessment of Climatic Changes Lecture presented at the Prime Minister's Seminar on global climate, 12 Downing Street, London 29 April, 1989.
79. Jager, J. (1988). Developing policies for responding to climatic change. Summary of discussions and recommendaions of workshop held in Villach and Bellagio, World Meteorological Organization.
80. Warrick, R.A. (1988). Carbon dioxide climatic changes and agriculture. The Geographical Journal 154: 221-233.
81. Kellogg, W.W., and Zhao, C. (1988). Sensitivity of soil moisture to doubling of carbon dioxide in climate model experiments. Part

I: North America, Journal of Climate 1: 348-366.

82. Zhao, Z.C. and Kellogg, W.W. (1988). Sensitivity of soil moisture to doubling of carbon dioxide in climate model experiments. part II: The Asian monsoon region. Journal of Climate 1: 367-378.

83. Schlesinger, M.J. and Mitchell, J.F.B. (1985). Model Projctions for equilibirium response to increased CO2 concentration. In: Projecting the Climatic Effects of Increased CO2, MacCracken, M C. and Luther, F.M., eds.,DOE/ER-2037, US. Department of Energy, Washington, DC., pp. 81-148.

84. Cure, J.D. (1985). Carbon dioxide doubling responses: a crop survey.In: Direct Effects of Increasing Carbon Dioxide on Vegetation, Strain, B.R.and Cure, J.D., eds., United States Department of Energy, Washington, pp. 19-116.

85. Parry, M.L., Carter, T.R. (1988). The assessment of effects of climatic variations on agriculture: aims, methods, and summary of results. In: The Impacts of climatic variation on agriculture, Vol. I: Assessment in Cool Temperate and Cold Regions, Parry, M.L. Carter, T.R.and Konijin, N.T.,eds., Kluwer, Dordrecht, Netherlands, pp. 11-95.

86. Brouwer, F. (1988). Determination of broad-scale latitude changes by climate and soils. Working Paper Wp-88-007, International Institute for Applied Systems Analysis, Laxemburg, Austr—ia, 21 pp.

87. Parry, M.L., Carter, T.R. and Konijin, N.T., eds., (1988). The Impact of Climatic Variations in Agriculture, Volume I, Assessment in Cool Temperate and Cold Regions, Kluwer, Dordrecht, Netherlands.

88. Kettunen, L., Mukula, J., Pohjonen, V., Rantanen, O. and Varjo, U. (1988). The effects of climatic variations on agriculture in Fin—land. In: The Impact of Climatic Variations on Agriculture, Volume I: Assessment in Cool Temperate and Cold Regions, Parry, M.L., Carter, T.R. and Konijin, N.T., eds., Kluwer, Dordrecht, Netherlands, pp. 55-614.

89. Bergthorsson., P., Bjornsson, H., Dyrmundsson, O., Gudmundsson, B., Helgadottir, A. and Jonmundsson, J.V. (1988). The effect of climatic variations on agriculture in Iceland. In: The Impact of Climatic Variations on Agriculture Volume 1: Assessment in cool Temperate and Cold Regions, Parry M.L., Carter, T.R. and Konijin, N.T., eds., Kluwer, Dordrecht, Netherlands, pp.383-509.

90. Parry, M.L., Carter, T.R. and Porter, J.H. (1989). The greenhouse effect and the future of UK agriculture. Journal of Royal Agricultural Society of England 150: 120-131.

91. Anon (1990). Intergovernmental Panel on climate change working Group II (Impacts). Section A, Report on Agriculture, Forestry and Land Use. 1. The Potential Impact of climatic change on Agriculture and Land use. In Press.

92. Rosenzweig, C. (1985). Potential CO2 Induced effects on North American wheat producing regions., climatic change 7, 367-389.

93. Pitovranov, S.E., Lakimets, V., Kiselev, V.I. and Sirotenko, O.D. (1888). The effects of climatic variations on agriculture in the subarctic zone of the USSR. In: The impact of Climatic Variations on Agriculture, Volume I: Assessment in cool temperatre and Cold Regions. Parry, M.L., Carter, T.R., and Konijin, N.T. eds., Kluwer, Dordrecht, Netherlands,

pp.617-724.
94. Zhang, J.C. (1989). The CO2 problem in climte and dryness in north China. Meteorological Magazine 15: 3-8.
95. Yoshino, M.M., Horie, T., Seino, H., Tsujii, H., Uchijima, T. and Uchijima, Z. (1988). In: The Impact of Climatic Variations on Agriculture volume 1: Assessment in cool temperate and cold Regions. Parry, M.L., carter, T.R. and Konijin, N.T., eds., Kluwer, Dordrecht, Netherlands, pp.725-863.
96. Parry, M.L. and Carter, T.R. (1989). An assessment of the effects of climatic change on agriculture. Climate change 15: 95-116.
97. Beran, M.A. and Arnell, W.M. (1989). Effect of climatic change on quantitative aspects of United Kingdom water resources. In: Contract Report to the Department of the Environment Water Directorate. Institute of Hydrology, Wallingford, 93 pp.
98. Williams, G.D.V., Fautley, R.A., jones, K.H., Stewart, R.B. and Wheaton,E.E. (1988). Estimating effects of climatic change on agriculture in Saskatchewan, Canada. In: The Impact of Climatic Variations on Agriculture, Volume I: Assessment in cool Temperate and Cold Regions, Parry, M.L., Carter, T.R. and Konijin, N.T., eds., Kluwer, Dordrecht, Netherlands, pp.121-382.
99. Imeson, A., Dumont, H. and Sekliziotis, S. (1987). Impact analysis of Climatic Change in the Mediterranean region, volume F, European Workshop on Inter-related Bioclimatic and Land Use changes, Noordwijkerhout, Netherlands.
100. Salinger, M.J. (1988). Climatic warming : Impact on the New Zealand growing season and implications for temperate Australia. In: Greenhouse Planning for climatic Change Pearman, G.H., ed., CSIRO, Australia, pp. 564-575.
101. Environmental Protection Agency (1989). The potential Effects of global climate change on the United States, Volume I: Regional Studies. Draft Report to congress, US Environmental Protection Agency, 91 pp.
102. Mackie, R.M. and Rycroft, M.J. (1988). Health and ozone layer. British Medical Journal 297: 369-370.
103. Rycroft, M.J. (1990). The Antarctic atmosphere: a hot topic in the cold cauldron. The Geographical Journal 156: 1-11.
104. Shanklin, J.D. and Gardner, B.G. (1989). The Antarctica ozone hole. British Antarctic Survey.
105. Kerr, R.A. (1988). Stratospheric ozone is decreasing. Science 239: 1489-1491.
106. Cochrane, J. (1990). Global warming and crop modeling. (Abstracts of communications, Audsley, E., ed.,) Journal of Agricultural Science, Cambridge. 115: 295-298.
107. Houghton, R.A. and Woodwell, G.M. (1989). Global climatic change. Scientific American 260: 36-44.
108. Singh, D.P., Singh, P., Pannu, R.K. and Sharma, H.C. (1990) Carbon dioxide enrichment, climate change and Indian agriculture. In: Proceedings of the Indo-US Workshop on Global Climatic Changes: Photosynthesis and Plant Productivity, Jan. 8-12, 1991, Delhi.

THE ORIGIN AND EVOLUTION OF C_4 METABOLISM OF THE CHENOPODIACEAE AS A RESULT OF GLOBAL ARIDIZATION OF CLIMATE

V.I.P'yankov, Ural State University, Sverdlovsk 620083, USSR

About 40 C_4 species from 5 tribes and 14 genera of Chenopodiaceae in the arid zone of Middle Asia were examined for their ecological distribution, anatomy and biochemical features (first products of photosynthesis and activity of main enzymes of carbon metabolism). The survey shows that in general evolutionary trend in carbon metabolism was the same in both the subfamilies (Chenopodioideae and Salsoloideae): C_3---NAD and NADP malic enzyme. Within the NAD- and NADP-ME sub-types biochemical groups were detected on the basis of the relation between activity of aspartate aminotransferase and NADP-malic enzyme, which changed from 3-5 to 200 times. These groups had not only biochemical features, but Kranz-anatomy types and life forms. We suppose that biochemical specialization of C_4 photosynthesis of Chenopodiaceae reflects the process of ecological differentiation of plants during aridization of climate in Afro-Asiatic region.

The evolution of plants has an adaptive character. A large geological and climatic process in the Earth History is reflected in the changes of plant cover. The expected change in global environment (rise of global temperature, destroying of ozone screen and increasing UV radiation, aridization of climate) may bring about a serious alterations of vegetation in future. In this context it is worth to understand the relation between the global climatic events and evolutionary processes, especially in the arid regions of the plants. To my mind the problem of the origin of C_4 photosynthesis and its evolution is a good example of climatic evolution in the world.

Among the 18 families including species with C_4 photosynthesis, a special role belongs to Chenopods (Chenopodiaceae). This family is second to the grasses in the abundance of C_4 species in world flora (1, 2) and hold the first place in the Afro-Asian desert region (3). We consider that among 1600 species of Chenopods (4) about 25 per cent have cooperative photosynthesis. C_4 species of this family have a broad ecological amplitude and inhabit both hot arid deserts and high mountains. That is why the investigation of cooperative photosynthesis of Chenopodiaceae reflects the common ways of adaptation of C_4 species to arid climate.

METHODS

Complex investigations of structural and functional characteristics of the assimilation apparatus of C_4 species of Chenopodiaceae in the arid zone of Middle Asia were conducted. Explorations were made in field conditions in different regions of the Soviet part of South-Turanian province of the Afro-Asian desert phytogeographical zone (Karakum and Kyzylkum deserts, Uzbekistan and Lajikistan arid territories). The climate of this region is desert continental and subtropical with annual precipitation of about 130-200 mm, average temperature in July 29-31°C and absolute maximum temperature as 47-49°C.

Assay of enzyme activity was done in crude extracts. The enzyme extraction medium contained: HEPES (Serva) - KOH (pH - 7.5) 100mM, EDTA (Serva) - 0, 25mM, $MgCl_2$ - 5mM, DTT (Reanal) - 5mM, 1% PVP (Polyclar AT, Serva). Volume of solution - 3 ml, weight of material 300 mg fresh weight. The assay of RuBP-carboxylase (EC 4.1.1.39) (RuBPC) and PEP-carboxylase (EC 4.1.1.31) (PEPC) was determined using ^{14}C incorporation into acid stable material (5), aspartate aminotransferase (EC 2.6.1.1) (AAT) (6) and NADP-malic enzyme (EC 1.1.1.40) (NADP-ME) (7). The chlorophyll content was measured in 80 per cent acetone extract spectrophotometrically. The equipment and methods of determination of the first products of photosynthesis are described by P'yankov and Vakhrusheva (8). Types of kranz-anatomy was determined under a light microscope using the typology suggested by Carolin et al. (9).

RESULTS AND DISCUSSION

The content of the short-term products of photosynthesis of C_4 Chenopods did not differ from the plants with cooperative photosynthesis from other families. After 10 seconds in atmosphere of $^{14}CO_2$ the main amount of label was included in dicarboxylic acids (60-90 %) and in the early C_3 products (phosphoglycerate and sugar phosphate esters 1.8-18 per cent of the total assimilated carbon). The main difference between various species of Chenopods was expressed in relation of formation of two C_4 acids - malate and aspartate. Figure 1 shows the types of primary CO_2 fixation in different taxons and anatomical groups of Chenopods. The close correlation between the types of kranz-anatomy and biochemical groups of C_4 photosynthesis was revealed exceptionally in the genus Salsola. Within C_4 species of Chenopodiaceae, Carolin et al. (9) detected four types of kranz-anatomy:atriplecoid-, suaedoid-, kochioid- and salsoloid-type. Their names correspond with the latin names of the genera Atriplex, Suaeda, Kochia and Salsola. These groups differ in anatomy, position of two assimilation tissues and some other features (9, 10).

All investigated species with suaedoid (genus Suaeda) and atriplecoid-types (genus Atriplex) had a strict aspartate NAD-ME type of CO_2 fixation. The plants with kochioid-type had the first products of photosynthesis both acid or mainly malate. Among the species with salsoloid-type all biochemical variant of C_4 photosynthesis were found.

The analysis of enzymes activities of C_4 Chenopods showed different types of CO_2 fixation (NAD- or NADP-ME) determining deep biochemical differentiation in this family (Figure 1, Table 1). These groups differ in the activity of AAT and NADP-ME, and possibly have different relation of malate and aspartate pathways of CO_2 translocation. The good test for the division of these groups is the relation AAT/NADP-ME activities. The other important biological pecularities of these groups suggest that these differences are not accidental and the biochemical variants reflect some steps of origin of C_4 syndrome in Chenopodiaceae.

We tried to find out the reason of such important structural-functional differentiation of C_4 Chenopods, and its importance. The present pattern of different ecological distribution of plants with

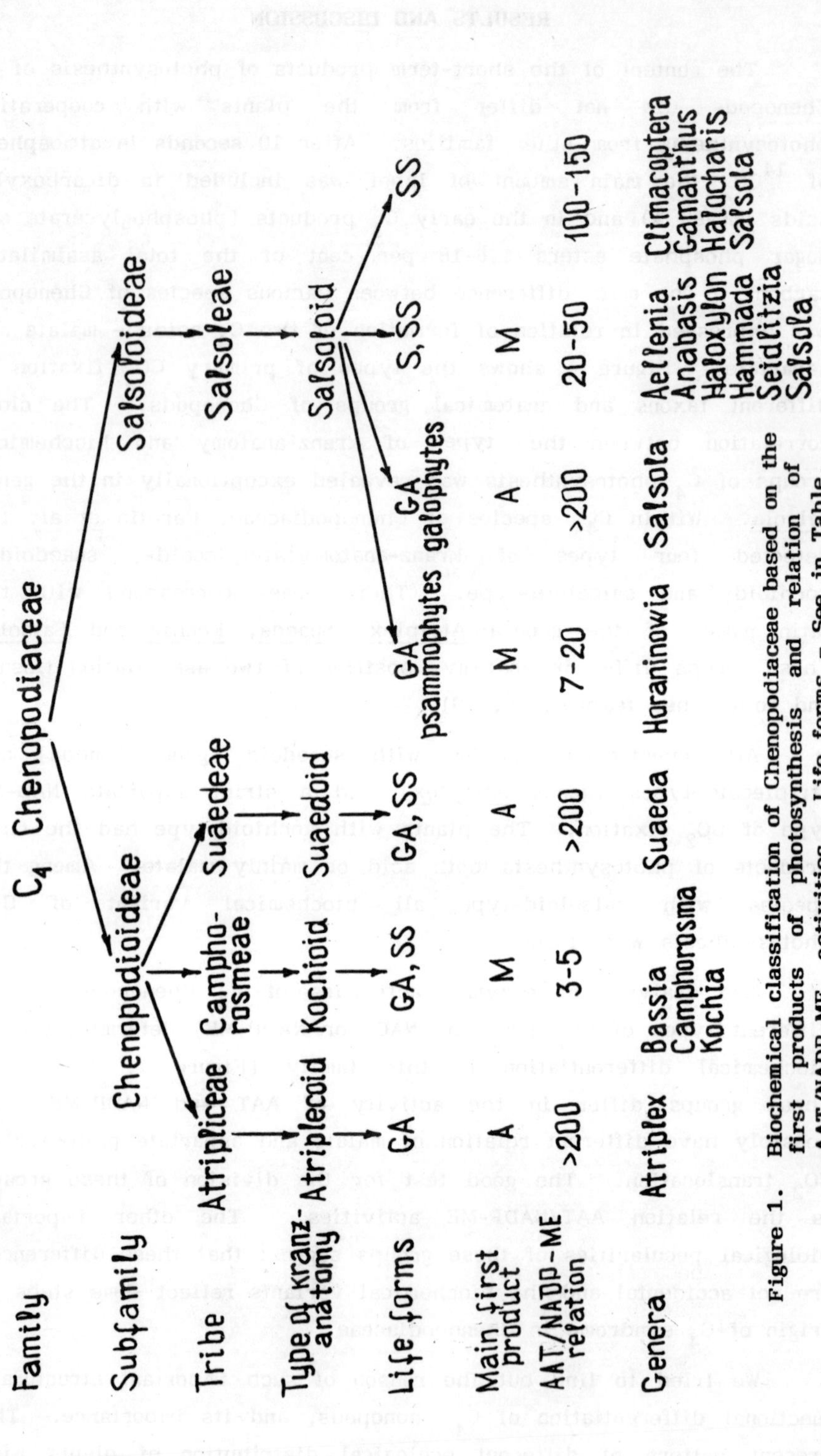

Figure 1. Biochemical classification of Chenopodiaceae based on the first products of photosynthesis and relation of AAT/NADP-ME activities. Life forms - See in Table.

Table 1. Activity of photosynthetic enzymes in the natural C_4 plants of different biological groups of Middle Asia

Groups[1]	Life[2] forms	Enzyme activity µmol min^{-1} mg^{-1} chlorophyll			
		RuBPC	PEPC	NADP-ME	AAT
Family Chenopodiaceae					
Atriplecoid NAD-ME	GA	2,1	12,9	0,08	50,0
Kochioid NADP-ME	GA,SS	0,7	29,1	9,26	43,2
Suaedoid NAD-ME	GA,SS	1,8	29,4	0,13	64,8
Salsoloid NAD-ME	GA	2,2	27,6	0,49	98,0
Salsoloid NADP-ME	GA	0,8	21,7	2,65	29,3
Salsoloid NAD-ME	SS	0,8	22,4	0,05	97,8
Salsoloid NADP-ME	S,SS	1,9	21,9	0,97	33,5
Family Poaceae					
Panicoid NADP-ME	GA,GP	1,3	19,2	4,83	9,9
Aristidoid NADP-ME	GP	2,2	30,5	21,07	19,2
Chloridoid NAD-ME	GP	2,1	28,1	0,42	59,2

[1]Types of kranz-anatomy was detected on typology (9)

[2]Life forms: GA - grass annaal, GP - grass perrenial, S - Shrubs, SS - semi-shrubs.

various types of photosynthesis C_3, C_4 and CAM plants (11-13) suggests the importance of biochemical characteristics of photosynthetic apparatus for adaptation and occupation of definite ecological niches. The analysis of phylogenetic selection in family Chenopodiaceae may indicate the ways for forming different stages of C_4-syndrome and show its connection with global climatic changes of the plant.

The Chenopodiaceae family has a comparatively late origin among angiosperms. Its exact absolute age is not known. However, some consider that it may vary from lower (14) to upper (15) Ceraceous period. It may be said that origin of this family was connected with the ancient supercontinent Gondwana or with the period when the close contacts between Africa, Australia and America existed which can be documented by the presence of Chenopods on all the continents (16). The presence of C_4-species in genera

Atriplex (subfamily Chenopodioideae tribe Atripliceae) and Suaeda (subfamily Salsoloideae tribe Susedeae) can suggest the origin of C_4 syndrome even at that early period (upper Creatous or beginning of tertiary). These taxons cantain C_3 and C_4 species and all C_4 plants have only NAD-ME type of metabolism (8, 17). That is why the first plants with C_4 photosynthesis were aspartate forming species with susaêdoid-and atriplecoid types of kranz-anatomy. We cannot know exactly the main reason of such evolutionary step (C_3---C_4 NAD-ME type), but possibly it is connected with climatic changes in upper Creatous. The origin of C_4 plants was connected with the seasonally arid areas (possibly in the continental interiors), and oligotrophic soils.

During the long period only aspartate forms of Chenopods existed. The first NADP-ME species in tribe Camphorosmeae originated from tribe Atripliceae. This step was connected with the appearance of new structure of the kranz-syndrome kochioid-type of anatomy. The broad geographical distribution of the genus Kochia on all continents proves the early origin of NADP-ME type of CO_2 fixation, but we have no data about the presence of NADP-ME mechanism in natural Kochia species from Australia and South America.

The more detailed information concerning the evolution of C_4 metabolism can be obtained during studies on the subfamily of Salsoloideae. One of the main steps in the evolution of C_4 syndrome was the appearance of the salsoloid-type anatomy in the primary section Caroxylon of genus Salsola. Botschantsev (15) considered that the first species of this genus appeared in South Africa during Oligocene or Miocene of the tertiary period. At that time, India was linked with Asia and Africa where climate in the region became more differentiated and arid. All explored species in the genus Salsola, section Caroxylon belong to the aspartate-forming type of C_4 photosynthesis (Figure 2). It seems that the appearance of the new structural type was the important precondition for the origin of the more complicated biochemical variants. During Miocene, the representatives of Section Caroxylon migrated to the region of North Africa and the Red Sea. There was rapid formation of the main sections of genus Salosola and main genera of tribe Salsoleae in that area. Probably this period can be considered as the time of the appearance of NADP-ME type of C_4 photosynthesis in the Salsoloideae

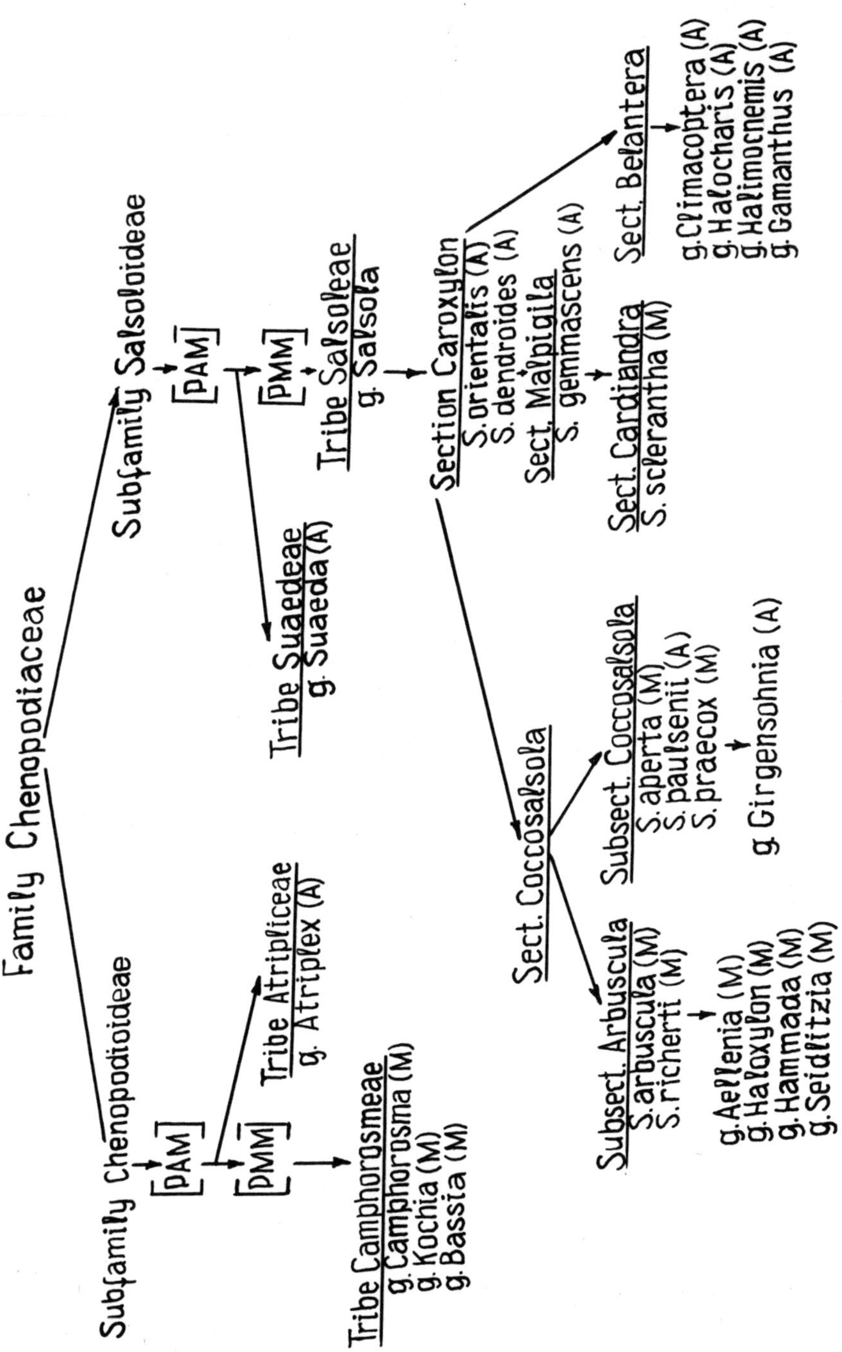

Figure 2. Evolution of photosynthetic pathway CO_2 fixation in Chenopodiaceae. M and A - malate (NADP-ME) and aspartate (NAD-ME) forming species. PAM - primary aspartate metabolism, PMM - primary malate metabol.

subfamily. At that time large sandy deserts were formed and plants with new biochemical pathways occupied new free ecological niches. High efficiency of NADP-ME mechanism promoted these plant successfully to compete with C_3 plants on the more favourable soil comparing with the usual substrates of NAD-ME species (saline and rocky places). All the various forms of tribe Salsoleae in the region of Red Sea were only shrubs. Perhaps, the primary malate-forming C_4 photosynthesis appeared first in the shrub's modification: CO_2 can be utilized through the malate channel but the significance of the aspartate way was still large. The high activity AAT and as a rule low NADP-ME activity in shrubs (Figure 1, Table 1) proves this mechanism. All malate-forming NADP-ME species had NAD-ME way of CO_2 fixation and translocation, but the perrenial NADP-ME Chenopods are "more aspartate" in comparison to the annual forms.

Further evolution of C_4 photosynthesis in the family Chenopodiaceae was connected with their broad distribution and occurred at the end of the tertiary and quaternary period in the arid regions of South Europe, Asia and Africa. The annual species originated from NADP-ME perrenial shrubs. They occupied favourable places with favourable edaphic conditions (sandy and semi-sandy non-saline soil). This species had higher NADP-ME and lower AAT activities in comparison with shrubs (Figure 1). Nevertheless their activity is considerably higher than in NADP-ME grasses (Table 1), and these plants also possess high lability malate/aspartate pathways of CO_2 fixation.

The group of halophyte annual plants (genera Climacoptera, Halocharis, Halimocnemis) have salsoloid-type kranz-anatomy and NAD-ME metabolism, but they have more active NADP-ME than primary aspartate-forming species from section Caroxylon (Table 1). These annual plants originated from section Belanthera (Figure 2) and formed during the quaternary period in the northern part of C_4 Chenopod's areas in saline conditions (15). We can suggest that inhabitance of this territories was done by psamophylous facultative NADP-ME species with gradual reduction of the malate way of CO_2 translocation in extreme conditions.

The analysis of this material showed a great importance of the photosynthetic function in the climatic evolution of plants. Global, geological and climatic events are reflected in vegetation. Increasing differentiation of climate leads to the formation of new soil-climatic conditions and induces the appearance and selection of new plant forms. The general trends of the evolution of higher plants are expressed in the widening of their climatic limits of survival (and they are achieved now) and increasing of ecological differentiation of their ecological niches. Expected rise of aridization may further bring about the expansion of C_4 species. Even now they are found in wild condition up to 65° N.L. in North America (18) and up to 4200 m in the Pamir mountains (Middle Asia) (19). They occupy, therefore, practically the same ecological interval as C_3 plants. Probably we have to register the changes of natural areas of C_3 and C_4 plants distribution in future.

REFERENCES

1. Downton, M.J.S. (1975). The occurrence of C_4-photosynthesis among plants. Photosynthetica 9: 96-105.

2. Raghavendra, A.S., Das, V.S.R. (1978)j. The occurrence of C_4-photosynthesis: a supplementary of C_4 plants reported during late 1974-mid 1977. Photosynthetica 12: 200-208.

3. Winter, K. (1981). C_4 plants of high biomass in arid regions of Asia: occurrence of C_4 photosynthesis in Chenopodiaceae and Polygonaceae from the Middle East and USSR. Oecologia 48: 100-106.

4. Taktajan, A. (1987). System of Magnolyphytes. Leningrad, Nauka 439 p.

5. Romanova, A.K. (1980). Biochemical methods studying of autotrophy in microorganism. Moscow, Nauka 158 p.

6. Hatch, M.D., Shaio-Lim Mau. (1973). Activity, location and role of aspartate aminotransferase and alanine aminotransferase isoenzymes in leaves with C_4-pathway of photosynthesis. Arch. Bioch. 156: 195-201.

7. Johnson, H.H., Hatch, M.D. (1970). Properties and regulation of leaf nicotinamide - adenine dinucleotide phosphate dehydrogenase and "malic" enzyme in plants with the C_4 dicarboxylic acid pathway of photosynthesis. Biochem. J. 119: 273-280.

8. P'yankov, V.I., Vakhrusheva, D.V. (1989). The pathways of primary CO_2 fixation in Chenopodiaceae C_4 plants of Central

Asian arid zone. Sov. Plant Physiol. 36: 178-187.

9. Carolin, R.C., Jacobs, S.W.L., Vesk, M. (1975). Leaf structure in Chenopodiaceae. Bot. Lahrb. Syst. Pflanzehgesch. Pflanzengeogr. 95: 226-255.

10. Vosnesenskaya, E.V., Gamaley, Yu. V. (1986). The ultra-structural characteristics of leaf types with kranz-anatomy. Bot. Zhurn. 71: 1291-1307.

11. Teeri, J.A., Stowe, L.G. (1976). Climatic pattern and distribution of C_4 grasses in North America. Oecologia 23: 1-12.

12. Hattersley, P.W. (1983). The distribution of C_3 and C_4 grasses in Australia in relation to climate. 57: 113-128.

13. Eickmeier, W.G. (1978). Photosynthetic pathway distribution along aridity gradient in Big Bend National Park, and implications for enhanced resource partirioning. Photosynthetica 12: 290-297.

14. Krasilov, V.A. (1989). Origin and early evolution of flowering plants. Moscow, Nsuka, 263 p.

15. Botschantzev, V.B. (1969). Genus Salsola L. (Composition, history of development and distribution). Leningrad Acad. Sci. Publishers, 45 p.

16. Lavrenko, E.M. (1962). Main characteristics of botanical geography deserts of Eurasia and North Africa. Moscow-Leningrad Acad. Sci. Publishers, 169 p.

17. Hatch, M.G., Kagawa, T., Craig, S. (1975). Subdivision of C_4-species based on differing C_4 acid decarboxylating system and ultrastructural features. Austral. J. Plant Physiol. 2: 111-128.

18. Schwarz, A.G., Redman, R.E. (1988). C_4 grasses from the boreal forest region of North-western Canada. Can. J. Bot. 66: 2424-2430.

19. P'yankov, V.I. (1990). Evolution and ecological importance of C_4-plants in arid zone of Middle Asia. In: Proceedings of the Soviet-Indian Simposium of Regulation of Photosynthesis. May 20-25, 1990. Puschino, USSR, pp 22-23.

MONITORING OF SALINIZATION AND DESERT ARID CLIMATE PRESSURE BY PHOTOSYNTHESIS ORGANIZATION IN PLANT COMMUNITIES

Bil' K.Ya. and Fomina I.R.
Institute of Soil Science and Photosynthesis, USSR Academy of Sciences, Pushchino, Moscow Region, 142292 USSR

While studying the peculiarities of plant photosynthesis in contrasting ecological regions it has been found that desert plants have the greatest variety of CO_2 fixation. Systematization of the structural-functional organization of photosynthetic apparatus indicates that there exists a certain correlation between the variants of carbon metabolism in plant communities and types of soils where the plants grow. In the desert flora C_3-xerophytes, C_3-succulents, C_4-xerophytes, C_4-succulents and a special group of halophytes with a unique C_4-CAM system were found. The discovery of all-the-day-round CO_2 assimilation mediated by the joint functioning of C_3-, C_4- and CAM systems and their location in specifically organized phototrophic tissues is a vivid example of plant communities existing in nature with the mechanisms stable to the pressure of extra arid conditions. We consider C_4-CAM system to be the most progressive stage of the evolution of plant photosynthetic apparatus: coordinated processes of CO_2 absorption and its storage as C_4- and relative acids, optimally requlated reactions of decarboxylation, balanced functioning of photosystems in chloroplasts, effective transport and accumulation of photosynthetic products rich in energy allow these plants to grow on solonchaks and heavy salinized soils and have a high productivity under extra arid conditions.

It is considered that C_4- biochemical "superstructures" of Benson-Calvin cycle appear due to the combined effect of high temperatures, insolation, water deficiency and increased salt content in soil. However, according to our data the active donation of CO_2 to the pentosophosphate reductive cycle adjusts the plants to water deficiency but it is absent in species which grow historically under conditions of sufficient water provision. In this case the interaction between the metabolism of C_4-acids and Benson-Calvin cycle in the plants without special phototrophic tissues results in the creation of specific carbon pathways all of which are characterized by the application of final products of photosynthesis as a substrate source for carboxylation reactions. Cooperation of heterotrophic and phototrophic tissues is one of the types of this mechanism. It was found by us in a number of succulents, sea grasses, macroalgae and Tradescantia fluminensis. This type provides the stability to a wide spectrum of unfavourable factors.

All the facts given above reveal new approaches in search for new ways of photosynthesis regulation, total productivity and stability of plants to global changes in the environment. In particular, C_4-CAM-system where the progressive aspects of the known processes are coordinated in the best way as well as the other variants of the cooperative interactions between the metabolism of C_4-acids and pentosophosphate cycle can serve as a criterion for selection and genetics to obtain new varieties and hybrids capable of using light energy, CO_2 and mineral nutrients with maximum efficiency. The experimental data obtained and their theoretical interpretation are the basis for the method for subsequent analysis of dynamic processes in deserts and semi-deserts developed by us. The degree of aridity and salinization pressure on the desert territories with different soil covers can be determined by the organization of photosynthetic apparatus of the dominating species of the plant cenosis. Comparing the rates of changing the areas occupied by certain soil types and vegetation growing on them one can evaluate the intensity of salinization and desert aridization processes.

EFFECT OF FOREST CONVERSION ON VEGETATION AND SOIL CARBON AND FUNCTIONAL TRAIT OF RESULTING VEGETATION

A.S. Raghubanshi, C.S. Jha, C.B. Pandey, Lalji Singh and J.S. Singh; Department of Botany, Banaras Hindu University, Varanasi-221005, India.

ABSTRACT

Major vegetation types and other landuses for an Indian dry tropical region were delineated using IRS 1A LISS I scene of 16 October 1988. This area has experienced rapid changes due to anthropogenic forcing. Carbon stored in vegetation and soil was estimated. Forest basal area was linearly related with C stored in vegetation, soil and total ecosystem. Total vegetation C (Tg) in the vegetated area (760385 ha) was 14.03 in forest, 3.03 in savanna and 0.66 in cropland. C stored in soil to a depth of 30 cm was 6.57 in forest, 5.30 in savanna and 11.71 Tg in cropland. CCTs of LANDSAT 4 MSS of 1982 and 1989 were used to estimate landuse change for 650339 ha area. The forest conversion during the past 7 years resulted into a net release of 6.62 Tg vegetation C and 1.11 Tg soil C. The resultant structurally simpler vegetation had a markedly greater proportion of C in rapidly turning over components, with increased flux relative to C storage. Functionally the derived vegetation was more tightly coupled to the rainfall variability, and would be expected to respond quickly to climate change.

INTRODUCTION

After fossil fuel combustion, forest clearing and shifting cultivation are most important causes of atmospheric build-up of

CO_2. The current release of carbon to the atmosphere from tropical deforestation could be 35-50 % of current emissions from world-wide combustion of fossil fuels. According to the analysis of Houghton et al. (1), changes in landuse over the past two centuries have caused a significant release of CO_2 to the atmosphere from the terrestrial biota and soils. Their estimates show that, between 1958 to 1980, there has been a net release of 38.76 x 10^{15}g C to the atmosphere. Deforestation in the tropics is responsible for nearly all of the flux because an annual net release of carbon to the atmosphere in the order of 0.9-2.5 x 10^{15}g is estimated from the tropics (Houghton et al. 1985). Researches by several workers (1, 3-7) suggest that the largest advance in determining the role of the biota in the global carbon cycle at present can be made by improved information on rate of conversion of forests. The greatest changes in terrestrial carbon results from the transformation of forest to non-forest and from the transformation of non-forest to the forest landuses (8).

Disturbances result in changes in species composition and vegetation structure. Overpeck et al. (9) observed that altered disturbance regimes tend to cause marked changes in simulated forest composition, and to accelerate the rate of forest response to climate change. In the dry tropical region, gradual forest destruction results in savannization. The functional importance shifts from the woody canopy to the herbaceous ground stratum where C-4 grasses predominate. This may lead to altered carbon storage/flux relationships and therefore will have implications for global carbon budget.

The purpose of this paper is to determine the effect of forest conversion on carbon storage in a dry tropical region. Present study also attempts to determine the net release of carbon from the study area using biomass inventory of the present and data of vegetation conversion over the 1982-89 period, and examines the functional trait of the converted vegetation.

STUDY AREA

Location

The study area covers the Sonbhadra district, Uttar Pradesh, as well as adjoining portions of Singrauli Coal Fields, in Sidhi district and a part of Sarguja district, Madhya Pradesh. It is located between 23°45' - 24°38' N lat. and 82°30' - 83°23' E long. The altitude varies between 315 to 415 m above mean sea level.

Climate and Geology

The study area represents a seasonally dry tropical environment. The mean maximum temperature varies from 23.2° in January to 40.5°C in May and mean minimum from 13.3°C in January to 30.5°C in June. The year is divisible into three seasons: summer (April to mid June), rainy (mid June to September) and winter (November to February). March and October represent transition periods. Annual average rainfall is 1035 mm of which 85% is received in the rainy season followed by an extensive dry period of 9 months in the annual cycle.

Geologically, four kinds of formations occur in the area: (i) the lower Vindhyans, (ii) the Bijawars, (iii) the Archeans, and (iv) the Gondwanas. Soils of the sites are largely residual ultisols.

MATERIALS AND METHODS

Present Landuse

The study area subscene was extracted from one IRS 1A LISS I scene (path 23 and row 51) acquired on 16 October 1988. Reconnaissance surveys were made with the help of False Colour Composite (bands, 4, 3 and 2) and Survey of India toposheets. For the extraction of better thematic details unsupervised classification was performed. It was difficult to provide training sets for the supervised classification, since the vegetation was quite patchy. 40 arbitrary classes were extracted and colour coded. The area was frequently visited for ground truth study.

With the help of field visits, and visual interpretation of FCC and enhanced products such as Principal Component Analysis and the Hue, Saturation and Intensity, 40 spectral classes of unsupervised classification were merged into seven forest classes and eight non forest classes. Area calculation was done by multiplying the number of pixels into each class by the spectral resolution of the LISS I sensor i.e., 72.5 x 72. 5 meters. Digital techniques were performed using SEPIMAGE VIPS-32 image processing software package.

Change in The Forest Cover from 1982-1989

CCTs of LANDSAT 4 MSS of December 1982 and LANDSAT 4 MSS of December 1989 were used (Path 142 and Row 043) for change detection studies. The data for these two dates are assumed to be exactly similar in all respects except the really changed pixels on the ground. The data of the same season were selected so that they have the least difference due to phenology of the vegetation. An accurate geometric registration of the images of interest is essential for change detection (10-11). To remove the nonsystematic errors, the map to image rectification was done. Prior to this, the subscene of interest was extracted from data of both dates. Transformation model was calculated between Survey of India toposheets (63L, 63P, 64I and 64M) and the December 1982 image . The December 1982 image rectification was done by selecting 32 GCPs (Ground Control points) on the toposheets and the corresponding points on the image with root mean square error less than 0.60 pixel.

To register the December 1989 image (the slave image) with reference to the rectified 1982 data (master image), 28 GCPs were generated on the master as well as the slave image and an image to image transformation model (3rd order polynomial) was calculated. Using this model, the slave was resampled for its pixel by pixel correspondence with the master image. By intensity interpolation technique, the brightness values of the shifted pixels were adjusted. The residual image was generated following Ingram et al. (12). Predicted values of the second date image bands, based on the regression between the four bands of the two dates, were subtracted from the corresponding bands of the first date. This subtraction procedure for the predicted values leads to the removal of the atmospheric haze. Thus four subtracted bands were generated. Principal Component Analysis was performed on these bands to emphasize the difference between the pixels. Since PC2 generally accounts for "greenness" component (13), it was chosen for further analysis. This PC2 image was colour sliced based on the ground experience of the changed areas as well as the comparison of the two date FCCs by visual interpretation. Ground truth for each category of changed areas was collected. Area under each category was calculated by multiplying the resolution of pixel with the pixel population of each colour slice (changed area category). The digital change detection analysis was done on Dipix system (14).

Measurement of Storage and Flux of Carbon

Sixty three sampling sites were selected to cover the entire range of forest variation. At each site ten randomly placed 10 x 10 m quadrats were examined for phytosociological data.

Woody plant biomass was measured using allometric equations relating circumference to biomass. Aboveground herbaceous biomass was sampled monthly through 50 x 50 cm harvest plots and fine root biomass ($<$5 mm diameter) through 15 x 15 x 15 cm monoliths.

Net primary productivity (NPP) for woody species both in the forest and savanna was estimated allometrically using girth increment, and that of herbaceous vegetation through peak-through analysis of biomass time-series using a decision matrix. Peak fine root biomass was added (assuming turnover of < 1 year) to the above estimates to obtain total net production. These methods are described in Rawat and Singh (15) and Singh et al. (16). Three forest sites and three savanna sites were selected for intensive study covering two annual cycles.

Carbon content in the different biomass components was determined using Perkin-Elmer Elemental Analyser. Per cent carbon in woody biomass was 45; in ground vegetation, 38; in foliage 45 and roots 43 %. Soil carbon was calculated using mean bulk density and organic carbon content to a depth of 30 cm. Long term vegetation C comprised woody (bole+branch) components, and coarse roots, and short-term vegetation C foliage, ground vegetation, litter and fine roots. The total vegetation C was sum of standing crops for different components. Stand C was the sum of vegetation C and soil C.

RESULTS AND DISCUSSION

Present Forest Cover and Other Landuses

The study area which was once covered with dense natural forests has experienced marked biotic stress particularly since approximately past three decades due to developmental activities, such as construction of reservoir, and establishment of thermal power stations, other factories, and mining for coal and limestone (17). Presently, out of the total area of 8883 sq. km studied, only 19.62% is under forest. A majority of the forest is of poor crown cover, with only 25.62% of the total forested area having crown cover >50%. Tree and grass savanna account for 23.26% barren or rocky exposures comprise 6.11% of the total area. A total of 42.61% of the area is under marginal cultivation (Table 1, Fig. 1a).

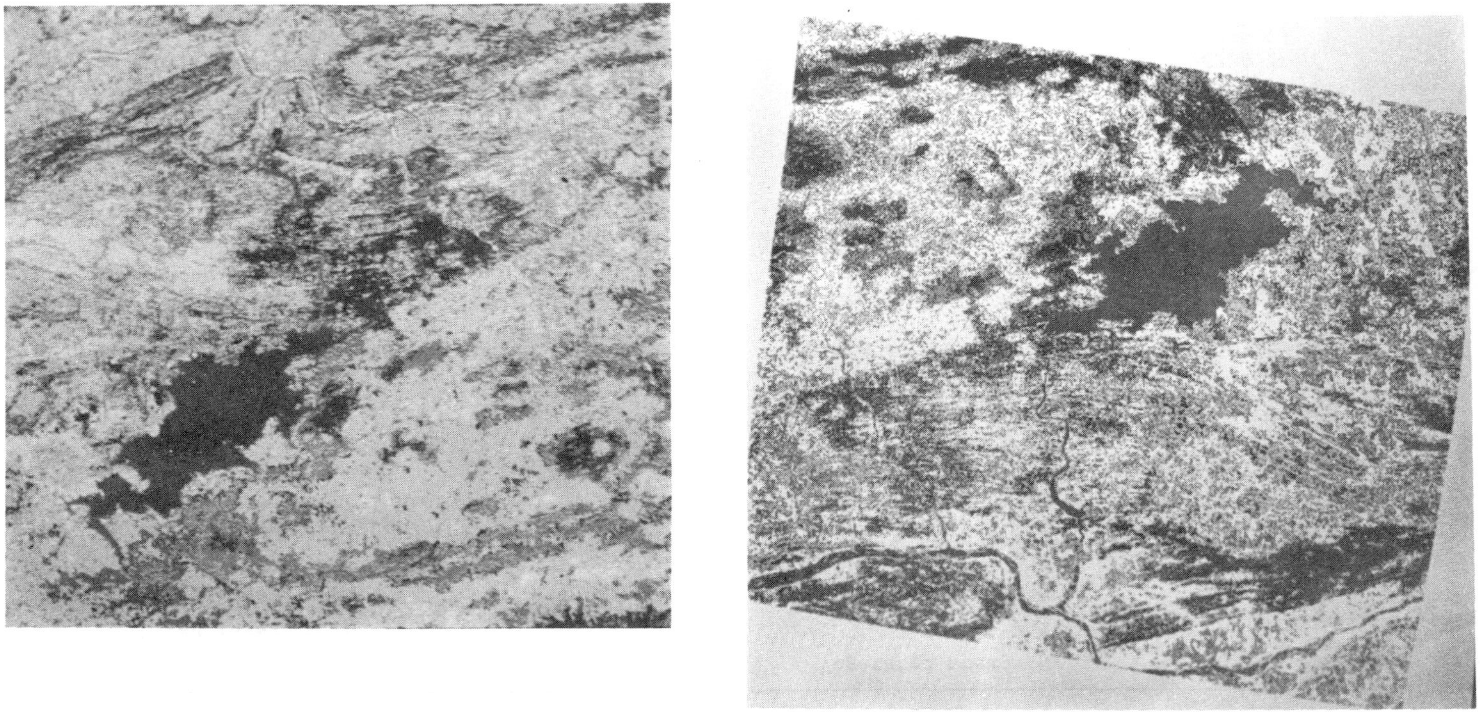

Fig. 1: (a) Unsupervised classification output showing five forest and eight eight non-forest classes. For colour codes see Table 1. (b) Colour sliced PC2 image based on subtraction of 1982 image and predicted 1989 image. For colour codes see Table 3.

Table 1: Landuse and forest cover of the study area determined on the basis of unsupervised classification of October 1988 LISS I Image (Total area of the subscene 8883 km² approx.).

Landuse class	Color in the FCC	% Area	Area (ha)
Forest	Vegetated area		
Mixed forest with			
Crown cover 50%	Parrot green	5.03	44681.49
Crown cover 50% to 40%	Dull green	4.68	41572.44
Crown cover 40% to 30%	Dark green	1.74	15456.44
Shorea dominated forest	Magenta	3.36	29846.88
Acacia forest	Yellow	4.82	42816.06
		19.63	
Non forest			
Tree savanna dominated with *Holarrhena* in the shrub layer	Pink	5.01	44503.83
Tree savanna dominated with *Ziziphus* in the shrub layer	Chocolate	2.74	24339.42
Grass savanna	Dark purple	5.66	50277.78
Degraded grass savanna	Cyan	9.95	88385.85
Cropland	Cream	42.61	378504.63
		65.97	
	Non vegetated area		
Water bodies	Blue	5.11	45392.13
Mines/Quarries	Black	3.1	28336.77
Barren/Rock exposures	White	6.11	54275.13
		14.32	

Relationship Between Vegetation Cover and Vegetation Carbon

In the forested area, tree basal cover was exponentially related with stand biomass (Fig. 2a), and appears to be a good predictor of the standing crop of carbon. Long term C (wood C + coarse root C) was linearly related (r^2= 0.923) with tree basal area (Fig. 2b). Short term C (foliage C+ fine root C) was also linearly related (r^2= 0.904) with tree basal area (Fig 2c). Total vegetation C (r^2=0.926), soil C (r^2=0.964) and stand C (r^2=0.942) were positively related with basal area (Fig 3a,b,c).

Standing Crop of Vegetation and Soil Carbon

Area-weighted mean standing crop values for short-term and long-term C and total vegetation and soil C are calculated for each vegetated landuse type in Table 2. Total vegetation C in the forested landuses ranged between 39.43 and 131.63 x 10^6 g ha . In the tree savanna this value ranged between 28.24 and 31.16 x 10^6 g ha^{-1}. The overall range of soil C was 26.44 to 40.22 x 10^6 g ha^{-1} in the forest and between 26.72 and 41.63 x 10^6 ha^{-1} in the tree savanna. Brown and Lugo (18) have reported vegetation C in undisturbed dry to seasonal tropical Asian forest from 60 to 150 x 10^6 g ha^{-1} and soil C from 50 to 80 x 10^6 g ha^{-1}. Based on data available on Asian tropical forests, Houghton et al. (19) reported high and low estimates of carbon stock in vegetation as 60-200 and 40-115 x 10^6 ha^{-1}, respectively. Thus our values are in agreement with other reports available on carbon storage of tropical forests. The total vegetation carbon in the present vegetated area (760385 ha) was 14.03 x 10^{12} g in forest, 3.03 x 10^{12} g in savanna and 0.66 x 10^{12} g in cropland. C stored in the soil to a depth of 30 cm was 6.57 x 10^{12} g in forest, 5.30 x 10^{12} g in savanna and 11.71 x 10^{12} g in cropland.

Change in Forest Cover in the Recent Past

The change (between 1982-1989)detection study could identify

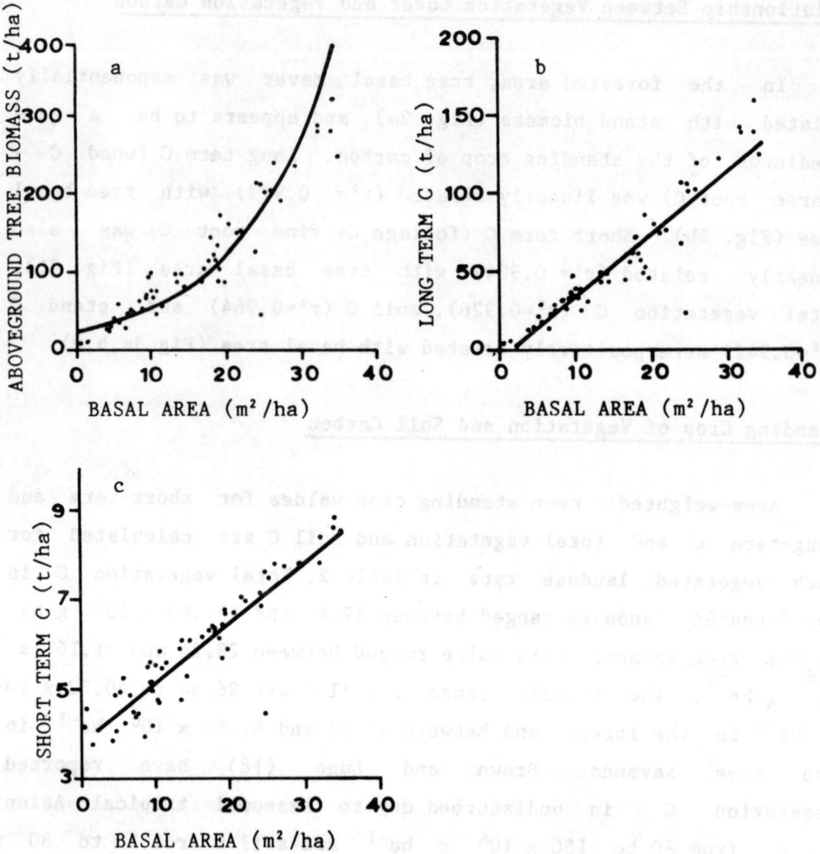

Fig. 2: (a) Relationship between tree basal area (X, m^2 ha^{-1}) and aboveground tree biomass (Y, t ha^{-1}). The regression equation was: Y= Exp (3.154+-.982 (X)) (r^2= 0.784; p 0.0001). (b) Relationship between tree basal area (X, m^2 ha^{-1}) and long term carbon (Y, t ha^{-1}). The regression equation was: Y = -6.858+4.15 (X) (r^2= 0.922; p 0.0001). (c) Relationship Between tree basal area (X, m^2 ha^{-1}) and short term carbon (Y, t ha^{-1}). The regression equation was: Y= 3.7504+0.1404 (X) (r^2= 0.903; p 0.0001).

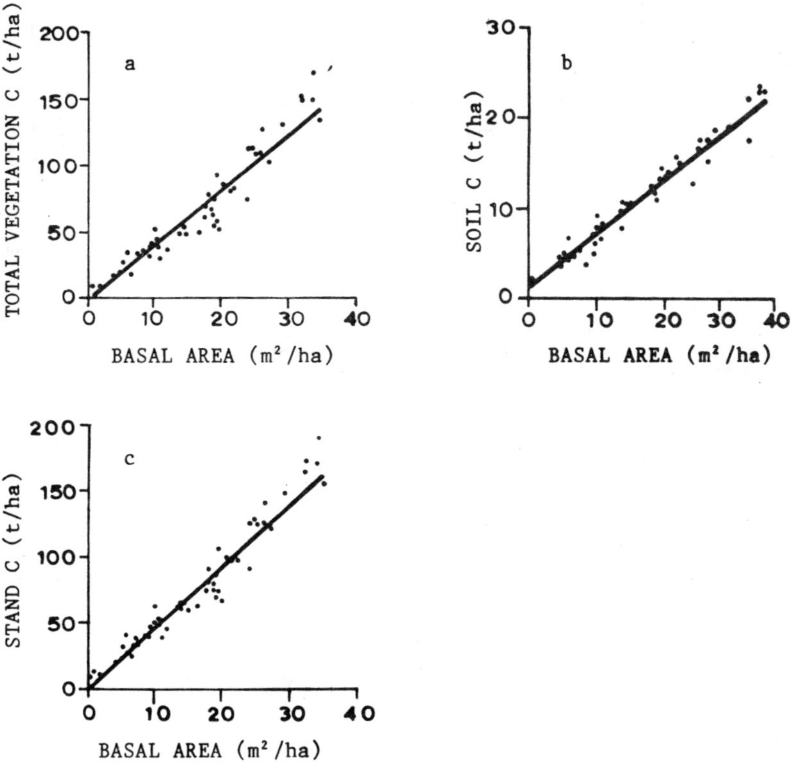

Fig.3: (a) Relationship between tree basal area (X, m^2 ha^{-1}) and total vegetation carbon (Y, t ha^{-1}). The regression equation was: Y = -3.107+4.297 (X) (r^2= 0.926; p 0.00001). (b) Relationship between tree basal area (X, m^2 ha^{-1}) and soil carbon (Y, t ha^{-1}). The regression equation was: Y= 0.981+6.22 (X) (r^2= 0.964; p 0.0001). (c) Relationship between tree basal area (X, m^2 ha^{-1}) and total stand carbon (Y, t ha^{-1}). The regression equation was: Y= -2.125+4.913 (X) (r^2= 0.942; p 0.0001).

Table 2: Area-weighted standing crop of vegetation and soil C of the vegetated portion of the study area.

Vegetation class	Area (ha)	Short term C	Long term C	Short term/ long term ratio	Vegetation C	Soil C	Stand C	Total vegetation C	Total soil C	Total stand C
Mixed forest with										
Crown cover 50%	44681.49	6.88	124.76	0.06	131.63	39.38	171.01	5.88	1.76	7.64
Crown cover 50% to 40%	41572.44	5.57	75.98	0.07	81.55	40.22	121.76	3.39	1.67	5.06
Crown cover 40% to 30%	1545.6	4.35	35.08	0.12	39.43	26.44	65.87	0.61	0.41	1.02
Shorea forest	29846.88	6.74	74.27	0.09	81.00	39.09	120.10	2.41	1.17	3.58
Acacia forest	42816.06	4.30	36.03	0.12	40.33	36.56	76.89	1.73	1.56	3.29
Tree savanna dominated with *Holarrhena* in the shrub layer	44503.83	5.37	25.79	0.21	31.16	41.63	72.79	1.39	1.84	3.23
Tree savanna dominated with *Ziziphus* in the shrub layer	24339.42	4.86	23.39	0.21	28.24	26.72	54.96	0.68	0.65	1.33
Grass savanna	50277.78	4.08	6.06	0.67	10.14	30.94	41.08	0.50	1.56	2.06
Degraded grass savanna	88385.85	2.03	3.03	0.67	5.06	14.06	19.12	0.44	1.25	1.69
Cropland	378504.63	1.75	0.00	-	1.75	30.94	32.69	0.66	11.71	12.37

Values for total vegetation C, total soil C and total stand C are in $\times 10^{12}$ g; the rest of the values are in $\times 10^{6}$ g ha^{-1}.

the conversion from mixed forest with crown cover > 50 to mixed forest with crown cover < 50 - 30%; from mixed forest with crown cover < 50% - 30% to tree savanna; and from grass savanna to degraded grass savanna (Fig. 1b). Only 30.76% of the forest area remained unchanged since 1982. 40.15% of the total forest area was converted from mixed forest with crown cover > 50% to mixed forest with crown cover < 50%- 30%. The good to poor forest conversion occurred at a rate of 6.59% of the forested area per year and savannization in the forested area took place at a rate of 3.29% per year (Table 3).

Several factors are responsible for such an accelerated pace of conversion of vegetation types into poorer crown cover types. Quarrying for limestone, establishment of cement factory, and thermal power projects and construction of the G.B. Pant Sagar reservoir, have resulted in a rapid build up of human population, and displacement of original population. The excessive fuel needs of the villagers and nearby townships result into felling and lopping of trees for firewood. The collection of tree-leaves for forage also involves excessive lopping of trees. According to Singh and Singh (20), in the present area the operation of the agro-ecosystems requires a considerable amount of subsidy from the surrounding forest ecosystems in terms of fodder and fuel wood. About 81 to 100% of the fuel needs, and 80-87 of the fodder needs are met with from the natural forest ecosystems and thus for each unit of energy obtained in agronomic yield (including milk), 3.1 units of energy are expended from the surrounding natural ecosystems in the form of fodder and fuel wood. These authors have estimated that fuel wood needed is 2,256.9 kJ x 10^5 per year per ha of cultivated land. Further labour population involved in quarrying activity harvests 5004 t of fire wood annually.

According to Houghton et al. (2), the degradation of the tropical open forests, primarily as a result of logging, harvest of fuelwood, deliberate burning and grazing, has occurred in many countries (21-22) but has not been properly quantified.

Table 3: Change in the landuse and forest cover between 1982 and 1989.

Area (ha)	Colour	% of the total area	From	To
78040.63	Red	12	Forest (cover 50%)	Forest (50-30%)
390020.317	Brown	6	Forest (cover 50-30%)	Tree savanna
52027.019	Green	8	Forest	Unchanged
52027.09	Yellow	8	Grass savanna	Degraded grass savanna
26013.54	Blue	4	Water body	Unchanged
78040.63	Orange	12	Cropland	Cropping pattern changed
32567.31	Grey	50	Cropland & grass savanna	Unchanged
650338.62		100		

The process of degradation is gradual and the associated reductions in the carbon stocks of vegetation and soil are variable, so the process is not as well documented as abrupt conversion of forest vegetation (2).

Net Release of Carbon due to Forest Conversion

The release of carbon or its accumulation depends on the standing stock of carbon in vegetation and soils and on the rates of deforestation and reforestation. Vegetation changed detection study involving a total area of 650339 ha indicated changes in 169088.03 ha. This involved changes in vegetation stock (as good forest converted into poor forest or tree savanna, and grass savanna converted into degraded grass savanna) and consequently in the standing crop of carbon. Thus the vegetation conversion during the past 7 years (1982-89) resulted into a net release of 6.6×10^{12} g vegetation C and 1.11×10^{12} g soil C (Table 4). This translates into release of 1.7×10^{12}g C ha^{-1} yr^{-1} due to land use change. For the entire Central Himalaya Singh et al. (23) estimated 3.2×10^{12} g $ha^{-1}yr^{-1}$ carbon injection to the atmosphere. If we assume that the rate of forest degradation observed in the present area is applicable to the entire dry tropical forest area of India (29.2 million ha, 24), forest degradation is releasing 49.6×10^{12} g C yr^{-1}. These values are much higher than the estimates of Houghton et al. (8) which is 33×10^{12} g yr^{-1} for whole of India. Of course, some of the vegetation carbon may not enter the atmosphere and may remain sequestered indefinitely in lumber, ash and charcoal.

Vegetation Structure and Short- vs Long-Term C Storage

Additionally, these conversions involved change from structurally more complex (high biomass) to structurally simpler (low biomass) vegetation. Consequently, the ratio of standing crop of carbon in short lived components of vegetation to standing crop of C in long lived components increased. Thus mixed forest with crown cover$>$ 50% had this ratio equal to 0.06,

Table 4: Vegetation C, soil C and total C injected into the atmosphere due to vegetation conversion during 1982-1989.

From	To	Changed area (ha)	Net release of vegetation C ($\times 10^{12}$ g)	Net release of soil C ($\times 10^{12}$ g)	Total release ($\times 10^{12}$ g)
Mixed forest with crown cover 50%	Mixed forest with crown cover 50%-30%	78040.63	4.795	0.226	5.021
Mixed forest with crown cover 50%-30%	Tree savanna	39020.31	1.562	0.003	1.565
Grass savanna	Degraded savanna	52027.09	0.264	0.878	1.142
Total		169088.03	6.622	1.107	7.729

the mixed forest with crown cover< 50 - 30% had a mean ratio of 0.10, the tree savanna 0.21 and the grass savanna 0.67. Figure 4 illustrates the inverse multiplicative relationship between percent short-term stored C and woodiness of vegetation. The increase in the short-term:long-term C ratio implies a change in the functional behaviour of the converted vegetation in the sense that long term storage of C declines relative to annual flux of C.

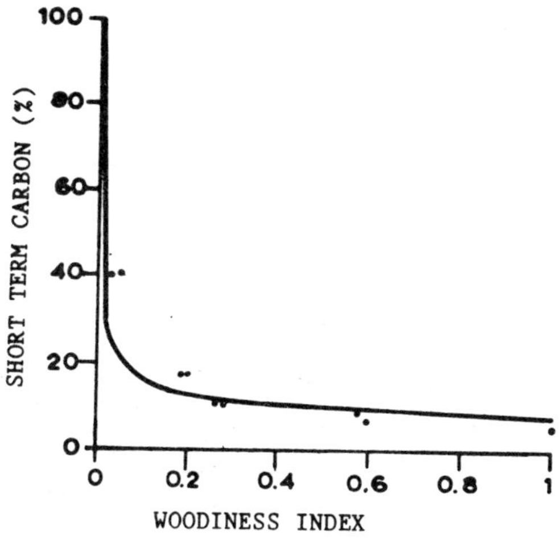

Fig. 4: Relationship between per cent short-term stored carbon and woodiness of vegetation (calculated as ratio of ecosystem wood carbon and wood carbon of best grade forest).

Carbon Storage and Flux in Forest and Derived Savanna

The total carbon storage in the intensively studied plots of forest and derived savanna vegetation averaged 28.68 t C ha^{-1} and 12.89 t C ha^{-1}, respectively (Table 5). The belowground biomass in the savanna contributed more (37%) to total C storage compared to forest (15%). Share of herbaceous vegetation plus fine roots in total stored vegetation C was only 4% in the forest compared to 49% in savanna. The net carbon input (total net production) in the forest and savanna was 6.27 t ha^{-1} yr^{-1} and 5.96 t ha^{-1} yr^{-1}, respectively (Table 6). Thus, although the carbon storage in savanna (B/P= 2.18) was half of the forest (B/P= 4.54) carbon flux was almost equal to that of the forest resulting into P/B ratio of 0.47 for savanna and 0.22 for forest. The contribution of

Table 5: Plant carbon mass ($t\ C\ ha^{-1}$) in tropical dry deciduous forest and derived savanna.

Components	Aboveground	Belowground	Total
	Forest (n=3)		
Woody	24.18	3.41	27.58
Herbaceous (+ tree fine roots)	0.12	0.93	1.05
Total	24.30	4.33	28.68
	Savanna (n=3)		
Woody	5.83	0.81	6.64
Herbaceous (+tree fine roots)	2.29	3.96	6.25
Total	8.11	4.77	12.89

Differences in the standing crops of carbon between forest and savanna were significant at p 0.001.

Table 6: Carbon input through net primary production ($t\ C\ ha^{-1}\ yr^{-1}$) in tropical dry deciduous forest and derived savanna.

Components	ANP	BNP	TNP
	Forest (n=3)		
Woody	5.70	0.37	6.07
Herbaceous (+ tree fine roots)	0.38	1.61	1.99
Total	6.08	1.98	8.06
	Savanna (n=3)		
Woody	0.86	0.08	0.94
Herbaceous (+ tree fine roots)	2.96	0.08	0.94
Total	3.82	2.13	5.96

Differences in carbon input values betweet forest and savanna were significant at p 0.001.

herbaceous vegetation plus fine roots to total net production averaged 84% for savanna and only 28% for forest. Evidently, drawing upon the local biodiversity resources the converted savanna system maintained the same level of the ecosystem functioning (productivity) as the forest through recruitment of fast growing species of shorter life span which replace the longer lived slow growing species of the forest at a cost of reduced C conservation in the biomass. Thus conversion of dry tropical forest into savanna increases carbon flux relative to storage.

Rainfall Variability and Savanna Response

Climate induced disturbance (increased frequency of forest fires due to global warming) could significantly alter the total biomass and compositional response of forests to future warming (9). Increased anthropogenic disturbances in the tropics are likely to accelerate savannization process. Most of the functioning roots in the derived savanna are located in the upper few centimetres (Fig. 5a) of soil profile making the savanna vegetation a pulsed system highly responsive to rainfall events. Our studies indicated that the monthly variation of green biomass was linearly related with rainfall variability (Fig. 5b). However, the green (photosynthetic) biomass of grazed savanna was not related with the variability of rainfall, rather it was exponentially related with grazing intensity (Fig. 5c). Evidently, the rainfall and grazing intensity are two most important factors explaining the variability of green biomass.

In the present study total net primary production (TnpP) of ungrazed savanna was related with the annual rainfall ($r^2 = 0.902$) (Fig. 6a). However, enhanced production in high rainfall years attracted greater consumption (Fig. 6b) with the result that the residual biomass was not related with rainfall, and the root production was inversely, though weakly, related with rainfall (Fig. 6c). Thus increased rainfall had no effect on carbon storage belowground.

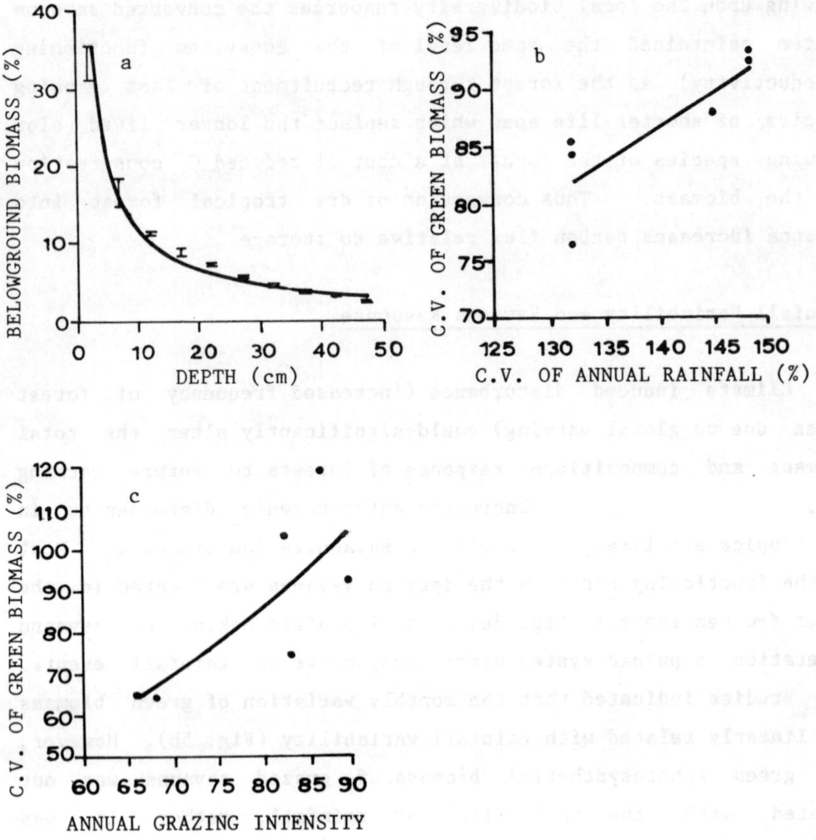

Fig.5: (a) Distribution of belowground biomass according to depth across three savanna sites. Bars represent ± 1 SE. (b) Relationship between coefficient of variation (c.v.) of green biomass and coefficient of variation of annual rainfall for ungrazed savanna. (c) Relationship between coefficient of variation (c.v.) of green biomass and annual grazing intensity for grazed savanna.

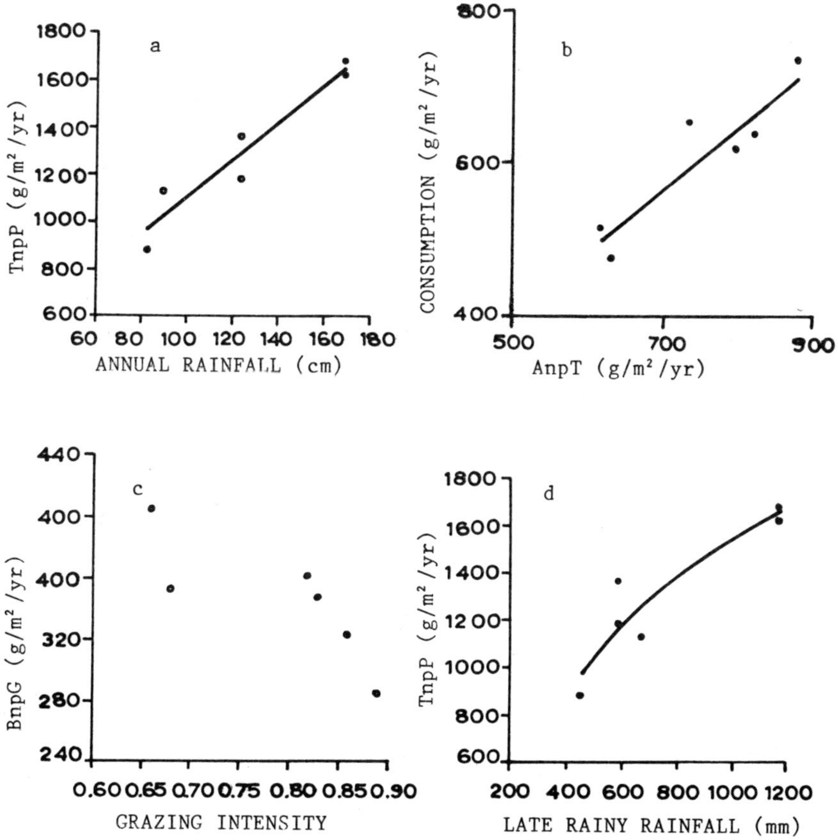

Fig. 6: Relationships between (a) total net primary production (TnpP) and annual rainfall for ungrazed savanna, (b) Annual consumption and annual aboveground net primary production (AnpT) for grazed savanna, and (c) belowground net primary production (BnpG) and grazing intensity for grazed savanna; and (d) TnpP and rainfall in late rainy season in ungrazed savanna.

TnpP was not related with early rainy rainfall (June+July), however, it was logarithmically related with late rainy rainfall ($r^2 = 0.861$) (Fig. 6d). While the rainfall in the early rainy season triggers growth, biomass accumulation is sustained by the late rainy season rainfall. Therefore, any variation in rainfall pattern due to future climate change is likely to affect more quickly the carbon flux and storage in the derived savanna vegetation.

The direction and magnitude of biotic flux in response to climate are among the most critical questions for the global carbon cycle and the earth's climate (8). The effect of climate on the biota could have either a positive or negative feedback on the interaction of CO_2 and global temperature. A possibility is expressed that the increased concentration of CO_2 in air enhances photosynthesis and leads to an increased storage of carbon in plants and soil (25-27). Plants and ecosystems may sequester carbon in response to increased levels of CO_2 despite nutrient limitations (28-29). Since the dominating vegetation in tropical savanna is C_4 type, CO_2-fertilisation may increase photosynthetic C fixation only marginally (30). However, increased C/N ratio in organic matter due to CO_2 fertilisation (31) could potentially lead to better carbon conservation on account of reduced decomposition (32). Nevertheless higher temperatures would still cause rapid C turnover. If climate change increases the intensity of monsoon circulation resulting into steady wet monsoon, net primary production may increase. And moister subsoil may ultimately result in an increase of woodiness (and hence C conservation) (33). However, higher production will lead to greater consumption, or in association with increased temperatures it could subsequently lead to more fire, and therefore decreased woodiness. Human and livestock population increases in future will further alter forest/savanna boundaries and rates of net release of C through gradual vegetation degradation. It will therefore be important to consider the forest/savanna patch dynamics in global carbon modeling.

REFERENCES

1. Houghton, R.A., Hobbie, J.E., Melillo, J.M., Moore, B., Peterson, B.J., Shaver, G.R. and Woodwell, G.M. (1983). Changes in the carbon content of terrestrial biota and soils between 1860 and 1980: A net release of CO_2 to the atmosphere. Ecol. Monogr. 53: 235-262.

2. Houghton, R.A., Schlesinger, W.H., Brown, S. and Richards, J.F. (1985). Carbon dioxide change between the atmosphere and terrestrial ecosystems. In: Atmospheric Carbon Dioxide and the Global Carbon Cycle, Trabalka, J.R. ed. U.S. Department of Energy, Washington, D.C. , pp. 113-140.

3. Woodwell, G.M., Whittaker, R.H., Reiners, W.A., Likens, G.E., Delwiche, C.C. and Botkin, D.B. (1978). The biota and the world carbon budget. Science 199: 141-146.

4. Woodwell, G.M. (1978). The carbon dioxide question. Sci. Am. 238: 34-43.

5. Moore, B. Boone, R.D., Hobbie, J.E., Houghton, R.A., Melillo, J.M., Peterson, B.J., Shaver, G.R., Vorosmarty, C.J. and Woodwell, G.M.. (1981). A simple model for analysis of the role of terrestrial ecosystems in the global carbon cycle. In: Modelling of Global Budget, Bolin, B. ed., John Wiley and Sons, New York, pp 365-385.

6. Woodwell, G.M., Hobbie, J.E., Houghton, R.A., Melillo, J.M., Moore, B., Peterson, B.J. and Shaver, G.R. (1983a). Global deforestation: contribution to atmospheric carbon-di-oxide. Science 1081-1086.

7. Woodwell, G.M., Hobbie, J.E., Houghton, R.A., Melillo, J.M., Moore, B., Park, A.B., Peterson, B.J., Shaver, G.R. and Stone, T.A. (1983b). Deforestation measured by LANDSAT: Steps Toward a Method. U.S. Department of Energy, Washington, D.C.

8. Houghton, R.A., Boone, R.D., Fruci, J.R., Hobbie, J.E., Melillo, J.M., Palm, C.A., Peterson, B.J., Shaver, G.R. and Woodwell, G.M. (1987). The flux of carbon from terrestrial ecosystems to the atmosphere in 1980 due to changes in land use: geographic distribution of the global flux. Tellus 396: 122-139.

9. Overpeck, J.T., Rind, D. and Goldberg, R. (1990). Climate-induced changes in forest disturbance and vegetation. Nature 343: 51-53.

10. Woodwell, G.M., Hobbie, J.E., Houghton, R.A., Melillo, J.M., Moore, B., Park, A.B., Peterson, B.J. and Shaver, G.R. (1984). Measurement of changes in the vegetation of the earth by satellite imagery. In: The Role of Terrestrial Vegetation in the Global Carbon cycle: Measurement by Remote Sensing, Woodwell, G.M., ed., John Wiley and Sons, Ltd.

11. Singh, A. 1986. Change detection in the tropical forest environment of northeastern India using Landsat. In: Remote Sensing and Tropical Land Management, Eden, M.J. and Parry, J.T., eds. John Wiley and Sons, Ltd.

12. Ingram, K., Knapp, E., and Robinson, J.W. (1981). Change-detection technique development for improved urbanised area deliniation. CSC/TM-81/6087, Report prepared for NASA, Computer Science Corporation, Springfield, Maryland.

13. Cochrone, G.R. (1986). The technology of remote sensing: System components, data processing and prospects. In: Remote Sensing and Tropical Land Management, Eden, M.J. and J.T. Parry eds., John Wiley and Sons Ltd.

14. Dipix System. (1983). ARIES-II Operators' Manual, Dipix System Ltd., Ottawa, Ontario.

15. Rawat, Y.S. and Singh, J.S. (1988). Structure and function of

Oak forests in central Himalaya. I. Dry matter dynamics. Ann. Bot. 62: 397-411.

16. Singh, J.S., Lauenroth, W.K. and Steinhorst, R.K. (1975). Review and assessment of various techniques for estimating net aerial primary productivity in grasslands harvest data. Bot. Rev. 41: 181-232.

17. Singh, J.S., Singh, K.P. and Agrawal, M. (eds.) (1990). Environmental Degradation of Obra-Renukoot-Singrauli Area and its Impact on Natural and Derived Ecosystems. MOEn Project Technical Report, Department of Botany, Banaras Hindu University, Varanasi, India.

18. Brown, S. and Lugo, A.E. (1982). The storage and production of organic matter in tropical forests and their role in global carbon cycle. Biotropica 14: 161-187.

19. Houghton, R.A. (1990). The future role of tropical forests in affecting the carbon dioxide concentration of the atmosphere. Ambio 19: 204-209.

20. Singh, V.P. and Singh, J.S. (1989). Man and Forests: A case study from dry tropics of India. Environ. Conserv. 16: 129-136.

21. Food and Agriculture Organisation of the United Nations (FAO. (1981). Tropical Forest resources Assessment Project: Forest resources of Tropical America, Forest Resources of Tropical Africa and Forest Resources of Tropical Asia. Technical reports Nos. 1, 2, and 3, United Nations 32/6: 1301-78-04, FAO, Rome.

22. Lanly, J.P. (1982). Tropical Forest Resources, FAO Forestry Paper 30, FAO, Rome.

23. Singh, J.S., Tiwari, A.K. and Saxena, A.K. (1985). Himalayan

forests: a net source of carbon for the atmosphere. Environ. Conserv. 12: 67-69.

24. Kaul, O.N. and Sharma, D.C. 1971. Forest types statistics. Ind. For. 97(7).

25. Grifford, R.M. (1980). Carbon storage by the biosphere. In: Carbon Dioxide and Climate: Australian Research, Pearman, G.I. ed., Australian Academy of Science, Canberra, p. 167-181.

26. Lemon, E.R. (ed.) (1983). The ratio of total to merchantable forest biomass and application to the global carbon budget. Can. J. For. Res. 13: 372-383.

27. Oechel, W.C. and Strain, B.R. (1985). Native species responses to increased atmospheric carbon dioxide concentration. In: Direct Effect of Increasing Carbon Dioxide on Vegetation, Strain, B.R. and Cure, J.D. eds., DOE/ER-0238, US Department of Energy, Washington, D.C. pp. 117-154.

28. Wong, S.C. (1979). Elevated atmospheric partial pressure of CO and plant growth. I. Interactions of nitrogen nutrition and photosynthetic capacity in C and C plants. Oecologia 44: 68-74.

29. Oechel, W.C., Hastings, S., Hilbert, D., Lawrence, W., Prudhomme, T.I., Riechers, G. and Tissue, D. 1984. Response of vegetation to carbon dioxide. Report no. 019, U.S. Department of Energy, Washington, D.C.

30. Carlson, R.W. and Bazzaz, F.A. (1980). The effects of elevated CO concentrations on growth, photosynthesis, transpiration, and water use efficiency of plants. In: Environmental and Climatic Impact of Coal Utilization, J.J. Singh and A. Deepak, eds., Academic Press, New York. pp. 609-623.

31. McElroy, M.B. and Moore III, B. (1988). Biogeochemical

dynamics. In: Toward an Understanding of Global Change, U.S. National Committee for the IGBP, National Academy Press, Washington, DC, USA. pp. 47-68.

32. Melillo, J.M., Aber, J.D. and Muratore, J.F. (1982). Nitrogen and lignin control of hardwood leaf litter decomposition dynamics. Ecology 63: 621-626.

33. Walker, B.H., Young, M.D., Parslow, J.S., Cocks, K.D., Flemming, P.M., Margules, C.K., Landsberg, J.J. (1989). Global climate change: Effects on renewable natural resources. In: Global Climate Change - Issues for Australia, Prime Minister's Science Council, Canberra, Australia, Commonwealth of Australia, 1989. pp. 31-76.

CAN MAJOR DEFORESTATION AFFECT CLIMATIC CHANGE?

V.M. MEHER-HOMJI
SALIM ALI SCHOOL OF ECOLOGY
PONDICHERRY UNIVERSITY
PONDICHERRY 605 001 - INDIA

Abstract

In the forest-hydrology link, distinction has to be maintained between rainfall and monsoon as rains could also be of cyclonic, orographic or convectional origin. In contrast to a deforested barren zone, a dense extensive forest-cover may influence convectional rainfall because of its lower albedo, aerodynamic surface roughness of its canopy, absence of dust particles and presence of organic debris in the atmosphere serving as effective condensation nuclei. Case studies on forest-rainfall relationship have been reviewed. As vegetation types and plant productivity depend on rainfall and length of dry season, major deforestation could shift the vegetation pattern towards drier side of the spectrum and reduce productivity.

Introduction

The present article analyses the putative relationship between forests and rainfall. It aims at putting the effects of deforestation on the climate in the right perspective. Not every shower is due to forest influence otherwise it would never even drizzle in the desert. The opposite of this is also true. Cherrapunji, one of the most rainy places in the world with over a metre of rain per annum, is totally devoid of forests. The reason is that rainfall depends on several factors, of which forest appears to be one. Rain-bearing phenomena, geographic location, topographic features, sea-surface temperature, land temperature and vegetation-cover, all have a role to play in the precipitation cycle.

All the arid zones are not due to the absence of forests. The deserts in general, owe their origin to their geographic location at the sub-tropical latitude on the western side of the continents. Deserts are the zones of large-scale descending air motion. A peculiar topographic situation may likewise render certain stations semi-arid. This would be the case of Bellary and Coimbatore located in the lee of the Western Ghats.

Delannoy (1982) has demonstrated the influence of the sea surface temperature on the coastal rainfall. Even the anomalies in the tropical sea-surface temperature of only a few degrees may induce large changes in circulation, cloudiness and rainfall (Julian and Chervin 1978; Rowntree 1978). Increase of ocean temperature by several degrees could increase evaporation rates by as much as 25-50 per cent. If the change in evaporation rate involves a very large area of several thousand kilometres, it can modify the climate over and downstream of the perturbed area.

In the normal monsoon-rainfall years it is the Indonesian region which experiences high temperature and consequently low pressure, but in the El-Nino years associated with weak monsoon and poor rainfall years, the eastern Pacific region off the coast of Peru (South America) is the area of high temperature and low pressure whereas the Indonesian region is under high pressure.

Micro-climates

What then is the role of forests and of deforestation in climatic changes? Among the most far-reaching and immediate effects of deforestation is the destruction of the micro-climate. The daily variation in ground temperature is much higher in the denuded areas compared to the land under forest; the soil is also less protected from the torrential rains. Studies from Singapore reveal higher temperatures at all depths from 3 to 50 cm in the bare soil, slightly lesser under grassland and the lowest and the least variable under mixed forest of Dipterocarpaceae (Hill 1966).

According to a UNESCO report (1978), higher soil temperature increases the rate of mineralisation of the organic matter, impairs the stability of soil crumb structure making the soil easily erodable.

In the Palni hills and the Nilgiris, during winter, the night temperature comes down as low as - 10°C in the open grassy area and whatever seeds of forest trees that may have germinated in the open, are destroyed by frost. At the same time, under the forest canopy, the temperature remains above 0°C (Legris and Blasco 1969). However, frost is not the only adverse factor here; there are rainless periods during which the relative humidity in the grasslands drops very low favouring the spread of fires. One to three months may pass without the region receiving a drop of rain-water. Again, the seedlings of forest species are eliminated by fire, as also hundreds of hectares of plantations of wattles and eucalyptus. The forest itself enjoys comparatively higher value of relative humidity. Thus, eco-climatic conditions are shown to be totally different under a forest and within a grassland though both may occur side by side. The extension of the forest is prevented by frost, dry spells and fire which wipe out the seedlings of trees.

The tree species of the sholas (small patches of tropical montane forests) are confined to valleys, hollows and depressions. Outside this habitat, their regeneration is very difficult. In the temperate countries, broad-leaved trees usually shed all their leaves in winter. In the tropical hill ranges, the forest trees are evergreen. The damage done by the winter cold to the seedlings of forest species in the open area is in fact a physiological drought effect rather than one of direct frost. The variation in night and day temperatures is high, of the order of 12°C, especially during winter. With the rising daytime temperature, the aerial parts of the young plants are subject to heavy loss of water due to transpiration while the soil rendered very cold during the night prevents absorption of water. As a result, the seedlings which have superficial root systems dry up. The successful establishment of the seedlings of shola species only under the forest cover may be explained by the fact that the forests stand on sites with higher moisture content. It has also been observed that the sholas mostly occupy the slopes protected from the morning sun. This may be another reason which accounts for the survival of the seedlings in the shola, as under this condition they will not be exposed to excessive transpiration immediately after the freezing cold of the night.

Herbarium specimens of leaves of the tree Palaquium grande collected a century ago from the Deltota estate near Peradeniya, Sri Lanka, show a lot of growth of a curious alga- like moss Ephemeropsis tjiboldensis which is a very good indicator of an ever-moist microclimate. However, a search of such epiphyllous mosses in the same locality in recent years by Professor W. Meijer of Kentucky University, USA, proved futile because of the destruction of forest. Now strong winds sweep over the barren land, eroded tea estates and a few remnants of forests near the crests of the hills.

Deforestation and decline in rains

Whether the rain in the Indian subcontinent arid zone has been on the wane has been a debated subject. Publications appearing till 1972 argued in favour of no climatic change (Pramanik et al. 1952; Rao 1958 and 1963). However, Winstanley (1973) has come out with the theory that since the late 1920s, the summer rainfall in the Sahel zone (south of the Sahara) and in north-western India has been showing a declining trend, with a corresponding increase in the winter-spring precipitation of the Middle East and of regions to the north of the Sahara along the Mediterranean coast. There was a temporary reversal in the 1950s but from the 1960s the summer-monsoon rainfall in the Sahel and Rajasthan is once again at a low point of irregular fluctuation with a decreasing trend, whereas the cold-season precipitation of the countries around the Mediterranean Sea is at a high point of an irregular fluctuation with an increasing trend.

One striking point in Winstanley's analysis is that he has combined the data of different situations, sometimes even of stations having totally different regimes. For example Marrakesh, Tunis, Tripoli, Jerusalem, Beirut, Mosul, Shiraz are in the Mediterranean region having a bulk of the rains in winter-spring, but Atar, Bikaner and Jodhpur are tropical, with rains concentrated in summer.

The stir caused by the statement of Winstanley (1973) that India was getting progressively drier, led to several studies ruling out this possibility (Agrawal 1976; Chowdhury and Abhyankar 1976; Legris and Meher-Homji 1976; Mukherjee and Singh 1978, Singh 1978).

However, trends are generally studied for the urbanised centres that have lost their forests long ago. Therefore it would be interesting to compare the trends for two groups of stations, both in the vicinity of forests - the surroundings of one group undergoing large-scale deforestation and destruction of its woodlands, and those of the other group being free of such devastation. Such an approach would eliminate the problem of periodic cycles of lean years. Within the framework of such a study, Meher-Homji (1980 a and b) selected two groups of stations-28 in western Karnataka and one (Munnar) in Kerala-the first group having suffered extensive deforestation within recent years and the other group having witnessed only slight reduction in their forest mantle. The trends of both rainfall and rainy days were assessed by applying six statistical criteria: (1) 20-year moving averages; (2) progressive averages of five years (each average is that of five years obtained by the formula $C' = (a+4b+6c+4d+e)/16$, where a,b,c,d, and e represent the mean rainfall in any five consecutive years and C' is the progressive average for the third year of the series); (3) comparison of the mean of 1966-1975 with the long-term average of 1906-1975; (4) comparison of decennial mean of 1966-75 with those of the earlier decades beginning from 1906; (5) comparison of a number of below average years during 1966-75 with the preceding decennial periods; and (6) as a below average year may fall short of the average value by a few milimetres, a comparison was also made taking into account the number of years showing 20 per cent deficit. Thus 12 criteria are considered in all-six of rainfall and six of rainy days. The deforested area within a radius of 16 kms around each station was measured by a planimeter by comparison of the earlier toposheets (from 1914 to 1954) of The Survey of India with the recent revised editions published from 1965 to 1973. In the case of Munnar, the Pre-investment Survey map of Kerala forests published in 1969 was also consulted.

As a rule, the larger the area of deforestation, more are the number of criteria showing diminishing tendency of rainfall and rainy days. Stations around which considerable forest devastation has occurred, as in the case of Hosanagara, Sampaji, Munnar, Chikmagalur, 9 to 11 criteria exhibit a declining trend. Hosanagara is located on the border of the Linganamakki reservoir in Karnataka. The reservoir gets filled up only once in four to five years. It got filled only thrice between 1959 and 1979. For the construction of this reservoir, deforestation was carried out over an area of 32,000 ha within a radius of 16 kms around Hosanagara.

Exceptions are the coastal stations Karwar, Kumta and Mangalore where in spite of considerable loss of forest, the number of criteria presenting declining trends in precipitation and rainy days is considerably smaller than expected. These peculiar cases may be accounted for by the high values of relative humidity due to the proximity of the sea that could compensate for the loss of forest-cover. The area presenting declining trends in more than five criteria is in the shape of a triangle with Hosanagara, Virajpet and Kollegal forming the three angles.

Where deforestation has been negligible (as at Kalkhatgi, Haliyal, Udipi, Shimoga, Mundgod, Hubli, Khanapur and Bhatkal) only one to three criteria reveal a declining trend.

Case of Udhagamandalam

A noteworthy case of a hill-station presenting a waning tendency of rainfall with the declining forest-cover is that of Udhagamandalam (Ootacamund) of the Nilgiri district. Average 20-years rainfall and number of rainy days of successive periods given in the Table 1 as well as 20-year moving averages (Meher-Homji 1984) show a net decline for the recent years, the record lowest rainfall being 800 mm in 1982. The average 20-year rainfall has declined from 1415 mm in 1902-1921 to 1200 mm for the 20-year period 1965-1984. The corresponding decline in rainy days is from 106 to 89, a decline of 16 per cent.

TABLE 1

Periodic fluctuations in rainy days and rainfall at Udhagamandalam

Period	Rainy days: Total	Rainy days: Excluding June to August	Rainfall (mm): Total	Rainfall (mm): Excluding June to August
1902-1906	522	312	7386	4226
1918-1922	505	326	6637	4483
1938-1942	540	303	7055	4412
1948-1952	417	242	5583	3389
1958-1962	590	351	7289	4340
1968-1972	455*	276*	6290	4400
1978-1982	448	271	5637	3871

* Data missing for 5th to 25th December, 1971

Padmavalli (1976) too noted a change in the pattern of the normal rainfall years of the Nilgiri district with the reduction in wooded land. Also Legris and Blasco (1969) observed a diminishing tendency of rainfall for Udhagamandalam from the turn of the century. The frequency of dry years having registered less than 1300 mm increased from eight years between 1902 and 1922 to 12 years between 1954 and 1964 with six consecutive years receiving below this mean value. Von Lengerke (1977) stated that the sub-normal rainfall of the years 1967 to 1970 at Udhagamandalam led farmers and planters to suggest a change of climate but he himself found no clear indication of a large-scale shift in the rainfall. All the same he noted a decrease from 1965 onwards.

Dr Voeckler in his report entitled "Improvement of Indian Agriculture" noted that the number of rainy days (excluding those of the months June, July and August when the rains are of monsoonic origin and not local) during the five-year period 1870-1874 was 374. The figures for the successive five-year periods presented in Table bring out the diminishing tendency of rainy days and rainfall, though the decline is not gradual. The percentage decline in rainy days from 1870-1874 to 1978-1982 is of the order of 28.

It would be an interesting exercise to seek a correlation between the percentage of wooded area at different periods and the oscillation of rainfall. It is doubtful whether the plantations of exotic trees play the same role as the indigenous trees in the water cycle, especially bearing in mind the fact that these exotics have xeromorphic features to reduce the rate of evapo-transpiration. Pines have needle-like leaves and in some species of wattles (Australian *Acacia*) the leaf stalk (phyllode) takes up the function of leaves. Whereas evapo-transpiration is reported to be very low in pines (Bonnemann and Rohrig n.d.), in *Eucalyptus* hybrid, the transpiration rate which is high under flooded conditions is reduced considerably under restricted soil moisture supply by stomatal closure (Rawat *et al*. 1985). However, these experiments have been conducted on seedlings in pots and comparative data from mature plantations and natural forests from a same eco-climate zone are missing.

Among the other case studies on the effect of deforestation on waning tendency of rainfall, mention may be made of Warren (1974) for the Ranchi plateau, Sarmah (1976) for Dibrugarh, Padmavalli (1976) for Nilgiri district, Biswas (1980) for Andaman-Car Nicobar Islands, Mishra and Dash (1984) for Sambalpur, Mukherjee *et al*. (1976) for Santa Cruz, Bombay and Singh *et al*. (Pers. Comm.) for Kumaun. Raju (1981) confirms certain observations of Meher-Homji (1980 a and b) on the effect of deforestation on rainfall for Uttara Kannada district of Karnataka.

Nicholson (cf. Ranganathan 1949) observed that the Chota Nagpur region which had a good area under forest towards the turn of the century, used to receive fairly frequent afternoon showers known as instability rain during summer which favoured tea plantations. Consequent upon the destruction of private forests, in spite of no apparent reduction in the monsoon rainfall, the instability rain has decreased so much that tea gardens have disappeared.

Warren (1974) attributes the decline in rainfall of the pre-monsoon months (May and June) over the Ranchi plateau to the degradation of forests to scrubs over an extensive area. The thunder-storm activity of the pre-monsoon season provides showers at the crucial moment when the water supply is rapidly dwindling towards the end of the long dry season. During the prevalence of long droughts in the (summer) monsoon season of weak monsoon years, conditions resemble those of the pre-monsoon months. Whereas the wooded areas are likely to benefit from such convective rains which may be low in amount but sufficient to maintain the crops and water supply for the local economy, the regions devoid of forests may not derive any benefit of thunder-storm showers.

Given the large-scale variation in rainfall over space and time, not always related to the forest-cover, some meteorologists are hesitant to admit the link between the changing pattern of precipitation and the manmade shifts in the land-use by consultation of even long-term statistics of both rainfall and the percentage of forests, cultivated-and waste-land.

Gadgil and Prasad (1986) have pointed out the utility of realistic models capable of simulating the various effects of deforestation like changes in albedo, soil moisture etc. Experiments showed that model-simulated climates are influenced by land-surface boundary conditions. The areas bordering the deserts were particularly prone to rapid desertification following deforestation in the wake of bio-geophysical feedbacks.

How does deforestation affect rainfall?

Charney et al. (1977) have emphasised the role of lower albedo (i.e. the proportion of the radiation reflected back to the amount striking a surface) for higher rainfall in a forested region. The albedo of deserts or dry soil is considerably greater (30 to 35 per cent) than that of vegetated of forested surfaces (15 to 25 per cent). So the actual net energy imparted to the atmosphere over a desert is less than that over green belts.

Forests act as an obstructing medium under orographic rainfall and increase the effective height of the land surface in providing an obstruction to air movement. Forests also reduce the wind speed through their aero-dynamically rough, undulating canopy; with the decrease in wind velocity, the air masses are forced to rise.

Sud and Smith (1985) have stressed the mechanical friction effect of forests in lifting the moist air and thereby enhancing the monsoon rainfall.

An essential condition for rainfall to take place is that the warm moisture-bearing air should be able to rise. But deserts are the zones of large-scale descending air motion. The large amount of dust over the desert increases the subsidence rate by as much as 50 per cent. Deforestation through erosion increases the dust content of the atmosphere. By cutting off a good portion of short wave solar radiation by scattering, the dust particles prevent it from reaching the ground affecting the albedo of the earth (Angstrom 1962). Another major effect of dust loading would be greater cooling and radiation divergence in the troposphere (Chakravarti 1978).

If the dust in the atmosphere could be eliminated, the radiation cooling would be less; in its turn, there would be a reduction in subsidence facilitating upward ascent of the air. As the air is already moisture laden, it would result in higher rainfall.

Yet another probable effect of deforestation on rainfall is that it diminishes the number of condensation nuclei in the atmosphere. These nuclei formed by plant debris and pollen grains seem to be more efficient than inorganic debris like the dust because in the clouds ice is formed over the inorganic debris at a much lower temperature (Glantz, 1987).

Concluding remarks

The ascension of the moisture-bearing air mass to a proper height to result in condensation of water vapour is linked to several meteorological phenomena after which the rain is named. Thus we have "convection rains or thunderstorms" when due to the excessive heating of some areas, mostly in the pre-monsoon months April-May, rising currents of air are set in motion carrying cloud moisture to great heights.

Rains of cyclonic or depression origin are caused by the differential heating of the sea and land masses resulting in converging winds. The principle of orographic rainfall is that when an air mass strikes a vertical obstacle it tends to ascend.

Whereas the forests do not seem to have control on the cyclonic or orographic types of rain, they do seem to influence the convection rains as shown by the study of Nicholson (cf. Ranganathan 1949) for the Chota Nagpur plateau of Bihar. Therefore, it may not be the absolute decline in the monsoon rainfall that affects the agriculture, flora, fauna or water supply but rather the lack of rains at critical stages. However marginal the increase may be due to the presence of the forest-cover, it makes a difference in sustaining the crops and maintaining the eco-systems.

The evidence from the Amazon basin of South America suggests that a rain-forest returns to the atmosphere, as much as 75 per cent of the moisture that it receives, through the process of evapo-transpiration forming new rain clouds (Salati et al. 1979).

Elsayam (1987) considering the data for the period 1953-1983 has also brought out how a change in the forest density in equatorial Africa causes a change in the air mass characteristics of Central Sudan; the surface temperature of the air increases and the dew point decreases; the south-westerlies monsoon air and the upper easterlies weaken.

Shukla and Mintz (1981) report that modifications in vegetation-cover due to the deforestation of large magnitude and large horizontal extent do influence the precipitation. The hypothesis is based on a correlation between the precipitation and the evapo-transpiration. However, the determinant factor is not only the vegetation but rather the relationship between the moisture content of the soil, the vegetation and the solar energy necessary for transforming water into atmospheric water vapour.

The forest soil, rich in humus, acts as a sponge, retaining the rain water and releasing it gradually to feed the streams, the rivulets and the rivers. But once the forest cover is gone, the shallow top-soil is soon washed away and then water is no longer retained by the substratum.

Dickinson (1980) states that in none of the numerical studies reviewed by him, the effect of deforestation on increasing temporal and spatial fluctuations between wet and dry conditions has been treated. However, he feels that such a change in surface conditions could in turn increase the intensity and decrease the duration of tropical rainfall, enhancing run-off even if the mean rainfall were unchanged.

If the intensity increases without a change in the annual quantum of rainfall, the result is lesser number of rainy days with long spells of dryness and much dreaded erratic distribution. Meher-Homji (1980 a and b) has already shown that large-scale deforestation reflects more through reduction in the number of rainy days than through the volume of rainfall. Soil erosion in its turn provokes two major disasters.

If rainfall continues to be "normal" but irregular with occasional torrential falls, the consequences are silting up of river beds and floods; if drought years prevail in succession, not only do the streams and rivulets depending on the gradual release of water from the forest-soil dry up but also the attendant problem of dust menace crops up, resulting in desertification of at least the marginally sub-humid zones. Increased dust particles in the atmosphere lead to desiccation and drought at least on the margins of the zones that are not so humid. Even the humid zones are in danger of getting progressively drier if droughts continue to recur over a series of years.

The density of the tree-cover and the leaf retention characteristics of trees are dependent on the amount and distribution of rainfall. Where the rainfall is very high and the length of the dry season short (two months or less), as at and around the equator, the result is a rain forest; with the increase in the length of dry season, the rain forest is replaced by tropical evergreen forest as in Kerala and western Karnataka. With subsequent decrease in rainfall and progressive increase in the number of dry months, the evergreen forest gives way to a semi-evergreen type with mixture of evergreen and deciduous species. A deciduous tree is one which loses all its leaves at one time while an evergreen tree loses its leaves gradually so that it is never completely bare. Along the gradient of lessening rainfall and increasing number of dry months, the semi-evergreen forest is followed in turn by moist deciduous forest, dry deciduous forest, thorn forest and eventually by desert vegetation (Meher-Homji 1982).

The net primary productivity figures (mean g/m^2/year) compiled by Lieth (1975) for the various vegetation types range from 2000 in the tropical evergreen rain forest to 600 in woodland (deciduous forest), 700 in tropical grasslands (Savanna types), 70 in Scrub (thorny type) and 3 in dry desert.

Total number of species according to the vegetation types in the various climatic zones of Peninsular India vary from 1700 to 2000 in the wet evergreen forests of the Western Ghats to 700-850 in the deciduous forests and 500 to 550 in the semi-desert to desert tracts

(Meher-Homji, 1990). As the vegetation types, their productivity and their species richness depend on rainfall amount and number of rainy days, it is most likely that major deforestation would shift the vegetation pattern towards the drier side of the spectrum, reducing their productivity and species richness.

At the United Nations University Workshop on Forest-Climate-Hydrology at the Commonwealth Forestry Institute, Oxford (March 1984), it was concluded that large-scale changes in vegetation induced by man, of which deforestation is the most extreme example, result not only in site impoverishment but also bring about a major change in regional heat and water balances. Most local changes in precipitation, over a few tens of kilometres due to forests are essentially caused by the redistribution of precipitation and as such are edge effects. Anthropogenic increase of atmospheric carbon dioxide and the accompanying global warming will have significant effects on forests and thus on the hydrological cycle. Thus from the current evidence it would seem that changes in vegetation are more likely to have important effects on climate at a regional scale.

REFERENCES

Agrawal, P.C. (1976) Variability and trend of annual rainfall in Chhatisgarh region, Madhya Pradesh. In: Proc. Symp. Tropical Monsoons. Indian Inst. Trop. Meteorol., Pune, pp. 465-470.

Angstrom, A. (1962) Atmospheric turbidity, global illumination and planetary albedo of the earth. Tellus 14: 435-454.

Biswas, B.C. (1980) Effect of forest cover on rainfall distribution at Andaman and Car Nicobar Islands. Mausam 31 (1): 178-180.

Bonnemann, A. and Rohrig, E. Der Wald als Vegetation styp und seine Bedeutung fur den Menslehn. Wald ban auf Oekologischen Grundlage Parey, Hamburg, Berlin.

Chakravarti, A.K. (1978) A case of droughts in the Sahara: The effect of the removal of natural vegetation on rainfall. Alternatives

1978: 55-56.

Charney, J.C., Quirk, W.J., Chow, S.H. and Kornfield, J. (1977) A comparative study of the effect of albedo change on drought in semi-arid region. J.Atmos.Sci. 34: 1366-1385.

Chowdhury, A. and Abhyankar, V.P. (1976) An attempt to determine climatic shift in rainfall over Gujarat State. In: Proc.Symp.Tropical Monsoons. Indian Inst. Trop. Meteorol., Pune, pp. 388-398.

Delannoy, H. (1982) Introduction à l'étude des relations entre les températures des eaux océaniques et les précipitations cotières marocaines. Norois 29(116): 535-545.

Elsayam, F.E. (1987) The effect of deforestation in equatorial Africa on rainfall in Central Sudan. Mausam 38(2): 233-238.

Gadgil, S. and Prasad, C.R. (1986) Forests and rainfall. Science Age (July): 25-28.

Glantz, M. (1987) La secheresse en Afrique. Pour la Science Aôut 1987, pp. 18-25.

Hill, R.D. (1966) Microclimatic observations in Bukit Timah Forest Reserve, Singapore. Malayan For. 29: 78-86.

Julian, P.R. and Chervin, R.M. (1978) A study of the southern oscillation and Walker circulation phenomenon. Mon. Wea Rev. 106: 1433-1451.

Legris, P. and Blasco, F (1969) Variabilité des facteurs du climat: Cas des montagnes du Sud de L'Inde et de Ceylan. Inst. Fr. Pondichéry. Tr. Sect. Sci. Tech. 8: 1-195.

Legris, P. and Meher-Homji, V.M. (1976) Is the rain in India on the wane? Bioclimatic consideration. Int. J. Biometeor. 20(3): 190-193.

Lieth, H. (1975) Primary production of the major vegetation units of the world. Primary Productivity of the biosphere, H. Lieth (ed.), Springer Verlag, Berlin, pp. 203-215.

Meher-Homji, V.M. (1980a) Repercussions of deforestation on precipitation in Western Karnataka. Archiv. fur Meteorol. Geophys. Biokl. 28B: 385-400.

Meher-Homji, V.M. (1980b) The link between rainfall and forest clearance: case studies from Western Karnataka. Transaction of the Institute of Indian Geographers. 2(1): 59-65.

Meher-Homji, V.M. (1982) Do forests control climate? Science Today 16(1): 37-42.

Meher-Homji, V.M. (1984) Udhagamandalam: A biogeographic perspective Indian Geogr. J. 59(2): 205-213.

Meher-Homji, V.M. (1990) Changes in phytomass cover in relation to hydrological cycle. National Symp. on Indian Geosphere and Biosphere Programme. Nat. Acad. of Sciences, India, Allahabad (In press)

Mishra, A. and Dash, M.C. (1984) Desertification around Hirakud dam. The Environmentalist 4: 51-58.

Mukherjee, A.K., Pradhan, S.K. and Rao, M.K. (1976) Effect of change of topography for development of Bombay airport on rainfall of Santa Cruz. In: Proc. Seminar on Man and His Environment. Institute of Science, Bombay, pp. 55-57.

Mukherjee, A.K. and Singh, B.P. (1978) Trends and periodicities in annual rainfall in monsoon areas over the northern hemisphere. Indian J. Met. Hydrol. Geophys. 29: 441-447.

Padmavalli, R. (1976) Utilisation of major natural resources. In: Proc. Symposium Resources Development and Planning. Geography Dept. Madras Univ., pp. 1-19.

Pramanik, S.K., Hariharan, S.P. and Ghosh, S.K. (1952) Analysis of the climate of the Rajasthan desert and its extension. Indian J. Met. Geophys. 3: 131-140.

Raju, K.V.S. (1981) Forest and rain report in Uttar Kannada. In: Major dams - A second look, Gadgil, M. and Sharma, L.T. eds. Gandhi Peace Foundation, New Delhi, pp. 99-101.

Ranganathan, C.R. (1949) Protective functions of forests. Proc. U.N. Conference on conservation and Utilisation of Resources.

Rao, K.N. (1958) Some studies on rainfall of Rajasthan with particular references to trends. Indian J. Met. Geophys. 9: 97-116.

Rao, K.N. (1963) Climatic changes in India. In: Changes of Climate, Proc. Rome Symp. Arid. Zone Res. 20, UNESCO, Paris.

Rawat, P.S., Negi, P.S. Rawat, J.S. and Gurumurti, K. (1985) Transpiration, stomatal behaviour and growth of *Eucalyptus* Hybrid seedlings under different soil moisture levels. Indian For. 3(12): 1097-1112.

Rowntree, P.R. (1978) Numerical prediction and simulation of the tropical atmosphere. Meteorology Over the Tropical Oceans, D.B. Shaw (ed.), Royal Meteorological Society, Bracknell, U.K., pp. 219-250

Salati, E., Dall Olio, A Natsui, E. and Gat, J.R. (1979) Recycling of water in the Amazon basin: An isotopic study. Water Resources Res. 15(5): 1250-1258.

Sarmah, S.K. (1976) Some observations on the variations of monsoon annual rainfall in Assam. In: Proc. Symposium Tropical Monsoon. Indian Inst. Trop. Meteorol. Pune, pp. 458-464.

Shukla, J. and Mintz, Y. (1981) Influence of land surface evapotranspiration on the earth's climate. Science 215: 1498-1501.

Singh, S.N. (1978) Is the climate of Hyderabad getting drier? Indian For. 104: 263-269.

Sud, Y.C. and Smith, W.E. (1985) Influence of local land surface processes on the Indian monsoon. Journal of Climate and Applied Meteorology Oct. 1985.

UNESCO (1978) Tropical Forest Ecosystems. UNESCO/UNEP/FAO Paris.

von Lengerke, H.J. (1977) The Nilgris - Weather and climate of a mountain area in South India. Franz Steiner Verlag, Wiesbaden.

Warren, W.D.M. (1974) A study of climate and forests in the Ranchi Plateau. Pt. II. Indian For. 100: 291-314.

Winstanley, D. (1973) Rainfall patterns and general atmospheric circulation. Nature, London. 245: 190-194.

PERSPECTIVES OF GREENHOUSE GASES IN CLIMATIC CHANGE AND PLANT PRODUCTIVITY

S.S.Hundal, Agrometeorologist, Department of Agricultural Meteorology, Punjab Agricultural University, Ludhiana 141 004 and I.P.Abrol, Deputy Director General (Soils, Agronomy & Agroforestry), Indian Council of Agricultural Research, New Delhi 110 001, India

Increasing concentrations of greenhouse gases such as carbon dioxide (CO_2), methane, nitrogen dioxide, ozone (O_3) and chlorofluorocarbons in the atmosphere warrant global warming and their buildup is likely to lead to surface temperature rise of 1.5-5.5°C and changes in precipitation pattern over the next 50-75 years. Reductions in crop yield and increase in crop water demands are projected due to unfavourable climatic changes. Adverse effects of global warming due to increase in CO_2 concentration may be outweighed by increased photosynthesis as doubling of CO_2 can increase photosynthesis in soybean, wheat and maize by 35, 25 and 10 percent, respectively. C_3 plants respond more favourably than C_4 plants to increasing CO_2. Higher levels of O_3, sulfur dioxide and oxides of nitrogen near the surface prove harmful to plant growth. Reductions of O_3 in the stratosphere lowers the protective effect in blocking incoming ultra-violet radiation which endangers plant and human life. The paper reviews the perspectives of greenhouse gases in relation to climatic change and crop productivity.

GREENHOUSE GASES AND CLIMATIC CHANGES

The main greenhouse gases are carbon dioxide (CO_2), water vapour, methane (CH_4), oxides of nitrogen (N_2X), ozone (O_3) and chlorofluorocarbons (CFCs). Out of these CO_2, CH_4 and nitrous oxide (N_2O) have significant manmade sources, whereas CFCs are purely manmade. These gases result in the well known greenhouse effect that arises because earth's atmosphere tends to trap heat near the surface. Water vapour, CO_2, and other trace greenhouse gases are relatively transparent to the visible and near infrared (IR) wavelengths that carry most of energy of sunlight, but they absorb more efficiently the longer IR wavelengths emitted by the earth. Hence an increase in the concentration of greenhouse gases tends to warm the surface by downward reradiation of IR wavelengths.

Global warming from the increase in greenhouse gases has become a major scientific and political issue in the past decade. It is argued that global warming will lead to rise in temperatures, changes in precipitation patterns and an average rise in sea level of about 6 cm a decade over the next century. Greenhouse gases cause a global climate forcing of 2-2.5 Wm^{-2}, as a result of change in solar irradiance. If the mean solar heating of the earth is taken as appx. 240 Wm^{-2}, a climate forcing of 2-2.5 Wm^{-2} corresponds to 0.8-1 per cent change in solar irradiance. For the period 1958-1989, the net climate forcing by CO_2, CH_4, N_2O and CFCs is about 1.17 Wm^{-2} (Fig. 1) and between 1850 and 1989 the corresponding forcing amounts to about 2 Wm^{-2}. The O_3 global climate forcing is probably less than 10-20 per cent of the net forcing by these other greenhouse gases (1). The UN working group has recently commented that the earth will be hotter by 1°C in just 35 years and by 3°C by the end of the 21st century. Thus the predicted rise of 0.3°C per decade is greater than that experienced over the past 10,000 years. Increasing levels of CFCs and other trace gases such as N_2O in the atmosphere are believed to have caused severe damage to ozone in the stratosphere. Ozone depletion results in increased ultraviolet radiation reaching the earth,thus endangering human and plant life. The rapid changes in climate will influence ecosystems and as a result some plants and wildlife may benefit while others will be

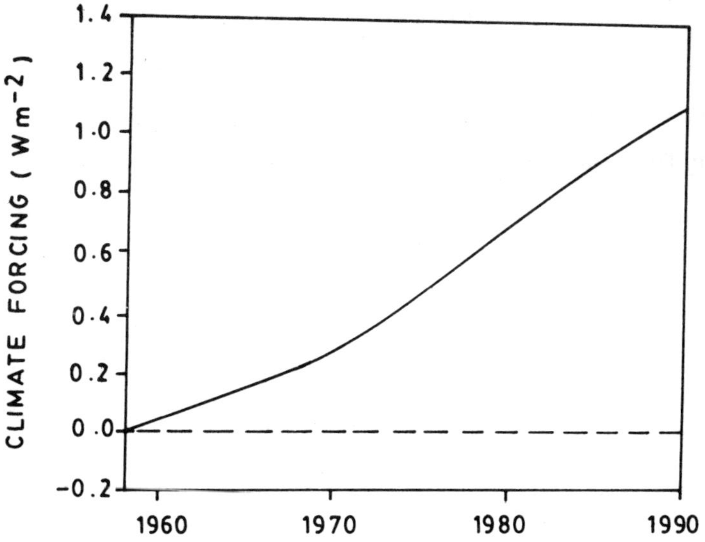

FIG. 1. CLIMATE FORCING IN PAST THREE DECADES OWING TO MEASURED CHANGES OF GREENHOUSE GASES (Sourse: Hansen & Lacis, 1990)

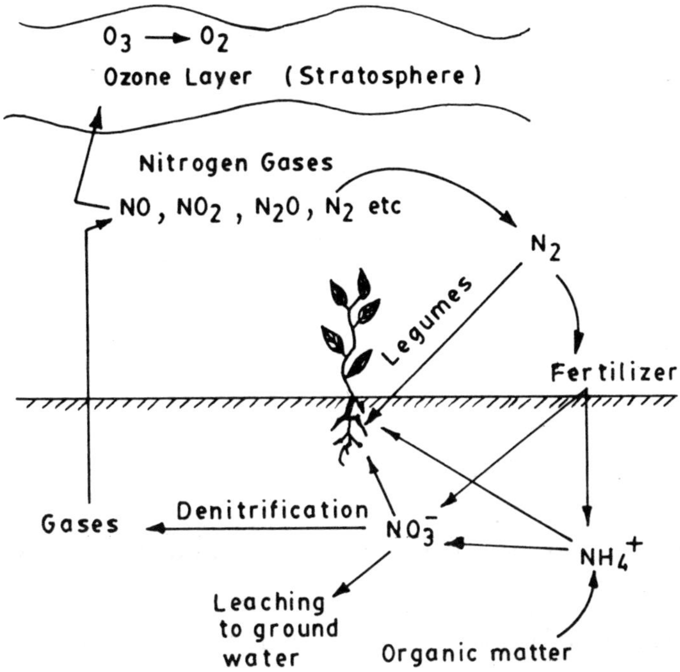

FIG. 2. THE NITROGEN CYCLE IN AGRICULTURE AS IT AFFECTS THE OZONE LAYER

unable to adapt fast enough and face extinction .

Carbon Dioxide Enrichment and Climatic Changes

The combustion of fossil fuels-coal, gas and oil has led to a steady increase in the quantity of CO_2 gas in the atmosphere. It is pointed out that CO_2 level has increased by some 25 percent since 1850 because of fossil fuel combustion and land use involving mainly deforestation (2). One-half or more of the increased supply of CO_2 finds its way into the sea or is absorbed by the plants but the remainder lingers in the air. As a result the concentration of CO_2 has risen from about 280 ppm before industrial revolution to nearly 350 ppm today. As the CO_2 concentration is increasing at an exponential rate,the next such increase could be in 20 years (3). Carbon dioxide blocks outgoing radiation from the earth more than incoming radiation from the sun and contributes to global warming. It is estimated that doubling the concentration of CO_2 in the atmosphere could lead to an eventual global warming anywhere in the range of 1.5-5.5 °C (4). In a modeling study(5) the effects of doubling the CO_2 concentration in a circulation model prediced a greater warming at the poles than at the equator. Carbon dioxide is believed to be potentially more effective at changing the climate and more than one-half of the enhanced greenhouse effect todate is ascribed to CO_2 enrichment. Some calculations(6) show that observed increase in CO_2 should have warmed the lower atmosphere by 0.2 to 0.3 °C. There is little doubt that CO_2 levels to 400 ppm would raise average global temperature by 1°C with greater or lesser regional variations(6).

Carbon Dioxide Enrichment and Plant Productivity

Carbon makes up about 40 percent of the dry matter weight of higher plants, therefore, CO_2 enrichment should increase photosynthesis and plant productivity significantly. Increased CO_2 has been reported to increase photosynthesis and decrease stomatal conductance in most crop plants resulting in reduced transpiration rate per unit area of leaf and overall increase in water use efficiency. Effects of CO_2 on photosynthesis of single leaves depends on irradiance and temperature. This interaction seems to be universal among different plant species (7). Within usually encountered ranges of temperature and irradiance, increasing CO_2 concentration in an atmosphere above 300 ppm increases photosynthetic rate of many plant species (8,9). The photosynthetic response of different crop species differs with increasing CO_2. Linear increases in CO_2 uptake were observed for sugarbeet, tomato and alfalfa whereas a more typical nonlinear response occurred with maize as the concentration of cO_2 increased between 55 to 575 ppm (10). Carbon dioxide enrichment of the atmosphere to 900 ppm has been reported to increase rice yields (11) considerably. For example, the rice grain yield increased from 9.0 to 11.6 t/ha in the dry season and from 5.7 to 7.7 t/ha in the wet season when CO_2 enrichment was provided for 30 days before heading in rice.

The CO_2 response of photosynthesis varies depending upon the C_3 and C_4 species, and there are striking similarities in CO_2 response among species having the same photosynthetic pathway. C_3 plants are known to respond somewhat more favourably to increasing CO_2 concentration than C_4 plants. C_4 plants appear to close their stomata in response to increasing CO_2 to a greater extent than C_3 plants (12). In maize, the stomata closure above a concentration of 400 ppm CO_2 was at a rate proportional to the increase in CO_2 concentration (13). About 40 percent increase in CO_2 uptake rate was observed in wheat (a C_3 plant) as CO_2 concentration was increased from 300 ppm to 800 ppm (14). A comparison of CO_2 uptake rate for sugarbeet (a C_3 plant) and maize (a C_4 plant) revealed 20 percent increase for sugarbeets whereas only 6 percent increase occurred for maize when CO_2 was increased from 315 ppm to 400 ppm over the whole day periods (15). The differential response of photosyntheis between C_3 and C_4 plants to CO_2 concentration are mainly attributed to the differences in diffusion resistance of stomata and photorespiration.

Increasing CO_2 inhibits photorespiration in C_3 plants as well as the photosynthetic enzymes in C_3 plants may be more responsive to higher internal concentrations of CO_2. The effects of possible CO_2-induced climate change on US agriculture were predicted in a modeling approach which incorporated global climate models, crop growth simulation models and an economic model (16). For doubled CO_2 concentration the climatic models prediced annual temperature increase of 4.3-5.1°C and 0.1-0.2 mm day^{-1} increase in annual water demand. These changes in temperature and precipitation lead to reductions in crop yield and increase in crop water demands. However, as a result of increased CO_2 the adverse effects of global warming were outweighed by increased photosynthesis which mitigated some or all of the climate induced yield changes. The model predicted an increase in photosynthesis of 35,25 and 10 per cent for soybean, wheat and maize,respectively upon doubling the concentration of CO_2 in the atmosphere.

Experimental data and simulation models predict a 0.5 percent increase in CO_2 uptake rate for 1 percent increase in CO_2 concentration at least over a range of 100-300 ppm CO_2 above the ambient. Thus we do have an inadvertent potential for increasing crop production as a result of global increase in CO_2 concentrations.

Other Greenhouse Gases and Plant Productivity

Oxides of nitrogen- N_2X

Oxides of nitrogen are found in many forms (NO, NO_2, N_2O, N_2 etc.) and they are emitted into the atmosphere from various industrial and combustion processes, automobile exhaust and agricultural fertilizers. Of the oxides of nitrogen, only NO_2 is considered to be particularly phytotoxic and that only in relatively high concentrations. Contact of NO_2 with wet cell walls or moisture on leaf surfaces will cause NO_2 to breakdown to form nitrous and nitric acids. Bleached and nacrotic tissues result from fumigations with high concentration of NO_2 but low level fumigations can reduce growth with no other visible symptoms.

Injury to plants depends upon numerous environmental factors and associated plant physiological processes.

Chemical oxides of nitrogen move into the stratosphere and contribute to depletion of O_3 which has attracted the criticism of environmentalists, who advocate that nitrogen fertilizers in agriculture are largely the origin of oxides of nitrogen. The nitrogen cycle in agriculture as it may affect the O_3 layer is depicted in fig.2. The fertilizer ammonium (NH_4^+) is changed to nitrate(NO_3^-) by bacteria in the process of nitrification. Denitrification in warm wet soils leads to formation of N_2 and N_2O gases. Nitrous oxide (N_2O) moves into the atmosphere and results into NO and NO_2 which affects O_3 destruction in the stratosphere. Some estimates point out that for every 5 to 15 parts of N_2O, 1 part of O_3 is destroyed. We know little about how much N_2O comes from agricultural practices. The best guess at present is that about 1 part in every 6 that is denitrified comes off as N_2O, the rest as N_2. To minimise damage to O_3 from contributions of oxides of Nitrogen from fertilizers it becomes more imperative to manage soil nitrogen more efficiently and reduce losses through denitrification. Use of ammonium fertilizers, improvement of soil drainage and nitrification inhibitors (such as nitrapyrin-N serve) needs to be encouraged.

Ozone

Ozone gas forms a wavy layer in the stratosphere and is confined to 10-15 and 50 km height with peak concentrations of 10 ppm (or 0.001 per cent) at 20-25 km. Ozone is constantly created by the action of sunlight on normal oxygen molecules and is also constantly destroyed through interaction with nitric oxide, CFCs and other molecules rising from the earth. Ozone absorbs much of the ultra violet radiation from the sun which would otherwise damage the DNA or reproductive molecule in all living systems. The O_3 layer also determines the temperature of the stratosphere, and thus indirectly affects conditions further down through the different rates at which it absorbs solar and terrestrial radiation. The effect of man made pollutants in damaging O_3 layer has gained much concern in recent years. Chlorofluoro carbons rise slowly into the stratosphere where they are turned into chlorine and fluorine atoms by action of sunlight, some of which in turn destory O_3.

Ozone as a greenhouse gas is believed to be decreasing in the stratosphere and increasing in much of the troposphere. Ozone is now considered to be the most damaging gaseous air pollutant in the US and causes 90 percent or more of the air pollution damage to the vegetation. A wide range of agronomic, horticultural and forest species are susceptible to O_3-induced leaf injury (18). Ozone injury develops if light is present only during the exposure period so that the stomata are open. Sensitivity of plants to O_3 is controlled by genetic and environmental factors, the physiological age of the leaves, maturity of the plants and interactions with other biotic pathogens (19). Environmental factors such as low soil and air moisture content or low light intensity influence sensitivity by altering stomatal opening. Effects of O_3 on sweet corn yields under field conditions where exposure of two varieties of sweet corn to 0,5, or 10 ppm O_3 for six hour per day from seedling emergence to harvest caused significant growth and yield loss in one of the varieties even with these low concentrations of the pollutant (20). In general, plants grown in humid area are more susceptible to O_3 damage than plants grown in arid areas.

Ultra-violet Radiation and Plant Productivity

Ozone, CO_2, water vapour and dust in the atmosphere absorb much of the solar radiation before it reaches the earth surface. Nearly 1.3 cal cm^{-2} min^{-1} of solar radiation reaches the earth surface and 52 per cent of this is in the infra-red region ($>$700 nm) and 4 percent in the ultra-violet region ($<$ 400 nm), thus about 44 percent or 0.6 Cal cm^{-2} min^{-1} of incident solar radiation is available for photosynthesis (21). Ultra-violet radiation is commonly divided into three regions: UV-A (320-400 nm);

UV-B (280-320 nm); and UV-C ($<$ 280 nm). An excess of UV radiation can act as an environmental stress since many cell constituents absorb radiation in the range from 250-400 nm. Effects of UV-C radiation are particularly damaging to living organisms, since nucleic acids and proteins absorb strongly in this region. These wavelengths, however, never reach the earth surface because of strong absorption of UV radiation in this region by stratospheric O_3. A normal cut off at 290 nm for solar UV radiation has been indicated to be presently reaching the earth surface (22). Due to reduction in stratospheric ozone, a significant change in the mean daily level of UV-B irradiation with a shift to shorter wave lengths in UV-B region can assume biological importance. In general, for every 1% reduction in stratospheric O_3 content, biologically effecive UV-B irradiation increases about 2 percent.

UV-B radiation and plant response

The increasing environmental concerns about potential depletion of O_3 layer have provided impetus to the role of UV-B radiation on plant productivity. Experiments have shown that greatly increased exposure to UV-radiation(more than would be likely from O_3 depletion for a long time), slows up the growth of plants, prevents ripening of crops in some cases, accelerates genetic mutation, and damages the process of photosynthesis. High levels of UV radiation can be injurious to horticultural and agronomic crops, particularly under conditions where photosynthetically active radiation and photorepair wave lengths are limiting . Seedlings of tomato, radish, cucumber, lettuce, green beans, cotton, soybean and pearl millet exhibit abnormal seedling growth under elevated UV-B irraidation (23). The inhibitory effects of increased UV-B radiation on net photosynthesis were reported to be non-linear in sensitive weed species, Rumex patientia L. when 38 percent depletion of atmospheric O_3 was simulated (24). Impairment of photosynthesis was related to total biologically effective UV-B radiation received and no apparent threshold of UV-B radiation could be established.

The ecological impact of increased UV-B radiation resulting from stratospheric O_3 reduction is difficult to assess. Current estimates suggest that over a single solar cycle of about 11 years, there is approximately a 20 per cent change in biologically effective UV radiation (25). Some shifts in competitive balance between higher plant species may be occurring from these natural fluctuations but documentation of such shifts is not readily available.

RESEARCH NEEDS

- Screening of species and varieties within species for tolerance to various air pollutants so as to minimize air pollution losses to crops.
- Research on adapting varieites, species and production techniques to increased temperature and drought stress resulting from climatic change.
- Research to understand likely crop yields, crop water demands and water supply as affected by regional climatic changes.
- The role of PAR and other wavelengthsin influencing plant responses to increased UV-B radiation need to be elucidated.
- Dose-response differences of plant species and cultivars to increased UV-B radiation need characterization.
- More information is needed regarding threshold injury levels for UV-B damage in native and economically important species.
- Collaboration with research institutions in other nations is required to monitor changes in climate on global basis and study impact on plant and human life.

REFERENCES

1. Hansen, J. E. and Lacis, A.A. (1990). Sun and dust Versus greenhouse gases: An assessment of their relative roles in global climatic change. Nature **346** : 713-719.

2. Schneider, S.H. (1989). The greenhouse effect: Science and policy. Science **243** : 771-781.

3. Keeling, C., Bacastow, R. and Whorf, T. (1982). In : Carbon dioxide review, ed. Clark, W, Oxford University Press, pp.377-384.

4. Dickinson, R.E. (1986). The greenhouse effect climatic change and ecosystems. Wiley, New York, 1986, pp 206-270.

5. Manabe, S. and Weatherald,R.T. (1975). The effects of doubling the CO_2 concentration on the climate of a general circulation model. J. Atmos. Sci.**32** : 3-15.

6. Tickell,C. (1978). Climate change and world affairs, Oxford, Pergamon Press, p 78.

7. Harper, L. A., Barker, D.N., Box, J.E. and Hesketh, J.D. (1973). Carbon dioxide and the photosynthesis of field crops: A metered carbon dioxide release in cotton under field conditions. Agron. J. **65**: 7-11.

8. Brun, W.A. and Cooper R.L. (1967). Effects of light intensity and carbon dioxide concentration on photosynthetic rate of soybean. Crop Sci. **1**: 451-454.

9. Ford, M. A. and Thorne, G. W. (1967). Effect of CO_2 concentration in growth of sugarbeet,barley, kale and maize. Ann. Bot. **31**: 629-644.

10. Thomas, M. D. and Hill, G. R. (1949). Photosynthesis under field conditions. In: Photosynthesis in plants, Frank, J. and Loomis, W.E. eds. The Iowa State College Press, Ames, pp.19-52.

11. Yoshida, S. (1976). Carbon dioxide and yield of rice. pp.211-221. In: Climate and rice, International Rice Research Institute, Los Banos, Philippines.

12. Akita, S. and Moss, D.N. (1972). Differential stomatal response between C_3 and C_4 species to atmospheric CO_2 concentration and light. Crop Sci. **12** : 789-793.

13. Gifford, R. M. (1970). Aspects of the variability and control of carbon dioxide exchange in maize leaves. Ph.D.Thesis, Cornell Univ.,Ithaca,NY. 241 pp. University Microfilms Diss. Abstr.No. 70-14, 383.

14. Akita, S. and Moss, D.N. (1973). Photosynthetic response to CO_2 and light by maize and wheat leaves adjusted for constant stomatal aperatures. Crop Sci. **13** : 234-237.

15. Lister, R. A. (1974). The effects of increasing atmospheric carbon dioxide and particles on plant photosynthesis. Ph.D. Thesis, Cornell University, Ithaca, NY. 159 pp. Univ. Microfilms Diss. Abstr. NO.75-4257.

16. Adams, R.M., Rosenzweig, C., Peart, R.M., Ritchie,J.T., McCarl, B.A., Glyer, J. D., Curry, R.B., Jones, J.W., Boote, K.J. and Allen Jr, L.H. (1990). Global climate change and US agriculture. Nature **345**: 219-224.

17. Taylor, O.C. and Eaton, F.M. (1966). Suppression of plant growth by nitrogen dioxide. Plant Physiol. **41**: 132-135.

18. Skelly, J.M., George, M.F., Heggestad,H.E. and Krizek, D.T. (1979). Air pollution and radiation stresses. In: Modification of the aerial environment of plants, Barfield,B.J. and Gerber, J.F., eds., American Society of Agricultural Engineers, pp. 114-138.

19. Heck, W.W., Dunning, J.A. and Hindawi, I.J. (1965). Interaction of environmental factors on the sensitivity of plants to air pollution. J. Air. Pollut. Control Assoc. **15**: 511-515.

20. Heagle, A.S., Body, E.D., and Pounds, E.K. (1972). Effect of ozone on the yield of sweet corn. Phytopathology **62**: 683-697.

21. Gates, D. M. (1965). Radiation energy, its receipt and disposal: pp. 1-26. In: P.E. Waggoner (ed.). Agricultural Meteorology. Am. Meteorol. Soc., Boston, MA..

22. Green, A. E.S., Sawada, T. and Shettle, E. P. (1974). The middle Ultraviolet reaching the ground. Photochem. Photobiol. **19**: 251-259.

23. Krizek, D. T. (1975). Influence of ultraviolet radiation on germination and early seedling growth. Plant Physiol. **34**: 182-186.

24. Sisson, W.B. and Caldwell, M.M. (1976). Photosynthesis, dark respiration, and growth of Rumex patientia L. exposed to ultraviolet irradiance (288 to 315 manometers) simulating a reduced atmospheric ozone column. Plant Physiol. **58** : 563-568.

25. Caldwell, M.M. (1977). Solar UV radiation and the growth and development of higher plants. pp.131-177. In: A.C. Giece (ed.). Photophysiology Vol. VI. Academic Press, NY.

GROWTH, PHOTOSYNTHESIS AND PRODUCTIVITY

V.I. Kefeli

Institute of Soil Science and Photosynthesis
USSR Academy of Sciences
puschino, 142292, USSR

The work was aimed at elucidating the role of growth and photosynthesis, as basic components of productivity, in the accumulation of phytohormones, inhibitors and biomass. The following pea mutants were used: dwarfs (inhibited stem development), albinos (chlorophyll deficiency). It was shown that the dwarfs contained very low quantity of phytohormones and increased amount of natural inhibitors (phenols). The albino plants had lost the ability for chlorophyll synthesis, had less carotenoids and a higher quantity of polyphenols.

INTRODUCTION

Biosynthesis of secondary substances is regulated by light either directly through phenolics or indirectly by endogenous phytohormones (indolics). Using different kinds of mutants it is feasible to separate the two effects (1-3). In the present study, dwarf and albino mutants of cotton (Gossypium hirsutum L.) and pea (Pisum sativum L.) were used to eliminate either growth or photosynthesis and thus investigate the biosynthesis and accumulation of phenolics and phytohormones (3). In some experiments, pea plants were grown under light of xenon lamps at various light intensities under phytotronic conditions. Analysis of phenolic and phytohormones was carried out adapting procedures described earlier (2).

Effect of high intensity light on stem and leaf growth of pea plant

Xenon arc lamps under phytotronic conditions block stem growth but did not inhibit photosynthesis (Figure 1). The plants having depressed stem growth produced thicker leaves with two-fold thick palisade parenchyma (Figure 2).

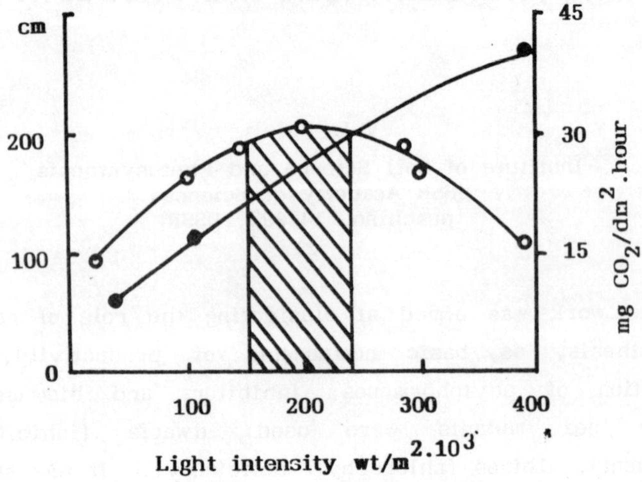

Figure 1. Growth (1) in cm and photosynthesis (2) in mg CO_2/dm^2.hour of pea plant "Torsdag" under light of xenon arc lamp under various intenties.

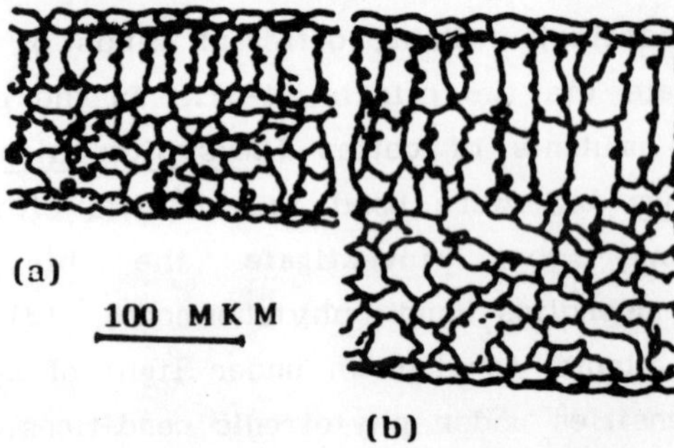

Figure 2. Leaf thickness (a) normal, (b) plants with inhibited growth of stem (400 wt/m^2. hour

Dwarf mutant (K 202) as well produced thicker leaf under high light intensity of 400 wt m^2.h (Table 1).

Table 1. Anatomical characteristics of leaves of 40 day old original and mutant pea plants

Thickness of (mm)	Light intensity, wt m^2.h		
	50	200	400
	Torstag		
Leaf	88.7±1.00	116.9±0.09	168.5±1.05
Palisade mesophyll	30.2±1.05	40.1±1.03	57.0±1.20
Spongy mesophyll	41.9±2.30	57.6±1.63	87.4±2.71
	Mutant K 202		
Leaf	190.5±3.21	270.0±4.49	303.5±4.10
Palisade mesophyll	82.6±1.72	106.7±1.99	120.4±2.30
Spongy mesophyll	91.3±2.27	140.7±3.90	161.2±1.40

Biosynthesis of flavanoids was proceeded from C^{14} PCA (para coumarate). Under the effect of high light intensity, accumulation of QGC (quercitin-glucosil-coumarate) was relatively higher in K 202 than in tall type 'Torsdag' (Table 2).

Table 2. Radioactivity of methanol-soluble metabolites of PCA* from pea

Variety Light intensity wt m^2	Radioactivity, % from methanol extract			
	Free PCA	Glucose ester of PCA	QGC**	Non-identified metabolite
Torsdag, 200 (tall)	58.48±2.9	24.77±1.2	4.15±0.2	12.54±0.6
Torsdag, 400	42.97±2.1	37.33±0.6	6.89±0.3	12.81±0.6
K 202, 200(dwarf)	37.96±1.9	40.50± 2.0	6.61±0.3	14.93±0.7

*PCA - para-coumarate ** QGC - quercetin glucosil coumarate

Albino mutants and phenolics

Normal and albino forms of cotton seedlings were grown in light for two weeks. Results indicate that levels of flavanoids were not affected whereas anthocyanins were depressed partially

(Table 3). Similar situation was noticed for pea seedlings. QGC accumulated in Xantha mutants (without chlorophyll) in light and its level was more or less at par in green leaves of normal plants or yellow leaves of the mutant (Table 4).

Table 3. Flavanoids and anthocyanins in cotton mutants

Forms	Flavanol glucoside	Anthocyanin	Stem length (cm)
	(mg g^{-1} of fresh weight)		
Green	13.0	20.0	10
Etiolated	0	0	21
Albino (grown in light)	12.5	10.0	10

Table 4. Phenolics in pea xantha mutant

Forma	Content of QGC in mg g^{-1}		PCA in mg g^{-1}		Growth length (cm)
	Fresh	Dry	Fresh	Dry	
Normal in light	5.0	36.3	0.04	0.30	8,0
Normal in darkness	1.3	17.1	0.30	0.40	17,0
Mutant in light	5.7	36.9	0.04	0.20	8,2
Mutant in darkness	1.2	17.5	0.02	0.30	16,5

CONCLUSIONS

Our study suggests clearly that the level of phenolics like QGC is not affected adversely in plants either with depressed stem growth or depressed chlorophyll synthesis. Under high light intensity, photosynthetic rate remains quite stable, though plants with depressed stem growth produced leaves with thicker palisade parenchyma. On the other hand, synthesis of flavanoids in dwarf plants is activated but does not depend directly on photosynthesis.

REFERENCES

1. Kefeli, V.I., Vutacek, M. (1979). Effects of phenolics on auxin metabolism. in: Regulation of secondary product and plant hormone metabolism, 12th Meeting, Dresden, Pergamon Press, pp 1-28.

2. Kefeli, V.I. (1978). Natural plant growth inhibitors and phytohormones. Junk Publ., The Netherlands, pp 278.

3. Kefeli, V.I. (1988). Abscisic acid and para-coumarate metabolism and functions. Intern. Congress of Biochemistry, Praha, Vol.V: 14 (Abstract).

IMPACT OF RISE IN TEMPERATURE ON THE PRODUCTIVITY OF WHEAT IN INDIA

Y.P.ABROL[1], A.K.BAGGA[1], N.V.K.CHAKRAVARTY[2] AND P.N.WATTAL[1]
Division of Plant Physiology[1] and Agricultural Physics[2], Indian Agricultural Research Institute, New Delhi-110 012, India

The increasing emissions of greenhouse gases such as carbon dioxide, chlorofluorocarbons, methane, nitrogen oxides and decrease in tropospheric ozone due to ongoing large scale burning of fossil fuel, deforestation, industrialization and intensive use of fertilizers, if unchecked, will raise the global temperature. The carbon dioxide concentration is expected to double and the ambient temperature is predicted to rise by 2-4°C by 2050-2075 which eventually would bring about changes in the intensity and spread of precipitation, vapour pressure deficit and evapotranspiration rates. In this paper, influence of changes in temperature on some aspects of growth and development relevant to grain yield of wheat are discussed.

Crop season and thermal environment

In India, wheat is grown under a wide range of soil and climatic conditions i.e. plains, river valleys or plateau lands as well as in the hills upto an elevation as high as ten thousand feet above mean sea level. The cultivation is restricted to cooler periods. Life cycle of the crop is short as the sowing is determined by gradual cooling of seedbed achieved after slowly receding warmer months to ensure optimal initial plant establishment. Maturation has to be completed before rapidly ascending temperatures combined with hot and dry winds which commence in summer. Thus, planting of the crop around mid-November is the established practice under irrigated conditions. In unirrigated areas, seeding is advanced by about four weeks so that the crop is able to utilize the moisture stored during the preceding monsoon. In heavily concentrated wheat growing areas, which lie between 26°-31° N and

74°-80°E and include the plains of Punjab, Haryana, Uttar Pradesh, Bihar, Madhya Pradesh and Rajasthan, the ripening of the crop terminates during March/April. Duration and severity of winter decreases from north to south. Correspondingly, the growing season becomes shorter and productivity declines. In eastern regions as well, the winter is warmer and yields are low. This is evident from the data compiled on the basis of trials conducted by All India Coordinated Wheat Improvement Project (AICWIP) at different centres under adequately fertilized and irrigated conditions (Table 1). It indicated that reduction in grain yield more or less corresponded with the increase in prevailing temperature. Crop duration and yields were reduced from 127 days and 45.2 q ha^{-1} (mean temperature 16.9°C at Delhi) to 108 days and 32.8 q ha^{-1} (mean temperature 20°C at Powarkheda). This is also true for seasonal fluctuations in temperature determining grain yields in a given tract (Table 2).

Growing period extends to as much as seven to eight months in Northern hills where the temperaturres are low. In some areas at high elevations, farmers prefer to sow a summer crop in May-June and harvest in September-October. With the advent of new technology and availability of improved varieties, farmers specially in the northern belt cultivate rice, sugarcane and potatoes on a large scale and fields get vacated late because of late harvest of the preceding crop. Under such situations, sowing of wheat is delayed towards last week of November to the end of December or even beginning of January and consequently later stages of the crop are pushed into rising temperatures in ensuing months. This causes reduction in growth period to less than four months.

Temperature, phenology and crop performance

The entire growth of wheat plant can be deliniated broadly into four phenological events: germination, double ridge or first spikelet initiation, terminal spikelet initiation, ear emergence or anthesis and grain development. These phenological events and duration of their phases are mainly controlled by temperature, photoperiod and irradiance (intensity of photosynthetically active radiation). In the following paragraphs, duration from first spikelet initiation to terminal spikelet initiation and from terminal spikelet initiation to ear emergence/anthesis are referred to as

Table 1. Growing season temperature (mean maximum and minimum °C), crop duration and yield in wheat at five locations in India (1984-85 season)

Place	Location		Temperature		Crop duration (days)	Grain ($q\ ha^{-1}$)
	Lat N	Long E	January	Growing season		
Powarkheda	22°24'	77°42'	19.0	20.0	108	35.8
Indore	22°30'	76°00'	18.4	19.9	113	32.5
Kanpur	26°28'	80°24'	14.1	19.8	116	32.4
Delhi	28°35'	77°12'	13.4	16.9	127	45.2
Ranchi	23°30'	85°30'	18.8	18.6	118	38.6

Data based on 16 varieties
Source: AICWIP Annual Report

Table 2. Mean maximum and minimum temperature °C during various phases of crop growth, their duration, grain number and grain yield in three seasons at three locations in India

Location	**Delhi** (28°35'N; 77°12'E)			**Kanpur** (26°28'N; 80°24'E)			**Indore** (22°30'N; 76°0'E)		
Season	1987-88	88-89	89-90	1987-88	88-89	89-90	1987-88	88-89	89-90
Mean temperature °C									
Sowing to anthesis	15.8	15.4	15.6	16.4	15.8	18.0	19.8	19.1	20.0
Sowing to maturity	17.7	17.0	17.6	18.9	18.3	19.5	20.7	20.8	21.1
Days to anthesis	93	99	87	82	86	84	74	82	78
Days to maturity	132	136	140	123	125	135	118	121	125
Grain number m^{-2}	13109	15120	14103	13081	14203	11978	10556	10962	10234
Grain yield m^{-2}	549	702	561	510	573	500	421	470	454

Data based on 33 (1987-88) and 10 varieties (1988-89 and 1989-90 seasons)
Source: AICWIP Annual Reports

spikelet initiation and spike growth phase, respectively. These phases also coincide with tillering and leaf area development.

a. Initial plant stand and tillering

Initial plant stand controls the magnitude of tillering and development of the leaf cover which eventually determines the crop growth rates and total dry matter per unit land area. While ambient temperatures in the range 20°-25°C (mean maximum and minimum) have been considered favourable for germination, seedling emergence and optimum plant establishment, mean daily temperature 16-20°C is good for tillering (Figure 1; Bhardwaj, 1978). Thus, mid-November temperatures are conducive for seeding to realize

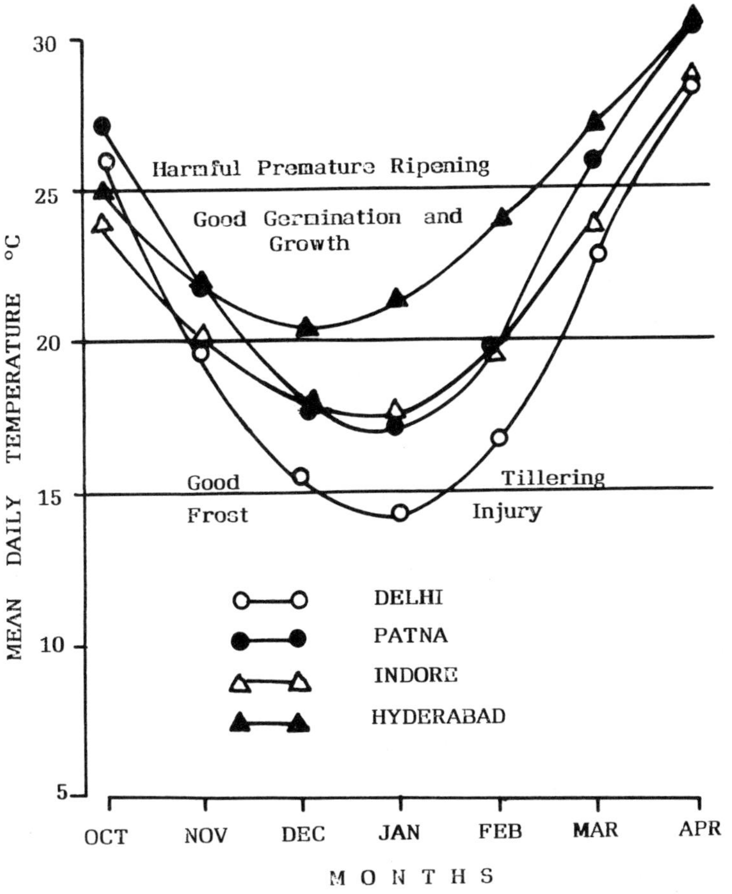

FIG.1. TEMPERATURE CURVES FOR DIFFERENT WHEAT ZONES.
(Adapted from Bhardwaj, 1978)

desirable plant population optimized approximately at 200 plants m^{-2} in irrigated areas (Saini, 1978). The level of plant stand is never adequate either in mid-October or in irrigated late December-January plantings. This is apparently due to supra-optimal or sub-optimal temperatures, respectively, prevailing during these months for establishment of proper plant stand. In the first case, rise in temperature is expected to prove drastic in unirrigated lands and more drastic at places where the temperature in the seeding zone of the soil exceeds by a few degrees relative to increase in atmospheric temperature. This is relevant in dark textured soils or in soils with low moisture retention capacity since both increasing temperature and soil moisture stress will operate simultaneously. In the second case, the increase in soil temperature may favour seedling emergence and initial plant stand.

Cooler temperatures required for normal tillering do not set in before fourth week of November in most of the wheat growing areas. Consequently, mid-November plantings are exposed during tillering phase to optimum range in December and January and achieve maximum tiller potential. In north India, early medium varieties (90-100 days flowering) show tiller number at its peak 50-55 days after sowing and subsequently decline to a constant number (Singh et al. 1972; Saini and Nanda, 1974; Nigam, 1977). Increase in temperatures is capable to change this time sequence and tillering behaviour. It may cause decrease in plant stand but not to the extent as to seriously affect crop growth rate since the initial effect of reduced plant density will eventually diminish with time due to compensatory mechanism of high tillering produced subsequently. On the other hand, both mid-October and late December-January plantings will suffer more for want of adequate tillering because increase in temperature during warmer months of November and rising hotter conditions from late January onwards coincide, respectively, with their tillering phase reducing its duration and number. This will also result in substantial reduction in crop growth rates, accelerated heading and in small plants.

b. Spikelet and grain number

Most of the reports confirm that duration of spikelet phase and subsequent phase of spike growth are most important for determining spikelet and floret number per spikelet (Rawson, 1970; Allison and Daynard, 1976; Rahman and Wilson, 1977; Saini et al., 1986). This is especially convincing for varieties, with almost negligible vernalization response, mostly grown in India. Thus, other environmental variables such as temperature and photoperiod are critical for the duration of these phases which determine accumulation of potential number of sites for the grains to develop prior to anthesis and for development of growth resources (LAI) and tillering.

Effect of temperature on duration of these phenological phases has been extensively studied on the basis of accumulation of heat units referred to as growing degree days (GDD). This is estimated as cumulative mean diurnal temperature above a base value of 4.5-5.0°C required to complete a specific phenological event. In some cases a proviso of photoperiodic need has also been incorporated and a photothermic unit adopted which is a product of GDD and mean diurnal day length during a particular phase. In almost all these studies it is well documented that a given variety has a definite temperature (GDD) or photothermic requirement before it attains a certain phenological stage in any environment. This would mean that increase in temperature or photoperiod or both will not disturb the plant time (i.e. GDD or photothermic days) but cause reduction in duration of the phase (calender time) aptly termed by Rawson (1988). Chinoy (1956) emphasized the importance of a definite energy requirement of a variety based on what he calls photothermic and vernalization quanta for predicting the time of its ear emergence in a known environment. Photoperiodic factor has limited relevance to controlling duration of these events in our major wheat growing areas which lie in a narrow latitudinal range where mean day lengths do not vary more than 10-15 minutes during different phases of growth of plant. Importance of this factor becomes less due to intensive cultivation of photoinsensitive varieties.

Chakravarty and Sastry (1983) observed almost similar GDD from sowing to anthesis for a variety Kalyansona i.e. 1086 and 1069 in November sowing, respectively, in warmer 1978-79 and cooler 1979-80 seasons but the time to anthesis was reduced by 9 days in first than in second season. GDD for this phase also did not show much variation in different plantings sown at different dates. Delayed sowing, however, reduced the time of maturation of the crop.

Interestingly, this variety had accumulated more or less similar units of GDD (almost 950) in three temperature regimes (15/10, 21/16, 27/22°C; maximum/minimum) under controlled conditions at Canberra (Rawson and Bagga, 1979). The duration to anthesis was reduced by 25 and 35 days, respectively, in the later two treatments than at 15/10°C.

Since there exists inverse relationship between spikelet number and duration of the developmental phase prior to anthesis, elevated temperature will invariably cause reduction in spikelet number (and/or floret number) and hence the decrease in potential grain number (Friend, 1965; Halse and Weir, 1974; Rahman and Wilson, 1978). We have worked out linear regression of grain number on the mean temperature from sowing to anthesis on the basis of data given in Table 2 on the assumption that the grain number is solely controlled by temperature prevailing during this period. It indicates reduction of 5.5 per cent in grain number for every 1°C rise in temperature (Figure 2). Saini and Nanda (1986) showed that there was a decline of approximately 600-650 grains m^{-2} (or about 4-5 per cent decrease) with every 1°C increase in mean temperatures above 17-17.7°C during the period from terminal spikelet initiation to anthesis.

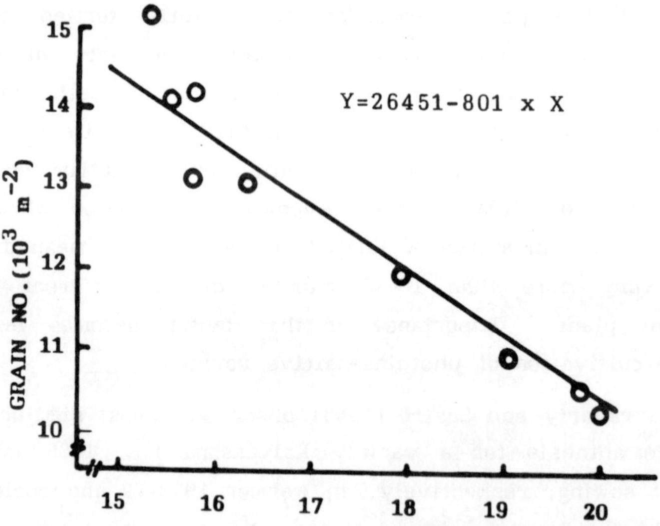

FIG.2.MEAN TEMPERATURE °C(SOWING TO ANTHESIS)

Temperature stress and grain development

Temperature has a pronounced effect on duration of grain filling. At low temperature, grain development continues for a longer period. This results in higher final grain weight (Wattal, 1965; Sofield *et al*. 1974; Spiertz, 1974; **Warrington** *et al*., 1977). Elevated temperatures after anthesis accelerate growth rate of grains but it is associated with greater respiratory losses, faster senescence of the foliage and ear which cause earlier termination of grain growth and lower grain yield. A number of studies have shown that mean maximum temperature above 25°C, during the grain filling period, tend to **depress** grain weight (Asana and Williams, 1965; Asana, 1966). They further observed that a rise of 5°C in mean maximum and minimum temperature (between approximately 16.8°C to 23.0°C) depressed one thousand grain weight by 13.9 and 17.8 per cent, respectively, in varieties PbC 281 and NP 720 under outdoor conditions at Delhi. Under controlled constant temperature conditions at Canberra they found that several other varieties responded more or less similarly to increase in 'day' temperature between 25 to 31°C and observed the mean reduction in grain weight by 16.4 per cent for 6°C rise in temperature. Wattal (1965) showed mean reduction of 17.3 per cent in grain yield with 5°C rise in mean maximum temperature (between 27.6 to 32.3°C) under open pot culture conditions.

From the foregoing analysis, it can be said that 1°C raise in global temperature will depress the grain yield by 8 to 10 per cent mediated through decline in grain number by 5.5 per cent and 1000 grain weight by about 3 per cent. The estimates of linear regression of grain yield on mean temperature of the whole growing season (Table 2; Figure 3) supports this contention. It shows losses in grain yield by about 9 per cent with a rise of 1°C in mean temperature.

Conclusion

To what extent the above estimates prove realistic will depend upon the benefits the crop may accrue due to expected increase in CO_2 levels concurrent with the increase in mean annual temperature above the existing normal seasonal fluctuations. Increase in CO_2 concentration will enhance the photosynthetic rates and

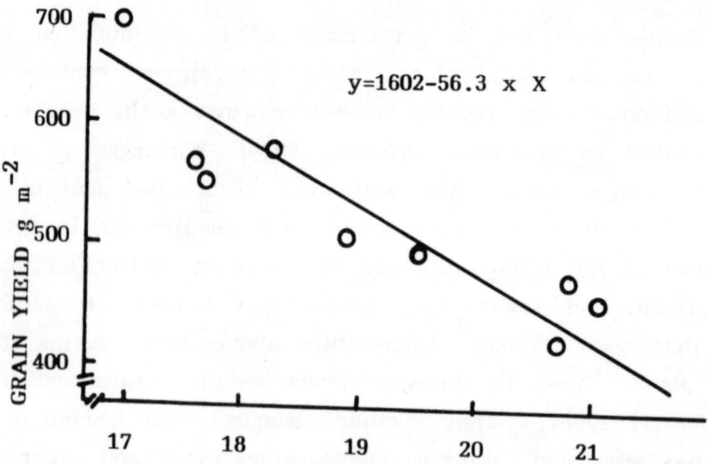

FIG.3.MEAN TEMPERATURE °C (SOWING TO MATURITTY)

productivity. A number of reports suggest that wheat yields may increase by about 35-40 per cent by doubling of CO_2 (Cure and Acock, 1986). It is likely that the temperature may rise to as much as 4°C at this level of CO_2. Thus the situation will not be grim since such an increase in CO_2 concentration would counterbalance the deleterious effect of temperature on grain yield. However, we may have to confront with the problem of shifting of wheat area more towards north as a result of global warming. This will not, however, acquire the dimensions which we perceive now as our conventional breeding methods have been able to isolate promising varieties tolerant to high temperatures existing in Central and Southern part of India.

REFERENCES

Allison, J.C.S. and Daynard, T.B. (1976). Effect of photoperiod on development and number of spikelets of a temperat and some low latitude wheats. Ann. Appl. Biol. 83: 93-120.

Asana, R.D. (1966). Physiological analysis of yield of wheat in relation to water stress and temperature. J. Post Graduate School, IARI, New Delhi 4: 17-31.

Asana, R.D. and Williams, R.F. (1965). The effect of temperature stress on grain development in wheat. Aust. J. Agric. Res. 16: 1-13.

Bhardwaj, R.B.L. (1978). Agronomic Wheat Practices. Indian Farming, Wheat Special No.27: 45-53.

Chakravarty, N.V.K. and Sastry, P.S.N. (1983). Phenology and accumulated heat unit relationships in wheat under different planting dates in the Delhi region. Agr. Sci. Progress 1: 32-42.

Chinoy, J.J. (1956). Determination of photothermic and vernalization quanta for the vegetative period of wheat. Physiol. Planta. 9: 1-18.

Cure, J.D. and Acock, B. (1986). Crop responses to carbon dioxide doubling: a literature survey. Agriculture and Forest Meteorology 38: 127-145.

Friend, D.J.C. (1965). Ear length and spikelet number of wheat grown at different temperatures and light intensities. Can J. Bot. 43: 345-353.

Halse, N.J. and Weir, R.N. (1974). Effects of temperature on spikelet number of wheat. Aust. J. Agric. Res. 25: 687-695.

Nigam, P.K. (1977). Physiology of tillering in wheat. Ph.D. thesis, Indian Agricultural Research Institute, New Delhi.

Rahman, M.S. and Wilson, J.H. (1977). Determination of spikelet number in wheat. I. Effect of varying photoperiod on ear developments. Aust. J. Agric. Res. 28: 565-574.

Rahman, M.S. and Wilson, J.H. (1978). Determination of spikelet number in wheat. III. Effect of varying temperature on ear development. Aust. J. Agric. Res. 29: 459-467.

Rawson, H.M. (1970). Spikelet number, its control and relation to yield per ear in wheat. Aust. J. Biol. Sci. 23: 1-15.

Rawson, H.M. (1988). Constraints associated with Rice-Wheat rotations. Effect of high temperatures on the development and yield of wheat and practices to reduce deleterious effects. In: 'Wheat Production Constraints in Tropical Environments'. Proceedings Int. Conference Chiang Mai, Thailand (Jan.19-23, 1987) ed. Klatt, A.R. pp 44-62.

Rawson, H.M. and Bagga, A.K. (1979). Influence of temperature between floral initiation and flag leaf emergence on grain number in wheat. Aust. J. Plant Physiol. 6: 391-400.

Saini, A.D. (1978). Physiology 'In: Wheat Research in India' Indian Council of Agricultural Research, New Delhi, pp 99-125.

Saini, A.D. and Nanda, R. (1974). Grain yield and tillering pattern of four wheat varieties. Ind. J. Agric. Sci. 44: 87-92.

Saini, A.D. and Nanda, R. (1986). Influence of temperature and light intensity during ear growth period on grain number in wheat. Ind. J. Agric. Sci. 56: 512-519.

Saini, A.D., Dadhwal, V.K.,Phadnawis, B.N. and Nanda, R. (1986). Influence of sowing dates on pre-anthesis phenology in wheat. Ind. J. Agric. Sci. 56: 503-511.

Singh, R.D., Chatterjee, B.M. and Sanyal, N.N. (1972). Tillering pattern in tall and dwarf wheat varieties under different levels of nitrogen and spacing. Ind. J. Agric. Sci. 42: 42-47.

Sofield, I., Evans, L.T. and Wardlaw, I.F. (1974). The effects of temperature and light on grain filling in wheat. In 'Mechanisms of Regulation of Plant Growth' ed. R.L. Bieleski, A.R. Ferguson and M.M. Cresswell. Bull. Roy. Soc. New Zealand. 12: 909-915.

Spiertz, J.H.J. (1974). Grain growth and distribution of dry matter in wheat plant as influenced by temperature, light energy and ear size. Neth. J. Agric. Sci. 22: 207-220.

Warrington, I.J., Dunstone, R.L. and Green, L.M. (1977). Temperature effects at three developmental stages on the yield of wheat ear. Aust. J. Agric. Res. 28: 11-27.

Wattal, P.N. (1965). Effect of temperature on the development of the wheat grain. Ind. J. Plant Physiol. 8: 145-159.

RECOMMENDATIONS

SESSION I: UV-B RADIATION EFFECTS

I Introduction

Increase in solar UV-B radiation adversely affects a number of important biological systems. Magnitude of the effects of enhanced levels of UV-B may vary among species and cultivars. Sensitive plants often show reduced photosynthesis, growth, flowering and yield . Important biological macromolecules such as DNA, proteins, phytohormones and secondary plant products are strongly influenced by increased UV-B radiation. Evidence is also available to suggest that deleterious effects of UV-B radiation on photosynthesis are accentuated when plants are concomitantly stressed by low levels of visible light, tropospheric ozone, heavy metals, herbicides and high temperatures. Plant resistance to UV-B radiation depends primarily upon protective mechanisms involving formation of UV-B absorbing pigments and repair processes such as photoreactivation. Many of these protective and repair processes are inducible by light *via* gene activation.

II Research priorities

Following areas have been identified:

1. *Primary molecular mechanisms*: A better understanding of the primary mechanism of UV-B induced damage on membranes, Photosystem II in chloroplast and DNA.

2. *Appropriate action spectra*: There is a need to characterize UV-B sensitivity by development of appropriate action spectra. This will provide information on photoactive species and include measurements on fluence-response relationships and lower limits to UV-B sensitivity. The biological effects of ozone depletion could only be meaningful if UV action spectra are steep i.e. increase sharply with decrease in wave length since only small increase of energy at shorter wave lengths would occur even with large ozone depletion.

3. *Sensitivity of natural eco-systems*: Long term information on UV-B sensitivity of natural eco-systems (including forests, fresh water and marine

eco-systems) is needed. Even plant communities which presently have limited economic value might act as a reservoir for genetic diversity for modern crop breeding programme and in future may provide new drugs, medicines and other natural products.

4. *Specific information from tropical environments*: Effects of UV-B radiation on low latitude tropical environments where plants are already exposed to a high UV-B radiation flux need to be studied. These investigations will help in understanding natural adaptations to UV-B in environments having the highest UV-B fluxes.

5. *Field variation studies*: We require more data from field studies of economically important crop species like rice, wheat, maize, sorghum etc.

6. *Interaction with other environment factors*: Factors such as CO_2, temperature, heavy metals viz. lead, mercury cadmium, nickel, chromium, zinc etc., are likely to increase, and the tropospheric ozone to decrease, in the future. It is important to study the interaction of UV-B radiation in concert with other changes.

SESSION II: CO_2 ENRICHMENT/BALANCE

I Introduction

It is presently well-documented that elevated CO_2 levels lead to increased photosynthetic capacity, decreased stomatal conductance, enhanced water use efficiency and altered photosynthetic partitioning to different organs such as reproductive structures and roots. Thus differences can be found among representatives in various plant communities (annuals, agricultural crops, forest trees, etc.). However, its effects in combination with other environment variables such as temperature, solar radiation and atmospheric precipitation need to be investigated. The aim is to isolate plant systems better suited to these concomitant changes. Both the short-term growth chamber experiments and field scale studies are required on agricultural crops and forest trees.

II Research priorities

Ambient CO_2 enhancement should involve studies on the following aspects:

1. *Photosynthesis*
   (i) Stomatal behaviour: relation between C_a and C_i.
   (ii) Mechanism for concentrating dissolved inorganic carbon (DIC): its operation in unicellular algae has been reported but its use for lowering atmospheric CO_2 and forming carbonate deposits (barrier reefs) is yet unexplored. Studies on the regulation of CO_2 by DIC need to be extended to higher plants.
2. *Source sink interaction*
   (i) Efficiency of dry matter production (g DW/g CO_2 assimilated), sink strength (harvest index).
   (ii) Regulation of feedback mechanism limiting photosynthesis with reference to sucrose-starch partitioning.
   (iii) Carbon partitioning (root : shoot ratio and other parameters).

3. *Respiration and photorespiration*
   (i) Whole plant respiration including maintenance and growth respiration components and carbon balance.
   (ii) Regulatory role of photorespiration.
   (iii) Mechanism and function of alternative respiration.

4. *Plant phenology*
   Growth and development patterns in relation to plant productivity.

5. *Rhizobial studies and plant microbial interaction*
   (i) Litter production in relation to carbon cycle.
   (ii) Role of cyanobacteria in fixing CO_2 and nitrogen.

6. *Water and nitrogen use efficiency*
   Species/cultivar response for water and nitrogen use efficiencies.

7. *Interaction with other environmental factors*
   (i) Light intensity (change in PAR levels).
   (ii) Temperature.
   (iii) Water Stress.

8. *Crop-weed interaction*
   Interaction between C_3-C_3, C_3-C_4, C_4-C_3, C_4-C_4 crop-weed species are required to be investigated in detail.

SESSION III: PHOTOSYNTHESIS AND ENVIRONMENTAL STRESSES

I Introduction

The global climatic change may accentuate and alter the geographical range of environmental factors which already constitute significant limitation to crop productivity e.g., high temperature, inadequate water supply and soil salinity due to non-judicious use of irrigation. The effect of enhanced levels of UV-B and steady increase in ambient CO_2 concentration in the atmosphere warrant detailed investigations on photosynthesis and environmental parameters limiting productivity.

II Research priorities

1. *Identification of direct effect of enhanced CO_2 and UV-B radiation levels on photosynthesis and plant productivity*
   - (i) CO_2 effects on:
     - (a) development of the photosynthetic apparatus
     - (b) feedback inhibition of photosynthesis
     - (c) assimilate transport
     - (d) sink demand ·
   - (ii) UV-B radiation effects on:
     - (a) mechanism for inhibition of PS II
     - (b) mechanisms and capacity for repair (unicellular green algae and cyanobacteria, as test material)
     - (c) protective mechanisms associated with PS II function — electron transport and pigment deexcitation processes
     - (d) impact on net photosynthesis and ultimately on productivity.
2. *Interaction between UV-B radiation and Co_2 influencing environmental factors affecting photosynthesis and plant productivity*
   - (i) Identification of mechanistic basis of moisture and temperature stresses.
   - (ii) Evaluation of stress proteins induced by moisture stress, heat and UV-B radiation.

3. *Development of physiological and genetical model systems*
   (i) Extensive genetic and molecular genetic information on *Arabidopsis* because of its close relationship to agronomically significant *Brassica* species.
   (ii) Chloroplast model systems: Study of DIC pumps in cyanobacteria, unicellular green algae such as *Chlamydomonas*, prochlorophytes would be of great promise in guiding specific model systems.

SESSION IV: CROP MODELLING

I Introduction

Crop modelling may help to predict the consequences of impending global climatic changes and integrate research on essential plant processes affecting productivity. Since water and nutrient supplies are finite, their efficiencies of utilization are of critical interest. Therefore, greater emphasis might be laid on the synthesis/construction of plant ideotypes suited to these environments.

II Research priorities

Some of the plant processes and production factors requiring immediate attention are as follows:

1. *Crop phenology*

(i) Relate components of yield to the vegetative, reproductive and grain filling periods.
(ii) Agromet data banks — break data out in terms of critical developmental period durations.
(iii) Conduct temperature typing to evaluate the relationship between metabolic and developmental paces.
(iv) Evaluate production/metabolic efficiencies as influenced by species, genotype and environment.
(v) Genetic manipulation and screening of field crops and wild populations — microcosm research.

2. *Developmental biology*

(i) Analyse cell differentiation and expansion pattern in reproductive organs (cereals and legumes).
(ii) Evaluate the environmental control on growth and differentiation.
(iii) Evaluate the role of stress shock proteins in seed stability, seed size and expansion, consider gene cloning.

3. *Cultural manipulations*

(i) Evaluate canopy carbon balance and water use efficiency at critical growth stages.

(ii) Nutrient optimization in low input systems — N fixation, intercropping, rotations, etc.

(iii) Water optimization — residue management effects on infiltration, erosion, evaporation and nutrient recycling.

(iv) Beneficial soil microorganisms — mycorrhizae, cyanobacteria etc.

SESSION V: AGRO AND NATURAL ECO-SYSTEMS

I Introduction

It is anticipated that global climatic changes will include alleviated levels of CO_2, temperature, radiation, precipitation, gaseous emissions (methane, nitrogen oxides, water vapour and Chlorofluoro-Carbons), dust, soil erosion and salinity. Strategies need to be developed to cope (or exist) with climate changes. This may be accomplished *via* altered crop management and/or improved crop cultivars in agricultural systems. Community composition in natural eco-systems might change depending on the competitive ability and reproductive success of each species under the changed environment. However, in agro-systems proper management practices such as (i) use of crop species which effectively store more CO_2 equivalents, (ii) developing useful land use strategies to utilise plant climatic resources in a more positive manner, (iii) wet land management (through crop cultures alongwith judicious fertilization), and (iv) irrigation practices which influence hydrologic cycles and potential salinity.

Plant eco-systems are directly involved with carbon and nitrogen cycles that regulate emission of gases. Direct effects of methane are not clear except through increased temperature and associated changes. The altered vapour pressure may have positive effects in water use efficiency but effects are negative where decreased radiation is critical. Dust might become a problem for radiation transmission especially under tropical environment. Furthermore, precise information is needed on sea rise, ozone levels (stratospheric and tropospheric), UV-B radiation, lightning events and cloud burst which also affect agro and natural eco-systems.

II Research priorities

Research on how global climatic change affects plant productivity should be multi-disciplinary. Of paramount significance is to have maps of soil productivity and climatic zonation for natural eco-systems which would help in extending the information collected both from short and long-term experiments at specific sites to other localities of the region. The following areas are identified for investigation.

1. *Rate and duration of photosynthesis of plant canopies*

In addition to having efficient photosynthetic conversion of radiation and CO_2 resources into biomass per unit land surface, it is important to sustain this activity throughout as much of the year as possible. Multiple cropping and related cropping systems need to be evaluated for improved productivity. Increased CO_2 levels are likely to enhance photosynthesis owing to diminished photorespiration. However, increase in temperature may partially offset this gain.

2. *Biomass production and partitioning*

Partitioning of biomass among plant tissues is known to be affected by CO_2 and altered temperature. New management strategies and cultivars need to be developed to optimize partitioning of photosynthate into economic yield. Higher partitioning to roots may enhance nitrogen fixation by the leguminous crops.

3. *Leaf resistance to CO_2 and H_2O*

Due to anticipated increase in water vapour in the atmosphere, it is critical to understand stomatal behaviour to cope with such a climatic change. Change in balance between photosynthesis and transpiration might affect the competitive ability of the plant in a given eco-system.

4. *Conversion efficiencies*

This involves evaluation of the impact of climate change on conversion of photosynthetically active radiation to biomass *per se* and chemical composition of the tissues. Changes in chemical composition of plant components might change pest and herbivore preference, thus affecting carbon and nitrogen recycling.

5. *Methane emissions*

There is likelihood of increase in acreage and cropping intensity of wetland rice in the sub-humid tropics. Quantification of emissions of methane and its impact on other environmental variables needs to be ascertained. Soil, water and crop management techniques will have to be modified to reduce emissions of CH_4.

6. *Interspecies competition*

Information on the effect of climatic factors on inter-species competition involving growth habit and reproductive success in agro-eco-system is lacking and needs detailed investigation, with particular reference to the effects of changes in temperature and water stress.

7. *Biotic and abiotic stresses*

Plants in agro and natural eco-systems must contend with a range of biological pests and environmental stresses. With alteration in precipitation and temperature fluxes, recurrence and virulence of pests might change. Research is needed to understand the short-term and long-term responses to these stresses.

GENERAL RECOMMENDATIONS

1. Development and standardization of newer methodology

(i) Standardization of techniques and instrumentation for accurate measurements of changes in ambient CO_2 and UV-B radiation.

(ii) Improvement of existing methods for monitoring physical and biochemical aspects of photosynthesis — leaf and other resistances, chloroplast morphology including excitation fluorescence of pigments and enzyme complexes.

(iii) Development of crop models to integrate essential plant processes and productivity based on data relating to increasing UV-B radiation, CO_2 levels and other factors of global climate change relevant to plant productivity.

(iv) Development of newer cropping systems interacting with changed resource strategies to minimize anticipated environmental constraints.

2. Interdisciplinary approach

Research facilities need to be developed in areas relating to meteorology, plant physiology, soil sciences and cognate disciplines for predicting the possible effects of global climatic change on plant productivity. To achieve this objective, it would be helpful to:

(i) Compile field and satellite data on soils, climate and canopy architecture.

(ii) Improve designs and statistical procedures for data interpretation.

3. International cooperation and technology transfer

(i) Formulation of protocol for transfer of information contained in project reports and agency documents.

(ii) Training of personnel and exchange of experts between developing and developed countries.

(iii) Organizing seminars and workshops to transfer users' expertise to the developing countries with the help of agencies like UNEP, IGBP, IPCC, WMO, USDA, FERRO and ICAR.

(iv) Collaboration in carrying out long term experiments on the impact of climate change on qualitative and quantitative changes in crop productivity.

4. Constitution of action programme committee

In India, the information available on global climatic change scenario is quite meagre and all the groups strongly felt the need to have updated data on emission of greenhouse gases by crops, incoming radiation fluxes both in terms of quality and intensity, temperature and precipitation changes in relation to their impact on agricultural systems in different agroclimatic zones. It is, therefore, absolutely necessary to establish a few main centres in the main agro-climatic zones having greater potential for productivity under the *aegies* of Indian Council of Agricultural Research. These centres should have the major responsibility to develop a data banks using ground and satellite information on soils, climate, photosynthesis and productivity of agricultural crops in close cooperation with other central agencies like Departments of Space, Science and Technology, Environment and Council of Scientific and Industrial Research and Meteorology.

Each of the centres, thus established, should have the facility of modern equipment and trained personnel. An expert committee at the national level may be constituted to formulate the programmes and monitor the progress of the different centres.

LIST OF PARTICIPANTS

I.P. Abrol
Deputy Director General (S&A)
Indian Council of Agril. Res.
Krishi Bhavan
New Delhi 110001
India

Y.P. Abrol
Div. of Plant Physiology
Indian Agril. Res. Inst.
New Delhi 110012
India

S.C. Adlakha
Agri. Research Specialist
FERRO, U.S. Dept. of Agril.
American Embassy
New Delhi 110021
India

P.K. Agarwal
Water Tech. Centre
Indian Agril. Res. Inst.
New Delhi 110012
India

R.S. Ambasht
Dept. of Botany
Banaras Hindu University
Varanasi 221005
India

K. Annamalainathan
Dept. of Plant Sciences
School of Biological Sciences
Madurai Kamraj University
Madurai 625021
India

C.R. Babu
Department of Botany
University of Delhi
Delhi 110007
India

T. Sudhakar Babu
School of Life Sciences
Jawaharlal Nehru University
New Mehrauli Road
New Delhi 110067
India

A.K. Bagga
Div. of Plant Physiology
Indian Agril. Res. Inst.
New Delhi 110012
India

P.N. Bahl
Asstt. Director General (Food)
Indian Council of Agril. Res.
New Delhi 110001
India

Neil R. Baker
Department of Biology
Wivenhoe Park
University of Essex
Colchester, C04 3SQ
England, UK

V. Balasubramanian
Central Research Institute for
Dryland Agril.
Santosh Nagar
Hyderabad 500659
India

B. Baldev
Pulse Laboratory
Indian Agril. Res. Inst.
New Delhi 110012
India

Paul W. Barnes
USEPA Environmental Res. Lab.
200SW 35th Street
Corvallis OR 97333
USA

Fakhri A. Bazzaz
Dept. of Organismic & Evolutionary Biology
Biological Laboratories
Harvard University
16 Divinity Avenue
Cambridge, Massachusetts 02138
USA

S.N. Bhardwaj
Div. of Plant Physiology
Indian Agril. Res. Inst.
New Delhi 110012
India

S.C. Bhargava
Div. of Plant Physiology
Indian Agril. Res. Inst.
New Delhi 110012
India

K. Ya. Bil
Institute of Photosynthesis and Soil Sciences
USSR Academy of Sciences
Puschino 142294, Moscow Region
USSR

Janet F. Bornman
Department of Plant Physiology
University of Lund
Box 7007, S 22007, Lund
Sweden

N.C. Chakraborty
School of Life Sciences
Jawaharlal Nehru University
New Mehrauli Road.
New Delhi 110067
India

N.V.K. Chakravarty
Div. of Agril. Physics
Indian Agril. Res. Inst.
New Delhi 110012
India

L. Karuna Chand
Div. of Plant Physiology
Indian Agril. Res. Inst.
New Delhi 110012
India

S. Chatterjee
Consultant
Ministry of Environment
CGO Complex, Lodi Road
New Delhi 110003
India

Renu K. Chopra
Water Technology Centre
Indian Agril. Res. Inst.
New Delhi 110012
India

Thomas P. Coohill
Dept. of Physics & Astronomy
Western Kentucky University
Bowling Green, KY 42101
USA

B.L. Deekshatulu
Director
National Remote Sensing Agency
Balanagar
Hyderabad 600037
India

R.K. Dixon
Team Leader, FTS
Ecological Effect
USEPA Environmental Res. Lab.
Corvallis, Oregon 97333
USA

Jerry D. Eastin
Department of Agronomy
Institute of Agriculture & Natural Resources
University of Nebraska
Lincoln, NE 68583-0817
USA

M.C. Ghildiyal
Div. of Plant Physiology
Indian Agril. Res. Inst.
New Delhi 110012
India

Yu.E. Giller
Institute of Plant Physiology and Biophysics
Academy of Sciences
Puschino, Moscow Region 142292
USSR

A. Gnanam
Vice-Chancellor
University of Pondicherry
R. Venkatraman Nagar
Pondicherry 605014
India

Susan Golden
Department of Biology
Texas A & M University
College Station, Texas 77843
USA

N.N. Goswami
Director
Indian Agril. Res. Inst.
New Delhi 110012
India

Govindjee
Department of Biophysics and Plant Biology
University of Illinois
Urbana Ill. 61801
USA

Arun Goyal
Dept. of Biochemistry
Michigan State University
East Lansing
Michigan 48824-1319
USA

Robert C. Hedlund
Deputy Director
FERRO, U.S. Dept. of Agri.
American Embassy
New Delhi 110021
India

J.D. Hesketh
U.S. Dept. of Agriculture-ARS
Department of Agronomy
8215 Turner Hall
University of Illinois
Urbana, Ill. 61801
USA

S.S. Hundal
Deptt. of Agro-Meteorology
Punjab Agricultural University
Ludhiana 141004
India

V.I. Kefeli,
Inst. of Soil Science & Photosynthesis, Puschino
Biocentre, Puschino 142292, Moscow
USSR

M.A.K. Khalil
Oregon Graduate Institute of Science & Technology
19600 N.W. Von Neumann Drive
Beaverton, Oregon 97006-1999
USA

T.N. Khoshoo
Tata Energy Research Institute
7, Jor Bagh
New Delhi 110003
India

D.T. Krizek
USDA Climate Stress Lab.
Room No. 206, Building 001
BARC-WEST
Beltsville, Md. 20105
USA

G. Kulandaivelu
School of Biological Sciences
Dept. of Plant Sciences
Madurai Kamraj University
Madurai 625021
India

H.D. Kumar
Department of Botany
Banaras Hindu University
Varanasi 221005
India

M. Udaya Kumar
Department of Crop Physiology
Univ. of Agril. Sciences
GKVK Campus
Bangalore 560065
India

P. Ananda Kumar
Nuclear Research Laboratory
Indian Agril. Res. Inst.
New Delhi 110012
India

Sushil Kumar
Head, HRDG
Council of Scientific & Industrial Res.,
Rafi Marg
New Delhi 110001
India

Santosh Kumari
Div. of Plant Physiology
Indian Agril. Res. Inst.
New Delhi 110012
India

A.N. Lahiri
Head, Div. of Arable Cropping System
Central Arid Zone Res. Inst.
Jodhpur 342003
India

A.K. Laisk
Inst. of Atmosphere Physics and Astrophysics
Estonian Academy of Science
202444 Observatory, Toravere
USSR

D. W. Lawlor
Crop Production Department
AFRC Inst. of Arable Crop Res.
Rothamsted Experimental Station
Harpenden, Hertz AL5 2JQ
UK

S.C. Maheshwari
Department of Plant Molecular Biology
University of Delhi
Delhi 110007
India

O.P. Makhija
Asstt. Director General (FC)
Indian Council of Agril. Res.
Krishi Bhavan,
New Delhi 110001
India

J.C. Malhotra
Asstt. Director General (PIU)
Indian Council of Agril. Res.
Krishi Bhavan
New Delhi 110001
India

C.P. Malik
Dept. of Basic Sciences
Punjab Agricultural University
Ludhiana 141004
India

V.M. Meher-Homji
Salim Ali School of Ecology & Research
French Institute , P.B.No. 33
Pondicherry 605001
India

M.G.K. Menon
72, Lodi Estate
New Delhi 110003
India

A.P. Mitra
Director-General
Council of Scientific & Industrial Research
Rafi Marg
New Delhi 110001
India

P. Mohanty
School of Life Sciences
Jawaharlal Nehru University
New Mehrauli Road
New Delhi 110067
India

S.D.S. Murthy
School of Life Sciences
Jawaharlal Nehru University
New Mehrauli Road
New Delhi 110067
India

K.S. Murty
Scientist Emeritus
Div. of Plant Physiology
Central Rice Research Inst.
Cuttack 753005
India

P.M. Nair
Food Technology & Enzyme Engineering Division
Bhabha Atomic Research Centre
Bombay 400085
India

Rajinder Nanda
Div. of Plant Physiology
Indian Agril. Res. Inst.
New Delhi 110012
India

Virendra Nath
Div. of Plant Physiology
Indian Agril. Res. Inst.
New Delhi 110012
India

C.J. Nelson
Department of Agronomy
University of Missouri
Columbia, Missouri 652111
USA

Donald R. Ort
Department of Plant Biology
University of Illinois
289 Morrill Hall
505 South Goodwin Avenue
Urbana, Ill. 61801
USA

William H. Outlaw, Jr.
Dept. of Biological Science
The Florida State University
Biology Unit 1, B-157
Tallahassee, Florida 32306-3-5
USA

S. Padhye
Department of Chemistry
University of Poona
Ganesh Kund Road
Pune 452002
India

Himadri Pakarasi
Plant Biology Program
Department of Biology
Washington University
Box 1137, St. Louis,
Missouri 63130 - 4899
USA

D.C. Parashar
Chemistry Division
National Physical Laboratory
Hillside Road
New Delhi 110012
India

R.S. Paroda
Deputy Director General (CS)
Indian Council of Agril. Res.
Krishi Bhavan
New Delhi 110001
India

Robert W. Pearcy
Dept. of Botany
University of California
Davis, California 95616
USA

Mary M. Peet
Dept. of Horticultural Science
North Carolina State University
Box 7609, Raleigh, NC 27695-7609
USA

John R. Porter
Long Ashton Res. Station
Dept. of Agril. Sciences
University of Bristol
Bristol BS 18 9AF
UK

T.G. Prasad
Dept. of Crop Physiology
University of Agril. Sciences
GKVK Campus
Bangalore 560065
India

A.N. Purohit
Director
Govind Ballabh Institute of
Humanity Studies & Development
Himalayan Ecology
Kosi, Almora
Distt. Nainital
India

V.I. P' Yankov
Deptt. of Biology
Urals State University
Sverdlovsk, 620083
USSR

P. Raghuveer
Div. of Plant Physiology
Indian Agril. Res. Inst.
New Delhi 110012
India

Mangala Rai
Asstt. Director General (Seeds)
Indian Council of Agril. Res.
New Delhi 110001
India

S. Ramachandran
Secretary, Department of Biotechnology
Block 2, 7-8 Floor
CGO Complex, Lodi Road
New Delhi 110003
India

V.S. Rama Das
Director, CSIR Complex
Palampur 176061
India

Ch. N. Rao
Div. of Plant Physiology
Central Rice Research Inst.
Cuttack 753005
India

G. Renger
Max-Volmer Inst.
fur Biophysikalische
und Physikalische Chemie
TU Berlin, Sekr. PC 14
Strasse des 17, Juni 135
D 1000 Berlin 12
Germany

Ann R. Rich (nee' Webb)
2, Earley Gate
Whiteknights
P.O. Box 239
Reading RG6 2AU
UK

S.C. Sabat
Bioenergetics Lab.
School of Life Sciences
Jawaharlal Nehru University
New Delhi 110067
India

U.K. Sengupta
Div. of Plant Physiology
Indian Agril. Res. Inst.
New Delhi 110012
India

Aruna Sharma
Div. of Plant Physiology
Indian Agril. Res. Inst.
New Delhi 110012
India

Manju Sharma
Department of Biotechnology
Block 2, 7-8th Floor
CGO Complex, Lodi Road
New Delhi 110003
India

I.S. Sheoran
Dept. of Chemistry and Biochemistry
Haryana Agricultural University
Hisar 125004
India

D.S. Shukla
Div. of Plant Physiology
Indian Agril. Res. Inst.
New Delhi 110012
India

D.R. Sikka
Director
Indian Institute of Tropical Meteorology
Pune
India

D.P. Singh
Department of Agronomy
Haryana Agricultural University
Hisar 125004
India

J.S. Singh
Department of Botany
Banaras Hindu University
Varanasi 221005
India

Phool Singh
Department of Agronomy
Haryana Agricultural University
Hisar 125004
India

Randhir Singh
College of Basic Sciences & Humanities
Haryana Agricultural University
Hisar 125004
India

G.S. Singhal
School of Life Sciences
Jawaharlal Nehru University
New Mehrauli Road
New Delhi 110067
India

S.K. Sinha
Water Technology Centre
Indian Agril. Res. Inst.
New Delhi 110012
India

G.C. Srivastava
Div. of Plant Physiology
Indian Agril. Res. Inst.
New Delhi 110012
India

H.K. Srivastava
Dept. of Biotechnology
Block 2, 7-8th Floor
CGO Complex, Lodi Road
New Delhi 110003
India

Ineke Stulen
Faculteit der Wiskunde en Natuuwetens-chappen
Rijksuniversiteit Groningen
Biologisch Centrum
Kerklaan 30, Postbus 14
9750 AA Haren (GN)
The Netherlands

A.R. Subramaniam
Saradamba
Ganjipet Junction
Vishakhapatnam 530002
India

Padmini Swain
Div. of Plant Physiology
Central Rice Research Inst.
Cuttack 753006
India

Alan H. Teramura
Department of Botany
University of Maryland
College Park, Md. 20742-5815
USA

James H. Thomas
Director
FERRO, U.S. Dept. of Agril.
American Embassy
New Delhi 110021
India

N.E. Tolbert
Department of Biochemistry
Michigan State University
East Lansing
Michigan 49924-1319
USA

D.P.S. Tomar
Div. of Plant Physiology
Indian Agril. Res. Inst.
New Delhi 110012
India

B.K. Tripathi
Asstt. Director General (Commercial Crops)
Indian Council of Agril. Res.
Krishi Bhavan
New Delhi 110001
India

Rashmi Tyagi
Department of Botany
Banaras Hindu University
Varanasi 221005
India

D.C. Uprety
Div. of Plant Physiology
Indian Agril. Res. Inst.
New Delhi 110012
India

P.N. Wattal
Div. of Plant Physiology
Indian Agril. Res. Inst.
New Delhi 110012
India

E. Wellman
Biologisches Institut II
Der Universitat Freiburg
Schanzlestrasse 1
D 7800 Freiburg
Germany

John Whitmarsh
U.S. Deptt. of Agril. ARS
Univ. of Illinois, 289, Morrill Hall
505 South Goodwin Avenue
Urbana, Ill. 61801
USA

AUTHOR INDEX